Models and Applications
of Chaos Theory
in Modern Sciences

Models and Applications of Chaos Theory in Modern Sciences

Elhadj Zeraoulia

Department of Mathematics
University of Tébessa, Tébessa, Algeria

CRC Press
Taylor & Francis Group
Boca Raton London New York

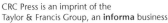

CRC Press is an imprint of the
Taylor & Francis Group, an **informa** business

Science Publishers
Jersey, British Isles
Enfield, New Hampshire

CRC Press
Taylor & Francis Group
6000 Broken Sound Parkway NW, Suite 300
Boca Raton, FL 33487-2742

First issued in paperback 2018

© 2011 by Taylor & Francis Group, LLC
CRC Press is an imprint of Taylor & Francis Group, an Informa business

No claim to original U.S. Government works

ISBN-13: 978-1-57808-722-8 (hbk)
ISBN-13: 978-1-138-11485-2 (pbk)

Library of Congress Cataloging-in-Publication Data

Zeraoulia, Elhadj.
    Models and applications of chaos theory in modern sciences / Elhadj Zeraoulia.
          p. cm.
    Includes bibliographical references and index.
    ISBN 978-1-57808-722-8 (hardcover)
    1. Chaotic behavior in systems. I. Title.
    Q172.5.C45Z47 2011
    003'.857--dc23

                                                              2011013329

Visit the Taylor & Francis Web site at
http://www.taylorandfrancis.com

and the CRC Press Web site at
http://www.crcpress.com

Preface

This unique book is a compilation of selected papers from the past decade presenting a perspective underlying the thematics and strategies related to the chaos theory and its applications in modern sciences, in particular, physical sciences and also human behavior, both individual and social (including the latest investigations in chaos theory and its interrelated problems in diverse theoretical and practical disciplines, incorporating the main engineering applications.) This collection of selected papers provide a comprehensive view on some models and applications of chaos theory in medicine, biology, ecology, economy, electronics, mechanical and human sciences...etc. The papers, written by many of the leading experts in the field, cover both the experimental and theoretical aspects of the subject. This volume presents a variety of fascinating topics of current interest and problems arising in the study of both discrete and continuous time chaotic dynamical systems modeling the several phenomena in nature and society. Exciting techniques stemming from the area of nonlinear dynamical systems theory are currently being developed to meet these challenges.

Models and Applications of Chaos Theory in Modern Sciences is devoted to setting an agenda for future research in this exciting and challenging field.

I would like to thank everyone who help me in writing this book.

Readership: Advanced undergraduates and graduate students in natural and human sciences and engineering such as physics, chemistry, biology or bioinformatics...etc; academics and practitioners in nonlinear physics and in various other areas of potential application; researchers, instructors, mathematicians, nonlinear scientists and electronic engineers interested in chaos, nonlinear dynamics and dynamical systems and all interested in nonlinear sciences.

Dr. Elhadj Zeraoulia
April, 2011

Contents

Preface   v

Section I: Models and Applications of Chaos Theory in Medicine   **1**

1. A Chaotic View of Behavior Change: A Quantum Leap for Health Promotion   3
   K. Resnicow and *R. Vaughan*

2. Crisis and Chaos in Behavioral Nutrition and Physical Activity   14
   T. Baranowski

3. Memory in Astrocytes: A Hypothesis   22
   R.M. Caudle

4. Nonlinear Dynamics and Chaos in a Fractional-Order HIV Model   37
   H. Ye and *Y. Ding*

5. Research on the Relation of EEG Signal Chaos Characteristics with   50
   High-Level Intelligence Activity of Human Brain
   X.Y. Wang, M. Juan, G.L.Tan and *L.X. Zou*

6. Nonlinear Filtering of Oscillatory Measurements in Cardiovascular   66
   Applications
   R.Vepa

7. Temporal Redistribution of Plantar Pressure Points in the Healthy and   84
   Diabetics: A Time Series Analysis of the Neuro-Capillary Chaos
   D.V. Oberoi, C. Pradhan, C.J. Kumar and *S.C. D'Souza*

Section II: Models and Applications of Chaos Theory in Biology   **93**

1. An Astrobiological Thought on Sustainable Life   95
   T. Naganuma

2. Matrix Frequency Analysis of Oryza Sativa (Japonica Cultivar-Group)   106
   Complete Genomes
   K. Manikandakumar, S.M. Kumaran and *R. Srikumar*

3. The Capabilities of Chaos and Complexity   123
   D.L. Abel

Section III: Models and Applications of Chaos Theory in Ecology   **167**

1. A Unified Approach of Catastrophic Events   169
   S. Nikolopoulos, P. Kapiris, K. Karamanos and *K. Eftaxias*

2. Evolutionary Geomorphology: Thresholds and Nonlinearity in   196
   Landform Response to Environmental Change
   J.D. Phillips

Section IV: Models and Applications of Chaos Theory in Economy   **215**

1. Physics and the Foundations of Economic Science: Comments in   217
   Memory of Ilya Prigogine
   Richard H. Day

2. Neumannian Economy in Multi-Agent Approach Investigation of   226
   Stability and Instability in Economic Growth
   Katalin Martinás

3. Theoretical Aspects of the Economic Transition: Case of Romania   236
   Cezar Scarlat and *Eugen I. Scarlat*

4. Complex Dynamics in a Nonlinear Cobweb Model For Real Estate   256
   Market
   Junhai Ma and *Lingling Mu*

5. Chaos Models in Economics   269
   Sorin Vlad, Paul Pascu and *Nicolae Morariu*

6. Complex Dynamics of an Adnascent-type Game Model   278
   Baogui Xin, Junhai Ma and *Qin Gao*

7. A Production-Inventory Model for Deteriorating Items with   290
   Production Disruptions
   Yong He and *Ju He*

8. Nonlinear Noise Estimation in International Stock Markets: 304
   Coarse-Grained Entropy Method
   Yong Fang

Section V: Models and Applications of Chaos Theory in Electronics   **315**

1. Simple Chaotic Oscillator: From Mathematical Model to Practical   317
   Experiment
   J. Petržela, Z. Kolka and *S. Hanus*

2. A Robust Chaos-Based True Random Number Generator Embedded in   328
   Reconfigurable Switched-Capacitor Hardware
   M. Drutarovský and *P. Galajda*

3. Conservative Chaos Generators with CCII+ Based on Mathematical   345
Model of Nonlinear Oscillator
J. Petržela and *J. Slezák*

Section VI: Models and Applications of Chaos Theory in Human Sciences   355

1. Ancient and Current Chaos Theories   357
G. Gündüz

2. Complex Freedom   376
D. Pećnjak

3. An Understanding of Language Development Models-Pidginization   384
from the Perspective of Chaos Theory
G. Zhao

4. Chaos and Natural Language Processing   393
M. Crisan

5. Discrete Phase-Locked Loop Systems and Spreadsheets   405
S. Abramovich, E. Kudryashova, G.A. Leonov and *S. Sugden*

6. Modeling Complex Spatial Dynamics of Two-Population Interaction in   429
Urbanization Process
Y. Chen and *F. Xu*

7. Characterizing Growth and Form of Fractal Cities with Allometric Scaling   449
Exponents
Y. Chen

8. Estimating the Distribution of Dynamic Invariants: Illustrated with an   473
Application to Human Photo-Plethysmographic Time Series
M. Small

9. Asymmetry, Symmetry and Beauty   487
H. Sabelli, A. Lawandow and *A.R. Kopra*

10. Nonlinear Dynamics in Psychology   520
S.J. Guastello

**Section VII: Models and Applications of Chaos Theory in Mechanical   545
Sciences**

1. Chemical Reactivity Dynamics and Quantum Chaos in Highly Excited   547
Hydrogen Atoms in an External Field: A Quantum Potential Approach
P.K. Chattaraj and *B. Maiti*

2. Regular and Chaotic Motion of a Bush-Shaft System with Tribological   569
Processes
J. Awrejcewicz and *Y. Pyryev*

3. Relaxed Plasma Equilibria and Entropy-Related Plasma Self-Organization
   Principles
   R.L.Dewar, M.J. Hole, M. MeGann, R. Mills and S.R. Hodson   582

4. Generalized Complexity and Classical-Quantum Transition
   A.M. Kowalski, A. Plastino and M. Casas   595

5. A New Mechanical Model for Particle Transport by Surface Waves and
   Applications
   M. Ragulskis, E. Sakyte, J.M. Seoane and M.A.F. Sanjuán   609

6. Chaotic Behavior of the Biharmonic Dynamics System
   V.S. Aslanov   627

7. Modeling, Chaotic Behavior and Control of Dissipation Properties of
   Hysteretic Systems
   J. Awrejcewicz and L. Dzyubak   645

8. Nonlinear Dynamics and Chaos of Microcantilever-Based TM-AFMs
   with Squeeze Film Damping Effects
   W.M. Zhang , G. Meng, J.B. Zhou and J.Y. Chen   667

9. Mathematical Identification of Homogenisation Processes in Argon
   Stirred Ladle
   K. Michalek and K. Gryc   689

10. Analysis of a Nonlinear Aeroelastic System with Parametric Uncertainties
    Using Polynomial Chaos Expansion
    A. Desai and S. Sarkar   697

11. Chaos Synchronization Criteria and Costs of Sinusoidally Coupled
    Horizontal Patform Systems
    J. Cai, X. Wu and S. Chen   717

Index   727

Section I

Models and Applications of Chaos Theory in Medicine

1. **A Chaotic View of Behavior Change: A Quantum Leap for Health Promotion**   3
   K. Resnicow and *R. Vaughan*

2. **Crisis and Chaos in Behavioral Nutrition and Physical Activity**   14
   T. Baranowski

3. **Memory in Astrocytes: A Hypothesis**   22
   R.M. Caudle

4. **Nonlinear Dynamics and Chaos in a Fractional-Order HIV Model**   37
   H. Ye and *Y. Ding*

5. **Research on the Relation of EEG Signal Chaos Characteristics with High-Level Intelligence Activity of Human Brain**   50
   X.Y. Wang, M. Juan, G.L. Tan and *L.X. Zou*

6. **Nonlinear Filtering of Oscillatory Measurements in Cardiovascular Applications**   66
   R. Vepa

7. **Temporal Redistribution of Plantar Pressure Points in the Healthy and Diabetics: A Time Series Analysis of the Neuro-Capillary Chaos**   84
   D.V. Oberoi, C. Pradhan, C.J. Kumar and *S.C.D'Souza*

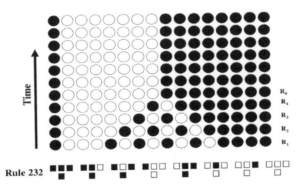

Rule 232

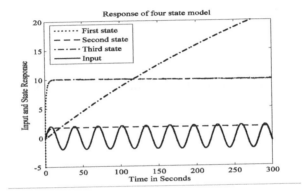

Chapter 1

A Chaotic View of Behavior Change: A Quantum Leap for Health Promotion[†]

Ken Resnicow[1] and *Roger Vaughan*[2]

[1]University of Michigan, Department of Health Education and Health Behavior, School of Public Health, Ann Arbor, MI, USA. Email: kresnic@umich.edu.
[2]Department of Biostatistics, Columbia University, 722 West 168th Street, 6th Floor, New York, NY 10032, USA. Email: rdv2@columbia.edu.

ABSTRACT

Background: The study of health behavior change, including nutrition and physical activity behaviors, has been rooted in a cognitive-rational paradigm. Change is conceptualized as a linear, deterministic process where individuals weigh pros and cons, and at the point at which the benefits outweigh the cost change occurs. Consistent with this paradigm, the associated statistical models have almost exclusively assumed a linear relationship between psychosocial predictors and behavior. Such a perspective however, fails to account for non-linear, quantum influences on human thought and action. Consider why after years of false starts and failed attempts, a person succeeds at increasing their physical activity, eating healthier or losing weight. Or, why after years of success a person relapses. This paper discusses a competing view of health behavior change that was presented at the 2006 annual ISBNPA meeting in Boston.

Discussion: Rather than viewing behavior change from a linear perspective it can be viewed as a quantum event that can be understood through the lens of Chaos Theory and Complex Dynamic Systems. Key principles of Chaos Theory and Complex Dynamic Systems relevant to understanding health behavior change include: 1) Chaotic systems can be mathematically modeled but are nearly impossible to predict; 2) Chaotic systems are sensitive to initial conditions; 3) Complex Systems involve multiple component parts that interact in a nonlinear fashion; and 4) The results of Complex Systems are often greater than the sum of their parts. Accordingly, small changes in knowledge, attitude, efficacy, etc. may dramatically alter motivation and behavioral outcomes. And the interaction of such variables can yield almost infinite potential patterns of

---

[†]Reused with permission from: Ken Resnicow, Roger Vaughan, A chaotic view of behavior change: a quantum leap for health promotion, *International Journal of Behavioral Nutrition and Physical Activity* 2006, 3: 25. http://www.ijbnpa.org/content/3/1/25.

motivation and behavior change. In the linear paradigm unaccounted for variance is generally relegated to the catch all "error" term, when in fact such "error" may represent the chaotic component of the process. The linear and chaotic paradigms are however, not mutually exclusive, as behavior change may include both chaotic and cognitive processes. Studies of addiction suggest that many decisions to change are quantum rather than planned events; motivation arrives as opposed to being planned. Moreover, changes made through quantum processes appear more enduring than those that involve more rational, planned processes. How such processes may apply to nutrition and physical activity behavior and related interventions merits examination.

Background

"What we call **chaos** is just patterns we haven't recognized. What we call random is just patterns we can't decipher. What we can't understand we call nonsense. What we can't read we call gibberish" Chuck Palahniuk.

The study of health behavior change, including nutrition and physical activity behaviors, has historically been rooted in a cognitive-rational paradigm. Extant models, such as Social Cognitive Theory, the Health Belief Model, the Theory of Planned Behavior, the Transtheoretical Model and others, have generally viewed change as an interaction of cognitive factors such as knowledge, attitude, belief, efficacy and intention [1,2]. Change is conceptualized as a linear, deterministic process where individuals weigh the pros and cons, and at the point at which the benefits outweigh the cost, "decisional balance" tips them toward change. An implicit assumption within this perspective is that change is a gradual process under conscious control. Consistent with this framework, the associated statistical models have almost exclusively assumed a linear relationship between psychosocial predictors and behavior (change); i.e., greater increases in knowledge, attitudes and intentions will lead to greater change in behavior.

However, the theoretical and statistical assumptions underlying this linear paradigm may be seriously flawed. In particular, such a perspective fails to account for nonlinear, quantum influences on human thought and action. The limitations of a rational-linear conceptualization of behavior change may in part (in addition to measurement error) explain the modest proportion of behavioral variance accounted for by such models; which typically has been in the range of around 10%–20% and rarely higher than 50% [3–11]. The fact that the majority of studies have employed cross-sectional designs and relied on self-report to measure behavior further suggests that the true variance accounted for by linear models may be even lower [12]. Below we provide an alternative model of health behavior change based on non-linear dynamics.

Discussion

An alternative view is that decisions to initiate (and possibly maintain) behavior change are quantum rather than linear events [13]. Such quantum leaps result from a surge of motivation or inspiration that is greater than the sum of its cognitive parts. It is not so much a planned decision, but something that arrives beyond cognition. The more dramatic form of quantum change is described by Miller [14]. "Buried in the statement 'I just decided', however can be another kind of experience that has been confused with ordinary decision making. It is the insightful type of quantum change. When people talk about such experiences in shorthand, they may say 'it just happened' or 'I just decided'. Inquire a little more closely, however, and it becomes apparent that the process is somewhat more complex." (page 37)

Miller delineates two types of quantum change, sudden insights and mystical epiphanies. Both kinds leave an indelible impact and often lead to lasting and pervasive change. Both usually involve a significant alteration in how the person perceives him/her self, others and the world. Although the cases described in Miller's book tend to involve an overwhelming transformation, less dramatic, less mystical "mini-epiphanies" may contribute to many behavior change decisions. From this perspective, behavior change can be understood through the lens of Chaos Theory and Complex Dynamic Systems. Four key principles from these theories relevant to understanding health behavior change are:

1. Chaotic systems can be mathematically modeled, usually in non-linear terms, but are nearly impossible to predict;
2. Chaotic systems are sensitive to initial conditions;
3. Complex Systems involve multiple component parts that interact in a nonlinear fashion; and
4. The results of Complex Systems are often greater than the sum of their parts.

Examples of chaotic systems include the weather, war, love, population growth, many epidemics and stock market prices. Chaos Theory has been used to explain psychologic health as well as specific health behaviors such as smoking and physical activity [15–17].

One of the first published works on Chaos Theory came from a meteorologist named Edward Lorenz. In the 1960's he was developing computer models of weather prediction. One day after running a predictive equation he decided to run the model a second time. But to save time he started the calculation in the middle of the sequence, plugging in manually some key numbers. But the predicted output diverged sharply from the original. He eventually discerned that in the original computation the number used was .506127 but in the simulation he had only entered the first three digits, .506 [18]. This phenomenon, eventually labeled "sensitivity to initial conditions", posits that a minor change at the beginning (or at various points) of a sequence of events can dramatically alter the long-term outcome of the system. This is commonly referred to as the butterfly effect.

The flapping of a single butterfly's wing today produces a tiny change in the state of the atmosphere. Over a period of time, what the atmosphere actually does diverges from what it would have done. So, in a month's time, a tornado that would have devastated the Indonesian coast doesn't happen. Or maybe one that wasn't going to happen, does. (Ian Stewart, Does God Play Dice? The Mathematics of Chaos, pg. 141) [19].

The weather is considered a classic chaotic system, as described in the text below. Yet, simple substitution of health behavior terminology for meteorological terminology reveals striking similarity. The weather (**BEHAVIOR CHANGE**) is an example of a chaotic system. In order to make long- term weather forecasts (**PREDICTIONS OF BEHAVIOR CHANGE**) it would be necessary to take an infinite number of measurements, which would be impossible to do. Also, because the atmosphere (**HUMAN BEHAVIOR**) is chaotic, tiny uncertainties would eventually overwhelm any calculations and defeat the accuracy of the forecast. Even if it were possible to fill the entire atmosphere of the earth with an enormous array of measuring instruments, e.g., thermometers, wind gauges, and barometers (**PSYCHOSOCIAL, BIOLOGIC, AND ENVIRONMENTAL MEASURES**) uncertainty in the initial conditions would arise from the minute variations in measured values between each set of instruments in the array. Because the atmosphere (**HUMAN BEHAVIOR**) is chaotic, these uncertainties, no matter how small, would eventually overwhelm any calculations and defeat the accuracy of the forecast (**PREDICTION**).

Another metaphor for sensitivity to initial conditions involves rolling two identical balls down a tall rocky mountain. Starting the balls even an inch or less apart at the top of the mountain could result in the two balls ending hundreds of feet apart at the bottom; having traversed vastly different courses. The different pathways created by slight differences in the impact point on a billiard ball is another example.

One additional concept from Chaos Theory, fractal patterns, may also be relevant to understanding human behavior. Fractals, which have been identified in natural science in the mapping of the microvascular system and snow flake geometry, are recurring patterns within larger systems that are self-similar, that is, a shape appears similar at all scales of magnification. In terms of human behavior, there may be common patterns of behavior change within and across individuals that follow certain complex, non linear patterns. Thus, although behavior change may unfold in an almost infinite combination of knowledge, attitude, efficacy, and intention, there may be recurrent patterns of change that may be used to identify audience segments which could be targeted by common interventions.

Linear	Quantum
Cognitive-Rational	Intuitive
Motivation is arrived at	Motivation arrives
Planned	Epiphany
Cortical	Limbic
Left Brain	Right Brain
Maintenance of Change	Initiation of Change
Engineers/Physicists	Artists

Figure 1 Continuum of Motivational Processes.

Application to Health Behavior

Health behavior may mirror other Complex Systems found in nature in that they involve multiple component parts that interact in a nonlinear fashion. Factors such as knowledge, attitude, belief, and efficacy no doubt exert some influence on health behavior change.

However, the interaction of these factors represent a complex system bound by chaotic regulation. For example, which particular bits of knowledge, attitude, belief, etc. and the amount of each required to "tip" the system for a particular individual is virtually impossible to predict, and the outcome is sensitive to initial conditions. Initial conditions within individuals, e.g., relevant prior experience with a particular disease (e.g., family history) or a genetic predisposition may alter the interaction in profound ways. And, the slightest change in the system, i.e., the addition of one more piece of information or persuasion could dramatically alter the outcome. Such complex relationships are well represented by the swirling patterns created by mixing multiple colors of dye with a stick. Given the non-linear nature of complex systems they are usually represented mathematically by quadratic or other non-linear models. In the linear framework unaccounted for variance is generally relegated to the catch all "error" term, when in fact such "error" may represent the chaotic component of the outcome. Stated otherwise, "error" may be the result of imposing a linear model on a non-linear phenomenon. Additionally, in complex dynamic systems the interaction of factors can yield almost infinite potential patterns. In linear terms, this may be analogous to higher order interaction terms that could involve 5, 10, or 15-way interactions. Although linear methods can be used to model such interactions, they are limited statistically and conceptually. First, the ability to detect such interactions would be underpowered, so unless the magnitudes of these interactions are pre-specified so that the study could be adequately powered, these analyses would generally lead one to assume, perhaps falsely, that no interaction exists. Second, untangling a 3-way or higher order interaction generally extends beyond our ability to map and interpret such a finding; a relatively simple two-way interaction states that the effect of one variable on the outcome is not constant, but depends upon the level or status of yet a second variable (e.g., the intervention effect on cholesterol reduction is not constant, but is greater for males that for females). The extension to a 3-way interaction says that that observed gender by treatment interaction is itself not always better for males than for females, but depends upon the status of a third variable (perhaps the intervention does better for tall males, but no better than it works for short females, etc.). And this is a reduced example where each variable in the interaction only has two levels. In complex systems the levels of interactions are copious. Finally, from a chaotic perspective the confluence of interactions both within and between individuals is highly variable and the system is sensitive to initial conditions making prediction of such complex interactions virtually impossible. From a chaotic perspective, rather than searching for main effects or simple 2-way interaction effects, behavior change is assumed to involve multiple levels of interaction that vary across individuals.

Linear models of behavior change are then both conceptually inappropriate and statistically futile. In traditional statistical terms this would equate to analyzing and reporting separate main effects for multiple independent variables when there are known interactions (non linear in nature) of these variables. The solution does not do justice to the complexity of the phenomena.

A potential important element of this model that should also be considered is the occurrence of random external and intrapsychic events. Chaotic systems are not synonymous with randomness, nonetheless, random events can significantly impact complex systems.

Consider why after years of false starts and failed attempts, a person succeeds at increasing their physical activity, eating healthier or losing weight. Or, why after years of success a person relapses. One explanation is that success or failure is determined by random events. The event may be external, such as hearing about someone they knew who lost weight, quit smoking, or perhaps passed away. This is similar to the "Cues" concept in the Health Belief Model [20,21].

The random event may also be intrapsychic. Without conscious thought, the person may experience a surge of motivation that they need to and/or are able to change or a craving may arise unexpectedly that triggers a relapse. Such feelings may be stimulated by associations created by classical conditioning about which the individual may not be conscious. Regardless, motivation and impulse arrives as opposed to being planned. Consistent with this perspective, West et al. recently reported an analysis of how smokers decided to quit.

Approximately half of the ex- and current smokers in their sample reported that their most recent quit attempt was unplanned and those who did quit this way were more likely to stay quit than those who made a specific plan to quit [22]. Another study of smokers found that more than half of quit attempts were spontaneous rather than planned [23]. West et al. explain their findings using "catastrophe theory" [22], which posits that dramatic outcomes can result from continuous pressure of a force on a system. An example often used to illustrate this concept is the result of gradually bending a plastic ruler until it snaps or the point at which water becomes vapor. So too, motivation may break or boil when enough pressure is applied to the system.

Chaotic patterns can stimulate behavior change in two distinct ways. In the first, single external random events such as a conversation, a public service announcement, newspaper article, word about the death of a friend or relative, etc. may serve as a tipping point for motivational change. Conversely, absent an external event, resident chunks of knowledge or attitude may randomly coalesce to form a perfect motivational storm. Miller also delineates two types of quantum change, with one being more a dramatic, mystical experience and the second being more a sudden insight or sense of finding one's truth. Common to both pathways is that they occur outside of conscious reasoning; that they happen to the person [13]. As Miller notes, the individual experiences a "fast forward to self actualization". Interestingly in a study of problem drinkers, those whose decision to quit drinking arose from a transformational experience (having experienced a negative/traumatic event such as hitting rock bottom or having a spiritual awakening) were twice as likely to be non-problem drinkers at followup whereas those who reported weighing the pros and cons of drinking were actually more likely to have drinking problems at followup [24]. The cognitive approach to behavior change in this study was associated with worse outcomes. Thus not only do there appear to be linear and quantum pathways to change, the two processes may impact behavioral outcomes differently. Another perspective that may be useful to include in this alternative paradigm is the concept of "Tipping Points". Tipping points are dramatic changes in social behavior that arise quickly and usually unexpectedly [25]. Whether it be a jingle or slogan; a political idea or mass purchase of a "fad" product, such tipping points are virtually impossible to predict, yet retrospectively coherent explanations for the phenomena are routinely offered. Similarly, each night after the stock market closes, pundants explain why certain events of the day or week "caused" the price fluctuations. Yet, a priori, few pundants could have predicted the impact of said events. If they possessed such prognostication ability they would be extremely wealthy. The stock market provides an excellent metaphor for chaos, as on an almost daily level, tipping points occur that lead to what has been called the random walk theory of wall street [26]. Additionally, just as our interventions often work, the stock market tends to rise. The former may be due to an inherent will to live and the latter inherent optimism of consumers. However, in both cases, there may be underlying human dynamics that predispose systems to moving in a particular direction.

Threshold effects or tipping points are commonly used in epidemiology. For example cutpoints for obesity, hyperlipidemia, and blood pressure are in part based on non-linear thresholds at which disease risk begins to rise at a faster rate [27]. In behavioral terms, the tipping point refers to the threshold at which individuals or groups of individuals adopt a particular idea or practice. Relating this to the obesity epidemic for example, there may be a societal tipping point at which a large percent of the population decides to alter their diet and activity patterns. A recent tipping point occurred in 2004–2005 when as much as 15% of the US population had tried the Atkins diet or some other low carbohydrate regimen [28], despite little scientific evidence demonstrating effectiveness [29–31]. Such non linear shifts have also occurred in the prevalence of smoking and illicit drug use [32,33]. However, they are difficult to predict let alone cause. It is important to note that the chaotic perspective of behavior change offered here focuses mostly on the individual intrapsychic dimension. Environmental factors such cost, availability, legal restrictions etc. also interact with intrapsychic determinants. In some cases, environmental determinants can overwhelm system constraints. For example, raising cigarette taxes by several dollars per pack, has a suppressing impact on individual smoking behaviors, whereas lack of availability of fruits and vegetables can constrain dietary choices.

Resistance to Chaos

Accepting randomness as a primal determinant of human behavior may be contrary to the deterministic view characteristic of western thought. Randomness may conflict with an innate tendency for humans to infer causality and a need for predictability. For example, when a punter wins the lottery, a completely random event, many individuals will assume that the winner used some replicable strategy that led to them to "earn" their prize or that some higher order "kharma" deemed the winner worthy. Accepting randomness requires that we relinquish the faith that reward and punishment; fortune and misfortune are doled out in an orderly, just fashion. Perhaps not surprisingly, Chaos theory and non-linear dynamics have met considerable resistance within the scientific community [18]. For public health professionals it requires a new conceptualization of health behavior as well as how and why we influence change.

In the complex system approach, the role of health communications may be analogous to the spinning of ping pong balls in a lottery machine. Say that each ping pong ball represents a chunk of knowledge, attitude, efficacy, or intention. On each ball lies a few strips of Velcro; the soft side. Inside the human psyche lies strips of the opposite, hard side of Velcro, which serve as potential motivational "receptors". Some of the motivational ping pong balls may have resided in the system for years while others may have been more recently implanted through a health education program, clinical counseling encounter, or health communication campaign. Rather than attempting to predict which piece or pieces of motivation may "tip" the individual, from the chaotic perspective, the role of the health professional is to ensure the balls are kept spinning at various intervals and velocities to maximize the chances that they adhere to their receptors. When sufficient balls have adhered a tipping point may occur. Which balls or combination of balls may trip the motivational switch as well as when and why they may stick, are chaotic events that defy accurate prediction. From a non-linear perspective, the goal of health professionals may be to encourage wing flapping.

The linear and chaotic paradigms are not necessarily mutually exclusive. Behavior change includes both chaotic and rational processes. As shown in the figure below, the Cognitive-Planned and Chaotic-Quantum aspects of motivation can be placed along a continuum. The continuum may be seen as a framework to both classify motivational styles (across individuals) or behavioral decisions (within individuals).

Some individuals may by their nature be prone to employ rationale decision making processes typically associated with left hemispheric function. On the other hand some may be more predisposed to quantum processes where change is more dramatic and less planned. Most individuals are likely influenced by both linear and quantum processes, perhaps depending on mood or other initial conditions. Another way to conceptualize the interaction of linear and quantum processes is that cognitive-rational factors may provide the fertile soil on which chaotic events may sprout. Thus, health promotion may be viewed as priming individuals so that when chaotic environmental or intrapsychic events occur, they have a greater likelihood of taking root. Whether individuals possess a predisposition to either style is an important issue with considerable implications for health communications. If valid, one implication is that program planners may need to tailor intervention content and delivery to match individual cognitive/motivational styles. Whereas quantum processes may be more operative at initiation of change, it is possible that cognitive-rational processes may be more relevant to maintenance of behavior change.

Summary and Implications for Practice and Future Research

The random component of health behavior change, though difficult to predict or control, can nonetheless be incorporated into practice and research. For example, using the "perfect storm" analogy, it may be important to provide individuals with periodic interventions so that the motivational ping pong balls are spun under varying "atmospheric" (i.e., psychologic and/or life circumstances) conditions. Periodic exposure is consistent with the approach used in many chronic disease management programs. Such program, from this new perspective can be viewed as providing repeated opportunities to produce the motivational storm. This approach is also consistent with counseling models such as motivational interviewing, which provide clients with considerable opportunity to explore life with and without their risk behavior; that is to spin the balls [34,35].

Another implication is that individually tailored interventions may be particularly promising as a means to maximize the likelihood of a perfect motivational storm [36–40]. Individually tailored communications increase both receiver attention and message salience, which together increase the chances that the "balls" are spun and that they have a optimal chance of sticking. There are also statistical implications. The potential variance in behavior accounted for by traditional cognitive factors should perhaps be assumed to have an upper limit far below 100%. Given prior studies, a reasonable upper limit may be in the 50% range. And rather than assuming unaccounted for variance simply reflects "error", non-linear models could be used to explore alternative mathematical relationships. And although the relationship of predictor variables may be complex and non-linear, there may be identifiable patterns, i.e., fractals in the parlance of chaos theory, that manifest across individuals that would allow for sophisticated audience segmentation and potentially powerfully tailored interventions.

We are not proposing that linear statistical models and linear-based health promotion interventions are of no value and need be discarded entirely. There is a vast scientific base indicating that our interventions can successfully change behavior. What we are proposing, however is that we need to rethink why our interventions work and for whom. Group interventions, we propose, work because they have spun the "balls" of motivation (or deactivated barrier balls) in a large group of individuals, and for a subset of these individuals the balls fit their motivational receptors and other psychologic and biologic settings. It is important to note that current theories and communication methodologies can greatly inform which "balls" we select to highlight in our interventions. Motivation is not random. Tailoring motivational messages to the audience remains a critical step in chieving positive outcomes, and our current theories can help select the most effective set of balls.

Additionally, although patterns of change likely follow unique, i.e., chaotic, patterns across individuals, it may nonetheless may be useful to know that, in aggregate, balls that have similar characteristic profiles tend to "pool" in a defined geographic area once rolled down the metaphorical intervention mountain, helping us to perhaps understand which ping pong balls to keep circulating and for whom. That is, there may be common pathways to change based on individual parameters that can be used to develop sophisticated audience segmenta ution analyses and more effective interventions that account for the chaotic element of change. A "mixture model" of both chaotic and linear progression may be one that helps us best understand change.

The proposition that a significant proportion of human behavior operates from a chaotic perspective, at first blush, may appear to defy empirical verification. However, with the advent of technologies such as Functional Magnetic Resonance Imaging (fMRI) and momentary psychologic assessment, it may be possible to examine where, neurologically different types of motivation arise, and even predict when and why quantum transformations occur. Theoretical and statistical research examining behavior change from a quantum perspective is encouraged. In particular, the degree to which transformational motivation observed in the addiction field operates in the nutrition and physical activity domains, and whether changes spurred by inspiration are more enduring than changes arrived at from the more cognitive, conscious pathway merits examination.

References

1. Glanz K., Rimer B.K. and Lewis F.M.: **Health Behavior and Health Education.** 3rd edition. San Fransisco, John Wiley & Sons; 2002.
2. Baranowski T., Cullen K.W., Nicklas T., Thompson D. and Baranowski J.: **Are current health behavioral change models helpful in guiding prevention of weight gain efforts?** *Obesity Research* 2003, **11 Suppl:** 23S–43S.
3. Godin G. and Kok G: **The theory of planned behavior: a review of its applications to health-related behaviors.** *American Journal of Health Promotion* 1996, **11:** 87–98.
4. Sjoberg S., Kim K. and Reicks M.: **Applying the theory of planned behavior to fruit and vegetable consumption by older adults.** *Journal of Nutrition for the Elderly* 2004, **23:** 35–46.
5. Astrom A.N.: **Validity of cognitive predictors of adolescent sugar snack consumption.** *American Journal of Health Behavior* 2004, **28:** 112–121.
6. Conn V.S., Burks K.J., Pomeroy S.H., Ulbrich S.L. and Cochran J.E.: **Older women and exercise: explanatory concepts.** *Womens Health Issues* 2003, **13:** 158–166.
7. Masalu J.R. and Astrom A.N.: **The use of the theory of planned behavior to explore beliefs about sugar restriction.** *American Journal of Health Behavior* 2003, **27:** 15–24.
8. Trost S.G., Pate R.R., Dowda M., Ward D.S., Felton G. and Saunders R.: **Psychosocial correlates of physical activity in white and African-American girls.** *Journal of Adolescent Health* 2002, **31:** 226–233.
9. Lien N., Lytle L.A. and Komro K.A.: **Applying theory of planned behavior to fruit and vegetable consumption of young adolescents.** *American Journal of Health Promotion* 2002, **16:** 189–197.
10. Kerner M.S. and Grossman A.H.: **Attitudinal, social, and practical correlates to fitness behavior: a test of the theory of planned behavior.** *Perceptual & Motor Skills* 1998, **87:** 1139–1154.
11. Conner M., Norman P. and Bell R.: **The theory of planned behavior and healthy eating.** *Health Psychology* 2002, **21:** 194–201.
12. Baranowski T., Lin L., Wetter D.W., Resnicow K. and Davis M.: **Theory as mediating variables: Why aren't community interventions working as desired?** Annals of Epidemiology 1997, **7:** 89–95.
13. Miller W.R.: **The phenomenon of quantum change.** *Journal of Clinical Psychology* 2004, **60:** 453–460.
14. Miller W.R. and C'De Baca J.: **Quantum change: When epiphanies and sudden insights transform ordinary lives.**, New York, NY, US: Guilford Press.; 2001:212.
15. Behrens D.A., Caulkins J.P. and Feichtinger G.: **A model of chaotic drug markets and their control.** Nonlinear Dynamics, Psychology, & Life Sciences 2004, **8:** 375–401.
16. Sprott J.C.: **Dynamical models of happiness.** *Nonlinear Dynamics, Psychology, & Life Sciences* 2005, **9:** 23–36.
17. Warren K., Hawkins R.C. and Sprott J.C.: **Substance abuse as a dynamical disease: evidence and clinical implications of nonlinearity in a time series of daily alcohol consumption.** *Addictive Behaviors* 2003, **28:** 369–374.
18. Glieck J.: **Chaos: Making a New Science.** New York, Penguin 1987.
19. Stewart I.: **Does God Play Dice?: The Mathematics of Chaos.** Cambridge MA, Blackwell 1989.
20. Rosenstock I.: **Social learning theory and the health belief model.** *Health Education Quarterly* 1988, **15:** 175–183.
21. Janz N.K. and Becker M.H.: **The health belief model: A decade later.** *Health Educaiton Quarterly* 1984, **11:** 1–47.

22. West R. and Sohal T.: **"Catastrophic" pathways to smoking cessation: findings from national survey.** *BMJ* 2006, **332:** 458–460.

23. Larabie LC: **To what extent do smokers plan quit attempts?** *Tobacco Control* 2005, **14:** 425–428.

24. Matzger H., Kaskutas L.A. and Weisner C.: **Reasons for drinking less and their relationship to sustained remission from problem drinking** [see comment]. *Addiction* 2005, **100:** 1637–1646.

25. Gladwell M.: **The Tipping Point**. Boston, Little, Brown, and Company 2000.

26. Malkiel B.: **A Random Walk Down Wall Street: Completely Revised and Updated**. 8th edition. New York, WW Norton 2003.

27. Norris J.C., van der Laan M.J., Lane S., Anderson J.N. and Block G.: **Nonlinearity in demographic and behavioral determinants of morbidity**. *Health Services Research* 2003, **38:** 1791–1818.

28. Opiniondynamics: **What Happened to the Low-Carb Craze?** 2005.

29. Foster G.D., Wyatt H.R., Hill J.O., McGuckin B.G., Brill C., Mohammed B.S., Szapary P.O., Rader D.J., Edman J.S. and Klein S.: **A randomized trial of a low-carbohydrate diet for obesity** [see comment]. *New England Journal of Medicine* 2003, **348:** 2082–2090.

30. Kushner R.F.: **Low-carbohydrate diets, con: the mythical phoenix or credible science?** *Nutrition in Clinical Practice* 2005, **20:** 13–16.

31. Truby H., Baic S., deLooy A., Fox K.R., Livingstone M.B., Logan C.M., Macdonald I.A., Morgan L.M., Taylor M.A. and Millward D.J.: **Randomised controlled trial of four commercial weight loss programmes in the UK: initial findings from the BBC "diet trials"**. *BMJ* 2006, **332:** 1309–1314.

32. Centers for Disease C, Prevention: **Cigarette smoking among adults—United States**, 2003. *MMWR— Morbidity & Mortality Weekly Report* 2005, **54:** 509–513.

33. Sloboda Z.: **Changing patterns of "drug abuse" in the United States: connecting findings from macro- and microepidemiologic studies**. *Substance Use & Misuse* 2002, **37:** 1229–1251.

34. Resnicow K., DiIorio C., Soet J.E., Borrelli B., Ernst D., Hecht J. and Thevos A.: **Motivational Interviewing in medical and public health settings. In** *Motivational interviewing: Preparing people for change* 2nd edition. Edited by: Miller W and Rollnick S. New York, Guildford Press 2002: 251–269.

35. Resnicow K., DiIorio C., Soet J.E., Ernst D., Borrelli B. and Hecht J.: **Motivational interviewing in health promotion: it sounds like something is changing**. *Health Psychology* 2002, **21:** 444–451.

36. Strecher V.J.: **Computer-tailored smoking cessation materials: A review and discussion**. *Pat Educ Couns* 1999, **36:** 107–117.

37. Brug J. and van Assema P.: **Differences in use and impact of computer-tailored dietary fat-feedback according to stage of change and education**. *Appetite* 2000, **34:** 285–293.

38. Oenema A., Brug J. and Lechner L.: **Web-based tailored nutrition education: results of a randomized controlled trial**. *Health Education Research* 2001, **16:** 647–660.

39. Kreuter M.W. and Wray R.J.: **Tailored and targeted health communication: strategies for enhancing information relevance**. *American Journal of Health Behavior* 2003, 27 Suppl **3:** S227–32.

40. Brug J., Oenema A. and Campbell M.: **Past, present, and future of computer-tailored nutrition education**. *Am J Clin Nutr* 2003, **77:** 1028S–1034S.m of Motivational Processes.

Chapter 2

Crisis and Chaos in Behavioral Nutrition and Physical Activity [†]

Tom Baranowski

Children's Nutrition Research Center, Department of Pediatrics, Baylor College of Medicine, 1100 Bates St., Houston, TX 77030, USA. Email: tbaranow@bcm.tmc.edu.

ABSTRACT

Resnicow & Vaughn challenged the field of behavioral nutrition and physical activity to conduct research in new ways. They challenged the predictiveness of our models, sensitivity to initial conditions, factors predisposing to change and measurement procedures. While the predictiveness of our models will reflect the sophistication of our thinking and research, and the sensitivity to initial conditions is subsumed under the sophistication of our models, research on conditions predisposing to change (e.g., epiphanies), more longitudinal designs, refined measurement procedures and testing of critical issues can only enhance the quality of our research. Improved research quality should lead to enhanced efficacy and effectiveness of our interventions, and thereby our making meaningful contributions to mitigating the chaos in our field and the crisis from the rising epidemic of obesity.

Background

Our field of behavioral nutrition and physical activity should be operating in crisis mode. The prevalence of obesity and overweight (an essentially nutrition and physical activity problem in its etiology and control) continues to increase at alarming rates in all age, demographic and gender groups in the US [1], Europe [2], and many other parts of the world [3,4]. There is concern that this will reverse the recent advances in chronic disease control [5]. In the face of this encroaching epidemic, obesity treatment programs have tended to have weak effects mostly for short periods of time [6]; and review after review have shown that obesity prevention programs also tend not to work [7–9]. Furthermore,

[†]Reused with permission from: Tom Baranowski, "Crisis and chaos in behavioral nutrition and physical activity," International *Journal of Behavioral Nutrition and Physical Activity,* 2006, vol. 3:27, doi:3/1/27. © 2006 Baranowski; licensee BioMed Central Ltd.

using the mediating variable model (see Fig.1) as a structured framework, it is not clear we know what changes in diet or physical activity behaviour have led to the current problems and thereby provide the best behavioral targets for change [link A in Fig.1] [10,11]; nor what mediating variables are most strongly related to these behaviors and thereby provide the best mechanisms for change [link B in Fig.1] [12]; nor how best to manipulate the mediating variables to obtain behavior change and lower obesity [link C in Fig.1] [13]. This is a frightful state of affairs. We should all be doing innovative theoretically guided, but high risk, research to quickly build a stronger knowledge base from which more effective interventions could be crafted. Yet, most of us appear to be acting in our usual way of doing things: "same old, same old,"

In this context, Resnicow & Vaughn [14] challenged our "same old" way of thinking about our field. They correctly specified the assumption of linearity in our predictive models, and proposed Chaos and Dynamic Systems Theories as iterative nonlinear models. They did not throw out all our theories per se, but challenged how we interrelated the variables, how we related them to behaviors, and offered some new variables predisposing to change. While Glass & McAtee [15] recently pointed out deficiencies in the social dimensions of our research, Resnicow & Vaughn targeted our thinking about behavior change. Some of the issues they raised are non-issues, but others deserve that we morph our basic methods to test the new ideas.

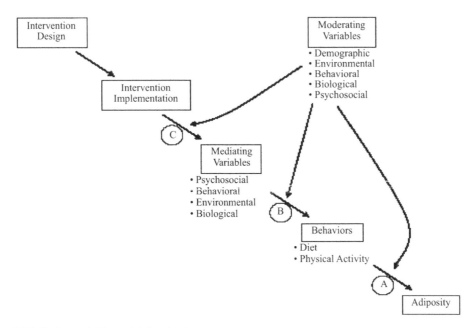

Figure 1 Mediating variable model for obesity.

Discussion

Nonissues

Since we use statistical methods, all our models are by definition probabilistic, rather than deterministic (unless we could account for 100% of the variance, which will not happen in our lifetimes).

A key issue in their article was the predictiveness of our current models. They correctly identified the very limited predictiveness of the current models. The key issue, however, is the level of predictiveness that could possibly be achieved in predicting behavior. For example, our biological research colleagues are not satisfied unless their models account for 90% or more of the variance in their phenomena of interest. We are well below that [16]. Resnicow & Vaughn have not taken into account the emerging research on environmental influences, e.g., home availability [17], neighbourhood characteristics [18,19]; biological influences, e.g., genes [20], sensitivity to tastes [21], the hormone rages of adolescent development [22,23]; emotional influences [23,24]; nor the likely interrelationships and interactions among these variables and our more usual psychosocial and behavioral predictors [25,26]. The higher the predictiveness of our models, the more we can engage in our logical approach to designing interventions based on these models. The larger number of and more diverse variables incorporated into these models, the more complex our interventions will need to be to address components of the model. And the interventions will need to both segment the population for differing types of interventions to different gender, age, ethnic, socioeconomic, and/or neighbourhood groups, and tailor the intervention to individual characteristics within these groups [27]. At this time, we need to build and test the more comprehensive models. This is a daunting, but exciting, challenge.

Resnicow & Vaughn proposed the principal of sensitivity to initial conditions, as if this were a new idea. All of our models of longitudinal relationships (as equations) have built into them sensitivity to initial conditions, i.e., the initial values of the variables. How diverse the outcomes depend on the nature of the relationships. As our models become more comprehensive and complex, fairly similar initial conditions could lead to quite divergent outcomes. In part this is a function of the sophistication of our knowledge base. We need to build more sophisticated predictive models.

The idea of a tipping point or when it might occur, is not well defined [28]. In some ways it reifies a change, as if there is something intrinsic to or magical about the change process. If a tipping point is nothing more than a critical point on a variable beyond which change occurs, it is not clear the concept adds much, but identifying those points would be helpful.

Issues Deserving Intensive Research

Investigators could take away from the Resnicow & Vaughn message that change is random and cannot be predicted, and thereby cannot be understood by our usual research methods on behavior or its change. This would be very unfortunate. Resnicow & Vaughn will have made a major contribution, only if it leads to innovative research and new insights.

Even in the vast complexities of molecular science, investigators are hammering away at delineating linear and nonlinear patterns to better understand the biology. Chances are we can do the same in behavior research.

Resnicow & Vaughn proposed that change does not occur in a linear "persuasion slowly overcoming resistance" manner, but rather in what they characterized as "quantum leaps," i.e., an epiphany or "aha!" event occurs from which the person decides to change. This is an interesting idea and should be testable. Innovative methods will be needed to identify people soon after the aha! experience to learn more about it. Perhaps interviewing new recruits to Weight Watchers™ or to fitness centers would accomplish this? A related issue would be what could we do to encourage aha! experiences? Are they a response to an overload of information (probably not, since we have done a lot of this already)? To repeated thinking about the issues (we could rogram prompts to thinking)? To setting off some emotional experience related to the behaviour (we might be able to tailor messages to issues people found emotionally charged)? Resnicow & Vaughn invoked the concept of "cues" from the Health Belief Model. There has been some research on cues [29–31], but this has not as yet led to substantial insights. Relating cues to aha! experiences could be an important avenue for research. Developing valid and reliable retrospective methods to identify and recall aha! experiences would be necessary to make much progress.

Resnicow & Vaughn correctly pointed out the cross sectional nature of most of our research. Dynamic Systems modelling proposes that dynamic research be done, and this would be focused on change over time which requires longitudinal designs [32]. The importance of longitudinal designs was emphasized when Nigg [33] found that physical activity predicted ensuing self efficacy, but not the other way around. If self efficacy is really caused by physical activity, but doesn't cause physical activity, it doesn't make sense to try to increase self efficacy in interventions.

While it is challenging to recruit and maintain longitudinal cohorts, such cohorts are required to address issues of direction of causality and thereby which variables should be targets for change in intervention programs. While ten year cohorts may not be necessary, perhaps 3 mo or 6 mo cohorts would provide tests for the changes we need. Longitudinal dynamic systems research has been initiated in other fields [34,35], which should provide a guide for our further development.

Whether behavior change can only be understood in retrospect instead of prospectively is an empirical issue. In part this is a function of how much variance our models will ultimately predict. Perhaps a few retrospective analyses will be necessary, perhaps using qualitative methods, to map out the processes occurring? But predictive science should be where we are headed, since predictive relationships clearly demonstrate what we know.

Resnicow & Vaughn correctly identified our current approaches to measurement as providing severe limitations to how we could understand our phenomena of interest. There have been limits on the extent to which existing measurement methods (e.g., classical test theory) have been used and reported [36], and limits on the predictiveness of existing measurement models [37]. One innovation in measurement theory that has recently drawn attention is Item Response Theory (IRT) [38]. IRT fits latent variables to items (and respondents) which identifies portions of the underlying variable being poorly measured [39], and assesses reliability across the range of the underlying variable [39,40]. Having items measuring specific locations on the underlying variable permits an assessment of whether the measures work differently after participation in an experiment [41]; differ by ethnic, gender or other groupings; and permit more efficient multidimensional modelling of the variable [42]. Use of IRT offers great promise for better understanding and minimizing the problems due to measurement of our constructs, and deserves much wider use.

An issue Resnicow addressed in his oral presentation in Boston (but not in his paper) was the falsifiability of a theory, and whether our current cognitive models are really theories. He correctly stated that in our current approach to research, no theories have been discarded (which would be considered a sign of progress and development in a field). Our best current research fits multivariate models to sets of variables [43], and determines which variables were significantly related to other variables in the model. This is useful for assessing the predictiveness of particular variables in certain situations, but does not necessarily address the usefulness of the larger theory. To move our field forward we need more attention to theoretical issues in our research, tests of clearer more specific predictions from theory applied to particular issues [44,45], and delineation of "critical issues" where two theories would make different predictions or model fitting research would need to test the fit of competing models [24], where the alternative models were predicated on different theories. More highly controlled experimental research on critical issues will also be necessary. Accumulation of findings across "critical" studies would enable the field to find more comprehensive and more predictive theoretical frameworks, and capitalize upon them in more likely to be effective interventions. There has been distaste for theory in our field [46], and some have proposed continuing conducting intervention research until randomly hitting on intervention procedures that work [47].

Alternatively, I believe highly predictive theory should guide the design of effective interventions. In a complicated set of many possible variables and relationships, a random search may never result in finding effective change techniques, and even if it did, we wouldn't have the conceptual framework to understand why it happened in order to exploit it.

Conclusion

Resnicow & Vaughn challenged the field of behavioural nutrition and physical activity to conduct research in new ways. While the predictiveness of our models will reflect the sophistication of our thinking and research, and the sensitivity to initial conditions is subsumed under the sophistication of our models, research on conditions predisposing to change (e.g., epiphanies), more longitudinal designs, refined measurement procedures and testing of critical issues can only enhance the quality of our research. Improved research quality should lead to enhanced efficacy and effectiveness of our interventions, and thereby our making meaningful contributions to mitigating the chaos in our field and the crisis from the rising epidemic of obesity.

Abbreviations

US   =   United States
IRT   =   Item Response Theory

Competing interests

The author(s) declare that they have no competing interests.

Acknowledgements

This work is in part a publication of the United States Department of Agriculture (USDA/ARS) Children's Nutrition Research Center, Department of Pediatrics, Baylor College of Medicine, Houston, Texas, and was funded in part with federal funds from the USDA/ARS under Cooperative Agreement No. 58-6250-6001. The contents of this publication do not necessarily reflect the views or policies of the USDA, nor does mention of trade names, commercial products, or organizations imply endorsement from the U.S. government.

References

1. Ogden C., Carroll M., Curtin L., McDowell M., Tabak C. and Flegal K.: **Prevalence of overweight in the United States, 1999–2004.** *JAMA* 2006, **295(13):** 1549–55.
2. Bostrom G. and Eliasson M.: **Chapter 5.3: Major public health problems—overweight and obesity.** *Scan J Public Health Suppl* 2006, **Jun(67):** 69–77.
3. Cheng T.O.: **Obesity is a global challenge.** *Am J Med* 2006, **119(6):** e11.
4. Malecka-Tendera E. and Mazur A.: **Childhood obesity: a pandemic of the twenty-first century.** *Int J Obes (Lond)* 2006: S1–3.
5. Jones A.P., Homer J.B., Murphy D.L., Essien J.D., Milstein B. and Seville D.A.: **Understanding diabetes population dynamics through simulation modeling and experimentation.** *Am J Public Health* 2006, **96(3):** 488–94.
6. Avenell A., Broom J., Brown T.J., Poobalan A., Aucott L., Stearns S.C., Smith W.C., Jung R.T., Campbell M.K. and Grant A.M.: **Systematic review of the long-term effects and economic consequences of treatments for obesity and implications for health improvement.** *Health Technol Assess* 2004, **8(21):** iii–iv. 1–182.
7. Flynn M.A., McNeil D.A., Maloff B., Mutasingwa D., Wu M., Ford C. and Tough S.C.: **Reducing obesity and related chronic disease risk in children and youth: a synthesis of evidence with 'best practice' recommendations.** *Obes Rev* 2006: 7–66.
8. Baranowski T., Cullen K.W., Nicklas T., Thompson D. and Baranowski J.: **School-based obesity prevention: A blueprint for taming the epidemic.** *Am J Health Behav* 2002, **26(6):** 486–93.
9. Flodmark C.E., Marcus C. and Britton M.: **Interventions to prevent obesity in children and adolescents: a systematic literature review.** *Int J Obes (Lond)* 2006, **30(4):** 579–89.
10. Bachman C.M., Baranowski T. and Nicklas T.A.: **Is there an association between sweetened beverages and adiposity?** *Nutr Rev* 2006, **64(4):** 153–74.
11. Nicklas T., Yang S.J., Baranowski T., Zakeri I. and Berenson G.S.: **Eating patterns and obesity in children: The Bogalusa Heart Study.** *Am J Prev Med* 2003, **25(1):** 9–16.
12. Baranowski T., Anderson C. and Carmack C.: **Mediating variable framework in physical activity interventions. How are we doing? How might we do better?** *Am J Prev Med* 1998, **15(4):** 266–97.
13. Baranowski T., Lin L.S., Wetter D.W., Resnicow K. and Hearn M.D.: **Theory as mediating variables: why aren't community interventions working as desired?** *Annals of Epidemiology* 1997, **7:** S89–S95.

14. Resnicow K. and Vaughn: **A chaotic view of behavior change: a quantum leap for health promotion.** *IJBNPA* in press.

15. Glass T.A. and McAtee M.J.: **Behavioral science at the crossroads in public health: extending horizons, envisioning the future.** *Soc Sci Med* 2006, **62(7):** 1650–71.

16. Baranowski T., Cullen K.W. and Baranowski J.: **Psychosocial correlates of dietary intake: Advancing intervention.** *Annual Review of Public Health* 1999, **19:** 17–40.

17. Jago R. and Baranowski T.: **Home food availability: A microenvironmental mediating variable?** *Public Health Nutrition* 2006 in press.

18. van der Horst K., Oenema A., Ferreira I., Wendel-Vos W., Giskes K., van Lenthe F. and Brug J.: **A systematic review of environmental correlates of obesity-related dietary behaviors in youth.** *Health Educ Res* 2006.

19. Jago R., Baranowski T. and Baranowski J.C.: **Observed, GIS, and selfreported environmental features and adolescent physical activity.** *Am J Health Promot* 2006, **20(6):** 422–8.

20. Rankinen T. and Bouchard C.: **Genetics of food intake and eating behavior phenotypes in humans.** *Annu Rev Nutr* 2006, **26:** 413–34.

21. Goldstein G.L. and Daun H., Tepper B.J.: **Adiposity in middle-aged women is associated with genetic taste blindness to 6-n-propylthiouracil.** *Obes Res* 2005, **13(6):** 1017–23.

22. Himes J.H., Obarzanek E., Baranowski T., Wilson D.M., Rochon J. and McClanahan B.S.: **Early sexual maturation, body composition, and obesity in African American girls.** *Obes Res* 2004: 64S–72S.

23. Richardson L.P., Garrison M.M., Drangsholt M., Mancl L. and LeResche L.: **Associations between depressive symptoms and obesity during puberty.** *Gen Hosp Psychiatry* 2006, **28(4):** 313–20.

24. Taylor S.D., Bagozzi R.P. and Gaither C.A.: **Decision making and effort in the self-regulation of hypertension: testing two competing theories.** *Br J Health Psychol* 2005, **10(Pt 4):** 505–30.

25. Kremers S.P., de Bruijn G.J., Visscher T.L., van Mechelen W., deVries N.K. and Brug J.: **Environmental influences on energy balance-related behaviors: A dual-process view.** *Int J Behav Nutr Phys Act* 2006, **3:** 9.

26. Cullen K.W., Baranowski T., Owens E., Marsh T., Rittenberry L. and de Moor C.: **Availability, accessibility and preferences for fruit, 100% juice and vegetables influence children's dietary behavior.** *Health Education & Behavior* 2003, **30:** 615–626.

27. Kroeze W., Werkman A. and Brug J.: **A systematic review of randomized trials on the effectiveness of computer-tailored education on physical activity and dietary behaviors.** *Ann Behav Med* 2006, **31(3):** 205–23.

28. Gladwell M.: **The Tipping Point—How Little Things Can Make a Big Difference.** New York, NY: Back Bay Books; 2002.

29. Jones T., Fowler M.C. and Hubbard D.: **Refining a tool to measure cues to action in encouraging health-promoting behavior-The CHAQ.** *American Journal of Health Promotion* 2000, **14:** 170–173.

30. Strychar I.M., Champagne F., Ghadirian P., Bonin A., Jenicek M. and Lasater T.M.: **Impact of receiving blood cholesterol test results on dietary change.** *American Journal of Preventive Medicine* 1998, **14:** 103–110.

31. Kip K.E., McCreath H.E., Roseman J.M., Hulley S.B. and Schreiner P.J.: **Absence of risk factor change in young adults after family heart attack or stroke.** *American Journal of Preventive Medicine* 2002, **22:** 258–266.

32. Howe M.L. and Lewis M.D.: **The importance of dynamic systems approaches for understanding development.** *Developmental Review* 2005, **25(3–4):** 247–251.

33. Nigg C.R.: **Explaining adolescent exercise behavior change; A longitudinal application of the transtheoretical model.** *Annals of Behavioral Medicine* 2001, **23:** 11–20.

34. van Geert P. and Steenbeek H.: **Explaining after by before: Basic aspects of a dynamic systems approach to the study of development.** *Developmental Review* 2005, **25 (3–4):** 408–442.

35. Smith L.B.: **Cognition as a dynamic system: Principles from embodiment.** *Developmental Review* 2005, **25(3–4):** 278–298.

36. Baranowski T., Klesges L.M., Cullen K.W. and Himes J.: **Measurement of outcomes, mediators and moderators in behavioral obesity prevention.** *Preventive Medicine* 2004, **38**: S1–S13.
37. Wilson M., Allen D. and Li H.: **Comparing item response modelling with the Classical Test Theory approach in measurement in the behavioral sciences.** *Health Educ Res* 2007 in press.
38. Wilson M.: **Constructing Measures: An Item Response Modeling Approach.** Mahwah, N.J.: Erlbaum 2005.
39. Watson K., Thompson D. and Baranowski T.: **Item response theory evaluation of the children's fruit and vegetable self efficacy scale.** *Health Education Research* 2007 in press.
40. Wilson M., Allen D.D. and Li J.C.: **Improving measurement in behavioural sciences using item response models: comparison with the classical test theory approach.** *Health Educ Res* . 2006: Jul 31.
41. Baranowski T., Allen D., Masse L. and Wilson M.: **Does participation in an intervention affect responses on self report questionnaires?** *Health Educ Res* 2006 in press.
42. Watson K., Baranowski T., Thompson D., Jago R., Baranowski J. and Klesges L.: **Innovative Application of a Multidimensional Item Response Model In Assessing the Influence of Social Desirability on the Pseudo-Relationship between Self-Efficacy and Behavior.** *Health Educ Res* in press.
43. Motl R.W., Dishman R.K., Saunders R.P., Dowda M. and Pate R.R.: **Perceptions of Physical and Social Environment Variables and Self-Efficacy as Correlates of Self-Reported Physical Activity Among Adolescent Girls.** *J Pediatr Psychol* . 2006: May 17.
44. Resnicow K. and Baranowski T.: **Are pre-contemplators less likely to change their dietary behavior? A prospective analysis.** *Health Education Research* 2003, **18(6):** 693–705.
45. Weinstein N.D., Sutton S. and Rothman A.J.: **Stage theories of health behavior: Conceptual and methodological issues.** *Health Psychology* 1998, **17**: 290–299.
46. Jeffery R.W.: **How can Health Behavior Theory be made more useful for intervention research?** *Int J Behav Nutr Phys Act* 2004, **1(1):** 10.
47. Robinson T.N. and Sirard J.R.: **Preventing childhood obesity: a solution-oriented research paradigm.** *Am J Prev Med* 2005, **28 (2Suppl 2):** 194–201.

Chapter 3

Memory in Astrocytes: A Hypothesis [†]

Robert M. Caudle

Department of Oral and Maxillofacial Surgery and Diagnostic Sciences, Universit of Florida College of Dentistry, Gainesville, Florida 32610, USA. Email: Caudle@ufl.edu.

ABSTRACT

Background: Recent work has indicated an increasingly complex role for astrocytes in the central nervous system. Astrocytes are now known to exchange information with neurons at synaptic junctions and to alter the information processing capabilities of the neurons. As an extension of this trend a hypothesis was proposed that astrocytes function to store information. To explore this idea the ion channels in biological membranes were compared to models known as cellular automata. These comparisons were made to test the hypothesis that ion channels in the membranes of astrocytes form a dynamic information storage device.

Results: Two dimensional cellular automata were found to behave similarly to ion channels in a membrane when they function at the boundary between order and chaos. The length of time information is stored in this class of cellular automata is exponentially related to the number of units. Therefore the length of time biological ion channels store information was plotted versus the estimated number of ion channels in the tissue. This analysis indicates that there is an exponential relationship between memory and the number of ion channels. Extrapolation of this relationship to the estimated number of ion channels in the astrocytes of a human brain indicates that memory can be stored in this system for an entire life span. Interestingly, this information is not affixed to any physical structure, but is stored as an organization of the activity of the ion channels. Further analysis of two dimensional cellular automata also demonstrates that these systems have both associative and temporal memory capabilities.

Conclusion: It is concluded that astrocytes may serve as a dynamic information sink for neurons. The memory in the astrocytes is stored by organizing the activity of ion channels and is not associated with a physical location such as a synapse. In order for this form of memory to be of significant duration it is necessary that the ion channels in the astrocyte syncytium be electrically in contact with each other. This function may be served by astrocyte gap junctions and suggests that agents that selectively block these gap junctions should disrupt memory.

[†]Reused with permission from: R.M. Caudle, Memory in astrocytes: a hypothesis, *Theoretical Biology and Medical Modelling* 2006, 3: 2, 1–10.

Background

Until recently astrocytes were considered to play no more than a supportive role for neurons in the central nervous system. This view has now been supplanted by a more active participation of astrocytes in information processing, where the astrocytes not only receive and respond to neuronal input, but also transmit signals to neurons [1–9]. These findings indicate that astrocytes contribute to the processing of information. In support of this concept it was recently demonstrated that spinal cord astrocytes are necessary to support hyperalgesia produced by peripheral injury [10–12]. Blocking gap junctions in the astrocytes suppressed hyperalgesia, which suggested that the astrocytes were processing the nociceptive information and regulating the function of spinal cord neurons [10]. These results are similar to work reported by Hertz et al. and Ng et al. who demonstrated that astrocytes are critical for the establishment of learned behaviors [13,14]. Furthermore, recent studies indicate that several general anesthetics suppress the function of astrocyte gap junctions at concentrations that are relevant for loss of consciousness [15,16]. These data suggest that the anesthetic properties of these agents may be mediated at least in part by their actions on astrocytes and may indicate some role for astrocytes in consciousness.

In a recent review Robertson outlined an astrocentric hypothesis of memory [17] as an alternative to the current neurocentric or synaptic based theories. In this hypothesis Robertson concludes that because astrocytes form large syncytium via gap junctions and that they are connected to neurons through synapses these cells can store and "bind" diverse information.

In this intriguing review Robertson hypothesizes that information is stored as a result of gap junctional plaques converting to a crystalline configuration that is a closed, high resistance, state of the gap junctions. As a result of these altered gap junctions ion flow between astrocytes is restricted resulting in a functional memory.

In examining the idea that astrocytes might play a major role in information processing it seemed prudent to examine other potential memory mechanisms that could support information processing in astrocytes. In experiments examining electrical potentials and calcium fluxes in astrocytes it was demonstrated that these cells can, on an individual basis, support potentials for several seconds [1,2,6,7]. These data suggest that ion channel activity in a group of gap junction linked astrocytes could retain information for substantial periods of time. Thus, the ion channels mediating the astrocyte potentials could function to store and process information in the central nervous system.

This paper examines the possible role of ion channels in storing information in astrocytes.

Results and Discussion

Similarity of Ion Channels to Cellular Automata

Ion channels communicate with each other via changes in voltage, changes in calcium concentrations or through other second messenger systems. In voltage gated ion channels, for example, the rules governing the relationship between channels specify that if neighboring channels alter the local membrane potential to some threshold the channel under observation will change state, i.e., open or close. Each ion channel functions as an independent unit that monitors information transmitted from its nearest neighbors. As a result of the information processing occurring at the single ion channel level ensembles of ion channels are capable of performing relatively complex functions, such as the generation of action potentials. This form of information processing by ion channels is remarkably similar to models known as cellular automata [18,19]. In cellular automata simple units that are capable of existing in a finite number of states are linked together using rules for the transfer of information between the units. The states occupied by the units and the rules of information transfer determine what state each unit will occupy in the next time period. These models have been extensively studied and demonstrate the emergence of complex behavior [20,21]. Some cellular automata have even demonstrated universal computation [22]. To illustrate how a cellular automata stores and processes information a one dimensional cellular automaton in which the units are binary (they are either in state 0 or state 1) is presented in Fig.1. The rule used was the mean of three units rounded to the nearest integer determines the state of the middle unit in the next iteration. This model was studied at length by Wolfram and this rule is Wolfram's rule number 232 [20,21].

In Fig.1 the initiating event (Representation 1(R_1)) was produced by randomly setting the states of the units in the automata. The time series was then calculated. In the figure it is evident that from R_1 to R_4 the automaton changes representations, but after R_4 the cellular automaton reaches a steady state and the representations no longer change. This stabile representation is the attractor R_0. The transition period from R_1 to R_0 is the memory of

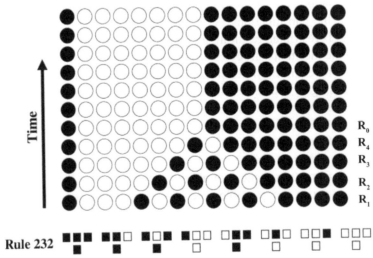

Figure 1 Memory in cellular automata. A sixteen unit one dimensional cellular automaton was constructed using binary units and Wolfram's rule number 232. This rule is illustrated at the bottom of the figure where the three squares on top are the current states of three adjacent units and the single square below is the resultant state of the middle unit during the next iteration. Open squares indicate state 0 and filled squares indicate state 1. The initial representation (R_1) was generated by randomly setting the state of each unit to either 0 (open) or 1 (filled). The time series was then calculated. Note that the memory of this system extends from R_1 to R_4 where the representations change with each iteration. Starting at R_0 the units no longer change state indicating that all information about R_1 is lost.

the automaton. At each iteration prior to R_0 the automaton retains information that can be used to determine something about the initial configuration. However, when the automaton reaches R_0 all information about the initial configuration has been lost. In astrocytes the ion channels in the membrane are distinct units with a finite number of states and they communicate with each other through a simple set of rules, i.e., a change in voltage or in Ca2+ concentration. Therefore, the astrocytes' membrane ion channels are acting as a two dimensional cellular automaton. As with the automaton presented in Fig.1 the initiating event can be inferred based on the configuration of the entire ensemble of ion channels up until the ion channel configuration returns to the attractor representation (R_0). At this point all information about the initiating event is lost. This concept suggests that ion channels working in collection can store information for at least brief periods of time. The remaining question is the maximum duration of memory in this type of system.

Memory in Cellular Automata

In a series of interesting experiments Langton examined the properties of cellular automata that optimize information storage and processing [23]. In these experiments he varied the rules by which the cellular automata operated and measured the resulting chaotic nature of the system. Langton found that automata whose rules made them operate at the junction between ordered and chaotic behavior were able to store information for the longest period of time. Memory dropped off markedly on either side of this phase transition. To illustrate how the chaotic nature of the cellular automata might influence memory a two dimensional cellular automaton with four different rule sets and a Moore neighborhood (8 neighbors) was set up (Fig. 2.A). The units in the automaton could occupy four different states, i.e., one open, one closed and two inactive. The cellular automaton was seeded with two units in the open state to invoke the initial representation R_1.

The left hand column illustrates a rule set that produces ordered behavior. Note that a signal cannot propagate in this cellular automaton. The second column demonstrates another form of ordered behavior where the behavior immediately becomes repetitive. This cellular automaton, like the one to the left of it, cannot process information due to the inability of the automaton to transition to novel representations. The third column is a rule set that produces behavior at the border between order and chaos. The net result is the smooth propagation of an "action potential" throughout the cellular automaton with the system eventually returning to the attractor representation R_0. The final column illustrates a chaotic system that evolves rapidly into a random pattern of channel openings. The nearly random behavior prevents proper processing of information since there is no relationship between successive representations. Figure 2.B illustrates the "potentials" produced by these different rule sets by plotting the number of open channels versus time. These models demonstrate that only the rule set with behavior at the transition between order and chaos produces a potential that is similar to an action potential observed in biological systems. Note that the rules that produce ordered behavior either returned to the attractor representation R_0 very rapidly or never returned to R_0, suggesting that the systems are incapable of supporting information storage. The chaotic rule set also never returns to the

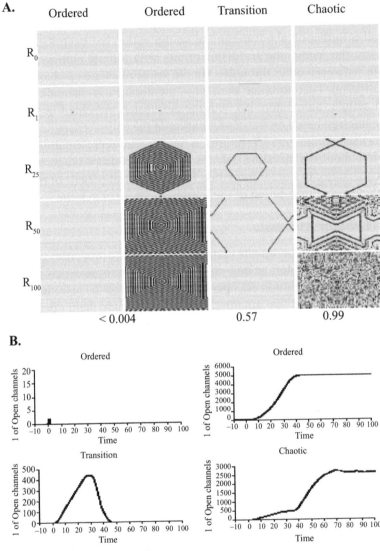

Figure 2 Two dimensional cellular automata operating between order and chaos behave like excitable membranes in biological cells. **A.** A two dimensional cellular automaton was constructed with the program CaSim using units with four states, i.e., one open, one closed and two inactive states. Four different rule sets were used to generate the four time series in the figure. The cellular automaton was seeded at R_1 by setting two units to the open state and the times series calculated. The configuration of the cellular automata at iterations 0, 1, 25, 50 and 100 are presented in the figure for the four rule sets. The entropy of the rule sets was determined by calculating the probability of each state (P_s) from 10 runs of 1000 iterations. For these calculations 10 percent of the units were set to the open state at R_1. Entropy was calculated using the equation: entropy $= -\sum P_s \ln(P_s)$. The entropy of each rule set was then expressed as a ratio of the calculated entropy to the maximum entropy (bottom of the figure). The maximum entropy is when all four states have a probability of 0.25. **B.** The "potentials" generated by the rule sets in A were graphed by plotting the number of open channels versus time. These plots indicate that only the transition rule set produces channel openings that are similar to action potentials in biological membranes.

attractor, which also indicates that the system cannot retain information for significant periods of time. Only the rule set that produced behavior between order and chaos could retain information about the initial event R_1 for a period of time and then return to the attractor representation. Based on the similarity of the potentials generated by the transition rule set these models suggest that the ion channels in the membranes of biological cells function as cellular automata with rules that set the behavior at the boundary between order and chaos. This region of the order to chaos spectrum balances information storage with transmission, which, in turn, supports information modification [23].

In addition to examining the length of memory in cellular automata relative to the chaotic nature of the automata, Langton [23] evaluated how the number of units in an automaton influenced memory. In these experiments Langton used rules that produced automata that operated in the order/chaos phase transition and then varied the number of units in the automata. He found that there was a log-linear relationship between the time that the cellular automata stored information and the number of units in the automata. This indicated that the addition of units to the automata exponentially increased the amount of time the automata stored information. This relationship is an extremely powerful property of cellular automata that has evolutionary significance for biological systems that process information with ion channels. The exponential relationship between memory and the number of units in an automaton indicates that a biological system simply has to add more units (ion channels) to its calculating device in order to dramatically increase its memory. With an increase in memory duration the complexity of the calculations that can be performed also increases [23].

The Human Cellular Automaton

The findings of Langton indicate that as a cellular automaton is increased in size the duration of memory increases. In the astrocentric hypothesis large numbers of astrocytes are connected through gap junctions [10,17,24–27], which suggests that astrocytes form extensive ion channel cellular automata. To examine the potential memory duration for a human brain sized cellular automaton data was collected from the literature for maximum ion channel open and closed times, duration of potentials evoked in single cells by very brief stimuli and the duration of potentials in brain slices and mollusk ganglia. The recordings in the slices and ganglia used for this analysis represented a large number of cells in the tissue rather than a single cell in the slice or a population response to a single synaptic event. Since data are limited for astrocytes, potentials from all forms of excitable cells were collected. In Fig. 3 the log maximum length of time reported for single ion channels to transition through an open and closed cycle and the log of the duration of evoked whole cell potentials were plotted versus the number of ion channels. For whole cells the number of ion channels was estimated to be 10^6. A regression line was fitted to these two sets of data. The duration of potentials from the slices and ganglia were then plotted on this line and the number of ion channels needed to produce these potentials was estimated by extrapolation. These potentials appeared to be generated by 10^7 to 10^8 ion channels. This finding suggests that Langton's relationship of the number of units to length of time that information is stored in cellular automata holds true for ion channel cellular automata.

Note that for convenience there was no attempt to limit the data collected to any one type of ion channel, cell type, or species. The assumption used here is that all biological systems evolved a similar mechanism to process information with ion channels and, as such, their ion channels have similar properties.

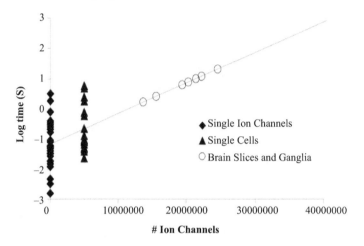

Figure 3 Memory as a function of the number of ion channels. Data was collected from the literature for the open/closed times for single ion channels, the length of potentials evoked in single cells and the length of potentials in groups of cells in brain slices or mollusk ganglia. The logs of the single ion channel and single cell data were graphed versus the number of ion channels. Cells were estimated to have 10^6 ion channels. The slope of the line defined by these two points was determined and the length of the potentials in the brain slices and mollusk ganglia were plotted onto the graph.

To generate an estimate of the total number of ion channels in a human astrocyte cellular automaton the number of astrocytes was approximated to be 10^{13} [28]. With 10^6 ion channels/cell this suggests 10^{19} ion channels in a human cellular automaton. Using the estimate of 10^{19} ion channels in the human cellular automaton the predicted duration of memory was extrapolated from the slope of the line in Fig. 3. The relationship between memory and the number of ion channels was estimated to be $t = e^{2.3 \times 10^{-7} N}$. Where t is time and N is the number of ion channels in the system. This calculation yielded a predicted maximum memory for a human sized astrocyte cellular automaton of $10^{10^{12}}$ years. Therefore, for all practical purposes, the predicted maximum duration of memory in human cellular automata is infinite. What is most notable about this memory is that it occurs without fixing the information to any physical structure such as a synapse or cell as predicted in Hebb's postulate [29]. The information is stored as a succession of representations, or ion channel configurations, with each individual representation lasting only a short period of time. The configuration of the ion channels is organized by the incoming information and then as this organization dissipates over time the information is lost. In thermodynamic terms the entropy of the system is decreased by the storage of information and, as the calculation presented above indicates, it takes a substantial amount of time for the entropy to return to baseline levels. Admittedly, the estimates for the number of ion channels and the number of astrocytes that make up a single syncytium are crude; however, even if the estimates are off by several orders of magnitude the overall conclusion that the potential duration of memory in a human ion channel cellular automaton is infinite, from a biological frame of reference, remains valid.

Another interesting comparison to be made between the astrocentric hypothesis and the neurocentric hypothesis is that there are $k^{10^{19}}$ distinct representations or unique configurations of the ion channels.

Using 10^{12} neurons each possessing 10^3 synapses we can estimate that there are 10^{15} synapses in a human brain [28] and a potential for $k^{10^{15}}$ distinct representations or unique configurations of the synapses. The term k is the number of states that an individual ion channels or synapse can take. These calculations demonstrate that the potential information processing capacity of the astrocytes using ion channels is many orders of magnitude larger than the capacity of neurons using synapses.

Associative Memory in Cellular Automata

An important component of memory is the ability to associate two or more events. In an ion channel cellular automata this is accomplished by the fact that the series of representations produced by a single event is significantly different from that produced by two events. Fig. 4 demonstrates the ability of a cellular automaton to associate information from two events. In the first column a single event produces a series of representations as the automaton progresses. In the second column two events occur simultaneously. The two events produce a series of representations that are distinct from the single event presented in the left column. This indicates that the two events have been associated to produce a unique memory.

Another interesting facet of ion channel cellular automata is that because they are dynamic systems they can readily store temporal differences between events. In the right hand column the two events are separated by ten units of time resulting in a series of representations that differs from either the single event in the left hand column or the two simultaneous events in the middle column. These observations suggest that the proposed astrocyte memory system can associate memories and that temporal information can be stored.

Research Supporting Astrocyte Cellular Automata as Memory Systems

In studies published over forty years ago Hyden demonstrated that glia were critical for memory [30–32]. More recent work using the one-trial aversive learning paradigm in chicks has confirmed Hyden's findings [13,14,33]. In these studies inhibitors of astrocyte function were found to block both short term and intermediate term memory, but, when administered later, had no effect on the long term retention of the learned behavior. During the short and intermediate periods it was demonstrated that ion fluxes in astrocytes are critical [13,33,34] for memory suggesting that the astrocyte ion channels may store information in the chicks for a brief period of time, approximately 60 minutes, while the appropriate rewiring of the neuronal circuitry takes place. It is important to note that this behavioral model involves

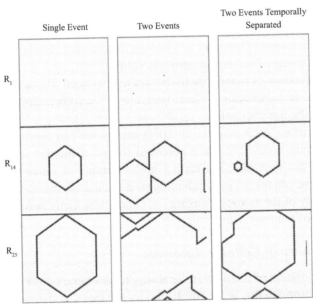

Figure 4 Associative memory in cellular automata. A cellular automaton operating at the transition between order and chaos was setup as described in Fig. 2 using the program CaSim. Three different stimuli were used. Iterations 1, 14 and 25 are presented in the figure. In the left column the cellular automaton was seeded by setting one unit to the open state at R_1 (Single Event). In the center column two units were seeded at R_1 (Two Events). In the right hand column the cellular automaton was seeded by setting one unit to the open state at R_1 and a separate unit to the open state at R_{10} (Two Events Temporally Separated). Note that each time series generates a different pattern of channel openings (representations) indicating that the two events in the second and third columns have produced unique memories by associating the events. Also note that the difference in representations produced by the automaton in the second and third columns indicates that the cellular automaton stores temporal information about the events. Therefore it is concluded that a two dimensional ion channel cellular automata is capable of associative memory.

both memory and learning, while the cellular automata hypothesis presented here is related purely to memory. Memory is the ability of an organism to store information about events in a retrievable format, whereas learning involves a change in behavior or potential behavior. Thus, a consolidated learned behavior, as occurs in the one-trial aversive learning paradigm, is likely to be the result of neuronal rewiring.

Furthermore, it does not require the organism to retain any specific memory of the event that precipitated the change in behavior beyond the length of time necessary to produce the rewiring. In this light, the chick in the aversive learning paradigm may actually recall the aversive stimulus for the short and intermediate term memory periods, which require astrocytes, but may not retain any recollection of the event once the aversive behavior has been established. It is enough for the chick to avoid certain objects without remembering why it needs to avoid them. The distinction between memory and learning is important because the two processes are likely mediated by different mechanisms. In the current hypothesis the ion channel cellular automata would be responsible for the specific memory of the event while changes in synaptic strength of the neurons would be responsible for learning and maintaining the new behavior. Astrocyte memory could support learning, but learning does not necessarily support the memory of events.

In addition to proposing that glia were involved in memory, Hyden predicted that mental diseases may involve glia [35] as reported in [34]. In the ion channel cellular automata hypothesis it is critical that the ion channels operate at the junction between order and chaos. Departure from this behavior is predicted to produce pathology. Deviation to the ordered side of the spectrum might produce depressive types of behaviors in the organism and memory deficits while deviation to the chaotic side might produce psychotic or manic types of behaviors that are also associated with memory deficits.

Several studies have demonstrated that long term treatment with antidepressant drugs at clinically relevant doses alters protein expression and function in astrocytes [36–41] and long term treatment with lithium ion results in suppression of mRNA for sodium-dependent inositol transporter in astrocytes [34]. The length of treatment required for the change in astrocyte proteins is consistent with the onset of the therapeutic effect of these agents. These studies suggest that these psychoactive agents may adjust the activity of astrocyte ion channel cellular automata toward the order/chaos border, thus improving the function of the memory system. Therefore, a number of studies, spanning over forty years, indicate that astrocytes are important for memory and possibly for the therapeutic effect of psychoactive drugs, which is consistent with the astrocyte ion channel cellular automata hypothesis.

Conclusion

In this study the hypothesis that astrocytes could store information in the central nervous system was considered. Based on the similarity of membrane ion channels to mathematical models known as cellular automata it seems reasonable to conclude that ion channels in astrocytes could store information for significant periods of time. This storage system does not rely on physically fixing information to any structure such as a synapse; rather information is stored by organizing the activity of the ion channels. If this concept is correct it suggests that neurons may use astrocytes as a dynamic information sink. In theory, this information would remain readily available to the neurons for extended periods of time.

Furthermore, this hypothesis indicates that to store information for significant periods of time the ion channels in the astrocyte syncytium must be in electrical contact with each other. This function could be served by the astrocytes' gap junctions. Thus, we can predict that agents that selectively block astrocyte gap junctions should disrupt memory. Clearly, further work is needed to verify this theoretical framework for memory in nervous systems.

Methods

One Dimensional Cellular Automaton

A 16 unit one dimensional cellular automaton was set up with each unit having 2 states. The rule used for this automaton was Wolfram's rule number 232 [20,21]. In this rule each unit is updated by averaging the states of the unit with its two nearest neighbors and then rounding to the nearest integer. The time series for this cellular automaton was calculated by hand.

Two Dimensional Cellular Automata

To examine the effects of different rule sets on 2 dimensional cellular automata the program CaSim [42] was used. A matrix of *100×100* units with a Moore neighborhood (eight neighbors) was set up with various rules. Each unit had 4 states. The entropy of the different rule sets was calculated using the equation entropy $= -\sum P_s \ln(P_s)$, where P_s is the probability of a unit occupying a particular state. The probabilities of the different states were determined from 10 runs of 1000 iterations for each cellular automaton. For these calculations the cellular automaton was seeded for each run by randomly setting ten percent of the units to the open state. The maximum entropy was calculated using the probability of 0.25 for each of the four states. The ratio of the calculated entropy of the rule set to the maximum possible entropy was used as an indicator of the chaotic nature of the system. Thus an entropy ratio of 0 is a completely ordered rule set and a ratio of 1 is a completely chaotic rule set.

For the examples presented in the figures the cellular automata where seeded with either 1 or 2 units set to the open state.

Duration of Memory Versus the Number of Ion Channels

To calculate the relationship between the number of ion channels in a system and the duration of information storage by the ion channels data was collected from published sources. The maximum open and closed times for various ion channels were obtained [43–54] and the open to closed cycle was used as the duration of memory in single ion channels. Similarly, potentials recorded in single cells were obtained [55–62] and used as an indication of the activity of multiple ion channels in concert. The log of the values for the duration of the responses in the ion channels and cells were plotted versus the number of ion channels. The number of ion channels in the cells was estimated to be 10^6. A line was then fitted to the two points and the log of the duration of potentials in slices and ganglia [63–71] were plotted on the line.

Acknowledgements

This work was supported by the University of Florida College of Dentistry and the McKnight Brain Institute.

References

1. Araque A., Carmignoto G. and Haydon P.G. Dynamic signaling between astrocytes and neurons. Annu Rev Physiol. 2001; 63: 795–813. doi: 10.1146/annurev.physiol. 63.1.795.
2. Araque A., Martin E.D., Perea G., Arellano J.I. and Buno W. Synaptically released acetylcholine evokes Ca2+ elevations in astrocytes in hippocampal slices. J Neurosci. 2002; 22: 2443–2450.

3. Carmignoto G. Astrocyte-neurone crosstalk: variants of the same language? Trends Pharmacol Sci. 2000; 21: 373–375. doi:10.1016/S0165-6147(00)01547-9.
4. Carmignoto G. Reciprocal communication systems between astrocytes and neurones. Prog Neurobiol. 2000; 62: 561–581. doi:10.1016/S0301-0082(00)00029-0.
5. Chiu S.Y. and Kriegler S. Neurotransmitter-Mediated Signaling Between Axons and Glial-Cells. Glia. 1994; 11: 191–200. doi:10.1002/glia.440110213.
6. Murphy T.H., Blatter L.A., Wier W.G. and Baraban J.M. Rapid communication between neurons and astrocytes in primary cortical cultures. J Neurosci. 1993; 13: 2672–2679.
7. Perea G. and Araque A. Properties of synaptically evoked astrocyte calcium signal reveal synaptic information processing by astrocytes. J Neurosci. 2005; 25: 2192–2203. doi: 10.1523/JNEUROSCI.3965-04.2005.
8. Zonta M., Angulo M.C., Gobbo S., Rosengarten B., Hossmann K.A., Pozzan T. and Carmignoto G. Neuron-to-astrocyte signaling is central to the dynamic control of brain microcirculation. Nat Neurosci. 2003; 6: 43–50. doi: 10.1038/nn980.
9. Zonta M. and Carmignoto G. Calcium oscillations encoding neuron-to-astrocyte communication. J Physiol Paris. 2002; 96: 193–198. doi:10.1016/S0928-4257(02)00006-2.]
10. Spataro L.E., Sloane E.M., Milligan E.D., Wieseler-Frank J., Schoeniger D., Jekich B.M., Barrientos R.M., Maier S.F. and Watkins L.R. Spinal gap junctions: potential involvement in pain facilitation. J Pain. 2004; 5: 392–405. doi:10.1016/j.jpain.2004.06.006.
11. Colburn R.W., Rickman A.J. and Deleo J.A. The effect of site and type of nerve injury on spinal glial activation and neuropathic pain behavior. Exp Neurol. 1999; 157: 289–304. doi: 10.1006/exnr.1999.7065.
12. Colburn R.W., Deleo J.A., Rickman A.J., Yeager M.P., Kwon P. and Hickey W.F. Dissociation of microglial activation and neuropathic pain behaviors following peripheral nerve injury in the rat. J Neuroimmunol. 1997; 79: 163–175. doi: 10.1016/S0165-5728(97)00119-7.
13. Hertz L., Gibbs M.E., O'Dowd B.S., Sedman G.L., Robinson S.R., Sykova E., Hajek I., Hertz E., Peng L., Huang R. and Ng K.T. Astrocyte-neuron interaction during one-trial aversive learning in the neonate chick. Neurosci Biobehav Rev. 1996; 20: 537–551. doi: 10.1016/0149-7634(95)00020-8.
14. Ng K.T., O'Dowd B.S., Rickard N.S., Robinson S.R., Gibbs M.E., Rainey C., Zhao W.Q., Sedman G.L. and Hertz L. Complex roles of glutamate in the Gibbs-Ng model of one-trial aversive learning in the new-born chick. Neurosci Biobehav Rev. 1997; 21: 45–54. doi: 10.1016/0149-7634(95)00079-8.
15. Mantz J., Cordier J. and Giaume C. Effects of general anesthetics on intercellular communications mediated by gap junctions between astrocytes in primary culture. Anesthesiology. 1993; 78: 892–901.
16. Wentlandt, Carlen P.L., Kushnir M., Naus C.C. and E.l. Beheiry H. General anesthetics attenuate gap junction coupling in P19 cell line. J Neurosci Res. 2005; 81: 746–752. doi: 10.1002/jnr.20577.
17. Robertson J.M. The Astrocentric Hypothesis: proposed role of astrocytes in consciousness and memory formation. J Physiol Paris. 2002; 96: 251–255. doi: 10.1016/S0928-4257(02)00013-X.
18. Ulam S.M. How to formulate mathematically problems of rate of evolution? Wistar Inst Symp Monogr. 1967; 5: 21–33.
19. Von Neumann J. and Burks A.W. Theory of self-reproducing automata. Urbana, University of Illinois Press; 1966.
20. Wolfram S. Cellular Automata. Los Alamos Science. 1983;9: 2–21. http: //www.stephenwolfram.com/publications/articles/general/83-cellular/index.html .
21. Wolfram S. Universality and Complexity in Cellular Automata. Physica D. 1984;10: 1–35. doi:10.1016/0167-2789(84)90245-8. http://www.stephenwolfram.com/publications/articles/ca/84-universality/index.html.
22. Perrier J.Y., Sipper M. and Zahnd J. Toward a viable, self-reproducing universal computer. Physica D. 1996; 97: 335–352. doi: 10.1016/0167-2789(96)00091-7.
23. Langton C.G. Computation at the Edge of Chaos- Phase-Transitions and Emergent Computation. Physica D. 1990; 42: 12–37. doi: 10.1016/0167-2789(90)90064-V.

24. Dere E., Souza-Silva M.A., Frisch C., Teubner B., Sohl G., Willecke K. and Huston J.P. Connexin30-deficient mice show increased emotionality and decreased rearing activity in the open-field along with neurochemical changes. Eur J Neurosci. 2003; 18: 629–638. doi:10.1046/j.1460-9568.2003.02784.x.

25. Frisch C., Souza-Silva M.A., Sohl G., Guldenagel M., Willecke K., Huston J.P. and Dere E. Stimulus complexity dependent memory impairment and changes in motor performance after deletion of the neuronal gap junction protein connexin36 in mice. Behav Brain Res. 2005; 157: 177–185. doi:10.1016/j.bbr.2004.06.023.

26. Frisch C., Theis M., Souza Silva M.A., Dere E., Sohl G., Teubner B., Namestkova K., Willecke K. and Huston J.P. Mice with astrocyte-directed inactivation of connexin43 exhibit increased exploratory behaviour, impaired motor capacities, and changes in brain acetylcholine levels. Eur J Neurosci. 2003; 18: 2313–2318. doi:10.1046/j.1460-9568.2003.02971.x.

27. Rose C.R. and Ransom B.R. Gap junctions equalize intracellular Na+ concentration in astrocytes. Glia. 1997; 20: 299–307. doi:10.1002/(SICI)1098-1136(199708)20: 4<299:: AID-GLIA3>3.0.CO; 2–1.

28. Kandel E.R., Schwartz J.H. and Jessell T.M. Principles of Neural Science. 4. New York, McGraw-Hill inc; 2000. pp. 180–182.

29. Hebb D. The organisation of behaviour: A neuropsychological theory. New York, Wiley; 1949.

30. Hyden H. A functional relationship between oligodendroglia and the nerve cell. Acta Morphol Neerl Scand. 1960; 3: 170–178.

31. Hyden H. and Egyhazi E. Nuclear RNA changes of nerve cells during a learning experiment in rats. Proc Natl Acad Sci USA. 1962; 48: 1366–1373.

32. Hyden H. and Egyhazi E. Glial Rna Changes During A Learning Experiment in Rats. Proceedings of the National Academy of Sciences of the United States of America. 1963; 49: 618.

33. Ng K.T., Gibbs M.E., Gibbs C.L., Sedman G., Sykova E., Svoboda J., Jendolova P., O'Dowd B., Rickard N. and Crowe S.F. Ion involvement in memory formation: the potential role of astrocytes. Prog Brain Res. 1992; 94: 109–115.

34. Hertz L., Hansson E. and Ronnback L. Signaling and gene expression in the neuron-glia unit during brain function and dysfunction: Holger Hyden in memoriam. Neurochem Int. 2001; 39: 227–252. doi: 10.1016/S0197-0186(01)00017-1.

35. Hyden H. A two-cell collaboration responsible for brain activity. Gèoteborg, Distr.: Almqvist & Wiksell, Stockholm; 1960.

36. Hertz L. and Richardson J.S. Acute and chronic effects of antidepressant drugs on beta-adrenergic function in astrocytes in primary cultures: an indication of glial involvement in affective disorders? J Neurosci Res. 1983; 9: 173–182. doi:10.1002/jnr.490090209.

37. Hertz L., Richardson J.S. and Mukerji S. Doxepin, a tricyclic antidepressant, binds to normal, intact astroglial cells in cultures and inhibits the isoproterenol-induced increase in cyclic AMP production. Can J Physiol Pharmacol. 1980; 58: 1515–1519.

38. Richardson J.S. and Hertz L. The effects of antidepressant drugs on adenylyl cyclase linked beta adrenergic binding sites on mouse astrocytes in primary cultures. Prog Neuropsychopharmacol Biol Psychiatry. 1983; 7: 675–680. doi:10.1016/0278-5846(83)90044-1.

39. Richelson E. The use of cultured cells in the study of mood-normalizing drugs. Pharmacol Toxicol. 1990; 66 Suppl 3: 69–75.

40. Sapena R., Morin D., Zini R., Morin C. and Tillement J.P. Desipramine treatment differently down-regulates beta-adrenoceptors of freshly isolated neurons and astrocytes. Eur J Pharmacol. 1996; 300: 159–162. doi:10.1016/0014-2999(96)00060-X.

41. Mercier G., Lennon A.M., Renouf B., Dessouroux A., Ramauge M., Courtin F. and Pierre M. MAP kinase activation by fluoxetine and its relation to gene expression in cultured rat astrocytes. J Mol Neurosci. 2004; 24: 207–216. doi:10.1385/JMN: 24: 2: 207.

42. Freiwald U. and Weimar J.R. The Java based cellular automata simulation system—JCASim. Future Generation Computer Systems. 2002; 18: 995–1004. doi: 10.1016/S0167-739X(02)00078-X.

43. Wang Z.F. and Shi Y.L. Modulation of inward rectifier potassium channel by toosendanin, a presynaptic blocker. Neuroscience Research. 2001;40: 211–215. doi: 10.1016/S0168-0102(01)00233-4.

44. Carabelli V., Hernandez-Guijo J.M., Baldelli P. and Carbone E. Direct autocrine inhibition and cAMP-dependent potentiation of single L-type Ca2+ channels in bovine chromaffin cells. Journal of Physiology-London. 2001; 532: 73–90. doi: 10.1111/j.1469-7793.2001.0073g.x.

45. Franciolini F., Hogg R., Catacuzzeno L., Petris A., Trequattrini C. and Adams D.J. Large-conductance calcium- activated potassium channels in neonatal rat intracardiac ganglion neurons. Pflugers Archiv-European Journal of Physiology. 2001; 441: 629–638. doi:10.1007/s004240000471.

46. Wu S.N., Li H.F. and Chiang H.T. Vinpocetine-induced stimulation of calcium-activated potassium currents in rat pituitary GH(3) cells. Biochemical Pharmacology. 2001; 61: 877–892. doi: 10.1016/S0006-2952(01)00553-6.

47. Niisato N. and Marunaka Y. Blocking action of cytochalasin D on protein kinase A stimulation of a stretch-activated cation channel in renal epithelial A6 cells. Biochemical Pharmacology. 2001; 61: 761–765. doi:10.1016/S0006-2952(01)00534-2.

48. Catacuzzeno L., Pisconti D.A., Harper A.A., Petris A. and Franciolini F. Characterization of the large-conductance Ca-activated K channel in myocytes of rat saphenous artery. Pflugers Archiv-European Journal of Physiology. 2000; 441: 208–218. doi:10.1007/s004240000414.

49. Pellegrini M., Menconi M.C. and Pellegrino M. Stretch-activated cation channels of leech neurons exhibit two activity modes. European Journal of Neuroscience. 2001; 13: 503–511. doi:10.1046/j.0953-816X.2000.01415.x.

50. Rodriguez-Contreras A. and Yamoah E.N. Direct measurement of single-channel Ca2+ currents in bullfrog hair cells reveals two distinct channel subtypes. Journal of Physiology-London. 2001; 534: 669–689. doi:10.1111/j.1469-7793.2001.00669.x.

51. Han J., Kim N. and Kim E. Trifluoroacetic acid activates ATP-sensitive K+ channels in rabbit ventricular myocytes. Biochemical and Biophysical Research Communications. 2001; 285: 1136–1142. doi:10.1006/bbrc.2001.5291.

52. Iftinca M., Waldron G.J., Triggle C.R. and Cole W.C. State-dependent block of rabbit vascular smooth muscle delayed rectifier and Kv1.5 channels by inhibitors of cytochrome P450-dependent enzymes. Journal of Pharmacology and Experimental Therapeutics. 2001; 298: 718–728.

53. Zhou Z., Hu S. and Hwang T.C. Voltage-dependent flickery block of an open cystic fibrosis transmembrane conductance regulator (CFTR) channel pore. Journal of Physiology-London. 2001; 532: 435–448. doi:10.1111/j.1469-7793.2001.0435f.x.

54. Wong A.Y.C., Burnstock G. and Gibb A.J. Single channel properties of P2X ATP receptors in outside-out patches from rat hippocampal granule cells. Journal of Physiology-London. 2000; 527: 529–547. doi:10.1111/j.1469-7793.2000.00529.x.

55. Baufreton J., Garret M., Dovero S., Dufy B., Bioulac B. and Taupignon A. Activation of GABA(A) receptors in subthalamic neurons *in vitro*: Properties of native receptors and inhibition mechanisms. Journal of Neurophysiology. 2001; 86: 75–85.

56. Barrett E.F. and Barrett J.N. Separation of 2 Voltage-Sensitive Potassium Currents, and Demonstration of A Tetrodotoxin-Resistant Calcium Current in Frog Motoneurons. Journal of Physiology-London. 1976; 255: 737–774.

57. Madison D.V. and Nicoll R.A. Cyclic Adenosine-3', 5'-Monophosphate Mediates Beta-Receptor Actions of Noradrenaline in Rat Hippocampal Pyramidal Cells. Journal of Physiology-London. 1986; 372: 245–259.

58. T.S. Action potentials in plant organs. Sympos Soc Exp Biol. 1966; 20: 49–74.

59. Mynlieff M. Identification of different putative neuronal subtypes in cultures of the superior region of the hippocampus using electrophysiological parameters. Neuroscience. 1999; 93: 479–486. doi: 10.1016/S0306-4522(99)00153-0.

60. Machemer H. and Ogura A. Ionic Conductances of Membranes in Ciliated and Deciliated Paramecium. Journal of Physiology-London. 1979; 296: 49–60.

61. Moser T. Low-conductance intercellular coupling between mouse chromaffin cells *in situ*. Journal of Physiology-London. 1998; 506: 195–205. doi:10.1111/j.1469-7793.1998.195bx.x.

62. Petruska J.C., Napaporn J., Johnson R.D., Gu J.G.G. and Cooper B.Y. Subclassified acutely dissociated cells of rat DRG: Histochemistry and patterns of capsaicin-, proton-, and ATP-activated currents. Journal of Neurophysiology. 2000; 84: 2365–2379.

63. Yuste R., Tank D.W. and Kleinfeld D. Functional study of the rat cortical microcircuitry with voltage-sensitive rye imaging of neocortical slices. Cerebral Cortex. 1997; 7: 546–558. doi:10.1093/cercor/7.6.546.

64. Schmidt J., Prinz P. and Deitmer J.W. Glial hyperpolarization upon nerve root stimulation in the leech Hirudo medicinalis. Glia. 1999; 27: 32–38. doi:10.1002/(SICI)1098-1136(199907)27: 1<32: : AID-GLIA4>3.0.CO; 2–9.

65. Deitmer J.W. and Kristan W.B. Glial responses during evoked behaviors in the leech. Glia. 1999; 26: 186–189. doi:10.1002/(SICI)1098-1136(199904)26: 2<186: : AID-GLIA9>3.0.CO; 2-E.

66. Caudle R.M. The demonstration of long latency potentials in the CA1 region of the rat hippocampal slice. Brain Res. 1993; 613: 247–250. doi:10.1016/0006-8993(93)90905-3.

67. Kojima S., Ogawa H., Kouuchi T., Nidaira T., Hosono T. and Ito E. Neuron-independent Ca2+ signaling in glial cells of snail's brain. Neuroscience. 2000; 100: 893–900. doi: 10.1016/S0306-4522(00)00338-9.

68. Komuro H., Sakai T., Momosesato Y., Hirota A. and Kamino K. Optical-Detection of Postsynaptic Potentials-Evoked by Vagal-Stimulation in the Early Embryonic Chick Brain-Stem Slice. Journal of Physiology-London. 1991; 442: 631–648.

69. Wu J.Y., Falk C.X., Cohen L., Tsau Y. and Zecevic D. Optical Measurement of Action-Potential Activity in Invertebrate Ganglia. Japanese Journal of Physiology. 1993; 43: S21–S29. doi:10.2170/jjphysiol.43.21.

70. Falk C.X., Wu J.Y., Cohen L.B. and Tang A.C. Nonuniform Expression of Habituation in the Activity of Distinct Classes of Neurons in the Aplysia Abdominal-Ganglion. Journal of Neuroscience. 1993; 13: 4072–4081.

71. Tsau Y., Wu J.Y., Hopp H.P., Cohen L.B., Schiminovich D. and Falk C.X. Distributed Aspects of the Response to Siphon Touch in Aplysia—Spread of Stimulus Information and Cross-Correlation Analysis. Journal of Neuroscience. 1994; 14: 4167–4184.

Nonlinear Dynamics and Chaos in a Fractional-Order HIV Model [†]

Haiping Ye[1] and Yongsheng Ding[2]

[1]Department of Applied Mathematics, Donghua University, Shanghai 201620, China and College of Information Sciences and Technology, Donghua University, Shanghai 201620, China.
[2]College of Information Sciences and Technology, Donghua University, Shanghai 201620, China.
Email: ysding@dhu.edu.cn.

ABSTRACT

We introduce fractional order into an HIV model. We consider the effect of viral diversity on the human immune system with frequency dependent rate of proliferation of cytotoxic T-lymphocytes (CTLs) and rate of elimination of infected cells by CTLs, based on a fractional-order differential equation model. For the one-virus model, our analysis shows that the interior equilibrium which is unstable in the classical integer-order model can become asymptotically stable in our fractionalorder model and numerical simulations confirm this. We also present simulation results of the chaotic behaviors produced from the fractional-order HIV model with viral diversity by using an Adams-type predictor-corrector method.

Introduction

An important part of the human immune response against viral infections is cytotoxic T Lymphocytes (CTLs) [1].They recognize and kill cells which are infected by virus. There are many immune models describing the virus dynamics with CTL immune response. Nowak and Bangham [2,3] proposed an ODE model which explores the relation among CTL immune responses, virus load, and virus diversity. In [2], a rate of specific CTL(Zj) proliferation in response to the corresponding specific infected cells (Ij) depends on the mass action law $cI_j Z_j$. This model has been important in the field of mathematical modelling of HIV infection. In their model, there is no interaction among different types of CTL(Zj). Iwami et al. [4] assumed that the correlation is incorporated as a function of the frequency

[†]Reused with permission from: Haiping Ye and Yongsheng Ding, "Nonlinear Dynamics and Chaos in a Fractional-Order HIV Model," *Mathematical Problems in Engineering*, vol. 2009, Article ID 378614, 12 pages, 2009. doi:10.1155/2009/378614.

that the specific CTLs(Z_j)-encounter in the specific infected cells (I_j). In a similar manner, they considered the rate of elimination of specific infected cells (I_j) by the specific CTLs(Z_j) to be proportional to this frequency. However, these models do not take into account the fractional order derivatives that have been extensively applied in many fields (e.g., [5–17]) and the reference cited therein). Recently many mathematicians and applied researchers have tried to model real processes using the fractional order differential equations (FODE) [16]. In biology, it has been deduced that the membranes of cells of biological organism have fractional order electrical conductance [13] and then, they are classified into group of noninteger order models. Also, it has been shown that modelling the behavior of brainstem vestibule-oculomotor neurons by FODE has more advantages than classical integer order modelling [8].

Particular emphasis is that a major difference between fractional order models and integer order models is that fractional order models possess memory [5,12], while the main features of immune response involve memory [18]. Hence, we attempt to model HIV infection with immune response using a fractional order system. Our presentation is based on the immune model of HIV infection which is developed by Iwami et al. [4]. For the one-virus model, we carry out a detailed analysis on stability of equilibrium. Our analysis shows that the interior equilibrium which is unstable in the classical integer order model can become asymptotically stable in our fractional order model. We also find that chaos does exist in the fractional order HIV model with viral diversity.

Model Derivation

We first give the definition of fractional order integration and fractional order differentiation [14,16]. For the concept of fractional derivative we will adopt Caputo's definition which is a modification of the Riemann-Liouville definition and has the advantage of dealing properly with initial value problems.

Definition 2.1. The fractional integral of order $\alpha > 0$ of a function $f : R^+ \to R$ is given by

$$I^\alpha f(x) = \frac{1}{\Gamma(x)} \int (x-t)^{\alpha-1} f(t) dt \qquad (2.1)$$

provided the right side is pointwise defined on R^+.

Definition 2.2. The Caputo fractional derivative of order $\alpha \in (n-1, n)$ of a continuous function $f : R^+ \to R$ is given by

$$D^\alpha f(x) = I^{n-\alpha} D^n f(x), \quad D = \frac{d}{dt}. \qquad (2.2)$$

Now we introduce fractional order into the ODE model by Iwami et al. [4]. The new system is described by the following set of FODE:

$$D^{q_1} T = \lambda - dT - \sum_{l=1}^{n} B_l' T V_l,$$

$$D^{q_{2j}} I_j = \beta_j^i TI_j - aI_j - bZ_j(I_j/T + \sum_{l}^{n} I_l'),$$

$$D^{q_{4j}} V_j = kaI_j - uV_j, (j = 1, 2, ..., n),$$

$$D^{q_{3j}} Z_j = cZ_j \frac{I_j}{T + \sum_{l}^{n} I_l'} - \delta Z_j, \quad (j = 1, 2, ..., n), \tag{2.3}$$

where $T(t)$ represents the concentration of uninfected cells at time t, $I_j(t)$ represents the concentration of infected cells with a virus particle of type j, $V_j(t)$ the concentration of free virus particle of type j, and $z_j(t)$ denotes the magnitude of the specific CTL response against variant j. Here, $0.95 \leq q_{1j}, q_{2j}, q_{3j}, q_{4j} \leq 1$ ($j = 1, 2, ..., n$) are restricted such that fractional derivative can be approximately described the rate of change in number.

Following [4], uninfected cells are assumed to be generated at a constant rate λ. Uninfected cells, infected cells, free viruses, and CTLs decline at rates d, a, u, and δ, respectively. The total number of virus particles produced from one cell is k. The rate of CTL proliferation in response to antigen is given by $cz_iI_i/(T + \sum_{l=1}^{n} I_l)$ and the specific infected cells are killed by specific CTLs at rate $bz_iI_i/(T + \sum_{l=1}^{n} I_l)$ while infected cells are produced from uninfected cells and free virus at rate $\beta_i'TV_l$, that is, a rate of specific CTL(Zj) proliferation in response to the corresponding specific infected cells (I_j) depends on the frequency, instead of the mass action law.

To simplify the model, it is reasonable to assume that the decay rate of free virus, u, is much larger than that of the infected cells, a, and this system describes the qualitative dynamics of the asymptomatic phase of HIV infection. Thus, we may introduce as a good approximation that the virus is in steady state (i.e., $Dq_{4j}V_j = 0$) and hence $V_j = kaI_j/u$ (see [4, 19]). This leads to the following simplified system of FODE:

$$D^{q_1} T = \lambda - dT - \sum_{l=1}^{n} \beta_l TI_l,$$

$$D^{q_{2j}} I_j = \beta_j^i TI_j - aI_j - bZ_jI_j/(T + \sum_{l}^{n} I_l'), \tag{2.4}$$

$$D^{q_{3j}} Z_j = cZ_jI_j / \left(T + \sum_{l}^{n} I_l'\right) - \delta Z_j, \quad (j = 1, 2, ..., n),$$

Where $\beta_j = \dfrac{ka\beta_j'}{u}$.

One-Virus Model

In this section, we discuss in detail an important special case of model (2.4) and perform an equilibrium and stability analysis for this special case.

We consider the one-virus model ($n = 1$) and assume that $q_1 = q_{21} = q_{31} = \alpha$ ($0.95 \leq \alpha \leq 1$). This one-virus model is described by the following system of FODE:

$$D^\alpha T = \lambda - dT - \beta_1 TI_1,$$

$$D^\alpha I_1 = \beta_1 TI_1 - aI_1 - \frac{bZ_1 I_1}{T + I_1}, \tag{3.1}$$

$$D^\alpha Z_1 = \frac{cZ_1 I_1}{T + I_1} - \delta Z_1.$$

To evaluate the equilibria, let

$$D^\alpha T = 0$$

$$D^\alpha I_1 = 0 \tag{3.2}$$

$$D^\alpha Z_1 = 0.$$

Then system (3.1) has three equilibria: the uninfected equilibrium $E_H = (\lambda / d, 0, 0)$ the boundary equilibrium $E_I = (T^*; I^*, 0)$ where $T^* = \dfrac{\alpha}{\beta_1}, I_1^* = \dfrac{\lambda}{\alpha} - \dfrac{d}{\beta_1}$ and the interior equilibrium $E_c = (\hat{T}, \hat{I}_1, \hat{Z}_1)$, where

$$\hat{T} = \frac{-d + \sqrt{d^2 + 4\lambda\hat{\beta}}}{2\hat{\beta}}, \hat{I} = \frac{\delta}{c - \delta} \hat{T}, \tag{3.3}$$

$$\hat{Z} = \frac{c\hat{T}}{b(c - \delta)}(\beta_1 T - \alpha), \hat{\beta} = \frac{\delta\beta_1}{c - \delta}. \tag{3.4}$$

Following the analysis in [4], we introduce a basic reproduction number which is defined by

$$R_0 = \frac{\lambda\beta_1}{\alpha d}. \tag{3.5}$$

Denote $R_+^3 = \{(T, I_1, Z_1) \in R^3 / T \geq 0, I_1 \geq 0, Z_1 \geq 0 \}$ and we always assume that $c > \delta$. Note that $D\alpha Z_1 < 0$ always holds true if $c \leq \delta$. By generalized mean value theorem [15], we get $Z_1(t)$ is decreasing if $c \leq \delta$.

Next we will discuss the existence and stability of the equilibria of the model (3.1).

Theorem 3.1. (a) *The uninfected equilibrium EH is locally asymptotically stable (LAS) if $R_0 \in (0,1)$ and unstable if $R_0 > 1$.*

(b) *If $R_0 > 1$, then the boundary equilibrium E_I exists. This equilibrium is LAS if $1 < R_0 < (\alpha\delta / d(c - \delta)) + 1$ and unstable if $R_0 > (\alpha\delta / d(c - \delta)) + 1$.*

(c) *If $R_0 > \alpha\delta / d(c - \delta)) + 1$, then E_c exists in $\text{Int}R_+^3$, where $\text{Int}R_+^3$ is the interior of R_+^3.*

Proof: (a) The Jacobian matrix $J(E_H)$ for system (3.1) evaluated at E_H is given by

$$J(E_H) = \begin{pmatrix} -d & -\dfrac{\beta_1\lambda}{d} & 0 \\ 0 & \dfrac{\beta_1\lambda}{d} - \alpha & 0 \\ 0 & 0 & -\delta \end{pmatrix}$$

(3.6)

E_H is locally asymptotically stable if all of the eigenvalues p of the Jacobian matrix $J(E_H)$ satisfy the following condition (6,17):

$$|\arg(p) > \frac{\alpha\pi}{2}$$

(3.7)

The eigenvalues of $J(E_H)$ are $-d$, $\beta_1\lambda\,(\,d-a)$, $-\delta$. It is clear that E_H is LAS if $R_0 < 1$ and is unstable if $R_0 > 1$.

(b) If $R_0 > 1$, then the existence of E_1 is obvious. The Jacobian matrix $J(E_1)$ for system (3.1) evaluated at E_1 is given by

$$J(E_H) = \begin{pmatrix} -d - \beta_1 I_1^* & -\beta_1 T^* & 0 \\ \beta_1 I_1^* & 0 & -\dfrac{bI_1^*}{T^* + I_1^*} \\ 0 & 0 & \dfrac{cI_1^*}{T^* + I_1^*} - \delta \end{pmatrix}$$

(3.8)

For $J(E_1)$ given by (3.8), the characteristic equation becomes

$$[p^2 - (d + \beta_1 I_1^*)p + \beta_1^2 I_1^*](p - \frac{cI_1^*}{T^* + I_1^*} + \delta) = 0$$

(3.9)

and hence all the eigenvalues

$$\begin{vmatrix} p_{1,2} = \dfrac{-(d + \beta_1 I_1^*)p \pm \sqrt{-(d + \beta_1 I_1^*)^2 - 4\beta_1^2 I_1^* T^*}}{2}, \\ p_3 = \dfrac{cI_1^*}{T^* + I_1^*} - d. \end{vmatrix}$$

(3.10)

If $R_0 > 1$, then $T^* > 0$, $I_1^* > 0$ and $p_{1,2}$ have negative real parts. Furthermore, if $1 < R_0 < (\alpha\delta\,/\,d(c-\delta)) + 1$, then $p_3 < 0$ and E_1 is LAS.

If $R_0 > (\alpha\delta\,/\,d(c-\delta)) + 1$, then $p_3 > 0$ and E_1 is unstable.

(c) If $R_0 > (\alpha\delta\,/\,d(c-\delta)) + 1$, then we obtain $\hat{Z}_1 > 0$. Thus, E_c exists in Int $\text{Int} R_+^3$. Therefore, the proof is complete.

To discuss the local stability of the interior equilibrium E_c, we consider the linearized system of (3.1) at E_c. The Jacobian matrix at E_c is given by

$$J(E_H) = \begin{pmatrix} -d - \beta_1 I_1^* & -\beta_1 \hat{T} & 0 \\ \beta_1 I_1^* + \dfrac{b\hat{Z}_1 \hat{I}_1}{(\hat{T}+\hat{I})^2} & \beta_1 \hat{T} - \alpha - \dfrac{b\hat{Z}_1 \hat{I}_1}{(\hat{T}+\hat{I})^2} & -\dfrac{b\hat{I}_1}{\hat{T}+\hat{I}} \\ -\dfrac{c\hat{Z}_1 \hat{I}_1}{(\hat{T}+\hat{I})^2} & \dfrac{c\hat{Z}_1 \hat{T}}{(\hat{T}+\hat{I})^2} & \dfrac{b\hat{I}_1}{\hat{T}+\hat{I}} - \delta \end{pmatrix} \qquad (3.11)$$

For convenience, we denote $\hat{T} = T, \hat{I}_1 = I, \hat{Z}_1 = Z$ and $\beta_1 = \beta$. In view of the above assumptions and using

$$\beta T - \alpha - \frac{bZ}{T+I} = 0, \lambda - dT - \beta TI = 0, \frac{cI}{T+I} = \delta, \frac{T}{T+I} = \frac{c-\delta}{c}, \qquad (3.12)$$

$J(E_c)$ can now be written as follows:

$$J(E_c) = \begin{pmatrix} -d - \beta I & -\beta T & 0 \\ \beta I + \dfrac{\delta}{c}(\beta T - \alpha) & \dfrac{\delta}{c}(\beta T - \alpha) & -\dfrac{b\delta}{c} \\ -\dfrac{\delta}{b}(\beta T - \alpha) & \dfrac{c-\delta}{b}(\beta T - \alpha) & 0 \end{pmatrix}. \qquad (3.13)$$

Then the characteristic equation of the linearized system of (3.1) is

$$\Phi(p) = p^3 + \alpha_1 p^2 + \alpha_2 p + \alpha_3 = 0, \qquad (3.14)$$

Where

$$\alpha_1 = d + \frac{\delta\alpha}{c} + \frac{\delta^2 \beta T}{c(c-\delta)},$$

$$\alpha_2 = \beta T(\beta I + \frac{\delta(\beta T - \alpha)}{c}) + \frac{\delta(c-\delta)(\beta T - \alpha)}{c} - \frac{\delta}{c}(d + \beta I)(\beta T - \alpha), \qquad (3.15)$$

$$\alpha_3 = \frac{\delta(c-\delta)(\beta T - \alpha)}{c} - (d + \beta I) + \frac{\beta\delta^2}{c}(\beta T - \alpha).$$

Proposition 3.2. *The interior equilibrium* E_c *is LAS if all of the eigenvalues p of* $J(E_c)$ *satisfy* $\left| \arg(p) \right| > \dfrac{\alpha\pi}{2}$.

Denote

$$D(\Phi) = \begin{vmatrix} 1 & \alpha_1 & \alpha_2 & \alpha_3 & 0 \\ 0 & 1 & \alpha_1 & \alpha_2 & \alpha_3 \\ 3 & 2\alpha_1 & \alpha_2 & 0 & 0 \\ 0 & 3_1 & 2\alpha_1 & \alpha_2 & 0 \\ 0 & 0 & 3 & 2\alpha_1 & \alpha_1 \end{vmatrix}$$

$$= 18\alpha_1 \alpha_2 \alpha_3 + (\alpha_1 \alpha_2)^2 - 4\alpha_3 \alpha_1^3 - 4\alpha_2^3 - 27\alpha_3^2. \qquad (3.16)$$

Using the results of [5,20], we have the following proposition:

Proposition 3.3. *One assumes that E_c exists in $IntR_+^3$.*

i) *If the discriminant of $\Phi(p)$ $D(\Phi)$ is positive and Routh-Hurwitz conditions are satisfied, that is, $D(\Phi) > 0$, $\alpha_1 > 0$, $\alpha_1\alpha_2 > \alpha_3$, then the interior equilibrium E_c is LAS.*

ii) *If $D(\Phi) < 0, \alpha_1 > 0, \alpha_3 > 0, \alpha_1\alpha_2 = \alpha_3, \alpha \in [0,1]$, then the interior equilibrium E_c is LAS.*

iii) *If $D(\Phi) < 0, \alpha_1 < 0, \alpha_3 < 0, \alpha > \dfrac{2}{3}$, then the interior equilibrium E_c is unstable.*

In our first example we set $\lambda = 10, d = 0.02$ which are chosen according to [21] and set $\alpha = \delta = 0.04, c = b, \beta_1 = 4 \times 10^{-4}$ which come from [4]. With these parameter values: $R_0 > 5/(\alpha\delta / d(c-\delta)) + 1$, $\alpha_1\alpha_2 < \alpha_3$, $D(\Phi) = -9.4073 \times 10^{-7} < 0$. By Pro.3.2., we obtain the interior equilibrium $E_c = (362.0335, 19.0544, 49.9289)$ is LAS when $\alpha < 0.9916$. Numerical simulations show that trajectories of system (3.1) approach to the interior equilibrium (see Figs. 1(a) and 1(b)) However, when $\alpha = 1$ that is the case of classical integer order, E_c is unstable by the Routh-Hurwitz criterion (see Figs. 2(a) and 2(b)).

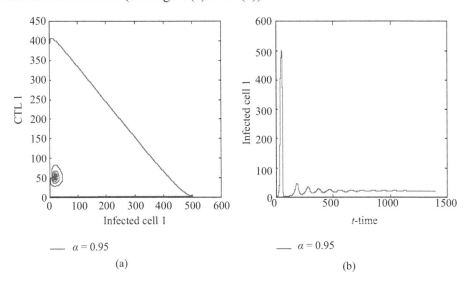

Figure 1 Numerical solutions of system (3.1). The plots show that trajectories of system (3.1) approach to the interior equilibrium for α=0.95.

Two-Virus Model

In this section, we consider viral diversity. We examine the two-virus model using numerical simulations. By examining the behavior of this simpler model we hope to get an idea as to how the more general models in system (2.4) may behave. The two-virus model is given by the following system of FODE:

$$D^{q_1}T = \lambda - dT - \beta_1 TI_1 - \beta_2 TI_2,$$

$$D^{q_{21}}I_1 = \beta_1 TI_1 - aI_1 - bZ_1 \frac{I_1}{T + I_1 + I_2},$$

$$D^{q_{22}}I_2 = \beta_2 TI_2 - aI_1 - bZ_2 \frac{I_2}{T + I_1 + I_2},$$

$$(4.1)$$

$$D^{q_{31}}Z_1 = cZ_1 \frac{I_1}{T + I_1 + I_2} - \delta Z_1,$$

$$D^{q_{32}}Z_2 = cZ_2 \frac{I_2}{T + I_1 + I_2} - \delta Z_2,$$

with initial value condition

$$T(0) = T_0, I_i(0) = I_{i0}; Z_i(0) = Z_{i0}, i = 1.2, \qquad (4.2)$$

where $0.95 \leq q_1, q_{2j}, q_{3j} \ (j = 1, 2) \leq 1$.

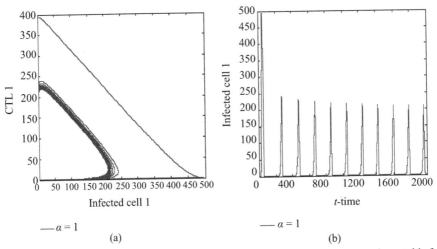

—— $\alpha = 1$ —— $\alpha = 1$

(a)   (b)

Figure 2 Numerical solutions of system (3.1). The plots show that the interior equilibrium is unstable for $\alpha = 1$.

To find numerical solution to (4.1) and (4.2) in the interval $[0, T]$, we reduce the systems (4.1) and (4.2) to a set of fractional integral equations, by using an equivalence (see [16, Theorem 3.24])

$$D^{\alpha} X = f(x) \Leftrightarrow X(t) = X(0) + I^{\alpha} f(x). \qquad (4.3)$$

Then we apply the generalized Adams-type predictor-corrector method or, more precisely, Predict, Evaluate, Correct, Evaluate (PECE) methods (see [22, 23]).

For notational convenience, we denote $\alpha = (q_1, q_{21}, q_{22}, q_{31}, q_{32})$.

We carry out numerical simulations for system, (4.1) and (4.2) with parameters $\lambda = 10$, $b = c = 0.08, a = 0.031, \delta = 0.03, d = 0.02, \beta_1 = 4 \times 10^{-4}$ and $\beta_2 = 2.08 \times 10^{-4}$ for the step size 0.07. Numerical solutions of systems (4.1) and (4.2) support that the system exhibits a chaotic behavior and systems (4.1) and (4.2) have a strange attractor in $IntR_+^5$ for $\alpha = [0.95, 0.95, 1, 1, 1]$ (see Figs. 3(a)–3(c)). It is clear that chaos does exist in our fractional order model with viral diversity as in the case of integer order model. The effect of viral diversity and the frequency dependence results in collapse of the immune system and make the behavior of the system dynamics complex [4]. However, as the value of some component or more components of the order α further decreases, for example, $\alpha = [0.95, 0.95, 0.95, 1, 1]$, the chaotic motion disappears and the systems (4.1) and (4.2) stabilize to a fixed point (see Figs. 4(a)–4(c)).

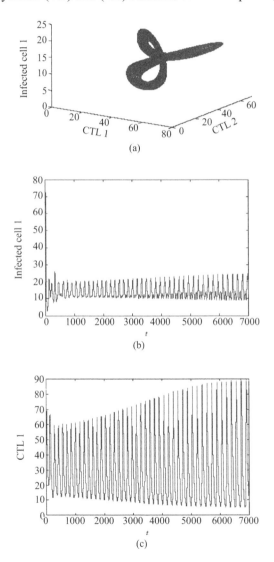

(a)

(b)

(c)

Figure 3 Numerical solutions of system (4.1) for $\alpha = [0.95, 0.95, 1, 1, 1]$. (a) A strange attractor in the Z_1-Z_2-I_1 phase. (b) Infected cell 1. (c) CTL 2.

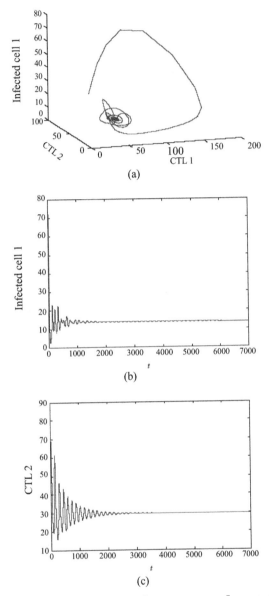

Figure 4 Numerical solutions of system (4.1) for $\alpha = [0.95, 0.95, 0.95, 1, 1]$. (a) Z_1-Z_2-I_1 phase. (b) Infected cell 1. (c) CTL 2.

Conclusions

In this paper, we have proposed a fractional order HIV model, as a generalization of an integer order model, developed by Iwami et al. [4]. The premise of the proposed model is the fact that fractional order models possess memory while the main features of immune response involve memory. It is an attempt to incorporate fractional order into the mathematical model of HIV-immune system dynamics and it is still an interesting exercise to determine, mathematically, how the order of a fractional differential system affects the dynamics of system.

In the case of one-virus model, the fractional order system has an interior equilibrium under some restriction. By using stability analysis on fractional order system, we obtain sufficient condition on the parameters for the stability of the interior equilibrium. Our analysis shows that the interior equilibrium which is unstable in the classical integer order model can become asymptotically stable in our fractional order model. Note that the interior equilibrium is globally asymptotically stable (GAS) (see [24]) if the terms associated with

immune reactions are given by cZ_1I_1 and bZ_1I_1 instead of $\dfrac{cZ_1I_1}{T+I_1}$ and $\dfrac{bZ_1I_1}{T+I_1}$ in (3.1). That is, the interior equilibrium of the one-virus model can become unstable because of the frequency dependence (see [4]). However, in our fractional order model with the frequency dependence, the interior equilibrium can also become asymptotically stable if the order $\alpha < 0.9916$.

We then consider viral diversity. If the terms associated with immune reaction depend on the mass action law instead of frequency, an interior equilibrium in [24] is GAS. Similar to the integer order model in [4], we find that strange chaotic attractors can be obtained under fractional order model with frequency dependence. That is, the effect of viral diversity and the frequency dependence results in collapse of the immune system and make the behavior of the system dynamics complex. However the chaotic motion may disappear and the fractional order system stabilizes to a fixed point if the value of the order α decreases. The specific biological meaning is deserved to further study.

Acknowledgments

The authors are very grateful to the referees for their valuable suggestions, which helped to improve the paper significantly. This work was supported in part by Specialized Research Fund for the Doctoral Program of Higher Education from Ministry of Education of China (no. 20060255006), Project of the Shanghai Committee of Science and Technology (no. 08JC1400100), and the Open Fund from the Key Laboratory of MICCAI of Shanghai (06dz22103).

References

1. M.A. Nowak, R.M. May, and K. Sigmund, "Immune responses against multiple epitopes," *Journal of Theoretical Biology*, vol. 175, no. 3, pp. 325–353, 1995.

2. M.A. Nowak and C.R.M. Bangham, "Population dynamics of immune responses to persistent viruses," *Science*, vol. 272, no. 5258, pp. 74–79, 1996.

3. M.A. Nowak and R.M. May, *Virus Dynamics: Mathematical Principles of Immunology and Virology*, Oxford University Press, Oxford, UK, 2000.

4. S. Iwami, S. Nakaoka, and Y. Takeuchi, "Frequency dependence and viral diversity imply chaos in an HIV model," *Physica D*, vol. 223, no. 2, pp. 222–228, 2006.

5. E. Ahmed and A.S. Elgazzar, "On fractional order differential equations model for nonlocal epidemics," *Physica A*, vol. 379, no. 2, pp. 607–614, 2007.

6. E. Ahmed, A.M.A. El-Sayed and H.A.A. El-Saka, "Equilibrium points, stability and numerical solutions of fractional-order predator-prey and rabies models," *Journal of Mathematical Analysis and Applications*, vol. 325, no. 1, pp. 542–553, 2007.

7. M.P. Lazarevi´c, "Finite time stability analysis of PDα fractional control of robotic time-delay systems," *Mechanics Research Communications*, vol. 33, no. 2, pp. 269–279, 2006.

8. T.J. Anastasio, "The fractional-order dynamics of brainstem vestibulo-oculomotor neurons," Biological Cybernetics, vol. 72, no. 1, pp. 69–79, 1994.

9. T.T. Hartley, C.F. Lorenzo and H.K. Qammer, "Chaos in a fractional order Chua's system," *IEEE Transactions on Circuits and Systems I*, vol. 42, no. 8, pp. 485–490, 1995.

10. B. Bonilla, M. Rivero, L. Rodr´ıguez-Germ´a, and J.J. Trujillo, "Fractional differential equations as alternative models to nonlinear differential equations," *AppliedMathematics and Computation*, vol. 187, no. 1, pp. 79–88, 2007.

11. A.M.A. El-Sayed, A.E.M. El-Mesiry, and H.A.A. El-Saka, "On the fractional-order logistic equation," *Applied Mathematics Letters*, vol. 20, no. 7, pp. 817–823, 2007.

12. R. Hilfer, Ed., *Applications of Fractional Calculus in Physics*, World Scientific, River Edge, NJ, USA, 2000.

13. K.S. Cole, "Electric conductance of biological systems," in *Proceedings of the Cold Spring Harbor Symposia on Quantitative Biology*, pp. 107–116, Cold Spring Harbor, NY, USA, January 1993.

14. I. Podlubny, *Fractional Differential Equations*, vol. 198 of Mathematics in Science and Engineering, Academic Press, San Diego, Calif, USA, 1999.

15. Z.M. Odibat and N.T. Shawagfeh, "Generalized Taylor's formula," *Applied Mathematics and Computation*, vol. 186, no. 1, pp. 286–293, 2007.

16. A.A. Kilbas, H.M. Srivastava, and J.J. Trujillo, *Theory and Applications of Fractional Differential Equations*, vol. 204 of *North-Holland Mathematics Studies*, Elsevier, Amsterdam, The Netherlands, 2006.

17. D. Matignon, "Stability results for fractional differential equations with applications to control processing," in *Computational Engineering in Systems Applications*, vol. 2, pp. 963–968, IMACS IEEESMC, Lille, France, 1996.

18. J. Velasco-Hern´andez, J. Garc´ıa, and D. Kirschner, "Remarks on modeling host-viral dynamics and treatment," in *Mathematical Approaches for Emerging and Reemerging Infectious Diseases: An Introduction to Models, Methods, and Theory*, vol. 125 of *The IMA Volumes in Mathematics and Its Applications*, pp. 287–308, Springer, New York, NY, USA, 2002.

19. R.R. Regoes, D. Wodarz, and M. A. Nowak, "Virus dynamics: the effect of target cell limitation and immune responses on virus evolution," *Journal of Theoretical Biology*, vol. 191, no. 4, pp. 451–462, 1998.

20. E. Ahmed, A.M.A. El-Sayed, and H.A.A. El-Saka, "On some Routh-Hurwitz conditions for fractional order differential equations and their applications in Lorenz, R¨ossler, Chua and Chen systems," *Physics Letters A*, vol. 358, no. 1, pp. 1–4, 2006.

21. A.S. Perelson, D.E. Kirschner, and R. De Boer, "Dynamics of HIV infection of CD4_ T cells," *Mathematical Biosciences*, vol. 114, no. 1, pp. 81–125, 1993.
22. K. Diethelm, N.J. Ford, and A.D. Freed, "A predictor-corrector approach for the numerical solution of fractional differential equations," *Nonlinear Dynamics*, vol. 29, no. 1–4, pp. 3–22, 2002.
23. K. Diethelm, N.J. Ford, and A.D. Freed, "Detailed error analysis for a fractional Adams method," Numerical Algorithms, vol. 36, no. 1, pp. 31–52, 2004.
24. Y. Iwasa, F. Michor, and M. Nowak, "Some basic properties of immune selection," *Journal of Theoretical Biology*, vol. 229, no. 2, pp. 179–188, 2004.

Research on the Relation of EEG Signal Chaos Characteristics with High-Level Intelligence Activity of Human Brain[†]

Wang Xing-yuan,[1,a] *Meng Juan,*[2] *Tan Gui-lin*[1] and *Zou Li-xian*[1]

[1]School of Electronic & Information Engineering, Dalian University of Technology, Dalian 116024, China.
[a]Email: wangxy@dlut.edu.cn.
[2]School of Information Engineering, Dalian Fisheries University, Dalian 116024, China.
Email: mjalbert@yahoo.cn.

ABSTRACT

Using phase space reconstruct technique from one-dimensional and multi-dimensional time series and the quantitative criterion rule of system chaos, and combining the neural network; analyses, computations and sort are conducted on electroencephalogram (EEG) signals of five kinds of human consciousness activities (relaxation, mental arithmetic of multiplication, mental composition of a letter, visualizing a 3-dimensional object being revolved about an axis, and visualizing numbers being written or erased on a blackboard). Through comparative studies on the determinacy, the phase graph, the power spectra, the approximate entropy, the correlation dimension and the Lyapunov exponent of EEG signals of 5 kinds of consciousness activities, the following conclusions are shown: (1) The statistic results of the deterministic computation indicate that chaos characteristic may lie in human consciousness activities, and central tendency measure (CTM) is consistent with phase graph, so it can be used as a division way of EEG attractor. (2) The analyses of power spectra show that ideology of single subject is almost identical but the frequency channels of different consciousness activities have slight difference. (3) The approximate entropy between different subjects exist discrepancy. Under the same conditions, the larger the approximate entropy of subject is, the better the subject's innovation is. (4) The results of the correlation dimension and the Lyapunov exponent indicate that activities of human brain exist in attractors with fractional dimensions. (5) Nonlinear quantitative criterion rule, which unites the neural network, can classify different kinds of consciousness activities well. In this paper, the results of classification indicate that the consciousness activity of arithmetic has better differentiation degree than that of abstract.

[†]Reused with permission from: X. Wang, J. Meng, G. Tan and L. Zou, Research on the relation of EEG signal chaos characteristics with high-level intelligence activity of human brain, *Nonlinear Biomedical Physics* 2010, 4: 2.

Introduction

EEG signal is a spontaneous bioelectricity activity that is produced by the central nervous system. It includes abundant information about the state and change of the neural system; therefore it is widely used in clinic and neural-electricity physiological research. In recent years, with the development of the nonlinear dynamics, more and more evidences indicate that the brain is a nonlinear dynamic system, and EEG signal can be regarded as its output [1,2]. In 1985, Babloyantz et al. first put forward that II and IV stage EEG signals of human sleep cycle are chaotic [3]. Hereafter, a large number of study results were reported that the EEG was derived from chaotic systems [4–8]. Therefore, people try to analyze EEG signals by way of nonlinear dynamics to get new knowledge of the brain. Lindenberg, Lehnertz and Ferri et al. researched several kinds of physiological and pathologic conditions; and computed the relevant data under various conditions. They point out finally, the nonlinear characteristic of the physiological EEG signals greatly differs from that of the pathology; when clear-headed, the brain has higher chaotic degree, processes information more quickly and can make more responses [9–12]. Chaos is unordered, but in some situations, it has organizing structures and high order and is the source of system information [13]. Therefore, in this paper, we study the relation of chaos characteristic of EEG signals with high-level intelligence activity of human brain through comparative studies of the nonlinear dynamic characteristic of the dynamic physiological EEG information of brain under different consciousness conditions.

Theory and Method

Chaotic system is described by strange attractors in the phase space [13]. In order to construct the phase space, we adopt the phase space reconstruct technique which was put forward by Packard et al. [14] and made reliable mathematical base by Takens [15]. Its principle is: Reconstruct m-dimensional phase space from EEG time series $\{x_n \mid n = 1, 2, \cdots, N\}$, then we get a group of phase space vectors

$$X_i = \{x_i, x_{i+\tau}, \cdots, x_{i+(m-1)\tau}\}, \; i = 1, 2, \cdots, M, X_i \in R^m;$$

where τ is the time-delay; $m \geq 2\delta + 1$, δ is the number of the system independent variables. M is less than N and they have the same order of magnitude. To reconstruct phase space, it is critical to analyze the phase graph, compute correlation dimension and Lyapunov exponent.

CTM Algorithm and the Determinism Computation of EEG Signals

Whether the brain is a deterministic system, determines the applicability of the nonlinear dynamic method of studying EEG signal [16]. Generally, the deterministic computation of the EEG signal requires much data; and supposes the spread of adjacent lines of EEG series in the phase space are similar. However, unstable data often generates false results. CTM algorithm is a method to express the second-order difference plot (SODP) characteristic of trajectory tangent vector quantificationally. It can be used in the deterministic computation of nonlinear time series effectively. This algorithm is real-time, stable and anti-noisy [17].

The tangent vector of trajectory in the reconstructing phase space is

$$Y(t) = x(t+1) - x(t).$$

The angle between the tangent vectors can be expressed by its cosine value

$$A(t) = \frac{Y(t+1) \cdot Y(t)}{\|Y(t+1)\| \|Y(t)\|}.$$

Compared with the angle itself, the cosine value can resist noises better. The SODP of signal expresses the change rate of the tangent vectors angle $A(n+2) - A(n+1)$ to $A(n+1) - A(n)$, its CTM value is

$$CTM = \frac{1}{N-2} \sum_{n=1}^{N-2} \sqrt{\left[A_{n+2} - A_{n+1}\right]^2 + \left[A_{n+1} - A_n\right]^2}.$$

The value of CTM reflects the smooth degree of the attractors' trajectory: the smaller the CTM value is, the less the changes of tangent vector angle, the smoother the trajectory is; and vice versa. The determinacy of the signal S can be measured by the ratio of the CTM value of the EEG series data and the surrogate data. The bigger S is, the stronger the randomicity of EEG signal is. The researches show: the deterministic signal S < 0.3; the random signal S > 0.7; as to part deterministic signal 0.3 < S < 0.7.

Approximate Entropy

In 1991, Pincus put forward a rule to measure the complexity and the statistic quantification of time series, i.e., approximate entropy [18]. The approximate entropy can weigh the probability of creating new pattern of time series. The bigger the probability is, the more complex the time series gets. Because only less data is needed to compute the stable estimated value of the approximate entropy, the approximate entropy is suitable for the classification of nonsteady consciousness EEG signal. For example, the sampling frequency for most EEG machines are between 100–1000Hz, but computing the approximate entropy needs 100–1000 data points, so the EEG data length used for classification can be taken as 0.5–1s. Although there are false mark disturbance and power frequency disturbance while gathering EEG signals, the EEG data needed is very short. So the approximate entropy has strong anti-chirp and antijamming ability. At present, there is still dispute on whether EEG is derived from chaotic systems or disorderly linear random systems [19]. The approximate entropy is suitable for deterministic and random signal, which further shows that the approximate entropy has better practicability.

The concrete algorithm for approximate entropy is described as follows: Suppose the initial data as $x(1), x(2), \cdots, x(N)$.

1. Form a group of m-dimensional vector according to the serial number order: $X(i) = [x(i), x(i+1), \cdots, x(i+m-1)]$ $(i = 1, 2, \cdots, N-m+1)$.

2. Define the distance between $X(i)$ and $X(j)$ as $d[X(i), X(j)] = \max_{k=0 \sim m-1} [|x(i+k) - x(j+k)|]$, and compute the distance $d[X(i), X(j)]$ between $X(i)$ and other vectors $X(j)$ $(j = 1, 2, \cdots, N-m+1; j \neq i)$ for every i value.

3. Given the threshold value r, count the number of $d[X(i), X(j)]$ which is smaller than r for every i value, and compute the ratio of this number to the total distance $N - m$:

$$C_i^m(r) = \frac{1}{N-m} \{\text{number of } [d[X(i), X(j)] < r]\} \ (i = 1, 2, \ldots, N-m+1).$$

4. The average value of i is computed according to logarithm of $C_i^m(r)$:

$$\phi^m(r) = \frac{1}{N-m+1} \sum_{i=1}^{N-m+1} \ln C_i^m(r).$$

5. Add the dimension by 1 again to $m+1$, repeat steps (1) to (4), and compute $C_i^m(r)$ and $\phi^m(r)$.

6. The theoretical value of the approximate entropy is

$$ApEn(m, r) = \lim_{N \to \infty} [\phi^m(r) - \phi^{m+1}(r)].$$

Generally speaking, the boundary value mentioned above exists by probability 1. N can't be ∞ in practice. When N is a finite value, the result is the estimated value of $ApEn$ when the series length is N, which is defined as $ApEn(m, r, N) = \phi^m(r) - \phi^{m+1}(r)$. Obviously, the value of $ApEn$ is related with the value of m and r. According to Pincus's work, $m = 2$ and $r = 0.1 : 0.25SD_x$ are suggested (SD_x is the standard deviation (SD) of initial data $x(i)$ ($i = 1, 2, \ldots, N$).

Multi-lead Correlation Dimension

In the study of nonlinear dynamics of EEG signals, the Takens's time delay reconstruction phase space method used EEG data of single channel record to reconstruct multi-dimensional EEG attractor, which reflects the time correlation of the system. In order to show the characteristic of the system from time and space, Eckmann and Ruelle proposed the multichannel reconstructing (multivariable embedding) method that can show the correlation of space and time simultaneously. When applied in time series with short-time noise, it can avoid problems such as the choice of delayed parameters and system errors with higher embedding dimension. Rombouts et al. thought the multichannel reconstructing method can provide more reliable results [20]. Take EEG signals as an example, recording variable of each lead is taken as a component of the reconstructing vector while reconstructing, the reconstructing dimension is decided by the electrode number of EEG signals.

Based on the multi-lead data, the principal step of computing the correlation dimension with GP algorithm [21] is: The m-dimensional embedding-space $\{X\}$ is got from m-lead observing time series.

1. Suppose $X(n) = \{x_1(n), x_2(n), \ldots, x_m(n)\} \ (n \le N, m \le M)$, here m is the number of the required variables.

2. For a given distance r, compute the correlation integral

$$C(r) = \frac{2}{N(N-1)} \sum_{j=1}^{N-1} \sum_{i=j+w}^{N} H(r - \|X_i - X_j\|), \text{ here } X \text{ is the vector in}$$

embedded space, N is the number of the vector, w is Theiler window, H is Heaviside function.

3. For an enough small r, the correlation integral approaches to the following formula:
   $LnC_m(r) = LnC + d_m Ln(r)$.
4. Evaluate the slope of the fitting straight line in the linearity range of $LnC_m(r) \cong Ln(r)$, namely the estimated value of the correlation dimension D_2.

Generally speaking, the correlation dimension of EEG represents the invariable measure for the self-similarity and the criterion irrelevance of the EEG signal, and shows the complex degree of the EEG signal.

Small Data Sets Method of Computing Lyapunov Exponent

The ordinary method of studying whether the actual observable series has chaotic characteristic or not, is to compute the biggest Lyapunov exponent λ_1 of the observable series. When $\lambda_1 > 0$, the observable system is believed to be chaotic. Since Wolf proposed and computed the Lyapunov exponent according to the observable series in 1985, there are some sophisticated methods in this respect, such as Jacobian method, p norm method and the small data sets method proposed by Rosenstein et al. [22]. The small data sets method is more robust than other methods to embedded dimension of the phase space, the reconstruction time delay, observable noises and so on.

Mark the constructed phase space as $\mathbf{X}=[\mathbf{X}_1, \mathbf{X}_2, ..., \mathbf{X}_N]$, phase point is $\mathbf{X}_j = \left[x_{j-(m-1)J}, x_{j-(m-2)J}, ..., x_j \right] (j=1,2,...,N)$, here N is the total number of the phase points, m is the embedding dimension of the phase space, J is the reconstructing time delay. Generally, $J = k\Delta t$, k is a positive integer, Δt is sampling interval. For $\forall \mathbf{X}_j \in \mathbf{X}$, define $d_j(0) = \inf_{\mathbf{X}_k \in} \|\mathbf{X}_j - \mathbf{X}_k\| = \|\mathbf{X}_j - \mathbf{X}_{\hat{j}}\|$, and $|j - \hat{j}| > p$, p is the average cycle of the time track. If $\exists \mathbf{X}_{j+i} \in \mathbf{X}$ and $\mathbf{X}_{\hat{j}+i} \in \mathbf{X}$, define $d_j(i) = \|\mathbf{X}_{j+i} - \mathbf{X}_{\hat{j}+i}\|$, then the advanced distance $d_j(i)$ has the following approximate relation

$$d_j(i) \approx d_j(0)e^{\lambda_1 i \Delta t} \tag{1}$$

here Δt is the sampling interval or the step length of the observable series; i is the sliding step ordinal of the phase point along the time track. Take natural logarithm to both sides of the formula (1), we can get $\ln d_j(i) \approx \ln d_j(0) + \lambda_1 i \Delta t$. When $d_j(i) = \|\mathbf{X}_{j+i} - \mathbf{X}_{\hat{j}+i}\|$ ($\|\cdot\|$ denotes the vector 2 norm), we get the empirical formula which Rosenstein et al. used to compute λ_1 [22]. In view of the influence of local computation, the last empirical formula is

$$\frac{1}{\Delta t} < \ln d_j(i) > \approx \frac{1}{\Delta t} < \ln d_j(0) > + \lambda_1 i \cdot$$

Here $< \cdot >$ is to get average.

Power Spectra

Using Auto-Regressive (AR) parameter model method to compute the self power spectra estimated value of the EEG signal [23]: The AR model of the EEG time series x_n is provided by the following formula

$$x_n = -\sum_{k=1}^{p} a_k x_{n-k} + w_n \qquad (2)$$

here p is the order of the AR model; $a_k (k = 1,2,\cdots p)$ is AR model parameter; w_n is the unpredictable part of x_n, namely residual error. If the model can well match the EEG time series, w_n should be white noise process. According to the AR model given by formula (2), we can get the estimated value of the AR spectra

$$P_x(\omega) = \frac{\sigma_\omega^2}{\left|A(e^{j\omega})\right|^2} = \frac{\sigma_\omega^2}{\left|1 + \sum_{k=1}^{p} a_k e^{-j\omega k}\right|^2} \qquad (3)$$

here σ_ω^2 is the variance of AR model residual error. From the formulas (2) and (3), we know the key to get the AR spectra estimation is to estimate the AR parameters $a_k (k = 1,2,\cdots p)$ through the EEG time series. Usually, Yule-Walker equation and Levinson-Durbin algorithm are used to estimate AR parameters. In this paper, we use Burg algorithm. Burg algorithm is an autoregression power spectra estimated method, on the premise of Levinson-Durbin recursion restraint, making the sum of the front and back forecast error energy smallest. Burg algorithm avoids the computation of self-correlation function. It can distinguish the extremely close sine signal in low noise signals, and may use less data record to estimate, and the result is extremely close to real values. Moreover, the forecasting error filter obtaining from Burg algorithm is minimum phase. The choice of the model order p is a critical problem in the AR model spectra estimate. If p is too low, it will cause smooth spectra estimate; while if p is too high, it will cause spectral line excursion and spectral line abruption and generate general statistic instability. In this paper, we adopt Akaike information criterion (AIC) to estimate the value of the order

$$AIC(p) = N \ln \hat{\rho}_p + 2p,$$

here N is the number of the data points, $\hat{\rho}_p$ is the estimated value of the white noise variance (forecasting error power) of p order AR model.

SOM Neural Network

The neural network is a highly nonlinear system; and it also shares similar characteristics with brain, so it is used in various classifications extensively. SOM neural network is composed of entire connection neuron array and it is a non-teacher, self-organizing and self-learning network. Its idea is that neurons in different areas of the space have different functions. When the neural network accepts an external input mode, it will be divided into different response areas, and each area has different response characteristics to the input mode.

A typical characteristic of SOM network is that it can generate the characteristic topology classification of input signal on one-dimensional or two-dimensional processing unit array, so the SOM network can extract the pattern characteristics of the input signal.

Generally, SOM network only includes one-dimensional array and two-dimensional array, but it can also be generalized into multidimensional processing unit array. This research uses two-dimensional array. SOM network is made up of the following four parts.

1. Processing unit array. Using to accept the input event and forming "discriminant function" of these signals.
2. Comparison and choice of mechanism. Using to compare the "discriminant function". And choosing one processing unit which has the biggest output value.
3. Partial interconnection action. Using to drive the chosen processing unit and the processing unit closest to it simultaneously.
4. Adaptive process. Using to revise the parameter of driven unit in order to increase its output value to the specific input "discriminant function".

Experiment and Result

EEG Data Source

The data used in this paper is the consciousness activities EEG data of 7 subjects that offered by the EEG research center of Colorado State University [24]. There are five kinds of human consciousness activities, i.e., relaxation, mental arithmetic of multiplication, mental composition of a letter, visualizing a 3-dimensional object being revolved about an axis, and visualizing numbers being written or erased on a blackboard [25]. The experimental process of data acquisition is: Subjects sit in the sound-insulated and light-weak room with the electrode cap and complete some consciousness tasks according to the indications. The corresponding electrical signals of the brain will be recorded. The electrode is laid in C3, C4,P3,P4,O1,O2 and EOG (Electro-Oculogram) altogether 7 leads according to international $10 \sim 20$ system standard. The sampling frequency is 250Hz, the simulative filtering range is $0.1 \sim 100$Hz. Signals polluted seriously by winks are excluded. Experimental data of each consciousness task last 10s. Figure 1 is the EEG signal of subject 1 while relaxing. It is obvious that even under relaxing conditions; healthy people's EEG signals fluctuate in a complicated way, which contains abundant nonlinear dynamic information.

Phase Graph Analysis

Using the phase space reconstruct technique from one-dimensional time series to determine the time delayτ: In the experimental system, it should be through repeated trial method to confirm choice of τ. If τ is undersize, the track of the phase space will approach to a straight line; per contra τ is oversize, the data point will centralize in a small range of the phase space, and we can't get the attractors' local structures from the reconstructed phase graph [13]. Testing repeatedly, we find that selecting $\tau = 3$, data point $N = 2000$, it can well reconstruct the EEG attractors. We construct the EEG attractors of all five kinds of consciousness activities of 7 subjects and find that EEG attractors of various patterns have similar characteristics.

Figure 2 is a representative one. As can be seen from Fig. 2, the attractors' track often rotate in an extremely complex way, even smear a group black in the plane, but there is

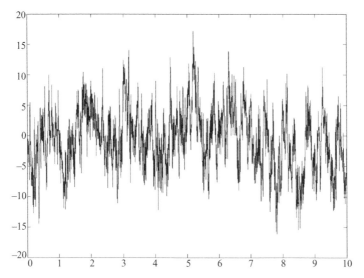

Figure 1 EEG signal waveform of subject 1 while relaxing.

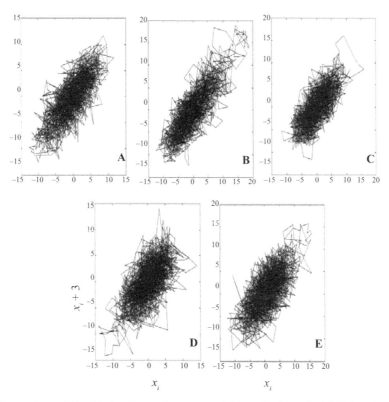

Figure 2 EEG attractors of five kinds of consciousness activities of subject 1. (a) Relaxation. (b) Mental arithmetic of multiplication. (c) Mental composition of a letter. (d) Visualizing a 3-dimensional object being revolved about an axis. (e) Visualizing numbers being written or erased on a blackboard.

still internal structure when the attractors is magnified. The attractors of relaxation, mental composition of a letter and visualizing a 3-dimensional object being revolved about an axis often distribute in a small ellipse region, while the point in the attractors of mental arithmetic of multiplication and visualizing numbers being written or erased on a blackboard centralize nearby the 45 degree line and there is a large distributing range along the 45 degree line. This is because while proceeding rational computation such as mathematics or imagination, the value of the adjacent sampling points of EEG signals are close, and the amplitude values of the whole EEG signals are great.

Power Spectra Analysis

Using the AR parameter model method, we select 250Hz sampling frequency to compute the power spectra of five kinds of tasks' EEG signal of 7 subjects. The parameters used in analysis are: the length of FFT M: 1024; the total number of the data N: 6000; order p: 320. By comparison of the power spectra of five kinds of tasks of 7 subjects, we find that the power spectra of five kinds of tasks for identical subject are similar and meet $1/f$ distribution. As can be seen from Fig. 3, although the attractors' difference is great (Fig. 2 (b) and 2(d)), their power spectra (Fig. 3(a) and 3(b)) show certain similarity. The peak in the high-frequency in Fig.3 is caused by the power frequency disturbance.

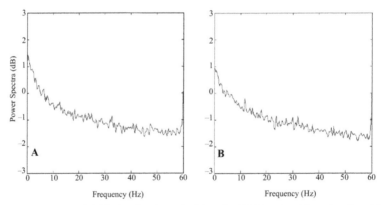

Figure 3 EEG power spectra of 2 kinds of tasks of subject 1. (a) Mental arithmetic of multiplication. (b) Visualizing a 3-dimensional object being revolved about an axis.

Practice prove: The EEG of human can be divided into four frequency sections: δ wave: the frequency is 1-4Hz, appears while sleeping, anaesthetizing deeply, oxygen deficit or the brain with organic disease; θ wave: the frequency is 4–8Hz, appears while feeling sleepy; α wave: the frequency is 8–13Hz, appears while closing eyes with clear-headed; β wave: the frequency is 14–30Hz, appears while opening eyes and looking at things or thinking. As can be seen from Fig. 3, although the spectral lines are similar, there are differences in the active frequency bands (8–30Hz) of different consciousness. So we add the energy of 8–13Hz and 14–30Hz separately in order to use it in SOM network to classify the consciousness.

CTM and the Deterministic Computation of the Signals

EOG signal is the main disturbance of each lead EEG signal, so we make a relevant analysis separately between the gathered EOG signal and another 6 leads in order to find several leads which are disturbed less. We choose $\tau = 3$ and $m = 16$ to compute the CTM. The method of surrogate data [26,27] is used to help detect nonlinear determinism. The surrogate data are linear stochastic time series that have the same power spectra as the EEG signal series. In this paper, we use "iteratively refined surrogate data", which have the same autocorrelation function, Fourier power spectrum, and probability distribution as the EEG time series. More detailed algorithms used in this study are present in the paper of Schreiber and Schmitz [27].

Figure 4 gives the statistic average histogram for each task of 100 times testing. It is obvious that the value of CTM accords with the phase graph 3 well. The statistic average results of the deterministic computations of the EEG signals are in the interval of $0.3 < S < 0.7$. It offers strong support that human brain which contains chaotic component is a highly nonlinear system. But while proceeding deterministic tests, we also find that its value's fluctuation is very big. As an empirical algorithm, when there is less data sample, its application also has certain limitation.

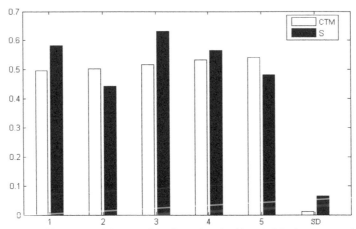

Figure 4 Statistic average histogram of CTM and S of 5 tasks of subject 1 (SD denotes standard deviation).

Approximate Entropy Computation

According to the characteristics of the processing data, we choose $r = 0.5SD_x$ and $r = SD_x$. The approximate entropy to 100 groups of data is computed separately. Because the data gathered from different electrodes may be asynchronous, we make interval eliminations to those unsuitable data. Figure 5 provides the statistic average histogram of the approximate entropy when $r = 0.5SD_x$ and $r = SD_x$.

Correlation Dimension Computation

According to the characteristics of the processing data, we precondition the EEG data first. Namely make a relevant analysis between the EOG and other leads, and sort them according to the order from weak to strong. Then carry through the phase space reconstruction. According to the discussion by Brandstater and Swinney [13]: The fluctuation of partial derivative in scale-free region should be less than 1%. Thus, the scale-free region can be determined. Then the least square method can be used to obtain the correlation dimension. After iterative trials, we found that the correlation dimension can be exactly determined with $m > 12$. Therefore, in these experiments, we choose $\tau = 3$ and $m = 16$ to compute the data of 4 subjects and each contains ten groups separately.

From Fig. 5, we can see, the consciousness activities (task 2 and 5), with more rational consciousness such as arithmetic, have relatively weaker ability to generate new pattern; while those consciousness activities (task 4), with more abstract consciousness such as visualizing graph rotating, have relatively stronger ability to generate new pattern, which means that the time series have more complexity. This also corresponds to the practice. Because mathematical computation is based on fixed rule, its ability to create new pattern ingredient is naturally lower.

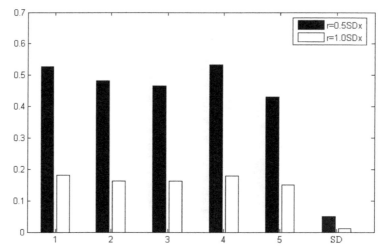

Figure 5 Statistic average histogram of the approximate entropy of subject 1 when $r = 0.5SDx$ and $r = 1.0SDx$.

Figure 6(a) is a representative curve $LnC(r)$ *vs* $Ln(r)$ of subject 1 while relaxing. Figure 6(b) provides the statistic results of the correlation dimension D_2 of 10 groups of data of five kinds of human consciousness activities (each vertical line represents the mean square error range of each task, the crossing point between the crosswise fold line and the vertical line is the mathematic expectation of the task). We can see from Fig. 6(b): For the same subject, do the same kind of tests in different time, its D_2 value may have great fluctuation, which means human brain has different excitable degree in different time slice. Figure 6(b) also shows that the error fluctuation of D_2 is minimum when implementing mathematical computation (task 2). This is because mathematical computation can make the spirit centralized more easily than other consciousness activities. In addition, we also compute the data of D_2 for other 3 subjects and each contains 10 groups of data. These D_2 will also be used in the ideology classification of the SOM.

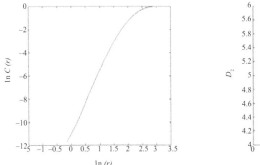 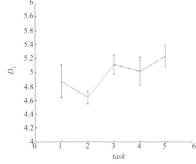

Figure 6 The result of the correlation dimension D_2 of subject 1. (a) $LnC(r)$ vs $Ln(r)$ curve while relaxing. (b) D_2 of 5 kinds of tasks and each contains 10 groups of EEG data.

Lyapunov Exponent Computation

Different consciousness activities stimulate different cerebrum regions, so the computation of single lead signal can't reflect the synthetic Lyapunov exponent of the brain consciousness activity well. The embedded dimension m is determined by iteratively trials. For the delay τ, the phase space of EEG signals is projected into the two-dimension plane. If τ is too small, the attractors will muster around the line $y = x$. If τ is too large, $m\tau$ will be much more than the average period. On this basis, τ is determined by iteratively trials. Furthermore, considering the fact that for different consciousness, different cerebrum region has different activity degree, implement sample splicing to the sampling data of each lead with $\hat{o} = 3$, $m = 16$ to reconstruct the phase space. Figure 7 is the biggest Lyapunov exponent λ_1 of 10 groups of EEG data with five kinds of human consciousness activities (each vertical line represents the result of mean square error range of each task, the crossing point between the histogram and the vertical line is the mathematic expectation of the task). It is obvious that the biggest Lyapunov exponents λ_1 of five human consciousness activities are all bigger than zero, which proves that human brain activity is chaotic.

SOM Network Consciousness Classification

The purpose of the investigation in this paper is to classify the intelligence consciousness activities. From the analysis above, we know that for the same subject, the methods described above may have better differentiation degree; but for different subjects, the above methods have difficulties to classify the consciousness activities, which also indicates that the brain is a highly complicated nonlinear system.

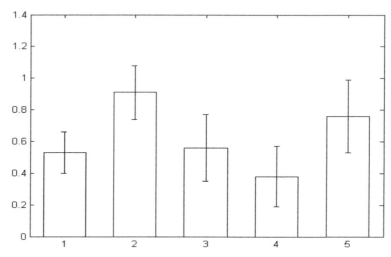

Figure 7 λ_1 of 5 kinds of tasks and 10 groups of EEG data of subject 1.

Therefore, we make the nonlinear criterions (mentioned above) into the prophase processing module, and input them to the input unit of the SOM network. That is, the SOM network has six inputs, including power spectra, CTM, S, approximate entropy, correlation dimension and Lyapunov exponents. According to the tests of the data, the competitive layer of the network is chosen as 8×6 structure. The predicted results of the network are shown in Fig. 8. In Fig. 8, the horizontal ordinate denotes the five outputs of the SOM network, and the vertical ordinate denotes the correct resolution. Figure 8(a) shows the correct resolution histogram of mixed tasks of single subject. In Fig. 8(a), the outputs of the SOM network are the mixed tasks which are combined in turn from the five kinds of human consciousness activities, i.e., relaxation, mental arithmetic of multiplication, mental composition of a letter, visualizing a 3-dimensional object being revolved about an axis, and visualizing numbers being written or erased on a blackboard. For example, "3" represents the combination of three tasks, i.e., relaxation, mental arithmetic of multiplication, and mental composition of a letter. Figure 8(b) shows the correct resolution histogram of four subjects. In Fig. 8(b), the outputs of the SOM network are the five individual tasks mentioned above. As can be seen from Fig. 8(b), the resolution of mathematical computation is relatively higher, while the resolutions of other tasks are about equivalent. The authors think that this is because the nonlinear quantitative parameters of the mathematical computation have great difference compared with other tasks. As can be seen from Fig. 8(a) and Fig. 8(b), the resolution of multi-individual drops obviously relative to single individual. This is because the nonlinear quantitative parameters of two subjects differ greatly, which makes the resolution of the network details drop. There will be better results if there are more individuals to train the network.

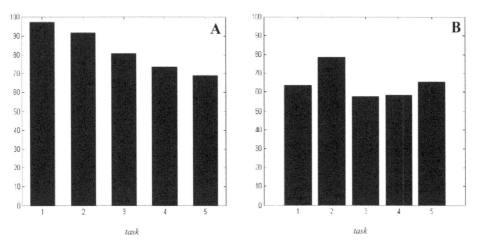

Figure 8 The predicted result of SOM network. (a) The correct resolution histogram of mixed tasks of single subject. (b) The correct resolution histogram of 4 subjects and 5 kinds of tasks'.

Discussion and Conclusion

1. In this paper, we use the determinacy, the phase graph, the power spectra, the approximate entropy, the correlation dimension and the Lyapunov exponent method etc. to study the EEG signal of 5 kinds of consciousness activities of 7 subjects. Although every method has merits and faults, the results show the nonlinear dynamic characteristics of the subject's brain from different perspective. Thereinto, from the deterministic computation we know that the EEG signal is between random signal and deterministic signal. This indicates that the brain may be a chaotic system. The analysis of the power spectra shows that various ideology of single subject is almost identical, but the activity frequency channels for different consciousness activities are different slightly. The analysis of the approximate entropy presents the degree of various consciousness activities on generating new pattern. The approximate entropy of different subjects exist discrepancy. The authors think that at the same state, the larger approximate entropy of the subject, the more innovational he has. The correlation dimension shows the change of chaos of different consciousness activities well, which can better indicate the activity degree of human consciousness, combining with the approximate entropy and the Lyapunov exponent. The above analyses indicate: Different consciousness activities have profound nonlinear dynamic differences. Some differences are difficult to perceive, and the nonlinear quantitative parameters of different individuals have great differences. So it is a critical problem to find a widely applicable criterion, which needs to be explored for a long time.

2. By analyzing the EEG signal of 5 kinds of human consciousness activities, the authors classify the EEG signal through SOM network. The result is almost satisfying. Because the neural network used in this research is classical SOM network, its self-applicability is rather bad. If it can be improved and applied to more samples, there will be better results.

3. The study on profound intelligence activity of human brain needs to integrate the achievements in the fields of life science, physics and modern mathematics. It needs multi-disciplinary cooperation in many aspects, especially the new branch in recent 20 years in mathematic-physics, i.e., the nonlinear theory, to stand on a new height to scan the intelligence activity problem of human brain to achieve the purpose of providing correct quantitative criteria for the intelligence activity of human brain. The task is very arduous. Therefore the theoretical and experimental works of this research should be furthered in the future.

Acknowledgements

This research is supported by the Chinese National Natural Science Foundation (No. 60973152, 60573172) and the Doctoral Program Foundation of Institution of Higher Education of China (No. 20070141014) and China Postdoctoral Science Foundation (No: 20090451268) and the Natural Science Foundation of Liaoning province (No: 20082165).

References

1. P. Faure and H. Korn. Is there chaos in the brain? I. Concepts of nonlinear dynamics and methods of investigation. Life Sciences, 2001, 324: 773–793.

2. H. Korn and P. Faure. Is there chaos in the brain? II. Experimental evidence and related models. Comptes Rendus Biologies, 2003, 326: 787–840.

3. A. Babloyantz, J.M. Salazar and C. Nicolis. Evidence of chaotic dynamics of brain activity during the sleep cycle. Phys Lett A, 1985, 3: 152–156.

4. W.S. Pritchard and D.W. Duke. Measuring "chaos" in the brain: a tutorial review of EEG dimension estimation. Brain and Cognition, 1995, 27: 353–397.

5. A. Babloyantz and A. Destexhe. Low-dimensional chaos in an instance of epilepsy. Proc Natl Ncad Sci, 1986, 83: 3513–3517.

6. H. Preißl, W. Lutzenberger, F. Pulvermüller and N. Birbaunrner. Fractal dimensions of short EEG time series in humans. Neuroscience Letters, 1997, 225: 77–80.

7. C.E. Elger, G. Widman, R. Andrzejak, J. Arnhold, P. David and K. Lehnertz. Nonlinear EEG analysis and its potential role in epileptology. Epilepsia, 2000, 41: 34–38.

8. S.N Sarbadhikari and K. Chakrabarty. Chaos in the brain: a short review alluding to epilepsy, depression, exercise and lateralization. Medical Engineering & Physics, 2001, 23: 445–455.

9. A.M. Lindenberg. The evolution of complexity in human brain development: an EEG study. Electroencephalography and Clinical Neurophysiology, 1996, 99: 405–411.

10. K. Lehnertz. Non-linear time series analysis of intracranial EEG recordings in patients with epilepsy-an overview [J]. International Journal of Psychophysiology, 1999, 34: 45–52.

11. R. Ferri, M. Elia, S.A. Musumeci and C.J. Stam. Nonlinear EEG analysis in children with epilepsy and electrical status epilepticus during slow-wave sleep (ESES). Clinical Neurophysiology, 2001, 112: 2274–2280.

12. R. Ferri, L. Parrino, A. Smerieri, M.G. Terzano, M. Elia, S.A. Musumeci, S. Pettinato and C.J. Stam. Nonlinear EEG measures during sleep: effects of the different sleep stages and cyclic alternating pattern. International Journal of Psychophysiology, 2002, 43: 273–286.
13. X.Y. Wang. Chaos in the complicated non-linear system. Beijing: Electronic Industry Press, 2003, 114–150.
14. N.H. Packard, J.P Crutchfield, J.D. Farmer and R.S. Shaw. Geometry from a time series. Phys Rev Lett, 1980, 45: 712–716.
15. F. Takens. Detection strange attractor in turbulence. In: Rand D.A., Young L.S., eds. Lecture Notes in Mathematica. New York: Springer-verlag, 1981, 366–381.
16. L.W. Salvino and R. Cawley. Smoothness implies determinism: A method to detect it in time series. Phys Rev Lett, 1994, 73 (8): 1091–1094.
17. J. Jeong, J.C. Gore and B.S. Peterson. A method for determinism in short time series, and its application to stationary EEG. IEEE Trans Biomed Eng, 2002, 49: 1374–1379.
18. S.M. Pincus. Approximate entropy as a measure of system complexity. Proc Natl Acad Sci USA, 1991, 88: 2297–2301.
19. L. Glass and M.C. Mackey. From clocks to chaos. Princeton: Princeton University Press, 1988, 24–62.
20. S.A.R.B. Rombouts, R.W.M. Keunen and C.J. Stam. Investigation a nonlinear structure in multichannel EEG. Phy lett A, 1995, 202: 352–358.
21. C. Subramanian, M.A. Cotter and E.S. Robertson. Epstein-Barr virus nuclear protein EBNA-3C interacts with the human metastasis. Nat Med, 2001, 7(3): 350–355.
22. M.T. Rosenstein, J.J. Collins and C.J. De Luca. A practical method for calculating largest Lyapunov exponents from small data sets. Physica D, 1993, 65: 117–134.
23. S.L. Marple. A new autoregressive spectrum analysis algorithm. IEEE Trans On Acoustic, Speech and Signal Processing. 1980, 28(3): 441–454.
24. http:// www. cs. colostate. edu/~anderson/res/eeg/.
25. Z.A. Keirn and J. I Aunon. Man-machine communications through brain-wave processing. IEEE Eng in Med and Bio Magazine, 1990, 28: 55–57.
26. J. Theiler, S. Eubank, A. Longtin, B. Galdrikian and J.D. Farmer. Testing for nonlinearity in time series: the method of surrogate data. Physica D, 1992, 58: 77–94.
27. T. Schreiber and A. Schemitz. Surrogate time series. Physica D, 2000, 142: 346–382.

Nonlinear Filtering of Oscillatory Measurements in Cardiovascular Applications[†]

Ranjan Vepa

School of Engineering and Material Science, Queen Mary, University of London, London, E14NS.
Email: r.vepa@ qmul.ac.uk.

ABSTRACT

An array of nonidentical and locally connected chaotic biological neurons is modelled by a single representative chaotic neuron model based on an extension of the Hindmarsh-Rose neuron. This model is then employed in conjunction with the unscented Kalman filter to study the associated state estimation problem. The archetypal system, which was deliberately chosen to be chaotic, was corrupted with noise. The influence of noise seemed to annihilate the chaotic behaviour. Consequently it was observed that the filter performs quite well in reconstructing the states of the system although the introduction of relatively low noise had a profound effect on the system. Neither the noise-corrupted process model nor the filter gave any indications of chaos. We believe that this behaviour can be generalised and expect that unscented Kalman filtering of the states of a biological neuron is completely feasible even when the uncorrupted process model exhibits chaos. Finally the methodology of the unscented Kalman filter is applied to filter a typical simulated ECG signal using a synthetic model-based approach.

Introduction

Oscillatory signals in the cardiovascular region either originate directly from the sino-atrial node or one of the neurons as an action potential traverses to the ventricle myocytes. Alternatively they are functions or weighted sums of action potentials arising at spatially distributed points. To consider a range of oscillatory measurements in the cardiovascular region it is important to consider the output of a typical neuronal cell.

[†]Reused with permission from: Ranjan Vepa, "Nonlinear Filtering of Oscillatory Measurements in Cardiovascular Applications," *Mathematical Problems in Engineering*, vol. 2010, Article ID 808019, 18 pages, 2010.doi:10.1155/2010/808019.

Neural information is mainly encoded in various firing patterns of a neuron, such as periodic spiking (or bursting) and chaotic spiking (or bursting), travelling among coupled neurons within a physiological domain of neurons such as the heart. The "action potential" is a spontaneously and rhythmically produced electrical impulse in a membrane of neuron cell that occurs during the firing of the neuron due to an exchange of charged ions inside and outside a neural cell. Although not a definition, a dynamic system may be considered chaotic if it exhibits i) sensitive dependence on the initial conditions and ii) a number of dense orbits with a multiplicity of periods for a range of parameters. Two nonlinear dynamic systems with chaotic responses can sometimes exhibit the phenomenon of synchronization when the responses of the two lock-in and seem to drive each other with a common feature such as the phase, phase-lag, amplitude, envelope or even some generalised property that can described in terms of a functional of the features of the response. Physiological observations have confirmed the existence of synchronous motion of neurons in different areas of the heart (Elson et al. [1], Pinto et al. [2] and Szücs, et al. [3]). Synchronization of neurons is possible when a single neuron faithfully encodes the timing of successive peaks, burst or spikes and a group of neurons can respond collectively to a common synaptic current. Moreover, a group of interacting coupled neurons can display various synchronous cardio-vascular rhythms. Several types of synchronization of coupled neurons have been studied under the influence of parameter changes and it is observed that when the coupling strength is above a critical value, certain synchronization mechanisms between neurons can be achieved. This applies both to bursting neurons as well as to neurons exhibiting periodic spikes. The presence of noise can have a profound effect and can enhance synchronization between neurons under certain conditions. Thus it was felt that one could employ a nonlinear filter such as the unscented Kalman filter (UKF) to estimate the states and parameters of an archetypal neuron.

In this paper the state and parameter estimation of an array of non-identical, locally connected chaotic biological neuronal models is considered. It is known that, under certain conditions, even a single biological neuron can exhibit chaotic behaviour. Chaos may be achieved by introducing the nonlinear effects of the chemical and electrical synapses. Alternately the chaotic behaviour of the single biological neuron is achieved by driving it with periodic excitations. The global behaviour of an array of biological neurons may then be investigated by considering a spatial distribution of identical neurons, where spatio-temporal chaos emerges, as well as in presence of spatial diversity, generated by a distribution law which could be stochastic or chaotic. In the latter case, it has been observed that the introduction of spatial disorder enhances the self-organization or synchronisation capability. In particular, in agreement with the results presented in Elson et al. [1], Pinto et al. [2] and Szücs et al. [3], the introduction of spatial diversity generated by such a distribution leads to an improvement in synchronization. While the phenomenon of synchronization in dynamical has been observed over a long time, two or more chaotic systems can be synchronized by linking them with mutual coupling or with a common signal or signals. Ideal synchronisation could be induced by mutually coupling a pair of identical chaotic systems when all trajectories converge to the same value and remain in step with each other during further evolution. Linking chaotic systems given by identical differential-dynamic models but with different system parameters can lead to practical synchronization involving phase synchronization. Initially unexcited biological neural models, subsequently externally excited by periodic oscillators can synchronize both in chaotic and periodic regimes. Provided the amplitudes and frequencies of certain modes are within certain limits it has been observed that a number of independent neurons can exhibit periodic or chaotic behaviour and achieve a regime of complete synchronization including phase synchronization.

In this paper we consider a typical extended four state Hindmarsh-Rose (HR) model (Hindmarsh and Rose, [4]) as a representation of an ensemble of biological neurons. This is preferred over the two-dimensional map model of Rulkov [5] and Shilnikov and Rulkov [6] although the map may be easier to implement in a filter. The neuron model was subjected to the same type of periodic forcing, as the biological neurons. The autonomous periodic bursting pattern of the four-dimensional neuron model was observed to be similar to a biological neuron. The fact that HR model represents an ensemble of biological models is accounted for by introducing low levels process noise. Thus both the process and measurement was assumed to be corrupted by the introduction of very low levels of white noise. The noise had a profound effect on the response of the model as it seemed to annihilate the chaos. The unscented Kalman filtering method was applied to estimate the states of the model. It was observed that the filter performs quite well in reconstructing the states the system, which was deliberately chosen to be chaotic. Neither the filter nor the noise corrupted process model gave any indications of chaos.

Finally the methodology is applied to the Electro Cardiogram (ECG) measurements which are modelled as oscillatory signals using a synthetic model first proposed by McSharry et al. [7]. Like the Hindmarsh-Rose model it exhibits limit cycle oscillations and chaos, and can represent the primary characteristic (P, Q, R, S, T) points in an ECG. The methodology of the UKF is used to filter and reconstruct a measured ECG signal and validated by simulation.

Chaotic Model of a Neuron

The analysis of biological neurons had shown they could be modelled with only three or four states, we chose initially to use a familiar simplified model put forward by Hindmarsh and Rose, [4]. The general form of this model contains three terms:

$$\begin{bmatrix} \dot{x} \\ \dot{y} \\ \dot{z} \end{bmatrix} = \begin{bmatrix} ax^2 - x^3 + y - z \\ bx^2 + y_0 - y \\ \mu(S(x + x_0) - z) \end{bmatrix} + \begin{bmatrix} I_0 + I_{\sin}(t) \\ 0 \\ 0 \end{bmatrix} \tag{1}$$

It is hard to establish a one to one correspondence between the states of the HR neuron and the states of a biological neuron. Yet the HR neuron model seems to reproduce the overall behaviour of the action potential fairly accurately. After appropriate scaling, the output of the HR neuron model can be made to lie within the same nominal limits as a biological neuron, -65 mV$<V<20$mV. Furthermore the other principal states of the HR neuron show the same behaviour as the principal compartmental currents and gating variables that can be established by considering the diffusion of ionic charge carriers from one compartment to the other. The net result of this type of diffusion is the generation of a potential difference, across the two compartments which can be described by the Nernst equation. For this reason the HR neuron may be employed as a representative model for constructing reduced order observers of the neuron dynamics. The three equations in (1) represent the original HR model where $x(t)$ corresponds to membrane voltage, $y(t)$ represents a 'fast' current and by making $\mu \ll 1$, $z(t)$ a 'slow' current. These three equations (the 3-state model) can produce several modes of spiking-bursting activity including a regime of chaos that appears similar to that seen in biological neurons. However the parameter space for the chaotic behaviour is much more restricted than we observe in real neurons. Following Szücs, et al. [3], the chaotic regime is greatly expanded by incorporation of the fourth term into the model:

$$\begin{bmatrix} \dot{x} \\ \dot{y} \\ \dot{z} \\ \dot{w} \end{bmatrix} = \begin{bmatrix} 3x^2 - x^3 + y - z \\ 1.8 + 7x^2 - y - w/80 \\ 0.0021(4(x+1.56)-z) \\ v(-w+0.88(y-0.9)) \end{bmatrix} + \begin{bmatrix} I_0 + I_{\sin}(t) \\ 0 \\ 0 \\ 0 \end{bmatrix} \qquad (2)$$

where, $I_{\sin}(t) = A\sin(2f)$ and $v = 0.0004$.

Adding the term $w(t)$ to introduce an even slower process ($v < \mu \ll 1$) is intended to represent the dynamics of intracellular *Calcium^{2+}* (Ca^{2+}) ions. To couple the additional equation to the original three state HR model a $- g * w(t)$ term is included in the second equation. When this term is taken into account, the model produces simulations of intracellular activity that are even more similar to the biological observations. However it is not known as yet if the $w(t)$ term actually represents Ca^{2+} ion kinetics in sino-atrial node (SAN) and ventricular neurons and numerical simulations are currently under way to compare Ca^{2+} transients in HR neurons using realistic biological models. Because of its relative simplicity, the extended HR model was extremely useful in constructing a simulation model that could perform the computations necessary to emulate SAN neurons in real time. A similar model has been employed by Mayer, Schuster and Claussen [8] to model thalamocortical circuits. Although this simplified model is difficult to compare with biological neurons which are made up of a multitude of individual conductance and compartments, we expect the model to provide us with the experience in estimating the states of a real biological model. Since the estimation of these states and parameters is crucial in establishing physiological mechanisms, we also developed several multi-compartmental type models that provide a more biologically realistic representation of the nonlinear voltage-current relations than that of Hindmarsh and Rose. Röbenack and Goel [9] and Goel and Röbenack [10], demonstrated that it was possible to reconstruct the currents and gate dynamics from measurements of the action potential, by using a "reduced order observer". An observer is an electronic circuit that is expected to reconstruct the internal dynamics of a system, whatever the nature of the dynamics may be, solely from the measurements, in such way that the error between the actual signal and its reconstruction is asymptotically stable. Goel and Röbenack employed a four and a six state model to construct their observer. While in this work the HR neuron model has been employed to demonstrate the viability of successfully observing the state of a neuron, the application of the methodology to a multi-compartmental biologically inspired model will presumably facilitate the reconstruction of the internal dynamics within the cell using measurements of the action potential. We accordingly employed a modified Hodgkin and Huxley type (Hodgkin and Huxley, [11]) seven state model to reconstruct all the states of the system. When this model, as well as several other biologically inspired models, was used to construct UKF based state estimators from a biological neural measurement in our first attempt, all of these models exhibited filter instability. Further analysis indicated that this could be due to one of three reasons: i) the chaotic nature of the dynamics ii) unobservability due to inadequate measurements and iii) the non-linear functions arising from the Nernst equations for the compartmental currents due to the ionic concentrations and the sigmoid like functions associated with the gate time constants and final values, which must lie within the prescribed final values. The question of unobservability was dealt with by including a range of simulated measurements. It was essential to identify which of the remaining reasons was the predominant cause for the filter instability. So it was decided to first eliminate the possibility of the chaotic dynamics being the primary factor in causing the filter instability. For this reason before employing these models it was decided to apply the UKF to the simplified extended HR model. In this context we note that observers and the extended Kalman filtering have been applied in the past to construct neural estimators by Cruz and Nijmeijer [12]. The reconstruction of the neural dynamics has been considered by Steur et al. [13] and Tyukin et al. [14] have considered the application of adaptive observers to neural systems. However our objective is to be to reconstruct the action potential and its features such as the duration, particularly of a group of spatially distributed neurons, over an extended time frame, and to ultimately extend the application to complex multi-compartmental models of biological neurons. An adaptive nonlinear observer wherein the gain of the observer is continually modified in accordance with the magnitude of the measurements and noise statistics by an appropriate adaption law would be more suitable in this case than a conventional nonlinear observer like the UKF. The neuron model described by (2), which represents a typical nonlinear oscillator, is described in the parameter plane with the coordinates as the amplitude and frequency of the forcing. (Glass and Mackey [15].) A study of the response characteristics of this model reveals sub-harmonic and super-harmonic synchronization or chaotic behaviour, depending on the amplitude and frequency of the forcing. In some cases the chaos occurs after a period-doubling bifurcation. For the parameter set considered in (2) the response is chaotic. However the addition of a relatively small level of noise to the initial conditions seemed to completely annihilate the chaos. A typical response of the model is shown in Fig.1a and the magnified plot in Fig. 1b and it illustrates the fact that the response is chaotic. In Fig.1a the second state is scaled down by 400 to plot it on the same figure. This can be demonstrated by a one-dimensional Poincaré plot (Abarbanel [16]). The Poincaré map corresponding to Fig.1b shows that the system is chaotic and is shown in Fig. 1c.

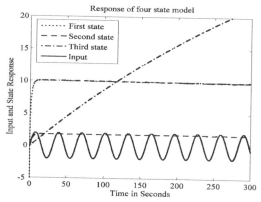

Figure 1a Extended HR neuron: State response plot comparing response of three states to input.

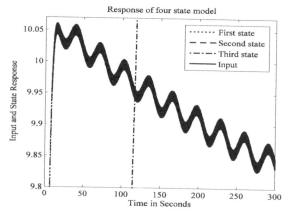

Figure 1b Extended HR neuron: Close-up of state response plot of first two states illustrating chaos.

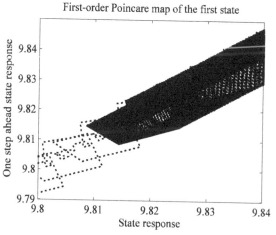

Figure 1c Extended HR neuron: Close-up of the Poincaré map of the chaotic first state corresponding to Fig. 1b.

The Unscented Kalman Filter

Most dynamic models employed for purposes of estimation neural action potential signals are generally not linear. To extend and overcome the limitations of linear models, a number of approaches such as the extended Kalman filter (EKF) have been proposed in the literature for nonlinear estimation using a variety of approaches. Unlike the Kalman filter, the EKF may diverge, if the consecutive linearizations are not a good approximation of the linear model over the entire uncertainty domain. Yet the EKF provides a simple and practical approach to dealing with essential non-linear dynamics.

The main difficulty in applying the EKF algorithm to problems related to the estimation of a neural action potential signal is in determining the proper Jacobian matrices. The UKF is a feasible alternative that has been proposed to overcome this difficulty, by Julier, Uhlmann and Durrant-Whyte [17] as an effective way of applying the Kalman filter to nonlinear systems. It is based on the intuitive concept that it is easier to approximate a probability distribution than it is to approximate an arbitrary nonlinear function or transformation, of a random variable.

The UKF gets its name from the unscented transformation, which is a method of calculating the mean and covariance of a random variable undergoing nonlinear transformation $y = f(w)$. Although it is a derivative free approach, it does not really address the divergence problem. In essence the method constructs a set of *sigma vectors* and propagates them through the same non-linear function. The mean and covariance of the transformed vector are approximated as a weighted sum of the transformed *sigma vectors* and their covariance matrices.

Consider a random variable w with dimension L which is going through the nonlinear transformation, $y = f(w)$. The initial conditions are that w has a mean $\overline{w}$ and a covariance $\mathbf{P}_{ww}$. To calculate the statistics of y, a matrix χ of $2L+1$ sigma vectors is formed. Sigma vector points are calculated according to the following equations:

$$\chi_0 = \overline{w} \tag{3a}$$

$$\chi_i = \overline{w} + \left(\sqrt{(L+\lambda_{us}) \, \mathbf{P}_{ww}} \, \right)_i, \, i = 1, 2, \ldots, L, \tag{3b}$$

$$\chi_i = \overline{w} - \left(\sqrt{(L+\lambda_{us}) \, \mathbf{P}_{ww}} \, \right)_i, \, i = L+1, L+2, \ldots, 2L, \tag{3c}$$

where, $\lambda_{us} = \alpha_{us}^2 (L + \kappa) - L$, α_{us} is a scaling parameter between 0 and 1 and κ is a secondary scaling parameter. $\left(\sqrt{(L+\lambda_{us}) \mathbf{P}_{ww}} \, \right)_i$ is the ith column of the matrix square root. This matrix square root can be obtained by Cholesky factorization. The weights associated with the sigma vectors are calculated from the following [18]:

$$W_0^{(m)} = \lambda_{us} / (L + \lambda_{us}) \tag{4a}$$

$$W_0^{(c)} = (\lambda_{us} / (L + \lambda_{us})) + 1 - \alpha_{us}^2 + \beta \tag{4b}$$

$$W_i^{(m)} = W_i^{(c)} = 1/2 \, (L + \lambda_{us}), \, i = 1, 2, \ldots, 2L, \tag{4c}$$

where β is chosen as 2 for Gaussian distributed variables. We have chosen to use the scaled unscented transformation proposed by Julier [18], as this transformation gives one the added flexibility of scaling the sigma points to ensure that the covariance matrices are always positive definite. The mean, covariance and cross-covariance of y calculated using the unscented transformation are given by,

$$y_i = \mathbf{f}(\chi_i) \tag{5a}$$

$$\overline{\mathbf{y}} \approx \sum_{i=0}^{2L} W_i^{(m)} \mathbf{y}_i \tag{5b}$$

$$\mathbf{P}_{yy} \approx \sum_{i=0}^{2L} W_i^{(c)} \left(\mathbf{y}_i - \overline{\mathbf{y}}\right)\left(\mathbf{y}_i - \overline{\mathbf{y}}\right)^T \tag{5c}$$

$$\mathbf{P}_{wy} \approx \sum_{i=0}^{2L} W_i^{(c)} \left(\chi_i - \overline{\chi}\right)\left(\mathbf{y}_i - \overline{\mathbf{y}}\right)^T \tag{5d}$$

where $W_i^{(m)}$ and $W_i^{(c)}$ are the set of weights defined in a manner so approximations of the mean and covariance are accurate up to third order for Gaussian inputs for all nonlinearities, and to at least second order for non-Gaussian inputs. The sigma points in the sigma vectors are updated using the nonlinear model equations without any linearization.

Given a general discrete nonlinear dynamic system in the form,

$$\mathbf{x}_{k+1} = \mathbf{f}_k\left(\mathbf{x}_k, \mathbf{u}_k\right) + \mathbf{w}_k , \quad \mathbf{y}_k = \mathbf{h}_k\left(\mathbf{x}_k\right) + \mathbf{v}_k \tag{6}$$

where $\mathbf{x}_k \in R^n$ is the state vector, $\mathbf{u}_k \in R^r$ is the known input vector, $\mathbf{y}_k \in R^m$ is the output vector at time k. $\mathbf{w}_k$ and $\mathbf{v}_k$ are, respectively, the disturbance or process noise and sensor noise vectors, which are assumed to Gaussian white noise with zero mean. Furthermore $\mathbf{Q}_k$ and $\mathbf{R}_k$ are assumed to be the covariance matrices of the process noise sequence, $\mathbf{w}_k$ and the measurement noise sequence, $\mathbf{v}_k$ respectively.

The unscented transformations of the states are denoted as,

$$\mathbf{f}_k^{UT} = \mathbf{f}_k^{UT}\left(\mathbf{x}_k, \mathbf{u}_k\right), \quad \mathbf{h}_k^{UT} = \mathbf{h}_k^{UT}\left(\mathbf{x}_k\right) \tag{7}$$

while the transformed covariance matrices and cross-covariance are respectively denoted as,

$$\mathbf{P}_k^{ff} = \mathbf{P}_k^{ff}\left(\hat{\mathbf{x}}_k, \mathbf{u}_k\right), \tag{8a}$$

$$\mathbf{P}_k^{hh-} = \mathbf{P}_k^{hh}\left(\hat{\mathbf{x}}_k^-\right) \tag{8b}$$

and

$$\mathbf{P}_k^{xh-} = \mathbf{P}_k^{xh-}\left(\hat{\mathbf{x}}_k^-, \mathbf{u}_k\right). \tag{8c}$$

The UKF estimator can then be expressed in a compact form. The state time-update equation, the propagated covariance, the Kalman gain, the state estimate and the updated covariance are respectively given by,

$$\hat{\mathbf{x}}_k^- = \mathbf{f}_{k-1}^{UT}\left(\hat{\mathbf{x}}_{k-1}\right) \tag{9a}$$

$$\hat{\mathbf{P}}_k^- = \mathbf{P}_{k-1}^{ff} + \mathbf{Q}_{k-1} \tag{9b}$$

$$\mathbf{K}_k = \hat{\mathbf{P}}_k^{xh-}\left(\hat{\mathbf{P}}_k^{hh-} + \mathbf{R}_k\right)^{-1} \tag{9c}$$

$$\hat{\mathbf{x}}_k = \hat{\mathbf{x}}_k^- + \mathbf{K}_k\left[\mathbf{y}_k - \mathbf{h}_k^{UT}\left(\hat{\mathbf{x}}_k^-\right)\right] \tag{9d}$$

$$\hat{\mathbf{P}}_\mathbf{k} = \hat{\mathbf{P}}_k^- - \mathbf{K}_k\left(\hat{\mathbf{P}}_k^{hh-} + \mathbf{R}_k\right)^{-1}\mathbf{K}_k^T. \tag{9e}$$

Equations (9) are in the same form as the traditional Kalman filter and the EKF. Thus higher order non-linear models capturing significant aspects of the dynamics may be employed to ensure that the Kalman filter algorithm can be implemented to effectively estimate the states in practice. For our purposes we adopt the both the UKF approach to estimate the neuron states in the process model.

The UKF is based on approximating the probability distribution function than to approximating a nonlinear function as in the case of EKF. The state distributions are approximated by a Gaussian probability density, which is represented by a set of deterministically chosen sample points. The nonlinear filtering using the Gaussian representation of the posterior probability density via a set of deterministically chosen sample points is the basis for the UKF. It is based on statistical linearization of the state dynamics rather than analytical linearization (as in the EKF). The statistical linearization is performed by employing linear regression using a set of regression (sample) points. The sigma points are chosen as the regression points. The mean and covariance at the sigma points then represent the true mean and covariance of the random variable with the particular Gaussian probability density. Thus when transformed to the nonlinear systems, they represent the true mean and covariance accurately only to the second order of the nonlinearity. Thus this can be a severe limitation of the UKF unless the nonlinearities can be limited to the first and second order in the process model.

UKF Estimation Applied to a Neuron Model

The success of the application of the UKF depends largely on the approximation to the covariance which is estimated as a weighted linear sum of the covariance at the sigma points. When this approximation is such that the covariance is not positive definite the UKF algorithm fails as the Cholesky decomposition is not possible. To ensure that this covariance it is essential adjust the scaling parameter α_{us}, if and when necessary. In the example illustrated, α_{us} was chosen to be very small positive number. First to see the need for the UKF, the traditional extended Kalman filter (EKF) is also applied to the same responses and the two sets of results compared. These comparisons are shown in Fig. 2. Figure 2 also illustrates the simulated neuron model states plotted to the same scale. While the state estimate obtained by the UKF and EKF are almost the same in the case of first state (which was measured) the EKF estimates of all the other states tend to zero. Although in the case of the third state z, the EKF seems to perform better than the UKF, the state estimate in this case as well tends to zero in steady state. This may be due to the inadequacy of the number of measurements but it is natural to assume that the internal states cannot be measured. Given that only the first state can be measured the UKF definitely tends to perform better than the traditional EKF.

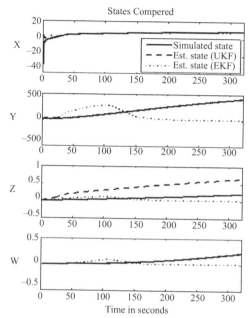

Figure 2 Neuron model states: Plots of the simulated and estimated states on the same scale versus the time in seconds.

In Fig. 3 are show the corresponding errors in the simulated states and UKF estimated states over the same time frame. Figure 4 shows the simulated measurement error of a typical sensor. Finally it must be said that the filter was run over a much longer time frame and the performance of the filter did not deteriorate in spite of this long term operation. Thus the implementation of an UKF based state estimator for the HR neuron is successfully demonstrated over a relatively long time frame.

In particular we observe the relatively large error in the third state, z. We also note this error does not significantly influence the error in the estimate of the first state. The addition of a relatively small level of noise to the initial conditions seems to have the effect of generating a response that completely shrouds and annihilates the chaotic behaviour and this appears to be a consequence of the sensitive dependence of the initial conditions as well.

What appears to be noise in the response may well be a combination of both noise and chaos, but it is not possible to distinguish between the two. This significant change in the response of z in the estimator, which can be recognized by comparing Figs.1 and 2, explains the reason for the chaos being annihilated as this state plays a key role in the appearance of the chaotic response in the first state. In fact it acts like a switch or gate and the addition of noise to the equation for z changes the dynamics of its mean value quite significantly which in turn is responsible for switching off the chaos. However we also observe that this not a feature of the estimator but is a result of the addition of noise to the Hiindmarsh-Rose model of the dynamics of the neuron. We had also observed that when no chaos was present and z was already well behaved, the introduction of noise was not so significant.

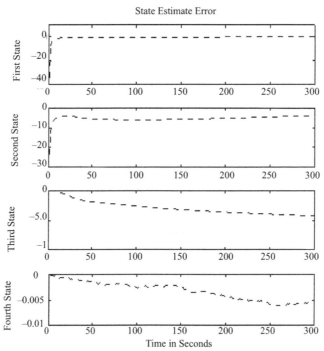

Figure 3 Neuron model states: Plots of the UKF estimate errors in the state variables versus the time in seconds.

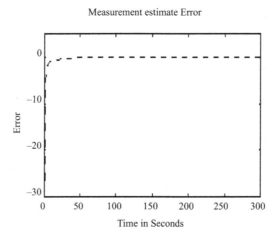

Figure 4 Neuron model states: Plot of the evolution of estimated measurement error versus the time in seconds.

Application to ECG Estimation

McSharry et al. [7] have proposed a theoretical non-linear dynamic model that is capable of emulating an ECG, which is characterised by several parameters that are adaptable to many measured ECG signals. A typical ECG signal, shown in Fig. 5, is characterised by six important points labelled as *P, Q, R, S, T* or *U*. These points define the "fiducial" points which are the landmarks on the ECG signal such as the isoelectric line (*PQ* junction), the onset of individual waves such as *QRS* complex and the *P* and *T* waves and the *PQ, QT* and *ST* time intervals. The ECG signal is periodic and the period is the elapsed time between two *R-R* peaks. The circular radian frequency $\omega = 2\pi/T_{R-R}$ is related to the *Heart Rate*.

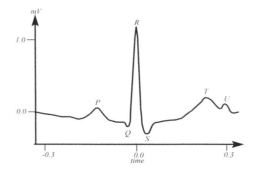

Figure 5 A typical R-R ECG signal.

The heart rate is by no means steady as several rhythmic variations are known to influence it. Coupling between the heart rate and the respiratory cycle causes oscillations in the heart rate at about 0.25Hz and is termed as the respiratory sinus arrhythmia. *Heart Rate Variability* (HRV) influences the fiducial points and is controlled by the baroreflex regulatory feedback. The baroreflex feedback mechanism is modelled by a nonlinear delay-differential equation by McSharry, McGuinness and Fowler [19] based a model by Fowler and McGuinness [20] to capture and to describe the interactions between the heart rate and blood pressure. The model gives rise to the oscillations in the blood pressure known as *Mayer waves* with a time period ranging from 10–25 seconds, due to the presence of a time delay. The model maintains an intrinsically stable heart rate in the absence of nervous control, and features baroreflex influence on both heart rate and peripheral resistance. Irregularities in the baroreflex feedback which can create disturbances in the blood pressure such as the *Mayer waves* that manifest themselves in some form in the ECG signal. The *Mayer waves* and the heart rate variability modelling have also been studied by Seydnejad and Kitney [21]. Analysis of Heart rate variability is also the basis for the assessment of the sympathetic and parasympathetic responses of the autonomic nervous system, with the sympathetic tone influencing the low frequency spectrum only while both the sympathetic and parasympathetic responses influence the high frequency component of the ECG spectrum. Consequently the heart rate estimation generally involves both ECG and additional measurements of the arterial blood pressure and/ or features associated with the respiratory system. For this reason in this paper the heart rate is assumed to be either known or independently estimated.

The original model proposed by McSharry et al. [7] generates a trajectory in a three-dimensional state space with co-ordinates (x, y, z). The ECG is plot of the z-co-ordinate with respect to time. An observation of the responses shows that they exhibit a limit-cycle behaviour and that it is not sinusoidal.

The dynamical equations of motion are given by a set of three ordinary differential equations,

$$\dot{x} = (1-r)x - \omega_1 y, \tag{10a}$$

$$\dot{y} = (1-r)y + \omega_1 x, \tag{10b}$$

$$\dot{z} = -\sum_{i=1}^{5} a_i \Delta\theta_i \exp\left(-\Delta\theta_i^2 / 2b_i^2\right) - (z - z_0) \tag{10c}$$

where $r = \sqrt{x^2 + y^2}$, $\Delta\theta_i = (\theta - \theta_i)\bmod 2\pi$, $\theta = \text{atan2}(y, x)$ is the four-quadrant inverse tangent (arctangent) given the sine (y) and cosine (x) of the angle θ defined in the range $-\pi < \theta \le \pi$, and ω_1 is the angular velocity of the trajectory as it moves around the limit cycle which is assumed to be either measured or estimated adaptively and hence is treated as a known parameter. The baseline value of z_0 in (10c) is assumed to be driven by the respiratory circular frequency ω_2 according to,

$$z_0(t) = A_0 \sin(\omega_2 t) \tag{11}$$

where the constant $A_0 = 0.15$ mV. These equations of motion may be integrated numerically using the MATLAB built-in m-file *ode45.m* which is based on an explicit Dormand-Prince Runge-Kutta (4,5) pair of formulae over each fixed time step $\Delta t = 1/f_s$ where f_s is the sampling frequency. Equation (10c) may be expressed as,

$$\dot{z} = -\sum_{i=1}^{5} a_i \Delta\theta_i \exp\left(-\gamma_i \Delta\theta_i^2\right) - (z - z_0), \; \gamma_i = 1/\left(2b_i^2\right). \tag{12}$$

The parameters of the modified representation of the (10c) given by (12) are defined in Table 1.

As rightly pointed by Sameni, Shamsollahi, Jutten and Babaie-Zadeh [22], the first two equations (10a) and (10b) could be transformed two other dynamic equations in terms of

$$r = \sqrt{x^2 + y^2} \tag{13a}$$

and

$$\theta = \text{atan2}(y, x). \tag{13b}$$

The r-dynamics takes the form, $\dot{r} = f(r)$ and is essentially unobservable. Consequently the equations (10a) and (10b) may be replaced by, $\dot{\theta} = \omega_1$. Thus equations (10a), (10b) and (12) may now be augmented by additional state equations and expressed as,

$$\dot{\theta} = \omega_1, \tag{14a}$$

$$\dot{z} = -\sum_{i=1}^{5} a_i \Delta\theta_i \exp\left(-\gamma_i \Delta\theta_i^2\right) - \left(z - A_0 \sin\phi_2\right).$$

(14b)

$$\dot{a}_i = 0, \ \dot{\gamma}_i = 0, \ i = 1, 2, \ldots, 5, \ \dot{\phi}_2 = \omega_2$$

(14c)

Table 1 Parameters of the ECG model given by (1).

Index (i)	1 (P)	2 (Q)	3 ®	4 (S)	5 (T)
Time (secs)	-0.2	-0.05	0	0.05	0.3
θ_i (radians)	$-\dfrac{1}{3}\pi$	$-\dfrac{1}{12}\pi$	0	$\dfrac{1}{12}\pi$	$\dfrac{1}{12}\pi$
a_i	1.2	-5.0	30.0	-7.5	0.75
γ_i	8.000	50.00	50.00	50.0	3.125

Equations (14a) and (14b) represent a classic pair of first order equations that exhibit both limit cycle and chaotic behaviour. The complete set of 13 equations characterised by eight parameters θ_i, i =1, 2, …, 5, ω_1, ω_2 and A_0 represents a dynamic model of the ECG with typical initial conditions as illustrated in Table 2. In addition one could assume that the state space dynamics includes a number of disturbances. The state space equations including the random White noise disturbances are given by (15a)–(15c) as,

$$\dot{\theta} = \omega_1 + w_1, \qquad \dot{z} = -\sum_{i=1}^{5} a_i \Delta\theta_i \exp\left(-\gamma_i \Delta\theta_i^2\right) - \left(z - A_0 \sin\phi_2\right) + w_2$$

(15a)

$$\dot{z} = -\sum_{i=1}^{5} a_i \Delta\theta_i \exp\left(-\gamma_i \Delta\theta_i^2\right) - \left(z - A_0 \sin\phi_2\right) + w_2.$$

(15b)

$$\dot{\phi}_2 = \omega_2 + w_3, \ \dot{a}_i = w_{i+3}, \ \dot{\gamma}_i = w_{i+8}, \ i = 1, 2, \ldots, 5,$$

(15c)

with the set w_j, i=1,2, …,13, are zero mean white noise process disturbances with a known covariance matrix.

Table 2 Typical initial conditions for the states in (5).

θ	1.5π	a_1	1.2	γ_1	50.0
Z	-0.0110	a_2	-5.0	γ_2	34.7222
ϕ_1	0	a_3	15.0	γ_3	55.4017
		a_4	-7.5	γ_4	78.1250
		a_5	0.75	γ_5	8.0

Given a set of continuously sampled ECG measurements, the measurements may be expressed by the equation,

$$z_m = z + v$$

(16)

with v is a zero mean White noise measurement disturbance with a known covariance. The UKF may be employed to estimate the states θ, z and the augmented states ϕ_1, a_i, γ_i, i=1,2, …,5. In Fig. 6 a typical set of simulated and estimated responses of the states θ,

z are compared. In Fig. 7 are shown the errors in the estimate over 10,000 time steps, $\Delta t = 0.0002s$.

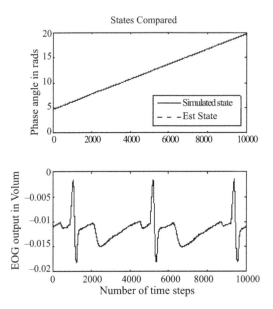

Figure 6 Comparison of simulated and estimated responses of the states θ, z plotted against the number of time steps.

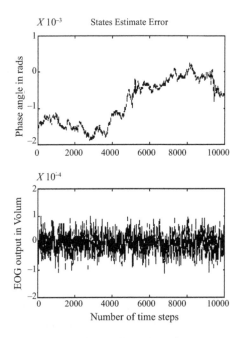

Figure 7 Errors in the estimated responses of the states θ, z, plotted against the number of time steps.

In Fig. 8 is shown a typical estimated error in the measurement. Thus the UKF is capable of performing extremely well given the measurements with well behave covariance characteristics. When the noise covariance matrices are unknown it is possible to estimate the states adaptively. The filter is currently undergoing extensive tests with actual measured ECG data and in this case the adaptive estimation appears not only to be more appropriate but also performs better than the non-adaptive UKF. A complete discussion of the application of the adaptive UKF to ECG measurements where the process and measurement noise covariance matrices are recursively updated, is beyond the scope of this paper and will be presented elsewhere.

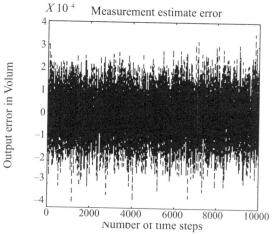

Figure 8 Measurement estimate error, plotted against the number of time steps.

Conclusions and Discussion

The unscented Kalman filtering method was applied to estimate the states of a HR like neuron model which in the absence of noise was deliberately chosen to be chaotic. The process and measurement was then corrupted by the introduction of very low levels of white noise. The noise had a profound effect on the response of the model as it seemed to annihilate the chaos. It was observed that the filter performs quite well in reconstructing the states the system. Neither the filter nor the noise corrupted process model gave any indications of chaos. Moreover the exercise gave us valuable experience in applying the UKF to a biological neuron.

Preliminary studies of the application of the UKF to a Hodgkin-Huxley type model indicated that the successful application of the unscented approach to an ensemble of biological neurons was feasible, provided the sigma points were scaled according certain scaling laws related to the gate constants. Finally the methodology of the unscented Kalman filter is successfully applied to filter a typical simulated ECG signal using a synthetic model based approach.

References

1. R.C. Elson, A.I. Selverston, R. Huerta, N.F. Rulkov, M.I. Rabinovich and H.D.I. Abarbanel, "Synchronous behavior of two coupled biological neurons," *Physical Review Letters*, vol. 81, no. 25, pp. 5692–5695, 1998.

2. R.D. Pinto, P. Varona, A.R. Volkovskii, A. Sz¨ucs, H.D.I. Abarbanel and M.I. Rabinovich, "Synchronous behavior of two coupled electronic neurons," *Physical Review E*, vol. 62, no. 2, pp. 2644–2656, 2000.

3. A. Szucs, R.C. Elson, M.I. Rabinovich, H.D.I. Abarbanel and A. I. Selverston, "Nonlinear behavior of sinusoidally forced pyloric pacemaker neurons," *Journal of Neurophysiology*, vol. 85, no. 4, pp. 1623–1638, 2001.

4. J.L. Hindmarsh and R. M. Rose, "A model of neuronal bursting using three coupled first order differential equations," *Proceedings of the Royal Society of London B*, vol. 221, no. 1222, pp. 87–102, 1984.

5. N.F. Rulkov, "Modeling of spiking-bursting neural behavior using two-dimensional map," *Physical Review E*, vol. 65, no. 4, Article ID 041922, 9 pages, 2002.

6. A.L. Shilnikov and N.F. Rulkov, "Subthreshold oscillations in a map-based neuron model," *Physics Letters A*, vol. 328, no. 2-3, pp. 177–184, 2004.

7. P.E. McSharry, G.D. Clifford, L. Tarassenko and L.A. Smith, "A dynamical model for generating synthetic electrocardiogram signals," *IEEE Transactions on Biomedical Engineering*, vol. 50, no. 3, pp. 289–294, 2003.

8. J. Mayer, H.G. Schuster and J.C. Claussen, "Role of inhibitory feedback for information processing in thalamocortical circuits," *Physical Review E*, vol. 73, no. 3, Article ID 031908, 15 pages, 2006.

9. K. R¨obenack and P. Goel, "Observer based measurement of the input current of a neuron," *Mediterranean Journal of Measurement and Control*, vol. 3, no. 1, pp. 22–29, 2007.

10. P. Goel and K. R¨obenack, "Observing the current input in neurons," Tech. Rep. 38, Mathematical Biosciences Institute, The Ohio State University, 2005.

11. A.L. Hodgkin and A.F. Huxley, "A quantitative description of membrane current and its application to conduction and excitation in nerve," *The Journal of Physiology*, vol. 117, no. 4, pp. 500–544, 1952.

12. C. Cruz and H. Nijmeijer, "Synchronization through extended Kalman filtering," in *New Directions in Nonlinear Observer Design*, H. Nijmeijer and T.I. Fossen, Eds., vol. 244 of *Lecture Notes in Control and Information Sciences*, pp. 469–490, Springer, London, UK, 1999.

13. E. Steur, I. Yu. Tyukin, H. Nijmeijer and C. van Leeuwen, "Reconstructing dynamics of spiking neurons from input-output measurements *in vitro*," in *Proceedings of the 3rd IEEE Conference on Physics and Control*, Potsdam, Germany, 2007, Paper no. 3–7.

14. I. Yu. Tyukin, E. Steur, H. Nijmeijer and C. van Leeuwen, "State and parameter estimation for systems in non-canonical adaptive observer form," in *Proceedings of the 17th IFAC World Congress on Automation Control*, vol. 17, Seoul, Korea, 2008, Paper no. 6–1.

15. L. Glass and M.C. Mackey, Eds., *From Clocks to Chaos: The Rhythms of Life*, Princeton University Press, Princeton, NJ, USA, 1988.

16. H.D.I. Abarbanel, *Analysis of Observed Chaotic Data*, Institute for Nonlinear Science, Springer, New York, NY, USA, 1996.

17. S. Julier, J. Uhlmann and H.F. Durrant-Whyte, "A new method for the nonlinear transformation of means and covariances in filters and estimators," *IEEE Transactions on Automatic Control*, vol. 45, no. 3, pp. 477–482, 2000.

18. S.J. Julier, "The scaled unscented transformation," in *Proceedings of the American Control Conference*, vol. 6, pp. 4555–4559, 2002.

19. P.E. McSharry, M.J. McGuinness and A.C. Fowler, "Confronting a cardiovascular system model with heart rate and blood pressure data," *Computers in Cardiology*, vol. 32, pp. 587–590, 2005.

20. A.C. Fowler and M.J. McGuinness, "A delay recruitment model of the cardiovascular control system," *Journal of Mathematical Biology*, vol. 51, no. 5, pp. 508–526, 2005.

21. S.R. Seydnejad and R.I. Kitney, "Modeling of mayer waves generation mechanisms: determining the origin of the low- and very low frequency components of BPV and HRV," *IEEE Engineering in Medicine and Biology Magazine*, vol. 20, no. 2, pp. 92–100, 2001.
22. R. Sameni, M.B. Shamsollahi, C. Jutten and M. Babaie-Zadeh, "Filtering noisy ECG signals using the extended Kalman filter based on a modified dynamic ECG model," *Computers in Cardiology*, vol. 32, pp. 1017–1020, 2005.

Temporal Redistribution of Plantar Pressure Points in the Healthy and Diabetics: A Time Series Analysis of the Neuro-Capillary Chaos[†]

Devesh V. Oberoi,[1] Cauchy Pradhan,[2] C. Jairaj Kumar[3]
and *Sidney C. D'Souza[4]*

[1]SGRR Institute of Medical & Health Sciences, Dehradun, UA, India. Email: devesh16_in@yahoo.co.in.
[2]National Institute of Mental Health and Neurosciences, Bangalore, KA, India.
[3]K S Hegde Medical Academy, NITTE University, Mangalore, KA, India.
[4]Kasturba Medical College, Mangalore, Manipal University, India.

ABSTRACT

Background: In diabetic individuals (DI), neuropathy hinders the redistribution of plantar pressure points thus leading to susceptible areas where there is constant capillary blanching which may develop into trophic ulcers. The redistribution of pressure points may precede evidence of clinical neuropathy. In this study we compare temporal redistribution of plantar pressure points (areas of capillary blanching) between normal subjects taken as controls and DI with no clinical signs of neuropathy.

Method: Four adults (45±4.55 years) diagnosed to have Type-2 Daiabetes, without signs of clinical neuropathy and age matched controls (43±3.74 years) were studied. The subjects were asked to stand on a glass slab and a 10 minute video recording of 10 selected plantar pressure points was made. Changes in the distance of these points with reference to a defined point on Mayer's line were measured at every 10 seconds. Standard deviation of difference of redistributed consecutive pressure point (SDPP) in cms., and fractal dimension (FD) was used to compare the two groups.

Results: Combined mean SDPP (DI =0.013 ± 0.008 cms, controls= 0.196±0.233 cms, P <0.001) and FD (DI =1.000 ± 0.000, controls= 1.010±0.017, P <0.001) of diabetic patients were significantly lower than controls. Pressure point at base of the 4th toe and the lower limit of blanching to the left Mayers line at the heel did not differ significantly between DI and controls.

[†]Reused with permission from: Oberoi D.V., Pradhan C., Kumar C.J., D'Sousa S.C. Temporal redistribution of plantar pressure points in diabetic and control subjects: A timeseries analysis of neuro-capillary chaos. AMJ 2010, 1, 2, 170–179. Doi. 10.4066/AMJ. 2010.221.

Conclusion: There is impaired redistribution of plantar pressure points in individuals with diabetes without signs of clinical neuropathy. This can be attributed to loss of chaos generating mechanisms in DI. Redistribution of pressure points may be essential in the prevention of trophic ulcers in susceptible individuals.

Keywords: Diabetes; neuropathy; trophic ulcers; fractal dimension; chaos.

Background

Diabetic foot syndrome is perhaps one of the most important complications of diabetes mellitus. The prevalence of diabetic foot ulcers ranges between 4–10%, while the lifetime incidence rates may be as high as 25% (1). A number of factors like the peripheral vascular disease, changes in foot architecture, peripheral sensory neuropathy and the plantar pressure are considered to be the prime etiological factors for the development of ulcers (2,3). Studies in the past have demonstrated that, the capillary flow is increased rather than decreased in the diabetic neuropathic foot and the high pressure areas are presumed to be associated with increased basal skin blood flow as compared to low pressure areas (4–6). Currently it is believed that the development of plantar pressure ulcers is associated with high amount of pressure exerted on certain regions of the foot (7–10).

Studies in the past have shown that the peak plantar pressure is a high risk for plantar ulcers yet it is a poor tool by itself to predict foot ulcers (11) Considering the minimal amount of pressure (< 30 *mm Hg*) required to occlude the capillary flow, any pressure greater than this is likely to cause an ulcer it would be more appropriate to suspect the prolonged sustenance of pressure at particular points causing unremitting tissue anoxia at these points rather than its magnitude as the prime cause for plantar ulcers. In our previous study we had demonstrated the role of redistribution of the pressure points or the weight bearing points in preventing the development of pressure ulcers in healthy individuals (12).

The possible disruption of this bio-mechanism in diabetics owing to peripheral neuropathy may lead to the development of pressure ulcers in susceptible individuals. As sub-clinical neuropathy is may exist for long periods in diabetic individuals (DI) (13), we hypothesize that the loss of redistribution of pressure points may precede evidence of clinical neuropathy. In our current pilot study we aim to we compare and quantify temporal redistribution of plantar pressure points (areas of capillary blanching) between normal subjects and DI with no clinical signs of neuropathy.

Method

This study was conducted at a premiere university teaching hospital in South India. The protocol of the study was approved by the Institute ethics committee. Written informed consent was obtained from each subject. Three adult male and 1 female diagnosed to have Type-2 Diabetes as per the Expert Committee on the Diagnosis and Classification of Diabetes Mellitus criteria were selected (12). Patients underwent routine outpatient neurological examination. None of the patients were found to have any abnormalities in sensation or muscle weakness or any evidence of claudication pain. Age matched healthy male individuals were included as controls.

The subjects were asked to stand on a transparent plexi glass slab of thickness 1 cm. A digital video camera (Sony DCR-SR87 with 690K pixel resolution) was placed 1 meter below the glass slab and a 10 minute video recordings of the plantar area was done for each of the study participants. Offline analysis of the images at every 10 second intervals was done. Ten pressure points (areas of capillary blanching) were selected from the right foot of all subjects. These pressure points were arbitrarily chosen from a previous study, analyzing pressure point changes in normal subjects (10). Figure 1 depicts the pressure points measured.

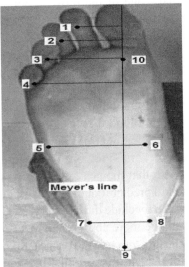

Figure 1 Image of plantar aspect of foot with points measured with reference to Meyer's line 1: The lowest point of the pale area on the 2nd toe. 2: The lowest point of the pale area on the 3rd toe. 3: The lowest point of the pale area on the 4th toe. 4: The lowest point of the pale area on the 5th toe. 5: A pressure point to the left of the mid-point of the Meyer's lines (an anatomical line passing through the middle of the great toe and the heel). 6: A pressure point to the right of the mid-point of the Meyer's line. 7: A point 2 cms above the base of heel, to the left of Meyer's line. 8: A point 2 cms above the base of heel, to the right of Meyer's line. 9: A point on the Meyer's line corresponding to the lower limit of the pale area on the foot. 10: A point on the Meyer's line corresponding to the upper limit of the pale area on the foot.

The points were:

(a) The lowest point of the pale area on the 2nd toe.
(b) The lowest point of the pale area on the 3rd toe.
(c) The lowest point of the pale area on the 4th toe.
(d) The lowest point of the pale area on the 5th toe.
(e) A point to the left of the mid-point of the Meyer's lines (an anatomical line passing through the middle of the great toe and the heel).
(f) A point to the right of the mid-point of the Meyer's line.
(g) A point 2 cm above the base of heel, to the left of Meyer's line.
(h) A point 2 cm above the base of heel, to the right of Meyer's line.

(i) A point on the Meyer's line corresponding to the lower limit of the pale area on the foot.

(j) A point on the Meyer's line corresponding to the upper limit of the pale area on the foot.

Measurement of perpendicular distances of the first eight points with reference to Mayer's line and of the last 2 points with reference to the lowest point on the heel (also the lowest point on the Meyer's line) was done at every 10 seconds. A total of 60 measurements were obtained per point per subject.

For each point the following parameters were calculated:

(a) Standard deviation of difference between consecutive redistribution of centre of the pressure point (SDPP) in cm. SDPP was used rather than raw distances since it is invariant to foot size.

(b) Fractal dimension (FD) as described by Katz, 1988 (13)

$$FD = \log(N-1) - [\log(N-1) + \log(d/L)]$$

N=Number of samples (i.e., 60), d=Maximum & L=Total 'distance' on waveform. *FD* (range 1–2) itself has no units.

Statistical Analysis

All statistical analysis were done using SPSS version 11. Continuous variables were expressed as mean ± standard deviation. Comparison of parameters between patients and controls was performed using Mann-Whitney U test for exact significance values.

Results

Demographics of the subjects are shown in Table 1. The age range for DI was 41–51 years and 38–47 years for controls. Duration of illness for DI at diagnosis ranged from 132–192 months. (BMI:25.1 ± 0.98kg/m^2 and 23.6 ± 0.49 kg/m^2 respectively) Fig. 2 depicts the mean distances of the centre of these points with reference to Mayer's line DI and control subjects Overall variability is higher in normal individuals when compared to diabetics.

Comparison of combined points show SDPP (DI =0.013 ± 0.008 cm, controls= 0.196 ± 0.233 cm, P <0.001) and FD (DI =1.000 ± 0.000, controls= 1.010 ± 0.017, P <0.001) of diabetic patients to be significantly lower than controls. On examination of individual

Table 1 Demographic data of controls and diabetic patients. DI: Diabetic individuals.

	Control (n=4)	DI (n=4)
Age (Year)	43±3.74	45±4.55
Gender (M: F ratio)	4:0	3:1
Body mass index (kg/m²)	23.6±0.49	25.1±0.98
Duration of illness (months)		156±25.92

pressure points, (Table 2) two pressure points on the plantar aspect of the foot in diabetics (point 3,8) (pressure point at base of 4th toe ,right of the midpoint of Meyer's line appear to have similar SDPP and FD in both patients and controls and do not reach significance levels. The FD of point 1 (base of second toe) is comparable between diabetics and controls.

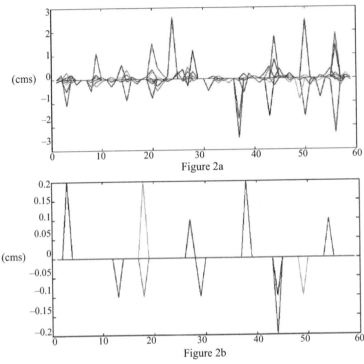

Figure 2a

Figure 2b

Figure 2 Mean distances of the centre of these points with reference to Mayer's line in controls (2a) and diabetic individuals (2b). The DI time series has fewer fluctuations and lower magnitude that that of controls.

Discussion

The present study is a pilot study with measurements done manually. This severely limited the number of patients, recordings and sampling frequency possible. The small numbers would have made any correlation analysis impossible with nerve conduction studies to determine relation with degree of neuropathy. The present study shows that even in a single individual there is variability in the magnitude of redistribution of weight bearing points (Fig. 2). Redistribution appears to form a complex time series in normal individuals. In DI this is lost giving rise to a monotonous time series. The resultant waveforms of DI has less fluctuations and lower magnitude that that of controls. The SDPP is a useful measure as it minimizes the affect of foot size and shape. The results suggest that point 3 and 8 (pressure point at base of 4th toe, right of the midpoint of Meyer's line, have the least variability of redistribution in both controls and diabetics. These points thus can be presumed to be the points through which maximum weight channelized and thus balance is obtained. These points indecently lie along the lateral osseous ridge and give rise to the observation that pressure ulcers occur under osseous pressure points (7–9). Pressure changes appear chaotic in nature and do not follow any set pattern. The overall frequency and magnitude of the redistribution of these points is significantly less in diabetics as compared to healthy individuals, demonstrated by both fractal dimension and statistical measures.

Table 2 Statistical and Fractal Dimensional comparison between Diabetic individuals and controls. SDPP: Standard deviation of difference between consecutive pressure points in cms; FD: Fractal dimension; DI: Diabetic individuals. Pressure points (1: The lowest point of the pale area on the 2nd toe. 2: The lowest point of the pale area on the 3rd toe. 3: The lowest point of the pale area on the 4th toe. 4: The lowest point of the pale area on the 5th toe. 5: A pressure point to the left of the mid-point of the Meyer's line (an anatomical line passing through the middle of the great toe and the heel). 6: A pressure point to the right of the mid-point of the Meyer's line. 7: A point 2 cms above the base of heel, to the left of Meyer's line. 8: A point 2 cms above the base of heel, to the right of Meyer's line. 9: A point on the Meyer's line corresponding to the lower limit of the pale area on the foot. 10: A point on the Meyer's line corresponding to the upper limit of the pale area on the foot).

Pressure points	SDPP			FD		
	DI (n=4)	Control(n=4)	p*	DI(n=d)	Control(n=4)	p*
1	0.017±0.013	0.184±0.129	0.029	1.000±0.000	1.009±0.009	0.057
2	0.016±0.006	0.291±0.294	0.029	1.000±0.000	1.021±0.024	0.029
3	0.010±0.007	0.302±0.323	0.114	1.000±0.000	1.023±0.029	0.114
4	0.019±0.000	0.316±0.258	0.029	1.000±0.000	1.018±0.018	0.029
5	0.020±0.006	0.081±0.022	0.029	1.000±0.000	1.002±0.001	0.029
6	0.014±0.012	0.115±0.082	0.029	1.000±0.000	1.004±0.004	0.029
7	0.011±0.008	0.079±0.058	0.029	1.000±0.000	1.002±0.002	0.029
8	0.008±0.009	0.051±0.025	0.057	1.000±0.000	1.001±0.001	0.114
9	0.008±0.009	0.113±0.123	0.029	1.000±0.000	1.006±0.010	0.029
10	0.007±0.008	0.429±0.456	0.029	1.000±0.000	1.024±0.027	0.029

As fractal dimension is related to other nonlinear measures of chaos like correlation dimension, maximal Lyapunov exponent, we expect that these measures too would be lower in diabetic individuals (DI). It would a much larger time series to calculate these measures and may require image processing tools to extract higher frequency changes. This is in line with the continuous wavelet transformation (CWT) plots that show the time dependent frequency changes of different pressure points in both healthy and diabetics. The CWT plot for healthy individual shows pronounced chaotic redistribution of pressure points displayed by high frequency color changes depicted by yellow and white (Fig. 3a) while those with diabetic sub clinical neuropathy with minimal to nil redistribution of weight bearing points produce a darker images depicted by maroon to completely black image on the CWT (Fig. 3b).

It may be concluded that the causation of diabetic ulcers or pressure ulcers may be the result of loss of mechanisms leading to generation of chaos in the pressure distribution in feet. This in contrast to the popular belief that the development of foot ulcers is related to the amount of pressure at a given point (8,9). We suggest a more intuitively plausible causality for foot ulceration. Ulceration may be linked to the sustenance of pressure rather than the degree as evidenced that very little pressure is required to cause capillary blanching. Tissue

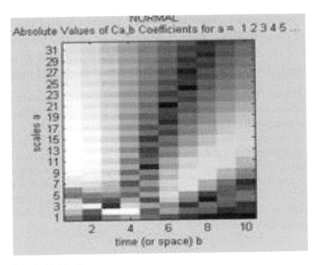

a: CWT plot for the healthy individual.

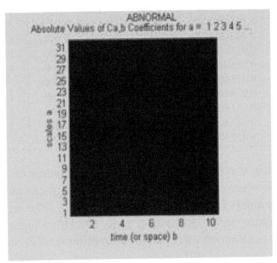

b: CWT plot for the chronic diabetic individual with subclinical neuropathy.

Figure 3 Continuous wavelet translation plots.

anoxia thus should occur if the pressure continues to cause blanching rather than a large pressure that disappears occasionally and does not cause sustained tissue anoxia.

The variation in the frequency of redistribution in different areas of the foot may pose a higher risk of ulcer formation in some specific parts of the foot. These pressure points in the diabetics which display a low frequency of redistribution may be at a higher risk for development of plantar ulcers as compared to other parts of the foot. Points over the heel and third toe seem especially vulnerable with lower frequency of re distribution.

Earlier theories, proposing high pressure point areas on the foot to be at a higher risk for developing plantar ulcers have led to several expensive diagnostic instruments for the recognition of such point (6,7). However, all such futuristic diagnostic technology may not be accessible to low income group patients in the in remote and economically challenged areas especially in the developing countries. Our method of analysis of such high risk points on the foot of diabetics is simple and inexpensive.

Conclusion

Our analysis of 10 randomly selected points shows that the sites of low to minimal frequency of chaotic redistribution of these points is specific for an individual and may vary from person to person. There is impaired redistribution of plantar pressure points in individuals with diabetes without signs of clinical neuropathy. This can be attributed to loss of chaos generating mechanisms in DI. Redistribution of pressure points may be essential in the prevention of trophic ulcers in susceptible individuals. The recognition of such points through such simple techniques may even help low-end shoe manufacturers to design comfortable footwear for sensory neuropathy patients specific to their need which shall provide them protection against the development of pressure ulcers.

Acknowledgements

We would like to thank the all the participants for consenting to be a part of this study.

References

1. Singh N., Armstrong D.G. and Lipsky B.A. (2005). Preventing foot ulcers in patients with diabetes. JAMA. Jan 12; 293(2): 217–28.
2. Richard J.L. and Schuldiner S. (2008).Epidemiology of diabetic foot problems. Rev Med Interne Sep; 29 2: S222–30.
3. Boulton A.J. (1992). The risk of foot ulceration in diabetic patients with high foot pressure: a prospective study. Diabetologia. Jul; 35(7): 660–3.
4. Netten P.M., Wollersheim H., Thien T. and Lutterman J.A. (1996) Skin microcirculation of the foot in diabetic neuropathy. Clin Sci.; 91: 559–565.
5. Flynn M.D., Edmonds M.E., Tooke J.E. and Watkins P.J. (1988) Direct measurement of capillary blood flow in the diabetic neuropathic foot. Diabetologia.; 31: 652–656.
6. Newton D.J., Bennett S.P., Fraser J., Khan F., Belch J.J.F., Griffiths G. and Leese G.P. (2005). Pilot study of the effects of local pressure on microvascular function in the diabetic foot. Diabetic Medicine 22; 11: 1487–1491.
7. Mueller M.J., Zou D., Bohnert K.L., Tuttle L.J. and Sinacore D.R. (2008) Plantar stresses on the neuropathic foot during barefoot walking. Phys Ther. Nov; 88(11): 1375–84.
8. Armstrong D.G., Lavery L.A., Vela S.A., Quebedeaux T.L. and Fleischli J.G. (1998) Choosing a practical screening instrument to identify patients at risk for diabetic foot ulceration. Arch Intern Med; 158: 289–92.
9. Stess R.M., Jensen S.R. and Mirmiran R. (1997) The role of dynamic plantar pressures in diabetic foot ulcers. Diabetes Care; 20: 855–8. X.
10. Veves A., Murray H.J., Young M.J. and Boulton A.J: The risk of foot ulceration in diabetic patients with high foot pressure: a prospective study. Diabetologia; 35: 660–663, 1992.
11. Lavery L.A., Armstrong D.G., Wunderlich R.P., Tredwell J. and Boulton A.J: Predictive value of foot pressure assessment as part of a population-based diabetes disease management program. Diabetes Care 26: 1069–1073, 2003.

12. Oberoi D.V., Kumar C.J., D'souza S., Kumar A. and Hegde B.M. (2007). Does non chaotic weight bearing foci cause foot ulceration in diabetics? Medical Hypotheses. 68(2), 468–9.
13. Hendriksen P.H., Oey P.L., Wieneke G.H., Bravenboer B. and van Huffelen A.C. (1993). Subclinical diabetic polyneuropathy: early detection of involvement of different nerve fibre types. J Neurol Neurosurg Psychiatry. May; 56(5): 509–514.
14. Gavin J.R., Alberti K.G.M.M., Davidson M.B., DeFronzo R.A., Drash A. and Gabbe S.G., et al. (2002) Report of the Expert Committee on the Diagnosis and Classification of DiabetesMellitus. Diabetes Care,;25 Suppl 1: S5–S20.
15. Katz M.Z.(1988) Fractals and analysis of waveforms. Comp Biol Med; 18: 145–156.

Section II

Models and Applications of Chaos Theory in Biology

1. **An Astrobiological Thought on Sustainable Life**
   T. Naganuma   95

2. **Matrix Frequency Analysis of Oryza Sativa (Japonica Cultivar-Group)**   106
   Complete Genomes
   K. Manikandakumar, S.M. Kumaran and *R. Srikuma*

3. **The Capabilities of Chaos and Complexity**   123
   D.L. Abel

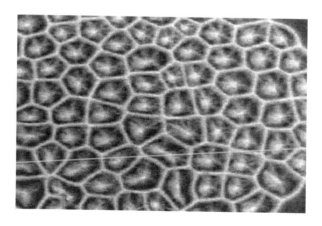

```
   1 gctagtgtag cttaagcaaa gcataacact gaagatgtta agatgggccg tagaaagccc
  61 cacgggcaca aaggtttggt cctgacttta ttatcagctt taacccaatt tacacatgca
 121 agcctccgca cccctgtgag gatgccctca atcccccgtc cggggacgag gagccggtat
 181 caggcacact tttagccca agacgccttg cttagccaca cccccaaggg aattcagcag
 241 tgatagacat taagccataa gtgaaaactt gacttagtca gggttaagag ggccggtaaa
 301 actcgtgcca gccaccgcgg ttatacgaga ggccctagtt gattcactcg gcgtaaagag
 361 tggttatgga gaataaaata ctaaagccga agacccctta ggccgtcata cgcacctagg
 421 ggctcgaatt atagacacga aagtagcttt acccttccc accagaaccc acgacagctg
 481 ggacacaaac tgggattaga taccccacta tgccccgccg taaacttaga tattccagta
 541 caacaaatat ccgccagggg actacgagcg ccagcttaaa acccaaagga cttggcggtg
 601 cttcagaccc ccctagagga gcctgttcta gaaccgataa ccccccgttca acctcactac
 661 tccttgcttt tcccgcctat ataccaccgt cgccagctta ccctgtgaag gtactacagt
 721 aagcagaatg agtaatactc aaaacgtcag gtcgaggtgt agcgtacgaa gtaggaagaa
 781 atgggctaca ttatctgatc cagattattc acggaaggtt gtctgaaacg acaatccgaa
 841 ggtggattta gcagtaaagg gggaatagag tgcccccttg aagccggctc tgaagcgcgc
 901 acacaccgcc cgtcactctc cccaacaacc gcctacacca aggtaaataa cacaacatcc
 961 gtcacaaggg gaggcaagtc gtaacatggt aagtgtaccg gaaggtgcac ttggaataat
1021 cagggtgtgg ctgagacagt taagcgactc ccttacaccg agaagacatc catgcaagtt
1081 ggatcaccct gaactaaaca gctagctcaa actataaaaa ccaaattaat gatatagata
```

Chapter 1

An Astrobiological Thought on Sustainable Life[†]

Takeshi Naganuma

Graduate School of Biosphere Science, Hiroshima University, Higashi-Hiroshima, 739-8528, Japan.
Email: takn@hiroshima-u.ac.jp.

ABSTRACT

Life on a global biosphere basis is substantiated as organics and organisms, and defined as the intermediate forms (briefly expressed as CH_2O) hovering between the reduced (CH_4, methane) and (CO_2, carbon dioxide) ends, different from the classical definition of life as complex organization maintaining ordered structure and information. Both definitions consider sustenance of life meant to be protection of life against chaos by input of external energy. The CH_2O-life is maintained as long as the supply of H and O lasts, which is in turn provided by the spilt of the water molecule H_2O. Water is split by electricity as well-known in school experiments and by solar radiation and geothermal heat on a global scale. In other words, the Sun's radiation and the Earth's heat as well as radioactivity split water to supply H and O for continued existence of life on the Earth. These photochemical, radiochemical and geothermal processes have influences on the evolution and current composition of the Earth's atmosphere, compared with those of Venus and Mars, and influences on the planetary climatology. This view of life may be applicable to "search-for-life in space" and to sustainability assessment of astrobiological habitats.

Keywords: split of water; photochemical; radiochemical; geothermal; reduction-oxidation; entropy.

Introduction

What is life? Erwin Schrödinger, Nobel laureate in physics 1933, tackled this old-lived question and defined life as the organization that maintains complex structure and heritable information in expense of "negentropy" (Schrödinger 1944). Negentropy is a useful conceptual tool to understand the physical basis underlying sustenance of biological machinery, and is the counter concept of entropy defined by the second principle of

---

[†]Reused with permission from: T. Naganuma, An Astrobiological Thought on Sustainable Life, *Sustainability,* 2009, *1*, 827–837; doi: 10.3390/su1040827.

thermodynamics, i.e., the time arrow theory. According to the principle, total amount of available energy, or energy, decreases irreversibly with time, and entropy is a measure of the ever-increasing unavailability partly as heat. Local entropy within a system may decrease in the expense of potential energy, i.e., negentropy, and such situation is substantiated in living organisms that expend chemical potential energy to maintain their structure and information. Heat, the energy-in-transit, does not directly support life by itself, but it may generate chemical potential energy via thermochemical reactions.

The question *what is life* is thus transformed as *what supports life*, and an answer for a living organism is the chemical potential energy that lowers local entropy despite increase of overall entropy. Then, the question is extended to a larger system where living organisms live, and revised as *what supports biosphere*. The Earth's biosphere receives heat *sens general* from both external and internal sources, i.e., the Sun and Earth's interior, respectively. Heat from the Sun derives ultimately from gravitational and nuclear potential energy of the hydrogen gas disk in the Hadean (Pregeologic) Eon, while heat from the Earth's interior originates from gravitational and nuclear potential energy of the silicate gas disk/microplanets and radioactive ^{40}K, respectively. Contradiction of gas disks and accumulation of microplanets liberated gravitational potential energy to yield heat. Heat of the ancient Sun ignited nuclear fusion to burn itself as a star, irradiating the Earth at the Solar constant of 1.4 kW m^{-2}. The profoundly underlying problem is how heat potentiates the Earth to host life.

Heat as energy-in-transit may form locally ordered structures in an open non-equilibrated system, as advocated by Ilya Prigogine, Nobel laureate in chemistry 1977 (Prigogine 1984). A visual example of the locally organized structure formed by heat is the Prigogine's hexagon, or Rayleigh-Benard convection (Getling 1998; Fig. 1). The heat used in the experiments to form convection cells derives from the chemical or electrical

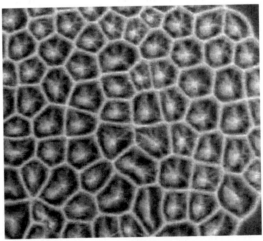

Figure 1 Rayleigh-Benard convection. Convection cells in a fluid are formed and maintained when heated from below. The upper surface of the heated fluid is unconstrained and is free to move and deform. This situation is sustained as long as heat energy is given to the fluid system and gives rise to the Prigogine's hexagons, or Rayleigh-Benard convection cells. Photo: Hideo Suzuki, Tokyo Metropolitan College of Technology.

potential energy, and therefore chemical potential energy indirectly (via heat) forms convection cells. The minimum unit of living organisms is coincidentally called "cell", and the biological cells are maintained by chemical potential energy contained in foods, or organic compounds. Animals eat organic compounds produced by others, while plants produce organics for themselves and others via photosynthesis.

These organic-eaters and organic-producers are collectively called heterotrophs and autotrophs, respectively. The non-photosynthetic mode of autotrophy is called chemolithoautotrophy, by which organics are produced in the expense of chemical energy liberated from the oxidation of inorganic compounds such as hydrogen and hydrogen sulfide. This mode of chemolithoautotrophic life is known to thrive in the deep-sea and deep sub-seafloor. This chapter tries to apply Schrödinger's negentropy concept to biosphere, and evaluates the sources of chemical potential energy for chemolithoautotrophic lives in deep-sea and deep sub-seafloor (Humphris et al. 1995; Fredrickson and Fletcher 2001) from a planetary point of view.

Vortex of Life

The Prigogine's hexagons, or Rayleigh-Benard convection cells, are generated by continuous flow of heat. Continuous flow of chemical potential energy maintains life. Similarly, continuous flow of water forms vortices, and therefore vortex serves as a key idea to understand life.

Flow of water sometimes form vortices. Vortex is only a temporal pattern of water movement, composed of different water molecules coming-in and going-out moment-by-moment. Kamo-no-Chomei, Japanese medieval essayist, expressed in *Hojoki* (1212) his view of transitory life as "The flowing river never stops and yet the water never stays the same. Foam floats upon the pools, scattering, re-forming, never lingering long. So it is with man and all his dwelling places here on earth" (translated by Moriguchi and Jenkins 1996). Chomei's transitory bubbles are parallel to vortices in my view of life. Although atoms and molecules of my body have replaced since my birth, I have never doubted my continuity and identity. That is, my identity is based more on a pattern like a vortex than materials, and I represent a tiny vortex of life (Fig. 2).

Vortices consist of water molecules flowing in and out every second, and are kept alive by constant current caused by gradient of water height. Water flows from highs to lows, and manifests gravitational potential energy as kinetic/mechanical energy and even electric energy at hydraulic power plants (Fig. 3), as well as vortices. Vortices are formed and maintained by the slopes between high and low water tables. If water tables become flat, there will be no flows and no vortices.

By what water is transported to high places? Water cycling, i.e., evaporation and precipitation, is mainly driven by solar heat energy. Therefore, hydraulic power plants are said to convert solar heat energy to electric energy via gravitational potential energy of water. Hydraulic high means the water on high that contain high gravitational potential energy. Then, what corresponds to chemical high for chemical potential energy? Taking examples of organic compounds (generalized as CH_2O), they manifest chemical potential energy through

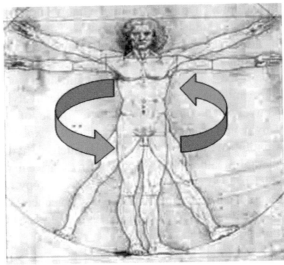

Figure 2 Leonardo Da Vinci's Vitruvian Man, showing his identity as a pattern like a vortex.

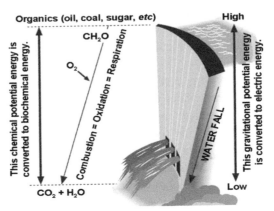

Figure 3 Manifestation of gravitational and chemical potential energy as electric and biochemical energy via water fall and oxidation, respectively.

combustion (parallel to water fall), or oxidation, to yield the most oxidized form of carbon (CO_2) on the lowest. Hence, chemical high and low correspond to more oxidizable (more reduced) and less oxidizable (less reduced) states, respectively (Fig. 3).

Life Vortex as Intermediate of CH_4 and CO_2

Since the life on the Earth is based on carbon, the reduced and oxidized ends of carbon, methane (CH_4) and carbon dioxide (CO_2), respectively, are mainly discussed. Organics and organisms are simply and collectively expressed here as CH_2O, instead of commonly used R for an organic functional group, because it is easily understandable that organics and organisms are substantial intermediates of CH_4 and CO_2. The merit of CH2O to represent organics/organisms would overcome potential confusion against the exact chemical formula of formaldehyde.

The Sun transports water to high places via evaporation, adding gravitational potential energy. Similarly, the Sun splits water into hydrogen (source of reducing power) and oxygen (source of oxidizing power). Split of water yields chemical highs and lows, or reducing and oxidizing ends, and thus forms chemical vortex of life. Water may split also by geothermal heat, including radioactivity of rock-borne ^{40}K, as described later.

Erwin Schrödinger defined life manifested by biological cells or individuals that maintain structure and information by eating negentropy (Schrödinger 1944). In contrast, I view life as "transitory intermediates on the balance of hydrogen and oxygen supplies", considering the Earth's biosphere. In this sense, organisms are only ephemeral and hovering between life and death. This view is depicted in Fig. 4.

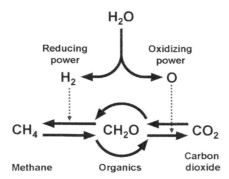

Figure 4 Transitory view of C-based life hovering between the reduced (CH_4) and oxidized (CO_2) ends. The reducing and oxidizing powers are derived from the split of water by heat or geothermal heat.

Methane is produced by the reaction $[CO_2+4H_2 \rightarrow CH_4+2H_2O]$ inorganically, as proposed by Paul Sabatier, Noble laureate in chemistry 1912 (Sabatier 1923). Methane is also produced by methanogenic microorganisms mediating the biological counterpart of the Sabatier reaction in deep-sea hydrothermal vents and deep subsurface (Humphris et al. 1995; Fredrickson and Fletcher 2001).

Organic compounds and organisms, generally expressed here as CH_2O or $[CH_2O]_n$, are intermediates of CH_4 and CO_2, and various organic forms are found in nature. Methane has the maximum number of hydrogen, i.e., four hydrogen atoms per carbon, and the greatest chemical potential energy (890 kJ mol^{-1}) among carbon compounds is manifested via full oxidation to yield carbon dioxide. Methane-oxidizing microorganisms, namely methanotrophs, utilize this chemical energy for metabolism for growth and reproduction, and are often found in deep-sea hydrothermal vents and deep subsurface (Bock and Goode 1996). That is, life vortex as realized by methanotrophs is manifested on the slope from CH_4 to CO_2 (Fig. 5).

All the organics and organisms are partially oxidized forms of CH_4, and are to be further oxidized to CO_2. Oil (petroleum) is the mixture of relatively less oxidized hydrocarbon chains (with more hydrogen per carbon on average), while formaldehyde (truly CH_2O) and acetaldehyde (CH_3CHO) are more oxidized forms and close to the CO_2-end. Life vortices may vary in numbers, sizes, features, etc., according to the amounts of manifested chemical potential energy (Fig. 6).

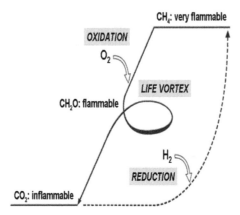

Figure 5 Life vortex in carbon cycling between CH_4 and CO_2. Chemical potential energy is manifested during the oxidation of CH_4 to generate life vortices as intermediates before full oxidation to CO_2, which is in turn re-potentiated via reduction by $H + e^-$ from the split of water.

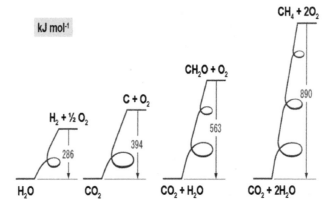

Figure 6 Different numbers and sizes of life vortices on the slope of oxidation of various carbon compounds. The vortices may have different characteristics according to the features of slopes.

Split of Water

Water, H_2O, may be the most common but miraculous molecule in the universe. The most abundant element in the universe is hydrogen, followed by helium, oxygen, carbon, nitrogen, and so on. As helium is chemically inert, hydrogen reacts with oxygen to form water, with carbon to form methane, and with nitrogen to form ammonia. Among these hydrogen compound, water shows a number of peculiarities that are not seen by other hydrogen compounds as follows:

1. Water molecule is held together by sharing electrons among two hydrogen and one oxygen atoms via two covalent bonds;
2. Water is a polar molecule. Although the net electrical charge on the molecule is zero, its structure causes the molecule to become polarized;

3. Electrostatic bonds (hydrogen bonds) form between the negatively charged oxygen side of one water molecule and the positively charged hydrogen of another molecule;
4. Existence of the hydrogen bonds explains many of the unique properties of water;
5. Ice has an orderly, open structure of water molecules held together by hydrogen bonding;
6. Water ice's crystal structure results in lower density of 0.92 g cm^{-3} than liquid water, 1.0 g cm^{-3};
7. Liquid water structure is intermediate between that of ice and water vapor, and consists of two types of aggregates of water molecules;
8. Structured water is composed of clusters of hydrogen-bonded water molecules that form and reform very quickly but slow enough to influence the physical behavior of water;
9. Unstructured water is composed of closely packed free water molecules, more dense than structured water;
10. If hydrogen bonding did not exist, water would only occur as a gas at the Earth's surface;
11. Water is the only naturally occurring substance on the Earth to exist at the surface in all three states: liquid, solid and gas;
12. Water dissolves more substances in greater quantity than any other common liquid;
13. Water has the highest surface tension of all liquids;
14. Water has the highest heat capacity of all common solids and liquids, which prevents extreme range in aquatic temperature;
15. Boiling and melting points of water are higher than those of other hydrogen compounds of similar size or oxygen-group ($_8$O, $_{16}$S, $_{34}$Se, $_{52}$Te and so on). For example, CH_4, NH_3, H_2S, H_2Se and H_2Te occur as gases at room temperature; and
16. Water has the highest heat conductivity of any common liquid.

A metaphysical consequence of the above-listed peculiarities of the water molecule is that hydrogen and oxygen atoms, i.e., H and O, attract each other. The most and third abundant elements in the universe bind by strong covalent bonding to contain chemical potential energy, and they may split by the input of external dissociation energy. The sources of external dissociation energy are the Sun (solar radiation) and the Earth (geothermal heat including ^{40}K radioactivity); and water molecules split via the light reaction of photosynthesis, photochemical reactions (photolysis), ^{40}K β-radiation (radiolysis) and the high-temperature water-rock interaction (thermolysis) (Fig. 7).

Organisms are realization of life, namely biological potential energy, and substantiated with organic compounds, CH_2O. The split products of water (H and O) bind again but intermediated by carbon (C) to form organics and organisms (CH_2O). Only continuous supplies of H and O maintain the ephemeral presence of CH_2O, otherwise CH_2O would shift to the reduced end (CH_4) or oxidized end (CO_2). The idea of split of water is a modern revision of the Schrödinger's negentropy for individual organisms, and may be extended to possible biospheres on other planetary bodies.

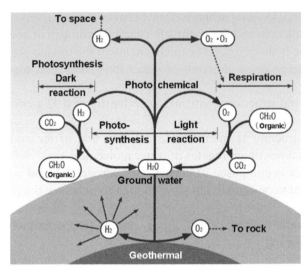

Figure 7 Split of the water molecule by solar radiation and geothermal heat (including radioactive decay) via the light reaction of photosynthesis and water-rock interaction, respectively.

Testing Planetary Biospheres

The Earth's biosphere is sustained by solar and geothermal heat via split of water. Even if the Sun ceases radiation, some part of the Earth's lives in the deep may continue as long as the Earth stays alive with liquid water and active plate/plume tectonics, or volcanism (Fig. 8a). Split of water maintains the existence of organics and organisms (CH_2O), and the levels of CH_4 and CO_2 are kept below 0.04%.

No liquid water but water vapor exists on Venus due to high surface temperature of about 500°C. Water vapor may split photochemically. The resultant hydrogen escapes to extra-Venus milieu, and the leftover oxygen accumulates in the Venus' atmosphere, with a CO_2 content of 98%. This biased split of water is unlikely to support life (Fig. 8b).

Hydrothermal activity may have existed or exist on ancient Mars (Bock and Goode 1996), although liquid water on ancient and recent Mars is still controversial (Bullock 2005). Regardless of liquid water, no plate tectonic activity is expected on modern Mars, due to short longevity of the planet as light as 1/10 of the Earth's mass. Therefore, only photochemical split, or photolysis, of water is presumed for Martian environment (Kransnopolsky and Feldman 2001), and accumulation of oxygen in the Martian atmosphere as presumed for the Venus' atmosphere results in a CO_2 content of 95%. Ancient Mars may have hosted geothermal split of water, and the remnant of ancient Martian life has been suggested in the Martian meteorite ALH84001 (McKay et al. 1996); however, the nanofossils are still controversially discussed in the scientific community (e.g., Golden et al. 2001; Thomas-Keprta et al. 2001).

In contrast, modern Mars is unlikely to capable of hosting a biosphere (Fig. 8c). The latest finding of ground ice of temporally melt-water (Byrne et al. 2009) suggests an icy/watery subterranean Mars. However, just occurrence-of-water on Mars, echoing with another

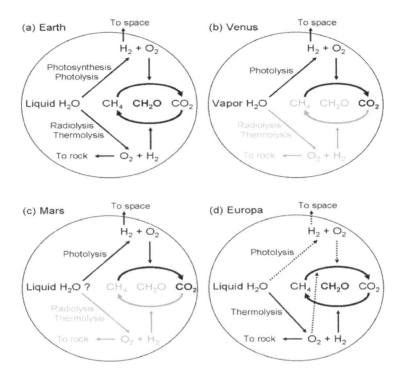

Figure 8 Possibility of carbon-based life (expressed as CH_2O) viewed from solar split of water (via photosynthesis and photolysis) and geothermal split of water (via radiolysis and thermolysis) in (a) Earth, (b) Venus, (c) Mars, and (d) Jovian satellite Europa. Black solid lines indicate existing and probable processes; black broken lines show possible pathways; and, gray lines suggest unlikely or negligible reactions. Recent finding of methane in Mars atmosphere (Formisano et al. 2004) may suggest occurrences of split-of-water and thus any form of life.

latest evidence for water on the Moon surface (briefed by Lucey 2009), does not imply life on modern Mars; split-of-water has more realistic relevance to life. In contrast, occurrence-of-methane has more implication for split-of-water, according to my scheme (Fig. 8c). It is interesting that methane (CH_4) is present in the Martian atmosphere (Formisano et al. 2004; Mumma et al. 2009), and that volcanic activity as recent as four million years ago is suggested (Neukum et al. 2004). This being so, a modern biosphere may be sustained in Martian subsurface that could store liquid water and remnant geothermal heat (Fisk and Giovannoni 1999). Another possibility is hydrogen (H_2) production via hydration of the most ancient volcanic rock of the Earth, komtiites, whose occurrence on Mars is also suggested (Nna-Mvondo and Martinez-Frias 2007). This is the split-of-water catalyzed by ultramafic komatiites, and a recent experiment confirmed that the komatiite-catalyzed H_2 produciton is robust enough to support H_2-based methanogenesis (Yoshizaki et al. 2009). Split-of-water catalyzed by komiites is likely to be a source of sustaining life forms on ancient Earth and Mars, and the possibility may extends to modern Mars.

Both liquid water and volcanism are postulated under the ice crust of Europa, the Jupiter's J2 satellite (Chyba and Phillips 2001). That is, geothermal split of water is likely to occur in

Europa, as well as photochemical split of water in the Europa's thin atmosphere (Chyba and Hand 2001). The resultant atmospheric oxygen may be incorporated to melt-and-refrozen surface ice and transported to interior ocean via tectonic ice convection of Europa (Gaidos and Nimmo 2000; Greenberg 2004). This planetary (or satellitary) setting may facilitate the formation of an extra-terrestrial biosphere (Fig. 8d), and thus provides a biospheric basis for search for life (Naganuma and Uematsu 1998).

Astrobiological Conclusion

Life of the Earth is carbon-based, and is substantiated as intermediate forms (expressed as CH_2O) hovering between the reduced end (CH_4) and the oxidized end (CO_2). The intermediate forms, organics and organisms, are ephemeral and eventually subject to full reduction or oxidation when the supplies of O or H cease, respectively. In other words, life is maintained only with the continuous supplies of H and O, which are in turn provided by the split of water. Solar radiation and geothermal heat would split water, and therefore it may not be too extreme to conclude that the Earth's lives are mainly sustained by either the Sun or the Earth, depending on the types of ultimate sources of nutrition, i.e., photosynthesis or chemolithoautotrophy.

In the Japanese language, the Sun is *hi*, and heat (fire) is also *hi* (originally *ho* or *fo*); water is *mi* or *mizu*; and, life is *i-no-chi* meaning energy of breath. The coincidence of two *hi* has impressed me, and I might say that split of *mi* by *hi* nourishes *chi*, at least, on the Earth. Both *hi*, that is the Sun's radiation and the Earth's interior heat, contribute to life. The degrees of contributions vary according to major modes of autotrophy, i.e., photosynthesis or chemolithoautotrophy.

Examples of chemolithoautotrophic communities that depend primarily on geothermal *hi* are found in deep-sea hydrothermal vents and deep subsurface, respectively (Humphris et al. 1995; Fredrickson and Fletcher 2001). The idea that non-solar split of water nourishes life thus derives from the studies of deep-sea and deep subsurface biospheres, and is extended to possible extra-terrestrial biospheres. The concept of planetary biospheres should accommodate more universal sense of life than traditional ones. The "non-solar split of water" idea is applicable to possible astrobiological biospheres.

References

1. Bock GR and Goode JA (eds) (1996) Evolution of Hydrothermal Ecosystems on Earth (and Mars?). John Wiley & Sons Ltd, Chichester.
2. Bullock MA (2005) The flow and ebb of water. Nature 438: 1087–1088.
3. Byrne S, Dundas CM, Kennedy MR, Mellon MT, McEwen AS, Cull SC, Daubar IJ, Shean DE, Seelos KD, Murchie SL, Cantor BA, Arvidson RE, Edgett KS, Reufer A, Thomas N, Harrison TN, Posiolova LV and Seelos FP (2009) Distribution of Mid-Latitude Ground Ice on Mars from New Impact Craters. Science 325: 1674–1676.
4. Chyba CF and Hand KP (2001) Life without photosynthesis. Science 292: 2026–2027.
5. Chyba CF and Phillips CB (2001) Possible ecosystems and the search for life on Europa. Proc Natl Acad Sci USA 98: 801–804.

6. Fisk MR and Giovannoni SJ (1999) Sources of nutrients and energy for a deep biosphere on Mars. J Geophys Res 104: 11805–11815.
7. Formisano V, Atreya S, Encrenaz T, Ignatiev N and Giuranna M (2004) Detection of methane in the atmosphere of Mars. Science 306: 1758–1751.
8. Fredrickson JK and Fletcher M (eds) (2001) Sursurface Microbiology and Biogeochemistry. Wikey-Liss, New York.
9. Gaidos EJ and Nimmo F (2000) Tectonics and water on Europa. Nature 405: 637.
10. Getling AV (1998) Rayleigh-Benard Convection. World Scientific Publishing, Singapore.
11. Golden DC, Ming DW, Schwandt CS, Lauer HV, Socki RA, Morris RV, Lofgren GE and McKay GA (2001) A simple inorganic process for formation of carbonates, magnetite, and sulfides in Martian meteorite ALH84001. Am Mineral 83: 370–375.
12. Greenberg R (2004) The evil twin of Agenor; tectonic convergence on Europa. Icarus 167: 313–319.
13. Humphris SE, Zierenberg RA, Mullineaux LS and Thomson RE (eds) (1995) Seafloor Hydrothermal Systems. American Geophysical Union, Washington, DC.
14. Kransnopolsky VA and Feldman PD (2001) Detection of molecular hydrogen in the atmosphere of Mars. Science 294: 1914–1917.
15. Lucey PG (2009) A Lunar Waterworld. Science Published Online September 24, 2009, DOI: 10.1126/science.1181471 (volume and page numbers to be announced).
16. McKay DS, Gibson EK Jr, Thomas-Keprta KL, Vali H, Romanek CS, Clemett SJ, Chillier XDF, Maechling CR and Zare RN (1996) Search for past life on Mars: possible relic biogenic activity in Martian meteorite ALH84001. Science 273: 924–930.
17. Moriguchi Y and Jenkins D (1996) *Hojoki* (text by Kamo-no-Chomei). Stone Bridge Press, Berkeley, California.
18. Mumma MJ, Villanueva GL, Novak RE, Hewagama T, Bonev BP, DiSanti MA, Mandell AM and Smith MD (2009) Strong Release of Methane on Mars in Northern Summer 2003. Science 323: 1041–1045.
19. Naganuma T and Uematsu H (1998) Dive Europa: A search-for-life initiative. Biol Sci Space 12: 126–130.
20. Neukum G, Jaumann R, Hoffmann H, Hauber E, Head JW, Basilevsky AT, Ivanov BA, Werner SC, van Gasselt S, Murray JB and McCord T, The HRSC Co-Investigator Team (2004) Recent and episodic volcanic and glacial activity on Mars revealed by the high resolution stereo camera. Nature 432: 971–979.
21. Nna-Mvondo D and Martinez-Frias J (2007) Review komatiites: from Earth's geological settings to planetary and astrobiological contexts. Earth Moon Planet 100: 157–179.
22. Prigogine I (1984) Order out of Chaos: Man's New Dialogue with Nature. Bantam Books, New York.
23. Sabtier P (1923) Catalysis in Organic Chemistry (translation of La Catalyse en Chimie Organique). Van Nostrand Co, New York.
24. Schrödinger E (1944) What is Life? The Physical Aspect of the Living Cell. Cambridge University Press, Cambridge.
25. Thomas-Keprta KL, Clemett SJ, Bazylinski DA, Kirschvink JL, McKay DS, Wentworth SJ, Vali H, Gibson EKJ and McKay MF, Romanek CS (2001) Truncated hexa-octahedral magnetite crystals in ALH84001: Presumptive biosignatures. Proc Natl Acad Sci USA 98: 2164–2169.
26. Yoshizaki T, Shibuya T, Suzuki K, Shimizu K, Nakamura K, Takai K, Omori S and Maruyama S (2009) H_2 generation by experimental hydrothermal alteration of komatiitic glass at 300°C and 500 bars: A preliminary result from on-going experiment. Geochem J 43: e17–e22.

Matrix Frequency Analysis of Oryza Sativa (Japonica Cultivar-Group) Complete Genomes [†]

K. Manikandakumar,[1] S. Muthu Kumaran[2] and R. Srikumar[3]

[1]Department of Physics, Bharathidasan University College (W), Orathanadu-614 625, Tanjavore District, Tamil Nadu, India. Email: bioinfokm@gmail.com.
[2]Department of Physics, Nehru Memorial College, Puthanampatti-621 007, Tiruchirappalli District, Tamil Nadu, India.
[3]Department of Microbiology, Bharathidasan University College (W), Orathanadu-614 625, Tanjavore District, Tamil Nadu, India.

ABSTRACT

The genome sequence information is essential to understand the function of extensive arrangements of genes. It is significant to combine all sequence information in a precise database to provide an efficient manner of sequence similarity search. The complete genome analysis, which is one of the essential steps to know their characteristics, is very important. Complete genome analysis is depends on matrix frequency of sequence residue calculation and CGR analysis. In this study, we select rice as the specimen for complete genome analysis. Rice is one of the most essential cereal crops providing food for more than half of the world's population. *Oryza sativa* (japonica cultivar-group) species is an important cereal and model monocot. We have generated a matrix frequency for genetic code analysis, which helps in the study of complete genome residues. Here we report the duplets and triplets codon for genetic code analysis of *O. sativa* chromosomes. We illustrate a new method of Chaos Game Representation, which produces the objects possessing self-similar structure. As per our findings, the average matrix frequency of stop codons is similar to the matrix frequency of start codon. This average is seems to be similar in the complete genome sequences of every *Oryza sativa* (japonica cultivar-group) chromosomes.

Keywords: *Oryza sativa* (japonica cultivar-group); Chaos game representation; chromosome; matrix frequency, fractal structure.

[†]Reused with permission from: K. Manikandakumar, S. Muthu Kumaran, R. Srikumar, Matrix Frequency Analysis of Oryza sativa (japonica cultivar-group) Complete Genomes, *Journal of Computer Science & Systems Biology*, Volume 2(2): 159–166 (2009).

Introduction

DNA is a double anti-parallel helix built by concatenating nucleotide blocks. Several physicochemical properties of DNA depends on the interactions between consecutive bases, thus, the classification of patterns from nearest neighbor bases could help in the description of nucleotide sequences (Mohanty A.K and A.V.S.S. Narayana Rao 2000). However, (Almeida J.S. et al. 2001) have followed the scale independence of CGR of genetic sequence method to investigate local and global homology. The two patterns identified from the analysis of whole genomes and the number of different dinucleotides are unequal frequencies of manifestation of some asymmetric pairs and preferences of certain nucleotides with specific nearest neighbors over equivalent dinucleotides (Nussinov R. 1980, 1981).

Small plant chromosomes, such as those in rice, often show irregular condensation at mitotic prometaphase. Thus, the condensation pattern appearing at prometaphase was only a morphological landmark to divide the rice chromosomes into sub-regions. Characteristics of each rice chromosome with uneven condensation have quantitatively been analyzed by using image analysis methods. (Fukui K. 1985; Iijima K. and Fukui K. 1991) developed a method for identifying rice chromosomes based on a flow chart that consists of 11 discriminates, which classify specific chromosome groups.

All rice chromosomes have identified and numbered by comparing the categories given by discriminates, one after another. The chromosomal spread is worth analyzing if, chromosomes 4,11, and 12 are distinguishable by visual inspection and if chromosomes 1,2, and 3 are completely recognized. If these six chromosomes are identified using discriminates 1 to 6 in order, then there is a great possibility of identifying all 12 chromosomes within the particular spread. The relevance of accessing the frequency of non-integer genomic sequences may not be apparent at first given that (Almeida J.S. et al. 2001) physically make all sequences of integer number of nucleotides.

The genome sequence information is indispensable in understanding the function of the wide array of genes that constitute the rice plant. Therefore, it is important to consolidate all sequence information in a specified database to provide an efficient method of sequence similarity search that eliminates artifactual matches be analysed by (Yoshiaki Nagamura et al. 2003). We have generated a matrix frequency for genetic code analysis with all available rice genome for *Oryza sativa* japonica-cultivar group. (Li-Zhi Gao et al. 2005) analysed that DNA shuffling is a direct evolution process which generates genetic diversity through the recombination of parental sequences in order to evaluate which pair of sequences could potentially produce the best result. *Oryza sativa* (japonica cultivar-group), a subspecies of rice, is an important cereal and model monocot (35884299 bp). The rice genome sequence provides a foundation for the improvement of cereals, our most important crops (Stephen A. et al. 2002). Experiments of direct evolution have successfully used to improve specific biological functions. (Jorge Luis Fuentes et al. 2005) analyses Genetic diversity of rice varieties (*Oryza sativa* L.) based on morphological, pedigree and DNA polymorphism data, Plant Genetic Resources and phenotypic, genealogical, RAPD and AFLP diversity groups.

Mathematical characterization of DNA sequences could help in the understanding of structural relationships among different whole genomes along the chromosomes. The degenerated translation of trinucleotide codons encode for 20 amino acids, and remaining three nonsense codons signal for the end of transcription. Base concentrations, stretches and patches are the main factors explaining the variability observed among sequences (Deschavanne P.J. et al. 1999). The genomic signature as expressed in terms of short nucleotide usage extends and generalizes the genomic signature and it takes advantages of whole genome data reveals genome wide trends (Karlin S. and Burge C. 1995). The measure of similarity using CGR can be the basis of a new set of algorithms to align sequences with considerable advantages over the conventional scoring methods (Almeida J.S. et al. 2001). The CGR is a formalism that bridges between sequence of discrete units and numeric coordinates in a continuous space. Consequently, basic statistic measures and techniques have applied to sequences and a wide range of new tools have devised for statistical analysis.

Here we report the genetic code analysis of complete genome of all chromosomes of *O. sativa*. We have generated a matrix frequency of the rice genome. We describe a new method of Chaos Game Representation applied to *O. Sativa* (japonica cultivar-group) species sequences, which produces fractal objects possessing self-similar structure.

Material and Methods

The *Oryza sativa* (japonica cultivar-group) Eukaryote complete genomes have downloaded from the GOLD (http://www.genomesonline.org/) database. The species have been totally 12 chromosomes. The details of the chromosomes are giving below.

[http://www.genomesonline.org/gold.cgi?want=Published+Co plete+Genomes]

Ch. No.	Web link
1	http://www.ncbi.nlm.nih.gov/entrez/viewer.fcgi?db=nucleotide&val=NC_008394
2	http://www.ncbi.nlm.nih.gov/entrez/viewer.fcgi?db=nucleotide&val=NC_008395
3	http://www.ncbi.nlm.nih.gov/entrez/viewer.fcgi?db=nucleotide&val=NC_008396
4	http://www.ncbi.nlm.nih.gov/entrez/viewer.fcgi?db=nucleotide&val=NC_008397
5	http://www.ncbi.nlm.nih.gov/entrez/viewer.fcgi?db=nucleotide&val=NC_008398
6	http://www.ncbi.nlm.nih.gov/entrez/viewer.fcgi?db=nucleotide&val=NC_008399
7	http://www.ncbi.nlm.nih.gov/entrez/viewer.fcgi?db=nucleotide&val=NC_008400
8	http://www.ncbi.nlm.nih.gov/entrez/viewer.fcgi?db=nucleotide&val=NC_008401
9	http://www.ncbi.nlm.nih.gov/entrez/viewer.fcgi?db=nucleotide&val=NC_008402
10	http://www.ncbi.nlm.nih.gov/entrez/viewer.fcgi?db=nucleotide&val=NC_008403
11	http://www.ncbi.nlm.nih.gov/entrez/viewer.fcgi?db=nucleotide&val=NC_008404
12	http://www.ncbi.nlm.nih.gov/entrez/viewer.fcgi?db=nucleotide&val=NC_008405

A simple model, which permits the simulation of these features of nucleotide residues, is discrete time Markov Chain (Goldman N 1993). In this model, a 4 X 4 matrix, P defines the probabilities with which subsequent bases follow the current base in a nucleotide residue. If the base labels A,T,G, and C are equated with the numbers 1,2,3 and 4; then P_{ij} is the jth element of the ith row of P which defines the probability that base j follows base i. The row sums of P must equal 1. Using this matrix, a simulated nucleotide residue may be obtained by selecting a first base randomly according to the frequencies of the bases in the nucleotide residue under study. If the base is i, then the probabilities will be P_{i1}, P_{i2}, P_{i3} and P_{i4}. These probabilities are used to select the next base, and so on until the simulated sequence is the same length as the original nucleotide residues.

This first-order Markov Chain model is in which successive bases in a residue depend only on the preceding base. The probabilities in the matrix *P* may be estimated by direct calculation from the residues dinucleotide frequencies. If the dinucleotide *XY* is observed n_{xy} times in the sequence, then probability P_{xy} is estimated by $n_{XY}/(n_{XA}+n_{XT}+n_{XG}+n_{XC})$. This permits a protein sequence to be simulated with both individual base frequencies and digroup frequencies matching those of the original sequence. Dinucleotide frequencies (n_{XY}) and Markov Chain probabilities (P_{XY}) for the *Oryza sativa* (japonica cultivar-group) genomes are given in Table 2.

The first-order Markov Chain model successfully recreates other genomes. The lack of banding suggests approximate equality of the frequencies of the bases A, C, G, and T, confirmed by direct calculation from the residues. The first-order Markov Chain model will not give the observed patterns, but a more complex second-order Markov Chain, in which each base depends on the previous two, does. Second-order Markov Chains have been used to describe both structure and with-in-structure of nucleotide residues. P_{XYZ}, the probability that base Z follows the trigroup XYZ, is estimated directly from the nucleotide residues trigroup frequencies n_{XYZ} using the formula $P_{XYZ}=n_{XYZ}/(n_{XYA}+n_{XYT}+n_{XYG}+n_{XYC})$. Trinucleotide frequencies (n_{XYZ}) and Markov Chain probabilities (P_{XYZ}) for the *Oryza sativa* (japonica cultivar-group) genomes are given in Table 3.

We apply the CGR method to *Oryza sativa* (japonica cultivar-group) species by considering the four different nucleotide residues into four groups namely Adenine, Thymine, Guanine and Cytosine. Using this distinctive way of CGR technique, the *Oryza sativa* (japonica cultivar-group) species produces the intrinsic fractal structure. The percentage values of nucleotide residues of the four groups available in the species under consideration has also computed and used for analysis. We find that some of the species of *Oryza sativa* (japonica cultivar-group) nucleotide sequences produce the similar kind of self-similar fractal structure. The CGR shows the characteristics of the *Oryza sativa* (japonica cultivar-group) genome.

To begin with, let us generate the typical fractal object namely the 'Square' possessing the self-similar structure using the Chaos Game Representation (CGR). Let us start with three vertices located at (0,0), (1,0), (0,1) and (1,1) labeled as A, T, C and G respectively. Now random sequences of 1, 2, 3 and 4 are obtained using a random number generator available in typical C compiler. In generating CGR, the nth point of the attractor is simply the mid-point between the $(n-1)^{th}$ point and the vertex corresponding to the nth value. Similarly, the successive application of this procedure for 100,000 points produces the 'Square' as shown in Fig. 1, which is a typical fractal object possessing self-similar structure.

We calculate the nucleotide contents of the above species into grouping of four types name as A, T, G and C. Used in computer algorithm the nucleotide contents are differentiating to each group. Then we calculate the A+G and T+C ratios of the above species. Finally, the average ratio of the each chromosome is calculating by the method A+G/T+C. All the results are given by percentage values. The Table 1 is given by all the chromosome details of the *Oryza sativa* (japonica cultivar-group) species. The CGR plots are drawn using Gnu plot method.

Table 1 Nucleotide content analysis of *Oryza sativa* (japonica cultivar-group) Total No. Base Pairs: **35884299.**

Sl. No.	Name of the species	Total No. of Residues	% of (A)	% of (T)	% of (G)	% of (C)	% of (A + G)	% of (T+C)	%of (A+G)/(T+C)
	chromosome 1	301936	26.99	28.15	22.70	22.16	49.69	50.31	1.0
	chromosome 2	662387	28.39	27.98	21.79	21.84	50.18	49.82	1.0
	chromosome 3	831805	27.50	28.34	22.05	22.11	49.55	50.45	1.0
	chromosome 4	17028043	27.88	27.95	22.10	22.08	49.98	50.03	1.0
	chromosome 5	476423	28.61	28.47	21.36	21.55	49.97	50.02	1.0
	chromosome 6	1949261	28.03	27.96	22.12	21.88	50.15	49.84	1.0
	chromosome 7	993326	28.67	28.24	21.47	21.63	50.14	49.87	1.0
	chromosome 8	8367279	28.49	28.37	21.57	21.57	50.06	49.94	1.0
	chromosome 9	2439243	28.06	27.81	22.06	22.07	50.12	49.88	1.0
	chromosome 10	306812	28.84	28.73	21.24	21.19	50.08	49.92	1.0
	chromosome 11	298736	28.50	28.66	21.63	21.21	50.13	49.87	1.0
	chromosome 12	2229048	28.54	28.21	21.60	21.65	50.14	49.86	1.0

Pictorial representation

Chaos Game Representation (CGR) for gene (or DNA) sequences was introduced by (Jeffrey H. J. 1990, 1992) and the essential structures of genome sequences of a few model organisms were obtained using CGR plots. Each chromosome has been taken in above 200,000 base pairs. Therefore, we did not represent the whole genomes. We have taken only first 100,000 base pairs nucleotide sequences for the above representation.

Results and Discussions

The structure of DNA is specific to each species and undergoes only slight variations along the whole genome (Deschavanne P.J. et al. 1999). Diversity among species is considerable and is primarily a consequence of base concentration, stretches of bases with unusual frequencies. The frequencies of occurrence are to point out the basis of the genome (Deschavanne P.J. et al. 1999). In our analysis is giving the matrix frequency calculation of every chromosome complete genome sequences. We analysed every chromosome and given in the Table 1 is shown by the individual nucleotide contents percentage. The Table 2 has shown by the first order Markov chain matrix frequency of all chromosomes and it is representing in dinucleotide codons. The Table 3 is show by the second order Markov chain matrix frequency of all chromosomes and it is representing in trinuclotide codons. The Table 4 is shown, the classification of triplets to occurring in which regions. The Table 5 is shown by the relations of the start and stop codons of all chromosomes.

Table 2 Nucleotide duplet Matrix frequency of *Oryza sativa* (japonica cultivar-group).

Name of the Duplets	Name of the Chromosome											
	C1	C2	C3	C4	C5	C6	C7	C8	C9	C10	C11	C12
AA	0.299	0.311	0.306	0.315	0.319	0.315	0.316	0.320	0.317	0.321	0.310	0.315
CA	0.289	0.314	0.301	0.293	0.317	0.300	0.308	0.304	0.299	0.302	0.315	0.316
GA	0.274	0.283	0.271	0.277	0.285	0.276	0.282	0.282	0.279	0.285	0.281	0.282
TA	0.223	0.233	0.228	0.233	0.232	0.232	0.245	0.237	0.231	0.249	0.241	0.234
AC	0.184	0.186	0.189	0.187	0.181	0.187	0.184	0.184	0.188	0.179	0.179	0.185
CC	0.245	0.232	0.239	0.249	0.237	0.238	0.237	0.238	0.248	0.239	0.234	0.233
GC	0.242	0.245	0.248	0.236	0.238	0.245	0.244	0.238	0.237	0.229	0.242	0.244
TC	0.223	0.219	0.217	0.220	0.217	0.215	0.212	0.213	0.220	0.213	0.207	0.215
AG	0.224	0.217	0.217	0.211	0.215	0.215	0.209	0.207	0.213	0.204	0.220	0.212
CG	0.184	0.175	0.180	0.192	0.160	0.192	0.182	0.186	0.185	0.182	0.163	0.173
GG	0.245	0.235	0.237	0.248	0.237	0.240	0.234	0.238	0.250	0.241	0.234	0.233
TG	0.248	0.240	0.242	0.232	0.236	0.236	0.231	0.230	0.233	0.221	0.239	0.240
AT	0.292	0.286	0.288	0.287	0.286	0.283	0.292	0.289	0.283	0.296	0.292	0.287
CT	0.282	0.279	0.279	0.266	0.287	0.270	0.273	0.271	0.268	0.276	0.289	0.278
GT	0.238	0.237	0.244	0.239	0.241	0.239	0.239	0.242	0.235	0.245	0.243	0.241
TT	0.306	0.308	0.313	0.314	0.315	0.317	0.313	0.320	0.316	0.317	0.313	0.311

We analyze the complete genome of *Oryza sativa* (japonica cultivar-group) species chromosomes nucleotide contents and the calculation is giving in Table 1. From the Table 1, the chromosome 4 has been largest residues (17028043 base pairs). The lowest residues have been chromosome 11 (298736 base pairs). Chromosome no. 1 is having the lowest Adenine residues (26.99%) and chromosome 10 is having the highest adenine residues (28.84%). The range of Thymine residues is 27.81% got the chromosome no. 9 and 28.73% of thymine residues are having chromosome no. 10. The range of Guanine ratio is 21.36% (chromosome 5) and 22.70 (chromosome 1). The range of Cytosine ratio is 21.19% (chromosome 10) and 22.16 (chromosome 1). The range of A+G content ratio is 49.55% (chromosome 3) and 50.18% (chromosome 2). The range of T+C content ratio is 49.82% (chromosome 2) and 50.45% (chromosome 3). The chromosome 8 has been representing in same nucleotide content ratio in Guanine and Cytosine residues (21.57%).

We generate and analyse the first order Markov chain matrix frequency for 16 (4x4) nucleotide doublet codons for *Oryza sativa* (japonica cultivar-group) species complete genome chromosomes and the matrix frequency for each doublet codon is given in Table 2. The Table 2 has shown, the AA codon minimum frequency range is 0.299 for chromosome-1

Table 3 Triplet codon Matrix frequency of *Oryza sativa* (japonica cultivar-group).

Name of the Triplet codon	Name of the Chromosome											
	C1	C2	C3	C4	C5	C6	C7	C8	C9	C10	C11	C12
AAA	0.339	0.351	0.350	0.347	0.354	0.357	0.361	0.359	0.346	0.360	0.346	0.352
AAC	0.320	0.344	0.328	0.321	0.348	0.330	0.346	0.336	0.323	0.329	0.343	0.346
AAG	0.289	0.303	0.286	0.297	0.304	0.302	0.303	0.305	0.299	0.307	0.302	0.302
AAT	0.254	0.265	0.258	0.266	0.267	0.267	0.283	0.270	0.260	0.281	0.273	0.264
CAA	0.294	0.300	0.292	0.313	0.312	0.302	0.304	0.311	0.317	0.319	0.309	0.308
CAC	0.291	0.312	0.298	0.294	0.314	0.297	0.303	0.304	0.303	0.305	0.311	0.314
CAG	0.258	0.260	0.247	0.258	0.262	0.251	0.256	0.257	0.255	0.267	0.255	0.254
CAT	0.203	0.217	0.211	0.225	0.214	0.220	0.231	0.226	0.228	0.246	0.233	0.223
GAA	0.283	0.295	0.291	0.299	0.306	0.300	0.294	0.302	0.298	0.300	0.293	0.302
GAC	0.280	0.302	0.290	0.275	0.306	0.283	0.286	0.284	0.282	0.288	0.303	0.304
GAG	0.284	0.296	0.286	0.281	0.293	0.283	0.292	0.287	0.284	0.290	0.300	0.294
GAT	0.229	0.240	0.237	0.223	0.240	0.226	0.239	0.231	0.218	0.232	0.242	0.239
TAA	0.270	0.284	0.275	0.289	0.287	0.287	0.286	0.293	0.293	0.290	0.280	0.285
TAC	0.273	0.300	0.289	0.282	0.302	0.293	0.296	0.295	0.289	0.288	0.303	0.302
TAG	0.264	0.268	0.260	0.268	0.272	0.263	0.272	0.274	0.270	0.270	0.263	0.271
TAT	0.204	0.211	0.208	0.215	0.207	0.213	0.221	0.216	0.215	0.229	0.217	0.212
ACA	0.173	0.178	0.179	0.178	0.175	0.178	0.172	0.175	0.179	0.165	0.170	0.177
ACC	0.238	0.225	0.230	0.246	0.232	0.230	0.227	0.233	0.249	0.231	0.225	0.226
ACG	0.230	0.236	0.232	0.225	0.226	0.231	0.230	0.227	0.226	0.216	0.227	0.229
ACT	0.210	0.206	0.205	0.204	0.207	0.201	0.197	0.200	0.209	0.195	0.197	0.202
CCA	0.200	0.198	0.207	0.212	0.191	0.208	0.207	0.207	0.212	0.211	0.200	0.201
CCC	0.223	0.217	0.225	0.225	0.223	0.217	0.221	0.216	0.227	0.223	0.217	0.211
CCG	0.260	0.250	0.281	0.247	0.258	0.260	0.266	0.255	0.243	0.236	0.270	0.265
CCT	0.268	0.259	0.260	0.263	0.255	0.257	0.264	0.258	0.262	0.262	0.259	0.257
GCA	0.177	0.179	0.174	0.183	0.169	0.181	0.180	0.178	0.183	0.178	0.166	0.176
GCC	0.262	0.236	0.253	0.272	0.239	0.257	0.254	0.261	0.264	0.257	0.247	0.249
GCG	0.252	0.262	0.258	0.258	0.251	0.267	0.260	0.259	0.259	0.248	0.248	0.258
GCT	0.214	0.208	0.202	0.216	0.200	0.212	0.204	0.208	0.214	0.215	0.193	0.204
TCA	0.191	0.193	0.199	0.179	0.191	0.185	0.182	0.179	0.179	0.168	0.180	0.189
TCC	0.254	0.248	0.247	0.253	0.251	0.245	0.244	0.243	0.253	0.246	0.244	0.244
TCG	0.233	0.239	0.236	0.221	0.229	0.229	0.232	0.222	0.223	0.220	0.236	0.233
TCT	0.208	0.210	0.206	0.208	0.210	0.202	0.196	0.200	0.205	0.196	0.189	0.204
AGA	0.218	0.204	0.203	0.205	0.206	0.199	0.194	0.196	0.208	0.199	0.209	0.202
AGC	0.163	0.158	0.165	0.169	0.143	0.173	0.161	0.164	0.162	0.170	0.146	0.154
AGG	0.245	0.232	0.242	0.242	0.237	0.236	0.232	0.230	0.244	0.235	0.237	0.229
AGT	0.252	0.239	0.245	0.233	0.236	0.236	0.228	0.230	0.233	0.222	0.243	0.244
CGA	0.213	0.215	0.213	0.187	0.207	0.208	0.194	0.193	0.187	0.174	0.203	0.203
CGC	0.206	0.191	0.202	0.221	0.171	0.219	0.207	0.215	0.210	0.204	0.182	0.197
CGG	0.263	0.277	0.262	0.282	0.265	0.275	0.263	0.276	0.296	0.270	0.253	0.263
CGT	0.235	0.234	0.239	0.207	0.225	0.228	0.217	0.214	0.203	0.193	0.224	0.229
GGA	0.261	0.257	0.260	0.252	0.255	0.254	0.259	0.249	0.253	0.252	0.268	0.250
GGC	0.192	0.197	0.195	0.200	0.184	0.207	0.196	0.198	0.199	0.189	0.175	0.184
GGG	0.232	0.208	0.218	0.225	0.221	0.213	0.219	0.216	0.227	0.223	0.211	0.213

Table 3 contd....

Table 3 contd....

Name of the Triplet codon	Name of the Chromosome											
	C1	C2	C3	C4	C5	C6	C7	C8	C9	C10	C11	C12
GGT	0.272	0.262	0.260	0.263	0.261	0.260	0.262	0.258	0.267	0.255	0.265	0.260
TGA	0.207	0.197	0.199	0.204	0.198	0.206	0.198	0.199	0.207	0.197	0.206	0.202
TGC	0.175	0.157	0.162	0.181	0.145	0.172	0.166	0.172	0.169	0.170	0.151	0.159
TGG	0.247	0.233	0.234	0.250	0.234	0.244	0.231	0.237	0.247	0.245	0.238	0.235
TGT	0.239	0.232	0.231	0.231	0.229	0.226	0.225	0.223	0.234	0.221	0.232	0.231
ATA	0.271	0.267	0.267	0.271	0.265	0.266	0.273	0.270	0.266	0.276	0.274	0.270
ATC	0.279	0.273	0.277	0.263	0.277	0.267	0.265	0.268	0.265	0.270	0.286	0.274
ATG	0.236	0.229	0.241	0.235	0.233	0.231	0.235	0.238	0.232	0.241	0.235	0.240
ATT	0.284	0.291	0.293	0.297	0.289	0.296	0.292	0.300	0.298	0.302	0.288	0.290
CTA	0.293	0.287	0.287	0.287	0.290	0.282	0.295	0.289	0.284	0.296	0.288	0.288
CTC	0.281	0.280	0.275	0.261	0.292	0.267	0.269	0.265	0.260	0.268	0.290	0.277
CTG	0.219	0.214	0.210	0.213	0.215	0.213	0.215	0.212	0.205	0.227	0.222	0.217
CTT	0.294	0.290	0.290	0.305	0.306	0.294	0.288	0.302	0.308	0.300	0.284	0.292
GTA	0.279	0.269	0.275	0.267	0.270	0.265	0.268	0.271	0.265	0.270	0.272	0.272
GTC	0.265	0.265	0.261	0.253	0.272	0.253	0.263	0.257	0.255	0.266	0.276	0.263
GTG	0.233	0.234	0.238	0.236	0.236	0.237	0.229	0.238	0.230	0.239	0.241	0.235
GTT	0.285	0.290	0.301	0.298	0.300	0.302	0.295	0.303	0.301	0.298	0.300	0.297
TTA	0.332	0.326	0.327	0.328	0.324	0.322	0.334	0.329	0.321	0.345	0.334	0.324
TTC	0.298	0.296	0.302	0.285	0.302	0.289	0.293	0.291	0.288	0.297	0.302	0.295
TTG	0.256	0.260	0.270	0.261	0.265	0.264	0.265	0.267	0.260	0.265	0.263	0.261
TTT	0.348	0.347	0.355	0.346	0.354	0.359	0.359	0.360	0.346	0.354	0.363	0.353

Table 4 Classification of triplet codon—complete genome of *Oryza sativa* (japonica cultivar-group).

Sl. No.	Name of the Chromosome	0.125–0.199	0.200–0.249	0.250–0.299	Above 0.300
	Chromosome 1	ACA, GCA, TCA, AGC, GGC, TGC	CAT, GAT, TAT, AC-C,G,T; CC-A,C; GC-G,T; TC-G,T; AG-A,G; CG-A,C,T; GGG, TG-A,G,T; ATG,CTG, GGG	AAG, AAT, CA-A,C,G; GA-A,C,G; TA-A,C,G; CC-G,T; GC-C,G; TC-C,G; AGT, CGG, GG-A,T; AT-A,C,T; CT-A,C,T; GT-A,C,T; TT-C,G	AAA, AAC, TTA, TTT
	Chromosome 2	ACA,CCA, GCA, TCA, AGC, CGC, GGC, TGA, TGC	CAT, GAT, TAT, AC-C,G,T; CCC, GC-C,T; TC-C,G,T; AG-A,G,T; CG-A,T; GGG, TG-G,T; ATG,CTG, GTG	AAT, CAG, GA-A,G; TA-A,G; CC-G,T; GCG, CGG, GG-A,T; AT-A,C,T; CT-A,C,T; GT-A,C,T; TT-C,G	AA-A,C,G; CA-A,C; GAC, TAC, TT-A,T
	Chromosome 3	ACA, GCA, TCA, AGC, GGC, TGA TGC	CA-G,T; GAT, TAT, AC-C,G,T; CC-A,C; GCT, TC-C,G,T; AG-A,G,T; CG-A,C,T; GGG, TG-G,T; ATG,CTG, GTG	AA-G,T, CA-A,C, GA-A,C,G; TA-A,C,G; CC-G,T; GC-C,G, CGG, GG-A,T; AT-A,C,T; CT-A,C,T; GT-A,C; TTG	AA-A,C; GTT, TT-A,C,T

Table 4 contd....

Table 4 contd....

Sl. No.	Name of the Chromosome	0.125–0.199	0.200–0.249	0.250–0.299	Above 0.300
	Chromosome 4	ACA, GCA, TCA, AGC, CGA, TGC	CAT, GAT, TAT, AC-C,G,T; CC-A,C,G; GCT, TC-G,T; AG-A,G,T; CG-C,T; GG-C,G, TG-A,T; ATG, CTG, GTG	AA-G,T, CA-C,G; GA-A,C,G; TA-A,C,G; CCT, GC-G,T; TCC, CGG, GG-A,T; TGG, AT-A,C,T; CT-A,C; GT-A,C,T; TT-C,G.	AA-A,C; CAA, CTT, TT-A,T
	Chromosome 5	ACA, CCA, GCA, TCA, AGC, CGC, GGC TGC	CAT, GAT, TAT, AC-C,G,T; CCC, GC-C,T, TC-G,T; AG-A,G,T; CG-A,T; GGG, TG-G,T; ATG, CTG, GTG	AAT, CAG; GAG; TA-A,G; CC-G,T, GCG, TCC; CGG, GG-A,T; AT-A,C,T; CT-A,C; GT-A,C; TTG	AA-C,G; CA-A,C; GA-A,C, TAC TTC, CTT, GTT, TT-A; AAA TTT
	Chromosome 6	ACA, GCA, TCA, AG-A,C; TGC	CAT, GAT, TAT, AC-C,G,T; CC-A,C, GCT, TC-C,G,T; AG-G,T; CG-A,C,T; GG-C,G, TG-A,G,T; ATG, CTG, GTG	AAT, CA-C,G; GA-C,G; TA-A,C,G; CC-G,T, GC-C,G; CGG, GG-A,T; AT-A,C,T; CT-A,C,T; GT-A,C; TT-C,G	AA-A,C,G; CAA, GAA, GTT, TT-A,T
	Chromosome 7	AC-A,T; GCA, TC-A,T; AG-A,C, CGA, GGC, TGA, TGC	CAT, GAT, TAT, AC-C,G; CC-A,C, GCT, TC-C,G; AG-G,T; CG-C,T; GGG, TG-G,T; ATG, CTG, GTG	AAT, CAG, GA-A,C,G; TA-A,C,G; CC-G,T, GC-C,G; CGG, GG-A,T; AT-A,C,T; CT-A,C,T; GT-A,C,T; TT-C,G	AA-A,C,G; CA-A,C; TT-A,T
	Chromosome 8	ACA, GCA, TCA, AG-A,C; CGA, GGC, TGA, TGC	CAT, GAT, TAT, AC-C,G,T; CC-A,C, GCT, TC-C,G,T; AG-G,T; CG-C,T; GG-A,G, TG-G,T; ATG, CTG, GTG	AAT, CAG, GA-C,G; TA-A,C,G; CC-G,T, GC-C,G; CGG, GGT, AT-A,C; CT-A,C; GT-A,C; TT-C,G	AA-A,C,G; CA-A,C; GAA, ATT, CTT, GTT, TT-A,T
	Chromosome 9	ACA, GCA, TCA, AGC, CGA, GGC, TGC	CAT, GAT, TAT, AC-C,G,T; CC-A,C,G; GCT, TC-G,T; AG-A,G,T; CG-C,T; GGG, TG-A,G,T; ATG, CTG, GTG	AA-G,T, CAG, GA-A,C,G; TA-A,C,G; CCT, GC-C,G; TCC, CGG, GG-A,T; AT-A,C,T; CT-A,C; GT-A,C; TT-C,G	AA-A,C; CA-A,C; CTT, GTT, TT-A,T
	Chromosome 10	AC-A,T; GCA, TC-A,T; AG-A,C, CG-A,T; GGC, TGA, TGC	CAT, GAT, TAT, AC-C,G; CC-A,C,G; GC-G,T, TC-C,G; AG-G,T; CGC, GGG, TG-G,T; ATG, CTG, GTG	AAT, CAG, GA-C,G; TA-A,C,G; CCT, GCC, CGG, GG-A,T; AT-A,C; CT-A,C; GT-A,C,T; TT-C,G.	AA-A,C,G; CA-A,C; GAA, ATT, CTT, TT-A,T
	Chromosome 11	AC-A,T; GCA, TC-A,T; AG-A,C, CG-A,T; GGC, TGC.	CAT, GAT, TAT, AC-C,G; CC-A,C,G; GC-G,T, TC-C,G; AG-G,T; CGC, GGG, TG-G,T; ATG, CTG, GTG	AAT, CAG, GA-C,G; TA-A,C,G; CCT, GCC, CGG, GG-A,T; AT-A,C; CT-A,C; GT-A,C,T; TT-C,G	AA-A,C,G; CA-A,C; GAA, ATT, CTT, TT-A,T
	Chromosome 12	ACA, GCA, TCA, AGC, CGC, GGC, TGC	CAT, GAT, TAT, AC-C,G,T; CC-A,C; GC-C,T; TC-C,G,T; AG-A,G,T; CG-A,T; GGG, TG-A,G,T; ATG, CTG, GTG	AAT, CAG, GAG; TA-A,G; CC-G,T, GCG, CGG, GG-A,T; AT-A,C,T; CT-A,C,T; GT-A,C,T; TT-C,G	AA-A,C,G; CA-A,C; GA-A,C; TAC, TT-A,T

Table 5 Analysis for Start codon and Stop codon of *Oryza sativa* (japonica cultivar-group).

Name of the Chromosome	Start Codon	Stop Codon			TAG+TGA/2
	ATG	TAA	TAG	TGA	
1.	0.236	0.270	0.264	0.207	0.236
2.	0.229	0.284	0.268	0.197	0.233
3.	0.241	0.275	0.260	0.199	0.230
4.	0.235	0.289	0.268	0.204	0.236
5.	0.233	0.287	0.272	0.198	0.235
6.	0.231	0.287	0.263	0.206	0.235
7.	0.235	0.286	0.272	0.198	0.235
8.	0.238	0.293	0.274	0.199	0.237
9.	0.232	0.293	0.270	0.207	0.239
10.	0.241	0.290	0.270	0.197	0.234
11.	0.235	0.280	0.263	0.206	0.235
12.	0.240	0.285	0.271	0.202	0.237

and the maximum frequency range is 0.321 for chromosome-10. The CA codon minimum frequency range is 0.289 for chromosome-1 and the maximum frequency range is 0.317 for chromosome-5. The GA codon minimum frequency range is 0.271 for chromosome-3 and the maximum frequency range is 0.285 for chromosome-5. The TA codon minimum frequency range is 0.223 for chromosome-1 and the maximum frequency range is 0.245 for chromosome-7. The AC codon minimum frequency range is 0.179 for chromosome-10 & 11 and the maximum frequency range is 0.189 for chromosome-3. The CC codon minimum frequency range is 0.232 for chromosome-2 and the maximum frequency range is 0.249 for chromosome-4. The GC codon minimum frequency range is 0.229 for chromosome-10 and the maximum frequency range is 0.248 for chromosome-3. The TC codon minimum frequency range is 0.223 for chromosome-1 and the maximum frequency range is 0.207 for chromosome-11. The AG codon minimum frequency range is 0.204 for chromosome-10 and the maximum frequency range is 0.224 for chromosome-1. The CG codon minimum frequency range is 0.160 for chromosome-5 and the maximum frequency range is 0.192 for chromosome-4 & 6. The GG codon minimum frequency range is 0.233 for chromosome-12 and the maximum frequency range is 0.250 for chromosome-9. The TG codon minimum frequency range is 0.221 for chromosome-10 and the maximum frequency range is 0.248 for chromosome-1. The AT codon minimum frequency range is 0.283 for chromosome-6 & 9 and the maximum frequency range is 0.296 for chromosome-10. The CT codon minimum frequency range is 0.266 for chromosome-4 and the maximum frequency range is 0.289 for chromosome-11. The GT codon minimum frequency range is 0.235 for chromosome-9 and the maximum frequency range is 0.245 for chromosome-10. The TT codon minimum frequency range is 0.306 for chromosome-1 and the maximum frequency range is 0.320 for chromosome-8.

We have generate and analyse the second order Markov chain matrix frequency for 64 (4x4x4) nucleotide triplet codons for *Oryza sativa* (japonica cultivar-group) species complete genome chromosomes and the matrix frequency for each tripet codon is given in Table 3. From Table 3, most of the highest nucleotide triplet codon is representing in AAA and TTT. The most of the lowest nucleotide triplets are AGC and TGC. All the above chromosomes, we identify the low sparseness regions are mostly played in species of triplets as AC-A,T; GCA, TC-A,T; AG-A,C; CG-A,T; GGC, TGA and TGC respectively. The highest sparseness regions are mostly played in species of triplets as AA-A,C,G; CA-A,C; GAA, ATT, CTT, GTT and TT-A,T respectively.

Table 4 has shown the frequency triplet codons have separated by four regions. The regions are classifying the frequency range of 0.125-0.199, 0.200-0.249, 0.250-0.299, and above 0.300 matrix frequency of triplet codons. This table is easy to analyse and to study, how many tripets are coming under particular range of frequency.

We have analysed the relations between the start codon and stop codon frequencies and it has given in Table 5. The genetic code is show in three types of stop codons. But the start codon is only one. Therefore, we tried to show one stop codon for every sequence. Our analysis has not succeeded. Nevertheless, the average of two-stop codon value is nearly equal to the start codon. This Table 4 is describes, separated and shown in start and stop codon for each chromosomes in *Oryza sativa* (japonica cultivar-group). This table shows every start codon frequency is equal to the average of two-stop codon frequency. So this analysis is used to finding and expressed the start codon is equal to stop codon for every *Oryza sativa* (japonica cultivar-group) chromosome complete genome sequences.

The Fig.1, is shown the CGR plot for the first 100,000 base pairs of 12 chromosomes of *Oryza sativa* (japonica cultivar-group) species, we identify the genomes are cross overlapping in A, G and T, C. Four triangles are connecting in the mid point of 0.5, 0.5, 0.5, and 0.5 respectively. The A-T region is keeping in more numbers of residues.

Analysis of Individual Chromosome

Chromosome 1

The chromosome 1 is, total of 301936 base pairs. From the Table 1, the highest percentage value is Thymine residue (28.15%). The lowest percentage value is Cytosine residue (22.16%). The Adenine and Guanine residues are 26.99% and 22.70%. The highest combination of nucleotide residues of T+C percentage is 50.31%. The ratio of A+G & T+C is 1.0. The triplet of chromosome 1, the highest tri-nucleotide is TTT (0.348%).

The lowest tri-nucleotide is AGC (0.163%). From the Table 3, the matrix frequency of chromosome 1, high frequency of tri-nucleotide sequence has been representing in AAA, AAC, TTA, and TTT (above 0.300%). Tri-nucleotide sequence of 0.250% to 0.299% has been represented in AAG, AAT, CA-A,C,G; GA-A,C,G; TA-A,C,G; CC-G,T; GC-C,G; TC-C,G; AGT, CGG, GG-A,T; AT-A,C,T; CT-A,C,T; GT-A,C,T; TT-C,G. Tri-nucleotide sequence of 0.200% to 0.249% has been represented in CAT, GAT, TAT, AC-C,G,T; CC-A,C; GC-G,T; TC-G,T; AG-A,G; CG-A,C,T; GGG, TG-A,G,T; ATG,CTG and GGG. Tri-nucleotide sequence of 0.150% to 0.199% is representing in ACA, GCA, TCA, AGC, GGC, and TGC.

Chromosome 2

The chromosome 2 is total of 662387 base pairs. The highest percentage value is Adenine residue (28.39%). The lowest percentage value is Guanine residue (21.79%).The Thymine and Cytosine residues are 27.98% and 21.84%. The highest combination of nucleotide residues of A+G percentage is 50.18%. The ratio of A+G & T+C content is 1.0. The triplet codon of chromosome 2, the highest triplet is AAA (0.351%). The lowest triplet codon is TGC (0.157%).

The matrix frequency of chromosome 2, the frequency of tri-nucleotide sequence has been represented in AA-A,C,G; CA-A,C; GAC, TAC, TT-A,T (above 0.300%).Tri-nucleotide sequence of 0.250% to 0.299% has been represented in AAT, CAG, GA-A,G; TA-A,G; CC-G,T; GCG,CGG,GG-A,T; AT-A,C,T; CT-A,C,T;GT-A,C,T; TT-C,G.Tri-nucleotide sequence of 0.200% to 0.249% has been represented in CAT, GAT, TAT, AC-C,G,T; CCC, GC-C,T; TC-C,G,T; AG-A,G,T; CG-A,T; GGG, TG-G,T; ATG,CTG and GTG. Tri-nucleotide sequence of 0.150% to 0.199% has been representing in ACA, CCA, GCA, TCA, AGC, CGC, GGC, TGA, and TGC.

Chromosome 3

The chromosome 3 is total of 831805 base pairs. The highest percentage value is Thymine residue (28.34%). The lowest percentage value is Guanine residue (22.05%). The Adenine and Cytosine residues are 27.50% and 22.11%. The highest combination of nucleotide residues of T+C percentage is 50.45%. The ratio of A+G & T+C content is 1.0.The triplet codon of chromosome 3, the highest triplet is TTT (0.355%). The lowest triplet codon is TGC (0.162%).

The matrix frequency of chromosome 3, the frequency of tri-nucleotide sequence has been represented in AA-A,C; GTT, TT-A,C,T (above 0.300%).Tri-nucleotide sequence of 0.250% to 0.299% has been represented in AA-G,T, CA-A,C, GA-A,C,G; TA-A,C,G; CC-G,T; GC-C,G, CGG, GG-A,T; AT-A,C,T; CT-A,C,T; GT-A,C; TTG. Tri-nucleotide sequence of 0.200% to 0.249% has been represented in CA-G,T; GAT, TAT, AC-C,G,T; CC-A,C; GCT, TC-C,G,T; AG-A,G,T; CG-A,C,T; GGG, TG-G,T; ATG,CTG and GTG. Tri-nucleotide sequence of 0.150% to 0.199% has been representing in ACA, GCA, TCA, AGC, GGC, TGA and TGC.

Chromosome 4

The chromosome 4 is total of 17028043 base pairs. The highest percentage value is Thymine residue (27.95%). The lowest percentage value is Cytosine residue (22.08%). The Adenine and Guanine residues are 27.88% and 22.10%. The highest combination of nucleotide residues of T+C percentage is 50.03%. The ratio of A+G & T+C is 1.0.The triplet codon of chromosome 4, the highest triplet is AAA (0.347%).The lowest triplet codon is AGC (0.169%). The matrix frequency of chromosome 4, the frequency of tri-nucleotide sequence has been represented in AA-A,C; CAA, CTT, TT-A,T (above 0.300%).Tri-nucleotide sequence of 0.250% to 0.299% has been represented in AA-G,T, CA-C,G; GA-A,C,G; TA-A,C,G; CCT, GC-G,T; TCC, CGG, GG-A,T; TGG, AT-A,C,T; CT-A,C; GT-A,C,T; TT-C,G. Tri-nucleotide sequence of 0.200% to 0.249% has been represented in CAT, GAT, TAT, AC-C,G,T; CC-A,C,G; GCT, TC-G,T; AG-A,G,T; CG-C,T; GG-C,G, TG-A,T; ATG, CTG and GTG. Tri-nucleotide sequence of 0.150% to 0.199% has been representing in ACA, GCA, TCA, AGC, CGA, and TGC.

Chromosome 5

The chromosome 5 is total of 476423 base pairs. The highest percentage value is Adenine residue (28.61%). The lowest percentage value is Guanine residue (21.36%). The Thymine and Cytosine residues are 28.47% and 21.55%. The highest combination of nucleotide residues of T+C percentage is 50.02%. The ratio of A+G & T+C content is 1.0. The triplet codon of chromosome 5, the highest triplets is AAA and TTT (0.354%). The lowest triplet codon is AGC (0.143%). The matrix frequency of chromosome 5, the frequency of tri-nucleotide sequence has been represented in AA-C,G; CA-A,C; GA-A,C(same ratio); TAC & TTC are the same ratio. CTT, GTT, TT-A; AAA and TTT are the same ratio of nucleotides (above 0.300%). Tri-nucleotide sequence of 0.250% to 0.299% has been represented in AAT, CAG; GAG; TA-A,G; CC-G,T, GCG=TCC; CGG, GG-A,T; AT-A,C,T; CT-A,C; GT-A,C; TTG. Tri-nucleotide sequence of 0.200% to 0.249% has been represented in CAT, GAT, TAT, AC-C,G,T; CCC, GC-C,T, TC-G,T; AG-A,G,T; CG-A,T; GGG, TG-G,T; ATG, CTG and GTG. Tri-nucleotide sequence of 0.150% to 0.199% has been representing in ACA, CCA, GCA, TCA, AGC, CGC, GGC and TGC.

Chromosome 6

The chromosome 6 is total of 1949261 base pairs. The highest percentage value is Adenine residue (28.03%). The lowest percentage value is Cytosine residue (21.88%). The Thymine and Guanine residues are 27.96% and 22.12%. The highest combination of nucleotide residues of A+G percentage is 50.15%. The ratio of A+G & T+C content is 1.0. The triplet codon of chromosome 6, the range of triplets is TTT (0.359%) and TGC (0.172%). The matrix frequency of chromosome 6, the frequency of tri-nucleotide sequence has been represented in AA-A,C,G; CAA, GAA, GTT, TT-A,T of nucleotides (above 0.300%). Tri-nucleotide sequence of 0.250% to 0.299% has been represented in AAT, CA-C,G; GA-C,G; TA-A,C,G; CC-G,T, GC-C,G; CGG, GG-A,T; AT-A,C,T; CT-A,C,T; GT-A,C; TT-C,G. Tri-nucleotide sequence of 0.200% to 0.249% has been represented in CAT, GAT, TAT, AC-C,G,T; CC-A,C, GCT, TC-C,G,T; AG-G,T; CG-A,C,T; GG-C,G, TG-A,G,T; ATG, CTG and GTG. Tri-nucleotide sequence of 0.125% to 0.199% has been represented in ACA, GCA, TCA, AG-A,C and TGC.

Chromosome 7

The chromosome 7 is total of 993326 base pairs. The highest percentage value is Adenine residue (28.67%). The lowest percentage value is Guanine residue (21.47%). The Thymine and Cytosine residues are 28.24% and 21.63%. The highest combination of nucleotide residues of A+G percentage is 50.14%. The ratio of A+G & T+C content is 1.0. The triplet codon of chromosome 7, the frequency of triplets is AAA (0.361%) and AGC (0.161%). The matrix frequency of chromosome 7, the frequency of tri-nucleotide sequence has been represented in AA-A,C,G; CA-A,C; TT-A,T of nucleotides (above 0.300%). Tri-nucleotide sequence of 0.250% to 0.299% has been represented in AAT, CAG, GA-A,C,G; TA-A,C,G; CC-G,T, GC-C,G; CGG, GG-A,T; AT-A,C,T; CT-A,C,T; GT-A,C,T; TT-C,G.

Tri-nucleotide sequence of 0.200% to 0.249% has been represented in CAT, GAT, TAT, AC-C,G; CC-A,C, GCT, TC-C,G; AG-G,T; CG-C,T; GGG, TG-G,T; ATG, CTG and GTG. Tri-nucleotide sequence of 0.125% to 0.199% has been represented in AC-A,T; GCA, TC-A,T; AG-A,C, CGA, GGC, TGA and TGC.

Chromosome 8

The chromosome 8 is total of 8367279 base pairs. The highest percentage value is Adenine residue (28.49%). The lowest percentage values are represented in Guanine and Cytosine residues (21.57%). The Thymine residue is 28.37%. The highest combination of nucleotide residues of A+G percentage is 50.06%. The ratio of A+G & T+C content is 1.0. The triplet codon of chromosome 8, the range of triplets is TTT (0.360%) and AGC (0.164%). The matrix frequency of chromosome 8, the frequency of tri-nucleotide sequence has been represented in AA-A,C,G; CA-A,C; GAA, ATT, CTT, GTT, TT-A,T of nucleotides (above 0.300%). Tri-nucleotide sequence of 0.250% to 0.299% has been represented in AAT, CAG, GA-C,G; TA-A,C,G; CC-G,T, GC-C,G; CGG, GGT, AT-A,C; CT-A,C; GT-A,C; TT-C,G. Tri-nucleotide sequence of 0.200% to 0.249% has been represented in CAT, GAT, TAT, AC-C,G,T; CC-A,C, GCT, TC-C,G,T; AG-G,T; CG-C,T; GG-A,G, TG-G,T; ATG, CTG and GTG. Tri-nucleotide sequence of 0.125% to 0.199% has been represented in ACA, GCA, TCA, AG-A,C; CGA, GGC, TGA and TGC.

Chromosome 9

The chromosome 9 is total of 2439243 base pairs. The highest percentage value is Adenine residue (28.06%). The lowest percentage value is Guanine residue (22.06%). The Thymine and Cytosine residues are 27.81% and 22.07%. The highest combination of nucleotide residues of A+G percentage is 50.12%. The ratio of A+T & G+C is 1.0. The highest triplets are AAA and TTT (0.346%). The lowest triplet codon is AGC (0.162%). The matrix frequency of chromosome 9, the frequency of tri-nucleotide sequence has been represented in AA-A,C; CA-A,C; CTT, GTT, TT-A,T (above 0.300%). Tri-nucleotide sequence of 0.250% to 0.299% has been represented in AA-G,T, CAG, GA-A,C,G; TA-A,C,G; CCT, GC-C,G; TCC, CGG, GG-A,T; AT-A,C,T; CT-A,C; GT-A,C; TT-C,G. Tri-nucleotide sequence of 0.200% to 0.249% has been represented in CAT, GAT, TAT, AC-C,G,T; CC-A,C,G; GCT, TC-G,T; AG-A,G,T; CG-C,T; GGG, TG-A,G,T; ATG, CTG and GTG. Tri-nucleotide sequence of 0.150% to 0.199% has represented in ACA, GCA, TCA, AGC, CGA, GGC and TGC.

Chromosome 10

The chromosome 10 is total of 306812 base pairs. The highest percentage value is Adenine residue (28.84%). The lowest percentage value is Cytosine residue (21.19%). The Thymine and Guanine residues are 28.73% and 21.24%. The highest combination of nucleotide residues of A+G percentage is 50.08%. The ratio of A+G & T+C content is 1.0. The highest triplet is AAA (0.360%). The lowest triplet codon is ACA (0.165%). The matrix frequency of chromosome 10, the frequency of tri-nucleotide sequence has been represented in AA-A,C,G; CA-A,C; GAA, ATT, CTT, TT-A,T (above 0.300%). Tri-nucleotide sequence of 0.250% to 0.299% has been represented in AAT, CAG, GA-C,G; TA-A,C,G; CCT, GCC, CGG, GG-A,T; AT-A,C; CT-A,C; GT-A,C,T; TT-C,G. Tri-nucleotide sequence of 0.200% to 0.249% has been represented in CAT, GAT, TAT, AC-C,G; CC-A,C,G; GC-G,T, TC-C,G; AG-G,T; CGC, GGG, TG-G,T; ATG, CTG, GTG. Tri-nucleotide sequence of 0.150% to 0.199% has been represented in AC-A,T; GCA, TC-A,T; AG-A,C, CG-A,T; GGC, TGA and TGC.

Chromosome 11

The chromosome 11 is total of 298736 base pairs. The highest percentage value of Thymine residue is 28.66%. The lowest percentage value of Cytosine residue is 21.21%. The Adenine and Guanine residues are 28.50% and 21.63%. The highest combination of nucleotide residues of A+G percentage is 50.13%. The ratio of A+G & T+C content is 1.0. The highest triplet codon frequency (TTT) is 0.363%. The lowest triplet codon frequency AGC is 0.146%. The matrix frequency of chromosome 11, the frequency of tri-nucleotide sequence has been represented in AA-A,C,G; CA-A,C; GA-C,G, TAC, GTT, TT-A,C,T (above 0.300%). Tri-nucleotide sequence of 0.250% to 0.299% has been represented in AAT, CAG, GAA, TA-A,G; CC-G,T, CGG, GG-A,T; AT-A,C,T; CT-A,C,T; GT-A,C; TTG. Tri-nucleotide sequence of 0.200% to 0.249% has been represented in CAT, GAT, TAT, AC-C,G; CC-A,C; GC-C,G; TC-C,G; AG-A,G,T; CG-A,T; GGG, TG-A,G,T; ATG, CTG, GTG. Tri-nucleotide sequence of 0.150% to 0.199% has been represented in AC-A,T; GC-A,T; TC-A,T; AGC, CGC, GGC and TGC.

Chromosome 12

The chromosome 12 is total of 2229048 base pairs. The highest percentage value of Adenine residue is 28.54%. The lowest percentage value of Guanine residue is 21.60%. The Thymine and Cytosine residues are 28.21% and 21.65%. The highest combination of nucleotide residues of A+G percentage is 50.14%. The ratio of A+G & T+C content is 1.0. The highest triplet codon frequency of TTT is 0.353%. The lowest triplet codon frequency of AGC is 0.154%. The matrix frequency of chromosome 12, the frequency of tri-nucleotide sequence has been represented in AA-A,C,G; CA-A,C; GA-A,C; TAC, TT-A,T (above 0.300%). Tri-nucleotide sequence of 0.250% to 0.299% has been represented in AAT, CAG, GAG; TA-A,G; CC-G,T, GCG, CGG, GG-A,T; AT-A,C,T; CT-A,C,T; GT-A,C,T; TT-C,G. Tri-nucleotide sequence of 0.200% to 0.249% has been represented in CAT, GAT, TAT, AC-C,G,T; CC-A,C; GC-C,T; TC-C,G,T; AG-A,G,T; CG-A,T; GGG, TG-A,G,T; ATG, CTG, GTG. Tri-nucleotide sequence of 0.150% to 0.199% has represented in ACA, GCA, TCA, AGC, CGC, GGC and TGC.

Conclusion

The new techniques just described identifying the rice chromosome and specifying the region using the first order Markov chain, second order Markov chain and CGR methods of the DNA sequence analysis. The probabilities defining these models can calculated directly and easily from the raw DNA sequences, implying that the CGR gives no further insight into the structure of the DNA sequence than is given by the dinucleotide and trinucleotide frequencies. In this paper, we have shown that simple Markov Chain models based solely on dinucleotide and trinucleotide frequencies can account for the complex patterns exhibited in CGR of *Oryza sativa* (japonica cultivar-group) chromosome sequences. The *Oryza sativa* (japonica cultivar-group) species chromosome sequences are more similar to each other. However, our analysis is visible of sequence pattern is similar in each other. Some high matrix frequency value (0.300) of tri-nucleotide codon is having by small number of tri-nucleotides, but the matrix values are different in each other. The low-resolution codon frequencies are having by small number of tri-nucleotides (0.125–0.199), but the matrix value is different. We observed the above results the low-resolution tri-nucleotides are very low. The chromosome 1 is having four high frequency (above 0.300%) tri-nucleotides. The chromosome 8 has been representing in same nucleotide content ratio in Guanine and Cytosine residues (21.57%). The frequency matrix values 0.200% to 0.299% are highly responsible for the *Oryza sativa* (japonica cultivar-group) species. It is representing more number of tri-nucleotide codons. Finally, we observed the frequency of start codon is equal to average of two stop codon frequencies. The *in silico* analysis of the matrix frequency study is used in *invitro/ invivo* studies for reassembling the particular repaired codon region which is modified by the gene tinkering (codon replacing) methods. The future analysis can integrate these procedures into one logical, individual gene.

References

1. Almeida JS, Carrico JA, Maretzek A, Noble PA, Fletcher M (2001) Analysis of genomic sequences by Chaos Game Representation. Bioinformatics 17: 429–437.
2. Deschavanne PJ, Giron A, Villain J, Fagot G, Fertil B (1999) Genomic Signature: Characterization and Classification of Species Assessed by Chaos Game Representation of Sequences. Mol Biol Evol 16: 1391–1399.
3. Fukui K (1985) Identification of plant chromosome by the image analysis method. The Cell 17: 145–149.
4. Fuentes JL, Cornide MT, Alvarez A, Suarez E, Borges E (2005) Genetic diversity analysis of rice varieties (*Oryza sativa* L.) based on morphological, pedigree and DNA polymorphism data, Plant Genetic Resources: Characterization and Utilization. Plant Genet Resour 3: 353–359.
5. Gao LZ, Zhang CH, Chang LP, Jia JZ, Qiu ZE et al. (2005) Microsatellite diversity within *Oryza sativa* with emphasis on indica–japonica divergence. Genet Res 85: 1–14.
6. Goff SA, Ricke D, Lan TH, Presting G, Wang R, et al. (2002) A Draft Sequence of the Rice Genome (*Oryza sativa* L. ssp. japonica). Science 296: 92–100.
7. Goldman N (1993) Nucleotide, dinucleotide and trinucleotide frequencies explain patterns observed in chaos game representation of DNA sequences. Nucleic Acids Res 21: 2487–2491.
8. Iijima K, Fukui K (1991) Clarification of the conditions for the image analysis of plant chromosomes. Bull Natl Inst Agrobiol Resour 6: 1–58.
9. Jeffrey HJ (1990) Chaos game representation of gene structure. Nucleic Acids Res 18: 2163–2170.
10. Jeffrey HJ (1992) Chaos game visualization of sequences. Computer Graphics 16: 25–34.
11. Karlin S, Burge C (1995) Dinucleotide relative abundance extremes: a genomic signature. Trends Genet 11: 283–290.

12. Mohanty AK, Narayana Rao AV (2000) Factorial Moments Analyses Show a Characteristic Length Scale in DNA Sequences. Phys Rev Lett 84: 1832–1835.
13. Nussinov R (1980) Some rules in the ordering of nucleotides in the DNA. Nucleic Acids Res 8: 4545–4562.
14. Nussinov R (1981) Nearest neighbor nucleotide patterns: Structural and biological implications. J Biol Chem 256: 8458–8462.
15. Yoshiaki NA, Baltazar A, Hisataka N, Ikuo H, Manabu A et al. (2003) A Comprehensive Homology Search for Rice Specific Sequences. Genome Inform 14: 533–534.

Chapter 3

The Capabilities of Chaos and Complexity†

David L. Abel

The Gene Emergence Project, The Origin of Life Science Foundation, Inc. 113-120 Hedgewood Dr.
Greenbelt, MD 20770-1610 USA. Email: life@us.net.

ABSTRACT

To what degree could chaos and complexity have organized a Peptide or RNA World of crude yet necessarily integrated protometabolism? How far could such protolife evolve in the absence of a heritable linear digital symbol system that could mutate, instruct, regulate, optimize and maintain metabolic homeostasis? To address these questions, chaos, complexity, self-ordered states, and organization must all be carefully defined and distinguished. In addition their cause-and-effect relationships and mechanisms of action must be delineated. Are there any formal (non physical, abstract, conceptual, algorithmic) components to chaos, complexity, self-ordering and organization, or are they entirely physicodynamic (physical, mass/energy interaction alone)? Chaos and complexity can produce some fascinating self-ordered phenomena. But can spontaneous chaos and complexity steer events and processes toward pragmatic benefit, select function over non function, optimize algorithms, integrate circuits, produce computational halting, organize processes into formal systems, control and regulate existing systems toward greater efficiency? The question is pursued of whether there might be some yet-to-be discovered new law of biology that will elucidate the derivation of prescriptive information and control. "System" will be rigorously defined. Can a low-informational rapid succession of Prigogine's dissipative structures self-order into bona fide organization?

Keywords: Complex adaptive systems (CAS); Complexity theory; Bio-cybernetics; Biosemiotics; Emergence; Non linear dynamics; Self-organization; Symbolic dynamicsanalysis; Systems theory.

†Reused with permission from: David L. Abel, The Capabilities of Chaos and Complexity, Int. J. Mol. Sci. 2009, 10(1) 247–291; doi: 10.3390/ijms10010247.

Introduction

Stand-alone chaos, complexity and catastrophe should never be confused with our theories and what we intelligent humans do using abstract conceptual nonlinear dynamic models. Life-origin science is not especially interested in:

1. Modern-day human applications of non linear dynamical systems theory.
2. Investigator involvement (artificial selection) in chaos, catastrophe, and complexity experimental designs.
3. Information defined in terms of the reduced uncertainty of subjective "observers" and "knowers", which did not exist for 99.9% of life's history.

Life origin science wants to know the capabilities of stand-alone chaos and complexity before any animal consciousness existed. If all known life depends upon genetic instructions, how was the first linear digital prescriptive genetic information generated by natural process? In the absence of human thought and involvement, can objective chaos, complexity and catastrophe generate either or both of two peculiar entities:

1. Prescriptive Information (PI) [1–3]? PI refers not just to intuitive or semantic information, but specifically to linear digital instructions using a symbol system (e.g., 0's and 1's, letter selections from an alphabet, A, G, T, or C from a phase space of four nucleotides). PI can also consist of purposefully programmed configurable switch-settings that provide cybernetic controls.
2. *Bona fide* Formal Organization [4]? By "formal" we mean function-oriented, computationally halting, integrated-circuit producing, algorithmically optimized, and choice-contingent at true decision nodes (not just combinatorial bifurcation points).

   Both PI and formal organization are abstract, conceptual, non physical entities [1–13]. Scientific endeavors to better understand cybernetic reality in nature are confronted with the uneasy suggestion of its transcendence over the physicality it controls. At the heart of all naturalistic life-origin models lies the presumption of self-organization of inanimate physicality into sophisticated formal utility. The notion of *emergence* can be traced back to Aristotle [14], but George H. Lewes was probably the first to define it in 1875: "The emergent is unlike its components insofar as these are incommensurable, and it cannot be reduced to their sum or their difference." [15. pg. 412]. The idea of emergence blossomed in the 1920's with contributions from C. Lloyd Morgan, C. D. Broad, Samuel Alexander, Henre Bergson, Alfred North Whitehead, and Arthur O. Lovejoy [16]. Weak and strong versions of emergence exist [17], but life-origin models of necessity require convincing models of strong emergence. The whole is greater than the sum of its parts [18]. Novel functional qualities are believed to arise spontaneously from inanimate physical components [19–22]. First, second, third and now fourth order (Types I–IV) emergence are said to exist [23]. Heritable linear digital genetic prescription can produce three-dimensional protein molecular machines that bind, transport and catalyze metabolic integration. Strong and Type IV emergent theory together attempt to explain the source of these phenomena. Admits Mark Bedau, "Although strong emergence is logically possible, it is uncomfortably like magic." [24].

If Pasteur and Virchow's First Law of Biology ("All life must come from previously existing life") is to be empirically falsified, direct observation of spontaneous generation is needed. In the absence of such empirical falsification, a plausible model of mechanism at the very least for both Strong and Type IV emergence (formal self-organization) is needed. Manfred Eigen [25–36] and Tibor Ganti [37–41] have been leaders in the search for mechanisms of biologic emergence from abiotic environments. Shuster joined with Eigen to hypothesize hypercycles [42–49].The Edge of Chaos [21,22,50–57] has been proposed as a possible source, though the description of all of the above models often seems more poetic or cartoon-like than real. Kauffman's and Dawkin's publications, for example, are often devoid of any consideration of the biochemical catastrophic realities that plague life-origin bench scientists [20–22,58–60] [61–63].

Attempts to define complexity are on-going [50,64–70]. Sequence complexity has been extensively studied, though far from exhaustively [1,71–77].

Much debate has occurred over the relation of linear complexity to semantic information [78–83] [84–92]. Some have attempted to reduce the information of linear digital prescription in genes to mere thermodynamics, combinatorial probabilism, and physicodynamic complexity [20,93–106]. Other investigators tend to view genetic information as literal and real [1,2,6, 107–112]. The special case of semiotic linear digital complexity has fostered the whole new field of Biosemiotics [2,113–133].

The cause and evolution of complexity are frequently addressed in the literature [10,134–141].How complexity relates to life has attracted innumerable papers [6, 142–148]. Systems Biology emphasizes the growing genomic and epigenetic complexity [149–151]. Attempts to deal with Behe's "irreducible complexity" [152] are appearing more often in scientific literature [153–157]. von Neumann [158] and Pattee [159–161] attempted to deal with the issue of Complementarity between the formal and physical aspects of complexity. Hoffmeyer and Emmeche have addressed the same basic problem with Code Duality [162,163]. Stein described the different sciences of complexity [164]. Norris has researched hypercomplexity [165]; Garzon dealt with bounded complexity [166]; and Levins the limits of complexity [167]. Bennett originated Logical Depth and its relation to physical complexity [168].

Attempts to relate complexity to self-organization are too numerous to cite [4,21, 169–171]. Under careful scrutiny, however, these papers seem to universally incorporate investigator agency into their experimental designs. To stem the growing swell of Intelligent Design intrusions, it is imperative that we provide *stand-alone natural process* evidence of non trivial self-organization at the edge of chaos. We must demonstrate on sound scientific grounds the formal capabilities of naturally-occurring physicodynamic complexity.

Evolutionary algorithms, for example, must be stripped of all artificial selection and the purposeful steering of iterations toward desired products. The latter intrusions into natural process clearly violate sound evolution theory [172,173]. Evolution has no goal [174,175]. Evolution provides no steering toward *potential* computational and cybernetic function [4,6–11].

The theme of this paper is the active pursuit of falsification of the following null hypothesis: "Physicodynamics alone cannot organize itself into formally functional systems requiring algorithmic optimization, computational halting, and circuit integration." At first glance the falsification of this hypothesis might seem like a daunting task. But a single exception of non trivial, unaided, spontaneous optimization of formal function by truly natural process would quickly falsify this null hypothesis.

Science celebrates positive and parsimonious descriptions of presumed objectivity. But we must never forget that our knowledge is only "best thus far." Even the most fundamental laws of physics technically must be viewed as "tentative." We rightly eschew diatribes of metaphysical pontifications. Science proceeds through open-mindedness and the falsification of null hypotheses, not through the rhetorical pronouncement of dogmas. Popper and many since have exposed the problems associated with trying to prove any positive hypothesis [176,177]. Neither induction nor deduction is foolproof. Theses that cannot be proven ought not to be proclaimed as positive statements of fact.

At the same time, we have spent much of the last century arguing to the lay community that we *have* proved the current biological paradigm. Unfortunately, very few in the scientific community seem critical of this indiscretion. One would think that if all this evidence is so abundant, it would be quick and easy to falsify the null hypothesis put forward above. If, on the other hand, no falsification is forthcoming, a more positive thesis might become rather obvious by default. Any positive pronouncement would only be labeled metaphysical by true-believers in spontaneous self-organization. Those same critics would disingenuously fail to acknowledge the purely metaphysical nature of the current Kuhnian paradigm rut [178]. A better tact is to thoroughly review the evidence. Let the reader provide the supposedly easy falsification of the null hypothesis. Inability to do so should cause pangs of conscience in any scientist who equates metaphysical materialism with science. On the other hand, providing the requested falsification of this null hypothesis would once-and-for-all end a lot of unwanted intrusions into science from philosophies competing with metaphysical materialism.

While proof may be evasive, science has an obligation to be honest about what the entire body of evidence clearly *suggests*. We cannot just keep endlessly labeling abundant evidence of formal prescription in nature "apparent." The fact of purposeful programming at multiple layers gets more "apparent" with each new issue of virtually every molecular biology journal [179–181]. Says de Silva and Uchiyama:

> "Molecular substrates can be viewed as computational devices that process physical or chemical 'inputs' to generate 'outputs' based on a set of logical operators. By recognizing this conceptual crossover between chemistry and computation, it can be argued that the success of life itself is founded on a much longer-term revolution in information handling when compared with the modern semiconductor computing industry. Many of the simpler logic operations can be identified within chemical reactions and phenomena, as well as being produced in specifically designed systems. Some degree of integration can also be arranged, leading, in some instances, to arithmetic processing. These molecular logic systems can also lend themselves to convenient reconfiguring. Their clearest application area is in the life sciences, where their small size is a distinct advantage over conventional

semiconductor counterparts. Molecular logic designs aid chemical (especially intracellular) sensing, small object recognition and intelligent diagnostics." [181]. What scientific evidence exists of physicodynamics ever having programmed a single purposeful configurable switch-setting? If we cannot present any such evidence, we should be self-honest enough to start asking ourselves, "How long are we going to try to maintain this ruse that the cybernetic programming we repeatedly observe is only 'apparent' rather than real?"

What Exactly is Complexity?

"Complexity" can tend to be a garbage-can catch-all term we use to explain everything we don't understand and cannot reduce. To define complexity, we need to start with one dimension and work up. When we progress from linear complexity into two and three dimensional complexity, quantifying the degree of complexity can quickly become intractable [182]. Thus, let us begin by precisely defining linear sequence complexity.

An unequivocal, pristine, mathematical definition of linear "complexity" already exists in scientific literature [1,71,183]: maximum complexity in a linear string, oddly enough, is randomness. Maximum complexity cannot be compressed because it lacks patterns and order [183,184]. A random string (Random Sequence Complexity, RSC) [1] is the most complex because its sequence cannot be enumerated using any algorithmically compressive string shorter than itself. Notice that this precise definition of linear complexity has nothing to do with meaning or function. Complexity in linear digital strings is fully measurable by the degree to which each string can be algorithmically compressed. This is true whether the string does anything useful or not. A string comprised of maximally uncertain elements will be the most complex string because it lacks order and pattern. The compressibility of that string is therefore extremely low. Uncertainty is measured in bits. The higher the number of bits of uncertainty, the greater the degree of complexity, and the closer we move toward a random string:

$$H = \sum_{i=1}^{M} p_i(-\log_2 p_i) \tag{1}$$

This of course is Shannon's basic measurement of uncertainty in linear sequence complexity.

We have invested so much confidence and anticipation in "complexity" as a potential source of spontaneous prescriptive information and organization that our senses are jolted by the pristine mathematical definition of sequence complexity reviewed above. We need to re-educate ourselves with the realization that maximum complexity is nothing more than randomness. The most complex of all strings is a random string. Random strings have never been observed to generate non trivial formal function of any kind. Complexity, therefore, has nothing to do with formal function. Complexity possesses no creative or computational talents. No justification exists for attributing exquisite formal organization to mere complexity.

Order, Structure and Pattern

Well, what about order and pattern? If complexity itself is not what produces utility in a linear digital string, surely order, structure and pattern can. But do they? The answer is no! To understand why, we must also define order and pattern.

What exactly is "order"? Starting with a single dimension, order in a sequence is defined by an increasing probability of occurrence of each structure, event, or alphabetical character in that string [183]. High probability is high order [185]. As the probability of an event approaches 1.0, its Shannon uncertainty approaches 0 bits [186]. 0 bits of uncertainty is maximum order. Maximum order is minimal complexity. Order and complexity are antithetical. They lie at opposite extremes of a bidirectional vector (Fig. 1) [71]. The relationship between order and complexity has been well-defined in the literature [1,3,71,187]. Figure 1 shows the antithetical relationship between order and complexity. Order lies on the opposite end of this bidirectional vector graph from complexity. The more complex a sequence is, the less ordered. The more ordered a sequence is, the less complex. The literature is filled with misunderstanding of the relationship between pattern and complexity.

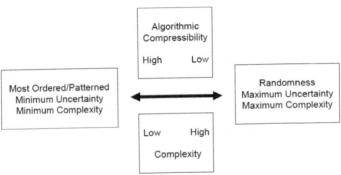

Figure 1 An antithetical relationship exists between linear sequence order and complexity. Randomness affords the greatest measure of complexity. The more ordered and patterned a sequence, the less uncertain are its components, and the less complex the sequence. Neither order nor complexity generates formal meaning or utility, both of which lie in a completely different dimension from order/complexity measures.

Ordered strings contain repeating patterns such as those found in sugar molecules. As we add dimensions, high order can be found in sine waves and inorganic crystals. Repeating patterns generate high order and low complexity. The probability of encountering the next element of a repeating pattern is high; the probability of coming across any uniqueness (e.g., a crystal impurity) is low. Highly ordered/patterned strings can be greatly compressed algorithmically. The most ordered string is exampled by a string of identical letters, or a DNA homopolymer consisting of all adenosines. A polyadenosine has maximum order, no uncertainty, and therefore no complexity. A polymer of 200 adenosines can be fully enumerated by the very short compression algorithm, "Give me an adenosine; repeat 200 times." This compression algorithm for a polyadenosine contains virtually no uncertainty, and therefore no information potential. It is an example of Ordered Sequence Complexity (OSC) [1] .

Note that statistical order and pattern have no more to do with function and formal utility than does maximum complexity (randomness). Neither order nor complexity can program, compute, optimize algorithms, or organize. A law of physics also contains very little information because the data it compresses is so highly ordered. The best way to view a parsimonious physical law is as a compression algorithm for reams of data. This is an aspect of valuing Occam's razor so highly in science. Phenomena should be explained with as few assumptions as possible. The more parsimonious a statement that reduces all of the data, the better [188,189]. A sequence can contain much order with frequently recurring patterns, yet manifest no utility. Neither order nor recurring pattern is synonymous with meaning or function.

Those trained in information theory will be quick to point out at this point that "information is always defined in terms of an observer or knower." They argue that information is not in the law's parsimonious statement or equation, but in the difference (R) between all of the uncertainty of the raw data, and the lesser amount of uncertainty generated by knowing the law. The problem with this concept of information is that for most of life's history, linear digital genetic instructions have been prescribing exquisite metabolic organization long before any observers or knowers existed. Observers and knowers themselves would not exist except for the extraordinary amount of cellular programming and organization that produced humans. Prescriptive Information (PI) [3] cannot be reduced to human epistemology. To attempt to define information solely in terms of human observation and knowledge is grossly inadequate. Such anthropocentrism blinds us to the reality of life's *objective* genetic programming, regulatory mechanisms, and biosemiosis using symbol systems [2,112,120,125,132, 190–194]. Well what about a combination of order *and* complexity? Doesn't that explain how prescriptive information comes into being?

Three subsets of linear complexity have been defined in an abiogenesis environment [1]. These subsets are very helpful in understanding potential sources of Functional Sequence Complexity (FSC) as opposed to mere Random Sequence Complexity (RSC) and Ordered Sequence Complexity (OSC) [1]. FSC requires a third dimension not only to detect, but to produce formal utility. Neither chance nor necessity (nor any combination of the two) has ever been observed to produce non trivial FSC [4]. Durston and Chiu at the University of Guelph developed a method of measuring what they call *functional uncertainty* (H_f) [195]. They extended Shannon uncertainty to measure a *joint variable* (X,F), where X represents the variability of data, and F its functionality. This explicitly incorporated the empirical knowledge of embedded function into the measure of sequence complexity:

$$H\left(X_f\left(t\right)\right) = -\sum P\left(X_f\left(t\right)\right)\log P\left(X_f\left(t\right)\right) \tag{2}$$

where X_f denotes the conditional variable of the given sequence data (X) on the described biological function f which is an outcome of the variable (F). The state variable t, representing time or a sequence of ordered events, can be fixed, discrete, or continuous. Discrete changes may be represented as discrete time states. Mathematically, the above measure is defined precisely as an outcome of a discrete-valued variable, denoted as $F = \{f\}$. The set of outcomes can be thought of as specified biological states.

Using this method allowed Durston and Chiu to compare quantifications of 2,442 aligned sequences of proteins belonging to the Ubiquitin protein family, among many other protein families evaluated. All of these sequences satisfied the same specified function *f*, which might represent the known 3-D structure of the Ubiquitin protein family, or some other function common to ubiquitin. The definition of functionality used by Durston and Chiu relates to the whole protein family. Thus this data can be inputted from readily available databases. Even subsets (e.g., the active sites) of the aligned sequences all having the same function can be quantified and compared. The tremendous advantage of using $H(X_f(t))$ is that *slight changes* in the functionality characteristics of biosequences can be incorporated and analyzed.

Subsequently, Durston and Chiu have developed a theoretically sound method of actually quantifying Functional Sequence Complexity (FSC) [77]. This method holds great promise in being able to measure the increase or decrease of FSC through evolutionary transitions of both nucleic acid and proteins. This FSC measure, denoted as ζ is defined as the change in functional uncertainty from the ground state $H(X_g(t_i))$ to the functional state $H(X_f(t_i))$, or

$$\zeta = \Delta H \left(X_g \left(t_i \right), X_f \left(t_i \right) \right) \tag{3}$$

The *ground state g* of a system is the state of presumed highest uncertainty permitted by the constraints of the physical system, when no specified biological function is required or present. Durston and Chiu wisely differentiate the ground state *g* from the *null state* H_ϕ. The null state represents the absence of *any* physicodynamic constraints on sequencing. The null state produces bona fide stochastic ensembles, the sequencing of which was *dynamically inert* (physicodynamically decoupled or incoherent [196,197]).

The FSC variation in various protein families, measured in Fits (Functional bits), is shown in Table 1 graciously provided here by Durston and Chiu. In addition to the results shown in Table 1, they performed a more detailed analysis of ubiquitin, plotting the FSC values out along its sequence. They showed that 6 of the 7 highest value sites correlate with the primary binding domain [77].

Table 1 FSC of Selected proteins. Supporting data from the lab of Kirk Durston and David Chiu at the University of Guelph [77] showing the analysis of 35 protein families.

	Length (aa)	Number of Sequences	Null State (Bits)	FSC (Fits)	Average Fits/Site
Ankyrin	33	1,171	143	46	1.4
HTH 8	41	1,610	177	76	1.9
HTH 7	45	503	194	83	1.8
HTH 5	47	1,317	203	80	1.7
HTH 11	53	663	229	80	1.5
HTH 3	55	3,319	238	80	1.5
Insulin	65	419	281	156	2.4
Ubiquitin	65	2,442	281	174	2.7
Kringle domain	75	601	324	173	2.3

Table 1 contd....

Table 1 contd....

	Length (aa)	Number of Sequences	Null State (Bits)	FSC (Fits)	Average Fits/Site
Phage Integr N-dom	80	785	346	123	1.5
VPR	82	2,372	359	308	3.7
RVP	95	51	411	172	1.8
Acyl-Coa dh N-dom	103	1,684	445	174	1.7
MMR HSR1	119	792	514	179	1.5
Ribosomal S12	121	603	523	359	3.0
FtsH	133	456	575	216	1.6
Ribosomal S7	149	535	644	359	2.4
P53 DNA domain	157	156	679	525	3.3
Vif	190	1,982	821	675	3.6
SRP54	196	835	847	445	2.3
Ribosomal S2	197	605	851	462	2.4
Viral helicase1	229	904	990	335	1.5
Beta-lactamase	239	1,785	1,033	336	1.4
RecA	240	1,553	1,037	832	3.5
tRNA-synt 1b	280	865	1,210	438	1.6
SecY	342	469	1,478	688	2.0
EPSP Synthase	372	1,001	1,608	688	1.9
FTHFS	390	658	1,686	1,144	2.9
DctM	407	682	1,759	724	1.8
Corona S2	445	836	1,923	1,285	2.9
Flu PB2	608	1,692	2,628	2,416	4.0
Usher	724	316	3,129	1,296	1.8
Paramyx RNA Pol	887	389	3,834	1,886	2.1
ACR Tran	949	1,141	4,102	1,650	1.7
Random sequences	1000	500	4,321	0	0
50-mer polyadenosine	50	1	0	0	0

Shown are sequence lengths (column 1), the number of sequences analyzed for each family (column 2), the Shannon uncertainty of the Null State H_ϕ (the absence of any physicodynamic constraints on sequencing: dynamically inert stochastic ensembles) for each protein (column 3), the FSC value ζ in Fits for each protein (column 4), and the average Fit value/site (FSC/length, column 5). For comparison, the results for a set of uniformly random amino acid sequences (RSC) are shown in the second from last row, and a highly ordered, 50-mer polyadenosine sequence (OSC) in the last row. All values, except for the OSC example, which was calculated from the constrained ground state required to produce OSC, were computed from the null state. The Fit values obtained can be discussed as the measure of the change in functional uncertainty required to specify any functional sequence that falls into the given family being analyzed. (Used with permission from Durston, K.K.; Chiu, D.K.; Abel, D.L.; Trevors, J.T. Measuring the functional sequence complexity of proteins. *Theor Biol Med Model* 2007, *4*, Free online access at: http://www.tbiomed.com/content/4/1/47).

In the pile of "Pick-up Sticks" seen in Fig. 2, very little order and patterning are present. Uncertainty as to how the sticks will fall is high. The pile of sticks is highly complex. In this three-dimensional model, it would probably be intractable to compute the complexity of relationships of each of these sticks to all of the other sticks. The degree of complexity would be staggering. But what exactly does this enormous degree of complexity DO? The pile of pick-up sticks achieves no utility of any kind. The imagined capabilities of stand-alone complexity are in reality miniscule at best. Attributing organization to chaos and complexity employs a combination of fallacious inferences involving category errors and non sequiturs. "The edge of chaos" [21,22,50–54] affords mesmerizing visions of potential accomplishment. While poetic and wonderfully inviting, the concept is sorely lacking in scientific content. The functional reality of "the edge of chaos" has been challenged [7,8,57,198].

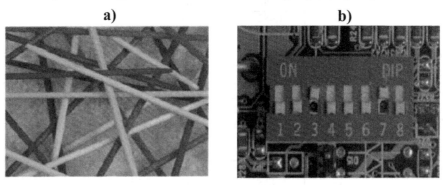

a)   **b)**

Figure 2 a) The degree of three-dimensional computational complexity within a pile of pick-up sticks is staggering. But what exactly does this enormous degree of complexity DO? What sophisticated formal function does this pile of objects generate? Mere combinatorial complexity must never be confused with formal utility. b) A row of dip switch settings depicts a different category of complexity-algorithmic, cybernetic programming complexity. Choice contingency is incorporated into purposeful configurable switch-settings that collectively prescribe formal function.

The association of complexity or patterns with most forms of bona fide organization should never be confused with causation [199]. Neither order nor complexity is a cause of organization or any other form of formal algorithmic optimization. We sling the words "chaos," "complexity," "order" and "pattern" around with vivid imagination and a great deal of blind faith in their capabilities. None of the latter states has ever been observed to produce the slightest amount of algorithmic organization. Stand-alone chaos and complexity have absolutely nothing to do with generating formal function. Neither do order and pattern. Self-ordering phenomena produce boring, unimaginative redundancy. Self-ordering phenomena, just like chaos and complexity, have never been observed to achieve 1) programming, 2) computational halting, 3) creative engineering, 4) symbol systems, 5) language, or 6) bona fide organization [4]. The latter are all formal processes, not physicodynamic processes.

Suppose stochastic ensembles of oligoribonucleotides were forming out of sequence space in an imagined "primordial soup." Since only 4 different nucleotides could be added next to a forming single positive strand, M in Equation 1 above would=4. Suppose next that the prebiotic availability p_i for adenine was 0.46, and the p_i's for uracil, guanine, and cytosine were 0.40, 0.12, and 0.02 respectively. This is being generous for cytosine, since cytosine would have been extremely difficult to make in any prebiotic environment [200]. Using these hypothetical base-availability probabilities, the Shannon uncertainty would have been equal to:

Adenine	$0.46 \, (-\log_2 0.46)$	$= 0.515$
Uracil	$0.40 \, (-\log_2 0.40)$	$= 0.529$
Guanine	$0.12 \, (-\log_2 0.12)$	$= 0.367$
Cytosine	$0.02 \, (-\log_2 0.02)$	$= 0.113$
	1.00	1.524 bits

Notice how unequal availability of the four nucleotides (*a form of ordering*) greatly reduces Shannon uncertainty at each locus, and in the entire sequence, of any biopolymeric stochastic ensemble (Fig. 1). Maximum uncertainty would occur if all four base availability probabilities were 0.25. Under these equally available base conditions, Shannon uncertainty would have equaled 2 bits per independent nucleotide addition to the strand. A stochastic ensemble formed under aqueous conditions of mostly adenine availability, however, would have had little information-retaining ability because of its high order [1].

As pointed out in the above reference, even less information-retaining ability would be found in an oligoribonucleotide adsorbed onto montmorillonite [201–206]. Clay surfaces would have been required to align ribonucleotides with 3' 5' linkages. The problem is that only polyadenosines or polyU's tend to form. Using clay adsorption to solve one biochemical problem creates an immense informational problem (e.g., high order, low complexity, low uncertainty, and low information retaining ability. See Fig. 1). High order means considerable compressibility. The Kolmogorov [207] algorithmic compression program for clay-adsorbed biopolymers (Fig. 2) would read: "Choose adenosine; repeat the same choice fifty times." Such a redundant, highly-ordered sequence could not begin to prescribe even the simplest protometabolism. Such "self-ordering" phenomena would not be the key to life's early algorithmic programming.

The RNA Word and pre-RNA World models [208,209] still prevail despite daunting biochemical problems. Life origin models also include clay life [210–213]; early three-dimensional "genomes" [214,215]; "Metabolism/Peptide First" [216–219]; "Co-evolution" [220–223]; "Simultaneous nucleic acid and protein" [224–226]; and "Two-Step" models of life-origin [227–229]. In virtually all of these life origin models, "self-ordering" is confused with "self-organizing." No mechanism is provided for the development of a linear digital prescription and oversight system to integrate metabolism. No known life form exists that does not depend upon such genetic instruction.

Autopoesis

Umberto Maturana and Francisco Varela [230–232] argue for a concept of autopoeisis that presupposes (or begins with) one-celled organisms and progresses evolutionarily all the way up to humans, their language, and their social structure. They used the term autopoiesis to characterize the nature of living systems more than to theorize how cellular life came into existence. An autopoietic system is self-sustaining, homeostatic and autonomous despite having a continuous flow of mass and energy through the cell. Maturana and Varela's basic contention is that organisms are inherently compelled to maintain their own inner nature and identity. Such a concept of "self-making" might better be described as "self-maintaining." It does not address the problem of abiogenesis-the spontaneous generation of life from non life at the molecular evolutionary level. Says Varela (who often writes jointly with Maturana),

> "If living systems are machines, that they are physical autopoietic machines is trivially obvious: they transform matter into themselves in a manner such that the product of their operation is their own organization. However, we deem the converse as also true: A physical system if autopoietic is living. In other words, we claim that the notion of *autopoiesis is necessary and sufficient to characterize the organization of living systems.*"

So far as abiogenesis research is concerned, such a statement seems circular or tautological in nature. Their publications offer no mechanisms or help in understanding the processes by which inanimate physics and chemistry wrote life's cybernetic programming. They do not address how an abiotic physicochemical environment organized chemical reactions into an 11-step biochemical pathway such as the Krebs Cycle. The latter yields no pragmatic benefit until the final biochemical step.

Pier Luigi Luisi, one of the world's leading experts in primordial membrane theory, points out that Maturana and Valari's concept of autopoiesis "is not a theory about the origin of life-but rather a pragmatic blueprint of life based on cellular life." [233] Luigi goes on to state that Maturana and Varela's theory of autopoiesis "had, and still has, a difficult time being accepted into the mainstream of life-science research." [233]

Margaret Boden [234–236] also tends to presuppose organization rather than to elucidate mechanisms of its abiotic derivation. She challenges Maturana's and Varela's use of cognitive language as being too liberal: "Life does not imply cognition." But many of Boden's publications themselves presuppose and incorporate human cognition and epistemology into her models. None of these authors purport to offer explanations for life origin. They are simply not on the forefront of abiogenesis research at the biochemical level. No physicochemical mechanism is provided for self-organization.

The term autopoesis has been used on occasion in a much broader sense than Maturana and Varela coined it. Some life-origin investigators use "autopoesis" to refer to the prebiotic "self-making from scratch" of life-of biochemical abiogenesis. But the use of this broader term in some publications has not afforded any new purely physical models of "self-organization" or "emergence."

Complex Adaptive Systems (CAS)

Complex adaptive systems (CAS) [137,237,238] are comprised of multiple interconnected (yet diverse) components. CAS readily undergo change. Healthy (CAS) are high-dimensional with respect to turbulence and potential for change. Rigid, low dimensional systems are said to be unhealthy. CAS are called "adaptive" because they are said to learn from experience. Cells, embryos, immune systems, central nervous systems, ecosystems, social insect colonies, human social systems, and economics, are all included in CAS studies. Either the CAS is itself alive, or it is a robot programmed (cybernetically determined) by life to "learn." It is not surprising that so many systems theory and CAS models presuppose and use life rather than explain the derivation of life. Already-existing cellular prescriptive information is incorporated into the model in virtually every CAS discipline. Self-organization is claimed, but never empirically demonstrated independent of experimenter steering.

Artificial life efforts generally pursue CAS models [6,146,239–246] in what amounts to an engineering context. So-called "evolutionary algorithms" and "directed evolution" strategies are often used to support an evolutionary paradigm. But both strategies amount to artificial selection, not natural selection. They have little or nothing to do with neoDarwinism.

Positive and negative feedback plays a major role in systems theory, especially in the social sciences [247–249]. Feedback mechanisms are characterized by a circularity of causation. Output components are fed back into the input. But feedback mechanisms need formal controls to generate sophisticated utilitarian results. Purely physicodynamic hypercycles [44,47,49], for example, consume all available resources in their redundant and unimaginative mutual replications. The result is catastrophic with regard to any formal self-organization. Empirical evidence is sorely lacking to support the hoped-for relentless growth in integrated protometabolic function [4]. The excitement that hypercycle theory generated in the 1980's has proven over time to be little more than Freudian wish fulfillment.

Trying to explain the spontaneous occurrence of CAS encounters early roadblocks. As with Shannon "information" theory, the *potential* information provided by uncertainty measures is not the same as intuitive information [250,251], semantic information [252–258], biological information [98,99,101,111,190,259–268], functional information [269,270], or the programmed prescriptive information (PI) that generates formal utility [1,3,12]. Combinatorial uncertainty is essential in any physical matrix for that matrix to be able to retain instantiated prescriptive information. But mere combinatorial uncertainty and the potential for change possess no programming talents [3,9,10].

For complex adaptive systems to progress in the direction of achieving formal utility, selection for potential function must take place at individual decision nodes, logic gates, and configurable switch settings prior to the realization of that function. An inanimate environment cannot do this. In addition, the syntax of such choices for utility must be integrated into programmable circuits to achieve computational halting. Combinatorial uncertainty does not provide any mechanism for steering physicality toward abstract, conceptual, formal computational success. Inanimate nature possesses no motivation, let alone formal skills, to pursue integration of pathways and cycles into a holistic metabolic scheme.

The illusion of abundant empirical support for the spontaneous generation of CAS arises from the conflation of physical combinatorial uncertainty with the agency of the experimenter. The experimenter invariably steers events toward desired function behind the scenes. Even before critiquing Materials and Methods of many CAS and systems theory papers, investigator involvement in experimental design is usually apparent right from the author's own words. For example, in a paper on adaptive feedback control for linearizable chaotic systems [271] we read: "A remarkable feature of *the proposed approach* is that *it can be used* for chaos control as well as chaos synchronization." [italics mine] Note the passive voice. Used by what or by whom? "*Numerical simulations* of two well-known chaotic systems are illustrated to show the effectiveness and robustness of the proposed adaptive *control strategy.*" [italics mine] Inanimate nature does not do "numerical simulations." Where exactly within natural physicodynamic interactions did "control *strategy*" come from? Such steering strategy toward utilitarian capability did not arise from physicodynamics. The steering arose purely from the experimenter's choice contingency-from the investigator's goals, experimental design, and artificial selection for what was wanted. Many papers actually acknowledge up front the role of "engineering," sometimes right in their titles [272–279] (e.g., "Simulation-Based Engineering Of Complex Adaptive Systems" [280]).

Attempts are usually made to attribute this acknowledged need for engineering to evolution [273,277,281,282]. But natural selection never works at the decision node programming level [10]. Evolution works only on already-programmed, already-living, already-fittest phenotypic organisms [175]. Selection pressure is nothing more than the differential survival and reproduction of the fittest small populations of living organisms [10]. An adequate selection mechanism for *potential* computational function has always been lacking in evolution theory. This is all the more painfully apparent at the molecular evolution level. No basis for preferring stand-alone function over non function exists in an inanimate prebiotic environment [10]. Worse yet, an inanimate environment has no ability to program for a *potential* function that does not yet exist [9]. Yet selection for potential function is exactly what genetic programming requires [12]. Genetic programming is "written in stone" into linear digital sequences bound by rigid 3'5' phosphodiester covalent bonds prior to transcription, translation, protein-folding, and three-dimensional metabolism [3].

The Big Three: Chance, Necessity and Selection

Selection must be included along with "chance and necessity" [174] as a fundamental category of reality. Why? First, biological science presupposes natural selection as its single most organizing paradigm. Without *selection*, evolution is impossible. Linear digital genetic instructions represent *selection-based* cybernetic programming. Life uses a symbol system as evidenced by the codon table. Symbols must be selected from an alphabet of symbols. Nucleotides must be selected from a phase space of four options at each locus in the DNA string. Second, the scientific method pre-assumes the reality and reliability of formal rationality, mathematics, cybernetic programming, and predictive computations. All of these operational tools depend upon decision theory [283–285]. The practice of science would be impossible without selection at bona fide decision nodes, logic gates and configurable switch settings. Chance and necessity, as Monod pointed out [174], are inadequate to describe everything we repeatedly observe, life especially. Science must acknowledge the reality and validity of selection as a fundamental, "properly basic" category.

A third dimension is required to see what relation, if any, order and complexity have to meaning and function. It is possible to write a highly functional computationally halting program that is non compressible (manifests no patterns). Each symbol selection represents an independent choice. A third dimension is required to distinguish what appears to be a random string from a functional program. That third dimension is the reality of selection, whether in the form of a) natural selection, or b) artificial selection:

a) Natural selection is a very special case indeed. Differential survival and reproduction of the fittest already-computed, already-living small populations of organisms is very indirect. Selection is not intended; it just happens secondarily. No purpose guides selection events. No true decision nodes are involved because evolution has no goal. In this sense, selection "pressure" is a misnomer. Differential survival is more happenstantial than pushed, more after-the-fact than pursued.

b) Artificial selection is the essence of formalism. Despite decades of concentrated research on consciousness and artificial intelligence, choice contingency remains elusive when approached from the direction of physicality. The mind/body problem is alive and well in the philosophy of science.

No known natural process exists that spontaneously writes meaningful or functional syntax. Only agents have been known to write meaningful and pragmatic syntax. Physicality cannot compute or make arbitrary symbol selections according to arbitrarily written rules. Physicality cannot compress. Physicality cannot value or pursue formal utility. Physicality is blind to pragmatic considerations, all of which are formally valued and pursued. No known mechanism exists in inanimate nature to steer physical events toward algorithmic optimization. Many epigenetic factors notwithstanding, genetics and genomics largely *program* phenotypes using a symbol system of linear digital prescription.

Do Symbol Systems Exist Outside of Human Minds?

Metabolism employs primarily proteins. The nucleotide sequences in mRNA prescribe the amino acid sequences that determine protein identity. DNA is largely inert. It plays no direct physicochemical role in protein binding, transport and catalysis. Molecular biology's two-dimensional complexity (secondary biopolymeric structure) and three-dimensional complexity (tertiary biopolymeric structure) are both ultimately determined by linear sequence complexity (primary structure; functional sequence complexity, FSC). The chaperone proteins that aid polyamino acid folding are also prescribed by the linear digital genetic programming instantiated into DNA sequencing.

Genetics not only utilizes a linear digital symbol system, but abstract Hamming block coding to reduce noise pollution in the Shannon channel (triplet codons to prescribe each amino acid). Anti-codons are at opposite ends of t-RNA molecules from amino acids. The linking of each tRNA with the correct amino acid depends entirely upon on a completely independent family of tRNA aminoacyl synthetase proteins. Each of these synthetases must be specifically prescribed by separate linear digital programming, but using the same MSS. These symbol and coding systems not only predate human existence, they *produced* humans along with their anthropocentric minds. The nucleotide and codon syntax of DNA linear digital prescription has no physicochemical explanation. All nucleotides are bound with the same rigid 3'5' phosphodiester bonds. The codon table is arbitrary and formal, not physical. The semantic/semiotic/bioengineering function required to make proteins requires dynamically inert configurable switch-settings and resortable physical symbol vehicles. Codon syntax communicates time-independent, non-physicodynamic "meaning" (prescription of biofunction). This meaning is realized only after abstract translation via a conceptual codon table. To insist that codon syntax only represents amino acid sequence in our human minds is not logically tenable.

Figure 3 shows the prescriptive coding of a section of DNA. Each letter represents a choice from an alphabet of four options. The particular sequencing of letter choices prescribes the sequence of triplet codons and ultimately the translated sequencing of amino acid building blocks into protein strings. The sequencing of amino acid monomers (basically the sequencing of their R groups) determines minimum Gibbs-free-energy folding into secondary and tertiary protein structure. It is this three-dimensional structure that provides "lock-and-key" binding fits, catalysis, and other molecular machine formal functions. The sequencing of nucleotides in DNA also prescribes highly specific regulatory micro RNAs and other epigenetic factors. Thus linear digital instructions program cooperative and holistic metabolic proficiency.

Not only are symbol systems used, but a bijection must occur between two independent symbol systems. Bijection (translation; a symbol system to symbol system correspondence) is rule-based, not physical law-based. No cause-and-effect necessity exists in the linking of anticodons, amino acids, tRNAs, and amino acyl tRNA synthetases with codons. The anticodon is located on the opposite end of tRNA from the amino acid. The correspondence between the two languages is arbitrary and abstract. By arbitrary, we do not mean random.

Arbitrary means free from physicodynamic determinism. Bijection rules are freely selected. Translation of this linear digital prescription into functionally specific polyamino acid chains cannot be explained by physicodynamics. It is not law-based, and it certainly is not random. If this were an empirical/inductive contention, "cannot" would have to be replaced with "has not yet been." The problem is that the statement is a valid inference of deductive logic. The conclusion is as unequivocal as that produced by balanced mathematical manipulations of any equation. Neither fixed/forced laws nor chance can logically make non trivial computationally halting programming decisions. It is a logical impossibility for chance and/or necessity to exercise bona fide choice contingency. They are in isolated categories (see Section 8). Neither unaided Markov chains nor physicodynamic determinism can select for *potential* formal function.

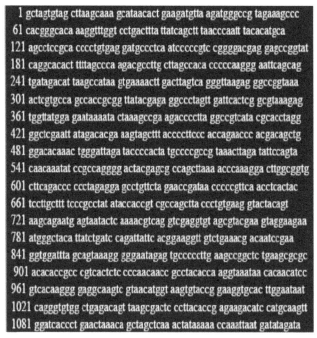

```
    1 gctagtgtag cttaagcaaa gcataacact gaagatgtta agatgggccg tagaaagccc
   61 cacgggcaca aaggtttggt cctgacttta ttatcagctt taacccaatt tacacatgca
  121 agcctccgca cccctgtgag gatgccctca atcccccgtc cggggacgag gagccggtat
  181 caggcacact ttttagccca agacgccttg cttagccaca cccccaaggg aattcagcag
  241 tgatagacat taagccataa gtgaaaactt gacttagtca gggttaagag ggccggtaaa
  301 actcgtgcca gccaccgcgg ttatacgaga ggccctagtt gattcactcg gcgtaaagag
  361 tggttatgga gaataaaata ctaaagccga agacccctta ggccgtcata cgcacctagg
  421 ggctcgaatt atagacacga aagtagcttt accccttccc accagaaccc acgacagctg
  481 ggacacaaac tgggattaga taccccacta tgccccgccg taaacttaga tattccagta
  541 caacaaatat ccgccagggg actacgagcg ccagcttaaa acccaaagga cttggcggtg
  601 cttcagaccc ccctagagga gcctgttcta gaaccgataa cccccgttca acctcactac
  661 tccttgcttt tcccgcctat ataccaccgt cgccagctta ccctgtgaag gtactacagt
  721 aagcagaatg agtaatactc aaaacgtcag gtcgaggtgt agcgtacgaa gtaggaagaa
  781 atgggctaca ttatctgatc cagattattc acggaaggtt gtctgaaacg acaatccgaa
  841 ggtggattta gcagtaaagg gggaatagag tgccccttg aagccggctc tgaagcgcgc
  901 acacaccgcc cgtcactctc cccaacaacc gcctacacca aggtaaataa cacaacatcc
  961 gtcacaaggg gaggcaagtc gtaacatggt aagtgtaccg gaaggtgcac ttggaataat
 1021 cagggtgtgg ctgagacagt taagcgactc ccttacaccg agaagacatc catgcaagtt
 1081 ggatcaccct gaactaaaca gctagctcaa actataaaaa ccaaattaat gatatagata
```

Figure 3 A section of *Alosa pseudoharengus* (a fish) mitochondrion DNA. This reference sequence continues on all the way up to 16,621 "letters." Each nucleotide is a physical symbol vehicle in a material symbol system. The specific selection of symbols and their syntax (particular sequencing) prescribes needed three-dimensional molecular structures and metabolic cooperative function *prior to* natural selection's participation. (Source: http://www.genome.jp/dbget-bin/www_bget?refseq+NC_009576).

The noise-reducing Hamming "block coding" of triplets of nucleotides to prescribe each specific amino acid is all the more abstract and formally conceptual. The triplet codon/ amino acid coding table has been shown to be conceptually ideal in a formal sense [286]. Block-coding greatly reduces the ill effects of a noisy channel on transmitted messages. Fewer prescriptive reading errors occur. Translation between the nucleotide and amino acid symbol systems is extraordinarily reliable. In addition, organisms possess amazing repair mechanisms to undo what noise pollution effects do occur to biomessages. Physics and chemistry provide no mechanisms to explain any of these sophisticated formal control and correction capabilities. They clearly traverse The Cybernetic Cut [9]-a great divide in nature between those phenomena that can be explained through the chance and necessity of natural process vs. those phenomena that can only be explained through formal steering and controls.

But the peculiarity of life over inanimate physics extends far beyond the above discussion. DNA requires editing in the course of its transcription to coding mRNA. And we have not even touched on the roles of many other independent players in the formal integration of transcription, translation, regulation, metabolism, and development. Epigenetic factors are a large part of overall holistic true organization [287–292]. Post-translational editing also plays a role [293–296].

As physicist Howard Pattee has demonstrated in many publications [191,192,297–302], open-ended evolution (OEE) is impossible without a linear digital genetic symbol system that can mutate independent of the real-time living of the phenotypic organisms that harbor them. Outwardly, the same relatively stable phenotypes exist and mate while tremendous modifications can be occurring in their genomes. Says Pattee, "A non-dynamic descriptive model evades an infinite regress by leaving time out of its rules and symbols. Self-describing models interact with dynamical systems by codes that we tacitly understand as writing, reading and interpreting."…"Separate description and construction components are necessary for complex systems that can adapt and evolve." [303]

Most mutations are silent. Genetic drift would be impossible without a genetic material symbol system (MSS) that can experience abundant variation within the same basic phenotype [196,304,305]. The phase space of potential new instructional sequences would be severely limited if genetic drift from successive point mutations, duplications, inversions, transpositions, crossings over, could not progress at the genetic level prior to phenotypic realization.

Literal genetic algorithms, not figurative ones, prescribe and control life. Nucleotides function in an objective, not just a human subjective symbolic capacity. The particular symbol selection at each decision node of nucleotide polymerization is isolated from physicodynamic causation by a *dynamic discontinuity* [196,304,305]. Although the instructions are physically instantiated into material symbol systems using physical symbol vehicles, the programming is fundamentally formal.

"Semantic/semiotic/bioengineering function requires dynamically inert, resortable, physical symbol vehicles that represent time-independent, non-dynamic "meaning." (e.g., codons)." [1] No empirical or rational basis exists for granting to physics or chemistry such non-dynamic capabilities of functional sequencing. Neither chance nor necessity (fixed law) can program configurable switches to integrate circuits or organize formal utility. Linear digital prescription in physical nucleic acid has thus far invariably been associated with life. A fully post modern anthropocentrism cannot argue a logically consistent macroevolutionary paradigm. If naturalistic/materialistic science believes anything, it believes that an objectively real "physical brain secretes mind as the liver secretes bile" [as Pierre Jean Georges Cabanis (1757–1808), Karl Vogt and many others since have phrased it]. Jakob Moleschott (1822–1893) is generally given credit for the renal version: "The brain secretes thought as the kidney secretes urine." For macroevolution theory to fly, a very real genetic symbol system must evolve through objectively real early eukaryotes, invertebrates, vertebrates, mammals and primates. A purely subjective or solipsistic view of nucleotides and codons-trying to deny that they are real physical symbol vehicles-totally compromises macroevolutionary theory.

Macroevolution theory of necessity presupposes a literal history of progressive adaptation of millions of objectively existent species through changes in objectively existent nucleotide symbol sequencing. The formal, representational codon table not only predates human minds, but humans themselves.

Symbolic Dynamics Analysis

It is important not to confuse objectively existent Random, Ordered and Functional Sequence Complexities in nature with symbolic dynamics analysis methodologies [306–309] created and applied by human experimenters. The symbolic dynamics models of human minds use abstract symbols to represent each state in discrete time intervals. Evolution is described by infinite sequences of symbols. A sophisticated shift operator must also be used. All aspects of symbolic dynamics, like the scientific method itself, is a formal enterprise, not a physicodynamic cause and effect chain of the inanimate physical world. As pointed out in the introduction of this paper, applied sciences such as symbolic dynamics analysis provide no help in explaining either gene emergence or spontaneous metabolic self-organization.

Both symbolic dynamics and objective genetic cybernetic programming [12] employ linear digital (discretized) symbol strings. In symbolic dynamics, if the state vector is not inherently discrete, it must be discretized to yield what is called a coarse-grained description of the system. But that is about where the similarities end.

In symbolic dynamics, we assign an arbitrary symbol to represent each discrete physicodynamic state. But inanimate nature cannot *represent* anything using symbols. The latter is a formal function, not a physicodynamic effect that would occur in a primordial environment. Second, using a symbol to represent a physicodynamically determined state is not a control function. It is merely a descriptive function similar to symbolizing initial conditions with formal units of measure. Although both are formal functions, neither is cybernetically determinative. The symbol does not represent a *prescriptive decision node choice* from among real options. Thus a sequence of symbols in symbolic dynamics serves no programming function. Linear digital genetic prescription does. The latter programming strings not only predate animal existence, they produced animals, their brain and minds. Such programming cannot be reduced to human epistemological models of information. Genetic cybernetics at the cellular level is objective, not subjective.

In symbolic dynamics, probability distributions of complexity measures are used to describe and analyze chaotic states. This places symbolic dynamics on similar footing with Shannon transmission engineering. Both systems measure probabilistic combinatorialism. Neither can address meaning, function, or the prescription of formal cybernetic function.

In symbolic dynamics (and most Monte Carlo simulations) time is measured in discrete intervals. Genetic symbol systems, like language, are time-independent in the sense that a seed's genome can remain in a state of suspended animation for centuries, yet still prescribe the same metabolic integration and life. The genome's messages are meaningful and functional in multiple time frames, environments, and with varying rates of catalysis.

The "words" in symbol sequences of symbolic dynamics analysis are derived and recognized through an arbitrary formal scheme generated by investigators' minds. This is artificial selection. It has no parallel in natural selection. Natural selection is nothing more than differential survival and reproduction of already-computed phenotypic organisms. Differential survival plays no role in molecular evolution or initial genetic programming.

Two Kinds of Contingency

Contingency means that events could have happened other than what unfolded [310]. Outcomes are not fully determined by prior cause-and-effect chains. Variability and degrees of freedom exist. Outcomes are not "necessary"-they are not mandated by natural laws working on initial conditions. But there are two kinds of contingency, 1) Chance contingency and 2) Choice contingency.

1. Chance contingency is exampled by heat agitation and Brownian movement of molecules in gas and fluid phases. We refer to chance contingency as "randomness." Chance contingency is statistically describable and predictable. Relative degrees of determinism and chance contingency can co-exist. Weighted means can be calculated for situations with seeming incomplete determinism. Some argue that all physical behavior is ultimately caused, and that chance contingency is only an illusion. Combinations of forces and their effects can be extremely complex. Yet-to-be-discovered forces and relationships may also be at work [199]. But functionally, on the macroscopic level especially, distinct advantages obtain from regarding chance contingency as real and for quantifying possible outcomes statistically.

2. Choice contingency obtains at true decision nodes. Decision nodes are much more than mere bifurcation points. Bifurcation points can be traversed by chance contingency. Any attempt to reduce decision nodes to mere bifurcation points results in rapid deterioration of any potential non trivial formal function. The existence of bifurcation points does not account for computational success. Organization and formal utility are achieved through the controlled opening and closing of logic gates. The latter requires bona fide choices made with steering and programming intent.

Configurable Switches

Figure 4 shows an old-fashioned binary configurable switch. Such a switch represents the simplest decision node. Everything computational and organizational stems back to binary decision nodes. Binary decision nodes are the basis of all formal function. Even analog and index systems are ultimately based on binary choices. An analog rheostat knob, for example, must be designed to increase power when turned in one direction (e.g., clockwise) and to decrease power when turned in the opposite direction (e.g., counterclockwise).

Can we describe any gradual "degrees of organization" that are possible in the flipping of each binary switch knob? Note that the pictured switch knob cannot be found in a neutral position. The switch is designed with a logical "excluded middle." It will always be found in either the on or off position. Such configurable switches are designed to record yes/no, on/off, 1/0 purposeful programming choices. There is no gradation of selection at each individual binary decision node. The switch knob will be found in either the right or left position.

Configurable switches are dynamically inert (dynamically incoherent; dynamically decoupled from physicodynamic causation) [196,197]. This means that on a horizontal switch board, the force of gravity works equally on all potential switch positions. Physicodynamics plays no role in which way the switch knob is pushed. This is the very meaning of "configurable" switches. Their setting is completely decoupled from physicodynamic causation. They can only be set by formal choice contingency, not by chance or law. It is the freedom of formal choice at configurable switches that makes all forms of formal sophistication possible in any physical system. Nonphysical formalism alone determines each switch setting. The switch is a "dynamically-inert configurable switch."

a)   b)

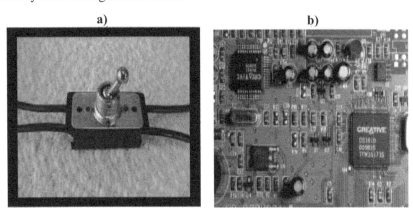

Figure 4 a) A binary configurable switch. Though physical, the switch-setting is nonetheless physicodynamically inert ("dynamically decoupled or incoherent" [196,197]). No physical force field determines the direction this knob is pushed. The vector of knob push is determined by formal choice contingency alone, not by chance or necessity, and not by order or complexity. b) An integrated circuit board arises only out of unified, coherent, purposefully cooperative, truly organized logic-gate switch-settings. The number of permutations of voluntary (choice-contingent; configurable) switch-setting combinations quickly becomes staggering. Often only one configuration achieves a certain functional computational halting.

The switch in Fig. 4 happens to be a binary switch. We could have just as easily photographed a quaternary switch. With a quaternary switch, the knob could be pushed away from you, pulled toward you, pushed to the right, or pushed to the left. A quaternary configurable switch represents 2 bits of uncertainty. The option space of equally available four possible nucleotides also represents 2 bits of uncertainty. Each potential add-on locus in a forming single-stranded oligoribonucleotide in an imagined primordial soup adds an additional 2 bits of uncertainty to the strand. The same is true of a single-stranded (positive, instructional) DNA polymer. Each locus corresponds to a four-way (tertiary) configurable switch. The high degree of uncertainty in a potential single-stranded DNA physical matrix is what allows DNA to retain such tremendous amounts of information. Spinelli & Mayer-Foulkes [311] found specific statistical differences between exon and intron DNA sequences, referrring to them as "linguistic DNA features." Large numbers of other researchers have found linguistic like properties in DNA prescriptive information as summarized by Searls [312].

Although statistical differences and patterns distinguish one linear digital prescriptive string from another, no prescriptive information exists because of probabilistic combinatorialism [77]. Prescriptive information only exists at the moment a particular choice for potential function is made [1]. When a nucleotide is rigidly (covalently) bound to the single-stranded string, the four-way configurable switch knob is actually pushed in one of four possible directions. At that moment all Shannon uncertainty is replaced with formal causation. The vector of the four-way switch knob is determined by choice contingency, not by physicodynamics. It is only when one of the four options is actually selected for potential function that prescriptive information comes into existence. It is only when that choice initiates movement of the physical switch knob in one of the four directions that formalism is instantiated into physicality.

Two Kinds of Selection

Two kinds of selection exist: 1) Selection *of existing* function (e.g., natural selection; differential survival) VS.2) Selection *for potential* function (e.g., artificial selection for formal function).

Selection *of existing* fitness is accomplished by selection pressure. Natural selection consists of differential survival and reproduction of the fittest already-computed phenotypes. It occurs only at the organismic level of already-living small populations of organisms. "Survival of the fittest" is environmental selection of the best existing breeds, varieties, and species.

Selection *for potential* fitness is always artificial rather than natural. Selection for potential fitness is a formal, not a physical enterprise. Selection for potential fitness occurs at decision nodes. Symbols systems and configurable switch settings are used to represent those decisions. Examples of formal selection include language, cybernetic programming, logic, math, computation, algorithmic optimization, design and engineering function, organization of any kind.

Linear digital genetic programming using a Hamming block code of 3 nucleotide selections to represent and prescribe each amino acid selection is a form of selection for potential fitness, not selection of existing fitness. Genetic programming cannot be explained by natural selection. The environment cannot select for potential function. Evolution has no goal or programming ability at the genetic level. As discussed above, the selection of each nucleotide corresponds to the setting of a four-way quaternary configurable switch. Three quaternary switch-settings in a row prescribe each amino acid "letter" of a very long protein "word." No fitness exists for the environment to favor or select at the level of 3'5' phosphodiester bond formation between nucleotides. These informational biopolymers must be sequenced prior to the realization of any prescriptive, enzymatic, or regulatory function. Selection at the level of nucleotide sequencing clearly falls within the category of "Selection for potential function" rather than the category of "Selection of existing function." This is called the GS (Genetic Selection) Principle [10]. The GS Principle states that selection must occur at the decision-node level of rigid covalent bond linkage of specific monomers to form functional syntax. After-the-fact selection of already computed phenotypic fitness is not sufficient to explain genetic programming or the metabolism it organizes.

We must also remember that natural selection does not favor function. Selection pressure favors only the survival of the fittest holistic, already-living organisms. No organism would be alive without thousands of cooperating molecular machines, integrated biochemical pathways and cycles, and the formal goal of maintaining a homeostatic metabolism. All of these algorithmic processes must be optimized and in place before any organism can come to life, let alone constitute the fittest selectable life. Chang et al. [313] state:

> 'Chemical evolution' should not be confused with Darwinian evolution with its requirements for reproduction, mutation and natural selection. These did not occur before the development of the first living organism, and so chemical evolution and Darwinian evolutions are quite different processes.

What Optimizes Genetic Algorithms?

Computational methods often employ genetic algorithms (GA's). The appeal of GAs is that they are modeled after biological evolution. The latter is the main motivation for tolerating such an inefficient awkward process. The GA search technique begins with a large random pool of representations of "potential solutions." Genetic algorithms are seen as a subset of evolutionary algorithms and as "evolutionary computation." The methodology is inspired by modeling a random beginning phase space, various kinds of mutations, inheritance and selection. The experimenter chooses the fittest solutions from each generation out of the "evolving" phase space of potential solutions. The goal of the process is optimization of a certain function.

All too many evolutionary computationists fail to realize the purely formal nature of GA procedures. GA's are not dealing with physicodynamic cause-and-effect chains. First, what is being optimized is *a formal representation* of meaning and function. A representation of any kind cannot be reduced to inanimate physicality. Second, "potential solutions" are formal, not merely physical entities. Third, at each iteration a (generation) a certain portion of the population of potential solution is deliberately selected by the agent experimenter (*artificial selection*) to "breed" a new generation. The optimized solution was purposefully pursued at each iteration. The overall process was entirely goal-directed (formal). Real evolution has no goal [172–175]. Fourth, a formal fitness function is used to *define* and *measure* the fittest solutions thus far to a certain formal problem. The act of defining and measuring, along with just about everything else in the GA procedure, is altogether formal, not physical [140,194,298,314,315].

Despite the appealing similarities of terms like "chromosomes," GA's have no relevance whatsoever to molecular evolution or gene emergence. Inanimate nature cannot define a fitness function over measures of the quality of representations of solutions. GAs are no model at all of natural process. GA's are nothing more than multiple layers of abstract conceptual engineering. Like language, we may start with a random phase space of alphabetical symbols. But no meaning or function results without deliberate and purposeful selection of letters out of that random phase space. No abiotic primordial physicodynamic environment could have exercised such programming prowess. Neither physics nor chemistry can dictate formal optimization, any more than physicality itself generates the formal study of physicality. Human epistemological pursuits are formal enterprises of agent minds. Natural process GAs have not been observed to exist. The GAs of living organisms are just metaphysically presupposed to have originated through natural process. We can liberally employ GAs and so-called evolutionary algorithms for all sorts of productive tasks. But GAs cannot be used to model spontaneous life origin through natural process because GAs are formal.

Order vs. Organization

Organization ≠ order. Disorganization ≠ disorder. Self-ordering of many kinds occurs spontaneously every day in nature in the absence of any organization. Spontaneous bona fide self-organization, on the other hand, has never been observed.

"Self-organization" is logically a nonsense term. Inanimate objects cannot organize themselves into integrated, cooperative, holistic schemes. Schemes are formal, not physical. To organize requires choice contingency, not just chance contingency and law-like necessity. Sloppy definitions lead to fallacious inferences, especially to category errors. Organization requires 1) decision nodes, 2) steering toward a goal of formal function, 3) algorithmic optimization, 4) selective switch-setting to achieve integration of a circuit, 5) choice with intent.

The only entity that logically could possibly be considered to organize itself is an agent. But not even an agent self-organizes. Agents organize things and events in their lives. They do not organize their own molecular biology, cellular structure, organs and organ systems. Agents do not organize their own being. Agents do not create themselves. They merely make purposeful choices with the brains and minds with which they find themselves. Artificial intelligence does not organize itself either. It is invariably programmed by agents to respond in certain ways to various environmental challenges in the artificial life data base.

Thus the reality of self-organization is highly suspect on logical and analytic grounds even before facing the absence of empirical evidence of any spontaneous formal self-organization. Certainly no prediction of bona fide *self*-organization from unaided physicodynamics has ever been fulfilled. Of course if we fail through sloppy definitions to discern between self-ordering phenomena and organization, we will think that evidence of self-organization is abundant. We will point to hundreds of peer-reviewed papers with "self-organization" in their titles. But when all of these papers are carefully critiqued with a proper scientific skepticism, our embarrassment only grows with each exposure of the blatant artificial selection that was incorporated into each paper's experimental design. Such investigator involvement is usually readily apparent right within Materials and Methods of the paper.

What Exactly is Chaos?

Chaos is a bounded state of *disorganization* that is extremely sensitive to the effects of initial conditions. Note that chaos is a disorganized state of matter, *not a disordered* state of matter. A considerable amount of order can arise spontaneously out of chaos. This is what chaos theory is about. Prigogine's dissipative structures are rapid successions of momentarily self-ordered states. Chaos theory deals with such spontaneously forming forms and order. All we have to do to observe spontaneous self-ordering is to pull the stopper out of our bathtub drain. Water molecules quickly self-order into a swirl-a vortex-from purely physicodynamic complex causation. We mistakenly call this self-organization. The vortex is not organized. It is only self-ordered [4]. What is the difference? No decision nodes are required for a bathtub swirl to self-order out of seemingly random Brownian movement. Proficient programming choices are not required for heat agitation of water molecules to self-order into a vortex. No configurable switches have to be purposefully set, each in a certain way, to achieve self-ordering. No pursuit of a goal is involved. No algorithmic optimization is required. In addition, Prigogine's dissipative structures do not DO anything formally productive. They possess no ability to achieve computational halting.

Chaos is capable of producing incredibly complex physicodynamic behavior. But we must never confuse this complexity with formal function. The shape of a candle flame is a spontaneously self-ordered shape or form. It is a rapid succession of dissipative structures that creates the illusion of a sustained structure. Order spontaneously appears out of disorder in the complete absence of any formal creative input or cybernetic management. But no algorithmic organization is produced by a candle flame.The sustained shape of a candle flame is self-ordered. It is not self-organized [4].

The dissipative structures of Prigogine arise out of high-order cause-and-effect "necessity." What seems to be a totally random environment is in fact a caldron of complex interaction of multiple force fields. The complexity of interactive causation can create the illusion of randomness, or of very real self-ordering. There may also be as-of-yet undiscovered physical causes. But dissipative structures self-order; they do NOT self-organize. The dissipative structures of chaos theory are unimaginative. Highly ordered structures contain very little information. Information retention in any physical medium requires freedom of selection of configurable switch settings. Switches must be "dynamically inert" with respect to their function as decision nodes.

Dissipative structures are

1. highly ordered
2. monotonous
3. predictable
4. regular (vortices, sand piles)
5. low informational
6. strings of momentary states

Dissipative structures are usually destructive, not cybernetically constructive (e.g., tornadoes, hurricanes). Trying to use "chaos" and "complexity" to provide mechanism for "self-organization" is like trying to use the Shannon transmission engineering to explain intuitive information, meaning and function. Shannon's equations define "surprisal" and "uncertainty," not semantic information. Just as we cannot explain and measure "intuitive information" using Shannon combinatorial uncertainty, we cannot explain a truly organized system appealing to nothing but a mystical edge of chaos. Reduced uncertainty ("mutual entropy") in Shannon theory comes closer to semantic information, but only because we mix in the formal elements of human knowledge. We measure the reduced uncertainty of *our knowledge*. At that point, we are no longer talking about objective information in nature. We are only talking about human epistemology. Human consciousness is highly subjective. The second we insist on defining information solely in terms of a human observer, we have destroyed all hope of elucidating the derivation of objective information in evolutionary history.

The disorganization of chaos is characterized by conceptual uncertainty and confusion. Disorganization lacks sophisticated steering and control. Disorganization pursues no purpose. Even if chaos had a purpose, it would lack all means of accomplishing that purpose. If chaos by definition is a bounded state of disorganization, how could we possibly attribute self-organization to chaos? No scientific basis exists for granting formal capabilities to chaos, complexity or catastrophe. None of these three has ever been observed to produce formal integration and algorithmic organization of any kind.

Scientists accomplish impressive feats using nonlinear dynamics. But our use of the phrase "nonlinear dynamics" all-too-easily starts referring to chaos as though chaos itself were capable of achieving formal function. We overlook the considerable degree of "investigator involvement" and artificial steering that went into nonlinear dynamic experiments. Formal mathematics was invariably employed by agents. No observers or knowers would exist were it not for a phenomenal amount of *objective* information instructing each cell. A great deal more objective prescriptive information is required to integrate cell systems, organs, organ systems, and holistic organisms. No observers or knowers were around when bacteria were being prescribed and their initial instruction sets being replicated and reproduced. Human observers are Johnny-come-lately *discoverers* of information. Human epistemology is not an essential component of what objective genetic prescriptive information *is* in nature.

Many scientists across a wide array of disciplines exercise a surprisingly blind faith in the amazing formal capabilities of spontaneous molecular chaos and combinatorial complexity. Empirical and rational support for this belief system is sorely lacking. Achieving sophisticated formal function consistently requires regulation and control. Control always emanates from choice contingency and intentionality, not from spontaneous molecular chaos.

The Edge of Chaos

If chaos is inadequate to explain self-organization, what about "the Edge of Chaos?" [7,8,21,22,50–57,198,316–328]. The edge of chaos is somehow much more appealing to us than just plain chaos. The edge of chaos is more poetic. It is terribly-inviting. It offers much more mystical allure. The question is, does the edge of chaos actually exist? If the edge of chaos is objectively real, what exactly is it? Where in time/space can we find it, and what can it independently do? Is the edge of chaos even scientifically addressable?

Let us first examine the potential interface of chaos with natural order-with the regularities of nature described by the physical laws. Can "order" program configurable switches? If "order" programmed configurable switches, they would all be programmed the same way. They would all be set to "On's," OR.. they would all be set to "Off's." Either way, the configurable switches would not be formally programmed into any algorithmic function. No more creativity would exist at the interface of bounded disorganization with forced order than in either single entity. No reason exists to expect any increased cybernetic potential at the edge of chaos than squarely in the middle of chaos (bounded disorganization). The fact that chaos is extremely sensitive to the effects of initial conditions adds no formal attributes. The latter certainly increases its changeability and the number of bits of uncertainty in the bounded state. But mere changeability and combinatorial uncertainty provide no optimization of formal function.

What about the interface of the bounded state of disorganization with heat agitation and Brownian movement? Maximum complexity would set all configurable switches randomly. What synergistic capabilities could emerge from the interface of disorganization with randomness? The two are not synonymous. But neither contributes anything to programming proficiency.

What scientific substance does the edge of chaos provide? What empirical support do we have of formal function arising spontaneously from the interface of chaos with chance OR necessity? What is the logic behind such anticipation? What empirical support do we have for the computational proficiency of the edge of chaos? Have we had any prediction fulfillments since it was first described in 1992 by Waldrop [50]? Is the notion of vast formal capabilities arising from the edge of chaos falsifiable? One has to wonder if the notion is worthy of serious discussion in a peer-reviewed science journal paper. It would not be were it not for the fact that so many peer-reviewed papers already cite this nebulous dream as an objective source of self-organization.

Systems Theory

Systems theory in the literature regularly presupposes the metaphysical belief of physicodynamic self-organization into formal function. One would think that systems theorists could readily offer a crystal-clear definition of "system." Sadly, this is not the case. It is not surprising, therefore, that chaos and such phenomena as weather fronts are referred to as systems with no eyebrows raised. Bona fide systems require organizational controls. True systems are cybernetic. Weather fronts are at best self-ordered by complex degrees of interactive physicodynamic causation. They are not formally controlled or organized to achieve sophisticated utility of any kind. A weather front is a physicodynamic interface complete with criticality and phase changes. It may become a highly self-ordered tornado or a hurricane. But it's not a true system because it is not formally organized or cybernetically programmed. No representational symbol system is used. No abstract conceptualizations are employed by weather fronts. They are simply physicodynamic interfaces totally lacking in algorithmic organization. We simply "murder the King's English" by referring to a weather front as a system. Such sloppy word usage leads to a great deal of confusion in understanding fundamental physics, the temporary and local circumvention of the 2nd Law, and the algorithmic processes that alone make the latter possible.

Chaos is neither organized nor a true system, let alone "self-organized." As pointed out above, organization is not the same as order. A bona fide *system* requires *organization*. Chaos by definition lacks organization. That's why we call it "chaos" even though it manifests extensive self-ordering tendencies. What could possibly be more self-ordered than a massive hurricane? But what formal functions does it perform? A hurricane doesn't DO anything constructive or formally functional because it contains no formal organizational components. It has no programming talents or creative instincts. A hurricane is not a participant in Decision Theory. A hurricane does not set logic gates according to arbitrary rules of inference. A hurricane has no specifically designed dynamically-decoupled configurable switches. No means exists to instantiate formal choices or function into physicality. A highly self-ordered hurricane does nothing but destroy organization. To call a hurricane "self-organized" constitutes one of the most egregious errors in science stemming from sloppy definitions, category errors, and non sequiturs.

In itself, chaos is NOT a

1. Calculus.
2. Algorithm.
3. Program that achieves computational halting.
4. Organizer of formal function.
5. A bona fide system.

Complexity is not a system, either, as we saw in the highly complex pile of pick-up sticks (Fig. 2). No programming is involved. No algorithms are optimized. No steering toward formal function occurs. A true system requires organization.

In physics, no empirical evidence exists, not even an anecdotal account, of Chaos, Catastrophe, maximum Complexity, order or pattern ever having produced sophisticated algorithmic function or cybernetic organization of any kind. A pulsar signal has abundant order and pattern. But it doesn't DO anything useful. It contains no meaningful or functional message. It knows nothing of decision nodes or choice contingency.

In biology, no rational or empirical justification exists for attributing linear, digital, encrypted, genetic recipes to stochastic ensembles OR to physical laws in *any* amount of time. Yet thousands of peer-reviewed papers exist in the literature on "self-organization." How can denial of self-organization possibly be correct? The answer is that all of these papers are universally misdefining what is being observed. Self-ordering phenomena are being observed, not self-organization. But self-ordering phenomena do not measure up to the task of genetic programming.

Formalism vs. Physicality

When it comes to life-origin studies, we have to address how symbol selection in the genetic material symbol system came about objectively in nature [2]. Life origin science must address the derivation of objective organization and control in the first protocells. How did prescriptive information and control arise spontaneously out of the chaos of a Big Bang explosion, primordial slime, vent interfaces in the ocean floor, or mere tide pools?

Self-ordering phenomena arise spontaneously out of phase space, but we have no evidence whatsoever of formal organization arising spontaneously out of physical chaos or self-ordering phenomena. Chance and necessity has not been shown to generate the choice contingency required to program computational halting, algorithmic optimization, or sophisticated function.

If chance and necessity, order and complexity cannot produce formal function, what does? *Selection for potential* utility is what optimizes algorithms, not randomness (maximum complexity), and not fixed law (highly patterned, unimaginative, redundant order). Utility lies in a third dimension imperceptible to chance and necessity. What provides this third dimension is when each token in a linear digital programming string is arbitrarily (non physicodynamically, formally) selected for potential function. The string becomes a cybernetic program capable of computation only when signs/symbols/tokens are arbitrarily *chosen* from an alphabet to *represent* utilitarian configurable switch settings. The choice represented by that symbol can then be instantiated into physicality using a dynamically inert (physicodynamically decoupled or incoherent) [196,197,329] configurable switch setting. At the moment the switch knob seen in Fig. 4 is pushed, nonphysical formalism is instantiated into physicality. Then and only then does algorithmic programming become a physical reality. Once instantiated, we easily forget the requirement of instantiation of *formal instructions and controls* into the physical system to achieve engineering function. It was the formal voluntary pushing of the configurable switch knob in a certain direction that alone *organized* physicality [1,3,4,7–9,12].

Degrees of integration are achieved through *a combination* of binary configurable switch-settings. The selection of any combination of multiple switch settings to achieve degrees of organization is called programming. But purposefully flipping the very first binary configurable switch is the foundation and first step of any form of programming. Programming requires choice contingency. The measure of algorithmic compression requires an added dimension. Only this extra dimension allows us to place a sequence on the unidimensional vector graph showing varying degrees of order and complexity (Fig. 1). No known natural process spontaneously compresses an informational message string. As Howard Pattee has repeatedly pointed out, any type of measurement is a formal function that cannot be reduced to physicodynamics [161,314,330,331]. We do not plug initial conditions into the formal equations known as "the laws of physics." We plug *symbolic representations* of those initial conditions into the laws of physics. Then we do formal mathematical manipulations of these equations to reliably predict physicodynamic interactions and outcomes. In this sense formalism governs physicality. The role that mathematics plays in physics is alone sufficient to argue for formalism's transcendence over physicality.

Just as it takes an additional dimension to measure the algorithmic compressibility of a sequence, it takes still another dimension to measure the formal utility of any sequence.

Formalisms are abstract, conceptual, representational, algorithmic, choice-contingent, non physical activities of mind. Formalisms typically involve steering toward utility. Formalisms employ controls rather than mere physicodynamic constraints. Formalisms require obedience to arbitrarily prescribed rules rather than forced laws. Physicodynamics cannot visualize, let alone quantify formal utility.

Formalisms cannot be produced by chance or necessity. Language, for example, uses arbitrary symbol selections from an alphabet of options. Logic theory uses rules, not laws, to judge inferences. Programming requires choice contingency at each decision node. Each logic gate and configurable switch must be deliberately set a certain way to achieve potential (not-yet-existent) computational halting. These are all formal functions, not spontaneous physicodynamic events. They are just as formal as mathematics. Decision nodes, logic gates, and configurable switches cannot be set by chance and/or necessity if sophisticated formal utility is expected to arise. They must be set with the intent to control and to program computational halting. Acknowledgement of the reality of formal controls was growing within the molecular biological community even prior to the now weekly new discoveries of extraordinarily sophisticated cybernetic mechanisms in cellular physiology [332].

The Cybernetic Cut

Formal function can invariably be traced back to the exercise of some form of decision theory. Achieving formal utility requires crossing The Cybernetic Cut [9]. The Cybernetic Cut is perhaps the most fundamental divide of scientifically addressable reality. A monstrous ravine runs through presumed objective reality. It is the great divide between physicality and formalism. On the one side of this Grand Canyon lies everything that can be explained by the chance and necessity of physicodynamics. On the other side lies those phenomena than can only be explained by formal choice contingency and decision theory-the ability to choose with intent what aspects of ontological being will be preferred, pursued, selected, rearranged, integrated, organized, preserved, and used. Physical dynamics includes spontaneous non linear phenomena, but not our formal applied-science called "non linear dynamics."

A configurable-switch (CS) Bridge traverses this great chasm. But this CS Bridge conveys one-way traffic only. Prescriptive information flows only from the formal side to the physical side of the ravine. Programming decisions can be instantiated into physical configurable switch settings. But physicality contributes no formal influence on those choices in reverse direction. The choices that set the physical configurable switches are themselves non physical. Physicodynamic forces have no influence upon non physical formalisms. Physicodynamics (the chance and necessity of physicality) cannot steer events toward computational halting.

Falsification of The Cybernetic Cut requires nothing more than demonstrating a single incident of two-way traffic across the CS Bridge. Thus far, no such incident of two-way traffic has even been observed. Logically, the chance and necessity of physicality cannot make purposeful choices (e.g., programming decisions). Physicality cannot plot and scheme. Physicality cannot prefer utility over non utility. It cannot even categorize formal function from non function. Stand-alone physicodynamics is blind to utility, and could care less whether anything "works" in a formal sense.

This is not to say that formalisms cannot employ elements of chance and/or physicodynamic determinism. Every day architects and engineers work around, depend upon, and deliberately employ the orderliness of physicodynamics. When we play the card game of poker, we incorporate stochastic reality and physical constraints into our formal scheming. These facts in no way threaten the reality of The Cybernetic Cut.

The Cybernetic Cut is logically, not empirically, absolute. Science does not expect induction to be absolute. But within any axiomatic deductive system, when the rules of inference are carefully adhered to, we have every right to draw as firm a conclusion as we do when predicting physical interactions with any mathematical law of physics. If the predictions fail, we have reason to question our initial axiomatic presuppositions. Thus far, we have no reason or empirical evidence that would cause us to doubt the axiom of The Cybernetic Cut and the one-way traffic across its CS Bridge.

In those fields relating to non linear dynamics, we tend to point to chaos theory, complexity theory, fractals, rugged fitness landscapes, Markov chains, evolutionary algorithms, and directed evolution as evidence for the self-organization of physicality. In reality, all of these fields and models serve only to reinforce the reality of The Cybernetic Cut. Investigator involvement (various forms of artificial selection, not natural selection) is readily identifiable in hundreds of these published experimental designs. A classic example is the body of published ribozyme engineering experiments [333–335]. Take away the experimenter's purposeful choosing of which iteration to pursue, and the desired ribozyme devolves every time toward either a non functional stochastic ensemble, or a self-ordered polymer such as a polyadenosine that also does nothing useful. Neither chance nor necessity can program the needed ribozyme.

Conclusions

The capabilities of stand-alone chaos, complexity, self-ordered states, natural attractors, fractals, drunken walks, complex adaptive systems, and other subjects of non linear dynamic models are often inflated. Scientific mechanism must be provided for how purely physicodynamic phenomena can program decision nodes, optimize algorithms, set configurable switches so as to achieve integrated circuits, achieve computational halting, and organize otherwise unrelated chemical reactions into a protometabolism. To focus the scientific community's attention on its own tendencies toward over-zealous metaphysical imagination bordering on "wish-fulfillment," we propose the following readily falsifiable null hypothesis, and invite rigorous experimental attempts to falsify it:

"Physicodynamics cannot spontaneously traverse The Cybernetic Cut [9]: physicodynamics alone cannot organize itself into formally functional systems requiring algorithmic optimization, computational halting, and circuit integration."

A single exception of non trivial, unaided spontaneous optimization of formal function by truly natural process would falsify this null hypothesis.

Acknowledgements

This work was supported by a grant from The Origin of Life Science Foundation, Inc. Many thanks to Kirk Durston and David Chiu at the University of Guelph for their willingness to share their impressive experimental data.

References

1. Abel, D.L.; Trevors, J.T. Three subsets of sequence complexity and their relevance to biopolymeric information. *Theoret. Biol. Med. Model.* 2005, *2*, Open access at http://www.tbiomed.com/content/2/1/29.
2. Abel, D.L.; Trevors, J.T. More than metaphor: Genomes are objective sign systems. *J. BioSemiotics* 2006, *1*, 253–267.
3. Abel, D.L. The BioSemiosis of Prescriptive Information. *Semiotica* 2009, In Press.
4. Abel, D.L.; Trevors, J.T. Self-Organization vs. Self-Ordering events in life-origin models. *Phys. Life Rev.* 2006, *3*, 211–228.
5. Abel, D.L. To what degree can we reduce "life" without "loss of life"? In *Workshop on Life: A satellite meeting before the Millenial World Meeting of University Professors*; Palyi, G., Caglioti, L., Zucchi, C., Eds.; University of Modena: Modena, Italy, 2000; Vol. Book of Abstracts, p. 4.
6. Abel, D.L. Is Life Reducible to Complexity? In *Fundamentals of Life*; Palyi, G., Zucchi, C., Caglioti, L., Eds.; Elsevier: Paris, 2002; pp. 57–72.
7. Abel, D.L. Life origin: The role of complexity at the edge of chaos. *Washington Science 2006*, Headquarters of the National Science Foundation, Arlington, VA 2006.
8. Abel, D.L. Complexity, self-organization, and emergence at the edge of chaos in life-origin models. *J. Wash. Acad. Sci.* 2007, *93*, 1–20.
9. Abel, D.L. The 'Cybernetic Cut': Progressing from description to prescription in systems theory. *Open Cybernet. Systemat. J.* 2008, *2*, 234–244; Open Access at http://www.bentham.org/open/tocsjopenaccess2.htm.
10. Abel, D.L. The GS (Genetic Selection) Principle. *Front. Biosci.* 2009, *14*, 2959–2969.
11. Abel, D.L. The capabilities of chaos and complexity. In *Society for Chaos Theory: Society for Complexity in Psychology and the Life Sciences*; Virginia Commonwealth University: Richmond, VA., Aug 8–10, 2008.
12. Abel, D.L.; Trevors, J.T. More than metaphor: Genomes are objective sign systems. In *BioSemiotic Research Trends*, Barbieri, M., Ed.; Nova Science Publishers, Inc.: New York, 2007; pp. 1–15.
13. Trevors, J.T.; Abel, D.L. Chance and necessity do not explain the origin of life. *Cell Biol. Internat.* 2004, *28*, 729–739.
14. Aristotle. *Metaphysics*, Book 8.6.1045a:8–10
15. Lewes, G.H. *Problems of Life and Mind (First Series)*. Trübner: London, 1875; Volume 2.
16. Lovejoy, A.O. The meanings of 'emergence' and its modes, with an introduction by Alicia Juarrero and Carl A. Rubino *E:CO* 2008, *10*, 62–78.
17. Chalmers, D.J. Strong and Weak Emergence. In *The Re-Emergence of Emergence*, Clayton, P., Davies, P., Eds.; Oxford Univeristy Press: Oxford, UK, 2006.
18. Steels, L. Towards a Theory of Emergent Functionality. In *Animals to Animats 1*; Meyer, J.-A., Wilson, S., Eds.; MIT Press: Cambridge, Mass, 1991.
19. Corning, P.A. The Re-Emergence of "Emergence": A Venerable Concept in Search of a Theory. *Complexity* 2002, *7*, 18–30.
20. Kauffman, S.A. *The Origins of Order: Self-Organization and Selection in Evolution*. Oxford University Press: Oxford, UK, 1993.
21. Kauffman, S. *At Home in the Universe: The Search for the Laws of Self-Organization and Complexity*. Oxford University Press: New York, 1995; p. 320.
22. Kauffman, S.A. *Investigations*. Oxford University Press: New York, 2000; p. 286.

23. Fromm, J. Types and Forms of Emergence. *arXiv:nlin* 2005, *0506028v1 [nlin.AO]*.
24. Bedau, M.A. Weak emergence. In *Philosophical Perspectives: Mind, Causation, and World*; Tomberlin, J., Ed.; Blackwell Publishers: Hoboken, N.J. USA, 1997; Volume 11, pp. 375–399.
25. Eigen, M. Self-organization of matter and the evolution of biological macromolecules. *Naturwissenchaften (In German)* 1971, *58*, 465–523.
26. Eigen, M. Molecular self-organization and the early stages of evolution. *Experientia* 1971, *27*, 149–212.
27. Eigen, M. Life from the test tube? *MMW Munch Med. Wochenschr* 1983, *Suppl 1.*, S125–135.
28. Eigen, M. New concepts for dealing with the evolution of nucleic acids. *Cold Spring Harb. Symp. Quant. Biol.* 1987, *52*, 307–320.
29. Eigen, M. The origin of genetic information: viruses as models. *Gene* 1993, *135*, 37–47.
30. Eigen, M. Selection and the origin of information. *Int. Rev. Neurobiol.* 1994, *37*, 35–46; discussion 47–50.
31. Eigen, M.; Biebricher, C.K.; Gebinoga, M.; Gardiner, W.C. The hypercycle. Coupling of RNA and protein biosynthesis in the infection cycle of an RNA bacteriophage. *Biochemistry* 1991, *30*, 11005–11018.
32. Eigen, M.; de Maeyer, L. Chemical means of information storage and readout in biological systems. *Naturwissenchaften* 1966, *53*, 50–57.
33. Eigen, M.; Winkler-Oswatitsch, R. Transfer-RNA: The early adaptor. *Naturwissenchaften* 1981, *68*, 217–228.
34. Eigen, M.; Winkler-Oswatitsch, R. Transfer-RNA, an early gene? *Naturwissenchaften* 1981, *68*, 282–292.
35. Eigen, M.; Winkler-Oswatitsch, R. Statistical geometry on sequence space. *Methods Enzymol.* 1990, *183*, 505–530.
36. Eigen, M.; Winkler-Oswatitsch, R.; Dress, A. Statistical geometry in sequence space: A method of quantitative comparative sequence analysis. *Proc. Natl. Acad. Sci. USA* 1988, *85*, 5913–5917.
37. Gánti, T. Organization of chemical reactions into dividing and metabolizing units: the chemotons. *Biosystems* 1975, *7*, 15–21.
38. Gánti, T. On the organizational basis of the evolution. *Acta Biol.* 1980, *31*, 449–459.
39. Gánti, T. Biogenesis itself. *J. Theor. Biol.* 1997, *187*, 583–593.
40. Gánti, T. On the early evolutionary origin of biological periodicity. *Cell Biol. Int.* 2002, *26*, 729–735.
41. Gánti, T. *The Principles of Life*. Oxford University Press: Oxford, UK, 2003; p. 200.
42. Eigen, M.; Gardiner, W.; Schuster, P.; Winkler-Oswatitsch, R. The origin of genetic information. *Sci. Am.* 1981, *244*, 88–92, 96, et passim.
43. Eigen, M.; Gardiner, W.; Schuster, P.; Winkler-Oswatitsch, R. The origin of genetic information, laws governing natural selection of prebiotic molecules have been inferred and tested, making it possible to discover how early RA genes interacted with proteins and how the genetic code developed. *Sci. Am.* 1981, *244*, 88–118.
44. Eigen, M.; Gardiner, W.C., Jr.; Schuster, P. Hypercycles and compartments. Compartments assists-but do not replace-hypercyclic organization of early genetic information. *J. Theor. Biol.* 1980, *85*, 407–411.
45. Eigen, M.; Schuster, P. The hypercycle. A principle of natural self-organization. Part A: Emergence of the hypercycle. *Naturwissenchaften* 1977, *64*, 541–565.
46. Eigen, M.; Schuster, P. *The Hypercycle: A Principle of Natural Self Organization*. Springer Verlag: Berlin, 1979.
47. Eigen, M.; Schuster, P. Comments on "growth of a hypercycle" by King (1981). *Biosystems* 1981, *13*, 235.
48. Eigen, M.; Schuster, P. Stages of emerging life-five principles of early organization. *J. Mol. Evol.* 1982, *19*, 47–61.
49. Eigen, M.; Schuster, P.; Sigmund, K.; Wolff, R. Elementary step dynamics of catalytic hypercycles. *Biosystems* 1980, *13*, 1–22.

50. Waldrop, M.M. *Complexity*. Simon and Schuster: New York, 1992.
51. Kauffman, S.A.; Johnsen, S. Coevolution to the edge of chaos: Coupled fitness landscapes, poised states, and coevolutionary avalanches. *J. Theor. Biol.* 1991, *149*, 467–505.
52. Bratman, R.L. Edge of chaos. *J. R. Soc. Med.* 2002, *95*, 165.
53. Ito, K.; Gunji, Y.P. Self-organisation of living systems towards criticality at the edge of chaos. *Biosystems* 1994, *33*, 17–24.
54. Munday, D. Edge of chaos. *J. R. Soc. Med.* 2002, *95*, 165.
55. Forrest, S. Creativity on the edge of chaos. *Semin. Nurse Manag.* 1999, *7*, 136–140.
56. Innes, A.D.; Campion, P.D.; Griffiths, F.E. Complex consultations and the 'edge of chaos'. *Br. J. Gen. Pract.* 2005, *55*, 47–52.
57. Mitchell, M.; Hraber, P.T.; Crutchfield, J.T. Dynamics, computation, and "the edge of chaos:" A re-examination. In *Complexity: Metaphors, Models, and Reality*; Cowan, G.P.D., Melzner, D., Eds.; Addison-Wesley: Reading, MA, 1994; pp. 1–16.
58. Kauffman, S. Behavior of randomly constructed genetic nets. In *Towards a Theoretical Biology Vol. 3*, Waddington, C.H., Ed.; Aldine Publishing Co.: Chicago, 1970; Vol. 3, p. 18.
59. Kauffman, S. Beyond Reductionism: Reinventing the Sacred. *Zygon* 2007, *42*, 903–914.
60. Kauffman, S.A. Prolegomenon to a general biology. *Ann. N.Y. Acad. Sci.* 2001, *935*, 18–36; discussion 37–38.
61. Dawkins, R. *The Selfish Gene*, 2nd Ed.; Oxford Univerisy Press: Oxford, UK, 1989.
62. Dawkins, R. *The Blind Watchmaker*. W. W. Norton and Co.: New York, 1986.
63. Dawkins, R. *Climbing Mount Impossible*. W. W. Norton and Co.: New York, 1996.
64. Gell-Mann, M. What is complexity? *Complexity* 1995, *1*, 16–19.
65. Ricard, J. What do we mean by biological complexity? *C.R. Biol.* 2003, *326*, 133–140.
66. van de Vijver, G.; van Speybroeck, L.; Vandevyvere, W. Reflecting on complexity of biological systems: Kant and beyond? *Acta Biotheor.* 2003, *51*, 101–109.
67. Edelman, G.M.; Gally, J.A. Degeneracy and complexity in biological systems. *Proc. Natl. Acad. Sci. USA* 2001, *98*, 13763–13768.
68. Simon, H.A. The architecture of complexity. *Proc. Am. Philos. Soc.* 1962, *106*, 467–482.
69. Nicolis, G.; Prigogine, I. *Exploring Complexity*. Freeman: New York, 1989.
70. Badii, R.; Politi, A. *Complexity: Hierarchical Structures and Scaling in Physics*. Cambridge University Press: New York, 1997; p. 318.
71. Yockey, H.P. *Information Theory and Molecular Biology*. Cambridge University Press: Cambridge, 1992; p. 408.
72. Yockey, H.P. *Information Theory, Evolution, and the Origin of Life*, 2nd Ed.; Cambridge University Press: Cambridge, 2005.
73. Lenski, R.E.; Ofria, C.; Collier, T.C.; Adami, C. Genome complexity, robustness and genetic interactions in digital organisms. *Nature* 1999, *400*, 661–664.
74. Lempel, A.; Ziv, J. On the complexity of finite sequences. *IEEE Trans. Inform. Theory* 1976, *22*, 75.
75. Konopka, A.K.; Owens, J. Complexity charts can be used to map functional domains in DNA. *Genet. Anal. Tech. Appl.* 1990, *7*, 35–38.
76. Adami, C.; Cerf, N.J. Physical complexity of symbolic sequences. *Physica D.* 2000, *137*, 62–69.
77. Durston, K.K.; Chiu, D.K.; Abel, D.L.; Trevors, J.T. Measuring the functional sequence complexity of proteins. *Theor. Biol. Med. Model.* 2007, *4*, Open access on-line at http://www.tbiomed.com/content/4/1/47.
78. Ebeling, W.; Jimenez-Montano, M.A. On grammars, complexity, and information measures of biological macromolecules. *Math. Biosci.* 1980, *52*, 53–71.
79. Gell-Mann, M.; Lloyd, S. Information measures, effective complexity, and total information. *Complexity* 1996, *2*, 44–52.
80. Zurek, W.H. *Complexity, Entropy, and the Physics of Information*. Addison-Wesley: Redwood City, CA, 1990.

81. Farre, G.L.; Oksala, T. *Emergence, Complexity, Hierarchy, Organization; Selected and Edited Papers from ECHO III.* Acta Polytechnia Scandinavica; Espoo: Helsinki, 1998.

82. Rosen, R. On information and complexity. In *Complexity, Language, and Life: Mathematical Approaches*, Casti, J. L., Karlqvist, A., Eds.; Springer: Berlin, 1985.

83. Zvonkin, A.K.; Levin, L.A. The complexity of finite objects and the development of the concepts of information and randomness by means of the theory of algorithms. *Russ. Math. Surv.* 1970, *256*, 83–124.

84. Konopka, A.K. Is the information content of DNA evolutionarily significant? *J. Theor. Biol.* 1984, *107*, 697–704.

85. Konopka, A.K. Theory of degenerate coding and informational parameters of protein coding genes. *Biochimie* 1985, *67*, 455–468.

86. Konopka, A.K. Sequences and Codes: Fundamentals of Biomolecular Cryptology. In *Biocomputing: Informatics and Genome Projects*; Smith, D., Ed. Academic Press: San Diego, 1994; pp. 119–174.

87. Konopka, A.K. Systems biology: Aspects related to genomics. In *Nature Encyclopidia of the Human Genome*; Cooper, D.N., Ed.; Nature Publishing Group Reference: London, 2003; Vol. 5, pp. 459–465.

88. Konopka, A.K. Information theories in molecular biology and genomics. In *Nature Encyclopedia of teh Human Genome*; Cooper, D.N., Ed. Nature Publishing Group Reference: London, 2003; Vol. 3, pp. 464–469.

89. Konopka, A.K. Sequence complexity and composition. In *Nature Encyclopedia of the Human Genome. Vol. 5*; Cooper, D.N., Ed. Nature Publishing Group Reference: London, 2003; pp. 217–224.

90. Koonin, E.V. Evolution of genome architecture. *Int. J. Biochem. Cell Biol.* 2009, *41*, 298–306.

91. Koonin, E.V.; Dolja, V.V. Evolution of complexity in the viral world: The dawn of a new vision. *Virus research* 2006, *117*, 1–4.

92. Koonin, E.V.; Wolf, Y.I. Genomics of bacteria and archaea: The emerging dynamic view of the prokaryotic world. *Nucleic Acids Res.* 2008, *36*, 6688–6719.

93. Toussaint, O.; Schneider, E.D. The thermodynamics and evolution of complexity in biological systems. *Comp. Biochem. Physiol. A Mol. Integr. Physiol.* 1998, *120*, 3–9.

94. Barham, J. A dynamical model of the meaning of information. *Biosystems* 1996, *38*, 235–241.

95. Stonier, T. Information as a basic property of the universe. *Biosystems* 1996, *38*, 135–140.

96. Boniolo, G. Biology without information. *Hist. Phil. Life Sci.* 2003, *25*, 255–273.

97. Sarkar, S. Biological information: A skeptical look at some central dogmas of molecular biology. In *The Philosophy and History of Molecular Biology: New Perspectives*; Sarkar, S., Ed. Kluwer Academic Publishers: Dordrecht, 1996; pp. 187–231.

98. Sarkar, S. Information in genetics and developmental biology: Comments on Maynard Smith. *Philos. Sci.* 2000, *67*, 208–213.

99. Sarkar, S. Genes encode information for phenotypic traits. In *Comtemporary debates in Philosophy of Science*; Hitchcock, C., Ed.; Blackwell: London, 2003; pp. 259–274.

100. Stent, G.S. Strength and weakness of the genetic approach to the development of the nervous system. *Annu. Rev. Neurosci.* 1981, *4*, 163–194.

101. Griffiths, P.E. Genetic information: A metaphor in search of a theory. *Philos. Sci.* 2001, *68*,394–412.

102. Godfrey-Smith, P. Genes do not encode information for phenotypic traits. In *Contemporary Debates in Philosophy of Science*; Hitchcock, C., Ed.; Blackwell: London, 2003; pp. 275–289.

103. Noble, D. Modeling the heart-from genes to cells to the whole organ. *Science* 2002, *295*, 1678–1682.

104. Mahner, M.; Bunge, M.A. *Foundations of Biophilosophy*; Springer Verlag: Berlin, 1997.

105. Kitcher, P. Battling the undead; how (and how not) to resist genetic determinism. In *Thinking About Evolution: Historical Philosophical and Political Perspectives*; Singh, R.S.; Krimbas, C.B.; Paul, D.B.; Beattie, J., Eds. Cambridge University Press: Cambridge, 2001; pp. 396–414.

106. Chargaff, E. *Essays on Nucleic Acids*. Elsevier: Amsterdam, 1963.

107. Jacob, F. *The Logic of Living Systems-a History of Heredity*. Allen Lane: London, 1974.

108. Alberts, B.; Bray, D.; Lewis, J.; Raff, M.; Roberts, K.; Watson, J.D. *Molecular Biology of the Cell*. Garland Science: New York, 2002.

109. Davidson, E.H.; Rast, J.P.; Oliveri, P.; Ransick, A.; Calestani, C.; Yuh, C.H.; Minokawa, T.; Amore, G.; Hinman, V.; Arenas-Mena, C.; Otim, O.; Brown, C.T.; Livi, C.B.; Lee, P.Y.; Revilla, R.; Rust, A.G.; Pan, Z.; Schilstra, M.J.; Clarke, P.J.; Arnone, M.I.; Rowen, L.; Cameron, R.A.; McClay, D.R.; Hood, L.; Bolouri, H. A genomic regulatory network for development. *Science* 2002, *295*, 1669–1678.

110. Wolpert, L.; Smith, J.; Jessell, T.; Lawrence, P. *Principles of Development*. Oxford University Press: Oxford, 2002.

111. Stegmann, U.E. Genetic information as instructional content. *Philos. Sci.* 2005, *72*, 425–443.

112. Barbieri, M. Biology with information and meaning. *Hist. Philos.Life Sci.* 2004, *25*, 243–254.

113. Deely, J. Semiotics and biosemiotics: Are sign-science and life-science coextensive? In *Biosemiotics: The Semiotic Web 1991*; Sebeok, T.A., Umiker-Sebeok, J., Eds.; Mouton de Gruyter: Berlin/N.Y., 1992; pp. 46–75.

114. Sebeok, T.A.; Umiker-Sebeok, J. *Biosemiotics: The Semiotic Web 1991*. Mouton de Gruyter: Berlin, 1992.

115. Hoffmeyer, J. Biosemiotics: Towards a new synthesis in biology. *Eur. J. Semiotic Stud.* 1997, *9*, 355–376.

116. Sharov, A. Biosemiotics. A functional-evolutionary approach to the analysis of the sense of evolution. In *Biosemiotics: The Semiotic Web 1991*; Sebeok, T.A., Umiker-Sebeok, J., Eds. Mouton de Gruyter: Berlin, 1992; pp. 345–373.

117. Kull, K. Biosemiotics in the twentieth century: A view from biology. *Semiotica* 1999, *127*, 385–414.

118. Kawade, Y. Molecular biosemiotics: molecules carry out semiosis in living systmes. *Semiotica* 1996, *111*, 195–215.

119. Barbieri, M. Life is 'artifact-making'. *J. BioSemiotics* 2005, *1*, 113–142.

120. Pattee, H.H. The physics and metaphysics of Biosemiotics. *J. BioSemiotics* 2005, *1*, 303–324.

121. Salthe, S.N. Meaning in nature: Placing biosemitotics within pansemiotics. *J. BioSemiotics* 2005, *1*, 287–301.

122. Kull, K. A brief history of biosemiotics. *J. BioSemiotics* 2005, *1*, 1–36.

123. Nöth, W. Semiotics for biologists. *J. BioSemiotics* 2005, *1*, 195–211.

124. Artmann, S. Biosemiotics as a structural science. *J. BioSemiotics* 2005, *1*, 247–285.

125. Barbieri, M. Is the Cell a Semiotic System? In *Introduction to Biosemiotics: The New Biological Synthesis*; Barbieri, M., Ed.; Springer-Verlag New York, Inc.: Secaucus, NJ, USA, 2006.

126. Barbieri, M. *Introduction to Biosemiotics: The New Biological Synthesis*. Springer-Verlag New York, Inc.: Dordrecht, The Netherlands, 2006.

127. Barbieri, M. Has biosemiotics come of age? In *Introduction to Biosemiotics: The New Biological Synthesis*; Barbieri, M., Ed.; Springer: Dorcrecht, The Netherlands, 2007; pp. 101–114.

128. Jämsä, T. Semiosis in evolution. In *Introduction to Biosemiotics: The New Biological Synthesis*, Barbieri, M., Ed. Springer-Verlag New York, Inc.: Dordrecht, The Netherlands; Secaucus, NJ, USA 2006.

129. Hoffmeyer, J. Semiotic scaffolding of living systems. In *Introduction to Biosemiotics: The New Biological Synthesis*; Barbieri, M., Ed.; Springer-Verlag New York, Inc.: Dordrecht, The Netherlands, 2006; pp. 149–166.

130. Kull, K. Biosemiotics and biophysics—The fundamental approaches to the study of life. In *Introduction to Biosemiotics: The New Biological Synthesis*; Barbieri, M., Ed. Springer-Verlag New York, Inc.: Dordrecht, The Netherlands, 2006.

131. Barbieri, M. *The Codes of Life: The Rules of Macroevolution (Biosemiotics)*. Springer: Dordrecht, The Netherlands, 2007.

132. Barbieri, M. Biosemiotics: A new understanding of life. *Naturwissenchaften* 2008, *95*, 577–599.

133. Hodge, B.; Caballero, L. Biology, semiotics, complexity: An experiment in interdisciplinarity *Semiotica* 2005, 477–495.

134. Adami, C.; Ofria, C.; Collier, T.C. Evolution of biological complexity. *P.N.A.S.* 2000, *97*, 4463–4468.

135. Goodwin, B. *How the Leopard Changed Its Spots: The Evolution of Complexity*. Simon and Schuster; Charles Scribner & Sons: New York, 1994.

136. Mao, C. The emergence of complexity: Lessons from DNA. *PLoS Biol.* 2004, *2*, e431.

137. Holland, J.H. *Hidden Order: How Adaptation Builds Complexity*. Addison-Wesley: Redwood City, CA, 1995.
138. Mikulecky, D.C. The emergence of complexity: Science coming of age or science growing old? *Computers Chem.* 2001, *25*, 341–348.
139. Salthe, S.N. *Development and Evolution: Complexity and Change in Biology*. MIT Press: Cambridge, MA, 1993.
140. Pattee, H.H. Causation, Control, and the Evolution of Complexity. In *Downward Causation: Minds, Bodies, and Matter*; Andersen, P.B.; Emmeche, C.; Finnemann, N.O., Christiansen, P.V., Eds.; Aarhus University Press: Aarhus, DK, 2000; pp. 63–77.
141. Szathmary, E.; Smith, J.M. The major evolutionary transitions. *Nature* 1995, *374*, 227–232.
142. Sole, R.; Goodwin, B. *Signs of Life: How Complexity Pervades Biology*. Basic Books: New York, 2000.
143. Stano, P.; Luisi, P.L. Basic questions about the origins of life: proceedings of the Erice international school of complexity (fourth course). *Orig. Life Evol. Biosh.* 2007, *37*, 303–307.
144. Homberger, D.G. Ernst Mayr and the complexity of life. *J. Biosci.* 2005, *30*, 427–433.
145. Pross, A. On the emergence of biological complexity: life as a kinetic state of matter. *Orig. Life Evol. Biosph.* 2005, *35*, 151–166.
146. Bedau, M.A. Artificial life: Organization, adaptation and complexity from the bottom up. *Trends Cogn. Sci.* 2003, *7*, 505–512.
147. Umerez, J. Howard Pattee's theoretical biology—a radical epistemological stance to approach life, evolution and complexity. *Biosystems* 2001, *60*, 159–177.
148. Branca, C.; Faraone, A.; Magazu, S.; Maisano, G.; Migliardo, P.; Villari, V. Suspended life in biological systems. Fragility and complexity. *Ann. N.Y. Acad. Sci.* 1999, *879*, 224–227.
149. Oltvai, Z.N.; Barabasi, A.L. Systems biology. Life's complexity pyramid. *Science* 2002, *298*, 763–764.
150. Rosen, R. Complexity and system description. In *Systems, Approaches, Theories, Applications*, Harnett, W.E., Ed.; Reidel Co.: Boston, MA, 1977.
151. Rosen, R. On Complex Systems. *Euro. J. Operational Rsrch.* 1987, *30*, 129–134.
152. Behe, M.J. *Darwin's Black Box*. The Free Press: New York, 1996.
153. Anderson, E. Irreducible complexity reduced: An integrated Approach to the complexity space. *PCID* 2004, *3.1.5 November*, 1–29.
154. Thompson, C. Fortuitous phenomena: On complexity, pragmatic randomised controlled trials, and knowledge for evidence-based practice. *Worldviews Evid. Based Nurs.* 2004, *1*, 9–17; discussion 18–19.
155. Pennock, R.T. Creationism and intelligent design. *Annu. Rev. Genomics Hum. Genet.* 2003, *4*, 143–163.
156. Aird, W.C. Hemostasis and irreducible complexity. *J. Thromb. Haemost.* 2003, *1*, 227–230.
157. Keller, E.F. Developmental robustness. *Ann. N.Y. Acad. Sci.* 2002, *981*, 189–201.
158. von Neumann, J.; Burks, A.W. *Theory of Self-Reproducing Automata*. University of Illinois Press: Urbana, 1966; p xix, 388 p.
159. Pattee, H.H. The complementarity principle in biological and social structures. *J. Soc. Biol. Struct.* 1978, *1*, 191–200.
160. Pattee, H.H. Complementarity vs. reduction as explanation of biological complexity. *Amer. J. Physiol.* 1979, *236*, R241–246.
161. Pattee, H.H. Evolving self-reference: Matter, symbols, and semantic closure. *Commun. Cog.* 1995, *12*, 9–28.
162. Hoffmeyer, J. Code-duality and the epistemic cut. *Ann. N.Y. Acad. Sci.* 2000, *901*, 175–186.
163. Hoffmeyer, J. Code duality revisited. *SEED* 2002, *2*, 1–19.
164. *Lectures in the Sciences of Complexity*. Stein, D.L., Ed.; Addison-Wesley: Redwood City, CA, 1988.
165. Norris, V.; Cabin, A.; Zemirline, A. Hypercomplexity. *Acta Biotheor* 2005, *53*, 313–330.
166. Garzon, M.H.; Jonoska, N.; Karl, S.A. The bounded complexity of DNA computing. *Bio. Systems* 1999, *52*, 63–72.

167. Levins, R. The limits of complexity. In *Biological Hierarchies: Their Origin and Dynamics*, Pattee, H., Ed.; Gordon and Breach: New York, 1971.

168. Bennett, D.H. Logical depth and physical complexity. In *The Universal Turing Machine: A Half-Century Survey*; Herken, R., Ed.; Oxford University Press: Oxford, 1988.

169. Grandpierre, A. Complexity, information and biological organization. *INDESC* 2005, *3*, 59–71.

170. Chandler, J.L. Complexity IX. Closure over the organization of a scientific truth. *Ann. N.Y. Acad. Sci.* 2000, *901*, 75–90.

171. Wimsatt, W.C. Complexity and organization. In *SA-1972 (Boston Studies in the Philosophy of Science)*; Reidel: Dordrecht, 1974; Vol. 20, pp. 67–86.

172. Mayr, E. Introduction, pp. 1–7; Is biology an autonomous science? pp. 8–23. In *Toward a New Philosophy of Biology, Part 1*; Mayr, E., Ed.; Harvard University Press: Cambridge, MA, 1988.

173. Mayr, E. The place of biology in the sciences and its conceptional structure. In *The Growth of Biological Thought: Diversity, Evolution, and Inheritance*; Mayr, E., Ed.; Harvard University Press: Cambridge, MA, 1982; pp. 21–82.

174. Monod, J. *Chance and Necessity*. Knopf: New York, 1972.

175. Mayr, E. *What Evolution Is*. Basic Books: New York, 2001.

176. Popper, K. *Conjectures and Refutations*. Harper: New York, 1963.

177. Popper, K. *The Logic of Scientific Discovery*. Hutchinson, London, 1968.

178. Kuhn, T.S. *The Structure of Scientific Revolutions*, 2nd Ed.; The University of Chicago Press: Chicago, 1970.

179. Dinger, M.E.; Pang, K.C.; Mercer, T.R.; Mattick, J.S. Differentiating protein-coding and noncoding RNA: Challenges and ambiguities. *PLoS computational biology* 2008, *4*, e1000176.

180. Banks, E.; Nabieva, E.; Chazelle, B.; Singh, M. Organization of Physical Interactomes as Uncovered by Network Schemas. *PLoS Computational Biology* 2008, *4*, e1000203.

181. de Silva, A.P.; Uchiyama, S. Molecular logic and computing. *Nat. Nano* 2007, *2*, 399–410.

182. Adami, C. What is complexity? *Bioessays* 2002, *24*, 1085–1094.

183. Li, M.; Vitanyi, P. *An Introduction to Kolmogorov Complexity and Its Applications*, 2nd Ed.; Springer-Verlag: New York, 1997; p. 637.

184. Chaitin, G.J. *Algorithmic information theory*, 1st paperback Ed.; Cambridge University Press: Cambridge, UK ; New York, 2004.

185. Yockey, H.P. Information theory, evolution and the origin of life. *Inform. Sci.* 2002, *141*, 219–225.

186. Shannon, C. Part I and II: A mathematical theory of communication. *Bell Sys. Tech. J.* 1948, *XXVII*, 379–423.

187. Chaitin, G.J. *Algorithmic Information Theory*. Revised Second Printing Ed.; Cambridge University Press: Cambridge, 1988.

188. Vitányi, P.M.B.; Li, M. Minimum Description Length Induction, Bayesianism and Kolmogorov Complexity. *IEEE Trans. Inform. Theory* 2000, *46*, 446–464.

189. Swinburne, R. *Simplicity as Evidence for Truth* Marquette University Press: Milwaukee, Wisconsin, 1997.

190. Barbieri, M. *The Organic Codes: An Introduction to Semantic Biology*. Cambridge University Press: Cambridge, 2003.

191. Pattee, H.H. The physical basis of coding and reliabiity in biological evolution. In *Prolegomena to Theoretical Biology*; Waddington, C.H., Ed.; University of Edinburgh: Edinburgh, 1968.

192. Pattee, H.H. How does a molecule become a message? In *Communication in Development; Twenty-eighth Symposium of the Society of Developmental Biology*; Lang, A., Ed.; Academic Press: New York, 1969; pp. 1–16.

193. Pattee, H.H. Physical problems of decision-making constraints. *Int. J. Neurosci.* 1972, *3*, 99–106.

194. Pattee, H.H. The physics of symbols: Bridging the epistemic cut. *Biosystems* 2001, *60*, 5–21.

195. Durston, K.K.; Chiu, D.K.Y. A functional entropy model for biological sequences. *Dynamics of Continuous, Discrete & Impulsive Systems, Series B*, 2005.

196. Rocha, L.M. Evolution with material symbol systems. *Biosystems* 2001, *60*, 95–121.

197. Rocha, L.M.; Hordijk, W. Material representations: From the genetic code to the evolution of cellular automata. *Artif. Life* 2005, *11*, 189–214.

198. Mitchell, M.; Hraber, P.T.; Crutchfield, J.T. Revisiting the edge of chaos: Evolving cellular automata to perform computations. *Complex Systems* 1993, *7*, 89–130.

199. Pearle, J. *Causation.* Cambridge University Press: Cambridge, 2000.

200. Shapiro, R. Prebiotic cytosine synthesis: A critical analysis and implications for the origin of life. *Proc. Natl. Acad. Sci. USA* 1999, *96*, 4396–4401.

201. Ferris, J.P.; Huang, C.H.; Hagan, W.J., Jr. Montmorillonite: A multifunctional mineral catalyst for the prebiological formation of phosphate esters. *Orig. Life Evol. Biosph.* 1988, *18*, 121–133.

202. Ferris, J.P.; Ertem, G. Oligomerization of ribonucleotides on montmorillonite: reaction of the 5'-phosphorimidazolide of adenosine. *Science* 1992, *257*, 1387–1389.

203. Ferris, J.P. Catalysis and prebiotic RNA synthesis. *Orig. Life Evol. Biosph.* 1993, *23*, 307–315.

204. Ferris, J.P.; Hill, A.R., Jr.; Liu, R.; Orgel, L.E. Synthesis of long prebiotic oligomers on mineral surfaces. *Nature* 1996, *381*, 59–61.

205. Miyakawa, S.; Ferris, J.P. Sequence- and regioselectivity in the montmorillonite-catalyzed synthesis of RNA. *J. Am. Chem. Soc.* 2003, *125*, 8202–8208.

206. Huang, W.; Ferris, J.P. Synthesis of 35–40 mers of RNA oligomers from unblocked monomers. A simple approach to the RNA world. *Chem. Commun. (Camb)* 2003, *12*, 1458–1459.

207. Kolmogorov, A.N. Three approaches to the quantitative definition of the concept "quantity of information". *Problems Inform. Transmission* 1965, *1*, 1–7.

208. Gilbert, W. Origin of life- the RNA World. *Nature* 1986, *319*, 618.

209. Gesteland, R.F.; Cech, T.R.; Atkins, J.F. *The RNA World.* 2nd Ed.; Cold Spring Harbor Laboratory Press: Cold Spring Harbor, 1999.

210. Cairns-Smith, A.G. *Seven Clues to the Origin of Life.* Canto Ed.; Cambridge University Press: Cambridge, 1990; p. 130.

211. Cairns-Smith, A.G. The origin of life and the nature of the primitive gene. *J. Theor. Biol.* 1966, *10*, 53–88.

212. Cairns-Smith, A.G. Takeover mechanisms and early biochemical evolution. *Biosystems* 1977, *9*, 105–109.

213. Cairns-Smith, A.G.; Walker, G.L. Primitive metabolism. *Curr. Mod. Biol.* 1974, *5*, 173–186.

214. Segre, D.; Ben-Eli, D.; Lancet, D. Compositional genomes: prebiotic information transfer in mutually catalytic noncovalent assemblies. *Proc. Natl. Acad. Sci. USA* 2000, *97*, 4112–4117.

215. Segre, D.; Lancet, D.; Kedem, O.; Pilpel, Y. Graded autocatalysis replication domain (GARD): Kinetic analysis of self-replication in mutually catalytic sets. *Orig. Life Evol. Biosph.* 1998, *28*, 501–514.

216. Guimaraes, R.C. Linguistics of biomolecules and the protein-first hypothesis for the origins of cells. *J. Biol. Phys.* 1994, *20*, 193–199.

217. Shapiro, R. A replicator was not involved in the origin of life. *IUBMB Life* 2000, *49*, 173–176.

218. Freeland, S.J.; Knight, R.D.; Landweber, L.F. Do proteins predate DNA? *Science* 1999, *286*, 690–692.

219. Rode, B.M. Peptides and the origin of life. *Peptides* 1999, *20*, 773–786.

220. Wong, J.T. A co-evolution theory of the genetic code. *Proc. Natl. Acad. Sci. USA* 1975, *72*, 1909–1912.

221. Wong, J.T. The evolution of a universal genetic code. *Proc. Natl. Acad. Sci. USA* 1976, *73*, 2336–2340.

222. Wong, J.T. Coevolution theory of the genetic code at age thirty. *Bioessays* 2005, *27*, 416–425.

223. Wong, J.T. Question 6: coevolution theory of the genetic code: a proven theory. *Orig. Life Evol. Biosph.* 2007, *37*, 403–408.

224. Zhao, Y.F.; Cao, P.-s. Phosphoryl amino acids: Common origin for nucleic acids and protein. *J. Biol. Phys.* 1994, *20*, 283–287.

225. Zhou, W.; Ju, Y.; Zhao, Y.; Wang, Q.; Luo, G. Simultaneous formation of peptides and nucleotides from N-phosphothreonine. *Orig. Life Evol. Biosph.* 1996, *26*, 547–560.

226. Nashimoto, M. The rna/protein symmetry hypothesis: Experimental support for reverse translation of primitive proteins. *J. Theor. Biol.* 2001, *209*, 181–187.

227. Dyson, F.J. *Origins of Life.* 2nd ed.; Cambridge University Press: Cambridge, 1998.

228. Dyson, F. In *Life in the Universe: Is Life Digital or Analog?* NASA Goddard Space Flight Center Colloquiem, Greenbelt, MD, 1999.

229. Dyson, F.J. A model for the origin of life. *J. Mol. Evol.* 1982, *18*, 344–350.

230. Maturana, H.; Varela, F. *Autopoiesis and Cognition: The Realization of the Living.* Reidel: Dordrecht, 1980.

231. Maturana, H.R. The organization of the living: a Theory of the Living Organization. International Journal of Human-Computer Studies, 51, 1999, 149–168

232. Maturana, H.R.; Varela, F.J. *Review of The Tree of Knowledge: The Biological Roots of Human Understanding.* Rev. Ed.; Shambhala; Distributed in the U.S. by Random House: Boston New York, 1992; p. 269

233. Luisi, P.L. Autopoiesis: A review and a reappraisal. *Naturwissenchaften* 2003, *90*, 49–59.

234. Boden, M.A. *The Philosophy of Artificial Life. Oxford Readings in Philosophy.* Publisher: New York, 1996.

235. Boden, M.A. Autopoiesis and life. *Cognitive Science Quarterly* 2000, *1*, 117–145.

236. Boden, M.A. *The Creative Mind: Myths and Mechanisms.* Routledge, New York 2004.

237. Holland, J.H. *Adaptation in Natural and Artificial Systems.* University of Michigan Press: Ann Arbor, MI, 1975.

238. Holland, J.H. *Emergence: From chaos to order.* Perseus Books: Reading, MA, 1998.

239. Johnson, J. *The Allure of Machinic Life: Cybernetics, Artificial Life, and the New AI* MIT Press: Bradford Books: Cambridge, 2008.

240. Rocha, L.M.; Yaeger, L.S.; Bedau , M.A.; Floreano, D.; Goldstone, R.L.; Vespignani, A. *Artificial Life X: Proceedings of the Tenth International Conference on the Simulation and Synthesis of Living Systems.* MIT Press: Cambridge, MA, 2006.

241. Rasmussen, S.; Chen, L.; Nilsson, M.; Abe, S. Bridging nonliving and living matter. *Artificial Life* 2003, *9*, 269–316.

242. Bedau, M.A.; McCaskill, J.S.; Packard, N.H.; Rasmussen, S.; Adami, C.; Green, D.G.; Ikegami, T.; Kaneko, K.; Ray, T.S. Open problems in artificial life. *Artif. Life* 2000, *6*, 363–376.

243. Adami, C. *Introduction to Artificial Life.* Springer/Telos: New York, 1998; p. 374.

244. Langton, C.G. Studying Artificial life with cellular automata. *Physica D.* 1986, *22*, 120–149.

245. Langton, C.G. Artificial life. In *Artificial Life*; Langton, C., Ed.; Addison-Wiley: Redwood City, CA, 1988; pp. 1–47.

246. Langton, C.G. Introduction. In *Artificial Life II*; Langton, C., Taylor, C.; Farmer, J.S., Rasmussen, E., Eds.; Addison-Wesley: Redwood City, California USA 1992; pp. 1–23.

247. Bateson, G. *Steps to an Ecology of Mind.* Chandler: New York, 1972.

248. Bateson, G. *Mind and Nature.* Bentam Books: New York, 1979.

249. Bateson, G.; Bateson, M.C. *Angels fear: Towards an epistemology of the sacred.* Hampton Press: Cresskill, New Jersey, USA 2005.

250. Carnap, R.; Bar-Hillel, Y. An outline of a theory of semantic information. *Technical Report #247, MIT Research Laboratory in Electronics; Also in Bar-Hillel, 1964, Language and Information, Chapter 15,* 1952.

251. Bar-Hillel, Y. In *Semantic Information and Its Measures*; Transactions of the Tenth Conference on Cybernetics, New York, 1952; Josiah Macy Jr. Foundation: New York, 1952; pp. 33–48.

252. Devlin, K. *Logic and Information.* Cambridge University Press: New York, 1991.

253. Barwise, J.; Perry, J. *Situations and Attitudes.* MIT Press: Cambridge, MA, 1983.

254. Floridi, L. Information. In *The Blackwell Guide to the Philosophy of Computing and Information*; Floridi, L., Ed.; Blackwell: Oxford, 2003; pp. 40–62.

255. Floridi, L. Open problems in the philosophy of information. *Metaphilosophy* 2003, *35*, 554–582.

256. Dretske, F. *Knowledge and the Flow of Information.* MIT Press: Cambridge, MA, 1981.

257. Bar-Hillel, Y. *Language and information: Selected essays on their theory and application.* Addison-Wesley Pub. Co.: Reading, MA, 1964.

258. Hintikka, J. On semantic information. In *Information and Inference*; Hintikka, J., Suppes, P., Eds. D. Reidel: Dorcrecht, 1970.

259. Godfrey-Smith, P. Theoretical role of "genetic coding". *Philos. Sci.* 2000, *67*, 26–44.

260. Godfrey-Smith, P. Information, arbitrariness, and selection: Comments on Maynard Smith. *Philos. Sci.* 2000, *67*, 202–207.

261. Khazen, A.M. [Origin and evolution of life and intellect from the point of view of information processing]. *Biofizika* 1992, *37*, 105–122.

262. Maynard Smith, J. The concept of information in biology. *Philos. Sci.* 2000, *67*, 177–194

263. Hoffmeyer, J.; Emmeche, C. Code-Duality and the Semiotics of Nature, (Forward to and reprinting of, with new footnotes). *J. BioSemiotics* 2005, *1*, 37–91.

264. Jablonka, E. Information: Its interpretation, its inheritance, and its sharing. *Philos. Sci.* 2002, *69*, 578–605.

265. Stegmann, U.E. The arbitrariness of the genetic code. *Bio. Philos.* 2004, *19*, 205–222.

266. Szathmary, E. From RNA to language. *Curr. Biol.* 1996, *6*, 764.

267. Szathmary, E. The origin of the genetic code: Amino acids as cofactors in an RNA world. *Trends Genet.* 1999, *15*, 223–229.

268. Szathmary, E. Biological information, kin selection, and evolutionary transitions. *Theor. Popul. Biol.* 2001, *59*, 11–14.

269. Szostak, J.W. Functional information: Molecular messages. *Nature* 2003, *423*, 689.

270. Hazen, R.M.; Griffin, P.L.; Carothers, J.M.; Szostak, J.W. Functional information and the emergence of biocomplexity. *Proc. Natl. Acad. Sci. USA* 2007, *104*, 8574–8581.

271. Bowong, S.; Kagou, A.T. Adaptive Control for Linearizable Chaotic Systems. *J. Vibrat. Cont.* 2006, *12*, 119–137.

272. Schimmel, P.; Soll, D. When protein engineering confronts the tRNA world. *Proc. Natl. Acad. Sci. USA* 1997, *94*, 10007–10009.

273. Benner, S.A.; Allemann, R.K.; Ellington, A.D.; Ge, L.; Glasfeld, A.; Leanz, G.F.; Krauch, T.; MacPherson, L.J.; Moroney, S.; Piccirilli, J.A.; *et al.* Natural selection, protein engineering, and the last riboorganism: rational model building in biochemistry. *Cold Spring Harb. Symp. Quant. Biol.* 1987, *52*, 53–63.

274. Gewolb, J. Bioengineering: Working outside the protein-synthesis rules. *Science* 2002, *295*, 2205–2207.

275. Yeung, M.K.S.; Tegner, J.; Collins, J.J. Reverse engineering gene networks using singular value decomposition and robust regression. *Proc. Natl. Acad. Sci. USA* 2002, *99*, 6163–6168.

276. Ohuchi, S.; Ikawa, Y.; Shiraishi, H.; Inoue, T. Modular engineering of a Group I intron ribozyme. *Nucleic Acids Res.* 2002, *Aug 1;30*, 3473–3480.

277. Csete, M.E.; Doyle, J.C. Reverse engineering of biological complexity. *Science* 2002, *295*, 1164–1169.

278. Luisi, P.L. Toward the engineering of minimal living cells. *Anat. Rec.* 2002, *268*, 208–214.

279. McCarthy, A.A. Microbia: engineering microbial network biology. *Chem. Biol.* 2003, *10*, 99–100.

280. Clymer, J.R. Simulation-Based Engineering Of Complex Adaptive Systems. *SIMULATION* 1999, *72*, 250–260.

281. Lewontin, R.C. Evolution as engineering. In *Integrative Approaches to Molecular Biology*, Collado-Vides, J.; Smith, T., Magasanik, B., Eds. MIT Press: Cambridge, MA, 1996.

282. Shapiro, J.A. A 21st century view of evolution: genome system architecture, repetitive DNA, and natural genetic engineering. *Gene* 2005, *345*, 91–100.

283. Kaplan, M. *Decision Theory as Philosophy*. Cambridge Univ. Press: Cambridge, 1996; p. 227.

284. Chernoff, H.; Moses, L.E. *Elementary Decision Theory*. 2nd Ed.; Dover Publications: Mineola, N.Y., 1986.

285. Resnik, M.D. *Choices: An Introduction to Decision Theory*. University of Minnesota Press: Minneapolis, Minn, 1987.

286. Bradley, D. Informatics. The genome chose its alphabet with care. *Science* 2002, *297*, 1789–1791.

287. Veening, J.-W.; Smits, W.K.; Kuipers, O.P. Bistability, Epigenetics, and Bet-Hedging in Bacteria. *Annu. Rev. Microbio.* 2008, *62*, 193–210.

288. Allis, D.C.; Jenuwein, T.; Reinberg, D.; Wood, R.; Caparros, M.-L. *Epigenetics*. Cold Springs Harbor Press: Woodbury, NY, 2007.

289. Qiu, J. Epigenetics: unfinished symphony. *Nature* 2006, *441*, 143–145.

290. Grant-Downton, R.T.; Dickinson, H.G. Epigenetics and its implications for plant biology. 1. The epigenetic network in plants. *Ann. Bot.* 2005, *96*, 1143–1164.

291. Jablonka, E.; Lamb, M.J. The changing concept of epigenetics. *Ann. N.Y. Acad. Sci.* 2002, *981*, 82–96.

292. Griesemer, J. What is "epi" about epigenetics? *Ann. N.Y. Acad. Sci.* 2002, *981*, 97–110.

293. Bachmair, A.; Novatchkova, M.; Potuschak, T.; Eisenhaber, F. Ubiquitylation in plants: A post-genomic look at a post-translational modification. *Trends Plant Sci.* 2001, *6*, 463–470.

294. Eisenhaber, B.; Bork, P.; Eisenhaber, F. Post-translational GPI lipid anchor modification of proteins in kingdoms of life: Analysis of protein sequence data from complete genomes. *Protein Eng.* 2001, *14*, 17–25.

295. Vaish, N.K.; Dong, F.; Andrews, L.; Schweppe, R.E.; Ahn, N.G.; Blatt, L.; Seiwert, S.D. Monitoring post-translational modification of proteins with allosteric ribozymes. *Nat. Biotechnol.* 2002, *20*, 810–815.

296. Mata, J.; Marguerat, S.; Bahler, J. Post-transcriptional control of gene expression: A genome-wide perspective. *Trends Biochem. Sci.* 2005, *30*, 506–514.

297. Pattee, H.H. Universal principles of measurement and language functions in evolving systems. In *Complexity, Language, and Life: Mathematical Approaches*; Casti, J.L., Karlqvist, A., Eds.; Springer-Verlag: Berlin, 1986; pp. 579–581.

298. Pattee, H.H. The physics of symbols and the evolution of semiotic controls. In *Proc. Workshop on Control Mechanisms for Complex Systems*, Coombs, M. e. a., Ed. Addison-Wesley: 1997.

299. Pattee, H.H. On the origin of macromolecular sequences. *Biophys. J.* 1961, *1*, 683–710.

300. Pattee, H.H. The nature of hierarchichal controls in living matter. In *Foundations of Mathematical Biology*; Rosen, R., Ed.; Academic Press: New York, 1971; Vol. 1, pp. 1–22.

301. Pattee, H.H. Laws and constraints, symbols and languages. In *Towards a Theoretical Biology*; Waddington, C.H., Ed. University of Edinburgh Press: Edinburgh, 1972; Vol. 4, pp. 248–258.

302. Pattee, H.H. Physical problems of the origin of natural controls. In *Biogenesis, Evolution, and Homeostasis*; Locker, A., Ed.; Springer-Verlag: Heidelberg, 1973; pp. 41–49.

303. Pattee, H.H. Dynamic and linguistic modes of complex systems. *Int. J. General Systems* 1977, *3*, 259–266.

304. Rocha, L.M. Selected self-organization and the semiotics of evolutionary systems. In *Evolutionary Systems: Biological and Epistemological Perspectives on Selection and Self-Organization*, Salthe, S.; van de Vijver, G., Delpos, M., Eds.; Kluwer: The Netherlands, 1998; pp. 341–358.

305. Rocha, L.M. Syntactic autonomy: Or why there is no autonomy without symbols and how self-organizing systems might evolve them. *Annals of the New York Academy of Science.* 2000, 901, 207–223.

306. Morse, M.; Hedlund, G.A. Symbolic Dynamics *Amer. J. Math.* 1938, *60*, 815–866.

307. Kitchens, B. *Symbolic dynamics. One-sided, two-sided and countable state Markov shifts*. Universitext, Springer-Verlag: Berlin, 1998; p. 252.

308. Lind, D.; Marcus, B. *An Introduction to Symbolic Dynamics and Coding*. Cambridge University Press: Cambridge 1995; p. 495.

309. Jiménez-Montaño, M.A.; Feistel, R.; Diez-Martínez, O. On the information hidden in signals and macromolecules. I. Symbolic time-series analysis *Nonlinear Dynamics Psychol. Life Sci.* 2004, *8*, 445–478.

310. Luisi, P.L. Contingency and determinism. *Phil. Trans. R. Soc. A* 2003, *361*, 1141–1147.

311. Spinelli, G.; Mayer-Foulkes, D. New Method to Study DNA Sequences: The Languages of Evolution. *Nonlinear Dynamics Psychol. Life Sci.* 2008, *12*, 133–151.

312. Searls, D.B. The language of genes. *Nature* 2002, *420*, 211–217.

313. Chang, S.; DesMarais, D.; Mack, R.; Miller, S.L.; Streathearn, G.E. Prebiotic organic syntheses and the origin of life. In *Earth's Earliest Biosphere: Its Origin and Evolution*; Schopf, J.W., Ed.; Princeton University Press: Princeton, NJ, 1983; pp. 53–92.

314. Pattee, H.H. Artificial Life Needs a Real Epistemology. In *Advances in Artificial Life*; Moran, F., Ed.; Springer: Berlin, 1995; pp. 23–38.

315. Pattee, H.H. Irreducible and complementary semiotic forms. *Semiotica* 2001, *134*, 341–358.

316. Albano, E.V.; Monetti, R.A. Comment on "Life at the edge of chaos". *Phys. Rev. Lett.* 1995, *75*, 981.

317. Baym, M.; Hubler, A.W. Conserved quantities and adaptation to the edge of chaos. *Phys. Rev. E. Stat. Nonlin. Soft Matter Phys.* 2006, *73*, 056210.

318. Bernardes, A.T.; dos Santos, R.M. Immune network at the edge of chaos. *J. Theor. Biol.* 1997, *186*, 173–187.

319. Bertschinger, N.; Natschlager, T. Real-time computation at the edge of chaos in recurrent neural networks. *Neural Comput.* 2004, *16*, 1413–1436.

320. Borges, E.P.; Tsallis, C.; Ananos, G.F.; de Oliveira, P.M. Nonequilibrium probabilistic dynamics of the logistic map at the edge of chaos. *Phys. Rev. Lett.* 2002, *89*, 254103.

321. Hiett, P.J. Characterizing critical rules at the 'edge of chaos'. *Biosystems* 1999, *49*, 127–142.

322. Legenstein, R.; Maass, W. Edge of chaos and prediction of computational performance for neural circuit models. *Neural Netw.* 2007, *20*, 323–334.

323. Melby, P.; Kaidel, J.; Weber, N.; Hubler, A. Adaptation to the edge of chaos in the self-adjusting logistic map. *Phys. Rev. Lett.* 2000, *84*, 5991–5993.

324. Mycek, S. Teetering on the edge of chaos. *Volunt. Leader* 1999, *40*, 13–16.

325. Mycek, S. Teetering on the edge of chaos. Giving up control and embracing uncertainty can lead to surprising creativity. *Trustee* 1999, *52*, 10–13.

326. Neubauer, J. Beyond hierarchy: working on the edge of chaos. *J. Nurs. Manag.* 1997, *5*, 65–67.

327. Schneider, T.M.; Eckhardt, B.; Yorke, J.A. Turbulence transition and the edge of chaos in pipe flow. *Phys. Rev. Lett.* 2007, *99*, 034502.

328. Stokic, D.; Hanel, R.; Thurner, S. Inflation of the edge of chaos in a simple model of gene interaction networks. *Phys. Rev. E. Stat. Nonlin. Soft Matter. Phys.* 2008, *77*, 061917.

329. Rocha, L.M. The physics and evolution of symbols and codes: Reflections on the work of Howard Pattee. *Biosystems* 2001, *60*, 1–4.

330. Pattee, H.H. The measurement problem in artificial world models. *Biosystems* 1989, *23*, 281–289; discussion 290.

331. Pattee, H.H. Laws, constraints, and the modeling relation—History and interpretations. *Chem. Biodivers.* 2007, *4*, 2272–2295.

332. Allweis, C. Proposal for APS-IUPS convention for diagraming physiological mechanisms. *Amer. J. Physiol.* 1988, *254*, R717–726.

333. Ellington, A.D.; Szostak, J.W. *In vitro* selection of RNA molecules that bind specific ligands. *Nature* 1990, *346*, 818–822.

334. Tuerk, C.; Gold, L. Systematic evolution of ligands by exponential enrichment-RNA ligands to bacteriophage—T4 DNA-polymerase. *Science* 1990, *249*, 505–510.

335. Robertson, D.L.; Joyce, G.F. Selection *in vitro* of an RNA enzyme that specifically cleaves single-stranded DNA. *Nature* 1990, *344*, 467–468.

Section III

Models and Applications of Chaos Theory in Ecology

1. **A Unified Approach of Catastrophic Events**   169
   S. Nikolopoulos, P. Kapiris, K. Karamanos and *K. Eftaxias*
2. **Evolutionary Geomorphology: Thresholds and Nonlinearity in
   Landform Response to Environmental Change**   196
   J.D. Phillips

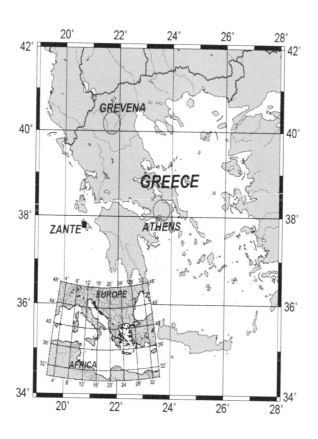

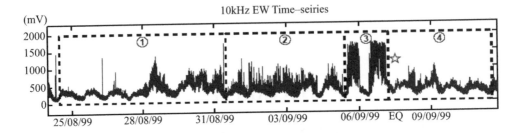

Chapter 1

A Unified Approach of Catastrophic Events[†,#]

S. Nikolopoulos,[1] P. Kapiris,[2] K. Karamanos[3] and K. Eftaxias[2,a]

[1]National Technical Univ. Athens, Dept. of Electrical & Computer Engineering Zografou Campus,
Zografou 15773, Greece.
[2]Dept. of Solid State Section, Faculty of Physics, Univ. of Athens Panepistimioupolis Zografou,
157 84 Athens, Greece.
[a]Email: ceftax@phys.uoa.gr.
[3]Centre for Nonlinear Phenomena and Complex Systems, Univ. Libre de Bruxelles, CP 231, Campus Plaine,
B-1050, Brussels, Belgium.

ABSTRACT

Although there is an accumulated charge of theoretical, computational, and numerical work, like catastrophe theory, bifurcation theory, stochastic and deterministic chaos theory, there is an important feeling that these matters do not completely cover the physics of real catastrophic events. Recent studies have suggested that a large variety of complex processes, including earthquakes, heartbeats, and neuronal dynamics, exhibits statistical similarities. Here we are studying in terms of complexity and non linear techniques whether isomorphic signatures emerged indicating the transition from the normal state to the both geological and biological shocks. In the last 15 years, the study of Complex Systems has emerged as a recognized field in its own right, although a good definition of what a complex system is, actually is eluded. A basic reason for our interest in complexity is the striking similarity in behaviour close to irreversible phase transitions among systems that are otherwise quite different in nature. It is by now recognized that the pre-seismic electromagnetic time-series contain valuable information about the earthquake preparation process, which cannot be extracted without the use of important computational power, probably in connection with computer Algebra techniques. This paper presents an analysis, the aim of which is to indicate the approach of the global instability in the prefocal area. Non-linear characteristics are studied by applying two techniques, namely the Correlation Dimension Estimation and the Approximate Entropy.

[†]Reused with permission from: S. Nikolopoulos, P. Kapiris, K. Karamanos, and K. Eftaxias, A unified approach of catastrophic events, Natural Hazards and Earth System Sciences (2004) 4: 615–631, SRef-ID: 1684-9981/nhess/2004-4-615.

[#] Part of Special Issue "Precursory phenomena, seismic hazard evaluation and seismo-tectonic electromagnetic effects"

These two non-linear techniques present coherent conclusions, and could cooperate with an independent fractal spectral analysis to provide a detection concerning the emergence of the nucleation phase of the impending catastrophic event. In the context of similar mathematical background, it would be interesting to augment this description of pre-seismic electromagnetic anomalies in order to cover biological crises, namely, epileptic seizure and heart failure.

Introduction

Prediction of natural phenomena has always been a well pondered problem. In physics, the predictability degree of a phenomenon is often measured by how well we understand it. Despite the large amount of experimental data and the considerable effort that has been undertaken by the material scientists, many questions about the fracture remain standing.

When a heterogeneous material is strained, its evolution toward breaking is characterized by the nucleation and coalescence of micro-cracks before the final break-up. Both acoustic as well as electromagnetic (EM) emission in a wide frequency spectrum ranging from very low frequencies (VLF) to very high frequencies (VHF), is produced by micro-cracks, which can be considered as the so-called precursors of general fracture. These precursors are detectable both at a laboratory and a geological scale. Several experimental results, which illustrate the connection between anomalous VLF-VHF electromagnetic phenomena and acoustic phenomena with earthquake preparation, were presented in a rather comprehensive collection of papers edited by (Hayakawa and Fujinawa 1994; Hayakawa 1999; Hayakawa and Molchanov 2002).

Aiming at recording VLF-VHF electromagnetic precursors, since 1994 a station was installed at a mountainous site of Zante island ($37.76°$ N–$20.76°$ E) in western Greece (Fig. 1). An important earthquake (Ms=5.9) occurred on 7 September 1999 at 11:56GMT at a distance of about 20 km from the center of the city of Athens, the capital of Greece. Very clear electromagnetic anomalies have been detected in the VLF band (Fig. 2), i.e., at 3 kHz and 10 kHz, before the Athens EQ (Eftaxias et al. 2000, 2001). The whole EM precursors were emerged from 31 August to 7 September 1999 (Fig. 2). It is characterized by an accelerating emission rate (Fig. 2), while, this radiation is embedded in a long duration quiescence period concerning the detection of EM disturbances at the VLF frequency band. These emissions have a rather long duration, (the data were sampled at 1 Hz), and thus it provides sufficient data for statistical analysis.

Recently, in a series of papers (Eftaxias et al. 2001, 2002; Kapiris et al. 2003, 2004b,a; Eftaxias et al. 2004), we attempt to establish the hypothesis that the pre-seismic electromagnetic emissions offer a potential window for a step by step monitoring of the last stages of earthquake preparation processes. However, it is difficult to prove association between any two events (possible precursor and earthquake) separated in times. As a major result, the present study indicates that it seems useful to combine various computational methods to enhance the association of the pre-seismic EM phenomena with micro-fracturing in the pre-focal area. The achievement of converging estimations would definitely improve the chances for an understanding of the physics behind the generation of earthquakes.

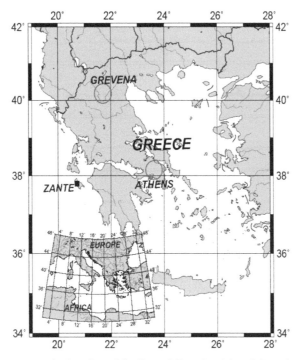

Figure 1 The map demonstrates the location of the Zante RF station (■) and the epicentres of the Athens and Kozani-Grevena earthquakes (○).

Background Information

In this section, we briefly describe the algorithms that were used and compared in this study. Their main characteristics as well as the reasons they were chosen are discussed.

The Delay Times Method

The Delay Times method is an important tool in non-linear analysis and gives both a qualitative and quantitative measure of the complexity of the time-series under examination. It was first established by (Grassberger and Procaccia 1983) and is based on the Takens Theorem (Takens 1981). A timeseries is constructed from a set of successive and experimentally derived values. From the original time-series we then construct a new series, which in this case is composed of vectors. For the construction of each of the vectors the estimation of two parameters, the embedding dimension, m, and the time lag τ, is required. The time lag represents the window that is used for the computation of the coordinates of these vectors. It is estimated from the decorrelation time, which is the window beyond which the signal ceases to present periodicities. The decorrelation time is calculated either from the first zero-value of the autocorrelation function, or from the first value of the mutual information function (Farmer and Swinney 1986) that is close to zero. The mutual information function is a widely acceptedmethod that computes non-linear and

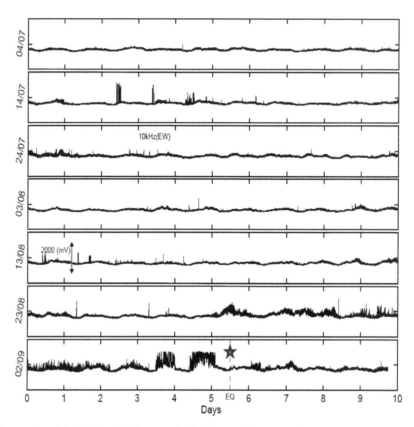

Figure 2 Time-series of the 10kHz (E-W) magnetic field strength between 4 July 1999 and 11 September 1999 in arbitrary units. The precursory accelerating emission is embedded in a long duration of quiescence period. The star indicate the time of the Athens earthquake occurrence.

linear correlation of a signal. The parameter m is assigned increasing integer values, in a range that satisfies both the Takens criterion and the maximum admitted window length, according to basic non-linear dynamics theory. Appendix A includes analytical information on these parameters, as well as a detailed description of the entire Delay Times method.

Once the above is completed, the correlation integral, $C(r)$, is computed for increasing values of r. This integral basically computes how many of the above vectors have a distance between them less than r, where r is a ray in the vector space. We are then able to plot $\ln(C)$ vs. $\ln(r)$, where $\ln$ is the natural logarithm function. From this plot, we select a scaling region and compute the slope of the curve in that region. This process is repeated for increasing values of the embedding dimension, m, and if the values of the slopes converge, then we have found the Correlation Dimension D_2 of the time-series $X(t)$. The convergence value of the slope is an estimation of the Correlation Dimension. A time-series that results from a complex non-linear dynamic system yields a larger value for the Correlation Dimension, as opposed to a time-series which results from a regular and linear dynamic system, lower Correlation Dimension values. Generally, the Correlation

Dimension, D_2, represents the independent degrees of freedom that are required for the proper description of a system or for the construction of its model.

The Approximate Entropy

There are various definitions of entropy, most of which usually arise from entropy computation such as the Shannon entropy or the Kolmogorov-Sinai entropy. From all known methods, Approximate Entropy (*ApEn*) is chosen, since it has been introduced as a quantification of regularity in data and as the natural information parameter for an approximating Markov Chain to a process (Pincus, 1991).

Given the original time-series $X(t)$, we construct a series of vectors, and then we find the heuristic estimation of an integer parameter, m, which in this case represents a window size. We then, one again, heuristically estimate a threshold, r, which arises from the product of the standard deviation of the time series and an arbitrary constant form 0 to 1, which is kept the same for all time-series. We then apply an iterative procedure which finally produces an approximation of *ApEn(m,r)*. Generally, random time-series produce increasing values of *ApEn(m,r)*, compared to regular time-series, a property which we exploit here. More details as well as a more analytical description of the method are included in Appendix B.

The Fractal Spectral Analysis

The concept of fractal is most often associated with irregular geometric objects that display self-similarity. Fractal forms are composed of subunits (and sub-sub-units, etc.) that resemble the structure of the overall object. The fractal analysis can be applied not just in irregular geometric forms that lack a characteristic (single) scale of length, but also to certain complex processes that generate irregular fluctuations across multiple time scales, analogous to scale-invariant objects that have a branching or wrinkly structure across multiple length scales. Earthquakes happen in self-organizing complex systems consisting of many non linear interacting units, namely opening micro-cracks. Self-organized complexity manifests itself in linkages between space and time producing fractal processes and structures. Herein, we concentrate on the question whether distinctive alterations in associated scaling parameters emerge as earthquakes are approaching.

We focus on the statistics of the detected electromagnetic fluctuations with respect to their amplitude, let's say $A(t_j)$. We attempt to investigate autocorrelation structures in these time-series. Any time series may exhibit a variety of autocorrelation structures; successive terms may show strong (brown noise), moderate (pink noise) or no (white noise) correlation with previous terms. The strength of these correlations provides useful information about the inherent "memory" of the system. The power spectrum, $S(f)$, which measures the relative frequency content of a signal, is probably the most commonly used technique to detect structure in time-series. If the time-series $A(t_j)$ is a fractal time series that series cannot have a characteristic time scale. But a fractal time series cannot have any characteristic frequency either. The only possibility is then that the power spectrum $S(f)$ has a scaling form:

$$S(f) \approx f^{-\beta} \tag{1}$$

where the power spectrum $S(f)$ quantifies the correlations at the time scale $\tau \approx 1/f$ and f is the frequency of the Fourier transform. In a $\ln S(f)–\ln(f)$ representation the power spectrum is a line with linear spectral slope β. The linear correlation coefficient, r, is a measure of the goodness of fit to the power law Eq. (1). Our approach is to calculate the fractal parameter β and the linear correlation coefficient r of the power law fit dividing the signal into successive segments of 1024 samples each, in order to study not only the presence of a power law $S(f) \approx f^{-\beta}$ but, mainly, the temporal evolution of the associated parameters β and r. The Continuous Wavelet Transform (CWT), using Morlet wavelet, is applied to compute the power spectrum, since being superior to the Fourier spectral analysis providing excellent decompositions within the maximum admitted window length (Kaiser, 1994).

Methods-Results

A convenient way to examine transient phenomena is to divide the measurements in time windows and analyze these windows. If this analysis yields different results for some precursory time intervals (epochs), then a transient behaviour can be extracted. We apply this technique for each of the methods used below.

We discriminate four epochs in the EM time series under study (Fig. 3). The first epoch refers to the electromagnetically quiescent period preceding the emergence of the EM anomaly. The second and third epochs include the precursory (possibly seismogenic) EM activity. We separate two time intervals during the detection of this EM anomaly, because we mainly search for the appearance of transient phenomena during the last preparation stage of the main shock. Finally, the fourth epoch refers to the period after the abrupt termination of the recorded EM anomaly.

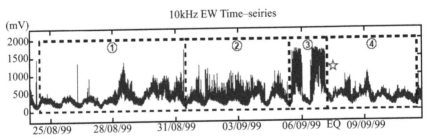

Figure 3 View of the time-series of 10kHz E-W. The four epochs in which the calculation are made are depicted.

Application of the Delay Times method

Through the use of the autocorrelation and mutual information functions, a value was determined for the time lag, τ, that is most suitable for this study and that was $\tau = 7$. The dimensions chosen for the phase space reconstruction started at $m = 3$ and went to $m = 20$. Both m and τ values were based on the fact that after several trials these values yield the best reconstruction and thus lead to more accurate results and subject discrimination. The correlation integral was then calculated for an extended range of r (up to 10^8, experimentally determined).

We calculate the correlation dimension, D_2, associated with successive segments of 3000 samples each and study the distributions of correlation dimension D_2 in four consecutive time intervals (Fig. 3). We recall that the recorded VLF EM anomaly of gradual increasing activity has been launched through a long duration kilohertz EM quiescence, while they ceased a few hours before the Athens earthquake. The first time interval corresponds to the quiescence EM period preceding the EM anomaly. The second and third time intervals correspond to the period of the recorded precursory anomaly; the third time interval includes the two strong impulsive bursts in the tail of the precursory emission. The fourth time interval refers to the quiescence period after the cessation of the precursory emission.

We underline the similarity of the distributions of the D_2-values in the first and fourth time intervals (Fig. 4). This almost common distribution characterizes the order of complexity in the background noise of the EM time series. The associated predominance D_2-values, from 7 up to 10, indicate a strong complexity and non linearity. Notice that a relevant lobe remains in the distributions of D_2-values in the second and third time interval, as it was expected.

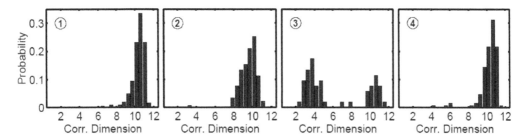

Figure 4 We first estimate the Correlation Dimension, D_2, in consecutive segments of 3000 samples each. Then, we trace the distribution of these D_2-values for four consecutive epochs. The four epochs are depicted in Fig. 3. The epochs 1 and 4 correspond to the EM quiescence that precedes and follows respectively the EM precursory activity. The almost similar distributions in the epochs 1 and 4 characterize the EM background (noise). In epoch 2, the little deformation of the distribution to the left side in respect to the distribution of the pure noise indicates that the initial part of the precursory emission is characterized by a little reduction of the complexity in respect to the high complexity of the pure noise. The right lobe that appears in the period 3 corresponds to the EM background, while the left lobe corresponds to the EM precursory activity. We observe a dramatic shift of the distribution of the D_2-values in epoch 3. This evidence indicates a strong reduction of complexity during the emergence of the two strong EM bursts in the tail of the precursory emission.

Now, we focus on the second and third time intervals, namely during the emergence of the precursory emission. We observe a significant decrease of the D_2-values as we move from the second to the third time window. The observed significant decrease of the D_2-values signals a strong loss of complexity in the underlying fracto-electromagnetic mechanism during the launching of the two strong EM bursts in the tail of the precursory emission. This evidence might be indicated by the appearance of a new phase in the tail of the earthquake preparation process, which is characterized by a higher order of organization. Sufficient experimental evidence seems to support the association of the aforementioned two EM bursts with the nucleation phase of the impending earthquake (Eftaxias et al. 2001; Kapiris et al. 2004b).

These findings suggest that there is important information in terms of correlation dimension hidden in the heterogeneities of the pre-seismic time series. The correlation dimension D_2 in the sequence of the precursory EM pulses seems to measure the distance from the global instability: the larger the D_2-values the larger the distance from the critical point.

Application of the Approximate Entropy method

The Approximate Entropy was computed for a variety of r-values proposed by previous researchers and it was found that the optimum value yielding clearest discrimination was the value $r = 0.65STD,$ where STD is the standard deviation of the time-series.

We calculate the Approximate Entropy associated with successive segments of 3000 samples each and study the distributions of the $ApEn$-values in four consecutive time intervals, as in the case of the study in terms of Correlation Dimension (Fig. 5).

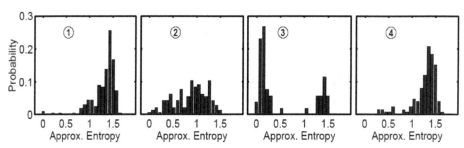

Figure 5 We first estimate the Approximate Entropy, $ApEn$, in consecutive segments of 3000 samples each. Then, we trace the distribution of these $ApEn$-values for four consecutive epochs that correspond exactly to the four epochs of Fig. 4. In the epoch 2, we observe an important shift of $ApEn$-values to lower values. This indicates that the emerged EM emission has a behavior far from this of the EM background. The right lobe that appears in epoch 3 corresponds to the EM background, while the left lobe corresponds to the EM precursory activity. We observe a dramatic reduction of complexity during the emergence of the two strong EM bursts in the tail of the precursory emission.

We observe again the similarity of the distributions of the $ApEn$-values in the first and fourth time intervals: this almost common distribution refers to the background noise of the EM time series. A relevant lobe remains in the distributions of the $ApEn$-values in the second and third time interval, as it was expected.

Now, we concentrate on the second and third time intervals, namely during the emergence of the precursory emission. We observe a significant decrease of the $ApEn$-values as we move from the second to the third time window. The observed considerable decrease of the $ApEn$-values in the third time interval reveals a strong loss of complexity in the underlying mechano-electromagnetic transduction during the launching of the two strong EM bursts in the tail of the precursory emission. In other words, the pre-focal area seems to be less responsive to the external stimuli when the pre-seismic EM signals are characterized by low *ApEn*-values.

In summary, in the pre-seismic EM time-series the values of the Correlation Dimension and Approximate Entropy are reduced as the main event is approached. This evidence indicates that the underlying fracto-electromagnetic mechanism exhibits a strong complexity and non-linearity far from the global failure. A significant loss of complexity and non-linearity is observed close to the global instability. This considerable alteration in both D_2-values and *ApEn*-values might be considered as candidate precursor of the impending event.

Remark

According to the appendixes, the method of Correlation Dimension (CorrDim) embeds the original time series into a phase space of dimension 3 to 20, examining thus the probability distribution of a norm defined in this phase space, contrary to the Approximate Entropy method which embeds the original time series in a 2-dimensional phase space only. As a result, the CorrDim method yields a more detailed description of a system's complexity, comparing to the *ApEn* method, which focuses mainly to coarse grained characteristics. When the examined time series is generated by a low dimensional process, it is better to use the *ApEn* method. On the other hand, when the complexity of the examined time series increases, the CorrDim method is more suitable as it is more sensitive to high complexity. In the cases of epochs 2 and 3 in Figs. 4 and 5, we are able to observe the above mentioned property. Focusing to epoch 2 in Fig. 5, the shift of *ApEn*-values to lower values witnesses the reduction of complexity, and thus the emergence of the precursor. In epoch 3 the complexity has further been diminished. Thus the *ApEn* method is the proper one to describe the associated grouping activity of the structures. Indeed, we observe that the probability distributions of *EQ-D_2* and Background-D_2 values seem to be similar (epoch 3 in Fig. 4), while in the case of *ApEn* method, the probability of *EQ-ApEn* values is larger than the Background *ApEn*-values.

Application of Fractal-dynamics

The spectral fractal analysis reveals that the pre-seismic electromagnetic fluctuations exhibit hidden scaling structure. We observe alterations in the associated dynamical parameters, which seem to uncover important features of the underlying earthquake preparation process (Kapiris et al. 2002; Eftaxias et al. 2003; Kapiris et al. 2003, 2004b,a; Eftaxias et al. 2004).

Figure 6 exhibits the temporal evolution of r as the main event is approached. We observe a gradual increase of the correlation coefficient with time: at the tail of the precursory activity the fit to the power law is excellent. The fact that the data are well fitted by the power-law (1) suggests that the pre-seismic EM activity could be ascribed to a multi-timescale cooperative activity of numerous activated fundamental units, namely, emitting-cracks, in which an individual unit's behavior is dominated by its neighbours so

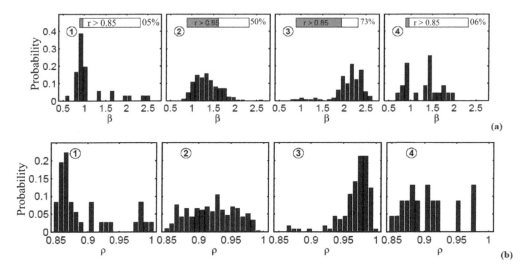

Figure 6 (a) We first estimate the exponent β, in consecutive segments of 1024 samples each. Then, we trace the distribution of these β-values for four consecutive epochs. The four epochs are depicted in Fig. 3 and correspond exactly to the four epochs of Fig. 4 and Fig. 5. Insets show percentage of segments with *r>0.85*. It is evident that the closer the final stage of seismic process, the larger the percentage of segments with *r* > 0.85 and the larger shift of β to higher values. Notice that in epoch 3 the signal becomes persistent. **(b)** The propability distributions of linear coef. *r* beyond *0.85*.

that all units simultaneously alter their behavior to a common large scale fractal pattern. On the other hand, the gradual increase of *r* indicates that the clustering in more compact fractal structures of activated cracks is strengthened with time.

Now we focus on the behavior of β-exponent. Two classes of signal have been widely used to model stochastic fractal time series (Heneghan and McDarby, 2000): fractional Gaussian noise (fGn) and fractional Brownian motion (fBm). These are, respectively, generalizations of white Gaussian noise and Brownian motion. The nature of fractal behavior (i.e. fGn versus fBm) provides insight into the physical mechanism that generates the correlations: the fBm represents cumulative summation or integration of a fGn. A formal mathematical definition of continuous fBm was first offered by (Mandelbrot and Ness, 1968).

For the case of the fBm model the scaling exponent β lies between 1 and 3, while, the range of β from −1 to 1 indicates the regime of fGn (Heneghan and McDarby, 2000). Figure 6 reveals that during the epochs 2 and 3 (Fig. 3) the β-values are distributed in the region from 1 to 3. This means that the possible seismogenic EM activity follows the fBm model. We concentrate on the quiescent EM period (first epoch in Fig. 3), preceding the emergence of the EM anomaly (second and third epochs in Fig. 3). We observe that only a very small number of segments, approximately 5%, follows the power law (1) (see inserts in Fig. 6). We can conclude that during the epoch 1 the associated time series do not behave as a temporal fractal. Moreover, if we concentrate on the 5% of the segments, the associated β-values range from 0 to 1, namely, this minority of segments may follow the fGn model.

We conclude that regime of the quiescent period is quite different from those of the possible seismogenic emission. The transition to the fractal structure and fBm class further identify the launch of the fracto-electromagnetic emission from the background (noise) of EM activity.

The distribution of β-exponents is also shifted to higher values (Fig. 6) during the precursory period.

The precursory shift of the distribution of both β-exponent and *r*-coefficient to higher values reveals important features of the underlying mechanism. The fractal-laws observed corroborate to the existence of memory; the system refers to its history in order to define its future. As the β-exponent increases the spatial correlation in the time-series also increases. This behaviour signals the gradual increase of the memory, and thus the gradual loss of complexity in the process. Maslov et al. (1994) have formally established the relationship between spatial fractal behaviour and long-range temporal correlations for a broad range of critical phenomena. By studying the time correlations in the local activity, they show that the temporal and spatial activity can be described as different cuts in the same underlying fractal. Laboratory results support this hypothesis: Ponomarev et al. (1997) have reported in phase changes of the temporal and spatial Hurst exponents during sample deformation in laboratory acoustic emission experiments. Consequently, the observed increase of the temporal correlation in the pre-seismic time-series may also reveal that the opening-cracks are correlated at larger scale length with time.

The following feature goes to the heart of the problem: first, single isolated micro-cracks emerge which, subsequently, grow and multiply. This leads to cooperative effects. Finally, the main shock forms. The challenge is to determine the "critical time-window" during which the "short-range" correlations evolve into "long-range" ones. Figure 6a indicates that the closer the global instability the larger the percentages of segments with *r* close to 1 and the larger the shift of β-exponent to higher values; the β-values are maximal at the tail of the pre-seismic state (Fig. 6b). This behaviour may reveal the "critical time-window".

The exponent β is related to the Hurst parameter, *H*, by the formula (Turcotte 1992)

$$\beta = 2H + 1 \text{ with } 0 < H < 1 \text{ and } 1 < \beta < 3 \tag{2}$$

for the fBm model (Mandelbrot and Ness 1968; Heneghan and McDarby 2000). Consequently, segments with Hurst exponents estimated by the previous formula out of the range $0 < H < 1$ do not follow the fBm model.

The exponent *H* characterizes the persistent/antipersistent properties of the signal according to the following scheme. The range $0.5 < H < 1$ ($2 < \beta < 3$) suggests persistence of the signal, i.e., if the amplitude of fluctuations increases in a time interval it is likely to continue increasing in the interval immediately following. The range $0 < H < 0.5$ ($1 < \beta < 2$) suggests anti-persistence of the signal, i.e., if the fluctuations increase in a period, it is likely to continue decreasing in the interval immediately following and vice versa. $H = 0.5$ ($\beta = 2$) indicates no correlation between the process increments, that is the system is characterized by random fluctuations (Mandelbrot and Ness 1968; Hristopulos 2003). Consequently, the particular value $\beta = 2$ takes on a special physical meaning: it signals the transition from anti-persistent to persistent behaviour in the time series.

Remark

As it was mentioned, the range $-1 < \beta < 1$ implies fGn behavior. For this model the exponent β is related to the Hurst parameter by the formula $\beta = 2H - 1$ (Heneghan and McDarby, 2000). Thus, H is also constrained to lie between 0 and 1 in epochs 1 and 4 (Fig. 6).

Physically, the Hurst exponent express the strength of the effect of excitation associated with the preceding event on succeeding events, or equivalently, the degree of negative/positive feedback in the dynamics. The range $0 < H < 0.5$ ($1 < \beta < 2$) during the first period of the EM precursor indicates a very large anti-persistency. This behaviour implies a set of fluctuations tending to induce a greater stability in the system. The observed shift of local H-exponents can be understood if we accept that the micro-heterogeneity of the system becomes less anti-correlated with time. The anti-persistent properties during this period are consistent with the existence of a non-linear feedback mechanism that "kicks" the cracking rate in the pre-focal area away from extremes. The systematic increase of the β-exponent (or Hurstexponent) indicates that the fluctuations become less anticorrelated with time, i.e., the nonlinear negative feedbacks gradually lose their ability to kick the system away from extremes. In other words, the decrease of heterogeneity appears to lead to a decrease in the ability to drive the system away from a persistent mode of opening-cracks evolution. It might be argued that "the first anti-persistent part of the precursory electromagnetic radiation is triggered by micro-fractures in the disordered system that surrounds the, almost, homogeneous backbones within the pre-focal area".

For times close to breakthrough, almost homogeneous backbones of high strength sustain the elastic strain energy. In the limit of a homogeneous system, once a crack nucleates in the rock, the stress is enhanced at its tip and therefore the next micro-crack almost surely develops at the tip. The appearance of persistence properties within the two strong impulsive signals at the tail of the precursory time series is thought to be clue to the fracture of the high strength homogeneous backbones (Kapiris et al. 2004b; Eftaxias et al., 2004; Kapiris et al. 2004a). This behaviour may witness that the system has been starting to self-organize by a positive feedback process, and thus, this acquires to a great degree the property of irreversibility. The concept that the launch of the persistence activity could give a significant hint of a considerable probability for a forthcoming global instability, namely, a significant event, can be accepted.

We have paid attention to the following experimental evidence: The accelerating EM precursor (Fig. 3) ends in two clear persistent signals with an energy ratio (second to first signal) ≈ 5. The radar interferometry analysis showed activation of two separate faults with corresponding energy release ratio (second to first signal) ≈ 5 (Eftaxias et al. 2001; Kapiris et al. 2004a). This surprising correlation in the energy domain enhances the consideration that the launch of persistent dynamics may signals the emergence of the nucleation stage of earthquake preparation.

It is worth mentioning that laboratory experiments by means of acoustic and electromagnetic emission also show that the main rupture occurs after the appearance of persistence behaviour (Ponomarev et al. 1997; Alexeev and Egorov 1993; Alexeev et al. 1993) in the time-series.

We conclude that the aforementioned three methods present coherent results: they clearly and accurately identify significant alteration in terms of D_2-values, *ApEn*-values, and Hurst-values, between the initial and terminal phase of the pre-seismic EM emission. The coherent results signal a significant loss of complexity in the tail of the precursory EM activity, while the underlying fracto-electromagnetic mechanism becomes persistent. The dynamical parameters, D_2, *ApEn*, and *H*, seems to represent a measure of the distance of the system from the global instability. The lower the D_2-values, the lower the *ApEn*-values, and the higher the Hurst-values, then, the smaller the distance of the system from the "critical point" (global instability).

Remark

A fundamental characteristic of probable EM precursors is their appearance in a wide frequency band, ranging from DC-ULF, ELF, VLF, and LF to VHF. The time elapsed from the detection of the DC-ULF pre-seismic EM activity to the occurrence of strong EQ (Varotsos et al. 1996; Hayakawa et al. 1999, 2000; Telesca and Lapenna 2001; Varotsos and Sarlis 2002; Ramirez-Rojas et al. 2004), is longer than the time for the VLF-VHF emissions, e.g., (Gershenzon and Bambakidis 2001; Eftaxias et al. 2001, 2002). Therefore, we can accept the concept that pre-seismic signals may arise from different mechanisms, which do not lie within the same time scales.

Authors have studied pre-seismic ULF geo-magnetic (Hayakawa et al. 1999, 2000) and geo-electrical signals (Telesca and Lapenna 2001) in terms of fractal spectral analysis. Regarding the behavior of the spectral-exponent observed, it seems that this is in contrast to the behavior observed during the present study. Characteristically, the β-values showed a tendency to gradually decrease during the process of the earthquake preparation. We think that this difference supports the hypothesis that the ULF signals on one hand and the VLF-VHF signals on the other hand may have originated on different mechanisms. Indeed, it has been suggested that the ULF geo-electical signals could be explained in these terms: (i) "Pressure Stimulated Currents" that are transient currents emitted from a solid containing electric dipole upon a gradual variation of pressure (Varotsos and Alexopoulos 1984a,b; Varotsos et al. 1996). (ii) The electro-kinetic effect, e.g., (Mizutani and Ishido 1976; Dobrovolsky et al. 1989; Gershenzon and Bambakidis 2001). Because electro-kinetic effect is controlled by the diffusion of water with the diffusion time comparable to the period of ULF emissions, more energy is provided to the ULF range (Gershenzon and Bambakidis 2001). We note that recently, (Surkov et al. 2002) have explained the logarithmic dependence of electric field amplitude E on the earthquake magnitude M that is indicated by experimental results (Varotsos et al. 1996).

From the Normal State to the Seismic Shock or Epileptic Seizure in Terms of Complexity

The world is made of highly interconnected parts on many scales, the interactions of which results in a complex behavior that require separate interpretation of each level. The laws that describe the behavior of a complex system are qualitatively different from those that govern its units. New features emerge as one moves from one scale to another, so it follows that the science of complexity is about revealing the principles that govern the ways in which these new properties appear.

A basic reason for our interest in complexity is the striking similarity in behaviour close to irreversible phase transitions among systems that are otherwise quite different in nature (Stanley 1999, 2000; Sornette 2002; Vicsek 2001, 2002; Turcotte and Rudle 2002). Recent studies have demonstrated that a large variety of complex processes, including earthquakes (Bak and Tang 1989; Bak 1997), forest fires (Malamud et al. 1998), heartbeats (Peng et al. 1995), human coordination (Gilden et al. 1995), neuronal dynamics (Worrell et al. 2002), financial markets (Mantegna and Stanley 1995) exhibits statistical similarities, most commonly powerlaw scaling behaviour of a particular observable. Stanley (2000) offer a brief and somewhat parochial overview of some "exotic" statistical physics puzzles of possible interest to biophysicists, medical physicists, and econophysics.

Interestingly, authors have suggested that earthquake's dynamics and neurodynamics could be analyzed within similar mathematical frameworks (Rundle et al. 2002). Characteristically, slider block models are simple examples of driven non-equilibrium threshold systems on a lattice. It has been noted that these models, in addition to simulating the aspects of earthquakes and frictional sliding, may also represent the dynamics of neurological networks (Rundle et al. 1995, and references therein). A few years ago, Bak et al. (1987) coined the term self-organized criticality (SOC) to describe the phenomenon observed in a particular automaton model, nowadays known as the sandpile-model. This system is critical in analogy with classical equilibrium critical phenomena, where neither characteristic time nor length scales exist. In general, the strong analogies between the dynamics of the "self-organized-criticality" (SOC) model for earthquakes and that of neurobiology have been realized by numerous of authors (Hopfield 1994; Herz and Hopfield 1995, and references there in; Usher et al. 1995; Zhao and Chen 2002, and references there in; Beggs and Plenz 2003).

Complexity does not have a strict definition, but a lot of work on complexity centers around statistical power laws, which describe the scaling properties of fractal processes and structures that are common among systems that at least qualitative are considered complex. The big question is whether there is a unified theory for the ways in which elements of a system organize themselves to produce a behavior that is followed by a large class of systems (Vicsek 2002). The aforementioned concepts motivated us to investigate whether common precursory patterns are emerged during the precursory stage of both epileptic seizure and earthquake (Li et al. 2004).

The brain possesses more than billions neurons and neuronal connections that generate complex patterns of behaviour. Electroencephalogram (EEG) provides a window, through which the dynamics of epilepsy preparation can be investigated. Figure 7 exhibits rat epileptic seizure.

As in the case of the pre-seismic EM emission, we monitor the evolution of fractal characteristics of pre-epileptic activities toward criticality in consecutive time windows. Our analysis reveals that numerous distinguishing features were emerged during the transition from normal states to epileptic seizures (Li et al. 2004): (i) appearance of long range power-

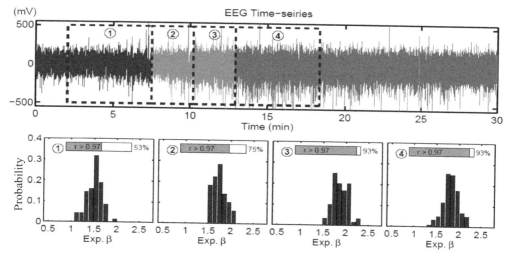

Figure 7 A rat epileptic seizure (red signal) in EEG time-series (upper part). Two electrodes were placed in epidural space to record the EEG signals from temporal lobe. EEG signals were recorded using an amplifier with band-pass filter setting of 0.5–100 Hz. The sampling rate was 200 Hz. Bicuculline i.p injection was used to induce the rat epileptic seizure. The injection time is at 7:49 (m-s) and the seizure at 12:55 (m-s), respectively. This pre-seizure period is depicted by the yellow part of the EEG time series. We estimate the exponent β, in consecutive segments of 1024 samples each. Then, we trace the distribution of these β-values for four consecutive epochs (lower part). The four epochs are depicted in the upper part with numbered dashed frames. Insets show percentage of segments with $r > 0.97$. Notice that at the last stage of the pre-ictal period (epoch 3) the signal emerges persistent behavior.

law correlations, i.e., strong memory effects; (ii) increase of the spatial correlation in the time-series with time; (iii) gradual enhancement of lower frequency fluctuations, which indicates that the electric events interact and coalesce to form larger fractal structures; (iv) decrease of the fractal dimension of the time series; (v) decrease with time of the anti-persistent behavior in the precursory electric time series; (vi) appearance of persistent properties in the tail of the pre-epileptic period. Figure 7 shows the aforementioned precursory behavior.

Notice that the aforementioned candidate precursors of the impending epileptic seizure or earthquake are launched in a way striking similar to those occurring just before the "critical point" of phase transition in statistical physics. Based on this similarity, it might be argued that the earthquake/epilepsy may be also viewed as "a generalized kind of phase transition" (Kapiris et al. 2004b; Contoyiannis et al. 2004).

Our results indicate that an individual firing neuron or an opening crack is dominated by its neighbours so that all activated biological or geological units simultaneously alter their behavior to a common fractal pattern as the epileptic seizure or the earthquake is approaching. Interestingly, common alterations in the associated parameters are emerged indicating the approach to the global instability in harmony with relevant theoretical suggestions (Hopfield 1994; Herz and Hop-field 1995; Usher et al. 1995; Zhao and Chen 2002; Rundle et al. 2002; Beggs and Plenz 2003). Consequently, the present analysis seems to support the concept that, indeed, a unified theory may describe the ways in which elements of a biological or geological system organize themselves to produce a catastrophic event.

From the Normal State to the Seismic Shock or Heartfailure in Terms of Correlation Dimension, Approximate Entropy and Multifractality

Recently, Fukuda et al. (2003) have investigated similarities between communication dynamics in the Internet and the autonomic nervous system. They found quantitative similarities between the statistical properties of (i) healthy heart rate variability and non-congested Internet traffic, and (ii) diseased heart rate variability and congested Internet traffic. The authors conclude that their finding suggest that the understanding of the mechanisms underlying the "human-made" Internet could help to understand the "natural" network that controls the heart. In the sense of this approach, we search for similarities from the normal state to the seismic shock or heart-failure.

Similarities in Terms of Multi-fractality

Mathematical analysis of both long-term heart-rate fluctuations (Ivanov et al. 1999; Stanley et al. 1999; Goldberger et al. 2002) and pre-seismic EM emissions (Kapiris et al. 2004b) show that they are members of a special class of complex processes, termed multi-fractals, which require a large number of exponents to characterize their scaling properties. In general, the detection of multi-fractal scaling may indicate that the underlying nonlinear mechanism regulating the system might interact as part as a coupled cascade of feedback loops in a system operating far from equilibrium (Meneveau and Sreenivasan, 1987).

Mono-fractal signals can be indexed by a single global exponent, i.e., the Hurst exponent H (Hurst 1951). Multi-fractal signals, on the other hand, can be decomposed into many subsets characterized by different local Hurst exponents h, which quantify the local singular behavior and thus relate to the local scaling to the time series. Thus, multi-fractal signals require many exponents to characterize their scaling properties fully (Vicsek 1993). The statistical properties of the different subsets characterized by these different exponents h can be quantified by the function $D(h)$, where $D(h_0)$ is the fractal dimension of the subset of the time series characterized by the local Hurst exponent h_0. Ivanov et al. (1999) have uncovered a loss of multi-fractality, as well as a loss of the anti-persistent behaviour, for a life-threatening condition, congestive heart failure. Following the method of multi-fractal analysis used by Ivanov et al. (1999), we examined multi-fractal properties in the VHF time series, namely, the spectrum of the fractal dimension $D(h)$, as a candidate precursor of the Kozani-Grevena earthquake (Kapiris et al. 2004b). Figure 8 shows that as the main event approaches, the EM time series manifest: a significant loss of multi-fractal complexity and reduction of nonlinearities, displaying a narrow (red) multi-fractal spectrum, and their fluctuations become less anti-correlated, as the dominant local Hurst exponents is shifted to higher values. These results reflect that for both the heart and pre-focal area at high risk the multi-fractal organization almost breaks down.

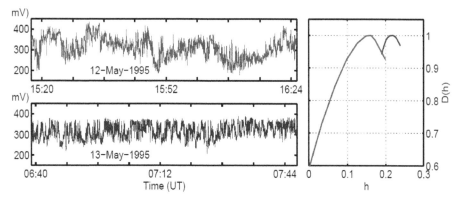

Figure 8 Two segments of the precursory 41MHz electromagnetic signal, recorded on 12 May 1995 (upper row) and 13 May 1995 (lower row) before the Kozani-Grevena earthquake *Ms=6.6* on May 13, 1995 at 08:47:12.9UTC. On the right part of the figure the corresponding fractal dimensions *D(h)* are presented.

In summary, the multi-fractality of the heart-beat time series and pre-seismic EM time series further enables us to quantify the greater complexity of the "healthy" dynamics compared to those of "pathological" conditions in both heart and pre-focal area.

Similarities in terms of Correlation Dimension and Approximate entropy

Recently, we have studied several methods which have been used for the categorization of two subjects groups, one which represents subjects with no prior occurrence of coronary disease events and another group who have had a coronary disease event (Nikolopoulos et al. 2003; Karamanos et al. 2004). It is worth mentioning that the Delay Times method and the computation of the Approximate Entropy present coherent results and succeed in clearly and accurately differentiating healthy subject ECGs from those of unhealthy subjects and coronary patients. Heart Rate Variability (HRV) time series coming from coronary patients exhibit more regular and periodical behaviour compared to ones coming from healthy subjects. The correlation dimensions of healthy time series are about $D2 \approx 9$ when the respective ones for the patients are about $D2 \approx 6$ (Nikolopoulos et al. 2003). Similarly, the mean *ApEn* value for the healthy time series was about ApEn ≈ 1.2 and for the patients ApEn ≈ 0.4 (Nikolopoulos et al. 2003). A similar reduction of complexity for heart failures has been observed in terms of Block-Entropy by some of the present authors (Karamanos et al. 2004). It is important to note that the D_2-values and *ApEn*-values associated with the second time interval of the pre-seismic EM time series are close to the ones coming from healthy subjects, while, the D_2-values and *ApEn*-values associated with the third time interval are close to the ones coming from patient subjects. Based on this analogy, we could say that the EM emissions in second and third time interval implies a kind of "healthy" and "patient" pre-focal area correspondingly. We focus on this analogy.

We recall that the EM time series in the second time interval is characterized by strong anti-persistence and multi-fractality. The multi-fractality indicates that the underlying nonlinear mechanism regulating the system might interact as part as a coupled cascade of feedback loops in a system operating far from equilibrium (Meneveau and Sreenivasan

1987). The anti-persistent properties during this period imply a set of fluctuations tending to induce a greater stability in the system. Thus, by the term "healthy pre-focal area", we mean a candidate focal area, which is consistent with a non-linear negative feedback system that "kicks" the cracking rate away from extremes.

By the term "patient pre-focal area", we mean a pre-focal area in which the system has been starting to self-organize by a non-linear positive feedback process, and thus, this acquires to a great degree the property of irreversibility. This behaviour may imply that the nucleation stage, the most interesting phase in the preparation process of the catastrophic fracture, has already been emerged. The study of pre-failure EM signals seems to provide a way for observing the Earth's crust ability to respond to stresses. Hallmarks of the "patient pre-focal area" are: the persistence behavior, the low multi-fractality, the low Correlation Dimension and the low Approximate Entropy. These hallmarks characterize a life-threatening condition for the human heart, too. Briefly, the "patient pre-focal area" and the "patient human heart" are characterized by a low complexity. On the other hand, signatures indicating the "healthy pre-focal area" are: the anti-persistence behavior, the high multi-fractality, the high Correlation Dimension, and the high Approximate Entropy. These signatures also characterize healthy human heartbeat. Briefly, the "healthy pre-focal area" and the healthy human heart are characterized by high complexity.

In a geometrical sense, the dynamical parameter β specifies the strength of the signal's irregularity as well. Qualitatively speaking, the irregularity of the signal decreases as the memory in the time-series increases. For the fBm model the fractal dimension d is found from the relation $d = \dfrac{5 - \beta}{2}$, which, after considering the aforementioned shift of β exponent to higher values, leads to a decrease of fractal dimension as the earthquake approaches. We recall the West-Goldberger hypothesis that a decrease in healthy variability of a physiological system is manifest in a decreasing fractal dimension (Goldberger et al.,2002, and references there in). Our results imply that this hypothesis could be extended to geological systems as well.

Conclusions

A method to asses the approach to the global instability has been applied in EM pre-seismic anomalies. The study of these pre-failure signals seems to provide a way for observing the Earth's crust ability to respond to stresses. The Delay Times method, the computation of the Approximate Entropy, and the monitoring of alteration of Fractal Spectral characteristics of pre-seismic EM activity toward global instability in consecutive time windows, present coherent results and succeed in a potential differentiation of the nucleation phase from previous stages of the earthquake preparation process. More precisely, the emergence of long-range correlations, i.e. appearance of long memory effects, the increase of the spatial correlation in the time series with time, the predominance of large events with time, as well as the gradual decrease of the anti-persistent behaviour may indicate the approach to the nucleation phase of the impending catastrophic event. The appearance of persistent properties in the tail of the precursory time series, the significant divergence of the energy release, the sharp significant decrease of the Approximate Entropy, and the quick reduction of the Correlation Dimension as well, all these, may hints that a new phase, probably the nucleation phase of the earthquake, has been started. This analysis may provide a useful way to the understanding of the fracture in the disordered media. The agreement of the "diagnostic" information given by each one of the methods indicates the necessity of further investigation, combined use, and complementary application of different approaches.

The performed analysis reveals that common precursory signs emerge in terms of fractal dynamics as the epileptic seizure and earthquake are approaching: common distinctive alterations in associated scaling dynamical parameters emerge as biomedical or geophysical shock is approaching. The experimental results verify relevant theoretical suggestions that earthquake dynamics and neural seizure dynamics should have many similar features and should be analyzed within similar mathematical frameworks (Hopfield 1994; Herz and Hopfield 1995; Usher et al. 1995; Zhao and Chen 2002; Rundle et al. 2002; Beggs and Plenz 2003).

In principle, it is difficult to prove associations between events separated in time, such as EQs and their precursors. The present state of research in this area requires a refined a definition of a possible pre-seismic anomaly in the record of EM radiation, and also the development of more objective methods of distinguishing seismogenic emissions from non-seismic EM events. A study in terms of complexity would seem to be useful in this regard. EEG time-series provide a window through which the dynamics of biological shock preparation can be investigated in the absence of non-biological events. We observe that both kinds of catastrophic events under investigation follow common behavior in their pre-catastrophic stage. This evidence may support the seismogenic origin of the detected EM anomaly.

We find also quantitative similarities between the properties of (i) healthy heart rate variability and initial antipersistence part of the pre-seismic EM time series, and (ii) diseased heart rate variability and terminal persistence part of the pre-seismic EM activity. These similarities have been emerged in terms of Correlation Dimension, Approximate entropy, and multifractal dynamics.

Fukuda et al. (2003) recall that very simple models of very complex systems in many cases provide deep insights. For example the Ising model and its simple variants as the Heisenberg model are sufficient to quantitatively describe a wealth of very complex systems in regions of their respective phase diagrams where scale invariant is displayed. The principle of "universality" in chemistry and physics, whereby diverse systems are described by the identical (simple) model, may have its counterpart in physiology (Stanley 1999). Even the numerical values of the critical-point exponents describing the quantitative nature of the singularities are identical for large groups of apparently diverse physical systems. It was found empirically that one could form an analog of the Mendeleev table if one partitions all critical systems into "universality classes". Two systems with the same values of critical-point exponents and scaling functions are said to belong to the same universality class. In the frame of this approach we have shown that the pre-seismic VHF emission belongs to the 3D-Ising-transition class (Contoyiannis et al. 2004). Fukuda et al.

(2003) argue that their finding suggest that the understanding of the mechanism underlying the "human-made" internet could help to understand the "natural" network that controls the heart. In this sense, it appears that the fracture in the disordered systems may provide another useful "model system" to investigate the mechanism responsible for the dynamics of the autonomic nervous system (ANS), which controls involuntary the heart or the epilepsy generation. In terms of complexity, this possibility is not implausible.

The science of complexity is in its infancy, and some research directions that today seem fruitful might eventually prove to be academic cul-de-sacs. Sethna et al. (2001) show that the seemingly random, impulsive events by which many physical systems evolve exhibit universal, and , to some extend, predictable behavior. Nevertheless, it is reasonable to believe that the results of the present study indicate that it is useful to transfer knowledge from the domain of biomedical shock preparation to the domain of earthquake generation and vice versa. This work could serve as an invitation to other specialists in these areas to transfer knowledge from the one field of research to the other.

Appendix A: Delay Times method

Given a timeseries, $X(t)$, t is an integer, $t \in (1, N)$ and N is the total number of timeseries points. The Delay Times method was first established by Grassberger and Procaccia (1983) and based on the Takens Theorem (Takens, 1981). According to this method, the timeseries $x(t)$ is a measure of a single coordinate of an m-dimensional system's underlying dynamics. Assuming m is the embedding dimension (the dimension of space in which the assumed system's trajectory is unfolded) and τ is the time lag, then phase space reconstruction (described below) is performed with time delays and the following m-dimensional vectors are constructed:

$$x(t) = \left[X(t), X(t-r), X(t-2r), ..., X(t-(m-1)\tau) \right] \tag{A1}$$

In this way, using the original timeseries, $X(t)$, we are able to construct a new vector timeseries, $x(t)$, which represents the trajectory from $x(0)$ up to and including $x(t)$ within the reconstructed phase space.

These vectors are defined in an m-dimensional phase space and are used in constructing the trajectory of the signal dynamics to this space. If the original phase space of the dynamics produce the attractor A, then the reconstruction of the phase space with the Delay Times method produces the reconstructed attractor, A'. If the reconstruction is accurate, then A' is the topological conjugate of the original attractor, A. Consequently, all dynamic properties of A are projected to A'. The criterion of the Takens Theorem (Takens, 1981) for a precise phase space reconstruction of an experimental trajectory dictates that m must be greater then [$2mc+1$], where mc is the estimated dimension of the attractor.

According to the Takens Theorem, this is efficient when the number of points of the timeseries, N, is infinite, meaning that for an infinite number of points, A and A' have the same properties. However, for most experimental methods, N is a finite number and in many cases is confined to 3000–4000 points. Therefore, only A' is estimated in the reconstructed space and retains only some of the properties of A (not all). Essential to phase space reconstruction, especially for the Delay Times method, is the estimation of the time lag, τ. There is a range of methods for estimating τ, the most popular being the calculation of the decorrelation time.

The decorrelation time is calculated either from the first zero-value of the Autocorrelation function or from the first minimum value of the mutual information function. The Autocorrelation function has been described in the previous section. The mutual information method is widely accepted and it computes the nonlinear and linear correlation of the Autocorrelation function. Once the Autocorrelation function has been normalized, the decorrelation time is found from the smallest time lag for which the function tends to zero. Similarly, the decorrelation time can also be found from the smallest time lag for which the mutual information function tends to zero.

According to Broomhead and King (1986); Albano et al. (1988); Kugiumtzis (1996) the results of a time-series analysis depends on the window length $(m-1)\tau$, which incorporates both the embedding dimension m and the time lag τ. Therefore, the constraint to the above methods is the limit on the size of the window, $(m-1)\tau$. A proper value for the window size provides good phase space reconstruction and ensures that all the points of the reconstructed phase space come from the same trajectory. As mentioned above, the Takens theorem dictates that proper phase space reconstruction is achieved when m is greater than $[2mc+1]$. This criterion is difficult to satisfy for increased values of τ due to the subsequently larger values of $(m-1)\tau$. A consistent window arises from the decorrelation time, seen as the time needed for the first decay of the Autocorrelation function. *A* time lag, τ, is chosen and the reconstructed dynamics are embedded in the m-dimensional phase space.

After the phase space reconstruction of the system's assumed dynamics, nonlinear dynamics algorithms are developed for the experimental analysis of a timeseries. The most popular algorithmic method is the Delay Times method, also known as the Algorithm of Grassberger and Procaccia (1983), which estimates the Correlation Dimension from the computation of the correlation integral.

The Grassberger Procaccia Algorithm (Grassberger and Procaccia 1983) assumes a time-series, *X(i)*, which is a measure over time *i* of a parameter of an m-dimensional dynamic system, for $i \in [1, N]$. The phase space reconstruction of this system is done according to the Takens theorem. Once again, the vector coordinates are constructed as in Eq. (8) and it is assumed that this vector is the trajectory vector of the ith time point of the reconstructed phase space of the dynamic system. The whole trajectory is $x(1), x(2),...,$ $x(i), P...., x(\rho)$ where $\rho = N - (m-1)\tau$. As mentioned and according to the Takens theorem, A' is the attractor to the reconstructed system dynamics and the topological conjugate to the original attractor *A*. Properties such as the Correlation Dimension are maintained after the projection of *A* to A'. The Correlation Dimension is defined as

$$D_2 = \lim_{N \to \infty} \frac{\log(C(m,r,\tau))}{\log(r)} \tag{A2}$$

where *r* is a distance radius in the reconstructed phase space. The index 2 in D_2 is used because the Correlation Dimension is a special case of the generalized dimension D_q where *q* integer. $C(m,r,\tau)$ is the correlation integral and is defined as Θ

$$C(m,r,\tau)=\frac{2}{N-1}\sum_{i=1}^{N}\sum_{j=i+1}^{N}\Theta\left[r-\left\|x_i-x_j\right\|\right],$$

(A3)

where x_i and x_j are as in Eq. (8). Θ is the Heavyside function:

$$\Theta=\begin{cases}1, & if\ i\geq 0\\ 0, & if\ i\leq 0\end{cases}$$

(A4)

The Euclidean norm used in the above equation states that the difference between xi and xj is the maximum difference among their coordinates:

$$\left\|x_i-x_j\right\|=\left\{x(i)-x(j)\right|^2+\left|x(i+\tau)-x(j+\tau)\right|^2+...+\left|x(i+(m-1)\tau)-x(j+(m-1)\tau)\right|^2\right\}^{\frac{1}{2}}$$

(A5)

The formula Eq. (10) simply says: for specific $m\ r,\tau$ find all pairs of x_i and x_j in the reconstructed time-series $x(t)$ for which the distance $\left\|x_i-x_j\right\|$ is smaller than r.

According to this algorithm the $C(m,r,\tau)$ is computed for increasing values of m and for a steady range of r. For each $\log(C)$ versus $\log(r)$plot a scaling region is been selected and the slope of the curve is calculated for this scaling region with a simple method (i.e., least squares). If the slope values estimated for each m converge in a steady value, then this steady value corresponds to the correlation dimension of the timeseries.

Appendix B Approximate Entropy

Given N data points, $X(1),X(2),X(3),.....,X(N)$, the $ApEn(m,r,N)$ is estimated, where r is a threshold and m a window size. The vector sequences necessary for phase space reconstruction, $x(i)$, are constructed with $x(N-m+1)$, defined by $x(i)=[X(i),...,,X(i+m-1)]$. These vectors represent m consecutive X values, using the ith point as the starting point. The distance $\left\|x(i)-x(j)\right\|$ is defined between the vectors $x(i)$ and $x(j)$ as the infinity norm

$$\left\|x(i)-x(j)\right\|=\max\left\{x(i)-x(j)\right|,\left|x(i+1)-x(j+1)\right|,...,\left|x(i+(m-1))-x(j+(m-1))\right|\right\}$$

(B1)

The probability that $\left|x(i+(m-1))-x(j+(m-1))\right|\leq r$given that $\left|x(i)-x(j)\right|\leq r$ and $\left|x(i+1)-x(j+1)\right|\leq r$ and $\left|x(i+2)-x(j+2)\right|\leq r$ and...is true is termed $C_r^m(i)$, where, once again, r is the a threshold and m the window size. For example, if $m=2$, $C_r^2(i)$ for $i=1,...,N$ is the probability that $\left|x(i+1)-x(j+1)\right|\leq r$ given that $\left|x(i)-x(j)\right|\leq r$.

The sequence in Eq. (14) is used to construct the $C_r^m(i)$ for each $i\leq N-m+1$ as in

$$C_r^m(i)=\frac{\left[no.of\ j\leq N-m-1,\ such\ that\ \left\|x(i)-x(j)\right\|\leq r\right]}{N-m-1}$$

(B2)

$\Phi^m(r)$is defined as

$$\Phi^m(r)=\frac{1}{N-m-1}\sum_{i=1}^{N-m-1}\ln C_i^m(r),$$

(B3)

where *ln* is the natural logarithm. Then Approximate Entropy is defined as

$$ApEn(m,r) = \lim_{N \to \infty} \left[\Phi^m(r) - \Phi^{m+1}(r) \right].$$

(B4)

It is therefore found that $-ApEn = \Phi^{m+1}(r) - \Phi^m(r)$ and is equal to the average over *i* of the natural *log* of the conditional probability that $|x(i+m) - x(j+m)| \le r|$, given that $|x(i+m) - x(j+m)| \le r$, for $k = 0,1,...,m-1$.

Several trials of this algorithm were run on the HRV data and it was adjusted accordingly in order to obtain a better distinction between the two subject groups. The first step in computing the Approximate Entropy is finding the length vector for *m=2*, which is $[x(i) \, x(i+1)]$, denoted *x(i)*. All vectors that are close to *x(i)*, $x(j) = [x(j) \, x(j+1)]$, are identified. As has already been stated, the vector *x(j)* is close to *x(i)* if $\|x(i) - x(j)\| \le r$. This, by definition, means that both $|x(i) - x(j)| \le r$ and $|x(i+1) - x(j+1)| \le r$ apply. A count of all the vectors *x(j)* close to *x(i)* is found and called *B*. The next step is to compute the rest of the *x(j)* vectors for which $|x(i+2) - x(j+2)| \le r$, and call it *A*. The ratio of *A/B* represents the conditional probability that *X(j+2)* is close to *X(i+2)*, given that the vector *x(j)* is close to *x(i)*.

The above process is repeated for each length 2 vector *x(i)*, calculating the conditional probability. The *ApEn* is found by calculating the average of the logarithm of these conditional probabilities and taking its negative (to make it positive), as seen in Eq. (18)

$$
\begin{aligned}
-ApEn(m,r) &= \Phi^{m+1}(r) - \Phi^m(r) \\
&= \left[\frac{1}{N-m} \sum_{i=1}^{N-m} \ln C_r^{m+1}(i) \right] - \left[\frac{1}{N-m} \sum_{i=1}^{N-m+1} \ln C_r^m(i) \right] \\
&\approx \frac{1}{N-m} \sum_{i=1}^{N-m} \left[\ln C_r^{m+1}(i) - \ln C_r^m(i) \right] \\
&= \frac{1}{N-m} \sum_{i=1}^{N-m} \frac{\ln C_r^{m+1}(i)}{\ln C_r^m(i)}
\end{aligned}
$$

(B5)

The calculation of the conditional probabilities will result in values between 0 and 1. If the timeseries is regular, the values *X(i), X(i+1), X(i+2)* are expected to be close to each other, as are *X(j), X(j+1), X(j+2)*. Therefore, the differences $|x(i) - x(j)|, |x(i+1) - x(j+1)|$ and $|x(i+2) - x(j+2)|$ will be close to each other for many values of *i,j*. This means that the conditional probabilities are expected to be closer to 1 for time-series coming from more regular processes. The negative logarithm of such a value will be closer to 0.

Conversely, random processes will produce conditional probabilities closer to 0, the negative logarithms of which will be closer to 1. The comparison of subsequent vectors in a random signal will result in different values in the successive vector distances. Thus, the *ApEn* values for signals coming from regular processes will be lower than the *ApEn* values coming from random signals. In this application, this implies that low *ApEn* values are to be clinically associated with cardiac pathology, while high values indicate a healthy and robust heart.

The previous algorithm calculates an estimation of the value of the Approximate Entropy, which is equal to the theoretical one, when the N tends to infinity. An examination of this algorithm reveals that it is analogous to the Grassberger & Procaccia Algorithm (the Delay Times method) for the Correlation Dimension estimation. Theoretical calculations by Wolf et al. (1965) indicate that reasonable estimations are achieved with an N value of at least 10m and preferably 30m. In the experimental analysis, $N=2000$ and $m = 2$ were used, producing satisfactory statistical *ApEn* validity.

However, the *ApEn* is a biased statistic. The expected value of *ApEn(m,r,N)* increases asymptotically with N to *ApEn(m,r)* for all processes. The choice of window for each vector x is also important for the *ApEn* estimation. However, the interest in this method is not in the reconstructed space, but rather in having a sufficient number of vectors in close proximity to each other, so that accurate conditional probabilities can be found. This work was partly supported by the PYTHAGORAS fellowships.

Edited by: M. Contadakis
Reviewed by: two referees

References

1. Albano, A., Muench, J., Schwartz, C., Mees, A., and Rapp, P.: Singular values decomposition and the Grassberger-Procaccia algorithm, Phys. Rev. A, 38, 3017–3026, 1988.
2. Alexeev, D. and Egorov, P.: Persistent cracks accumulation under loading of rocks and ccncentration criterion of failure, Reports of RAS 333, 6, 769–770, (in Russian), 1993.
3. Alexeev, D., Egorov, P., and Ivanov, V.: Hurst statistics of time dependence of electromagnetic emission under rocks loading, Physical-Technical problems of exploitation of treasures of the soil, 5, 27–30, (in Russian), 1993.
4. Bak, P.: How nature works, Oxford: Oxford UP, 1997.
5. Bak, P. and Tang, C.: Earthquakes as a self-organized critical phenomenon, J. Geophys. Res., 94, 15 635–15 637, 1989.
6. Bak, P., Tang, C., and Weisenfeld, K.: Self-organized criticality: an explanation of 1/f noise, Phys. Rev. A, 38, 364–374, 1987.
7. Beggs, J. and Plenz, D.: Neuronal avalanches in neocortical circuits, J. Neurosci, 23, 11 167–11 177, 2003.
8. Broomhead, D. and King, G.: Extracting qualitative dynamics from experimental data, Physica D, 20, 217–236, 1986.
9. Contoyiannis, Y., Diakonos, F., Kapiris, P., and Eftaxias, K.: Intermittent dynamics of critical pre-seismic electromagnetic fluctuations, Phys. Chem. Earth, 29, 397–408, 2004.
10. Dobrovolsky, I., Gershenzon, N., and Gokhberg, M.: Theory or Electrokinetic Effects occurring at the Final State in the preparation of a Tectonic Earthquake, Physics of the Earth and Planetary Interiors, 57, 144–156, 1989.
11. Eftaxias, K., Kopanas, J., Bogris, N., Kapiris, P., Antonopoulos, G., and Varotsos, P.: Detection of electromagnetic earthquake precursory signals in Greece, Proc. Japan Acad., 76(B), 45–50, 2000.
12. Eftaxias, K., Kapiris, P., Polygiannakis, J., Bogris, N., Kopanas, J., Antonopoulos, G., Peratzakis, A., and Hadjicontis, V.: Signatures of pending earthquake from electromagnetic anomalies, Geophys. Res. Lett., 28, 3321–3324, 2001.
13. Eftaxias, K., Kapiris, P., Dologlou, E., Kopanas, J., Bogris, N., Antonopoulos, G., Peratzakis, A., and Hadjicontis, V.: EM anomalies before the Kozani earthquake: A study of their behavior through laboratory experiments, Geophys. Res. Lett., 29, 69/1–69/4, 2002.

14. Eftaxias, K., Kapiris, P., Polygiannakis, J., Peratzakis, A., Kopanas, J., and Antonopoulos, G.: Experience of short term earthquake precursors with VLF-VHF electromagnetic emissions, Natural Hazards and Earth System Sciences, 3, 217–228, 2003, **SRef-ID: 1684-9981/nhess/2003-3-217.**

15. Eftaxias, K., Frangos, P., Kapiris, P., Polygiannakis, J., Kopanas, J., Peratzakis, A., Skountzos, P., and Jaggard, D.: Review and a Model of Pre-Seismic electromagnetic emissions in terms of fractal electrodynamics, Fractals, 12, 243–273, 2004.

16. Farmer, A. and Swinney, H.: Independent coordinates for strange attractors from mutual information, Phys. Rev. A, 33, 1134–1140, 1986.

17. Fukuda, K., Nunes, L., and Stanley, H.: Similarities between communication dynamics in the Internet and the automatic nervous system, Europhys. Lett., 62, 189–195, 2003.

18. Gershenzon, N. and Bambakidis, G.: Modelling of seismoelectromagnetic phenomena, Russ. J. Earth Sci., 3, 247–275, 2001.

19. Gilden, D., Thornton, T., and Mallon, M.: 1/f noise in human cognition, Science, 267, 1837–1839, 1995.

20. Goldberger, A., Amaral, L., Hausdorff, J., Ivanov, P., and Peng, C.-K.: Fractal dynamics in physiology: Alterations with disease and aging, PNAS, 2466–2472, 2002.

21. Grassberger, P. and Procaccia, I.: Characterization of strange attractors, Phys. Rev. Lett., 50, 346–349, 1983.

22. Hayakawa, M.: Atmospheric and Ionospheric Electromagnetic Phenomena Associated with Earthquakes, Terrapub, Tokyo, 1999.

23. Hayakawa, M. and Fujinawa, Y.: Electromagnetic Phenomena Related to Earthquake Prediction, Terrapub, Tokyo, 1994.

24. Hayakawa, M. and Molchanov, O.: Seismo Electromagnetics, Terrapub, Tokyo, 2002.

25. Hayakawa, M., Ito, T., and Smirnova, N.: Fractal analysis of ULF geomagnetic data associated with the Guam eartquake on August 8, 1993, Geophys, Res. Lett., 26, 2797–2800, 1999.

26. Hayakawa, M., Itoh, T., Hattori, K., and Yumoto, K.: ULF electromagnetic precursors for an earthquake at Biak, Indonesia on February 17, 1996, Geophys. Res. Lett., 27(10), 1531–1534, 2000.

27. Heneghan, C. and McDarby, G.: Establishing the relation between detrended fluctuation analysis and power spectral density analysis for stohastic processes, Phys. Rev. E, 62, 6103–6110, 2000.

28. Herz, A. and Hopfield, J.: Earthquake cycles and neural reverberations: collective oscillations in systems with pulse-coupled threshold elements, Phys. Rev. Lett., 75, 1222–1225, 1995.

29. Hopfield, J.: Neurons, dynamics and computation, Physics Today, 40, 40–46, 1994.

30. Hristopulos, D.: Permissibility of fractal exponents and models of band-limited two-point functions for fGn and fBm random fields, Stohastic Environmental Research and Risk Assessment, 17, 191–216, 2003.

31. Hurst, H.: Long term storage capacity of reservoirs, Trans. Am. Soc. Civ. Eng., 116, 770–808, 1951.

32. Ivanov, P., Amaral, L., Goldberger, A., Havlin, S., Rosenblum, M., Struzik, Z., and Stanley, H.: Multifractality in human heartbeat dynamics, Nature, 399, 461–465, 1999.

33. Kaiser, G.: A. Friendly Guide to Wavelets, Birkhauser, 1994.

34. Kapiris, P., Polygiannakis, J., Nomicos, K., and Eftaxias, K.: VHFelectromagnetic evidence of the underlying pre-seismic critical stage, Earth Planets Space, 54, 1237–1246, 2002.

35. Kapiris, P., Eftaxias, K., Nomikos, K., Polygiannakis, J., Dologlou, E., Balasis, G., Bogris, N., Peratzakis, A., and Hadjicontis, V.: Evolving towards a critical point: A possible electromagnetic way in which the critical regime is reached as the rupture approaches, Nonlin. Proc. Geophys., 10, 1–14, 2003.

36. Kapiris, P., Balasis, G., Kopanas, J., Antonopoulos, G., Peratzakis, A., and Eftaxias, K.: Scaling similarities of multiple fracturing of solid materials, Nonlin. Proc. Geophys., 11, 137–151, 2004a, **SRef-ID: 1607-7946/npg/2004-11-137.**

37. Kapiris, P., Eftaxias, K., and Chelidze, T.: The electromagnetic signature of prefracture criticality in heterogeneous media, Phys. Rev. Lett., 92, 065 702/1–4, 2004b.

38. Karamanos, K., Nikolopoulos, S., and Hizanidis, K.: Block Entropy analysis of long recorded Electrocardiograms as a good way for discrimination between Normal subjects and Coronary patients,

International Journal of Computing and Anticipatory systems (IJCAS, Liege, Proceedings of CASYS 2003), 2004.

39. Kugiumtzis, D.: State space reconstruction parameters in the analysis of c haotic time series – the role of the time window length, Physica D, 95, 13–28, 1996.

40. Li, X., Polygiannakis, J., Kapiris, P., Balasis, G., Peratzakis, A., Yao, X., and Eftaxias, K.: Evolving towards a biological or geophysical catastrophic event: emergence of isomorphic precursory alterations in scaling parameters in terms of intermitent criticality, in 4th International Conference on Fractals and Dynamics Systems in Geoscience, Kloster Seeon, Germany, May 19–22, Book of Abstract, 55–57, 2004.

41. Malamud, B., Morein, G., and Turcotte, D.: Forest fires: an example of self-oganized critical behavior, Science, 281, 1840–1842, 1998.

42. Mandelbrot, B. and Ness, J.: Fractional Brownian motions, fractional noises and applications, SIAM Rev., 10, 422–437, 1968.

43. Mantegna, R. and Stanley, H.: Scaling behavior in the dynamics of an economic index, Nature, 376, 46–49, 1995.

44. Maslov, S., Paczuski, M., and Bak, P.: Avalanches and $1/f$ Noise in Evolution and Growth Models, Phys. Rev. Lett., 73, 2162, 1994.

45. Meneveau, C. and Sreenivasan, K.: Simple multifractal cascade model for fully developed turbulance, Phys. Rev. Lett., 59, 1424–1427, 1987.

46. Mizutani, H. and Ishido, T.: A new interpretation of magnetic field variation associated with the Matsushiro earthquake occurrence, J. Geomagn. Geoelectr., 28, 179–188, 1976.

47. Nikolopoulos, S., Alexandridi, A., Nikolakeas, S., and Manis, G.: Experimental Analysis of Heart Rate Variability of Long-Recording Electrocardiograms in Normal Subjects and Patients with Coronary Artery Disease and Normal Left Ventricular Function, J. Biomed. Inf., 36, 202–217, 2003.

48. Peng, C., Havlin, S., Stanley, H., and Goldberger, A.: Quantification of scaling exponents and crossover phenomena in nonstationary heartbeat timeseries, Chaos, 5, 82–87, 1995.

49. Pincus, S.: Approximate Entropy: A Complexity Measure for Biologic Time series Data, in The proceedings of IEEE 17th Annual Northeast Bioengineering Conference, pp. 35–36, New York: IEE Press, 1991.

50. Ponomarev, A., Zavyalov, A., Smirnov, V., and Lockner, D.: Physical modelling of the formation and evolution of seismically active fault zones, Tectonophysics, 277, 57–81, 1997.

51. Ramirez-Rojas, A., Pavia-Miller, C., and Angulo-Brown, F.: Statistical behavior of the spectral exponent and the correlation time of electric self-potential time series associated to the Ms=7.4 September 14, 1995 earthquake in Mexico, Phys. Chem. Earth, 29, 305–312, 2004.

52. Rundle, J., Klein,W., Gross, S., and Turcotte, D.: Boltzmann fluctuations in numerical simulations of nonequilibrium lattice threshold systems, Phys. Rev. Lett., 75, 1658–1661, 1995.

53. Rundle, J., Tiampo, K., Klein, W., and Martins, J. S.: Selforganization in leaky threshold systems: the influence of nearmean field dynamics and its implications for earthquakes, neurology, and forecasting, PNAS, 99, 2514–2521, 2002.

54. Sethna, J., Dahmen, K., and Myers, C.: Crackling noise, Nature, 410, 242–250, 2001.

55. Sornette, D.: Predictability of catastrophic events: Material rupture, earthquakes, turbulence, financial crashes, and human birth, PNAS, 99, 2522–2529, 2002.

56. Stanley, H.: Scaling, universality, and renormalization: Three pillars of modern critical phenomena, Rev. Mod. Phys., 71, S358–S366, 1999.

57. Stanley, H.: Exotic statistical physics: Applications to biology, medicine, and economics, Physica A, 285, 1–17, 2000.

58. Stanley, H., Amaral, L., Goldberger, A., Havlin, S., Ivanov, P., and Peng, C.: Statistical physics and physiology: Monofractal and multifractal approaches, Physica A, 106, 309–324, 1999.

59. Surkov, V., Uyeda, S., Tanaka, H., and Hayakawa,M.: Fractal properties of medium and seismoelectric phenomena, J. Geodynamics, 33, 477–487, 2002.

60. Takens, F.: Detecting strange attractors in turbulence, in Dynamical systems and turbulence, Warwick 1980: Proceedings of symposium held at the university of Warwick 1979–1980, Springer-Verlag, Berlin, 366–381, 1981.
61. Telesca, L. and Lapenna, V.: A new approach to investigate the correlation between geoelectrical time fluctuations and earthquakes in a seismic area of southern Italy, Geophys. Res. Lett., 28, 4375–4378, 2001.
62. Turcotte, D.: Fractals and chaos in geology and geophysics, Cambridge University Press, 1992.
63. Turcotte, D. and Rudle, J.: Self-organized complexity in the physical and social sciences, PNAS, 99, 2463–2465, 2002.
64. Usher, M., Stemmler, M., and Olami, Z.: Dynamic pattern formation leads to 1/f noise in neural populations, Phys. Rev. Lett., 74, 326–329, 1995.
65. Varotsos, P. and Alexopoulos, K.: Physical properties of the variation of the electric field of the earth preceding earthquakes, I, Tectonophysics, 110, 73–98, 1984a.
66. Varotsos, P. and Alexopoulos, K.: Physical properties of the variation of the electric field of the earth preceding earthquakes, II, Tectonophysics, 110, 99–125, 1984b.
67. Varotsos, P. and Sarlis, N.: A review of the recent VAN efforts: The explanation of the SES physical properties, in Lithosphere-Atmosphere-Ionosphere Coupling, Terrapub, Tokyo, 131–140, 2002.
68. Varotsos, P., Lazaridou, M., Eftaxias, M., ANtonopoulos, K., Makris, G., and Kopanas, J.: Short term earthquake prediction in Greece by seismic electric signals, in: A Critical Review of VAN, edited by: Lighthill, J., World Scientific, Singapore, 29–76, 1996.
69. Vicsek, T.: Fractal Growth Phenomena, World Scientific, Singapore, 1993.
70. Vicsek, T.: A question of scale, Nature, 411, 421, 2001.
71. Vicsek, T.: The bigger picture, Nature, 418, 131, 2002.
72. Wolf, A., Swift, J., Swinney, H., and Vastano, J.: Determining Lyapunov exponents from a time series, Physica D, 16, 285–317, 1965.
73. Worrell, G., Cranstoun, S., Litt, B., and Echauz, J.: Evidence for self-organized criticality in human epileptic hippocampus, Neurophysiology, Basic and Clinical, 13, 1–6, 2002.
74. Zhao, X. and Chen, T.: Type of self-organized criticality model based on neural networks, Phys. Rev. E, 65, 026 114-1–026 114-6, 2002.

Evolutionary Geomorphology: Thresholds and Nonlinearity in Landform Response to Environmental Change[†]

J.D. Phillips

Tobacco Road Research Team, Department of Geography, University of Kentucky, Lexington, KY 40506-0027, USA. Email: jdp@uky.edu.

ABSTRACT

Geomorphic systems are typically nonlinear, owing largely to their threshold-dominated nature (but due to other factors as well). Nonlinear geomorphic systems may exhibit complex behaviors not possible in linear systems, including dynamical instability and deterministic chaos. The latter are common in geomorphology, indicating that small, short-lived changes may produce disproportionately large and long-lived results; that evidence of geomorphic change may not reflect proportionally large external forcings; and that geomorphic systems may have multiple potential response trajectories or modes of adjustment to change. Instability and chaos do not preclude predictability, but do modify the context of predictability. The presence of chaotic dynamics inhibits or excludes some forms of predictability and prediction techniques, but does not preclude, and enables, others. These dynamics also make spatial and historical contingency inevitable: geography and history matter. Geomorphic systems are thus governed by a combination of "global" laws, generalizations and relationships that are largely (if not wholly) independent of time and place, and "local" place and/or time-contingent factors. The more factors incorporated in the representation of any geomorphic system, the more singular the results or description are. Generalization is enhanced by reducing rather than increasing the number of factors considered. Prediction of geomorphic responses calls for a recursive approach whereby global laws and local contingencies are used to constrain each other. More specifically a methodology whereby local details are embedded within simple but more highly general phenomenological models is advocated. As landscapes and landforms change in response to climate and other forcings, it cannot be assumed that geomorphic systems progress along any particular

[†]Reused with permission from: J.D. Phillips, Evolutionary geomorphology: thresholds and nonlinearity in landform response to environmental change, *Hydrol. Earth Syst. Sci.*, 10, 731–742, 2006.

pathway. Geomorphic systems are evolutionary in the sense of being path dependent, and historically and geographically contingent. Assessing and predicting geomorphic responses obliges us to engage these contingencies, which often arise from nonlinear complexities. We are obliged, then, to practice evolutionary geomorphology: an approach to the study of surface processes and landforms which recognizes multiple possible historical pathways rather than an inexorable progression toward some equilbribrium state or along a cyclic pattern.

Introduction

Geomorphologists have long made reference to landform and landscape *evolution*, usually using the latter word as a general term referring to change over time. Traditional chronological models of landscape evolution such as those of Davis, Penck, and King postulate a deterministic cycle or progression of forms. Process-based "equilibrium" models also postulate a specific developmental pathway, towards some final steady-state. To the extent these models apply, they simplify efforts to predict the response of earth surface processes and landforms to climate and other environmental changes. In equilibrium-based theory, a given set of boundary conditions produces a given outcome, indicating that we should be able to work out a one-to-one correspondence between changes in boundary conditions and geomorphic reponse. In cyclical models, exogenous changes can be treated as interruptions, accelerations, or decelerations of the prescribed cycles.

Though existing models of landscape evolution and geomorphic response to disturbance are all applicable in some situations, none provides a general framework applicable to all (or even a majority of) geomorphic systems. Further, geomorphic change over time is often characterized by pathways more complex than progression toward some endstate, be the latter a planation surface, equilbrium form, mature zonal soil, or other hypothesized destination. Accordingly, several geomorphologists have espoused an explicitly evolutionary approach that distinguishes between complex, nonlinear, historically contingent, path-dependent evolution and classically deterministic development over time. This paper explores the links between evolutionary geomorphology, thresholds, and nonlinear dynamics, in the context of predicting effects of environmental change on geomorphic systems. Special attention is given to implications regarding geographical and historical contingency.

Evolutionary Geomorphology

Instead of seeking universal theories (be they based on equilibrium notions, cycles, or otherwise), Ollier (1979) suggested, it might be more useful to see how landscapes actually evolve. Ollier's evolutionary geomorphology emphasizes dates, ages, and history, and stresses the consistent internal (to the landscape or system under study) evidence rather than a priori theoretical notions. Ollier (1979) does not advocate anecdotal, a theoretical approaches, but rather adapting or devising conceptual frameworks to fit the evidence rather than imposing conceptual frameworks at the beginning. The evolutionary geomorphology of Ollier can be interpreted as working out the pathway or trajectory of change in a multidimensional space encompassing multiple possibilities.

Thornes' (1983) vision of evolutionary geomorphology is also concerned with the long-term behavior of landforms. Thornes (1983) laid out a blueprint for evolutionary geomorphology based on complex dynamical systems. Defining an area dominated by a particular landform or process as a domain, process geomorphology is chiefly concerned with behavior determining the character and configuration of the domains. Evolutionary geomorphology, by contrast, is concerned with "the initiation and development of the structure giving rise to the domains" (Thornes 1983:227). Structure here refers to the structural relationships among processes, geological controls, climate, relief, and other factors rather than geological structure per se, and evolutionary geomorphology is portrayed as being more analytical than chronological approaches which essentially describe particular historical pathways. Thornes explicitly (rather than implicitly as in Ollier's case) advocated a concern with defining geomorphic system trajectories through a multidimensional phase or state space defined by the key variables or components of the system.

In soil geomorphology, Johnson and colleagues (Johnson and Watson-Stegner 1987; Johnson et al. 1990) challenged the view of pedogenesis as an inexorable (though perhaps occasionally interrupted) pathway of increasing pedological development toward a steady-state climax soil. The evolutionary model of pedogenesis (Johnson and Watson-Stegner, 1987) allows for the possibilities of both progressive and regressive pedogenesis, and for complex changes in the state of the soil landscape. This model was explicitly linked to dynamical systems by Johnson et al. (1990), and to complex nonlinear dynamics by Phillips (1993).

Huggett (1995, 1997) is concerned with geoecosystems, including geomorphic systems, more generally. He contrasted an evolutionary viewpoint with a "developmental" view characterized by progress along a predetermined path, whether a Davisian cycle or progress toward a single steadystate equilibrium. Huggett's evolutionary approach emphasizes inconstancy, based on the unlikelihood of sufficient time for full developmental sequences to occur, the likelihood of nonlinearity and complexity, and dependence on initial conditions. An evolutionary view thus recognizes that at any instant earth surface "systems are unique and constantly changing, and are greatly influenced by historical events (owing to the relevance of initial conditions)" (Huggett, 1997:315). The historical path of an earth surface system is interpreted in an evolutionary context as changes in the state of the system rather than as progression (or retrogression) along a particular developmental pathway (Huggett, 1995:268).

The notions of evolutionary geomorphology outlined above are consistent in several regards. All are concerned with change over time in landforms and landscapes, emphasizing historical and geological time scales. All recognize multiple possible historical pathways for such changes, rather than an inevitable progression toward some final equilibrium state or along a cyclic pattern. The conceptions of evolutionary geomorphology explicitly acknowledge historical contingency, whether in the form of inheritance, path dependence, or dependence on initial conditions.

Applications of nonlinear and complex systems analysis in the geosciences has often been (accurately) characterized as the importation of ideas from systems theory, mathematics, and theoretical physics and chemistry into a new domain, particularly with respect to relatively new constructs such as chaos, fractals, and self-organized criticality. However, the notions of evolutionary geomorphology show that threads of inquiry within geomorphology also lead to the consideration of nonlinear complexity in earth surface systems, independently of ideas transferred from other fields. Most of the fundamental implications of nonlinear and complexity science as they apply to earth surface processes and landforms are entirely consistent with existing and well-known (though certainly not necessarily universally accepted) concepts in geomorphology developed via geographical and geological reasoning (Phillips 1992).

Nonlinearity in Geomorphic Systems

Notwithstanding the comments above, nonlinear dynamics and complexity have been widely discussed in geography, geology, and geomorphology with an emphasis on abstractions of theory rather than concrete aspects of surface forms and processes, and on imported rather than home-grown methods and terminology. As a consequence several widely held (mis) perceptions exist in the earth science community about complex nonlinear dynamics. One is that this type of complexity, readily generated by equation systems, simulation models, and controlled experiments, has not been convincingly demonstrated in real-world earth surface processes and landforms. Another is that some forms of nonlinear complexity, such as deterministic chaos, imply hopelessly innate complexity and an inability to predict. Earth scientists are also often put off by claims on behalf of some strains of nonlinear theory (for example self-organized criticality) that they represent meta-explanations for nature.

These perceptions, while pervasive, are inaccurate. Nonlinear complexity is not necessarily pathological, and may enhance some modes of understanding and predictability. Most scientists working in nonlinear dynamics do not make claims of meta-explanation. Complex nonlinear dynamics are not (merely) an artifact of models, equations, and experiments, but have been observed and documented in many geomorphic phenomena and are not rare or isolated phenomena. These points have been addressed more fully elsewhere (Phillips 2003a), along with arguments that the identification of and engagement with nonlinear dynamics in earth surface systems has profound implications for prediction, explanation, and application.

Geomorphology is dominantly (and appropriately) an empirical discipline where ground truth is paramount and "field relations are the final court of appeal" (Bretz 1962). Thus, while acknowledging the critical roles of theory, modeling, and experimentation, geomorphologists ultimately find work with a field component most convincing, and understanding nonlinear dynamics (and applying the lessons there from to practical problems) requires linking complex system behaviors to histories, relationships, and phenomenologies in real landscapes. This further implies a need to problematize based on principles and conceptual frameworks of the earth and environmental sciences, as opposed to those of the mathematical and laboratory sciences.

Causes of Nonlinearity

A system is nonlinear if the outputs (or responses or outcomes) are not proportional to the inputs (or stimuli, changes, or disturbances) across the entire range of the latter. Nonlinearity creates possibilities for complex behavior not possible in linear systems. However, nonlinear systems may be simple and predictable, and complexity may have causal roots other than nonlinearity. Geomorphic systems are overwhelmingly nonlinear, owing to a number of general phenomena summarized in Table 1 and discussed in detail by Phillips (2003a). These phenomena are mostly common to ecosystems and to earth surface systems in general (Phillips 2004).

Table 1 Sources of nonlinearity in geomorphic systems (adapted from Phillips 2003a).

Source of Nonlinearity	Why is it nonlinear ?	Examples(see text for Details)
Thresholds	Where thresholds exist, outputs or responses by definition cannot be proportional to inputs or stimuli across the entire range of the inputs	Force vs. resistance ; relative rates of linked processes
Storage effects	Addition or removal of mass from storage creates lags and discontinuities in mass balances and input-output relationships	Sediment transport , storage , and deposition
Saturation and depletion	Effects of a unit change in an input or forcing varies with respect to some optimum	Effects of moisture availability on weathering rates; effect of soil / regolith thickness on bedrock weathering
Self -reinforcing, positive feedback	Changes or disturbances promote their own growth and development independently of external forcings	Growth of solutional depressions and nivation hollows ; enhancement of variations in weathering/ erosion resistance; preferential flow phenomena
Self-limiting processes	Developmental pathways are limited by internal factors independently of external forcings	Weathering limited by depletion of weatherable minerals; diffusional slope degradation; floodplain vertical accretion
Opposing or competitive interactions or feedbacks	Opposing interactions or competitive feedbacks may cause systems to tip or switch abruptly	Vegetation-erosion interactions; indirect geomorphic effects of ecological competition
Multiple modes of adjustment	Multiple possible configurations in response to a single forcing or set of boundary conditions	Fluvial hydraulic geometry
Self-organization	May involve complex adaptations independent of external forcings	Flow-bedform interactions; periglacial patterned ground; channel networks
Hysteresis	A dependent variable may have two or more values associated with a single value of an independent variable	Discharge vs . sediment transport relationships ; river channel changes ; solute concentrations

Thresholds are of particular significance in geomorphology, as discussed by Chappell (1983), Schumm (1979, 1991), Coates and Vitek (1980) and any geomorphology textbook published in the last 20 years. In simple terms a threshold is the point at which a system's behavior changes. Geomorphic thresholds may be either intrinsic, and associated with the inherent structure or dynamics of the geomorphic system, or extrinsic, associated with external factors such as climate, tectonics, and base level. Most commonly geomorphic thresholds are of two general types: the ratio of force or power (or a surrogate thereof) to resistance, or the relative rates of linked processes. Examples of force: resistance thresholds include shear strength vs. shear stress in slope stability, and critical stream power or wind velocity in sediment transport and deposition. Examples of linked process thresholds include relative rates of regolith formation and erosion, and glacial accumulation vs. ablation.

Recently it has been argued that some nonlinear systems evolve to a "critical" state, generally characterized by proximity to a threshold. Schumm (1979) argued that due to the predominance of thresholds, landforms typically evolve to a condition of incipient instability. Schumm's work thus anticipates recent studies of self-organized criticality, but arrives at similar basic conclusions based on geological reasoning.

Other key sources of nonlinearity in geomorphology include storage effects, saturation and depletion relationships, self-reinforcing positive feedbacks, self-limiting negative feedbacks, "competitive" relationships (for example between soil erosion and vegetation cover), multiple modes of adjustment, self-organization, and hysteresis. These are summarized in Table 1, and geomorphic examples given in Phillips (2003a). These general sources of nonlinearity are overlapping and interrelated, and despite the generality of the list, it is undoubtedly not exhaustive.

Self-organization deserves further comment, as the term has various and often conflicting definitions, some of which are unrelated to complex nonlinear dynamics, and some of which are subsumed in the categories above (Phillips 1999b). Some forms, such as self-organized criticality, involve nonlinearities as systems evolve toward critical states (e.g., Dearing and Zolitschka 1999; Gomez et al. 2002). Others, such as dynamically unstable self organization (Phillips 1999b) are an outcome rather than a cause of nonlinearity. In the most general sense self organization refers to the formation of patterns attributable to the internal dynamics of a geomorphic system, independently of external controls or inputs. Because this may offset or intensify the effects of external forcings and boundary conditions, self-organization may be a source of nonlinearity in a system.

Implications of Nonlinearity

Nonlinearity implies landforms and landscapes are likely to vary in their sensitivity to environmental change. Systems near a threshold, approaching saturation, or characterized by strong positive or overcompensating negative feedbacks, for example, are much more sensitive to a given disturbance than would otherwise be the case. Landscape sensitivity in this sense is discussed at length by Brunsden (1980), Begin and Schumm (1984), Downs and Gregory (1995), and Thomas (2001).

Nonlinearity admits the possibility of dynamical instability and chaos (equivalent in the case of nonlinear dynamical systems). While the significance of this in geomorphology is contested, the evidence that geomorphic systems can be, and often are, chaotic is now overwhelming, even when work based strictly on models is excluded. Several available reviews should suffice to make this point (Baas 2002; Christofoletti 1998; Hergarten 2002; Phillips 1999a, 2003a, 2005; Sivakumar 2000, 2004a; Thomas 2001). The implications are discussed below.

Dynamical Instability and Chaos

Geomorphic systems are not all, or always, chaotic. Indeed, many appear to have both stable, non-chaotic modes and unstable, chaotic modes (Phillips 1999a, 2003a, 2005). Implications for long-term landscape evolution are discussed elsewhere (Phillips, 2003b, 2005). Here the focus is on predicting and responding to effects of environmental change on geomorphic processes and forms.

Geomorphic systems are conceptualized as n-dimensional systems with components x_i, $i=1, 2, ..., n$, such that

$$dx_i / dt = f(dx / dt) \tag{1}$$

where x indicates the vector of all x_i. Thus the components of the system potentially effect, and are potentially effected by, each other. The system state at time t is given by

$$x(t) = C.x(0).e^{\lambda t} \tag{2}$$

where $x(0)$ is the initial state (at the onset of landscape evolution or at the time of a change or disturbance) and C is a vector constant related to the initial conditions. The λ are the n Lyapunov exponents of the system (equivalent to the real parts of the complex eigenvalues of a Jacobian interaction matrix of the system), where $\lambda_1 > \lambda_2 > .... > \lambda_n$.

If randomly selected pairs of locations in a landscape are compared in terms of some indicator of system state (elevation or regolith thickness, for example), the mean difference or separation at time t is given by

$$\delta(t) = ke^{\lambda_1} \tag{3}$$

where the constant k normalizes the initial separation and λ_1 is the largest Lyapunov exponent.

Stable, non-chaotic geomorphic systems have all $\lambda < 0$, while any positive exponent ($\lambda_1 > 0$) indicates instability and chaos. Methods for detecting and analyzing chaos in geomorphic and hydrologic systems are discussed elsewhere (Phillips 1999a; Sivakumar 2000, 2004a). The key point here is the (finite) exponential divergence that occurs in unstable, chaotic geomorphic systems.

Because the effects of minuscule initial variations and small disturbances are exaggerated over time, the implications for geomorphic response to environmental change are that

1. Small changes may produce disproprotionately large results.
2. Short disturbances may have dispoportionately longlived effects.
3. Evidence of landform change may not reflect proportionally large environmental changes or events. In the absence of perfect isotropy, initial conditions vary locally. This sensitivity to initial conditions leads to a fourth implication
4. Geomorphic systems may have multiple response trajectories or modes of adjustment to changes.

For instance, the nonlinear dynamical systems models of Thornes (1985) and Kirkby (1995) indicate that the relationship between vegetation and soil erosion in semiarid environments is unstable. When disturbed, the system will "tip" to either a maximum vegetation/no erosion or maximum erosion/ no vegetation state. These predictions have been validated by subsequent field studies (Abrahams et al. 1995; Puigdefabregas and Sanchez 1996). Stratigraphic, morphological or other evidence of erosional episodes or vegetation changes therefore may not imply a major change in climate, land use, or other forcings, but the "tipping" of the unstable system in response to a small, short-term perturbation such as a storm, a fire, or the grazing of a cattle herd.

Other examples include Dearing and Zolitschka (1999), who addressed the implications of nonlinear complexities in interpreting lake sediment archives, demonstrating how complex internal dynamics rather than external forcings account for some observations in the sediment record. Gaffin and Maasch (1991) showed that multiple equilibria associated with nonlinear feedbacks can result in large coastal onlap shifts associated not with accordingly large sea level change, but rather arising from small perturbations. The behavior of glacial feeder systems has been reconstructed from steep faced glaciodeltaic progradational successions, but Richards and others (2000) showed that such glacier-fed successions in Ireland and Scandinavia have evidence of complex nonlinear dynamics, leading to sedimentation patterns that reflect internal interactions involving delta front steepness and sediment texture, rather than external forcings.

Predicting or interpreting geomorphic responses to climate change thus requires that chaotic or potentially chaotic systems be identified. More specifically, as many systems have both stable and unstable modes, and as both stability and instability are emergent behaviors which appear and disappear as temporal and spatial scales are changes, the scales or circumstances under which chaos and instability are relevant need to be determined. Chaos detection methods are discussed in a separate paper (Phillips 2006).

Prediction

Due to (among other things) nonlinear complexity, predicting the response of landforms and surface processes to climate change cannot rely uncritically on "equilibrium" frameworks based on the notion of a new steady-state configuration. Neither can it be safely assumed that responses will be quantitatively or even qualitatively similar to those in the historical record. Where does this leave us?. One approach is probabilistic. Stochastic forecasting methods work equally well whether the phenomenon is truly random or merely apparently so (as in a chaotic sequence). Probabilistic methods can be improved on in a chaotic system, as the pseudo-random behavior occurs within well-defined boundaries. A possible analog to some problems of geomorphic predictability is the field of demographics, where individual human behavior is inherently unpredictable, but characteristic aggregate behaviors can be probabilistically predicted. A second possibility is to exploit chaotic dynamics. Chaos may preclude deterministic long-term prediction, but does not preclude iterative, short-term predictions. A number of studies in geomorphology, sedimentology, and hydrology have shown that where chaos exists nonlinear prediction models give better results than either traditional deter ministic or stochastic models (Barton et al. 2003; Jaffe and Rubin 1996; Lall et al. 1996; Porporato and Ridolfi 2001; Sangoyami et al. 1996). Sometimes the unstable growth of small perturbations, but with finite and well-defined limits and aggregate statistical regularity, is reflected in a syndrome of chaotic instability at one scale resolved into orderly, even regular patterns at a broader scale. Studies based on this approach have led to improved models of fluvial, coastal, and aeolian bedforms (e.g., Nelson 1990; Rey et al. 1995; Rubin 1992; Werner 1995). A third approach to prediction exploits the emergent properties of chaos. A deterministically chaotic system, by definition, has some underlying deterministic dynamics, which may be (though are not necessarily) quite simple. Likewise, at broader scales the complex irregularities are bounded, and exhibit some degree of irregularity. Turbulent flows are a canonical example (Escultura 2001; Tsinober 1998), where the basic underlying physics are well known and deterministic predictions are straightforward where particle interaction is insignificant. The complex interactions of more than a few particles, however, is chaotic, and the location and velocity of any given particle is unpredictable in any deterministic sense more than a few instants into the future. At still broader scales, however, the aggregate fluid lows are quite adequately predicted from gravitational and pressure gradients. In some cases it may be possible to restrict or expand spatial or temporal scales to get into a non-chaotic mode.

In meteorology, despite vast improvements in deterministic modeling and atmospheric physics and chemistry, the backbone of forecasting is still synoptic meteorology and climatology—the study of weather maps, though now largely automated and embedded in numerical models. By examining situations in spatial and temporal context, behavioral typologies are developed. The atmospheric equations of motion are a classic example of chaos, but with general physical "global" laws and "local" synoptic observations constraining each other, reasonable predictions are possible. This suggests a useful analog for geomorphology, where avowedly synoptic, event-based, or situationally-constrained forecasts have been shown to be effective in several recent cases (Knighton and Nanson 2001; Knox 2000; Miller et al. 2003; Slattery et al. 2006).

Geography and History

The implications of instability and chaos in predicting geomorphic responses could be summed up as: Geography matters, and history matters. Geography matters because local variations and disturbances result in increasing divergence over time. History matters because geomorphic systems "remember" initial variations and perturbations. Because geography and history matter, factors and controls specific to place and time (local factors) are irreducibly significant—a source of frustration to many geomorphologists, who like other scientists strive for explanation and prediction based on "global" laws (or principles or generalizations) which are independent of time and place.

Several recent developments in the earth and environmental sciences support the emerging view that historical and spatial contingencies are ubiquitous and must be engaged on their own terms-that is, the contingencies cannot be subsumed under global laws by simply collecting more and better data or constructing more involved models. These developments include a shift away from a search for global generalizations within spatial data to efforts to explain spatial variability by explicitly incorporating local factors (so-called *local forms of spatial analysis*). This shift is most evident in quantitative geography (Fotheringham and Brunsdon 1999), and successful applications in geomorphology include Atkinson et al. (2003) and Nelson (2001). Landscape ecology and soil geography have also focused on explaining spatial variability rather than extracting global laws, with the dominant conceptual frameworks based on the search for applicable process laws within local and regional contexts (Christakos 2002; Goovaerts 1999; Haines-Young and Chopping 1996; Ibanez et al. 1995; Walsh et al. 1998).

Studies of effects of high-magnitude, low-frequency events further underscores the inescapable elements of historical and geographical contingency in geomorphology. Impacts of floods, hurricanes, and other large events may be influenced or controlled by event timing, sequence, and initial conditions in addition to (or rather than) event magnitude and force-resistance relationships governed by generally applicable laws. Because timing, sequence, and initial conditions are inherently contingent, effects of such events cannot be (entirely) addressed via global laws. Examples include Carling and Beven (1989), Lecce et al. (2004), Magilligan (1992), Magilligan et al. (1998), Miller et al. (2003); Phillips (1999c); and Pickup (1991). The critical role of place- and time-based explanation is also indicated by a cumulative and repeated inability to extract generalizations. For example, Schumm et al.'s (2000) book on tectonics and alluvial rivers relies heavily on four case studies, but generalizations are still hard to come by: "Because the four rivers are subjected to different types of active tectonism and each river is different, the only firm conclusion that can be reached is that deformation causes river variability" (p. 151). Similarly, even in a relatively restricted geographical context no generalizations about downstream geomorphic effects of dams on large rivers could be discerned (Friedman et al. 1998). A multi-investigator, multinational effort to link landslides to climate change in Europe was no more successful in producing generalizations: "...the complexity of the relationship between climate and landsliding seems to make it not feasible to establish 'universal laws' all over Europe" (Dikau and Schott, 1999:1).

In geomorphology and the earth sciences more broadly, the undeniable role of history has repeatedly defeated efforts to understand landscape entirely on the basis of reductionist global laws, and concern over global change has rejuvenated palaeoenvironmental reconstructions. The recognition that landscape evolution has irreducible elements of contingency and path-dependency leads to acknowledgement that in many cases geomorphology calls for an approach to science fundamentally different from that of the reductionist laboratory science ideal (e.g., Baker 1996; Bishop 1998; Harrison 1999; Spedding 1997). Several recent studies explicitly address the necessity of dealing with historical contingency in specific field problems (Bishop 1998; Brierly and Fryirs 2005; Fryirs 2002; Lane and Richards, 1997; Sauchyn 2001; Vandenbergehe, 2002).

Contingency can arise from a number of different phenomena, and would be an issue even without complex nonlinear dynamics. However, the fact that geomorphic systems in many cases are dynamically unstable indicates that initial variations matter, local disturbances matter, and history matters. This indicates that the components of a geomorphic system as described in Eq. (1) can be represented as

$$x = x_g + x_l \tag{4}$$

where x_g represents the components governed or represented by laws, regularities, or relationships which are widely applicable and not place- or time-specific. Components x_l are associated with local, contingent factors.

Geomorphic problems can be defined so that only global factors are relevant, or so that local factors have negligible influence. The response of landforms and landscapes to climate or other changes, however, is not ultimately concerned with changes in the stability of a simplified or idealized slope, or the erosion of a modelled field. Eventually, the fate of specific landforms and landscapes must be addressed, involving both x_g and x_l.

Returning to the notions of evolutionary geomorphology as the trajectory of system states through time, the state of a geomorphic system (combining Eqs. and 3) is

$$x(t) = f\left(C, x_g(0), x_l(0), \lambda\right) \tag{5}$$

A beach, for instance, is determined partly by global laws and general relationships pertaining to the physics of wave generation, propagation, shoaling, and breaking; sediment entrainment and transport; wave-nearshore-beach interactions; etc. The state of the beach (defined, for example, on the basis of its morphology or erosion/accretion status) is also determined on the basis of a number of local, contingent factors such as recent storm, wind and wave history, underlying geologic controls, sea level history, vegetation, proximity to sediment sources and sinks, and human (or other animal) effects. With $i=1, 2,...,n$ general or global controls $x_{g,i}$, and $j=1, 2,...,m$ local or contingent controls $x_{l,j}$, the probability of a specific state $p(S)$ is a function of the joint probabilities:

$$p(S) = \prod_n p(x_{g,i}) \cdot \prod_m p(x_{l,j}) \tag{6}$$

where probabilities $p(G_i).p(G_j) < 1$. The $p(x_{g,i})$ may approach unity in some cases–this is certainly the ideal, though in practice even universal laws are conditioned by uncertainty associated with parameterization and the form of the relevant law. However, $p(x_{ij}) < 1$, and often < 1. Accordingly, $p(S) << 1$, assuring (alas) at least some elements of uniqueness in every landscape.

Equation (5) shows that the key to increasing generality of landscape decriptions and analyses comes from reducing the number of components, variables, or controls considered, as including more $x_{g,i}$ or x_j can only reduce $p(S)$. The more variables and parameters included, or the more processes modelled, the more singular the outcome.

The Way Forward

This analysis should not discourage the search for generalizations, or be interpreted as advocating a purely idiographic approach. It should be clear that both global and local factors are critical in geomorphic systems, and that approaches exclusively based on one or the other, while perhaps successful in particular problems or applications, cannot ultimately explain landscape evolution and response.

In general, the way forward involves dealing with the mutual constraints of local and global factors on each other. Understanding changes in karst processes and landforms in response to climate change, for example, will require addressing the particular combination of lithologic, structural, topographic, and biotic (at least) controls in an area, and many potential specific outcomes are possible. However, general principles of karst geomorphology should allow one to rule out some possibilities, and to further identify lower- and higher-probability responses. Conversely, general principles of fluvial reponse to sea-level changes can inform predictions of responses to climate change, but river-specific predictions must be made in the context of the particular geologic and hydrologic controls and recent geomorphic history of the river. More specifically, we may seek generalizations in pareddown, more generalized models–recognizing that the more pared-down, the more general-and then embed within these specific field problems.

Hergarten (2002) and Werner (1999) have argued that the fundamental qualitative behavior of geomorphic systems is more important than the quantitative details. This is

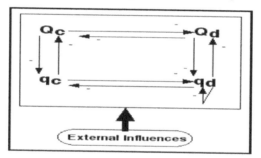

Figure 1 Flow partitioning model for fluviokarst landscapes in central Kentucky, after Phillips and Walls (2004). The model is based on a mass balance partition of a unit of effective precipitation among surface (Q) and subsurface (q) flow, in each case allocated into concentrated (subscript c) or diffuse (d) flow. Included links are based on field observations in the study area.

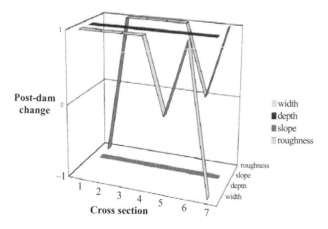

Figure 2 General pattern of increases (1), decreases (−1), or negligible change (0) at seven cross-sections in a 55 km reach downstream of Livingston Dam on the Trinity River, Texas, following dam construction.

a persuasive argument in an applied context, as questions such as whether or not gully erosion may be initiated on rangeland as a consequence of environmental change are far more important than predicted rates of soil removal or gully incision. The types of analyses recommended and illustrated, while rigorous and mathematical, are essentially phenomenological and qualitative or semi-quantitative. This approach has been successful in modeling and explaining (among other things) landslides, aeolian dunes, soil erosion, beaches, glaciers, channel networks, and periglacial patterned ground (Hergarten 2002; Werner and Fink 1994; Werner 1995; Favis-Mortlock 1998; Masselink 1999; Bahr and Meier 2000; De Boer 2001). Methodologically distinct but conceptually similar is qualitative modeling based on the set of positive, negative, or negligible interrelationships among the key components of a geomorphic system. Originally conceived (or at least perceived) as an expedient in the absence of data or knowledge necessary to fully specify the quantitative relationships, a number of authors have pointed out that qualitative models actually increase the generality of the results (Escultura 2001; Harrison 1999; Phillips 1992, 1999a; Phillips and Walls 2004; Slingerland 1981; Trofimov and Moskovkin 1984). While the quantitative aspects of many processes and relationships are highly variable, the qualitative features may be universal (for example fully developed turbulence; Escultura 2001; Tsinober 1998; weathering and erosion; Phillips 2005). Specific quantitative relationships between vegetation cover and erosion, for instance, are strongly variable in space and time, while the qualitative link (more vegetation cover = greater resistance) applies always and everywhere. Qualitative stability models have been particularly successful in ecology (see reviews by Logofet 1993; Pahl-Wostl 1995), but there are also several examples of fruitful applications in geomorphology (see reviews by Phillips 1999a, 2005).

Hydrology faces closely related problems of contingency (Beven 2000). The dominant processes concept (DPC) is a recognition that there are difficulties in trying to model all potentially relevant processes. These difficulties, along with field observations that often only a few processes dominate hydrological responses in any watershed, and the cumulative experience of modellers, which suggests that simple models with a few dominant factors can capture the essential features of hydrologic response (Sivakumar 2004b). Hydrological analysis should therefore be based on simpler models and fewer processes, but with the included processes tied to local conditions in individual watersheds.

Examples

Below, I will highlight two attempts by myself and co-workers to implement the general approach described above. This is in the spirit of practicing what one preaches; I am not suggesting these as exemplars. For the latter, at least in an applied framework, I recommend Brierly and Fryirs (2005) recent book on geomorphology and river management.

Michael Walls and I (Phillips and Walls 2004), used an approach similar to the DPC in our study of divergent evolution of fluviokarst landscapes in central Kentucky. A qualitative model of flow partitioning (Fig.1) between surface and subsurface, and between concentrated and diffuse, flow was used to explain the tendency of the most eroded portions of the study area to diverge into either strongly karstified zones with few or no channels, or fluviallydissected zones with few solutional landforms. The qualitative model is very general in that it is based on a universal mass balance principle, whereby all effective precipitation (precipitation minus evapotranspiration) must flow or be stored either at the surface or subsurface, and in either case is either concentrated or diffuse. Even as applied to the study area this partitioning does not depend on specific, necessarily local, parameterizations.

Conversely, some of the links in the model are not universal, and the sign of those links in our model was based on conditions and field observations within the study area. The model is entirely "competitive" in the sense that once moisture is in one of the four states it remains there. While clearly true on an instantaneous basis (water cannot be in two places simultaneously), in some systems and situations transfers such as subsurface to surface return flow or exfiltration or channelization of diffuse flow are important. This particular system construction is based on our study area-specific observations that the initial disposition of runoff is the critical factor for landform development (Phillips et al. 2004; Phillips and Walls 2004). Results are not applicable to all fluviokarst landscapes, but are potentially relevant to those where the links in the flow partitioning model are the same as in the inner Bluegrass region of Kentucky. The second example involves downstream geomorphic effects of a dam, viewed as an opportunistic experiment to assess what happens if (in this case) sediment load is drastically reduced without significant change in the discharge regime. The interrelationships between width, depth, velocity, roughness, and slope at a cross-section are dynamically unstable, indicating multiple modes of adjustment, complex responses, and an inability to predict even qualitative responses without specific information at each cross-section. Thus the response of the Trinity River channel (southeast Texas) is characterized by qualitatively different combinations of increases, decreases and relative constancy of channel width, depth, slope, and roughness following construction of Livingston Dam (Phillips et al. 2005).

The qualitative system model of the interaction of the hydraulic variables, the instability of these interrelationships, and the implications thereof, are universal and applicable to any fluvial system. However, the range of possible responses is constrained not only by fundamental flow resistance hydraulics, but also by the systematic qualitative relationships between discharge, slope, sediment load, and grain size that underpin essentially all hydraulic geometry models (Phillips et al. 2005). These in turn depend on site-specific and historically contingent factors such as substrate and bank material, geologic setting and history, vegetation, flood history, and many other factors. For example, adjustments of depth and slope via incision in our study reach are limited at present because the Trinity River has already incised to resistant bedrock. In this example, the applicable global laws constrain the possible responses, guide the interpretation of observed responses, and allow for some generalization to other cases. However, these laws cannot (by themselves) explain changes at specific cross-sections, and allow for multiple outcomes. The specific place- and time-specific details of the study area can, in conjunction with the laws, explain observations and allow predictions (Phillips et al. 2005). However, by themselves these local details have little or no relevance beyond the study area. The combination of the two, as mandated by an evolutionary approach to geomorphology, as necessary to explain the system under study.

Conclusions

Thresholds, nonlinearity, and complex dynamics in geomorphic systems suggest that we are quite limited in discerning universal laws applicable to predicting geomorphic response to environmental change. Rather, the suggestion is to refocus on a search for lessons—typologies, patterns, and synoptic situations we can learn from. In that spirit, the major proposed lessons of this paper can be summarized as follows:

- Geomorphic systems are typically nonlinear, owing largely to their threshold-dominated nature (but due to other factors as well).
- Nonlinear geomorphic systems are capable of complex behaviors not possible in linear systems, including dynamical instability and deterministic chaos.
- Dynamical instability and chaos are common in geomorphic systems, indicating that small, short-lived changes may produce disproportionately large and long lived results; that evidence of geomorphic change may not reflect proportionally large external forcings; and that geomorphic systems may have multiple potential response trajectories or modes of adjustment to change.
- Instability and chaos do not preclude predictability, but do modify the context of predictability. The presence of chaotic dynamics inhibits or excludes some forms of predictability and prediction techniques, but does not preclude, and enables, others.
- Geography matters.
- History matters.
- While the geographical and historical contingency indicated above would occur independently of complex nonlinear dynamics, instability and chaos dictate that such contingency is important.
- Geomorphic systems are thus governed by a combination of "global" laws, generalizations and relationships that are largely (if not wholly) independent of time and place, and "local" place and/or time-contingent factors.
- The more components, variables or processes included in the representation of any geomorphic system, the more singular the results or description are.
- Generalization is enhanced by reducing rather than increasing the number of factors considered.
- Prediction of geomorphic responses calls for a recursive approach whereby global laws and local contingencies are used to constrain each other. More specifically, I advocate a methodology whereby local details (be they process mechanical, historical, or both) are embedded within simple but more highly general phenomenological models. There are examples of successful applications of the advocated approach.

Landscapes and landforms change over time, and in response to changes in climate and other external forcings. It cannot be assumed that geomorphic systems progress along any particular pathway, whether that pathway leads to a steady-state form, a peneplain, or any other predordained endpoint. Geomorphic systems are evolutionary in the sense of being path dependent, and historically and geographically contingent. Assessing and predicting geomorphic responses obliges us to engage these contingencies, which often arise from nonlinear complexities. We are obliged, then, to practice evolutionary geomorphology.

Acknowledgements

The invitation from M. Sivapalan, C. Hinz, and G. Hancock to present these ideas at a conference in 2005 was instrumental in leading to the crystallization of thoughts expressed in this paper. Helpful review comments from A. James are greatly appreciated.

Edited by: M. Sivapalan.

References

1. Abrahams, A.D., Parsons, A.J., and Wainwright, J.: Effects of vegetation change on interrill runoff and erosion, Walnut Gulch, Arizona, in: Biogeomorphology, Terrestrial and Freshwater Systems, Proceedings of the 26th Binghamton Geomorphology Symposium, edited by: Hupp, C.R., Osterkamp, W., and Howard, A., Amsterdam, Elsevier, 37–48, 1995.
2. Atkinson, P.M., German, S.E., Sear, D.A.,and Clark, M.J.: Exploring the relations between riverbank erosion and geomorphological controls using geographically weighted logistic regression, Geogr. Anal., 35, 58–83, 2003.
3. Baas, A.C.W.: Chaos, fractals, and self-organization in coastal geomorphology: simulating dune landscapes in vegetated environments, Geomorph., 48, 309–328, 2002.
4. Bahr, D.B. and Meier, M.F: Snow patch and glacier size distributions, Water Resour. Res., 36, 495–501, 2000.
5. Baker, V.R.: Hypotheses and geomorphological reasoning, in: The Scientific Nature of Geomorphology, edited by: Rhoads, B.L. and Thorn, C.E., Wiley, New York, 57–86, 1996.
6. Barton, C.C., Dismukes, J.S., and Morton, R.A.: Complexity analysis of the change in shoreline position at Duck, North Carolina, in: Coastal Sediments '03, edited by: Davis, R.A., American Society of Civil Engineers, New York, 1–7, 2003.
7. Begin, Z.B. and Schumm, S.A.: Gradational thresholds and landform singularity: significance for Quaternary studies, Quat. Res., 27, 267–274, 1984. Beven, K.J.: Uniqueness of place and process representations in hydrological modeling, Hydrol. Earth Syst. Sci., 4, 203–213, 2000, http://www.hydrol-earth-syst-sci.net/4/203/2000/.
8. Bishop, P.: Griffith Taylor and the southeast Australia highlands: issues of data sources and testability in interpretations of longterm drainage history and landscape evolution, Austral. Geog. 29, 7–29, 1998.
9. Bretz, J.H.: Dynamic equilibrium and the Ozark land forms, Am.J. Sci., 260, 427–438, 1962. Brierly, G.J. and Fryirs, K.A: Geomorphology and River Management. Blackwell, Oxford, UK, 2005.
10. Brunsden, D.: Applicable models of long term landform evolution, Z. Geomorph., suppl. 36, 16–26, 1980.
11. Carling, P. and Beven, K.: The hydrology, sedimentology, and geomorphological implications of floods: an overview, in: Floods: Hydrological, Sedimentological, and Geomorphological Implications, edited by: Beven, K. and Carling, P., Wiley, Chichester, 1–9, 1989.
12. Chappell, J.: Thresholds and lags in geomorphologic changes, Aust. Geog., 15, 358–366, 1983.
13. Christakos, G.: On the assimilation of uncertain physical knowledge bases: Bayesian and non-Bayesian techniques, Adv. Water Resour., 25, 1257–1274, 2002.
14. Christofoletti, A: Noc ̧ ̃oes b ́asica para a compreens ́ao da establidade e mudanc ̧as nos sistemas geograficos. Cîencia Geografica 4, 13–19, 1998.
15. Coates, D.R. and Vitek, J.V. (Eds.): Thresholds in Geomorphology, Unwin Hyman, London, 1980.
16. Dearing, J.A. and Zolitschka, B.: System dynamics and environmental change: an exploratory study of Holocene lake sediments at Holzmaar, Germany, The Holocene, 9, 531–540, 1999.

17. DeBoer, D.H.: Self-organization in fluvial landscapes: sediment dynamics as an emergent property, Comp. Geosci., 27, 995–1003, 2001.

18. Dikau, R. and Schrott, L.: The temporal stability and activity of landslides in Europe with respect to climate change (TESLEC): main objectives and results, Geomorph., 30, 1–12, 1999.

19. Escultura, E.E.: Turbulence: theory, verification, and applications, Nonlinear Anal., 47, 5955–5966, 2001. Favis-Mortlock, D.: A self-organizing dynamic systems approach to the simulation of rill initiation and development on hillslopes, Comp. Geosci., 24, 353–372, 1998.

20. Fotheringham, A.S. and Brunsdon, C.: Local forms of spatial analysis, Geogr. Anal., 31, 340–358, 1999.

21. Friedman, J.M., Osterkamp, W.R., Scott, M.L., and Auble, G.T.: Downstream effects of dams on channel geometry and bottomland vegetation: regional differences in the Great Plains, Wetlands, 18, 619–633, 1998.

22. Fryirs, K.: Antecedent landscape controls on river character, behavior, and evolution at the base of the escarpment in Bega catchment, New South Wales, Australia, Z. Geomorph., 46, 475–504, 2002.

23. Gaffin, S.R. and Maasch, K.A.: Anomalous cyclicity in climate and stratigraphy and modeling nonlinear oscillations, J. Geophys. Res., 96B, 6701–6711, 1991.

24. Gomez, B., Page, M.J., Bak, P., and Trustrum, N.A.: Selforganized criticality in layered lacustrine sediments formed by landsliding, Geol., 30, 519–522, 2002.

25. Goovaerts, P.: Geostatistics in soil science: state-of-the-art and perspectives, Geoderma, 89, 1–45, 1999.

26. Haines-Young, R. and Chopping, M.: Quantifying landscape structure: a review of landscape indices and their application to forested landscapes, Prog. Phys. Geog., 20, 418–445, 1996.

27. Harrison, S.: The problem with landscape, Geogr., 84, 355–363, 1999. Hergarten, S.: Self-organized Criticality in Earth Systems, Springer, Berlin, 2002.

28. Huggett, R.J.: Geoecology: An Evolutionary Approach, Routledge, London, 1995.

29. Huggett, R.J.: Environmental Change. The Evolving Ecosphere, Routledge, London, 1997.

30. Ibanez, J.J., De-Alba, S., Bermudez, F.-F., and Garcia-Alvarez, A.: Pedodiversity: concepts and measures, Catena 24, 215–232, 1995.

31. Jaffe, B.E. and Rubin, D.M.: Using nonlinear forecasting to learn the magnitude and phasing of time-varying sediment suspension in the surf zone, J. Geophys. Res., 101C, 14 283–14 296, 1992.

32. Johnson, D.L., Keller, E.A., and Rockwell, T.K.: Dynamic pedogenesis: new views on some key soil concepts, and a model for interpreting Quaternary soils, Quat. Res. 33, 306–319, 1990.

33. Johnson, D.L. and Watson-Stegner, D.: Evolution model of pedogenesis, Soil Sci., 143, 349–366, 1987.

34. Kirkby, M.J.: Modelling the links between vegetation and landforms, in: Biogeomorphology, Terrestrial, and Aquatic Systems, edited by: Hupp, C.L., Osterkamp, W.R., and Howard, A., Elsevier, Amsterdam, 319–335, 1995.

35. Knighton, A.D. and Nanson, G.C.: An event-based approach to the hydrology of arid zone rivers in the channel country of Australia, J. Hydrol., 254, 102–123, 2001.

36. Knox, J.C.: Sensitivity of modern and Holocene floods to climate change, Quat. Sci. Rev. 19, 439–451, 2000.

37. Lall, U., Sangoyami, T., and Abarbanel, H.D.I.: Nonlinear dynamics of the Great Salt Lake: nonparametric short-term forecasting, Water Resour. Res., 32, 975–985, 1996.

38. Lane, S.N. and Richards, K.S.: Linking river channel form and process: time, space, and causality revisited, Earth Surf. Proc. Landf., 22, 249–260, 1997.

39. Lecce, S.A., Pease, P.P., Gares, P.A., and Rigsby, C.A.:Floodplain sedimentation during an extreme flood: the 1999 flood on the Tar River, eastern North Carolina, Phys. Geog., 25, 334–346, 2004.

40. Logofet, D.O.: Matrices and Graphs: Stability Problems in Mathematical Ecology, CRC Press, Boca Raton, FL, 1993.

41. Magilligan, F.J.: Thresholds and the spatial variability of flood power during extreme floods, Geomorph., 5, 373–390, 1992.

42. Magilligan, F.J., Phillips, J.D., Gomez, B., and James, L.A.: Geomorphic and sedimentological controls on the effectiveness of an extreme flood, J. Geol., 106, 87–95, 1998.

43. Masselink, G.: Alongshore variation in beach cusp morphology in a coastal embayment, Earth Surf. Proc. Landf., 24, 335–347, 1999.

44. Miller, D., Luce, C., and Benda, L.,: Time, space, and episodicity of physical disturbance in stream, For. Ecol. Manage., 178, 121–140, 2003.

45. Nelson, J.M.: The initial instability and finite-amplitude stability of alternate bars in straight channels, Earth-Sci. Rev., 29, 97–115, 1990.

46. Nelson, A.: Analyzing data across geographic scales in Honduras: detecting levels of organization within systems, Agric., Ecosys. Environ., 85, 107–131, 2001.

47. Ollier, C.D.: Evolutionary geomorphology of Australia and Papua New Guineau, Trans. Inst. Brit. Geog. NS4, 516–539, 1979.

48. Pahl-Wostl, C.: The Dynamic Nature of Ecosystems, Chaos and Order Entwined, John Wiley, Chichester, UK, 1995.

49. Phillips, J.D.: Qualitative chaos in geomorphic systems, with an example from wetland response to sea level rise, J.Geol., 100, 365–374, 1992.

50. Phillips, J.D.: Stability implications of the state factor model of soils as a nonlinear dynamical system, Geoderma, 58, 1–15, 1993.

51. Phillips, J.D.: Earth Surface Systems, Complexity, Order, and Scale, Basil Blackwell, Oxford, UK, 1999a.

52. Phillips, J.D.: Divergence, convergence, and self-organization in landscapes, Ann. Assoc. Am. Geog., 89, 466–488, 1999b.

53. Phillips, J.D.: Event timing and sequence in coastal shoreline erosion: Hurricanes Bertha and Fran and the Neuse estuary, J. Coas. Res., 15, 616–623, 1999c.

54. Phillips, J.D.: Sources of nonlinear complexity in geomorphic systems, Prog. Phys. Geog., 26, 339–361, 2003a.

55. Phillips, J.D.: Alluvial storage and the long term stability of sediment yields, Bas. Res., 15, 153–163, 2003b.

56. Phillips, J.D.: Divergence, sensitivity, and non-equilibrium in ecosystems, Geogr. Anal., 36, 369–383, 2004.

57. Phillips, J.D.: Weathering, instability, and landscape evolution, Geomorph., 67, 255–272, 2005.

58. Phillips, J.D.: Deterministic chaos and historical geomorphology: a review and look forward, Geomorph., 76, 109–121, 2006.

59. Phillips, J.D., Martin, L.L., Nordberg, V.G., and Andrews, W.A.: Divergent evolution in fluviokarst landscapes of central Kentucky, Earth Surf. Proc. Landf., 29, 799–819, 2004.

60. Phillips, J.D., Slattery, M.C., and Musselman, Z.A.: Channel adjustments of the lower Trinity River, Texas downstream of Livingston Dam, Earth Surf. Proc. Landf., 30, 1419–1439, 2005.

61. Phillips, J.D. and Walls, M.D.: Flow partitioning and unstable divergence in fluviokarst evolution in central Kentucky, Nonlin. Processes Geophys., 11, 371–381, 2004, http://www.nonlin-processes-geophys.net/11/371/2004/.

62. Pickup, G.: Event frequency and landscape stability on the floodplain systems of arid central Australia, Quat. Sci. Rev., 10, 463–473, 1991.

63. Puigdefabregas, J. and Sanchez, G.: Gemorphological implications of vegetation patchiness on semi-arid slopes, in: Advances in Hillslope Processes, edited by: Anderson, M.G. and Brooks, S.M., John Wiley, Chichester, UK, 1027–1060, 1996.

64. Rey, V., Dames, A.G., and Belzons, M.: On the formation of bars by the action of waves on an erodible bed: A laboratory study, J. Coas. Res., 11, 1180–1194, 1995.

65. Richards, A., Phipps, P., and Lucas, N.: Possible evidence for underlying non-linear dynamics in steep-faced glaciodeltaic progradational successions, Earth Surf. Proc. Landf., 25, 1181–1200, 2000.

66. Rubin, D.M.: Use of forecasting signatures to help distinguish periodicity, randomness, and chaos in ripples and other spatial patterns, Chaos 2, 525–535, 1992.

67. Sangoyomi, T.B., Lall, U., and Abarbanel, H.D.I.: Nonlinear dynamics of the Great Salt Lake: Dimension estimation, Water Resour. Res., 32, 149–159, 1996.
68. Sauchyn, D.J.: Modeling the hydroclimatic disturbance of soil landscapes in the southern Canadian plains: the problems of scale and place, Environ. Mon. Assess., 67, 277–291, 2001.
69. Schumm, S.A.: Geomorphic thresholds: the concept and its applications, Trans. Insti. Brit. Geogr., NS4, 485–515, 1979.
70. Schumm, S.A.: To Interpret the Earth. Ten Ways to Be Wrong, New York, Cambridge University Press, New York, 1991.
71. Schumm, S.A., Dumont, J.F., and Holbrook, J.M.: Active Tectonics and Alluvial Rivers, Cambridge University Press, New York, 2000.
72. Sivakumar, B.: Chaos theory in hydrology: important issues and interpretations, J. Hydrol., 227, 1–20, 2000.
73. Sivakumar, B.: Chaos theory in geophysics: past, present, and future, Chaos, Sol. Fract., 19, 441–462, 2004a.
74. Sivakumar, B.: Dominant processes concept in hydrology: moving forward, Hydrol. Proc., 18, 234–2353, 2004b.
75. Slattery, M.C., Gares, P.A., and Phillips, J.D.: Multiple modes of runoff generation in a North Carolina coastal plain watershed, Hydrol. Proc., in press, 2006.
76. Slingerland, R.: Qualitative stability analysis of geologic systems with an example from river hydraulic geometry, Geol., 9, 491–493, 1981.
77. Spedding, N.: On growth and form in geomorphology, Earth Surf. Proc. Land., 22, 261–265, 1997.
78. Thomas, M.F.: Landscape sensitivity in time and space—an introduction, Catena, 42, 83–98, 2001.
79. Thornes, J.D.: Evolutionary geomorphology, Geogr., 68, 225–235, 1983.
80. Thornes, J.B.: The ecology of erosion, Geogr., 70, 222–235, 1985.
81. Trofimov, A.M. and Moskovkin, V.M.: The dynamic models of geomorphological systems, Z. Geomorph., 28, 77–94, 1984.
82. Tsinober, A.: Turbulence: beyond phenomenology, in: Chaos, Kinetics, and Nonlinear Dynamics in Fluids and Plasmas, edited by: Benkadda, S. and Zaslavsky, G.M., Springer, Berlin, 85–143, 1998.
83. Vandenberghe, J.: The relation between climate and river processes, landforms, and deposits during the Quaternary, Quat. Internat., 91, 17–23, 2002.
84. Walsh, S.J., Butler, D.R., and Malanson, G.P.: An overview of scale, pattern, process relationships in geomorphology: a remote sensing perspective, Geomorph., 21, 183–205, 1998.
85. Werner, B.T.: Eolian dunes: computer simulation and attractor interpretation, Geol., 23, 1107–1110, 1995.
86. Werner, B.T.: Complexity in natural landform patterns, Science, 284, 102–104, 1999.
87. Werner, B.T. and Fink, T.M.: Beach cusps as self-organized patterns, Science, 260, 968–971, 1994.

Section IV

Models and Applications of Chaos Theory in Economy

1. **Physics and the Foundations of Economic Science: Comments in Memory of Ilya Prigogine**
   Richard H. Day   217

2. **Neumannian Economy in Multi-Agent Approach Investigation of Stability and Instability in Economic Growth**
   Katalin Martinás   226

3. **Theoretical Aspects of the Economic Transition: Case of Romania**
   Cezar Scarlat and *Eugen I. Scarlat*   236

4. **Complex Dynamics in a Nonlinear Cobweb Model for Real Estate Market**
   Junhai Ma and *Lingling Mu*   256

5. **Chaos Models in Economics**
   Sorin Vlad, Paul Pascu and *Nicolae Morariu*   269

6. **Complex Dynamics of an Adnascent-type Game Model**
   Baogui Xin, Junhai Ma and *Qin Gao*   278

7. **A Production-Inventory Model for Deteriorating Items with Production Disruptions**
   Yong He and *Ju He*   290

8. **Nonlinear Noise Estimation in International Stock Markets: Coarse-Grained Entropy Method**
   Yong Fang   304

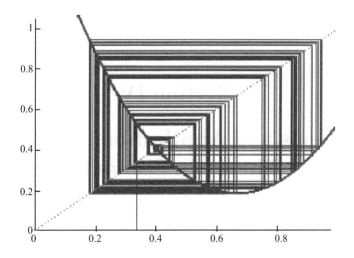

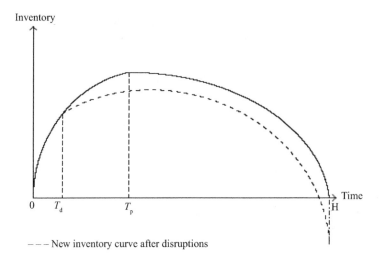

‒ ‒ ‒ New inventory curve after disruptions

Physics and the Foundations of Economic Science: Comments in Memory of Ilya Prigogine[†]

Richard H. Day

Department of Economics, University of Southern California, Los Angeles, CA, 90089-0253, USA.
Email: rday@usc.edu.

ABSTRACT

This paper discusses how the concepts of nonlinearity, discrete dynamics, chaos, and emergent order that Ilya Prigogine advanced correspond to similar ideas involved in the adaptive, evolutionary branch of economics.

Ilya Prigogine and Interdisciplinary Science

Ilya Prigogine must always have had his eye out for kindred spirits, individuals who loved science-all science-and who were contributing to some science in particular. That may be why he appeared so often at multidisciplinary conferences and why he sponsored so many such events himself. I first met him at one such conference, one devoted to nonlinear dynamics sponsored by the U.S. Department of Transportation under the leadership of Robert Crosby. That institution's immediate concern was the growth of cities and the complex, developing transportation networks that cities require. Crosby had already contracted with Ilya's Solvay Institute to study the phenomenon. Bob's conference brought mathematicians, physicists, and regional scientists together to consider the methodological problems involved, to inform one another of the various approaches being taken already, or that seemed to have potential, and to expose commonalities among them: nonlinearities of causal structure and complex patterns of system development over time and space (see [4]).

There followed a series of even more inclusive multidisciplinary conferences sponsored by Prigogine at his centers in Brussels and Austin with specialists in mathematics, physics, biology, and economics who, in spite of their varied interests and expertise, shared

[†]Reused with permission from: Richard H. Day, "Physics and the foundations of economic science: comments in memory of Ilya Prigogine," *Discrete Dynamics in Nature and Society*, vol. 2004, no. 1, pp. 91–99, 2004. doi:10.1155/S1026022604312045.

common scientific visions. These conferences provided an outlet for research that was pressing beyond the boundaries of "standard" science. They helped establish a critical mass of intellectual work on nonlinear dynamics and evolutionary processes that has itself become a main stream. Many who might have remained in their lonely quests thus found a congenial intellectual home. Prigogine successfully created a main stream into which his own work could flow and in which many lesser fish could swim (see [1,4,5]).

These informal remarks consider the nature of physics, mathematics, and economics vis-à-vis the complex, evolutionary world which was Ilya's life work to comprehend.

Time, Instability and Novelty

How can we relate these various meanings of time-time as motion, time related to irreversibility as in thermodynamics, time as history as in biology and sociology?

Prigogine [12]

Prigogine shared with many philosophers an obsession with time. What is *it*? Is *it* reversible or irreversible? Is *it* something real, something that could itself be explained by or inferred from more fundamental phenomena? Is *it* simply another dimension like space, so that its passing is merely an illusion?

Whatever *it* is, it is an ordering of events that are experienced and remembered or recorded. Without *it* we should not carry in our minds the partially remembered story of our lives, no joys to be savored, no lessons to be learned, no regrets to be endured: we should have no past. We should not be able to orient our actions toward any anticipated goal, for there could be no anticipation, no problems to solve, no choices to be made; in short, no future. So time exists for us and it has value, like any good in short supply.

Given a finite chain of time-ordered events associated with a distinguishable process and given quantitative indexes that describe the process at each "time," one can inquire about the past or the future of the quantitative measures or states. Are they growing or shrinking, or some growing and some shrinking? Are they heading for a steady state where all the indexes would be changing at exactly the same rate, proportional to one another? Are they heading for a stationary state where the indexes become constant? Or, in between these extremes, are they approaching some intertemporal pattern that repeats itself at regular intervals-or fails ever to repeat itself, always generating states that form no identifiable pattern, that do not converge to stationary, steady, or periodic states? These are the mathematical questions posed in the dynamic analysis of observable, quantifiable processes.

If a process converges to a periodic, steady, or stationary state, it is asymptotically stable, becoming ever more predictable with the passage of time. If on the other hand a process is irregular (chaotic), its trajectories are unstable in that its states do not converge to predictable patterns, yet are stable in the sense of being confined, that is, bounded and predictable in a statistical, probabilistic sense. Finally, quantum processes are defined in terms of "waves" or distributions of conceptual events, distributions that change their "shape" over time so that the chance of experiencing (at the human scale) any given event may itself change over time.

The instability that arises in human experience would seem to be pervasive: it is far more difficult to predict human events that will take place in the future than to explain those that occurred in the past. Wassily Leontief, in a perceptive essay, implied that human history was essentially unstable and suggested that, like geologists and paleontologists, the historian interested in isolating the cause of changes should start with a given set of known recent events, then identify their immediate antecedents, rather than start with an early situation and try to explain what happened years later as a result (see [9,10]). Thus, to provide a variation on Leontief's parable, if the historian wanted to explain why Alexander's empire split into three soon after his death, it would be futile to derive an explanation based on the battle where Alexander received his fateful injury. Rather, going backwards from the split through the sequence of events that preceded it, finally arriving at the injury that led to his demise before the structure of empire could be solidified, one would see that one event inevitably precedes another, even if the ultimate "cause" was Alexander's hubris in scaling the Mallian fortress wall and assaulting the defenders by himself (see [2, page 311]). The successive swallowing up of Greece in the Roman, Byzantine, and Ottoman Empires follows in no obvious way from that isolated event in Asia that preceded them. Yet, proceeding backwards, an understanding might emerge as to why history churned out that sequence of events and not another.

As a kind of mathematical "outline of history," I constructed a discrete time "regular mountainous" map that generates chaotic paths of population numbers through the mega stages of demoeconomic development. These time paths mimic in a qualitative way the progression of the world's population through the hunting and food collecting band, settled agriculture, city states, trading empires, the nation state, and the global information economy with growth and fluctuations in population numbers within stages, and with switching to more advanced or reverting to less advanced stages, more or less as it actually occurred in the records archaeologists and historians have constructed. Defining a scenario as a particular sequence of stages and using *backwardly* iterated, multivalued maps, it is possible in principle to derive the set of initial conditions in the first stage that would lead to any given hypothetical scenario. Contrastingly, picking initial conditions that would generate any specified scenario would have a finite but very small probability of success (Day [7, Chapters 21–24]).

Thus, by proceeding backwards as Leontief advised, a sequence of phases through which history passed can be identified, from which the causal dynamics can be inferred. Looking ahead we can dimly perceive the fundamental instabilities that will eventually drive us from our current system of relationships into a new phase of history. What momentous decisions will be made, what new forms of organization and technology will emerge, and what will remain of those now existing can only be guessed.

Here is something just as baffling as the events that hint at subatomic structure: the emergence of novelty. Modern cosmology describes the universe as a process of change in which successive structures emerge from their predecessors: particles, atoms, molecules, complex compounds, the various stellar bodies and systems, our solar system, the emergence of living things, culminating with humans as the most intelligent and creative among the sentient beings. With humans, novelty follows after novelty with each individual, a unique, complex biochemical, evolutionary process that begins and ends, that is manifestly stable in its persistence, manifestly unstable in its growth, transformation, and death.

Scale and Theory

If there is simplicity somewhere in physics and chemistry, it is not in the microscopic models. It lies more in idealized macroscopic representation.

Prigogine [12]

In the explanations of process, scale is everything. Nature appears to us as a hierarchy of nested systems: a solar system in the cosmos, our world and its geological dynamics, its biological covering of interacting species; each a physiology of organs and processes, ultimately made up of molecules, in turn comprised of atoms, themselves a baffling quantum manifestation of matter and energy. The behavior of phenomena at a given scale exhibits patterns of movement and transformation that seem to be comprehensible *at that scale*. The manner in which patterns of movement of systems at one scale depend upon the hierarchy of constituent processes within these systems is often obscure.

Lucky for us! Thus, a thermometer tells a physician what needs to be known about the movement of atoms within a sick patient, and Newton's equations, appropriately augmented with perturbation terms to make up for the departures of nature from mathematical assumptions, guide astronauts to the moon with no knowledge whatsoever of curvature in space-time or of the manner in which electrons distribute their paths through slits in a screen. At each level in the hierarchy, physical systems work according to laws based on quantitative measures that emerge from phenomena *at that level*-not all aspects of behavior, but some-those that can be represented by quantitative indexes. Those physicists who think themselves building a single theory of everything will have to find a way to bridge the gaps between the theories that explain events at different scales. As far as I understand the situation at present, not all of the bridges have been built.

As one passes from physical to social science, one finds similar issues confronting the theorist. How can the thought and behavior of individuals at one level be related to the play of chemicals in the brain below and to the discourse among people in a group above? How is the rationality of the individual related to the goals adopted by a family, a business firm, or a unit of government?

In economics, general equilibrium theory characterizes the interaction of individuals or families and business firms, each of whom thoughtfully pursues wants and needs or profits. Using prices, they allocate their resources and operate their production capabilities in the best way, which-given equilibrium prices-leads to compatible choices by everyone. Macroeconomic theorists initially leapt from this characterization to aggregate demand and supply relationships among indexes of goods, money, and prices, leaving implicit any connection to component organizations specified at the microeconomic level (Keynes [8] originated the theory; Tinbergen [15] originated its empirical implementation). The economy so represented is viewed as a macroscopic phenomenon with the hope that laws (or statistical relationships) of aggregate demand and supply will emerge at that level. It is analogous to the way Boyles law describes a relationship between temperature and pressure with no explicit connection to the underlying dynamics of the constituent molecules, or much as Newton's equations of motion applied to the earth and its moon are based solely on the aggregate index of mass with no reference to the complex structures and the dynamics of their movement that make up the two aggregate bodies.

An alternative approach represents the macroscopic economy as a set of identical individuals or, equivalently, as a single person with assumed preferences that order all possible trajectories of aggregate output, consumption, investment, and capital accumulation according to the statistical expectation of their discounted values. Such a radically reductionist approach models the growth of the GDP index remarkably well but at the cost of squeezing out any room for differences in the behavior of savings and investment, between labor demand and supply, or in the role of government agencies, or a monetary authority; in short, squeezing out all semblance of coordination or lack thereof (see [3]). It is analogous to a theory of an ideal gas based on a single representative molecule which at a speed near that of light could follow an ergodic trajectory that during any tiny interval of time visited every tiny area of surface a sufficient number of times to produce a pressure that could float a balloon. Has a theorem of that kind been proved? And if so, should we be convinced that it was the right characterization of the macroscopic behavior of an ensemble of gas molecules in a container? Analogously, is it reasonable to characterize an economy as if it were a single omniscient person? Certainly, on the face of it, the answer is obviously "No!" Nonetheless, the estimated "time-preference" parameter can be thought of as a cultural constant that characterizes the average propensity to save in an economy as a whole.

Mathematics and Experience

In continuous dynamics, time is represented by the real numbers, and in discrete dynamics by the integers with any initial condition associated with zero and "now" represented by t, a real or integer number, as the case may be. The simplest possible example of reversibility of the index t is in a linear difference equation,

$$x_{t+1} = ax_t \tag{1}$$

The variable x after s periods have passed is

$$x_{t+s} = a^s x_t. \tag{2}$$

If we think of "t" as "now," then $a^s x_t$ is what x is expected to be s periods in the future. After those s periods have passed, we can say that x_t "occurred" s periods in the past. Thus,

$$x_t = a^{-s} x_{t+s}. \tag{3}$$

This mathematical process, however, does not reverse the order of time. Such an equation *represents* in a convenient notation a simple growth or decay process but the equation itself *is not a simple growth or decay process*. It is *just symbols on a page* that can be manipulated in a way that characterizes a mental image of something salient about an ordering of events that behave in a very specific way over time and that enables us to tell where it is going and whence it came. But when I think of my childhood-no matter how vivid the memory-I have not reversed the passage of time for I continue to age as every moment passes. It is the ordering of memory that makes possible each mental history; it is the ordering of the physiological process of living matter which guarantees that history will cease to exist.

Philosophically minded physicists sometimes appear to confuse the mathematics of a theory with the subject matter of the theory itself (the reversibility of the time index in an equation with the reversibility of a physical process of some kind). On the other hand, they sometimes seem to want material things to explain everything such as jumping gravitons holding things together. It is the irreversibility of the *physical process* that intrigued Prigogine. As I see it, mathematics is *never* more than a symbolic model of something or some process. It is *always* abstract, always approximate, always resting ultimately on untested and untestable assumptions.

Mind Over Matter

We do not find theories in nature *outside the mind*, but we can understand and talk to each other about our understanding of how things work. Still more wondrous, the human mind-by thinking-can conjure the image of objects that do not exist, stories of events that might happen in the future, then fashion and follow sequences of actions that make those objects exist, that turn those stories into history. Imagine a molecule in Alexander's white plume, the one he always wore in battle so his troops could always know where he was. Is there any law of physics, or can there ever be a law of physics that could explain the trajectory of that molecule through time? Alexander's example proves that in history human volition has the potential to make things happen, to *cause* events ("Wilken...pointed out this moment of decision [to move on Egypt instead of pursuing Darius] is one of history's great proofs that individuals, not mere economic forces, can change the destinies of mankind" [13, page 116]).

This capacity of mind is not just wondrous, it is also ordinary. Every sane person possesses it to some degree, for every person thinks of and chooses to do things *they* have never previously done. Every person recognizes degrees of freedom that provide scope for choice, is aware of constraints that bound potential action, and considers preferences that connect potential actions with the anticipated satisfaction of needs and wants. Every sane person exercises self-control that guides action along chosen paths, yet possesses the flexibility to change course in midstream in response to developing external situations that induce a reconsideration in the perceived possibilities and anticipated satisfactions.

Economists have formalized these attributes of mind in an abstract mathematical theory of rational choice, progressively refining and generalizing it. Incorporated into the mathematical model of a general equilibrium, it represents in an abstract and idealized way a decentralized market economy in a state of perfect coordination, mediated by equilibrium prices. It has the practical significance of hinting how and why market economies have flourished relative to the nonmarket alternatives. I say "hinting" because, though it captures the rational aspect of individual human action, it does so at the expense of inadequately accounting for the boundedness and fallibility of rational thought and the evolutionary character of economic life. Rational thought is fallible because knowledge of options and the processes that must be used to execute plans is incomplete; it is fallible because choosing is a problem of constrained optimization, a problem that can usually be solved only approximately, or not at all, the first often most difficult step being to find a solution that is feasible.

The result is that human action cannot normally be based on formal thinking but on other modes for determining a course of action: imitating others, trial and error search or experimentation, consulting an authority, doing what you did before as long as it "works," or acting on a "hunch" or intuition, or even thoughtless impulse. Economic science is rapidly expanding its inquiry into these alternative modes of behavior. Such inquiries are the economist's equivalent of looking inside the atom.

We are a social species so many, if not most of our actions are *interactions* with others. We begin in some sort of "family," go to schools, play in sports and music groups, work under the direction of others and/or in cooperation with others in some sort of hierarchical structure, or in competition with other individuals or groups. The competitive equilibrium defined by the general economic equilibrium model takes account of interaction of individuals in terms of flows of goods but without any other interchange among them. The theory of games has focused on the interaction of individuals using strategies that are optimal in the context of rather narrowly drawn competitive situations. Organization theorists, however, have investigated the inner workings of large scale organizations comprised of individuals performing varied tasks, arranged in a cooperative hierarchy, arriving in this way at a sort of political science of the business enterprise involving structures of authority and power, that is, of *governance* (Commons (1950), [11,14,16]).

The point to be emphasized here is that economic phenomena introduce volition in a fundamental way. The atoms (people), governed by volition, form molecules and compounds (households, firms, government agencies) according to mentally emergent laws that evolve: laws governed by legal or quasilegal processes with new institutions being invented and with existing ones undergoing continual revision and transformation.

All what we want to say is that the operation of *mind* governs the flow of material in the economic world. It is as if atoms could decide what kind of atom they want to be: what valence, what atomic weight; could decide with whom they want to form molecules; could invent completely new molecules that never existed before; or could set off motion into orbits of their choosing and return to some other trajectory at will. Those who hope to discover laws of human behavior will have to deal with the dynamics of organizations and the mechanisms that lead people to obey or violate the laws they create for themselves.

Development

> Far from equilibrium...new processes set in . . . .

Prigogine [12, page 67]

To encompass fundamental aspects of physical and human nature, economic theorists must commit themselves to a dynamic, evolutionary point of view. In so doing, it is unlikely that a single unified framework will serve the purpose. As in physics, distinct theoretical relationships have to be developed for the distinct levels at which action can be measured. If we take the individual to be the basic atom of economic phenomena, then the micro theory is concerned with what, why, and how people choose. At the macro level is the economics of a nation based on indexes of aggregated micro variables: GDP, total investment and consumption, measures of total capital and productivity, population, working age population, the employed and unemployed, total government spending, and the total money supply. In between lies the theory of organizational structures, households, business firms, and government agencies. The explanation of behavior at this level is in between the micro and the macro, thus, sometimes referred to as meso-economics.

Mechanical phenomena-both physical mechanisms and mathematical theories- have played fundamental roles in economic development. The direct result of creative intelligence is that their progeny have accumulated to such an extent as to dominate the environment within which further human development takes place. Alluded to briefly above was an explanation of how economic choice operates on alternative technologies and drives human societies through phase transitions that take place one after another. The process is modeled using the mathematical tools of multiple phase dynamics (see [6,7]). Examples can be specified in which given phase structures can be repeated in a cyclic or irregular sequence reminiscent of the Hindu paradigm, a history repeating its former structures again and again. In reality, given the irreversibility of chemical processes and "a fortiori biological processes," cycles, if indeed they exist at all, must inevitably be interrupted (see [12, page 212]). Even if the number of possible phase structures that exist at any one time is finite, the creative intelligence can always be counted on to add more to the repertoire. We will always be pioneers, each generation opening up a previously inexperienced new age. Our phase progressions will form unique sequences; the future perpetually producing uncertainty and novelty.

On the basis of reflections like these, it has been observed that physics, chemistry, economics,..., science in general is not so far removed from literature and poetry. Ilya had a great love of the arts as well as the sciences. His work as a physicist seems to have been less-and more-than a quest for a theory of everything. His approach may not have built bridges between all the levels of physical and social reality. He did build a bridge between the various ways the human mind attempts to comprehend experience: an understanding of the evolutionary and uncertain character that is common to the material world and to all of life that has emerged from it.

References

1. P.M. Allen, W.L. Schieve, and R.N.Adams (eds.), *Special volume on modeling complex systems*, European Journal of Operations Research **30** (1987), no. 3.
2. Arrian (Flavius Arrianus Xenophon), *The Campaigns of Alexander*, Penguin Books, Harmondsworth, 1971, translated by Aubrey de Selincourt.
3. T.F. Cooley and E.C. Prescott, *Economic growth and business cycles*, Frontiers of Business Cycle Research (T.F. Cooley, ed.), Princeton University Press, Princeton, 1995, Chapter 1.
4. R. Crosby (ed.), *Cities and Regions as Nonlinear Decision Systems*, Westview Press, Colorado, 1983.
5. R.H. Day, *Nonlinear dynamics and evolutionary economics*, Nonlinear Dynamics and Evolutionary Economics (P. Chen and R. Day, eds.), Oxford University Press, New York, 1993, Chapter 3.
6. R.H. Day, *Multiple-phase economic dynamics*, Nonlinear and Convex Analysis in Economic Theory (Tokyo 1993), Lecture Notes in Economics and Mathematical Systems, vol. 419, Springer-Verlag, Berlin 1995, pp. 25–45.
7. R.H. Day, The Divergent Dynamics of Economic Growth: Studies in Adaptive Economizing, Technological Change, and Economic Development, Cambridge University Press, Cambridge, 2004.
8. J.M. Keynes, The General Theory of Employment, Interest and Money, Harcourt Brace and Company, New York, 1936.
9. W.W. Leontief, *When should history be written backwards*, Economic History Review Second Series **16** (1963).
10. W.W. Leontief, *Essays in Economics: Theories and Theorizing*, Oxford University Press, London, 1966, Chapter 2.
11. J.G. March and H.A. Simon, *Organizations*, JohnWiley & Sons, New York, 1958.
12. I. Prigogine, *From Being to Becoming: Time and Complexity in the Physical Sciences*, W. H. Freeman, New York, 1980.
13. M. Renault, The Nature of Alexander, Pantheon Books, New York, 1975.
14. H.A. Simon, Administrative Behavior: A Study of Decision Making Processes in Administrative Organization, The Free Press, New York, 1957.
15. J. Tinbergen, *Statistical Testing of Business Cycle Theories*, League of Nations, Geneva, 1939, reprinted in 1962 by Agathon Press, New York.
16. O.E. Williamson, *Economic Organizations: Firms, Markets and Policy Control*, New York University Press, New York, 1986.

Neumannian Economy in Multi-Agent Approach Investigation of Stability and Instability in Economic Growth[†]

Katalin Martinás

ELTE, Atomfizikai Tanszak, Budapest, Hungary.

ABSTRACT

Axiomatic foundation of non-equilibrium microeconomics is outlined. The economic activity is modeled as transformation and transport of commodities (materials) owned by the agents. Rate of transformations (production intensity), and the rate of transport (trade) are defined by the agents. Economic decision rules are derived from the observed economic behavior. The non-linear equations are solved numerically for a Neumannian economy. The emergence of the equilibrium market structure appears as an order out of chaos process.

Keywords: economics, irreversibility, growth, chaos.

Introduction

Neumann, in his 1937 paper introduced a model economy to study the conditions for equilibrium growth [1]. In the present paper the Neumannian economy will be investigated in an irreversible (thermodynamic like) multi-agent approach.

There is a basic similarity between chemistry and economics, namely that they are concerned with transformation and transport of material. Formal description of the material flows in economics and in chemistry lead to equations of the same structures, with some important differences. In thermodynamics the laws of transport and transformations are formulated as the First and Second Law of Thermodynamics. The analogous formulation in economics for the First Law is trivial [2]. Economic activity can be described via the

[†]Reused with permission from: Martinas Katalin (2004), Neumannian Economy in Multi-agent Approach. Investigation of Stability and Instability in Economic Growth, *Interdisciplinary Description of Complex Systems—scientific journal*, 2, issue 1, p. 70–78.

changes of the quantities of goods. Second Law states that the purpose of economic activity to be better off. (Later we give a more precise definition.) The First Law, in its direct or indirect form, was already articulated by economists. The classical economists accepted it. Neoclassical approach considers it trivial and non-important. Second Law is applied in a stronger form. We always choose the best possibility.

Organization of the paper is as follows. In Section 2 the thermodynamic description of chemical systems will be outlined. First and Second Law will be given in such form, which can be applied also in economics. In Section 3 a thermodynamic (or irreversible) microeconomics is outlined. In Section 4 we give numerical solutions for a Neumannian economy [1]. The results show a chaos-order-chaos behavior.

Thermodynamics

A Quick Summary of the History of Economics and Thermodynamics

Analogous properties of chemical and economic processes are well known for a long time. Kalman Szily (a physicist) gave a lecture in 1871 with the title *Communistic state of the physical world*, where he used some economic examples to explain the First and Second Law of Thermodynamics. In 1903, Jankovich [3] published a paper on the mechanical foundation of value. Imre Fenyes (professor of thermodynamics) emphasized the similarity between the economic goods and the thermodynamic quantities. Here we can cite hints by [4], Samuelson [5], Moffat [6], Berry and Andresen [7] among others. In an excellent summary on the use of physical analogies by early neoclassic economists [8]. Mirowski writes: "The metaphor of energy/utility which neoclassical economics appropriated was derived from the physics of a specific historical period: the years of the mid nineteenth century just prior to the elaboration of the second law of thermodynamics" [9]. Yet pre-entropic physics has been basically a theory based on mechanical considerations, without a "time-arrow", without irreversibility.

There were several attempts to exploit the analogous properties of the thermodynamic entropy and the utility concept of neoclassical economics [10], Rosonoer [11–12], Bródy et al. [13], or utility and internal energy [14], Stepanic et al [15], Mironova, Amelkin and Tsirlin [16]. These analogies are useful but different from the irreversible approach.

Laws of Thermodynamics

There are two basic approaches to understand the equilibration process: a macro (thermo-dynamics) and a micro (statistical mechanics) approach. Nowadays the statistical approach is more popular. Statistical mechanics, developed initially by Ludwig Boltzmann, defines entropy as the logarithm of the number of "microstates". Boltzmann's definition of entropy rather suggests the order/disorder metaphor. Boltzmann's statistical approach is a useful tool in a wide range of applications. Classical (or phenomenological) thermodynamics starts from the observed properties of our material world.

First Law

In thermodynamic investigations it is worthwhile to distinguish between the extensive variables (volume, energy) and intensive variables (e.g., temperature, pressure). An extensive variable must satisfy two conditions. First, its time dependence is described by the generic balance equation:

$$dX / dt = J + G \tag{1}$$

where J is the flux, and G is a generalized source term. Second, it must be *additive*, in the sense that if X^a and X^b are values of the variables for two systems 'a' and 'b', the variable has the value $X^a + Y^b$ for the combined system, consisting of the union of the two. A general rule of thermodynamics states that all the interactions of the thermodynamic system with its environment take place through the flows of extensive variables. Mass, energy, the number of molecules and total volume are examples of extensive properties. That property is summarized in the First Law.

First Law: Any simple system has particular states that are characterized completely by extensive quantities.

The postulate reflects an important feature of thermodynamic systems. The evolution of the system is governed by the balance equations of extensive quantities, in the form:

$$dX_{ki} / dt = \sum_{k,l,i} J_{kli} + G_{ki} \tag{2}$$

where index i and l identify the subsystem, X_{ki} is the quantity of the extensive parameter k, $J_{il,k}$ denotes the flow from the system l to the system i.

The independent set of extensive variables necessary to describe a given system is determined essentially by trial and error. The choice is not unique. It depends on one's purpose or (from another perspective) on the accuracy of one's measurements. For example, consider the air in a room. For certain purposes it is sufficient to treat it as an equilibrium gas. But for a more precise measurement one has to take all the different types of molecules present in the air into account. In still more precise calculations one might also consider the different isotopes. Going to extremely fine details it might be necessary to consider the internal structures of the atoms. In real calculations it is necessary that we take into account only those details what are important for the problem in question.

Second Law

The Second Law of thermodynamics is essentially different from the First Law, since it deals with the direction in which a process takes place in nature. It expresses the preferences of Nature. Not every change which is consistent with the balance equations is a possible change. The conservation laws do not suffice for a unique determination of natural processes. As for instance, in the previous example, the water equation offers no information, whether hydrogen and oxygen actually combine to form water, or water decomposes into hydrogen and oxygen or whether such a process can go into both directions.

The essence of the Second law is that all the independent elementary (infinitesimal) processes that might take place may be divided into three types: natural processes, unnatural (forbidden) processes, and reversible processes:

- Natural processes are all such that actually do occur. (Example: heavy body falls down.)
- Unnatural process (the reverse of a natural process): such a process never occurs, only as forced processes.
- As a limiting case between natural and unnatural processes are the reversible processes. They do not actually occur, but they are important for mathematical causes.

The Second Law implies a relation between the quantities connected with the initial and final states of any natural process. The final state of a natural process has to be discriminated from the initial state, while in a reversible process they have to be in some sense equivalent.

Non-equilibrium Thermodynamics—Dynamics

The balance equations describe the time evolution of the systems, the relation of flows and the distribution of stocks has to be defined empirically. These relations are empirical, material dependent, but not arbitrary. The flows must obey the Second Law. That condition has a great power. It allows us to introduce the concept of thermodynamic forces, and the force law which connects the flows with the state variables.

The uni-directionality of natural processes is formulated in the form of a dynamic law in non-equilibrium thermodynamics. The change (flow) is proportional to the force (difference in intensive parameters, δY_k)

$$J_i = \sum L_{ik} \delta Y_k \tag{3}$$

where L is the so called conductivity matrix. It is positive definite. That property of matrix L follows from the Second Law. That is the "time's arrow". Thermodynamic based microeconomics.

Irreversible Economics

For the mathematical structure of irreversible microeconomics see [17]. Here we summarize the most important definitions and concepts [18].

An economic agent (EA) is defined for our purposes as the smallest entity with an implicit or explicit decision-making rule. An EA would normally be either a firm or an individual. EA are characterized by their scope of activities, by their knowledge, experiences and by their stocks of goods and money. Our primary interest is the change of stocks and its economic effect. Every stock which can be affected by the economic activity of an agent can be listed, and those also, which affect the economic activity of the agent. The list of goods may contain the money, but it is not necessary.

An agent consumes produces or exchanges the goods. In *consumption* the quantity change is always negative. *Production* is a transformation of stocks from the initial form to a final form. Here the change is positive for the products and byproducts, and negative for the input materials. Total quantity increases only when there is an input from the nature. Otherwise the obligatory losses decrease the total quantity of goods. *Trade* modifies only the owner of stocks. Economic actions are trade and production. Trade and production are described as decisions. EA selects or rejects the offers provided by other EAs or (in case of production) by his/her internal state.

Decisions are tantamount to selections from a limited set of possibilities for immediate action. The set of possibilities is constrained by the external environment (for example, the legal framework) and by the assets of the agent, including financial assets, physical assets and intangibles such as knowledge, know-how, reputation, and so on.

Proposition 1 (First Law): Evolution of an economic system is described by the balance equation for stocks of goods and money.

Proposition 2 (Second Law of Economics): The purpose of economic action is to increase the expected economic welfare.

There is no economic action decreasing the expected economic welfare. It is the no loss rule. That rule implies that every agent has a measure to rank her/his belongings as a measure of economic welfare. The economic welfare of an economic agent is a function of the stocks of goods and money belonging to the economic agent: $Z=Z(X)$ The proof is given in [17]. Sign convention is selected so that $dZ > 0$ for allowed (no-loss) processes, and $dZ < 0$ for forbidden (loss-making) transactions.

Assuming the function Z is continuous and differentiable, the partial derivatives in respect of the stocks can be interpreted as the marginal Z-value of the good i. It is measured in welfare/quantity units. Let M is for the quantity of the money owned by the agent. The marginal Z-value of money is the marginal change of the welfare, that is: $w_M = \partial Z / \partial M$. The marginal value of the good i is: $v_i = \dfrac{\partial Z / \partial X_i}{\partial Z / \partial M}$. The expected change of economic welfare is:

$$dz = w_M \left(\sum v_i dX_i + dM \right) \tag{4}$$

The expected gain (profit) in trade of a unit of good I for price p_i is $F = (v_i - p_i)$, the expected gain for production $F = vC$. C_i is the change of the ith good in a unit production. It is positive for the products and byproducts, and negative for the input materials.

The technology defines the production vector. The agent defines the production level. Technology (capital) gives an upper limit (y_{max}), but real systems work with less efficiency.

Assumption: Production level is proportional to expected profit, that is

$$y_n = L_n vC \tag{5}$$

In trade the unit processes are given by the market institutions. Traded quantity when agent k trades with the agent m at price p_i will be proportional to the expected gain, that is

$$y_{km,i} = L_{km,i} (v_i - p_i) \tag{6}$$

Trade is viable only if the agent *m* agrees the same quantity with opposite sign. Price, p_i comes form the solution of equation

$$\sum_{nm} y_{nm,i} = \sum_{nm} L_{mn,i} \left(v_i^n - p_i \right) = 0 \tag{7}$$

where the summation is for the agents participating in the bargaining process.

The equation system is closed. (We omit the indices for the sake of clarity and brevity.)

Balance equations:

$$X(t+1) = X(t) + L(v-p) + L_n(vC)C \tag{8}$$

$$M(t+1) = M(t) - pL(v-p) \tag{9}$$

$$L(v-p) = 0 \tag{10}$$

It is a non-linear coupled mapping. The properties can be investigated with numerical solutions. A numerical solution needs the following data:

- identification of the agents
- the production vectors
- initial stocks
- welfare function and the coupling parameters (L)

Technological innovations and monetary policy of the economic system may also have to be specified. Further, exogenous effects such as consumption not connected with production, taxation, depreciation, and natural constraints, if any, must be specified.

Neumann Economy

Our minimum model of an economy has 3 economic agents, corresponding to sectors, namely: agriculture, industry, and households and we apply the simplifying assumptions of Neumann, namely:

- C vector is constant.
- Nature is infinite (there are free goods of nature in quantities without limit).
- Consumption consists of two parts. A fix part and a part which is proportional to the production.
- Labor is considered as a normal stock (the households produce it).
- Agents get interest payment for their money stock.

Selection of the Welfare Function

For the present investigations a logarithmic welfare function was selected:

$$Z_n = \sum_i X_{in} \log\left(C_n M_n / x_{in}\right) \tag{11}$$

where X_{in} is the stock i of the agent n, M_n is the money stock of the agent n, C_n is constant.

Production vectors: Agriculture (2,–.13–0.08), Industry (–.4,2,–.36), Households (–1.83,–.1, 2).

The coupling parameter L for all trades is assumed to be unity, viz. $L = 1$. The production coupling parameters are: $L1 = 0.352, L2 = 0.288$ and $L3 = 0.352$. The interest rate is given for the money. After each cycle the money of the agents is multiplied with α.

Initial stocks were selected as:

	Money	Food	Tools	Labor
Agriculture	1000	22.98	18.51	14.07
Industry	1000	22.05	19.34	14.16
Households	1000	21.97	18.86	14.73

Numerical Solutions

Figure 1 shows the intensity of production as a function of time. Time means here the number of completed cycles. The interest rate is $\alpha = 1.0005$. The system starts from chaos. The initial period (in a magnified form it is on Fig. 2) is the non-equilibrium part. The initial stock distribution did not correspond to an equilibrium. Figure 2. demonstrates the working of the invisible hand. Order develops from chaos. Fluctuations decrease, quasi-periodic oscillations emerge with decreasing amplitudes. The system finds the equilibrium path.

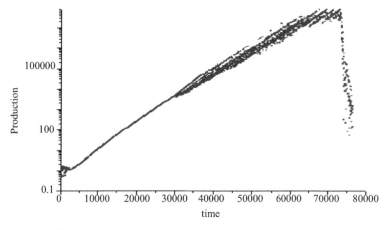

Figure 1 Production as a function of time.

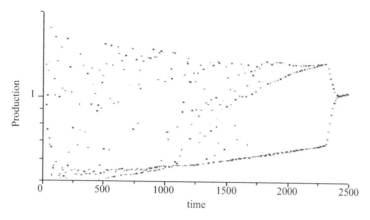

Figure 2 Working of the Invisible Hand.

There is a near-exponential growth of the production intensity. The equilibrium growth is stable only for $\alpha=1.0000$. Nevertheless the agents' welfare increases unequally. This inequality leads to the loss of stability, a new chaos appears. The time needed to reach the instability (or the collapse of economy) is called as the lifetime of the economy. Life time is defined by the technology (production vectors) and the interest rate. In Fig. 3 the lifetime is plotted as a function of interest rate. The figures confirm Neumann's result, the golden result, for a given technology there is only one interest rate, which gives a stable growth. All the other interest rates give growth with limited stability, but in the nearly stable growth regimes the price ratio is nearly constant.

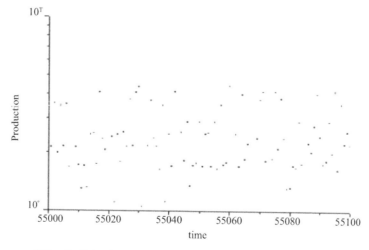

Figure 3 Emergence of Chaotic Behaviour.

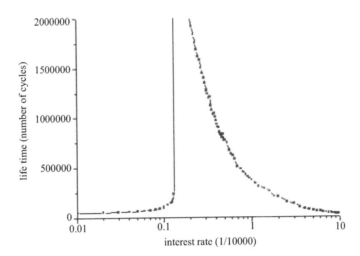

Figure 4 Life time versus interest rate

Acknowledgment

This paper is based on the results of a research program T: 043522, supported by the Hungarian Scientific Research Fund.

References

1. Zalai, E.: *Matematikai közgazdaságtan*, KJK–Kerszöv Jogi és Üzleti Kiadó Kft., Budapest, 2000.
2. Martinás, K.: *Is the utility maximum principle necessary?*, In: *Crisis in Economics*. Editor: E. Fullbrook, Routledge, London, 2003.
3. Jankovich, B.: *Mechanical foundation of value theory* (in Hungarian: Az értékelmélet mechanikai alapelveiről) Közg. Szemle, **35**, 10-23,1903.
4. Von Neumann, J.: *Über ein ökonomisches Gleichungssystem und eine Verallgemeinerung des Brouwerschen Fixpunktsatzes*, Ergebnisse eines mathematischen Kolloquiums, **8**, pp. 73–83, 1937.
5. Samuelson, P.A.: *Thermodynamics and Economic Isomorphisms. in:Global Econometrics*, ed. by F.G.Adams-Hickman, B.G.Hickman, The MIT Press, Cambridge., 1983.
6. Moffat, M.J.: *A Dynamic Model of Adaptive Economic Behaviour*, in: A. Taton (ed.): Sadi Carnot et l`essor de la thrmodynamiqe, CNRS, Paris, 1976.
7. Berry R.S.: Andresen B. *Thermodynamic constraints in economic analysis. Self-organization and dissipative structures: Applications in the physical and social sciences*. Ed. By W.C. Schieve and P.M. Allen. University of Texas Press, Austin, Texas, 1982.
8. Mirowski, P.: *Physics and Marginalist Revolution*, J. Economic History, 44, 345–354, 1984.
9. Mirowski, P.: in: *The reconstruction of economic theory*, ed. by P. Mirowski, Kluwer-Nijhoff Publ. Co., Boston, 1986.
10. Lihnierowicz M.: *Un modele d`echange economique (Economie et thermodinamique)*, Annales de l`Institut Henry Poincare, nouvelle serie. Section B, **4**(2), 159–200, 1970.
11. Rosonoer, L.I.: *Exchange and resource distribution (A generalized thermodynamic approach)* I, II, III, Avtomatika i telemekhanika 5, 115–133, 6, 65–80, 8, 82–104, 1973.
12. Rozonoer L.I., Tsirlin A.M.: *Optimal control of thermodynamic processes*. I–III. Automation and Remote Control, 1983, Nos, **1,2,3**.
13. Bródy, A., Martinás, K., Sajó, K.: *Essay on Macroeconomics*, Acta Oec. **36**, 305, 1985.

14. Saslow, W.M.: *Economic analogy to thermodynamics*, Am. J. Phys. **67**, 1239, 1999.
15. Stepanic, J., Stefancic, H, Zebec, M.S. and Perackovic, K.: *Approach to a Quantitative Description of Social Systems Based on Thermodynamic Formalism* , Entropy **2**, 98–105, 2000.
16. Mironova V.A., Amelkin S.A. and Tsirlin A.M.: *Mathematical methods of finite-time thermodynamics.* Moscow: Khimia, 2000. (in Russian).
17. Martinás, K.: *Irreversible Microeconomics*, Complex Systems in Natural and Economic Sciences, Eds: K. Martinás, M. Moreau, ELFT, Budapest, 114–123, 1996.
18. Ayres, R.U. and Martinás, K.: *Wealth Accumulation and Economic Progress*, J. Evolutionary Economics, **6**, 347–360, 1996.

Chapter 3

Theoretical Aspects of the Economic Transition: Case of Romania[†]

Cezar Scarlat[1,a] and *Eugen I. Scarlat*[1,b]

[1]University Politehnica of Bucharest, Romania.
[a]Email: cezarscarlat@yahoo.com.
[b]Email: eugen.scarlat@mct.ro.

ABSTRACT

For Romania, like for all other ex-communist countries from Eastern Europe, the transition from the rigid centrally planned economic system to the free-market economy, fair competition based, was an amazing experience. From the academic standpoint, the economic reform was a huge research opportunity, as well-with extremely important practical consequences. Based on the case of Romania, the authors have developed an original, bi-dimensional matrix model of this transition process (Scarlat Model), emphasizing the typology of four basic economic systems. Managerial aspects are underlined-both for economic systems and transition process-as well as some stability considerations. Two features of the transition strategy are presented: the transition path and duration of the process. Special attention was paid to assessing the moment by when the economic transition ends. Analysis of the transition path—based on the theory of deterministic chaos (i.e., short-run predictability)-has led to interesting results: a comprehensive research on the evolution of the Romanian currency exchange over a period of sixteen years (1990–2005) revealed three intervals in the Romanian recent history of economic transition and confirmed the diagnostic of transition end. The general model is applied in case of Romania and some interesting findings are presented, but it is fully applicable to all Eastern European countries and not only. The EU accessing process is a different type of transition-rigorously planned, regulated and monitored.

Keywords: economy model, economic transition, transition path, deterministic chaos, EU accession.

[†]Reused with permission from: C. Scarlat, E.I. Scarlat, Theoretical Aspects of the Economic Transition: The Case of Romania, *Managing Global Transitions* 5 (4): 307–331, 2007.

Introduction

In 1989, the pretended solid system of centrally planned economy has collapsed. The revolutionary changes that started in Poland demolished the whole Soviet-style command economic system of Eastern Europe and the Soviet Union itself. Dramatic changes in the economic systems of these countries followed but the scientists were not prepared for such a moment: there was no valid theory of economic reform to rely on-as there was no such precedent. Within the empty space of the economic theory in this respect, the fall of communism has given way to a variety of strange, archaic, alternate and even "informal" and "parallel" forms of management and enterprise—as the entrepreneurial energy existed but it was not properly channelled (Dana and Dana 2003). As Aligica describes (2006), the raise of the *new institutionalism* is one of the most significant reactions of the economic reform experience. A comprehensive World Bank Report (World Bank Report 2004) evaluates the World Bank assistance in 26 countries in Europe and Central Asia. After 1989, the transition countries "have undertaken massive reforms of their economic systems, transforming institutions, processes, attitudes, and fundamental concepts of individual and organisational behaviour". Considering its complexity, the development of an "easy-to-understand" but working model of economic transition is a considerable challenge.

The economic reforms have had significant impact in all sectors and important research efforts have been made to clarify, explain or solve problems related to the transition processes. Even gender issues were such a subject (Homlong, Springler 2006)—in order to identify quality and quantity of female labour participation in such processes.

Based on the authors' experience and research during over fifteen years of transition from a centrally planned economy toward a free-market economic system, an *original model of the economic systems and transition* was developed. More or less authorised voices claim sentences about "privatisation", "economic reform" or "transition toward market economy" and—more recently—European Union. In spite of their actuality, they often generate questions and are even confusing.

The proposed model allows clear definitions and possible transition strategies are analyzed. The role of government is emphasized, as well as the key-role of the public administration in the process of accession to the European Union.

The economic and transition model, which is proposed by the authors as a conceptual framework, is bi-dimensional. Types of ownership (state/private) and management (centralized/decentralized) are considered. Consequently, the resulting "two by two matrix" reveals four types of economic systems—among them the centrally-planned economy and the free-market economic system.

The major advantage of this model is its analytical potential. The different transition strategies are presented and the position of a certain economic system, on its path toward the free-market economy, at a certain moment, can be easily identified. The topic of this paper is very actual, extremely important and relevant to the current state of knowledge for—at least—two reasons:

- The proposed transition model is both theoretical and practical (explanatory) as well as investigation tool (Ardelea and Scarlat 1991; Scarlat 1994, 1999, 2001, 2003a; Scarlat and Curaj 2004).
- The research conducted in Romania, the analysis of time series of certain macroeconomic indicators (as the exchange rate) during the transition period or the transition path, from the standpoint of deterministic chaos (i.e., short-term predictability), has revealed interesting characteristics (Scarlat 2005; Scarlat E.I. et al. 2007a, 2007b) and was in accordance with the other estimations about the end of transition.

Some of the conclusions of the research might be applicable to other transition economies as important decision support tools.

Economy Model

Using a political economy approach, and based on different criteria (political structures, 'what is a good society?' and the pace of implementing the transition policies), Marangos (2005, 2006) has identified and developed five alternative models of transition: Shock Therapy, the Neoclassical Gradualist model, the Post-Keynesian model of transition, the Pluralistic Market Socialist, and the Non-Pluralistic Market Socialist model of transition (the Chinese model). These models are associated with alternative institutional development processes—i.e., the decision depends on what institutions are considered to be more efficient: institutions produced by state intervention or market-generated. The neoclassical gradualist model maximizes the social welfare under the given internal and external constraints (Marangos 2006). Unfortunately, this meritorious and complex typology presents definition elements of both economic transition (as shock therapy) and economic system (others), which might be confusing.

Definitely, an economic system is described by a certain number of features: more features, more information and—finally—the system is better described. The issue is to investigate the possibility to characterise (any) economic system by a *minimum number* of parameters, in order to simplify the analysis as much as possible (but they still describe the system completely). The point is to find the *most important features*. The basic assumption is that any economic system can be characterised by two major features, considered as determining features:

- The type of ownership that could be, basically, state ownership or private ownership (the intermediate or mixed ownership is accepted);
- The type of management adopted by businesses/organisations active within the economic system.

We agree to define the management type as centralised if all the vital decisions are made at the macroeconomic level (government), and the decentralised management, where decisions are made at the microeconomic (organisation) level (intermediate forms of management are also accepted).

The result of these assumptions is the two-dimension (matrix) model presented in Fig. 1. Validity of the model's assumptions is proved by the fact that all the important types of economic systems are described pretty completely. While the association "private ownership" & "decentralised management" is typical to the democratic countries' "market economy" (quarter I), the association "state ownership" & "centralised management" defines "the command economy" or centrally planned economy of the communist/socialist countries (quarter III). Analysing the model with combinatorial techniques, two more associations are shown: "private ownership" & "centralised management" (quarter II), defining the economy of monopoly and "state ownership" & "decentralised management" (quarter IV), introducing the so-called "social-market" economy.

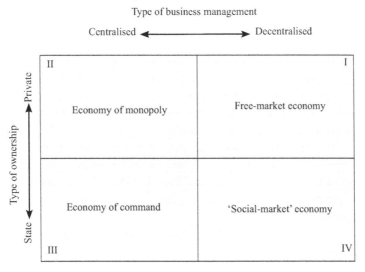

Figure 1 The matrix model of four basic types of economic systems.

It is important to avoid the confusion between "centrally planned economy" and "planned business activity"-core of the modern free-market economy. Unfortunately, such confusions were frequently presented in Romanian newspapers and mass media, right after 1990. Another trap to be avoided: the modern free-market economy *is not* the old "market" economy dominated by powerful monopolies (economy of monopoly).

Free-market economic system is superior to the centrally planned economy for two reasons, at least: *motivation* (as a result of private ownership) and *flexibility* (as a result of decentralized management.

Stability and Management Considerations

From the management standpoint, the description of the proposed model (Scarlat's Model) starts from two basic prerequisites:

(i) private ownership, opposed to the public one, generates motivation;
(ii) decentralisation, determines the flexibility of management systems and processes.

On the other hand, the basic management principle states that *the owner makes the decisions (regardless who are the owners—individuals, businesses or the government)*. This situation is met in quarters I and III, which corresponds to relatively stable systems. Any other situation means conflicts or potential conflicts between the owner and decision maker (quarters II and IV). This explains the relatively short life of economic (and political) systems corresponding to the quarters II and IV. The failure of centrally planned economies (quarter III) is not because of management conflicts but mostly for rigidity, poor motivation and low efficiency. It is interesting to underline the unstable character of the economic models belonging to quarters II and IV. Contradiction between the private ownership and centralised management (quarter II) will have serious effects like economic and social unbalance, excessive and immoral revenue sharing and, finally, the danger of significant unemployment.

The opposition between state ownership and decentralised management (as much as it is possible) lead to the property's depreciation through poor efficiency, as a result of the back interest, with a major negative consequence, inflation (quarter IV).

A different situation, wherein the frames of private ownership and free competition principles are broken and/or replaced by the centralised command mechanisms, is that of the war economy (Germany, Italy, and Romania, during the World War II). By laws and/or decrees, the economy's direction is accomplished by the state both at macro and, somehow, microeconomic levels. The state targets all the nation's resources in order to sustain the war efforts. During the crisis this strategy can lead to results that are superior to those from quarter I. But in normal times it hinders initiative and dynamism, placing the economy behind the free markets.

It is also important to analyze the relation between company management and ownership, by quarters.

Quarter I. Interest aroused by private ownership and managerial flexibility results in elastic and efficient management. Under the necessity of free-market circumstances, governed by the principles of the demand-supply balance and the fair competition, the necessity of a fast market information transfer appears. Extensive and intensive use of information technology becomes a management tool.

Quarter II. There is private ownership's motivation but centralised leadership that creates over-sized structures with slow reaction time and a long information circuit. Quite rigid and disadvantageous management is the result. This can be improved by using information systems.

Quarter III. Management is imposed from outside the system. The economy is of "command" type but the company is deprived of a basic managerial function: "decision-making". That is ultra-centralised, the communication channel is long, and information is distorted and weakened through a bureaucratic chain with non-existent feedback. In the few situations where it appears, it doesn't contribute to the decision making. The lack of motivation increases the system's immobility.

Quarter IV. Accepting that a company is allowed to make decision, state ownership rejects that kind of motivation. The result is a formal management. The information systems are used in an inefficient way, as well.

Transition process

A successful transition process from communism to democracy and/or market economy may be a tool for the economic development of a country (Bitzenis 2007).

The free market economy's superiority (through motivation, adaptability and flexibility, and progress, and finally through productivity and efficiency) is practically proved as well, confirmed by the social and historical dynamics of economic systems. At least for economic reasons, as the free-market economic system is more efficient than the centrally planned one, the transition to free-market economy is a must.

Since late 90s, a rich literature has been developed, screening the various aspects of the transition. The prestigious journal *Post-Communist Economies* (formerly *Communist Economies and Economic Transformation,* Routledge) has published, systematically, interesting articles on various economic sectors, in different transition countries as such: Azerbaijan (Sabi 1997), Bulgaria (Mihaylova & Howe 1998), Croatia (Cengic 1996), (East) Germany (Hölscher 1997), Hungary (Mihalyi 1996), Poland (Kaminski 1998), Romania (Hunya 1998), Russia (Magomedov 1998), Kazahstan and Uzbekistan (Abazov 1997), Ukraine (Ishaq 1997, von Hirschhausen 1998). The critical issue of economic development is important for each individual country as well as for an entire region—it could be either Central Asia or Eastern Europe (Brzeski & Colombatto 1999).

Over the last decade, comparative and multi-country studies were conducted. Such a research completed in four countries from Central and Eastern Europe (Romania, Bulgaria, Czech Republic, Hungary) has identified several areas of concern for the very dynamic sector of small and medium-sized enterprises (SMEs), active internationally: lack of appropriate financial instruments, difficult access to market intelligence, lack of professionalism within the business support infrastructure (Lloyd-Reason et al. 2005). At the other extreme, there are numerous examples of successful economic development, SMEs-based. Analysing the Vietnamese economy, Harvie (2004) conclude that SMEs are job creators, contribute to sustainable economic development by efficient allocation of resources, expand the exports, achieve a more equal distribution of incomes, and assist the rural and regional development. Actually, the "Scarlat Model" demonstrates that SMEs are the ideal actors to play on the stage of free-market economy: SMEs are totally private and decentralized decision makers.

The Scarlat Model can be used to analyze the economic transition as well. The transition process—from centrally planned economy to the free-market economy—means simultaneous privatisation and decentralisation of the business management (Fig. 2). The adopted legal acts in Romania *(the Reorganisation of the State Owned Enterprises Act, the Corporations Act, the Act of Land)* target all the transition aspects mentioned before.

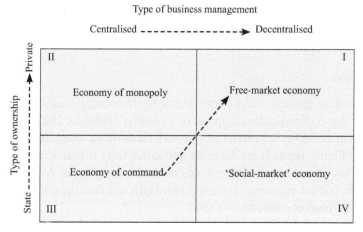

Figure 2 The transition model-from centrally planned economy to the free-market economy.

The fundamental problem of the transition is choosing and justifying the optimal strategy, defined by several components—each of them assuming critical decision making:

- Duration and speed of transition ("shock" or "gradual" transition);
- Transition path (trajectory);
- Privatisation techniques—as an essential component of the transition and economic reform;
- Transition management and strategy—as a result of all the above.

As far as duration/pace of transition—either "shock therapy" or "gradual transition"—the champions of each of them might have their arguments. However, ten years of transition offer enough hard evidences—as some countries have opted for "shock" while most of them (as Romania) have decided to follow the "gradual" path (Giannaros 2000).

The proposed model allows analysing all strategy aspects. For example, as far as transition path, the model's analysis reveals three types of trajectories, at least:

a) The direct transition (quarters III → I); theoretically the most advantageous but requires an outstanding managerial effort.

b) Transition through quarter 2 (quarters III → II → I), which is potentially high unemployment generative.

c) Transition through quarter IV, threatened by inflation (quarters III → IV → I).

It is important to mention that these three types are just basic trajectories. Bitzenis (2007) has introduced nine general reform paths of transition. If the economic criterion (inflation avoidance) is overwhelming, variant (b) will be chosen meaning prioritising the privatisation. If the social criterion (lowest possible unemployment) prevails, (c) will be followed and companies will become fully autonomous. From the theoretical, managerial standpoint, transition should be as short and straight as possible. Prolonging the transition through unstable statuses (quarters II and IV) induces stresses and leads to a drop in economic efficiency or in the speed of the economic reform process a decrease. As mentioned before, the split between private/state or centralized/decentralized is pretty rigid. The model can be improved, considering both ownership and decentralisation as having continuous variation: ownership percentage and degree of decentralisation. The proposed model—slightly modified (Fig. 3)—allows investigating the real path followed by Romania on its way to free-market economy. The overall survey on the Romanian economy and business environment allows us to conclude that the most of the legal framework is in place and the business is very active. According to a detailed survey conducted and published back in 2004: "Romania does not request a transition period or impairment of law and states that it will be possible to fully enforce the acquis after the accession" (Fuerea et al. 2004, 29).

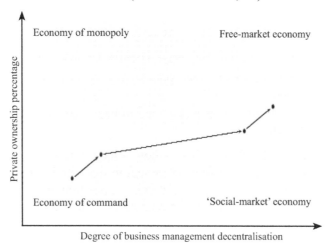

Figure 3 The model of the transition path of the Romanian Economy.

Role of the Government

As have seen, the match between ownership and management in quarter I is only the background for an efficient economic system. For countries in transition-as Romania-the government is less and less an economic actor but a crucial rule maker. After the transition management role, the government must play another essential role: the guard and the guarantor of the free-market economic system. In order to maintain the viability of the free-market economy system, its borders should be safely defended by (Fig. 4):

- "Anti-trust" regulations (antimonopoly) in order to discourage the movement back to quarter II,
- Laws for property protection and limitation of nationalisation processes that shouldn't allow the access towards quarter IV,

as well as measures of stability within the system:

- Fair competition legislation;
- Laws for consumers' protection.

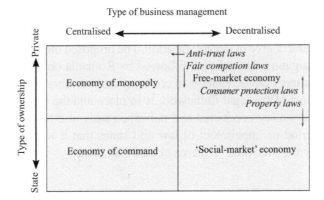

Figure 4 The role of government-the legal environment development and protection.

To all these, the legal framework for environment protection is a *sine-qua-non* condition for sustainable development.

During its EU accession process, Romania had to pay attention to all these aspects as follows (Fuerea et al. 2004):

- Fair competition: "In this field … legislation is harmonized with similar European legislation. There is a high proportion of compatibility with EU legislation regarding the regulation of corporate agreements, dominant position abuse and the control of economic concentrations." (Chapter 6. Competition policy, 30).
- Consumer protection: "In conclusion, the degree of concordance between the Romanian legislation and *acquis* … is estimated to be relatively high since all measures taken by Romania are fully compatible" (Chapter 23. The consumer and health protection, 88).
- Environment protection: in this area, Romania still has "to promote a normative measure to support the local budgets in creating/improving the environment infrastructure" (Chapter 22. Environment protection, 85).

Justification of those measures lies in the fundamental principles of the free-market economy: open competition, balance between the demand and supply, and company autonomy.

As far as private sector, the government plays an important role, too. The "privatization" process is only one of the ways to create a significant private sector in economy:

- Privatisation of the former state-owned enterprises;
- Creation of new private firms—small business start-ups, mostly (which started right after December 1989);
- Restitution of the properties confiscated by the communist regime between 1945 and 1985.

This is why the "creation of the private sector" syntagm—which is strongly recommended to be used—reflects the real-life processes better than the word "privatization".

The Romanian government took action in all the above directions. Only the pace was different. As example, the major steps for property restitutions are chronologically displayed in Table 1. Similar investigations can be performed for other means to create the private sector. The Romanian Statistics Authority periodically provides data in this respect.

Table 1 Property restitution acts, in Romania (Source: Romanian Official Gazette (Monitorul Oficial) 1990–2005).

Year	Act number	Restitution object
1991	Law No. 18 / 1991	Land for agriculture (less than 10 ha)
1992	Law No. 82 / 1992	Urban pieces of land
2000	Law No. 1 / 2000	Forests. Land larger than 10 ha
2001	Law No. 10 / 2001	Buildings (restrictive)
2005	Law No. 247 / 2005	Most of the properties

It has to be emphasized that the government has to be the referee of the economy game, not a player. But its role and implication are sometimes rather difficult to assess, mainly in transition economies. As example, in Czech Republic, the influence of the government in the economy was higher than what the government officially conceded during the 1990s (Kreuzbergova 2006). So-called "banking socialism" has meant the indirect control of the recent privatized companies by the government-controlled banks.

Too much involvement of the government in the economy means less economic freedom. The Heritage Foundation and Dow Jones & Company, Inc. is annually publishing-since 1995—the world ranking of economic freedom, according to the "index of economic freedom", which is based on some tens of independent variables, grouped in 10 broad factors of economic freedom, strongly influenced by the government policy.

The economic freedom "encompasses all liberties and rights of production, distribution, or consumption of goods and services. The highest form of economic freedom provides an absolute right of property ownership, fully realized freedoms of movement for labour, capital, and goods, and an absolute absence of coercion or constraint of economic liberty beyond the extent necessary for citizens to protect and maintain liberty itself. In other words, individuals are free to work, produce, consume, and invest in any way they please, and that freedom is both protected by the state and unconstrained by the state" (Beach and Kane 2007, 38). To note that the absence of government coercion or constraint must also include a sense of liberty as distinct from anarchy.

The 2007 Index of Economic Freedom ranks Romania on 67th place (compared to the 92nd place in 2006) in a global list of 161 countries (Kane, Holmes, and O'Grady 2007). The position corresponds to the pretty large group of "moderately free-countries"—where the majority of former communist countries are ranked. The lower the position, the greater the level of government interference in the economy and less economic freedom the respective country enjoys.

End of Transition

The question of "when the transition process ends" (Scarlat 1999) is twice important: not only for the theory sake but for practical reasons, as well (just think about the status of "functional market economy" according to which the candidate countries for EU accession were assessed). On the other hand, Kumar (2006) finds that there is a strong correlation between world trade and a well-established market economy such as the USA, Japan and countries in EU, while there is small correlation between the world trade and a transitional economy such as Russia. In other words, the "functional market economies" have better established import and export transactions and, consequently, economic welfare. The Scarlat model is able to offer a reliable answer to the sensitive question "when the transition process ends". In theory, by the time when the private ownership is prevailing and the business management is dominantly decentralized, then, the free-market economy system is in place (Fig. 2). In practice, during the transition process, it is necessary to measure both the private ownership percentage and degree of business management decentralisation (Fig. 3). When more than 50% at both of them (or a different but higher than 50% privatisation and/or decentralisation target) is reached, then the process of *economic transition towards the free-market economy* is considered over. While the percentage of private ownership is relatively easy to calculate, based on regular statistics (see, as example, Fig. 5, which depicts the Romanian case for its privatization target), the degree of business management decentralisation is rather difficult to assess. In any case, a coherent set of criteria for assessing the degree of decentralisation has to be developed. Such a set of criteria should include vital decisions at the company level—as decisions on: company mission and strategy, budget, company suppliers and clients, pricing policy, personnel policy (number, structure, salaries, hiring and firing), investments, acquisitions and mergers, insolvency and

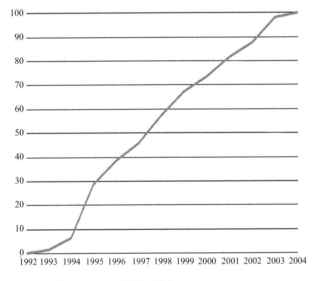

Figure 5 Privatisation dynamics in Romania, 1992–2004.

bankruptcy. Further research on this issue is to be completed and eventually published.

There are also other standpoints, methods and/or information sources that could be used to assess, conventionally, the point in time when the economic transition gets—conventionally—to its end. The question is, somehow, rhetorical—because ... the transition never ends: the economic system—even after reaching the "free-market economy" quadrant —is not rigid, it is continuously evolving (within the same quadrant, hopefully).

Several approaches mostly used to diagnose and identify that moment when the economic transition is conventionally over are mentioned below.

- The *political* approach (it considers a complex of elements although): the transition is over when the EU sentences it—in Country Report—and considers the economic system as "functional market economy". For Romania, this happened by the end of 2003: "Romania can be considered as a functional market economy once the good progress made has continued decisively" (European Commission 2003, 7).
- The *economic* approach: the economic transition is over when the country's yearly GDP reaches the pre-transition maximum level. The Romanian economy has reached this value in the years 2002–2003, according to WB and UN estimates, respectively, at market prices, current prices, in USD (United Nations Database 2007). It is significant that countries, which were forced to follow reform paths in belligerent and/or violent forms, have reported limited economic growth and "not only lag behind the other transition countries in the region, but also, as of today, have not even reached the levels they had attained in 1989" (Bitzenis 2007).
- The *management* approach: the economic transition is completed at that point in time when the strategic objectives are reached (in terms of GDP, private sector dimension, etc.). For example: the privatisation target was achieved by 2003 (Fig. 5).
- The *econo-statistics* approach: based on time-frequency series analysis, which will be further discussed.

As all the above may present cross influences, a correlated approach is recommended. In case of Romania, the different approaches converge to indicate that economic transition got to its end by 2002–2003. This conclusion will be supported by the conclusions of the next chapter.

Analysis Tools for the Transition Path

The transition process was not linear. Besides its trajectory and duration, it is interesting to assess when the transition ends and identify if other elements have occurred on the transition path. There are also more sophisticated tools to analyze the transition path, behavioural patterns and specific features. For example, from the point of view of the deterministic chaos (i.e., short-run predictability), the analysis of time series might lead to interesting results. A time series might be anything between randomness and deterministic; in turn, the determinism might be chaotic (i.e., sensitive to initial conditions) or classic (non sensitive).

The research methodology is based on the analysis of the time series of macroeconomic indicators, by statistical methods; the macroeconomic indicator chosen: the US dollar (USD) exchange rate. The figures were collected based on the daily values reported by the Romanian central bank (Banca Nationala a Romaniei) over a period of sixteen years of Romanian recent history: 1990–2005. This period includes the interval used previously for presenting the privatisation process: 1992–2004 (Fig. 5). A comprehensive research on the evolution of the Romanian currency exchange rate (ROL) with respect to the US dollar (USD) was conducted. Figure 6 depicts the overall evolution. The logarithm scale was used because of large variations of the indicator (up to 10^4).

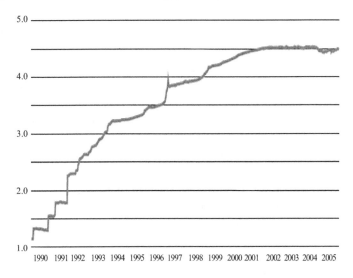

Figure 6 Time evolution of the ROL/USD exchange rate, at logarithm scale (Source of data: National Bank of Romania, at: www.bnr.ro).

The analysis was performed taking into consideration three criteria:

- System complexity (this is not necessarily the whole economy, but part of it, as monetary system)
- Sensitivity to initial conditions (as a pointer of the predictability)
- Persistence of the evolutionary trend (indicating the similitude with the random-walk behaviour, i.e., the lack of deterministic predictability in favour of probabilistic predictability).

In the nonlinear dynamics theory, the corresponding quantities for the above mentioned criteria are the correlation dimension, the largest Lyapunov exponent (Bask 1996) and the Hurst coefficient (Tsonnis, Heller, Takayasu, Marumo, Shimizu 2001).

The degree of complexity of the underlying system is quantitatively given by the correlation dimension referring to the minimum number of variables that is needed to replicate the dynamic system. The last one is the same with the dimension of the (strange) attractor characterizing which one of the topological-equivalent systems (Takens 1981).

It is worth noting that the number of variables that replicates the system is fractional, accordingly to the theory of the strange attractors (Kantz and Schreiber 1997).

The sensitivity to initial conditions could explain the spreading diversity of the nowadays economic systems, even they were starting from quite similar prerequisites. This is a basic feature of the chaotic systems. Hence, one way of revealing the existence of deterministic chaos in a time series is to measure the degree of divergence of nearby orbits in the phase-space. Such divergence $\delta(t)$ can be measured by the Lyapunov exponents λ, and the presence of at least one positive exponent is taken as indication to what extent the evolution is predictable over a specific time range (Rosenstein 1993):

$$\delta(t) \cong \delta(0) \cdot e^{\lambda t}$$

As far as *the persistence of the evolutionary trend*, the Hurst exponent is an additional pointer for making distinction between the randomness and causality; besides, its value of 0.5 separates the persistent and anti-persistent evolutionary trends (Stanley, Mantegna 2004). It is defined by the time evolution of the root-mean-square fluctuation of the averaged displacement over all possible positions t_0 (the bars indicate the average):

$$\left\{ \overline{\left[x(t_0 + t) - x(t) \right]^2} - \overline{\left[x(t_0 + t) - x(t) \right]}^2 \right\}^{\frac{1}{2}} \sim t^H .$$

The values of the Hurst exponent (H) range between 0 and 1. A value of 0.5 indicates a true random walk (a Brownian time series). In a random walk there is no correlation between any element and a future element. A Hurst exponent value $0.5 < H < 1$ indicates "persistent behaviour" (e.g., a positive autocorrelation). If there is an increase from time step $i-1$ to i, there will probably be an increase from i to $i+1$. The same is true for decreases: a decrease tends to follow a decrease. A Hurst exponent value $0 < H < 0.5$ indicates a time series with "anti-persistent behaviour" (or negative autocorrelation): an increase tends to be followed by a decrease, or conversely, a decrease will be followed by an increase. This behaviour is sometimes called "mean reversion". Some results of the research conducted in Romania are further presented. The analysis of the transition path—using all the above criteria—has revealed three intervals (Scarlat, E.I. et al. 2005) as follows (Fig. 7):

i) 1990–1997
ii) 1998–2001
iii) 2002–2005.

Obviously, the three intervals are splits of the diagram depicted in Figure 6.

The interval no. 1 (1990–1997) is characterized by simple underlying dynamics and low fractal dimension, weak sensitivity to initial conditions, and very aggressive, positive long-run trend. After the social and political events that provoked the fall of the totalitarian regime, the inertia of the total deterministic evolution of the economy of command that has been legally operated until the end of 1989 prolonged the effect for several more years, when the economy diminished to work as a "national holding", and both the lack of an adequate institutional feed-back and the social pressure lead to political decisions of maintaining the exchange rate at artificial constant values. Thus, the accumulation of the demand pressure upon the slow-developing supply lead to an increased import of goods; the insufficient feed-forward of a weak monetary system could not balance the long response time of production, and explosive corrections occurred in the form of stepped jumps of the exchange rate ROL/USD (see Fig. 7a). The exchange rate is deterministic and "very" predictable from the long-run perspective (short-run understood), with a lot of angular points. The main features of this time interval are summarized in Table 2.

The interval no. II (1998–2001), as Fig. 7b shows, is characterized by increased correlation dimension, i.e., more complex dynamics, and clear positive values for the

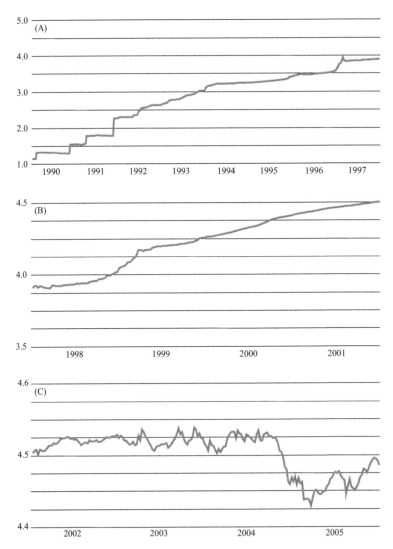

Figure 7 The dynamics of the three intervals of the transition period (1990–2005)—the evolution of the USD exchange rate against Romanian ROL.

Table 2 The three time intervals and their characteristics—based on the evolution of USD exchange rate against Romanian ROL.

Characteristics	Interval no. I	Interval no. II	Interval no. III
Time range	1 January 1990–31 December 1997	1 January 1998–31 December 2001	1 January 2002–31 October 2005
Complexity (system dimension)	1.74	3.09	4.45
System sensitivity to initial conditions	weak	strong	strong
Trend	persistent	persistent	anti-persistent
Type of predictability	long-run, deterministic	long-run, chaotic deterministic	short-run, chaotic deterministic

largest Lyapunov exponent. All the usual hypotheses for chaotic dynamics are fulfilled, and this is the domain with significant functional changes toward the open market economy. The positive long-run trend continues, but the averaged relative variation per year of the exchange rate is hundred times lower (72.27% compared to 7264.29% for the interval no. I). The exchange rate remains predictable both from short- and long-run perspective (Table 2), exhibiting a much smoother curve.

The interval no. III (2002–2005)—see Fig. 7c—is closer to the random walk behaviour, as pointed out by the anti-persistence of the short run trend. It might be a chaotic deterministic, with a higher volatility than the previous one, or stochastic, or mixed; in order to decide in this matter, further analysis is necessary. To simplify, we only mention here that the evolution remains of causal type. For the purpose of this paper, it is important the fact that there are arguments to state that the transition seems to come to its end by the end of 2001 due to at least two reasons: the first is the disappearance of the long-run trend and, consequently, the beginning of the steady state regime, and the second is the complexity of the structure revealed by the fractal dimension of the interval no. III. As stated in literature (Schwartz and Yousefi 2003), a free market economy in a steady state regime contains a strange attractor with a correlation dimension around 4.5, and that is the case for Romania (Table 2). The exchange rate predictability is weak, with small deviations from the nearly horizontal baseline (excepting a significant fall in March 2005—see Fig. 7c). This is the consequence of the new settled macroeconomic equilibriums, very sensitive to the environmental stimuli, but also with high capacity to recover the functional "working points".

Significantly low predictability is consistent with the long run behaviour, which in fact has no trend and the curve approaches a random walk appearance (Fig. 7c). In a steady-state-running economy, as mentioned by other authors, a positive real part of the largest Lyapunov coefficient seems to be normal. Moreover, the greater the correlation dimension, the more complex the economic system (and the monetary policy of the Central Bank). A smaller Lyapunov exponent is consistent with a higher degree of short run predictability in an economic system; for the first period, the exchange rate is predictable from the long run perspective (positive trend). The smaller Lyapunov exponent characterizing the structural changes period indicates smaller sensitivity to the initial condition together with a non-randomly evolution toward a more complex system characterizing the steady-state-regime.

Overall, it is important to note that time series analysis indicates the end of 2001 as the end of the transition process—which confirms the previous conclusion (end of transition process in Romania: 2002–2003).

The transition trajectory in case of Romania seems to be of the type III → IV → I.

Open Closing: EU Accession Process—A Different Type of Transition

As the EU accession is a transition process itself, Romania—like other Eastern European countries—has been "in transition process" twice. Fortunately, the transition to the "functional" free-market economy is naturally connected to the EU accession process (Scarlat and Richevaux 2006). Several impact studies were conducted in Romania under PAIS (Pre-Accession Impact Studies): PAIS I (2001–2002), PAIS II (2003–2004), PAIS III (2005–2006). Overall, the PAIS studies were intended to provide recommendations to the Romanian Government to support negotiations with the European Commission. Under PAIS I (Scarlat, Popescu, and Warner 2002) primary and secondary research (desk research and formal face-to-face interviews) was conducted. The main objective of the study was to provide an overview of the institutional requirements for implementing the *acquis communautaire* in Romania and to provide a set of recommendations. Most of the above issues and recommendations were presented to the RIA International Conference in Sofia (Scarlat 2003b). As these recommendations were acknowledged by the Conference participants coming from ex-communist countries, it means that *this set of recommendations are applicable not only to Romania but to most of the ex-communist, EU accession countries.*

The studies conducted under PAIS II (Fuerea et al. 2004) confirmed that Romania-like other EU candidate countries in economic transition—is aiming a "mobile target": harmonization of the Romanian legislation with the *acquis communautaire* is based on two processes:

- transposition of the Community legislation into the Romanian legislation and
- rendering compatible of the national normative measures which transpose the *acquis* with the provisions of the Community legislation). The research revealed that the concordance degree of the Romanian legislation with the EU legislation (in force in 2002) was approximately 75%. The problem is that the *acquis* covers more than 90,000 pages (out of which about 20,000 are continuously being amended!).

The EU accession process was a different type of transition—not only planned but rigorously regulated and monitored. Following to the EU accession, the EU integration process is a different topic, quantitatively and qualitatively different.

Conclusions

- The general economy model that was presented ("Scarlat Model") is a useful analysis framework to discuss the transition features—mainly its path but also transition duration, delays, priorities (privatisation vs. decentralisation)—the transition strategy ultimately.

- The sophisticated research tools applied for transition path analysis, based on the theory of deterministic chaos, lead to revealing results: in case of Romania, three neatly distinct intervals were identified and described. According to its transition policy and development strategy, the Romanian transition trajectory seems to be of the type III (centrally planned economy) → IV ("social market" economy) → I (free-market economy).

- Despite of several improvements still expected to be done, the basic structural changes towards a functional market economy in Romania were implemented up to the end of 2001. It is important to note that time series analysis indicates the end of 2001 as the end of the transition process—which confirms the previous conclusion (end of transition process in Romania: 2002–2003).

- This model was applied in case of Romania but it is fully applicable to all Eastern European countries and not only. In this case, interesting comparative analyses might be performed.

References

1. Abazov, R. 1997. Formation of the Non-state Sector and Privatisation in Kazahstan and Uzbekistan. *Post-Communist Economies,* 9(4): 431–48.
2. Aligica, P.D. 2006. Learning in time: new institutionalism and the Central and Eastern European economic reform experience. *Global Business and Economics Review,* 8(1/2): 25–43.
3. Ardelea, D. and C. Scarlat. 1991. De la economia de comanda la economia de piata. *Sisteme logistice,* 1 (July): 33–5.
4. Bask, M. 1996. Dimensions and Lyapunov Exponents from Exchange Rate Series. *Chaos, Solitons & Fractals,* 7(12): 2199–214.
5. Beach, W.W. and T. Kane. 2007. Methodology: Measuring the 10 Economic Freedoms. In *2007 Index of Economic Freedom*, ed. Kane, T., K.R. Holmes, and M.A. O'Grady, 37–55. Washington, D.C.: The Heritage Foundation and Dow Jones & Company, Inc.
6. Bitzenis, A. 2007. Political and economic alternatives for the Central and East European Region and China. *Global Business and Economics Review,* 9(1): 101–22.
7. Brzeski, A. & E. Colombatto. 1999. Can Eastern-Europe Catch-Up? *Post-Communist Economies,* 11(1): 5–25.
8. Cengic, D. 1996. Privatisation and Management Buy-out: The Example of Croatia. *Post-Communist Economies,* 8(4): 549–64.
9. Dana, L.P., T. Dana. 2003. Management and enterprise development in post-communist economies. *International Journal of Management and Enterprise Development*, 1(1): 45–54.
10. European Commission. 2003. *Conclusions of the 2003 Regular Report on Romania's progress towards accession: Romania—Main Results.* Bucharest: Delegation of the European Commission in Romania, 5 November 2003 At: http://www.infoeuropa.ro/.
11. Fuerea, A. et al. 2004. *A chapter-by-chapter assessment of the conformity of the Romanian legislation with the acquis communautaire.* Bucharest: European Institute of Romania.
12. Giannaros, D. 2000. Did the "shock therapy" approach work in the economic restructuring of Eastern Europe? Some evidence from Poland and Russia: a brief review. *Global Business and Economics Review,* 2(1): 53–66.
13. Harvie, C. 2004. The contribution of SMEs in the economic transition of Vietnam. *Journal for International Business and Entrepreneurship Development,* 2(2): 1–16.

14. Homlong, N., E. Springler, 2006. Gender aspects of economic transition: attitude towards female labour participation in the new member states of the European Union. *Global Business and Economics Review,* 8(1/2): 6–24.
15. Hölscher, J. 1997. Economic Dynamism in Transition Economies: Lessons for Germany. *Post-Communist Economies.* 9(2): 173–82.
16. Hunya, G. 1998. Romania 1990–2002: Stop-go Transformation. *Post-Communist Economies,* 10(2): 241–58.
17. Ishaq, M. 1997. The Ukrainian Economy and the Process of Reform. *Post-Communist Economies,* 9(4): 501–18.
18. Kaminski, B. 1998. Poland's Transition from the Perspective of Performance in EU Markets. *Post-Communist Economies,* 10(2): 217–40.
19. Kane, T., K.R. Holmes, and M.A. O'Grady. 2007. *2007 Index of Economic Freedom.* Washington, D.C.: The Heritage Foundation and Dow Jones & Company, Inc. Also available at: www.heritage.org/index.
20. Kantz, H. and T. Schreiber. 1997. *Nonlinear time series analysis.* Cambridge: University Press.
21. Kreuzbergova, E. 2006. Banking socialism in transition: the experience of the Czech Republic. *Global Business and Economics Review,* 8(1/2): 161–77.
22. Kumar, S. 2006. Challenges and benefits of international trade on developed and transitional economies and individual companies: a case study. *International Journal of Management and Enterprise Development,* 3(6): 558–78.
23. Lloyd-Reason, L. et al. 2005. Internationalisation process, SMEs and transitional economies: a four-country perspective. *International Journal of Entrepreneurship and Innovation Management,* 5(3/4): 206–26.
24. Magomedov, A. 1998. Krasnodar-krai: A 'Growth Pole' in the Transitional Economy of Russia? *Post-Communist Economies,* 10(3): 363–74.
25. Marangos, J. 2005. Alternative models of transition and institutional development. *Global Business and Economics Review,* 7(4): 390–408.
26. Marangos, J. 2006. Was there an optimum model of transition? *Global Business and Economics Review,* 8(1/2): 133–60.
27. Mihalyi, P. 1996. Privatisation in Hungary: Now Comes the 'Hard Core'. *Post-Communist Economies,* 8(2): 205–16.
28. Mihaylova, M. & K.S. Howe. 1998. An Analysis of Milk Processing in Bulgaria: Policy Implications of Market Structure, Price and Production Trends During Economic Transition. *Post-Communist Economies,* 10(4): 539–56.
29. Rosenstein, M.T., J.J. Collins and C.J. De Luca 1993. A practical method for calculating largest Lyapunov exponents from small data sets. *Physica D,* 65: 117–34.
30. Sabi, M. 1997. Banking in Transition: Development and Current Problems in Azerbaijan. *Post-Communist Economies,* 9(4): 491–500.
31. Scarlat, C. 1994. Strategic model on the transition process to the free-market economy. Proceedings of the *International Simulation and Gaming Association—ISAGA'93 Conference,* ed. E. Radaceanu, 43–6. Bucharest: IROMA.
32. Scarlat, C. 1999. Romanian Business Environment: Did the Transition Process to the Free-Market Economy Get to an End in Romania? *Research in Business Management* No. 2: *European Cooperation and Expertise,* ed. C. Scarlat and I. Ilinca, 13–20. Bucharest: Center for Business Excellence from University "Politehnica" of Bucharest.
33. Scarlat, C. 2001. *Initierea, dezvoltarea si managementul afacerilor mici si mijlocii.* Rm. Valcea: Conphys.
34. Scarlat, C. 2003a. *Antreprenoriat si Managementul intreprinderilor mici si mijlocii.* Bucharest: Printech.
35. Scarlat, C. 2003b. On impact of the EU accession on Romanian public institutions management. In *Politics of Regulatory Impact Assessment. Best practices and lesson-drawing in Europe,* ed. O. Borissova, 94–102. Sofia: American University in Bulgaria.

36. Scarlat, C. 2005. Aspects of the Romanian Transition. Proceedings of the 6th International Conference *Managing the Process of Globalisation in New and Upcoming EU Members*, Portoroz, 24–26 November 2005, 225–234. Koper: Faculty of Management.
37. Scarlat, C. and A. Curaj. 2004. SME management in contemporary knowledge society, in Romania. In *Knowledge Society-Challenges to Management. Globalisation, Regionalism and EU Enlargement process. Proceedings of the 4th International Conference of the Faculty of Management Koper*, ed. E. Zizmond, 133–48. Koper: Faculty of Management.
38. Scarlat, C., D. Popescu, M. Warner. 2002. *An overview of the implications of EU accession for Romanian public sector institutions*. Bucharest: European Institute of Romania.
39. Scarlat, C., M. Richevaux. 2006. Aspects de la transition roumaine. *Humanisme & Entreprise*, 275– Fevrier 2006: 73–86.
40. Scarlat, E.I. et al. 2005. Time-frequency series analysis of the exchange rate USD-ROL. In *Proceedings of the National Physics Conference*. Bucharest: University "Politehnica".
41. Scarlat, E.I. et al. 2007a. Self-similar characteristics of the currency exchange rate in an economy in transition. *Physica A*, 379(1): 188–98.
42. Scarlat, E.I. et al. 2007b. Chaotic features in Romanian transition economy as reflected onto the currency exchange rate. *Chaos, Solitons and Fractals*, 33: 396–404.
43. Schwartz, B. and S. Yousefi. 2003. On complex behavior and exchange rate dynamics. *Chaos, Solitons and Fractals*, 18: 503–23.
44. Stanley, H.E., R. Mantegna. 2004. *Introduction to Econophysics*. Cambridge: University Press.
45. Takens, F. 1981. Detecting strange attractors in turbulence. In *Dynamical systems and turbulence*, eds. D. Rand, L. Young, 366–81. Berlin: Springer.
46. Tsonnis, A.A., F. Heller, H. Takayasu, K. Marumo, T. Shimizu. 2001. A characteristic time scale in dollar-yen exchange rates. *Physica A*, 291: 574–82.
47. United Nations Common Database (UNCDB). 2007. At: http://unstats.un.org/unsd/cdb.
48. von Hirschhausen, C. 1998. Industrial Restructuring in Ukraine Seven Years After Independence: From Socialism to Planning Economy? *Post-Communist Economies*, 10(4): 451–66.
49. World Bank Report Number 30264. 2004. *Economies in Transition: An OED Evaluation of World Bank Assistance*, ed. Galenson, A. Washington, D.C.: The World Bank. At: http://www.worldbank.org/oed.

Chapter 4

Complex Dynamics in a Nonlinear Cobweb Model for Real Estate Market[†]

Junhai Ma[1,a] and Lingling Mu[1,b]

[1]School of Management, Tianjin University, Tianjin, PR China, 300072.
[a]Email: mjhtju@yahoo.com.cn.
[b]Email: lingmu1020@163.com.

ABSTRACT

In this paper we establish a nonlinear real estate model based on cobweb theory, where the demand function and supply function are quadratic. The stability conditions of the equilibrium are discussed. We demonstrate that as some parameters varied, the stability of Nash equilibrium is lost through period doubling bifurcation. The chaotic features are justified numerically via computing maximal Lyapunov exponents, sensitive dependence on initial conditions. The delayed feedback control (DFC) method is applied to control the chaos of system.

Keywords: Discrete systems, real estate, bifurcation, chaos, delay feedback control.

Introduction

Cobweb models describe the price dynamics in a market of a nonstorable good that takes one time unit to produce [3]. In economic modeling, many examples of cobweb chaos have been demonstrated. Some of the most famous examples include [4–11]. Hommes [7] applies the concept of adaptive expectations in a cobweb model with a single producer, to investigate the occurrence of strange and chaotic behavior. Finkenstädt [5] applied linear supply and nonlinear demand functions; Hommes [6], Jensen and Urban [8] used linear demand functions with nonlinear supply equations. These findings indicate that the nonlinear cobweb model may explain various irregular fluctuations observed in real economic data. In this study, we go one step further to study the cobweb model with nonlinear demand

[†]Reused with permission from: Junhai Ma and Lingling Mu, "Complex Dynamics in a Nonlinear Cobweb Model for Real Estate Market," *Discrete Dynamics in Nature and Society*, vol. 2007, Article ID 29207, 14 pages, 2007. doi:10.1155/2007/29207.

and supply function. A possible source of such an evolutionary market dynamics is an interaction between government and real estate developer.

Traditional cobweb models usually describe a dynamic price adjustment in agricultural markets with a supply response lag [4]. Consider, for instance, the supply of housing. The time of housing construction guarantees a finite lag between the time the production decision is made and the time the housing is ready for sale. The real estate developer's decision about how many houses should be built and sale is usually based on current and past experience. This principle is the same as that of agricultural product. So it is feasible to introduce cobweb model into real estate market.

The present paper attempts to establish a nonlinear model for the real estate market, and introduce adjustment parameters of housing prices and land price into the model, which can denote the game behavior of players. The system stability with the variation of parameters is analyzed. Numerical simulations verify the complexity of system evolvement. Finally, time-delayed feedback control method is used to keep the system from chaos and bifurcation.

Nonlinear Models for Real Estate Market

In this paper we assume that all real estate developers in the market are belong to one benefit group and have a common benefit target. Usually the price p is characterized by the nonlinear inverse demand function of $p = a - b\sqrt{Q}$. Where a and b are positive constants. a is the maximum price in the market, and Q is the total quantity in the market. This kind of form has been used in other oligopoly models and in the experimental economics dealing with learning and expectations formation (see e.g., [1,2,12]). The transformation of this formula is as follows:

$$D_1(t) = b_0 - b_1 p_1(t) + b_2 p_1^2(t), \quad D_2(t) = c_0 - c_1 p_2(t) + c_2 p_2^2(t) \tag{2.1}$$

where $b_0, b_1, b_2, c_0, c_1, c_2$ are positive constants. $p_1(t)$ is the land price at time period t, $p_2(t)$ is the housing prices at time period t, $D_1(t)$ is the land demand at time period t, $D_2(t)$ is the housing demand at time period t. Due to the law of demand that the slope of demand curve is negative, the price $p_1(t)$ and $p_2(t)$ must respectively satisfy the inequality: $2b_2 p_1(t) - b_1 < 0$ and $2c_2 p_1(t) - c_1 < 0$, $4b_2 b_0 - b_1^2 > 0$, $4c_2 c_0 - c_1^2 > 0$, must hold, thus the signs of demand equations in formula (2.1) are positive.

In this case, the land market and housing market are interrelated. Though the housing market does not directly affect land market, the land price impacts the housing supply which decreases with increasing land price. This rule is the same as that of hog and corn as stated by Waugh [16]. Real estate companies adjust the housing supply according to the relative policies and the situation of housing price and land price. The formula of supply can be supposed as follows:

$$S_1(t) = e_0 + e_1 p_1(t) + e_2 p_1^2(t), \quad S_2(t) = d_0 + d_1 p_2(t) + d_2 p_2^2(t) - d_3 p_1(t) \tag{2.2}$$

where $d_0, d_1, d_2, d_3, e_0, e_1, e_2$ are positive constants. $S_1(t)$ is the land supply at time period t, $S_2(t)$ is the housing supply at time period t. Because $2e_2 p_1(t) + e_1 > 0$ and $2d_2 p_2(t) + d_1 > 0$, so we can affirm that the slope of supply curve is positive and it is in accordance with the law of supply. Providers begin to supply the products only when

$$p_1(t) > \frac{-e_1 + \sqrt{e_1^2 - 4e_2 e_0}}{2e_2} \text{ and } p_2(t) > \frac{-d_1 + \sqrt{d_1^2 - 4d_2 d_0}}{2d_2} \tag{2.3}$$

Define

$$Z(p) = D(p) - S(p) \tag{2.4}$$

$Z(p)$ is excess demand function descending with price, which denotes the gap between demand and supply. When the price is low, excess demand exists and when the price is high excess supply exists, thus p^* that satisfies the equation $Z(p^*) = 0$ is called equilibrium point.

Substitute (1) and (2) into (3), we obtain:

$$\begin{cases} Z(p_1(t)) = b_0 - e_0 - (e_1 + b_1)p_1(t) + (b_2 - e_2)p_1^2(t) \\ Z(p_2(t)) = c_0 - d_0 - (d_1 + c_1)p_2(t) + (c_2 - d_2)p_2^2(t) + d_3 p_1(t) \end{cases}, \quad t = 0,1,2,... \tag{2.5}$$

Since $Z(p)$ follows the law of demand, the following conditions must hold:

$$\begin{cases} b_2 - e_2 > 0 \\ c_2 - d_2 > 0 \\ 2(c_2 - d_2)p_2(t) - (d_1 + c_1) < 0 \\ 2(b_2 - e_2)p_1(t) - (e_1 + b_1) < 0 \end{cases} \tag{2.6}$$

where α_1 is the adjustment parameter of land price, which denotes the adjustment degree of benchmark land price controlled by government through the land supply plan. α_2 is the adjustment parameter of housing price, the dynamic model of land price and housing price can be established as follows:

$$\begin{cases} p_1(t) = p_1(t-1) + \alpha_1 Z(p_1(t-1)) \\ p_2(t) = p_2(t-1) + \alpha_2 Z(p_2(t-1)) \end{cases}, \quad t = 0,1,2,... \tag{2.7}$$

where α_1 and α_2 are positive parameters. It's clear that the excess functions of land and housing with adjustment parameters are two dimensional nonlinear map, which can be regarded as a discrete dynamic system.

Stability Analysis

Bifurcation and Chaos

Expansion formula of (2.7) is:

$$
\begin{cases}
p_1(t) = p_1(t-1) + \alpha_1 \left(b_0 - e_0 - (e_1 + b_1)p_1(t-1) + (b_2 - e_2)p_1^2(t-1) \right) \\
p_2(t) = p_2(t-1) + \alpha_2 \left(c_0 - d_0 - (d_1 + c_1)p_2(t-1) + (c_2 - d_2)p_2^2(t-1) + d_3 p_1(t-1) \right)
\end{cases}
$$

$$t = 0,1,2,\ldots \tag{3.1}$$

let $u = \dfrac{\alpha_1(e_2 - b_2)}{1 + \alpha_1 \sqrt{(e_1 + b_1)^2 - 4(e_2 - b_2)(e_0 - b_0)}}$, $\ U = \dfrac{1}{2} - \dfrac{1}{2} \cdot \dfrac{1 - \alpha_1(e_2 - b_2)}{1 + \alpha_1 \sqrt{(e_1 + b_1)^2 - 4(e_2 - b_2)(e_0 - b_0)}}$

$$x(t) = u p_1(t) + U, \tag{3.2}$$

the transform of the first equation in formula (3.1) is:

$$x(t) = \lambda x(t-1)(1 - x(t-1)) \tag{3.3}$$

where $\lambda = 1 + \alpha_1 \sqrt{(e_1 + b_1)^2 - 4(e_2 - b_2)(e_0 - b_0)}$. The stability of $x(t)$ varies along with variety of λ according to Logistic rule.

If $\alpha_1 < 0$, then $\lambda < 1$ implies that one fixed point exists in system (3.1), however, $\alpha_1 < 0$ is insignificant so we do not give consideration.

If $0 \leq \alpha_1 < 2 / \sqrt{(e_1 + b_1)^2 - 4(e_2 - b_2)(e_0 - b_0)}$, then $1 < \lambda < 3$ implies that two fixed points exist in system (3.1) and bifurcation appears.

If $2 / \sqrt{(e_1 + b_1)^2 - 4(e_2 - b_2)(e_0 - b_0)} \leq \alpha_1 \leq \sqrt{6} / \sqrt{(e_1 + b_1)^2 - 4(e_2 - b_2)(e_0 - b_0)}$, then $3 \leq \lambda \leq 1 + \sqrt{6}$ implies that four fixed points exist in system (3.1) and period doubling bifurcation appears.

As λ increases, the amount of fixed point continue to grow until $\lambda = 3.5699$, when $\alpha_1 = 2.5399 / \sqrt{(e_1 + b_1)^2 - 4(e_2 - b_2)(e_0 - b_0)}$, the value of $x(t)$ is unequal to any point that appeared before, system enters chaos from period doubling bifurcation.

The same argument holds for the second equation in formula (3.1). Let $\alpha_1 = 0$: When $0 < \alpha_2 < 2\sqrt{(d_1 + c_1)^2 - 4(d_2 - c_2)(d_0 - c_0)}$, bifu]rcation occurs.

When

$$2 / \sqrt{(d_1 + c_1)^2 - 4(d_2 - c_2)(d_0 - c_0)} \leq \alpha_2 \leq \sqrt{6} / \sqrt{(d_1 + c_1)^2 - 4(d_2 - c_2)(d_0 - c_0)},$$ period doubling bifurcation occurs.

When $\alpha_2 = 2.5699 / \sqrt{(d_1 + c_1)^2 - 4(d_2 - c_2)(d_0 - c_0)}$, system enters chaotic state from period doubling bifurcation.

Stability Analysis

Now we discuss the stability of fixed points of the discrete dynamic system (3.1) through analyzing the eigenvalues of asymptotic linear equation of formula (3.1). Four fixed points of difference function (3.1) are obtained:

$$E: \begin{cases} p_1 = \dfrac{b_1 + e_1 \pm \sqrt{(b_1 + e_1)^2 - 4(e_2 - b_2)(e_0 - b_0)}}{2(b_2 - e_2)}, \\[4mm] p_2 = \dfrac{d_1 + c_1 \pm \sqrt{(d_1 + c_1)^2 - 4(c_2 - d_2)(c_0 - d_0 + d_3 p_1)}}{2(c_2 - d_2)}, \end{cases} \tag{3.4}$$

Provided that:

$$\begin{cases} (b_1 + e_1)^2 - 4(e_2 - b_2)(e_0 - b_0) \geq 0 \\ (d_1 + c_1)^2 - 4(c_2 - d_2)(c_0 - d_0 + d_3 p_1) \geq 0 \end{cases} \tag{3.5}$$

Lemma 3.1 *The equilibrium*

$$E_1 \left(\frac{b_1 + e_1 \pm \sqrt{(b_1 + e_1)^2 - 4(e_2 - b_2)(e_0 - b_0)}}{2(b_2 - e_2)}, \frac{d_1 + c_1 \pm \sqrt{(d_1 + c_1)^2 - 4(c_2 - d_2)(c_0 - d_0 + d_3 p_1)}}{2(c_2 - d_2)} \right) \tag{3.6}$$

is an unstable equilibrium point.

Proof: In order to prove this result, we find the eigenvalues of the Jacobian matrix J. In fact at E_1, the Jacobian matrix becomes a triangular matrix:

$$J(E_1) = \begin{pmatrix} 1 + \alpha_1 \sqrt{(b_1 + e_1)^2 + 4(b_2 - e_2)(b_0 - e_0)} & 0 \\ \alpha_2 d_3 & 1 + \alpha_2 \sqrt{(d_1 + c_1)^2 + 4(d_2 - c_2)(d_0 - c_0 - d_3 p_1)} \end{pmatrix} \tag{3.7}$$

whose eigenvalues are given by the diagonal entries. They are:

$$\begin{cases} \lambda_1 = 1 + \alpha_1 \sqrt{(b_1 + e_1)^2 + 4(b_2 - e_2)(b_0 - e_0)} \\ \lambda_2 = 1 + \alpha_2 \sqrt{(d_1 + c_1)^2 + 4(d_2 - c_2)(d_0 - c_0 - d_3 p_1)} \end{cases} \tag{3.8}$$

It is clear that when condition (3.5) holds, then $|\lambda_1| > 1$ and $|\lambda_2| > 1$. Then E_1 is an unstable equilibrium point of the system (3.1). This completes the proof of the proposition.

The stability of other points can also be judged by the above method.

The Stable Region of Equilibrium Point

In this subsection we analyze the asymptotic stability of the equilibrium point for the two-dimensional map (3.1). We determine the region of stability in the plane of the parameters (α_1, α_2). The Jacobian matrix at $E^*(p_1^*(t), p_2^*(t))$ takes the form

$$J = \begin{pmatrix} 1-\alpha_1\left(b_1 + e_1\right)+2\alpha_1\left(b_2 - e_2\right)p_1^*(t) & 0 \\ \alpha_2 d_3 & 1-\alpha_2\left(d_1 + c_1\right)+2\alpha_2\left(d_2 - c_2\right)p_2^*(t) \end{pmatrix}$$

(3.9)

The characteristic equation of the matrix (3.9) has the form $F(\lambda) = \lambda^2 - Tr\lambda + Det$, where "$Tr$" is the trace and "$Det$" is the determinant of the Jacobian matrix (3.9) which are given by

$$Tr = 2 - \alpha_1\left(e_1 + b_1\right) + 2\alpha_1\left(b_2 - e_2\right)p_1^*(t) - \alpha_2\left(d_1 + c_1\right) + 2\alpha_2\left(c_2 - d_2\right)p_2^*(t)$$

$$Det = 1 - \alpha_1\left(e_1 + b_1\right) + 2\alpha_1\left(b_2 - e_2\right)p_1^*(t) - \alpha_2\left(d_1 + c_1\right) + 2\alpha_2\left(c_2 - d_2\right)p_2^*(t)$$
$$+ \left[\alpha_1\left(e_1 + b_1\right) - 2\alpha_1\left(b_2 - e_2\right)p_1^*(t)\right]\left[\alpha_2\left(d_1 + c_1\right) - 2\alpha_2\left(c_2 - d_2\right)p_2^*(t)\right]$$

(3.11)

Since

$$\left(Tr\right)^2 - 4Det = \left[\alpha_2\left(d_1 + c_1\right) - \alpha_1\left(e_1 + b_1\right) + 2\alpha_1\left(b_2 - e_2\right)p_1^*(t) - 2\alpha_2\left(c_2 - d_2\right)p_2^*(t)\right]^2 > 0$$

(3.12)

we deduce that the eigenvalues of equilibrium are real. The local stability of equilibrium point is given by Jury's conditions [13] which are:

a) $1 - Tr + Det > 0$

Lemma 3.2 *The condition (a) is always satisfied*

Proof: Because

$$1 - Tr + Det = \left[\alpha_1\left(e_1 + b_1\right) - 2\alpha_1\left(b_2 - e_2\right)p_1^*(t)\right]\left[\alpha_2\left(d_1 + c_1\right) - 2\alpha_2\left(c_2 - d_2\right)p_2^*(t)\right]$$

(3.12)

moreover, the last two conditions in Eq.(2.6) are hold, α_1 and α_2 are positive parameters, so the sign of "$1 - Tr + Det$" is positive and the Lemma is proven.

b) $1 + Tr + Det > 0$

$$1 + Tr + Det = 4 - 2\alpha_1\left(e_1 + b_1\right) + 4\alpha_1\left(b_2 - e_2\right)p_1^*(t) - 2\alpha_2\left(d_1 + c_1\right) + 4\alpha_2\left(c_2 - d_2\right)p_2^*(t)$$
$$+ \left[\alpha_1\left(e_1 + b_1\right) - 2\alpha_1\left(b_2 - e_2\right)p_1^*(t)\right]\left[\alpha_2\left(d_1 + c_1\right) - 2\alpha_2\left(c_2 - d_2\right)p_2^*(t)\right]$$

(3.13)

c) $Det - 1 < 0$

$$Det - 1 = \alpha_1\left(e_1 + b_1\right) + 2\alpha_1\left(b_2 - e_2\right)p_1^*(t) - \alpha_2\left(d_1 + c_1\right) + 2\alpha_2\left(c_2 - d_2\right)p_2^*(t)$$
$$+ \left[\alpha_1\left(e_1 + b_1\right) - 2\alpha_1\left(b_2 - e_2\right)p_1^*(t)\right]\left[\alpha_2\left(d_1 + c_1\right) - 2\alpha_2\left(c_2 - d_2\right)p_2^*(t)\right]$$

(3.14)

The conditions (b) and (c) define a bounded region of stability in the parameters space (α_1, α_2). Then the second and third conditions are the conditions for the local stability of equilibrium point which becomes:

$$\begin{cases} 1 + Tr + Det = 4 - 2\alpha_1 \left(e_1 + b_1\right) + 4\alpha_1 \left(b_2 - e_2\right) p_1^*\left(t\right) - 2\alpha_2 \left(d_1 + c_1\right) + 4\alpha_2 \left(c_2 - d_2\right) p_2^*\left(t\right) \\ + \left[\alpha_1 \left(e_1 + b_1\right) - 2\alpha_1 \left(b_2 - e_2\right) p_1^*\left(t\right)\right] \left[\alpha_2 \left(d_1 + c_1\right) - 2\alpha_2 \left(c_2 - d_2\right) p_2^*\left(t\right)\right] > 0 \\ Det - 1 = \alpha_1 \left(e_1 + b_1\right) + 2\alpha_1 \left(b_2 - e_2\right) p_1^*\left(t\right) - \alpha_2 \left(d_1 + c_1\right) + 2\alpha_2 \left(c_2 - d_2\right) p_2^*\left(t\right) \\ + \left[\alpha_1 \left(e_1 + b_1\right) - 2\alpha_1 \left(b_2 - e_2\right) p_1^*\left(t\right)\right] \left[\alpha_2 \left(d_1 + c_1\right) - 2\alpha_2 \left(c_2 - d_2\right) p_2^*\left(t\right)\right] < 0 \end{cases} \quad (3.15)$$

The stability region is bounded by the portion of hyperbola with positive values of α_1 and α_2, whose equations are given by the vanishing of the left-hand side $1 + Tr + Det = 0$ and $Det - 1 = 0$. For the values of (α_1, α_2) inside the stability region (see Fig.3.1), the equilibrium point is stable node and looses its stability through a period doubling bifurcation. The bifurcation curve intersects the axes α_1 and α_2, respectively, whose coordinates are given by

$$A\left(0, \frac{2}{d_1 + c_1 - 2\left(c_2 - d_2\right) p_2 *\left(t\right)}\right) \quad and \quad B\left(\frac{2}{e_1 + b_1 - 2\left(b_2 - e_2\right) p_1 *\left(t\right)}, 0\right). \quad (3.16)$$

Numerical Simulations

In order to study the complex dynamics of system (3.1), it is convenient to take the parameters values as follows:

$b_0 = 1.2$, $b_1 = 2$, $b_2 = 1.6$, $c_0 = 4$, $c_1 = 1.6$, $c_2 = 0.04$, $d_0 = 0$, $d_1 = 3$, $d_2 = 0.02$, $d_3 = 0.4$, $e_0 = 0.5$, $e_1 = 0.3$, $e_2 = 0.2$.

$$(4.1)$$

Figure 3.1 shows the region of stability of Nash equilibrium. Eq. (3.15) defines the region of stability in the plane of (α_1, α_2). Figure 4.2 shows the map of f_{α_1, α_2}. Dynamics of land price in the cobweb model is given by system $p1(t) = f_{\alpha_1, \alpha_2}\left(p_1\left(t - 1\right)\right)$ with two model parameters. A graphical analysis in Fig.2 shows that the map f_{α_1, α_2} is non-monotonic with one critical point, where the graph has a (local) minimum, and that initial state $p_1(0)$ does not converge to a low periodic orbit. Since the graphical analysis in this case does not converge, it suggests that the dynamical behavior is chaotic.

Figures 4.2 and 4.3 show the bifurcation diagrams with respect to the parameter α_1 and for $\alpha_2 = 0.2$ and 0.4. In both figures the Nash equilibrium $E* = (0.4, 0.9)$ is locally stable for small values of the parameter α_1. If α_1 increases, the Nash equilibrium point becomes unstable, and one observes complex dynamic behavior occurs such as cycles of higher order and chaos. Also the maximal Lyapunov exponent is plotted in Figs. 4.2 and 4.3.

Figures 4.4, 4.5, 4.6 and 4.7 show the graph of strange attractors for the different values of α_2. The parameter α_2 takes the values as 0.07, 0.1, 0.2 and 0.3, which exhibit fractal structure in both cases.

We compute the difference of two orbits with initial points $\left[p_1(0), p_2(0)\right]$ and $\left[p_1(0) + 0.0001, p_2(0)\right]$, as well as $\left[p_1(0), p_2(0)\right]$ and $\left[p_1(0), p_2(0) + 0.0001\right]$, to demonstrate the sensitivity to initial conditions of the system (3.1). The parameters take values as $(\alpha_1, \alpha_2) = (2.3, 0.6)$ and $\left[p_1(0), p_2(0)\right] = (1, 2)$. The results are shown in Figs. 9–10, where $\Delta p_1(t) = p_1(t) - p_1'(t)$ and $p_1'(t)$ is the value of land price at time period t with

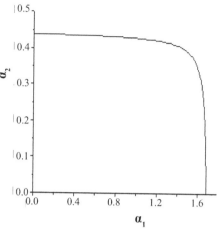

*Figure 1** Stability region of Nash equilibrium.

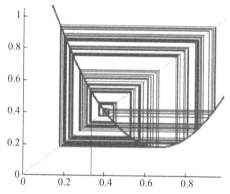

Figure 2 The graph of map f_{α_1, α_2} .

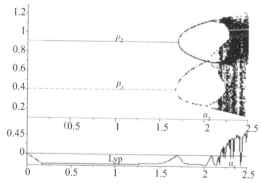

†**Figure 3** Bifurcation diagram for $\alpha_2 = 0.4$.

---

*Figure 3.1 in original source
**Figure 4.1 in original source
†Figure 4.2 in original source

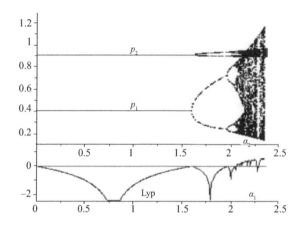

*Figure 4** Bifurcation diagram for $\alpha_2 = 0.2$.

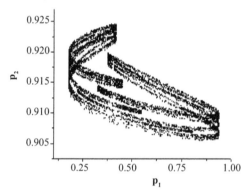

Figure 5 Strange attractor for $\alpha_2 = 0.07$.

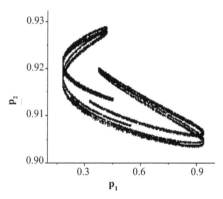

†**Figure 6** Strange attractor for $\alpha_2 = 0.1$.

---

*Figure 4.3 in original source
**Figure 4.4 in original source
†Figure 4.5 in original source

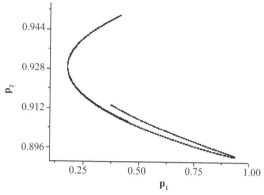

*Figure 7** Strange attractor for $\alpha_2 = 0.2$.

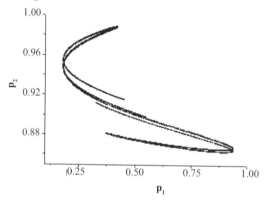

Figure 8 Strange attractor for $\alpha_2 = 0.3$.

initial value of $p_1(0) + 0.0001$; $\Delta p_2(t) = p_2(t) - p_2'(t)$ and $p_2'(t)$ is the value of housing price at time period t with initial value of $p_2(0) + 0.0001$; In both figures, initial condition of one coordinate differs by 0.0001, the other coordinate keeps equal. At the beginning, the difference is indistinguishable but after a number of iterations the difference between them builds up rapidly. From Figs. 4.8 and 4.9 we show that the time series of the system Eq.(3.1) is sensitive dependence on initial conditions, i.e., complex dynamics behaviors occur in this model.

Chaos Control

Delay feedback control (DFC) method was brought forward by K. Pyragas [14]. The method allows a noninvasive stabilization of unstable periodic orbits (UPOs) of dynamical systems [15]. It feeds back part of system output signals as exterior input to the system after a time delay. $u(\bullet)$ is control signal gained by self-feedback coupling between output and input signals in chaotic system. $x(t) = f(x(t-1)) + u(t)$ is the form of DFC, where

---

*Figure 4.6 in original source
**Figure 4.7 in original source

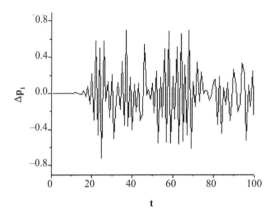

Figure 9 Sensitivity to initial conditions of P_1.

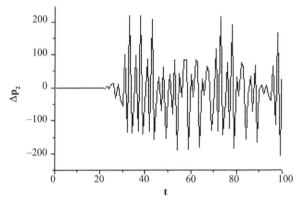

Figure 10 Sensitivity to initial conditions of P_2.

$u(t) = k(x(t)) - x(t - \tau), (t > \tau), \tau$ is time delay, k is controlling factor. Though delay feedback control only carries out on one variable, it enables other variables in the system to achieve stability simultaneously. Our goal is to control the system in such way. The system with controlling factor is shown as follows:

$$\begin{cases} p_1(t) = p_1(t-1) + \alpha_1 Z(p_1(t-1)) - k(p_1(t-1) - p_1(t-\tau)) \\ p_2(t) = p_2(t-1) + \alpha_2 Z_2(p_2(t-1)) \end{cases}, \quad t = 0,1,2,... \tag{5.1}$$

From Fig. 3.1 we know that chaos exists in system (3.1) when $\alpha_2 = 0.4, \alpha_1 = 2.3$, therefore we carry out control under this condition .Choosing $\tau = 1$, first inspect the relation of k and system stability. The Jacobian matrix of system (5.1) is:

$$J = \begin{bmatrix} (1 - \alpha_1(e_1 + b_1) + k + 2\alpha_1(b_2 - e_2)p_1(t))/(1+k) & 0 \\ \alpha_2 d_3 & 1 - \alpha_2(d_1 + c_1) + 2\alpha_2(c_2 - d_2)p_2(t) \end{bmatrix} \tag{5.2}$$

Substitute equilibrium point (0.4,0.9) into (5.2), we obtain eigenvalues $\lambda_1 = -0.83, \lambda_2 = \dfrac{k-1.7}{1+k}$. So when $k > 0.35$, absolute values of both eigenvalues are less than 1, which means the system is stable.

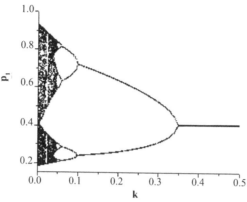

*Figure 11 Relation graph of p_1 and k.

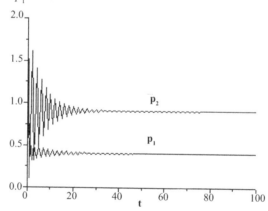

**Figure 12 Time series of p_1 and p_2 with $k=0.4$.

As shown in Fig.4.10 land price is controlled from chaotic state to stable state when k is greater than 0.35, so we select $k=0.4$. Housing price and land price are also controlled to equilibrium point (0.4,0.9) as shown in Fig.4.11.

Conclusion

A nonlinear model for real estate market has been presented based on the cobweb theory. It is a simple dynamic model with nonlinear demand and supply function. From numerical simulations we deduce that the land supply system has the remarkable influence on real estate market. Therefore policy makers who intervene in one market should recognize

---

*Figure 4.10 in original source
**Figure 4.11 in original source

that what they do not may also influence other relative markets. We showed that the fast adjustment cause a market structure to behave chaotically. Therefore the dynamics of market is changed when players apply different adjustment speed. Attempts are also made to stabilize the chaotic system with the delay feedback method. Combining with this method the land price and housing price evolve from chaotic to stable.

Acknowledgements

The authors are supported by National Nature Foundation of China (71271071). We are grateful to Professor Chen Yu-shu (Academician of Chinese Academy of Engineering) and Mrs. Liang jiao-jie for their helpful comments. The authors thank three anonymous referees for their valuable suggestions and remarks. Any errors or shortcomings are our own.

References

1. W.A. Brock and C.H. Hommes, "A rational route to randomness," *Econometrica*, vol. 65, no. 5, pp. 1059–1095, 1997.
2. C. Chiarella, "The cobweb model, its instability and the onset of chaos," Economic Modeling, vol. 5, no. 4, pp. 377–384, 1988.
3. B. Finkenst"adt, *Nonlinear Dynamics in Economics: A Theoretical and Statistical Approach to Agricultural Markets*, Lecture Notes in Economics and Mathematical Systems no. 426, Springer, Berlin, Germany, 1995.
4. C.H. Hommes, "Adaptive learning and roads to chaos: the case of the cobweb," *Economics Letters*, vol. 36, no. 2, pp. 127–132, 1991.
5. C.H. Hommes, "Dynamics of the cobweb model with adaptive expectations and nonlinear supply and demand," *Journal of Economic Behavior & Organization*, vol. 24, no. 3, pp. 315–335, 1994.
6. R.V. Jensen and R. Urban, "Chaotic price behavior in a nonlinear cobweb model," *Economics Letters*, vol. 15, no. 3-4, pp. 235–240, 1984.
7. A. Matsumoto, "Ergodic cobweb chaos," *Discrete Dynamics in Nature and Society*, vol. 1, no. 2, pp. 135–146, 1997.
8. A. Matsumoto, "Preferable disequilibrium in a nonlinear cobweb economy," *Annals of Operations Research*, vol. 89, pp. 101–123, 1999.
9. H.E. Nusse and C.H. Hommes, "Resolution of chaos with application to a modified Samuelson model," *Journal of Economic Dynamics & Control*, vol. 14, no. 1, pp. 1–19, 1990.
10. H.N. Agiza, A.S. Hegazi, and A.A. Elsadany, "Complex dynamics and synchronization of a duopoly game with bounded rationality," *Mathematics and Computers in Simulation*, vol. 58, no. 2, pp. 133–146, 2002.
11. A.K. Naimzada and L. Sbragia, "Oligopoly games with nonlinear demand and cost functions: two boundedly rational adjustment processes," *Chaos, Solitons & Fractals*, vol. 29, no. 3, pp. 707–722, 2006.
12. T. Offerman, J. Potters, and J. Sonnemans, "Imitation and belief learning in an oligopoly experiment," *Review of Economic Studies*, vol. 69, no. 4, pp. 973–997, 2002.
13. F.V. Waugh, "Cobweb models," *Journal of Farm Economics*, vol. 46, no. 4, pp. 732–750, 1964.
14. T. Puu, Attractors, Bifurcations, and Chaos: Nonlinear Phenomena in Economics, Springer, Berlin, Germany, 2000.
15. X. Li, G. Chen, Z. Chen, and Z. Yuan, "Chaotifying linear Elman networks," *IEEE Transactions on Neural Networks*, vol. 13, no. 5, pp. 1193–1199, 2002.
16. K. Pyragas, "Continuous control of chaos by self-controlling feedback," *Physics Letters A*, vol. 170, no. 6, pp. 421–428, 1992.
17. V. Pyragas and K. Pyragas, "Delayed feedback control of the Lorenz system: an analytical treatment at a subcritical Hopf bifurcation," *Physical Review E*, vol. 73, no. 3, Article ID 036215, 10 pages, 2006.

Chapter 5

Chaos Models in Economics[†]

Sorin Vlad,[1,a] *Paul Pascu*[1,b] *and Nicolae Morariu*[1,c]

[1]University of Suceava, 13 University Street, Suceava ,Romania, 720229.
[a]Email: sorinv@seap.usv.ro.
[b]Email: paulp@seap.usv.ro.
[c]Email: nmorariu@eed.usv.ro.

ABSTRACT

The paper discusses the main ideas of the chaos theory and presents mainly the importance of the nonlinearities in the mathematical models. Chaos and order are apparently two opposite terms. The fact that in chaos can be found a certain precise symmetry (Feigenbaum numbers) is even more surprising. As an illustration of the ubiquity of chaos, three models among many other existing models that have chaotic features are presented here: the nonlinear feedback profit model, one model for the simulation of the exchange rate and one application of the chaos theory in the capital markets.

Index Terms: chaos, nonlinear systems, complex behavior, bifurcation diagram.

Introduction

One of the axioms of the modern science asserts that if an accurate description of a physical system can be identified then the possibility of a deeper understanding of the system and the prediction of the system evolution is possible. These assertions are not always correct. For instance, if one applies the laws of motion stated by Newton, then there is possible to predict exactly the orbit of the Moon around the Earth if the influence of other planets is not considered. These predictions were verified and proved to be accurate. If the third planet is included, the mathematical model of the interaction of the two bodies becomes "the three bodies problem", solved by Newton but for a limited set of cases and unsolved for the general case. Today by means of a computer, "the tree bodies problem" can be solved, but one can observe that the prediction of the orbit of the third planet is often impossible.

[†]Reused with permission from: Sorin Vlad, Paul Pascu and Nicolae Morariu, Chaos Models in Economics, *Journal of Computing*, 2 (1), 2010, 79–83.

A large number of real systems have a nonlinear behavior despite the idealized linear behavior used in modeling. The development of a new way of dealing with nonlinear systems is obvious. This "new way of dealing" exists already despite the fact that the study of the nonlinearity is still at the beginning. Some changes in nonlinear systems can lead to a complex and erratic behavior called chaos. The nonlinearity is one of the conditions needed by a system in order to develop chaos. The term chaos is used to describe the behavior of a system that is aperiodic and apparently random. S.H. Strogatz defines chaos as an aperiodic long time behavior developed by a deterministic system highly sensitive on initial condition. [1] Behind this apparently random behavior lies the deterministic character determined by the equations describing the system. Most of the systems that are used as examples to explain the concepts of chaos theory are deterministic.

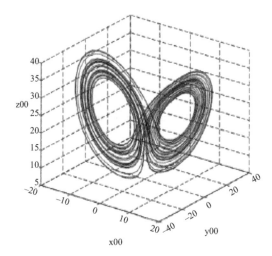

Figure 1 Lorentz attractor-the symbol of chaos theory.

There are two types of chaos: deterministic and nondeterministic. The deterministic chaos represents the chaotic motion of the nonlinear systems whose dynamic laws determines uniquely the evolution of the system's state based on the previous evolution. The deterministic chaos represents only one particular case of what is called nondeterministic chaos that exhibits a superexponential divergence of the trajectories. In this case the equations describing the evolution of the system are not known. The both ways of chaos manifestations are short-term predictable but long term unpredictable. The chaos and the concepts related to the dynamics of the systems and their modeling using differential equations is named the chaos theory and is tightly related with the notion of nonlinearity [4]. The nonlinearity implies the loss of the causality correlation between the perturbation and effect propagated in time. The study of the nonlinearity is named nonlinear dynamics—a captivating domain using a mathematical apparatus still under development.

Despite the fact that the ideas leading to the emergence of the chaos theory existed before longtime, Lorenz (1963) created a mathematical model of the convection currents circulation in atmosphere and observed that when the systems begins with initial conditions slightly changed from the previous ones, the results are completely different. This phenomenon lies at the basis of a very popular paradigm of chaos named "the butterfly effect", that states that if the flapping of a butterfly slightly modifies the atmospherically conditions in the Amazonian jungle, this fact can have an impact, at the end of a complex cause—effect chain in setting off a tornado in Texas. The butterfly effect paradigm contains the essence of the phenomenon characterizing the chaos: first, the sensitive dependence on initial conditions and second—the fact that to predict the future state of a chaotic system, the current state need to be known with infinite prediction. The manifestation of chaos can be found everywhere in the real world, for instance: the propagation of the avalanches, epidemics spreading, climate evolution, heart beats, lasers, electronic circuits, etc.

A legitimate question that rises is that the chaos is the rule or the exception from the rule. Taking into account that most of the systems of the real world are nonlinear (the basic condition for the emergence of chaos), seems that chaos could be one of the not so obvious features of the nature. The importance of studying chaos is that chaos offers an alternate method that explains the apparently random behavior of the complex systems. The chaos plus the specific mathematical tools is a framework of studying different models from different fields, models that can be reduced to elementary models with known chaotic behavior for some values of the parameters. The way to chaos begins with the phenomenon of period doubling. The period doubling evolves in 2, 4, 8, 16 and so on periods and the system evolution can abruptly fall into chaotic regime. In the case of unimodal function there is an interesting symmetry in the parameter values for what the period doubling occurs. If A_1 is the value of the control parameter for what the first period doubling occurs and A_n is the value for what the nth period doubling occurs, then:

$$\delta = \lim_{n \to \infty} \frac{A_n - A_{n-1}}{A_{n+1} - A_n} = 4.66920$$

(1)

where δ is the Feigenbaum number valable for all unimodal functions [5].

Nonlinear Models Chaos in Exchange Rates

For the simulation of the volatile behavior of the exchange rates were created models that treat the exchange rates as being prices of the financial assessments traded on efficient markets. The current exchange rate contains the currently available information and the changes observed reflect the effect of the new events that are unpredictable by definition. The theory states that an accurate a priori prediction of the exchange rate evolution is impossible to be made but the subsequent explanation of the changes is possible. In order to eliminate these difficulties, the chaos theory and the nonlinear models are extensively used. The first researches have been carried out starting from 1980.

In the majority of situations these models are highly nonlinear and result in a wide range of dynamic behavior, including chaotic dynamics. There is a dispute over the manifestation of chaotic dynamics in exchange rates. There are many studies that are positive to the chaotic dynamics (Federici 2001, Westerhoff, Darvas 1998, Hommes 2005, Vandrocicz 2006) and also a number of studies that are rejecting the chaos in exchange rate (Brooks, Serletis). The chaos theory demonstrates that even the simplest dynamical systems can exhibit at some point a very complex behavior. If the exchange rates variation is caused due to the chaotic nature of the system, this should lead to the fact that the smallest influences should have the effect of a nonlinearity over the exchange rates—exactly what happens in reality. The first model presented demonstrates the fact that even the simplest models can exhibit chaotic behavior. [3]. The demand of foreign currency is determined as percentage of the deviation of current exchange rate towards the expected one.[2]

$$S_t = \alpha \left(\frac{e^e}{e_t} - 1 \right), \alpha \geq 0$$

(2)

where

e_t is the domestic price of the foreign currency.

e^e is the future estimated exchange rate.

α is the sensitivity parameter.

Figure 2 illustrates the evolution of the system for two initial slightly different values: 0.2 and 0.2005 (the dotted line). The values of the two time series are identical for a short period of time (the first 10 iterations) and then the trajectories of the systems are diverging. The scatterplots for the two time series are provided to demonstrate the independence of the two time series after 10 iterations. The scatterplots presented in Fig. 3 and Fig. 4 one of the fingerprints of chaos: the distance between two trajectories starting from nearby points in the state space diverge over time.

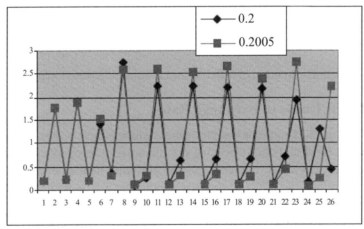

Figure 2 The influence of the initial conditions.

When the sensitivity parameter is varied, the same effects can be observed. Figure 4 presents the trajectories of the system for two very near values of α. The apparently irrelevant changes can affect the longtime behavior of the exchange rate modeled using the Ellis model and some of these small shocks can determine the system to fall into the chaotic regime.

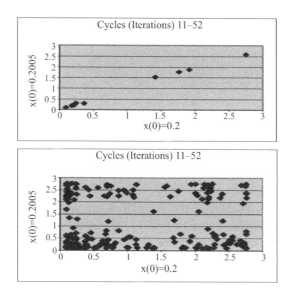

Figure 3 The scatterplot for the first 10 iterations and the scatterplot for the last 41 iterations.

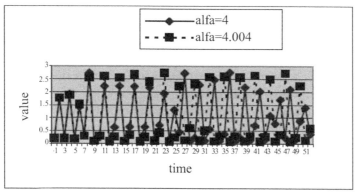

Figure 4 The influence of the initial conditions.

The trade balance (T_i) is a linear function depending on the current exchange rates and the corresponding exchange rate for the last period, written as deviation from the expected values and is given by the equation: The expected exchange rate represents the stable state at which the speculators on the market do not wish to sell nor buy.

$$T_t = \beta\left(e_t - e^e\right) + \gamma\left(e_{t-1} - e^e\right) \quad \beta, \gamma > 0 \tag{3}$$

The clearing of the exchange markets writes as:

$$\Delta S_t = T_t \tag{4}$$

After replacing equations (2) and (1) in (4), equation (4) becomes:

$$\beta e_{t-1} e_t^2 - \left[\left(\beta + \gamma\right)e^* e_{t-1} - \gamma e_{t-1}^2 - ae^*\right]e_t - \alpha e^* e_{t-1} = 0 \tag{5}$$

The equation 5 has two roots, the positive one being considered for obvious reasons. The resulting nonlinear equation is:

$$e_t = \frac{\left[(\beta + \gamma) e^* e_{t-1} - \gamma e_{t-1}^2 - ae^* \right]}{2\beta e_{t-1}} +$$

$$\frac{\sqrt{\left[(\beta + \gamma) e^* e_{t-1} - \gamma e_{t-1}^2 - ae^* \right]^2 + 4 * \beta e_{t-1} * \alpha * e_{t-1}}}{2\beta e_{t-1}} \tag{6}$$

for $\alpha = \beta = 4$ and $\gamma = 26$.

The graphical representation of the solution e_t show that the graph presents a peak value of 2.76 and a minimum value of 0.091. Any other value from outside the interval represented by these two values is attracted. The evolution of the system with the specified parameters is chaotic because it satisfies the Li-Yorke condition [3].

The Model of the Nonlinear Feedback Mechanism of the Profit

The current spending of a firm can influence the value of the profit obtained at the end of the reference period. The profit will influence the spending over the next period. The dependence between the previous value of the profit and the current value is nonlinear because an increase of the spending does not reflect in an increase of the profit. The law of the decrease of the efficaciousness asserts that a certain mean value reaches minimum or maximum value when its magnitude equals the marginal value. One can invest in a certain production capability but this doesn't guarantee an unlimited increase of the production but the increase up to a certain point. Beyond that point the increase of the investment does not generates a corresponding increase of the production.

The dependence between the current profit and the previous profit can be modeled by using the equation:

$$\Pi_{t+1} = A\Pi_t - B\Pi_t^2 \tag{7}$$

The maximum profit Π_{max} is supposed that it can be determined.
Dividing the equation (6) with Π_{max} the following result is obtained:

$$\frac{\Pi_{t+1}}{\Pi_{max}} = A \frac{\Pi_t}{\Pi^{max}} - B \left(\frac{\Pi_t}{\Pi^{max}} \right)^2 \Pi^{max} \tag{8}$$

Let $\pi_t = \dfrac{\Pi_t}{\Pi^{max}}$ and the equation (7) becomes:

$$\pi_{t+1} = A\pi_t - B\pi_t \Pi^{max} \tag{9}$$

If we take $\Pi^{max} = \dfrac{A}{B}$ the equation above becomes the logistic equation:

$$\pi_{t+1} = A\pi_t - A\pi_t^2 = A(1 - \pi_t)\pi_t \tag{10}$$

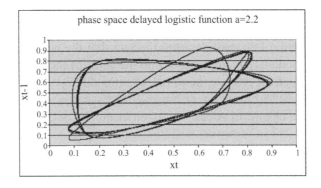

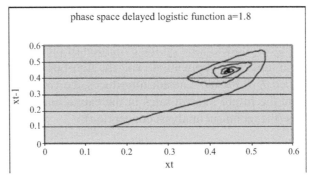

Figure 5 Phase space portrait of the logistic delayed function $xt = ax_{t-1}(1 - x_{t-2})$. For small values of the parameter a the non-chaotic attractor is a point. For a larger value of the parameter the non-chaotic attractor is a limit cycle.

The logistic map exhibits the same dependence on the initial condition: the slightest change of the initial condition causes a completely different evolution. The complex behavior of the apparently simple functions can be observed using the bifurcation diagram. The bifurcation diagram (Fig. 5) is an excellent tool allowing analyzing the behavior of a function by varying a control parameter (in the case of logistic function, the control parameter is A). The logistic function is known to have a chaotic behavior with small isles of periodicity for a value of the parameter A greater that 3.57. For $A \in [3.57, 4]$ there are small areas of periodicity, the white stripes that can be observed in the figure. For $A > 4$ the behavior is completely chaotic.

The K-Z Model of Larrain

The theory and the models regarding the functioning of the capital markets initially developed on the hypothesis according to which these markets are efficient. The efficient-market hypotheses comprise a series of conditions which basically say that the prices of the assets and their turnover can be determined based on the supply and demand in the competitive market where there are rational agents. These rational agents quickly assimilate any piece of information that is relevant to determining the prices of the assets and their turnover, adjusting the price in accordance with this information. In other words, the agents do not have different comparative advantages in acquiring information.

That is to say that such a market does not provide opportunities to obtain a profit on an asset whose turnover is superior to the risk undertaken by the agent. Thus, the normal profits will be nil, taking into consideration the fact that the agents procure this piece of information and immediately incorporate it into the price of the assets. If the last piece of information and the current one are incorporated immediately into the price of the assets, then only a new piece of information, or however else we may call it—the "novelty", will be able to determine a change in prices. As novelties are, by definition, unpredictable, then the changes in prices (or in turnover) will be unpredictable, too: no piece of information at time t or previous to this time will be able to help improve the forecast concerning the prices and turnovers (nor contribute to the decrease in the forecast errors made by the agents in this respect). This forecast error independence towards the previous information is called the feature of orthogonality and is widely used in testing the efficient market hypothesis. M.Larrain elaborated a model combining a classical description of a Keynesian economy with a non-linear model based on the evolution of the interest rates. Because he named the behavioural model "Z application" and the non-linear model "K application", this led to the so-called *K-Z model*. Larrain introduces the two components separately. Thus, he observes that the future interest rates in the capital markets depend both on the previous interest rates (the technical analysis conception):

$$r_{t+1} = f(r_{t-n}), \ t = 0, 1, 2, ... n \tag{11}$$

where f is a non-linear function, and on a series of fundamental economic variables (fundamentalist conception):

$$r_{t+1} = g(Z) \tag{12}$$

Where $Z = (y, M, P, ...)$, y - being the real GNP, M - the money offer, P - the consumer price index, etc.

The component (1a) shows that the future interest rate depends on its previous rates up to a certain lag n. This dependence of the future sizes on the previous ones is specific to the conception of the technical analysis of approaching the capital markets.

The exact form of $f(r_{t-m})$ is unknown, it may differ from one analyst to another. Larrain chooses for this function the expression: $r_{t+1} = a + br_t^n - cr_t^{n+1}$ where we can notice that if $c = b$ becomes the logistic equation, which is known to have a chaotic behaviour for certain values of the control parameter b (or c):

$$r_{t+1} = a - br_t^n(1 - r_t) \tag{13}$$

This represents the *K* component (application) of the model.

In what follows we shall present the way in which is built the Z component:

$$r_{t+1} = dy_t + eP_t - f \cdot M_t - g\sum(Y_t - c_t) \tag{14}$$

where d, e, f, g are constants and y_t represents the real GNP, M_t—the money offer (expressed through the aggregate M_t), p_t the—the consumer price index, Y_t the real personal and c_t—the real personal consumption. This component reflects the fundamentalist conception according to which the interest rates in the capital markets depend on the evolution of fundamental sizes.

Larrain combines the two components, K and Z, in one single expression as follows:

$$r_{t+1} = a - br_t^n(1-r_t) + dy_t + eP_t - f \cdot M_t - g\sum_t (Y_t - c_t)$$

(15)

This expression shows that the future interest rates are a combined function of technical and fundamental factors. Meanwhile, the former or latter of the two components can dominate the other one. Thus, during the stability periods, the capital markets are efficient and the interest rates will depend on the Z component, to a larger extent. In the unstable periods of the markets in question, the investors lose their trust in fundamental variables, making decisions by extrapolating tendencies. Thus, it is the *K* component that becomes dominant. In this situation, under certain circumstances, the c control parameter can take values in intervals for which the logistic equation has a chaotic behaviour, thus inducing crises and chaos episodes in the markets in question. The tests made with Larrain's model led to a series of interesting conclusions. Thus, for the stable capital markets, such as the bond market or the security market, the obtained forecasts covered quite well the evolution of the interest rates observed in reality, which for such a model represents a success. Still, for the estimation of the equation parameters (14), the model used techniques of linear regression, which annulled the premise that one or another of the two components can be dominant in one period or another. In order to introduce such an alternation of the dominance of the *K-Z* application components it is necessary that the parameters of the function in question should be variable in time, which the model in its initial form cannot allow. Improving such a model could reconcile the two big tendencies in the analysis of the capital markets, technical and fundamentalist, offering a powerful instrument of forecasting these markets.

Conclusion

Chaos is can be found almost everywhere in the nature. Chaos theory and fractals are currently applied in the study of the natural phenomenon. An essential condition needed in order that chaos to emerge is to have nonlinear systems. In fact very few of all models are purely linear, the vast majority of the systems are nonlinear. The paper emphasizes two of the features of the chaotic systems: dependence to initial conditions and the divergence of nearby trajectories. The chaos theory has a significant impact on economy and especially on capital markets. If the behavior of one economic system is proved to be chaotic this guarantees that, using appropriate methods, a short-term prediction can be made.

Refferences

1. E. Peters, "Chaos and Order in the Capital Markets", New York: John Wiley & Sons, 1996.
2. J. Ellis, *Non-linearities and chaos in exchange rates*, in "Chaos and Non-Linear Models in Economics: Theory and Applications", pp.187–195, Edward Elgar Publishing, 1994.
3. R.C. Hilborn, "Chaos and Nonlinear Dynamics", Oxford University Press, 1999.
4. S. Kuchta, "Nonlinearity and Chaos in Macroeconomics and Financial Markets".
5. S. H. Strogatz, "Nonlinear dynamics and chaos with applications to Physics, Chemistry and Engineering", Perseus Books, 1994.
6. V. Mărăcine, E. Scarlat, *Aplicaţii ale teoriei haosului în economie*, "Informatică Economică", no.1(21), 2002.

Chapter 6

Complex Dynamics of an Adnascent-type Game Model[†]

Baogui Xin,[1,a] Junhai Ma[1] and Qin Gao[1]

[1]Nonlinear Dynamics and Chaos Group, School of Management, Tianjin University, Tianjin 300072, China.
[a]Email: xin@tju.edu.cn.

ABSTRACT

The paper presents a nonlinear discrete game model for two oligopolistic firms whose products are adnascent. (In biology, the term adnascent has only one sense, "growing to or on something else," e.g., "moss is an adnascent plant." See Webster's Revised Unabridged Dictionary published in 1913 by C. & G. Merriam Co., edited by Noah Porter.) The bifurcation of its Nash equilibrium is analyzed with Schwarzian derivative and normal form theory. Its complex dynamics is demonstrated by means of the largest Lyapunov exponents, fractal dimensions, bifurcation diagrams and phase portraits. At last, bifurcation and chaos anticontrol of this system is studied.

Keywords: Discrete dynamical system; Triangular map; Normal forms; Flip bifurcations; Schwarzian derivative; Chaos anticontrol.

Introduction

Economic thought has had some significant influence on the development of ecological theory [1]. (Worster claimed that Darwin was influenced in his development of the theory of evolution of species by the views of Malthus.) In the opposite direction, many scientists such as Marshall [2] and Lotka [3], have stated that biology can be a source of inspiration for economics. (Marshall [2] suggested that "The Mecca of the economist lies in economic biology rather than in economic dynamics;" Lotka [3] said that "Man's industrial activities are merely a highly specialized and greatly form of the general biological struggle for existence,...the analysis of the biophysical foundations of economics, is one of the problems coming within the program of physical biology.") Thus further analogies

[†]Reused with permission from: Baogui Xin, Junhai Ma and Qin Gao, "Complex Dynamics of an Adnascent-Type Game Model," *Discrete Dynamics in Nature and Society*, vol. 2008, Article ID 467972, 12 pages, 2008. doi:10.1155/2008/467972.

between biology and economics can be discovered as both disciplines adopt concepts such as competition, mutualism and adnascent relation. Such ideas have greatly influenced a good many researchers in economics, for example, Barnett and Glenn [4] investigated competition and mutualism among early telephone companies; Hens and Schenk-Hoppé [5] studied evolutionary stability of portfolio rules in incomplete markets; Levine [6] Compared products and production in ecological and industrial systems. In addition, there are a lot of phenomena with adnascent relation in economics, for example, a car key ring is adnascent to a car. In this paper, the definition of adnascent will be applied into economics to investigate a novel game model with two oligopolistic firms *X* and *Y*, where product *B* of the firm *Y* is adnascent to product *A* of the firm *X*, and the output of product *B* is determined by the output of product *A*, but not vice versa.

In 1838, Cournot proposed the classical oligopoly game model. In 1883, Bertrand reworked Cournot's duopoly game model using prices rather than quantities as the strategic variables. In 1991, Puu [7] introduced chaos and bifurcation theory into duopoly game models. Over the past decade, many researchers, such as Tramontana et al. [8], Ahmed and Agiza [9] and Ahmed et al. [10], Agiza and Elsadany [11], Bischi et al. [12], Kopel [13] and Den Haan [14], have paid a great attention to the dynamics of games. As mentioned above, if one draws an analogy between species in biology and products in economics, it is easy to find that some of relationships among different products are substitutable or parasitic, and others are supportive or adnascent. But all the models cited above are based on the assumption that all players (firms) produce goods which are perfect substitutes in an oligopoly market. In this paper, we assume that the relationship of two players' products is not substitutable but adnascent.

This paper is organized as follows. In Section 2, a nonlinear discrete adnascent-type game model is presented. In Section 3, local stability of the Nash equilibrium of this system is studied. In Section 4, the bifurcation is studied with Schwarzian derivative and normal form theory. In Section 5, bifurcation and chaos anticontrol of the model is considered with nonlinear feedback anticontrol technology. In Section 6, the model's complex dynamics is numerically simulated by the largest Lyapunov exponents, fractal dimensions, bifurcation diagrams and phase portraits.

An Adnascent-type Dynamical Game Model

Assumptions

This model is based on these following assumptions.

Assumption 2.1. There are two heterogeneous firms X and Y producing adnascent products. The production decision of firm Y must depend on firm X, but not vice versa.

Assumption 2.2. Each firm is a monopoly of its products market.

Assumption 2.3. Firms have respective nonlinear variable cost functions [15] and nonlinear inverse demand functions [16]. (The linear cost function $C(x) = x$ or $C(x) = a + bx$ is usually adopted in the classical economics. Indeed, quadratic cost functions are often met in many applications (see [17–19]).)

Assumption 2.4. Firm X can compete solely on price and then make its output decision, which can have effect on firm Y.

Assumption 2.5. Firms always make the optimal output decision for the maximal margin profit in every period.

Nomenclature

The following is a list of notations that will be used throughout the paper.

(i) x_t, y_t are outputs of firms X and Y in period t, respectively, and they must be positive for any $t > 0$. (ii) $P_{xt} = a_1 - b_1 x_t^2, P_{yt} = a_2 - b_2 y_t^2$ are nonlinear inverse demand functions [16] for firms X and Y in period t, respectively, where $a_1, b_1, a_2, b_2 > 0$. (iii) $C_{xt} = c_1 x_t^2, C_{yt} = c_2 y_t^2$ are nonlinear variable cost functions [15] for firms X and Y in period t, respectively, where $c_1, c_2 > 0$. (The nonlinear variable cost function $C(x) = cx^2$ can be derived from a Cobb-Douglas-type production function (see [19–21]). (iv) $\Pi_{xt} = P_{xt}x_t - C_{xt} = x_t(a_1 - b_1 x_t^2) - c_1 x_t^2$, $\Pi_{yt} = P_{yt}y_t - C_{yt} = y_t(a_2 - b_2 y_t^2) - c_2 y_t^2$ are single profits of firms X and Y in period t, respectively. (v) $\alpha_1, \alpha_2 > 0$ are respective output adjustment parameters of firms X and Y, which represent the fluctuation of two firms' output decisions. Generally speaking, the two parameters should be very small.

Model

With Assumptions (2.5), the margin profits of firms X and Y in period t are given respectively by

$$\frac{\partial \Pi_{x_t}}{\partial x_t} = a_1 - 3b_1 x_t^2 - 2c_1 x_t,$$

$$\frac{\partial \Pi_{y_t}}{\partial y_t} = a_2 - 3b_2 y_t^2 - 2c_2 y_t. \tag{2.1}$$

One of the methods to find out the Nash equilibrium is to let Eqs. (2.1) be equal to 0. Thus one can get firms' reaction functions, i.e., the optimal outputs x_t^* and y_t^*. Under Assumptions (2.1) and (2.4), the dynamic adjustment of the adnascent-type game can be written as follows:

$$\begin{cases} x_{t+1} = x_t + \alpha_1 x_t \dfrac{\partial \Pi_{x_t}}{\partial x_t}, \\ \\ y_{t+1} = y_t + \alpha_2 y_t \dfrac{\partial \Pi_{y_t}}{\partial y_t}. \end{cases} \tag{2.2}$$

The game model with bounded rational players has the following nonlinear form:

$$\begin{cases} x_{t+1} = x_t + \alpha_1 x_t (a_1 - 3b_1 x_t^2 - 2c_1 x_t), \\ y_{t+1} = y_t + \alpha_2 y_t (a_2 - 3b_2 y_t^2 - 2c_2 y_t). \end{cases} \quad (2.3)$$

Note that the model has a particular form, it is a so-called triangular map which is the class of maps in which one dynamic variable is independent on the other, that is of the type $x' = f(x), y' = g(x, y)$ while the other, y, strongly depends on the first. A peculiarity of this class of maps is that the eigenvalues in any point of the phase plane are always real, and that many bifurcations are explained via the one-dimensional map $x' = f(x)$.

Nash Equilibrium and its Local Stability of System (2.3)

A Nash equilibrium, named after John Nash, is a solution concept of a game involving two or more players, such that no player has incentive to unilaterally change his or her action. In other words, players are in equilibrium if a change in strategies by any one of them would lead that he(she) to earn less than if he(she) remained with his(her) current strategy.

System (2.3) is a two-dimensional non-invertible map that depends on eight parameters. The Nash equilibrium point of the system (2.3) is the solution of the following algebraic system.

$$\begin{cases} \alpha_1 x(a_1 - 3b_1 x^2 - 2c_1 x) = 0, \\ \alpha_2 y(a_2 - 3b_2 y^2 - 2c_2 y) = 0. \end{cases} \quad (3.1)$$

Note that system (3.1) does not depend on the parameters α_1 and α_2. By simple computation of the above algebraic system it was found that there exists one interesting positive Nash equilibrium as follows:

$$E^*(x^*, y^*) = (\frac{A - c_1}{3b_1}, \frac{B - c_2}{3b_2}), \quad (3.2)$$

where $A = \sqrt{c_1^2 + 3a_1 b_1}$, $B = \sqrt{c_2^2 + 3a_2 b_2}$.

The Jacobian matrix of system (2.3) at the Nash equilibrium $E^*(x^*, y^*)$ has the following form.

$$J(E^*) = \begin{pmatrix} 1 - 2\alpha_1(a_1 - c_1 x^*) & 0 \\ 0 & 1 - 2\alpha_2 B x^* \end{pmatrix}. \quad (3.3)$$

Thus, its eigenvalues can be expressed as $\lambda_1 = 1 - 2\alpha_1 A x^*$ and $\lambda_2 = 1 - 2\alpha_2 B x^*$. Then the condition $\lambda_1 < 1$ is always satisfied while $\lambda_1 > -1$ holds if

$$\alpha_1 < \frac{1}{Ax^*} = \frac{3b_1}{A(A - c_1)} = C, \quad (3.4)$$

and the condition $\lambda_2 < 1$ is always satisfied while $\lambda_2 > -1$ holds if

$$\alpha_2 < \frac{1}{Bx^*} = \frac{3b_1}{B(A-c_1)} = D.$$

As a result, the following proposition holds.

Proposition 3.1 *The Nash equilibrium* $E^*(x^*, y^*)$ *is called*

(i)   *a sink if* $\alpha_1 < C$ *and* $\alpha_2 < D$, *so the sink is locally asymptotically stable;*
(ii) *a source if* $\alpha_1 > C$ *and* $\alpha_2 > D$, *so the source is locally unstable;*
(iii) *a saddle if* $\alpha_1 < C$ *and* $\alpha_2 > D$ *or* $\alpha_1 > C$ *and* $\alpha_2 < D$;
(iv) *non-hyperbolic if either* $\alpha_1 = C$ *or* $\alpha_2 = D$.

Bifurcation Analysis

Due to the fact that the map is triangular, the stability of the variable x is independent on the other, thus the bifurcation analysis for this variable can be easily performed with the one-dimensional map $x' = f(x)$, which is a cubic, and the interest is only in the positive part.

The best known and most popular projective differential invariant is the Schwarzian derivative. The map's Schwarzian derivative [22] is

$$S_f(x) = \frac{f'''(x)}{f'(x)} - \frac{3}{2}\left(\frac{f''(x)}{f'(x)}\right)^2 = -\frac{6\alpha_1(3b_1 + 3\alpha_1 a_1 b_1 + 54\alpha_1 b_1^2 x^2 + 24\alpha_1 b_1 c_1 x + 4\alpha_1 c_1^2)}{(1+\alpha_1 a_1 - 9\alpha_1 b_1 x^2 - 4\alpha_1 c_1 x)^2} \tag{4.1}$$

Obviously $S_f(x) < 0$ for $x > 0$, so that all the flip bifurcations are supercritical [23]. An example of supercritical flip bifurcation will be presented with normal form theory as follows. Generally speaking, for given firms X and Y, their parameters a_1, b_1, a_2, b_2, c_1 and c_2 are invariable, and their output adjustment parameters α_1 and α_2 are changeable.

In what follows, for convenience of studying the bifurcation parameter α_1 and α_2, we let $a_1 = 10, b_1 = 0.5, a_2 = 9.75, b_2 = 0.182, c_1 = 5$ and $c_2 = 4$. Then we can get the following system:

$$\begin{cases} x_{t+1} = x_t + \alpha_1 x_t (10 - 1.5x_t^2 - 10x_t), \\ y_{t+1} = y_t + \alpha_2 y_t (9.75 - 0.546y_t^2 - 8y_t). \end{cases} \tag{4.2}$$

The system (4.2) exists a Nash equilibrium point $E^*(0.883, 1.132)$ which is independent of the parameters α_1 and α_2. The Jacobian matrix at $E^*(0.883, 1.132)$ is

$$A = \begin{pmatrix} 1 - 11.17\alpha_1 & 0 \\ 0 & 1 - 8.1557\alpha_2 \end{pmatrix}. \tag{4.3}$$

Obviously, its eigenvalues satisfy (i) $\lambda_1 = -1$, if $\alpha_1 = 0.179$; $\lambda_2 = -1$ if $\alpha_2 = 0.245$. Thus system (4.2) may undergo flip bifurcation at $\alpha_1 = 0.179$ or $\alpha_2 = 0.245$.

Lemma 4.1 (Topological norm form for the flip bifurcation [24]). *Any generic, scalar, one-parameter system* $x \mapsto f(x,\alpha)$, *having at* $\alpha = 0$ *the fixed point* $x_0 = 0$ *with* $\mu = f_x(0,0) = -1$, *is locally topologically equivalent near the origin to one of the following normal forms:* $\eta \mapsto -(1+\beta)\eta \pm \eta^3$.

The following system can be obtained with $\alpha_2 = 0.2$,

$$\begin{cases} x_{t+1} = x_t + \alpha_1 x_t (10 - 1.5x_t^2 - 10x_t), \\ y_{t+1} = y_t + 0.2y_t (9.75 - 0.546y_t^2 - 8y_t). \end{cases} \tag{4.4}$$

Proposition 4.2 (Critical norm form for flip bifurcation). *System* (4.4) *can be written as following critical normal form for flip bifurcation:*

$$\xi_{t+1} = -\xi_t + c\xi_t^3, \tag{4.5}$$

where $c = 12.23$

Proof. To compute coefficients of normal form, we translate the origin of the coordinates to this Nash equilibrium $E^*(0.883, 1.132)$ by the change of variables as by the change of variables

$$x = 0.883 + u, \quad y = 1.132 + v \tag{4.6}$$

This transforms system (4.2) with parameters $\alpha_1 = 0.179$ into

$$\begin{cases} u_{t+1} = 0.883 - u_t - 2.5u_t^2 - 0.269u_t^3, \\ v_{t+1} = 1.131 - 0.001u_t - 0.63v_t - 1.85u_t v_t - 0.1v_t^2 - 0.11u_t v_t^2. \end{cases} \tag{4.7}$$

This system can be written as

$$X_{n+1} = AX_n + \frac{1}{2}B(X_n, X_n) + \frac{1}{6}C(X_n, X_n, X_n) + O(X_n^4), \tag{4.8}$$

where

$$A_0 = A(E^*) = \begin{pmatrix} -1 & 0 \\ 0 & -0.63 \end{pmatrix}. \tag{4.9}$$

and the multilinear functions $B : R^2 \times R^2 \to R^2$ and $C : R^2 \times R^2 \times R^2 \to R^2$ are also defined, respectively, by

$$B_i(x,y) = \sum_{j,k=1}^{2} \left. \frac{\partial^2 X_i(\xi,0)}{\partial \xi_j \partial \xi_k} \right|_{\xi=0} x_j y_k,$$

$$C_i(x,y,z) = \sum_{j,k,l=1}^{2} \left. \frac{\partial^3 X_i(\xi,0)}{\partial \xi_j \partial \xi_k \partial \xi_l} \right|_{\xi=0} x_j y_k z_l. \tag{4.10}$$

For system (4.7),

$$B(\xi,\eta) = \begin{pmatrix} -5\xi_1\eta_1 \\ -1.85\xi_1\eta_2 - 0.2\xi_2\eta_2 \end{pmatrix},$$

$$C(\xi,\eta,\zeta) = \begin{pmatrix} -1.61\xi_1\eta_1\zeta_1 \\ -0.22\xi_1\eta_2\zeta_2 \end{pmatrix}. \tag{4.11}$$

The eigenvalues of the matrix J are $\lambda_1 = -1$ and $\lambda_2 = -0.63$. Let $q, p \in R^2$ be eigenvectors corresponding to λ_1, λ_2, respectively:

$$q = \begin{pmatrix} 1 \\ 0 \end{pmatrix}, \quad p = \begin{pmatrix} 1 \\ 0 \end{pmatrix}. \tag{4.12}$$

satisfy $A_0 q = -q, A_0^T p = -p$ and $< p, q > = 1$. So the coefficient of the normal form of system (4.7) can be computed by the following invariant formula:

$$c = \frac{1}{6}\langle p, C(q,q,q)\rangle - \frac{1}{2}\langle p, B(q,(A - I_n)^{-1} B(q,q))\rangle = 12.23 \tag{4.13}$$

The proposition is proved.

The bifurcation type is determined by the stability of the Nash equilibrium as at the critical parameter value. According to the above Proposition 4.2, for system (4.7), the critical parameter $c = 12.23 > 0$, so the flip bifurcation at the Nash equilibrium $E^*(0.883, 1.132)$ is supercritical.

Bifurcation and Chaos Anticontrol

A government may pay attention to chaos anticontrol on the game system. Its motivations are as follows. Chaos exhibits high sensitivity to initial conditions, which manifests itself as an exponential growth of perturbations in the initial conditions. As a result, two firms' decision behaviors of the anticontrolled chaotic game systems appear to be random. So it can weaken the negative effect of excessive monopoly at least. In addition, Huang [25] has proved that, in some sense, chaos is beneficial not only to all oligopolistic firms but also to the economy as a whole.

There are various methods can be used to control or anticontrol bifurcations and chaos, for example, impulsive control [26], adaptive feedback control [27], linear and nonlinear feedback control [28–30]. In this section, the nonlinear feedback technique will be employed to anticontrol system (4.4). As mentioned above, system (4.4) is an adnascent-type game model, that is, firm Y must depend on firm X, but not vice versa. In other words, the production decision of firm X is independent. Since firm X of system (4.4) undergoes bifurcation and chaos, one may merely anticontrol firm Y. Considering the principle of simplification and maneuverability, one may choose a generalized nonlinear feedback anticontroller (e.g., production tax rebate) on firm Y as follows:

$$u = \sum_{i=1}^{n} k_i y^i,$$ (5.1)

where the linear terms in the anticontroller are used to shift the location of the equilibrium and bifurcation because only the linear part affects the Jacobian matrix of the linearized system, the nonlinear terms are used to change the property of the bifurcation and chaos. But it is not necessary to take too much components unless one wants to preserve all equilibria of the original system. In this paper, since it is unnecessary to preserve all equilibria of system, the anticontroller can be greatly simplified as

$$u = ky^2,$$ (5.2)

Then the anticontrolled system can be represented as

$$\begin{cases} x_{t+1} = x_t + \alpha_1 x_t (10 - 1.5x_t^2 - 10x_t), \\ y_{t+1} = y_t + 0.2y_t (9.75 - 0.546y_t^2 - 8y_t) + ky^2, \end{cases}$$ (5.3)

for system (5.3), it is easy to get its Nash equilibria

$$E_1 \left(0.88, \frac{106 - \sqrt{14964 - 38743k}}{150k - 14.47} \right), \quad E_2 \left(0.88, \frac{106 + \sqrt{14964 - 38743k}}{150k - 14.47} \right)$$ (5.4)

and Jacobian matrix

$$J(E^*) = \begin{pmatrix} 1 + \alpha_1 (10 - 20x - 4.5x^2) & 0 \\ 1.95 - 0.11y^2 - 1.6y & 1 - 0.22xy - 0.6x + 2ky \end{pmatrix}.$$ (5.5)

As mentioned above, system (4.4) undergoes a flip bifurcation at $\alpha_1 = 0.179$ and $x = 0.88$. Like system (5.3), after an anticontroller $u = ky^2$ is put on firm Y of system (4.4), firm X is uninfluenced. As a result, in system (5.3), when $x = 0.88$ and $\alpha_1 = 0.179$, the two conditions of flip bifurcation at Nash equilibria can be expressed as follows:

$$1 + \alpha_1 (10 - 20x - 4.5x^2) = -1, \quad |1 - 0.22xy - 0.6x + 2ky| < 1$$ (5.6)

hold with $0.1 < k < 0.39$.

Numerical Simulations

In this section, some numerical simulations are presented to confirm the above analytic results and to demonstrate added complex dynamical behaviors. To do this, one will use the largest Lyapunov exponents, fractal dimensions, bifurcation diagrams and phase portraits to show interesting complex dynamical behaviors.

In system (4.2), the largest Lyapunov exponents, fractal dimensions and bifurcation diagrams with two parameters α_1 and α_2 are shown in Fig. 1.

Figure 1(a) is the outputs bifurcation diagram of firm X with the parameters $\alpha_1 \in [0, 0.27]$ and $\alpha_2 \in [0, 0.2]$. When the output adjustment parameter α_1 increases, the outputs of firm X present complex dynamics as follows. Its outputs change from Nash equilibrium to bifurcation till chaos. Obviously the output adjustment parameter α_2 of firm Y has no effect on firm X, which just verifies the adnascent relationship between firms X and Y.

Figure 1(b) is the outputs bifurcation diagram of firm Y with the parameters $\alpha_1 \in [0, 0.27]$ and $\alpha_2 \in [0, 0.2]$. It is obviously that there is no bifurcation and chaos in Figure 1(b).

Figure 1(c) is the largest Lyapunov exponents diagram of system (4.2) with the parameters $\alpha_1 \in [0, 0.27]$ and $\alpha_2 \in [0, 0.2]$. The Lyapunov exponent of a dynamical system is a quantity that characterizes the rate of separation of infinitesimally close trajectories. A positive Lyapunov exponent is usually taken as an indication that the system is chaotic [31].

Figure 1(d) is a fractal dimensions diagram of system (4.2) with the parameters $\alpha_1 \in [0, 0.27]$ and $\alpha_2 \in [0, 0.2]$. A fractal dimension is taken as a criterion to judge

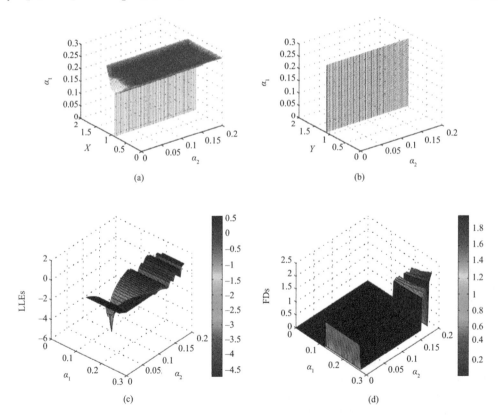

Figure 1 For system (4.2) with $\alpha_1 \in [0, 0.27]$ and $\alpha_2 \in [0, 0.2]$ (a) bifurcation diagram of firm X; (b) bifurcation diagram of firm Y; (c) largest Lyapunov exponents (LLEs); (d) fractal dimensions (FDs).

whether the system is chaotic. There are many specific definitions of fractal dimension and none of them should be treated as the universal one. This paper adopts the following definition of fractal dimension [32].

$$d_L = k - \frac{1}{\lambda_{k+1}} \sum_{i=1}^{k} \lambda_i,$$

(6.1)

where $\lambda_1 \geq \lambda_2 \geq \cdots \geq \lambda_n$ are the Lyapunov exponents and k is the largest integer for which $\sum_{i=1}^{k} \lambda_i \geq 0$ and $\sum_{i=1}^{k+1} \lambda_i < 0$. If $\lambda_i \geq 0$ for all $i = 1, 2, \cdots, n$ then $d_L = n$. If $\lambda_i < 0$ for all $i = 1, 2, \cdots, n$ then $d_L = 0$.

In system (4.4), firm X has supercritical flip bifurcation at $\alpha_1 = 0.179$ shown in Fig. 2(a), while firm Y undergoes neither bifurcation nor chaos.

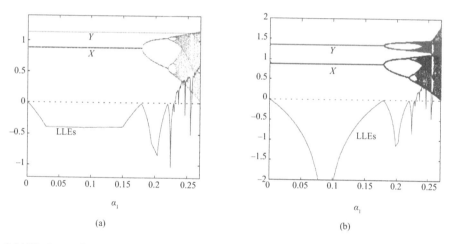

(a)   (b)

Figure 2 (a) The largest Lyapunov exponents and bifurcation diagram of system (4.4) versus $\alpha_1 \in [0, 0.27]$; (b) the largest Lyapunov exponents and bifurcation diagram of system (5.3) versus $k=0.2$ and $\alpha_1 \in [0, 0.27]$.

In system (5.3), when one fixes $k = 0.2$, he can get the largest Lyapunov exponents and bifurcations diagram shown in Fig. 2(b) and chaotic attractor portrait shown in Fig. 3. Obviously firms X and Y undergo synchronously bifurcations and chaos with $k = 0.2$.

The government can anticontrol the synchronization of bifurcation and chaos by varying the anticontrol parameter k.

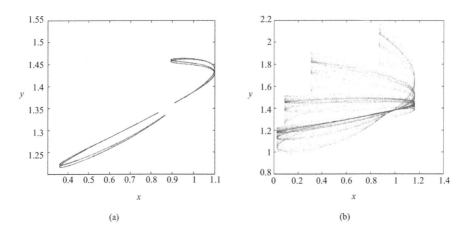

(a)   (b)

Figure 3 For system (5.3) with k=0.2. (a) Chaotic attractor versus $\alpha_1 = 0.24$; (b) chaotic attractor versus $\alpha_1 = 0.27$.

Conclusion

In this paper, we have presented a nonlinear adnascent-type game dynamical model with two oligopolistic firms, and emphatically reported its some complex dynamics, such as Nash equilibrium, bifurcation, chaos and their anticontrol. By means of the largest Lyapunov exponents, fractal dimensions, bifurcation diagrams and phase portraits, we have demonstrated numerically its complex dynamics. For the system, other complexity anticontrol theory and methodology will be considered in future work.

Acknowledgments

The authors acknowledge partial financial support by the National Natural Science Foundation of China (Grant No. 60641006). They also would like to express sincere gratitude to anonymous referees for their valuable suggestions and comments.

References

1. D. Worster, *Nature's Economy: A History of Ecological Ideas*, Cambridge University Press, Cambridge, UK, 1985.
2. A. Marshall, *Principles of Economics*, Macmillan, London, UK, 4th edition, 1898.
3. A.J. Lotka, *Elements of Mathematical Biology*, Dover, New York, NY, USA, 1996.
4. W.P. Barnett and R.C. Glenn, "Competition and mutualism among early telephone companies," *Administrative Science Quarterly*, vol. 32, no. 3, pp. 400–421, 1987.
5. T. Hens and K.R. Schenk-Hoppé, "Evolutionary stability of portfolio rules in incomplete markets," *Journal of Mathematical Economics*, vol. 41, no. 1-2, pp. 43–66, 2005.
6. S.H. Levine, "Comparing products and production in ecological and industrial systems," *Journal of Industrial Ecology*, vol. 7, no. 2, pp. 33–42, 2008.
7. T. Puu, "Chaos in duopoly pricing," *Chaos, Solitons & Fractals*, vol. 1, no. 6, pp. 573–581, 1991.
8. F. Tramontana, L. Gardini, and T. Puu, "Cournot duopoly when the competitors operate multiple production plants," *Journal of Economic Dynamics and Control*, vol. 33, no. 1, pp. 250–265, 2009.

9. E. Ahmed and H.N. Agiza, "Dynamics of a Cournot game with *n*-competitors," *Chaos, Solitons & Fractals*, vol. 9, no. 9, pp. 1513–1517, 1998.

10. E. Ahmed, H.N. Agiza, and S.Z. Hassan, "On modifications of Puu's dynamical duopoly," *Chaos, Solitons & Fractals*, vol. 11, no. 7, pp. 1025–1028, 2000.

11. H.N. Agiza and A.A. Elsadany, "Nonlinear dynamics in the Cournot duopoly game with heterogeneous players," *Physica A*, vol. 320, pp. 512–524, 2003.

12. G.I. Bischi, C. Mammana, and L. Gardini, "Multistability and cyclic attractors in duopoly games," *Chaos, Solitons & Fractals*, vol. 11, no. 4, pp. 543–564, 2000.

13. M. Kopel, "Simple and complex adjustment dynamics in Cournot duopoly models," *Chaos, Solitons & Fractals*, vol. 7, no. 12, pp. 2031–2048, 1996.

14. W.J. Den Haan, "The importance of the number of different agents in a heterogeneous asset-pricing model," *Journal of Economic Dynamics and Control*, vol. 25, no. 5, pp. 721–746, 2001.

15. J. Friedman, "Oligopoly theory," in *Handbook of Mathematical Economics*, K. J. Arrow and M.D. Intriligator, Eds., vol. 2, pp. 491–534, North-Holland, Amsterdam, The Netherlands, 1982.

16. Y. Zhang, X. Luo, and J. Xue, "Adaptive dynamic Cournot model in oligopoly market," *Journal of Industrial Engineering and Engineering Management*, vol. 18, no. 3, pp. 78–81, 2004.

17. H.-X. Yao and F. Xu, "Complex dynamics analysis for a duopoly advertising model with nonlinear cost," *Applied Mathematics and Computation*, vol. 180, no. 1, pp. 134–145, 2006.

18. A.K. Naimzada and L. Sbragia, "Oligopoly games with nonlinear demand and cost functions: two boundedly rational adjustment processes," *Chaos, Solitons & Fractals*, vol. 29, no. 3, pp. 707–722, 2006.

19. G.I. Bischi, F. Lamantia, and L. Sbragia, "Competition and cooperation in natural resources exploitation: an evolutionary game approach," in *Game Practice and the Environment*, C. Carraro and V. Fragnelli, Eds., pp. 187–211, Edward Elgar, Cheltenham, UK, 2004.

20. C. W. Clark, *Mathematical Bioeconomics: The Optimal Management of Renewable Resources*, Pure and Applied Mathematics, JohnWiley & Sons, New York, NY, USA, 2nd edition, 1990.

21. F. Szidarovszky and K. Okuguchi, "An oligopoly model of commercial fishing," *Seoul Journal of Economics*, vol. 11, no. 1, pp. 321–330, 1998.

22. V. Ovsienko and S. Tabachnikov, *Projective Differential Geometry Old and New: From the Schwarzian Derivative to the Cohomology of Diffeomorphism Group*, vol. 165 of *Cambridge Tracts in Mathematics*, Cambridge University Press, Cambridge, UK, 2005.

23. J. Guckenheimer and P. Holmes, *Nonlinear Oscillations, Dynamical Systems, and Bifurcations of Vector Fields*, vol. 42 of *Applied Mathematical Sciences*, Springer, New York, NY, USA, 1983.

24. Y.A. Kuznetsov, *Elements of Applied Bifurcation Theory*, vol. 112 of *Applied Mathematical Sciences*, Springer, New York, NY, USA, 2nd edition, 1998.

25. W. Huang, "The long-run benefits of chaos to oligopolistic firms," *Journal of Economic Dynamics and Control*, vol. 32, no. 4, pp. 1332–1355, 2008.

26. L. Qian and Q. Lu, "Bifurcation of the competitive system with impulsive control," *Dynamics of Continuous Discrete and Impulsive Systems. Series B*, vol. 14, supplement 5, pp. 24–28, 2007.

27. G. Qi, Z. Chen, and Z. Yuan, "Adaptive high order differential feedback control for affine nonlinear system," *Chaos, Solitons & Fractals*, vol. 37, no. 1, pp. 308–315, 2008.

28. Z. Chen and P. Yu, "Controlling and anti-controlling Hopf bifurcations in discrete maps using polynomial functions," *Chaos, Solitons & Fractals*, vol. 26, no. 4, pp. 1231–1248, 2005.

29. X.-Y. Wang and M.-J. Wang, "Chaotic control of the coupled Logistic map," *Acta Physica Sinica*, vol. 57, no. 2, pp. 731–736, 2008.

30. X. F. Wang and G. Chen, "Chaotification via arbitrarily small feedback controls: theory, method, and applications," *International Journal of Bifurcation and Chaos in Applied Sciences and Engineering*, vol. 10, no. 3, pp. 549–570, 2000.

31. Q. Lu, *Bifurcation and Singularity*, Shanghai Scientific and Technological Education, Shanghai, China, 1995.

32. J.L. Kaplan and J.A. Yorke, "Preturbulence: a regime observed in a fluid flow model of Lorenz," *Communications in Mathematical Physics*, vol. 67, no. 2, pp. 93–108, 1979.

A Production-Inventory Model for Deteriorating Items with Production Disruptions[†]

Yong He[1] and *Ju He*[2]

[1]School of Economics and Management, Southeast University, Nanjing 210096, China.
Email: heyong@126.com.
[2]School of Management and Engineering, Nanjing University, Nanjing, 210093, China.

ABSTRACT

Disruption management has recently become an active area of research. In this study, an extension is made to consider the fact that some products may deteriorate during storage. A production-inventory model for deteriorating items with production disruptions is developed. Then the optimal production and inventory plans are provided, so that the manufacturer can reduce the loss caused by disruptions. Finally, a numerical example is used to illustrate the model.

Keywords: supply chain management; deteriorating items; disruption management; inventory management.

Introduction

In real life, the effect of decay and deterioration is very important in many inventory systems. In general, deterioration is defined as decay, damage, spoilage, evaporation, obsolescence, pilferage, loss of utility, or loss of marginal value of a commodity that results in decreasing usefulness [1]. Most of the physical goods undergo decay or deterioration over time, the examples being medicine, volatile liquids, blood banks, and others. Consequently, the production and inventory problem of deteriorating items has been extensively studied by researchers. Ghare and Schrader [2] were the first to consider ongoing deterioration of inventory with constant demand. As time progressed, several researchers developed inventory models by assuming either instantaneous or finite production with different assumptions on the patterns of deterioration. In this connection, researchers may refer to

[†]Reused with permission from: Yong He and Ju He, "A Production Model for Deteriorating Inventory Items with Production Disruptions," *Discrete Dynamics in Nature and Society*, vol. 2010, Article ID 189017, 14 pages, 2010. doi:10.1155/2010/189017.

[3–7]. Interested readers may refer to review [8,9]. Recently, several related articles were presented, dealing with such inventory problems [10–17].

At the beginning of each cycle, the manufacturer should decide the optimal production time, so that the production quantity should satisfy the following two requirements: one, it should meet demand and deterioration; second, all products should be sold out in each cycle, that is, at the end of each cycle, the inventory level should decrease to zero. Some researchers have studied such production model for deteriorating items under different condition. For example, Yang and Wee [18] derived the optimal production time for a single-vendor, multiple-buyers system. Liao [19] derived a production model for the lot-size inventory system with finite production rate, taking into consideration the effect of decay and the condition of permissible delay in payments. Lee and Hsu [20] developed a two-warehouse inventory model with time-dependent demand. He et al. [21] provided a solution procedure to find the optimal production time under the premise that the manufacturer sells his products in multiple markets.

The above papers all assume that production rate is known and keeps constant during each cycle. They do not consider how to adjust the production plan once the production rate is changed during production time. However, after the plan was implemented, the production run is often disrupted by some emergent events, such as supply disruptions, machine breakdowns, earthquake, H1N1 epidemic, financial crisis, political event, policy change, and so on. For example, the Swedish mining company Boliden AB suffered the production disruptions at its Tara zinc mine in Ireland due to an electric motor breakdown at one of the grinding mills. As a result of the breakdown the production of zinc and lead concentrates is expected to fall by some 40% over the next six weeks [22]. These production disruptions will lead to a hard decision in production and inventory plans. Recently, there is a growing literature on production disruptions. For example, some researchers studied the production rescheduling problems with the machine disruptions [23–26]. Some researchers analyzed the optimal inventory policy with supply disruptions [27–30]. In most of the existing literature, products are assumed to be no deterioration when the production disruptions are considered. But, in real situation the deterioration are popular in many kinds of products. Hence, if the deterioration rate is not small enough, the deterioration factor cannot be ignored when the production system is disrupted.

Therefore, in this paper, we develop a production-inventory model for deteriorating items with production disruptions. Once the production rate is disrupted, the following questions are considered in this paper:

i) Whether to replenish from spot markets or not?
ii) How to adjust the production plan if the new production system can still satisfy the demand?
iii) How to replenish from spot markets if the new production system no longer satisfies the demand?

The paper is organized as follows: Section 2 is concerned with the mathematical development and the method for finding the optimal solutions. In Section 3, we present a numerical example to illustrate the model. In Section 4, conclusions and topics for further research are presented.

Mathematical Modeling and Analysis

Suppose a manufacturer produces a certain product and sells it in a market. All items are produced and sold in each cycle. The following assumptions are used to formulate the problem:

a) A single product and a single manufacturer are assumed.
b) Demand rate are deterministic and constant.
c) Normal Production rate is greater than demand rate.
d) Lead time is assumed to be negligible.
e) Deterioration rate is deterministic and constant.
f) Shortages are not allowed.
g) Time horizon is finite.
h) There is only one chance to order the products from spot markets during the planning horizon.

Let the basic parameters be as follows:
p normal production rate
d demand rate
θ constant deterioration rate of finished products
H planning horizon
T_p the normal production period without disruptions
T_d the production disruptions time
T_p^d the new production period with disruptions
T_r the replenishment time from spot markets once shortage appears
Q_r the order quantity from spot markets once shortage appears
$I_i(t)$ inventory level in the ith interval ($i = 1,2,...,n$), n can be different in different scenario.

The Basic Model-without Disruptions

At first, the manufacturer makes decisions about the optimal production time T_p under the normal production rate. The inventory model for deteriorating items with normal production rate can be depicted as Fig. 1.

The instantaneous inventory level at any time $t \in [0, H]$ is governed by the following differential equations:

$$\frac{dI_1(t)}{dt} + \theta I_1(t) = p - d \ , \ 0 \le t \le T_p \tag{2.1}$$

$$\frac{dI_2(t)}{dt} + \theta I_2(t) = -d \ , \ T_p \le t \le H \tag{2.2}$$

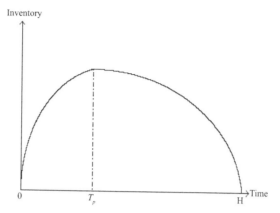

Figure 1 Inventory system without disruptions.

Using the boundary condition $I_1(0)=0$ and $I_2(T_d)=0$, the solutions of above differential equations are

$$I_1(t)=\frac{p-d}{\theta}\left(1-e^{-\theta t}\right),\ 0\le t\le T_p \tag{2.3}$$

$$I_2(t)=\frac{d}{\theta}\left[e^{\theta(H-t)}-1\right],\ T_p\le t\le H \tag{2.4}$$

The condition $I_1(T_p)=I_2(T_p)$ yields

$$\frac{p-d}{\theta}\left(1-e^{-\theta T_p}\right)=\frac{d}{\theta}\left[e^{\theta(H-T_p)}-1\right] \tag{2.5}$$

From (2.5), the production time T_p satisfies the following equation:

$$T_p=\frac{1}{\theta}\ln\frac{p-d+de^{\theta H}}{p} \tag{2.6}$$

In order to facilitate analysis, we do an asymptotic analysis for $I_i(t)$. Expanding the exponential functions and neglecting second and higher power of θ for small value of θ, equations (2.3) and (2.4) becomes

$$I_1(t)\approx(p-d)\left(t-\frac{1}{2}\theta t^2\right),\ 0\le t\le T_p \tag{2.7}$$

$$I_2(t)\approx d\left[(H-t)+\frac{1}{2}\theta(H-t)^2\right],\ T_p\le t\le H \tag{2.8}$$

and T_p approximately satisfies the equation

$$(p-d)\left(T_p - \frac{1}{2}\theta T_p^{\,2}\right) = d\left[(H-T_p) + \frac{1}{2}\theta(H-T_p)^2\right] \tag{2.9}$$

From Misra [31], we have

$$T_p \approx \frac{d}{p-d}(H-T_p)\left[1 + \frac{1}{2}\theta(H-T_p)\right] \tag{2.10}$$

Since

$$\frac{dT_p}{d\theta} = \frac{1}{2}\frac{d(H-T_p)^2}{p+\theta d(H-T_p)} > 0 , \tag{2.11}$$

We can get the following corollary:

Corollary 1. Assuming $\theta \ll 1$, then T_p is increasing in θ.

Corollary implies that the manufacturer has to produce more products when deterioration rate increases. Hence, decreasing deterioration rate is an effective way to reduce the product cost of manufacture.

The Production-inventory Model Under Production Disruptions

In the above model, the production rate is assumed to be deterministic and known. In practice, the production system is often disrupted by various unplanned and unanticipated events. Here, we assume the production disruptions time is T_d. Without loss of generality, we assume that the new disrupted production rate is $p + \Delta p$, where $\Delta p < 0$ if production rate decreases suddenly, or $\Delta p > 0$ if production rate increases.

Proposition 1. If $\Delta p \geq \dfrac{-(p-d)(1-e^{-\theta H})}{1-e^{-\theta(H-T_d)}}$, then the manufacturer can still satisfy the demand after production disruptions. Otherwise, i.e., $-p \leq \Delta p < \dfrac{-(p-d)(1-e^{-\theta H})}{1-e^{-\theta(H-T_d)}}$, there will exist shortages due to the production disruptions.

Proof. Without considering the stop time of production or replenishment, the inventory system with production disruptions can be depicted as Fig. 2.
From subsection 2.1, we know

$$I_1(t) = \frac{p-d}{\theta}\left(1 - e^{-\theta t}\right), \ 0 \leq t \leq T_d \tag{2.12}$$

The inventory system after disruptions can be represented by the following differential equation:

$$\frac{dI_2(t)}{dt} + \theta I_2(t) = p + \Delta p - d , \ T_d \leq t \leq H \tag{2.13}$$

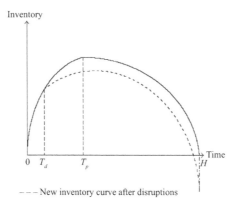

--- New inventory curve after disruptions

Figure 2 Inventory system with production disruptions.

Using $I_1(T_d) = I_2(T_d) = \dfrac{p-d}{\theta}\left(1 - e^{-\theta T_d}\right)$, we have

$$I_2(t) = \frac{1}{\theta}\left[p + \Delta p - d - \Delta p e^{-\theta(t-T_d)} - (p-d)e^{-\theta t}\right], \; T_d \le t \le H \tag{2.14}$$

Hence, we know that

$$I_2(H) = \frac{1}{\theta}\left[p + \Delta p - d - \Delta p e^{-\theta(H-T_d)} - (p-d)e^{-\theta H}\right] \tag{2.15}$$

Hence, if $I_2(H) \ge 0$, i.e., $\Delta p \ge \dfrac{-(p-d)(1-e^{-\theta H})}{1-e^{-\theta(H-T_d)}}$, this means that the manufacturer can still satisfy the demand after production disruptions. But if $I_2(H) < 0$, i.e., $-p \le \Delta p < \dfrac{-(p-d)(1-e^{-\theta H})}{1-e^{-\theta(H-T_d)}}$, we know that the manufacturer will face shortage since the production rate decreases deeply. The proposition is proved. From proposition 1, we know that if $\Delta p \ge \dfrac{-(p-d)(1-e^{-\theta H})}{1-e^{-\theta(H-T_d)}}$, the production-inventory problem is to find the new optimal production period T_p^d. If $-p \le \Delta p < \dfrac{-(p-d)(1-e^{-\theta H})}{1-e^{-\theta(H-T_d)}}$, the production-inventory problem is to find the optimal replenishment time T_r and replenishment quantity Q_r.

Proposition 2. If $\Delta p \ge \dfrac{-(p-d)(1-e^{-\theta H})}{1-e^{-\theta(H-T_d)}}$, then the manufacturer's production time with production disruptions is

$$T_p^d = \frac{1}{\theta}\ln\frac{p + d(e^{\theta H} - 1) + \Delta p e^{\theta T_d}}{p + \Delta p}.$$

Proof. From Proposition 1, we know that the new production time $T_p^d \in [T_d, H]$ if

$\Delta p \ge \dfrac{-(p-d)(1-e^{-\theta H})}{1-e^{-\theta(H-T_d)}}$. The inventory model can be depicted as Fig. 3. So if

$\Delta p \ge \dfrac{-(p-d)(1-e^{-\theta H})}{1-e^{-\theta(H-T_d)}}$, the inventory system after disruptions can be represented by

the following differential equations:

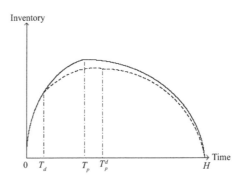

--- New inventory curve after disruptions

Figure 3 Inventory system, ($T_P^d \in [T_d, H]$).

$$\frac{dI_2(t)}{dt} + \theta I_2(t) = p + \Delta p - d , \; T_d \le t \le T_p^d \tag{2.16}$$

$$\frac{dI_3(t)}{dt} + \theta I_3(t) = -d , \; T_p^d \le t \le H \tag{2.17}$$

Using the boundary conditions $I_1(T_d) = I_2(T_d) = \dfrac{p-d}{\theta}\left(1 - e^{-\theta T_d}\right)$ and $I_3(H) = 0$, we know

$$I_2(t) = \frac{1}{\theta}\left[p + \Delta p - d - \Delta p e^{-\theta(t - T_d)} - (p-d)e^{-\theta t} \right], T_d \le t \le T_p^d \tag{2.18}$$

$$I_3(t) = \frac{d}{\theta}\left[e^{\theta(H-t)} - 1 \right], \; T_p^d \le t \le H \tag{2.19}$$

Using the boundary condition $I_2(T_p^d) = I_3(T_p^d)$, we have

$$T_p^d = \frac{1}{\theta} \ln \frac{p + d(e^{\theta H} - 1) + \Delta p e^{\theta T_d}}{p + \Delta p} \tag{2.20}$$

The proposition is proved.

Since $\dfrac{dT_p^d}{dT_d} = \dfrac{(p + \Delta p)\Delta p e^{\theta T_d}}{p + d(e^{\theta H} - 1) + \Delta p e^{\theta T_d}}$, we can easily get corollary 2.

Corollary 2. (I) If $\dfrac{-(p-d)(1-e^{-\theta H})}{1-e^{-\theta(H-T_d)}} \le \Delta p < 0$, then T_p^d is decreasing in T_d. (II) If

$\Delta p > 0$, then T_p^d is increasing in T_d.

Expanding the exponential functions and neglecting second and higher power of θ for small value of θ, equations (2.18) and (2.19) becomes:

$$I_2(t) \approx \Delta p(t-T_d)\left[1-\frac{1}{2}\theta(t-T_d)\right]+(p-d)t(1-\frac{1}{2}\theta t)\, , T_d \le t \le T_p^d \tag{2.21}$$

$$I_3(t) \approx d(H-t)\left[1+\frac{1}{2}\theta(H-t)\right],\ T_p^d \le t \le H \tag{2.22}$$

and T_p^d approximately satisfies the equation

$$\Delta p(T_p^d - T_d)\left[1-\frac{1}{2}\theta(T_p^d - T_d)\right]+(p-d)T_p^d\left[1-\frac{1}{2}\theta T_p^d\right]$$

$$= d(H-T_p^d)\left[1+\frac{1}{2}\theta(H-T_p^d)\right] \tag{2.23}$$

From Misra [31], we have

$$T_p^d \approx \frac{\Delta p T_d + d(H-T_p^d)\left[1+\frac{1}{2}\theta(H-T_p^d)\right]}{p+\Delta p - d}. \tag{2.24}$$

According to (2.24) we know

$$\frac{d\,T_p^d}{d\theta} = \frac{1}{2}\frac{d(H-T_p^d)^2}{p+\Delta p + d\theta(H-T_p^d)} > 0. \tag{2.25}$$

Hence, we can get the following corollary:

Corollary 3. Assuming $\theta < 1$, then T_p^d is increasing in θ.

If $-p \le \Delta p < \dfrac{-(p-d)(1-e^{-\theta H})}{1-e^{-\theta(H-T_d)}}$, there will exist shortage.

The manufacturer will have to produce products during the whole planning horizon, i.e., $T_p^d = H$. In order to avoid shortage, the manufacturer need order products from spot markets to satisfy the demand. The inventory model can be depicted as Fig. 4.

Proposition 3. If $-p \le \Delta p < \dfrac{-(p-d)(1-e^{-\theta H})}{1-e^{-\theta(H-T_d)}}$, then the replenishment time and quantity are:

$$T_r = \frac{1}{\theta}\ln\frac{p-d+\Delta p e^{\theta T_d}}{p+\Delta p - d},\ Q_r = \frac{p+\Delta p - d}{\theta}\left[1-e^{\theta(H-T_r)}\right].$$

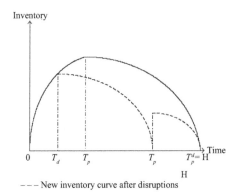

--- New inventory curve after disruptions

Figure 4 Inventory system ($T_p^d = H$).

Proof. First, we need to determine the order time point T_r. Let

$$I_2(t) = \frac{1}{\theta}\left[p + \Delta p - d - \Delta p e^{-\theta(t-T_d)} - (p-d)e^{-\theta t} \right] = 0 \text{, we have}$$

$$T_r = \frac{1}{\theta}\ln\frac{p - d + \Delta p e^{\theta T_d}}{p + \Delta p - d} \tag{2.26}$$

So

$$\frac{dI_3(t)}{dt} + \theta I_3(t) = p + \Delta p - d \text{, } T_r \le t \le H \tag{2.27}$$

Using the boundary condition $I_3(H) = 0$, we have

$$I_3(t) = \frac{p + \Delta p - d}{\theta}\left[1 - e^{\theta(H-t)} \right]; \ T_r \le t \le H \tag{2.28}$$

Hence, the order quantity is

$$Q_r = I_3(T_r) = \frac{p + \Delta p - d}{\theta}\left[1 - e^{\theta(H-T_r)} \right] \tag{2.29}$$

The proposition is proved.

If $-p \le \Delta p < \dfrac{-(p-d)(1-e^{-\theta H})}{1 - e^{-\theta(H-T_d)}}$, according to (2.26), we have

$$\frac{dT_r}{dT_d} = \frac{p + \Delta p - d}{p - d + \Delta p e^{\theta T_d}}\Delta p e^{\theta T_d} < 0 \tag{2.30}$$

$$\frac{\mathrm{d}Q_r}{\mathrm{d}T_d} = (p + \Delta p - d)e^{\theta(H-T_r)}\frac{\mathrm{d}T_r}{\mathrm{d}T_d} < 0 \qquad (2.31)$$

Hence, we can obtain the following corollary:

Corollary 4. If $-p \le \Delta p < \dfrac{-(p-d)(1-e^{-\theta H})}{1-e^{-\theta(H-T_d)}}$, then T_r and Q_r is decreasing in T_d.

A Numerical Example

Our objective in this section is to gain further insights based on a numerical example. We use the following numbers as the base values of the parameters: $p = 350$, $d = 200$, $\theta = 0.03$, $H = 20$, $T_d = 8$ and $\Delta p = -200$. Using Eq. (2.6) we obtain $T_p = 12.8$. Next, we observe how T_p^d, T_r, and Q_r would change as θ and T_d. Figure 5 and 6 depict T_p^d

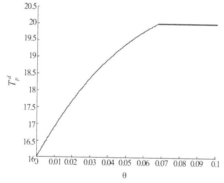

Figure 5 T_p^d with respect to θ.

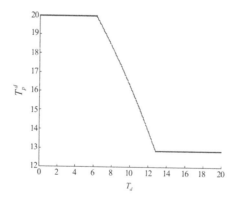

Figure 6 T_p^d with respect to T_d.

with respect to θ and T_d respectively. Figure 7 and 9 depict T_r with respect to θ and T_d respectively. Figure 8 and 10 depict Q_r with respect to θ and T_d respectively.

From Fig. 5, we can find that T_p^d is increasing in θ when $\theta \leq 0.068$. When $\theta > 0.068$, since the deterioration rate is so high that the manufacturer cannot satisfy the demand by self-producing. He has to buy products from spot markets in order to avoid shortage. From Fig. 7 and Fig.8, we can see that T_r is decreasing in θ and Q_r is increasing in θ when $\theta > 0.068$. From Fig. 6, we can find that T_p^d is decreasing in T_d when $6.2 \leq T_d \leq 12.8$. If $0 \leq T_d < 6.2$, the manufacturer will have to replenish inventory from spot markets. From Fig. 9 and Fig. 10, we can see that T_r is increasing in T_d and Q_r is decreasing in θ when $0 \leq T_d < 6.2$.

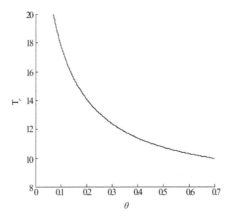

Figure 7 T_r with respect to θ.

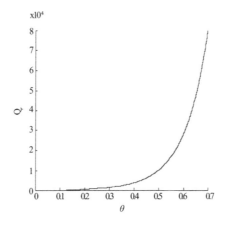

Figure 8 Q_r with respect to θ.

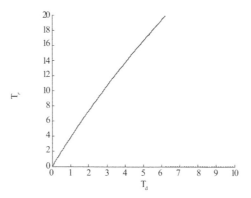

Figure 9 T_r with respect to T_d.

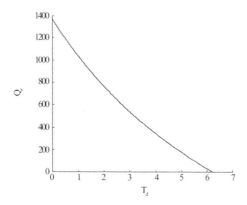

Figure 10 Q_r with respect to T_d.

Conclusions

In this paper, we propose a production-inventory model for a deteriorating item with production disruptions. Here, we analyze this inventory system under different situations. We have showed that our method helps the manufacturer reduce the loss caused by production disruptions. In this study, the proposed model considers the deterioration rate as constant. In real life, we may consider the deterioration rate as a function of time, stock, and so on. This will be done in our future research.

Acknowledgements

This research is supported by the National Natural Science Foundation of China (No.71001025). Also, this research is partly supported by a grant from the Ph.D. Programs Foundation of Ministry of Education of China (No.200802861030), the Project Sponsored by the Scientific Research Foundation for the Returned Overseas Chinese Scholars, the China Postdoctoral Science Foundation (No.20070411043), and the Postdoctoral Foundation of Jiangsu Province of China (No.0701045C).

References

1. H.M. Wee, "Economic production lot size model for deteriorating items with partial back-ordering," *Computers & Industrial Engineering*, vol. 24, no. 3, pp. 449–458, 1993.
2. P.M. Ghare and S.F. Schrader, "A model for exponentially decaying inventory," *Journal of Industrial Engineering*, vol. 14, no. 5 (1963) 238–243.
3. R.P. Covert, G.C. Philip, An EOQ model for items with Weibull distribution deterioration, AIIE Transactions 5(1973) 323–326.
4. T. Chakrabarti, B.C. Giri, K.S. Chaudhuri, An EOQ model for items Weibull distribution deterioration shortages and trended demand—an extension of Philip's model, Computers and Operations Research 25(1998) 649–657.
5. U. Dave, On a discrete-in-time order-level inventory model for deteriorating items, Operational Research 30(1979) 349–354.
6. E.A. Elsayed, C. Terasi, Analysis of inventory systems with deteriorating items, International Journal of Production Research 21(1983) 449–460.
7. S. Kang, I. Kim, A study on the price and production level of the deteriorating inventory system, International Journal of Production Research 21(1983) 899–908.
8. F. Raafat, Survey of literature on continuously deteriorating inventory models, Journal of the Operational Research Society 42(1991) 27–37.
9. S.K. Goyal, B.C. Giri, Recent trends in modeling of deteriorating inventory, European Journal of Operational Research 134(2001) 1–16.
10. K.J. Chung, J.J. Liao, The optimal ordering policy in a DCF analysis for deteriorating items when trade credit depends on the order quantity, International Journal of Production Economics 100(2006) 116–130.
11. A.K. Maity, K. Maity, S. Mondal, M. Maiti, A Chebyshev approximation for solving the optimal production inventory problem of deteriorating multi-item, Mathematical and Computer Modelling 45(2007) 149–161.
12. M. Rong, N.K. Mahapatra, M. Maiti, A two warehouse inventory model for a deteriorating item with partially/fully backlogged shortage and fuzzy lead time, European Journal of Operational Research 189(2008) 59–75.
13. M.S. Chern, H.L. Yang, J.T. Teng, S. Papachristos, Partial backlogging inventory lot-size models for deteriorating items with fluctuating demand under inflation, European Journal of Operational Research 191(2008) 127–141.
14. S.S. Mishra, P.P. Mishra, Price determination for an EOQ model for deteriorating items under perfect competition, Computers & Mathematics with Applications 56(2008) 1082–1101.
15. L.Y. Ouyang, J.T., Teng, S.K. Goyal, C.T. Yang, An economic order quantity model for deteriorating items with partially permissible delay in payments linked to order quantity, European Journal of Operational Research 194(2009) 418–431.
16. Chih-Te Yang, Liang-Yuh Ouyang, and Hsing-Han Wu, Retailer's Optimal Pricing and Ordering Policies for Non-Instantaneous Deteriorating Items with Price-Dependent Demand and Partial Backlogging, Mathematical Problems in Engineering, Volume 2009 (2009), Article ID 198305, 18 pages.
17. K. Skouri and I. Konstantaras, Order Level Inventory Models for Deteriorating Seasonable/Fashionable Products with Time Dependent Demand and Shortages, Mathematical Problems in Engineering, Volume 2009 (2009), Article ID 679736, 24 pages.
18. P.C. Yang, H.M. Wee, A single-vendor and multiple-buyers production-inventory policy for a deteriorating item, European Journal of Operational Research 143(2002) 570–581.
19. J.J. Liao, On an EPQ model for deteriorating items under permissible delay in payments, Applied Mathematical Modelling 31(2007) 393–403.
20. C.C. Lee, S.L. Hsu, A two-warehouse production model for deteriorating inventory items with time-dependent demands, European Journal of Operational Research 194(2009) 700–710.

21. Y. He, S.Y. Wang, K.K. Lai, An optimal production-inventory model for deteriorating items with multiple-market demand, European Journal of Operational Research 203(2010) 593–600.

22. Boliden AB reports production disruptions at Tara mine in Ireland. Nordic Business Report, June 13, 2005.

23. S.W. Chiu, Robust planning in optimization for production system subject to random machine breakdown and failure in rework, Computers & Operations Research 37(2010) 899–908.

24. G.C. Lin, D.E. Kroll, Economic lot sizing for an imperfect production system subject to random breakdowns, Engineering Optimization 38(2006) 73–92.

25. B.C. Giri, T. Dohi, Exact formulation of stochastic EMQ model for an unreliable production system, Journal of the Operational Research Society 56(2005) 563–575.

26. B. Liu, J. Cao, Analysis of a production-inventory system with machine breakdowns and shutdowns, Computers & Operations Research 26(1999) 73–91.

27. A.M. Ross, Y. Rong, L.V. Snyder, Supply disruptions with time-dependent parameters, Computers & Operations Research 35(2008) 3504–3529.

28. D. Heimann, F. Waage, A closed-form approximation solution for an inventory model with supply disruptions and non-ZIO reorder policy, Journal of Systemics, Cybernetics and Informatics 5(2007) 1–12.

29. Z. Li, S.H. Xu, J. Hayya, A periodic-review inventory system with supply interruptions, Probability in the Engineering and Informational Sciences 18(2004) 33–53.

30. H.J. Weiss, E.C. Rosenthal, Optimal ordering policies when anticipating a disruption in supply or demand, European Journal of Operational Research 59(1992) 370–382.

31. R.B. Misra, Optimal production lot size model for a system with deteriorating inventory, International Journal of Production Research 15(1975) 495–505.

Nonlinear Noise Estimation in International Stock Markets: Coarse-Grained Entropy Method[†,#]

Yong Fang

Post-doctoral Research Center for Applied Economics, Shanghai University of Finance and Economics, Shanghai 200433, China. Email: yongf72@163.com.

ABSTRACT

With the step-by-step opening of China Stock Market and gradual strengthening of international linkage, how to efficiently measure and manage risk, evaluate and improve market operation efficiency is an important project in present financial research. According to nonlinear dynamics and chaos and fractal theory, we apply phase space reconstruction technique and coarse-grained entropy method to estimate the nonlinear noise levels in stock markets of Chinese Mainland, Hong Kong, US, UK and Japan, and we emphasize on discussing the standard deviation of nonlinear noise σ and noise-to-signal ratio NSR which are two important indexes about risk measurement and efficiency evaluation, and further we make a comprehensive comparison analysis on the risk and operation efficiency of stock markets of above countries or areas.

Keywords: Coarse-grained entropy, International stock markets, Market efficiency, Noise estimation, Nonlinear time series, Phase space reconstruction.

Introduction

Traditional financial models are linear equilibrium models based on rational expectation. Excluding the influence of psychological factors such as greed and fear, these models depict the negative feedback leading to the stabilization of financial system and indicate that the response to exogenous disturbance of financial system is reverting to the equilibrium in a continuous and linear way.

[†]Reused with permission from: Yong Fang, Nonlinear Noise Estimation in International Stock Markets:Coarse-grained Entropy Method, *International Journal of Economics and Finance,* Vol. 2, No. 1, 97–104, 2010.

[#]The research is financed by China Postdoctoral Science Foundation (Grant No. 20090450075), Shanghai Municipal Natural Science Foundation (Grant No. 09ZR1421900) and Innovation Program of Shanghai Municipal Education Commission (Grant No. 10YZ184). (Sponsoring information).

Nevertheless, what traditional financial models depict is only an ideal state and is widely divergent from the truth. Subjected to noise trading, positive feedback trading, overreaction and herding behavior which are universal in financial market (see De Long, Shleifer, Summers, and Waldmann (1990a, 1990b), De Bondt and Thaler (1985), Banerjee (1992)), the fluctuation of asset price presents strong nonlinearity. Moreover, the evolution of financial market is driven by numerous heterogeneous investors with bounded rationality, so financial market is a complex dynamic system which is essentially characterized by intrinsic randomness of deterministic system, nonrepetitive aperiodic cycle, equilibrium which is far from balance, sensitivity to initial conditions, tendency, self-similarity and mutation. Therefore, nonlinear dynamics and chaos and fractal theory which have developed rapidly in recent years provide a new perspective for us to study complex financial system. Among numerous financial research projects, estimating the nonlinear noise level of complex financial dynamic system is of great practical significance. It can provide financial supervision and management departments with efficient decision supports such as evaluating the operation efficiency of financial market, guarding against and managing financial risk and exporting so-called policy function developed by Boldrin and Montrucchio (1986). Nonlinear noise of complex financial system has two kinds of sources, and one kind is measurement noise from outside of system while another kind is dynamical noise from inside of system. Some scholars have put forward a series of methods on estimating nonlinear noise level, among which the methods developed by Cawley and Hsu (1992), Schreiber (1993), Diks (1996) and Oltmans and Verheijen (1997) only apply to the estimation of low level measurement noise while the coarse-grained entropy method developed by Urbanowicz and Holyst (2003) can efficiently estimate high level measurement noise as well as dynamical noise.

In recent years, many scholars have been drawn to the nonlinearity of stock markets. Longbing Xu and Rong Lu (1999) studied the nonlinearity of Chinese stock market by use of R/S method, and their empirical results indicated that both Shanghai and Shenzhen stock market were characterized by nonlinearity, long memory and volatility clustering. By use of R/S analysis and the calculation of correlation dimension, auto correlation function and Lyapunov index, Haihua Wu and Daoye Li (2001) found that Shanghai stock market was distinctly characterized by fractal and chaos, and was a nonlinear system where strange attractors existed. By implementing a series of tests such as normality test, ADF unit root test, BDS test and ARCH test and calculating correlation dimension and Lyapunov index, Xusong Xu and Yanbin Chen (2001) found that both nonlinearity and chaos existed in Chinese stock market. But existing literatures are in large limited to testing the nonlinearity and chaos in stock markets, and few literatures focus on how to measure nonlinear risk, how to evaluate the efficiency of a stock market, how to form an efficient investment strategy and how to manage systematic risk. These problems are worthy of deeper research. In this paper, we will make an exploration on studying these problems. We apply phase space reconstruction technique and coarse-grained entropy method to estimate the nonlinear noise levels in stock markets of Chinese Mainland, Hong Kong, US, UK and Japan, and further a comprehensive comparison analysis on risk and operation efficiency of above stock markets is made.

Sample Data and Descriptive Statistics

We choose the daily return time series of SSE Composite Index (Chinese Mainland), Hang Seng Index (Hong Kong), S&P 500 (US), FTSE 100 (UK) and Nikkei 225 (Japan) as our research objects. The period studied is from December 20, 1990 to June 21, 2007 and the sample data is from RESSET (http://www.resset.cn) and Yahoo Finance (http://finance.yahoo.com).

To be exempt from the influence of outliers on statistical results, we take values which distances from the mean of time series exceed four times standard deviation as outliers, and then get rid of them from the time series. The descriptive statistics of daily return time series of various stock indexes after getting rid of outliers are given in Table 1.

As shown from Table 1 and line graph and histogram of time series, the above five daily return time series of stock indexes all significantly deviate from normal distribution and are all characterized by leptokurtic, fat tail and volatility clustering.

Table 1 Descriptive statistics of daily return time series of various stock indexes.

	SSE Composite Index	Hang Seng Index	S&P 500	FTSE 100	Nikkei 225
Number of observations	4027	4065	4142	4148	4062
Mean	0.000489	0.000572	0.000364	0.000294	0.000023
Standard deviation	0.020895	0.014261	0.009335	0.009556	0.014206
Maximum	0.1191	0.062	0.0393	0.0406	0.0796
Minimum	−0.1118	−0.0607	−0.0391	−0.0397	−0.0698
Skewness	0.01118	−0.037123	−0.026864	−0.100655	0.147949
Kurtosis	8.742128	4.892077	4.788616	4.540808	5.249315
Jarque-Bera	5532.517	607.2887	552.6178	417.3261	871.1254
P-value	0.000000	0.000000	0.000000	0.000000	0.000000

Preliminary Tests

Nonlinearity Tests

We apply BDS test to detect the nonlinearity of various time series. At first, linear correlation in the time series is filtered by the use of ARMA model, and then BDS test is conducted for residual series. BDS statistics with various thresholds r and embedded dimensions m are given in table 2 (σ_τ denotes the standard deviation of residual series). As shown from Table 2, the null hypothesis that residuals are i.i.d. is significantly rejected (the critical value of normal distribution is 1.96 under the significance level of 5%), and it indicates that the stock markets of above five countries or areas are all of significant nonlinearity.

Determinism Tests

So-called determinism of system is that the future states can be determined by the past states. The system with determinism usually presents some degree of tendency and self-similarity. We apply the method of recurrence plot developed by Eckmann, Kamphorst

Table 2 BDS statistics of daily return time series of various stock indexes.

		SSE Composite Index	Hang Seng Index	S&P 500	FTSE 100	Nikkei 225
$r = 0.5\sigma_\varepsilon$	$m = 2$	18.62	8.68	7.73	9.66	5.21
	$m = 3$	25.56	10.79	12.30	13.14	8.43
	$m = 4$	34.07	13.63	15.81	17.25	10.89
	$m = 5$	44.73	16.69	20.15	21.00	14.17
$r = \sigma_\varepsilon$	$m = 2$	20.86	9.55	7.80	11.09	5.84
	$m = 3$	25.92	12.70	12.99	14.93	9.11
	$m = 4$	30.42	15.85	16.60	18.88	11.54
	$m = 5$	34.38	18.47	20.63	22.25	14.13
$r = 1.5\sigma_\varepsilon$	$m = 2$	22.36	10.17	8.49	12.59	6.67
	$m = 3$	26.47	13.95	13.57	16.67	9.99
	$m = 4$	29.40	17.03	16.82	20.27	12.19
	$m = 5$	31.40	19.38	20.13	23.16	14.12

and Ruelle (1987) to test the determinism of various time series. At first, phase space reconstruction is conducted by the use of nonlinear dynamics, and the time delays τ and embedded dimensions m (the maximum embedded dimension is 15) of various time series which help to reconstruct phase space are given in Table 3, and they are determined by average mutual information method and false nearest neighborhood method.

Table 3 Time delays and embedded dimensions of daily return time series of various stock indexes which help to reconstruct phase space.

	SSE Composite Index	Hang Seng Index	S&P 500	FTSE 100	Nikkei 225
τ	2	1	1	2	1
m	9	11	9	12	13

After phase space reconstruction, recurrence plots of various time series can be made (thresholds are the standard deviations of various time series), and recurrence quantitative analysis (RQA) can be further conducted. By observing the recurrence plots of above five time series, we find that there scatter some small bands parallel with the diagonal, which indicates obvious determinism. Results of RQA on original data and on the new data after randomly disturbing the original sequence of various time series are respectively given in Table 4. As contrasted to original data, the determinism of the new data after randomly disturbing the original sequence significantly declined and this indicates that our determinism tests are robust.

Chaos Tests

According to G-P algorithm (see Grassberger and Procaccia (1983)), double logarithmic scatter plots of correlated integral $C(r)$ with respect to threshold r with various embedded dimensions (as shown in Fig. 1) are made and we can find the following things. On one hand,

Table 4 RQA results of daily return time series of various stock indexes.

		SSE Composite Index	Hang Seng Index	S&P 500	FTSE 100	Nikkei 225
Original data	%recurrence	8.62%	1.06%	2.37%	0.63%	0.41%
	%determinism	37.27%	92.55%	92.59%	3.83%	93.66%
	Longest diagonal line segment	271	45	55	23	51
New data after randomly disturbing the original sequence	%recurrence	1.55%	0.22%	0.61%	0.11%	0.06%
	%determinism	1.64%	82.58%	81.85%	0.10%	80.95%
	Longest diagonal line segment	13	19	16	4	12

with the gradual increase of embedded dimension m, the slope of linear part in $\ln C(r) - \ln r$ graph (correlated dimension) gradually goes to stabilization, which indicates that there exists strange attractor in the system. On the other hand, before arriving at stabilization correlated dimension presents obvious jump (SSE Composite Index acts significantly). From these findings, we may believe that various time series are all characterized by chaos in some degree.

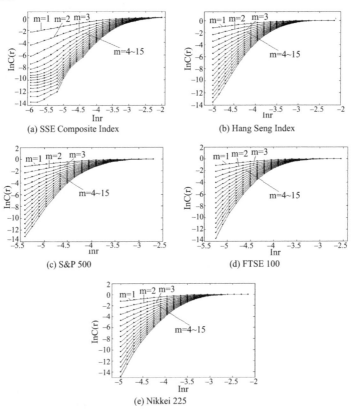

Figure 1 $\ln C(r) - \ln r$ graphs of daily return time series of various stock indexes.

Nonlinear Noise Estimation

Coarse-grained Entropy Method

Let $\{x_i\}, i = 1, 2, \cdots, N$ be a nonlinear time series. After selecting suitable time delay τ and embedded dimension m we reconstruct phase space, and then get m dimension vector sequence

$$\vec{y}_i = (x_i, x_{i+\tau}, \cdots, x_{i+(m-1)\tau}), i = 1, 2, \ldots, M \tag{1}$$

where $M = N - (m-1)\tau$. The correlated integral of embedded phase space is defined as the follows

$$C_m(r) = \frac{2}{M(M-1)} \sum_{1 \le i < j \le M} H(r - \|\vec{y}_i - \vec{y}_j\|) \tag{2}$$

where $H(\cdot)$ is Heaviside function, and r is threshold and $\|\cdot\|$ is maximum norm. Therefore, coarse-grained entropy can be estimated as follows

$$K_2(r) \approx -\frac{d\ln[C_m(r)]}{dm} \tag{3}$$

Let σ be the standard deviation of the noise (including measurement noise and dynamical noise) of nonlinear time series $\{x_i\}, i = 1, 2, \ldots, N$, and the observation values of coarse-grained entropy $K_{noisy}(r)$ can be fitted by the following formula

$$K_{noisy}(r) = -\frac{d\ln[C_m(r)]}{dm}$$

$$= -cg(\frac{r}{2\sigma})\ln r + [\kappa + b\ln(1-ar)] \times (1 + \sqrt{\pi} \frac{\sqrt{r^2/3 + 2\sigma^2} - r/\sqrt{3}}{r}) \tag{4}$$

where k, a, b, c, σ are parameters under estimation, and the function $g(z)$ is defined as follows

$$g(z) = \frac{2}{\sqrt{\pi}} \frac{ze^{-z^2}}{erf(z)} \tag{5}$$

where $erf(\cdot)$ is error function.

In formula (4), $g(\cdot)$ shows the influence of noise on correlated dimension, and $\kappa + b\ln(1-ar)$ is coarse-grained entropy of signal uncontaminated by noise while $[\kappa + b\ln(1-ar)] \times (\sqrt{\pi} \frac{\sqrt{r^2/3 + 2\sigma^2} - r/\sqrt{3}}{r})$ is increment of coarse-grained entropy caused by noise. After estimating the standard deviation of noise σ by the use of fitting formula (4), finally we can calculate noise-to-signal ratio (NSR) of time series as follows

$$NSR = \frac{\sigma}{\sigma_{data}} \times 100\% \tag{6}$$

and we take it as a measurement of noise level of time series.

Estimation Results

Taking the essentiality that correlated integral $C_m(r)$ varies with embedded dimension m and threshold r into account, we choose m from 1 to 20, and r takes 100 different values in interval $\left[\dfrac{\sigma_{data}}{2}, 2\sigma_{data}\right]$. Observation value of coarse-grained entropy $K_{noisy}(r)$ can be estimated by the use of OLS method for the following linear regression

$$\ln C_m(r) = k \cdot m + e \tag{7}$$

where the negative of slope k is just $K_{noisy}(r)$.

Then we apply Levenberg-Marquardt method and general global optimization method to fit $K_{noisy}(r)$ according to (4). The estimation value of σ, NSR, decision coefficient of fitting R^2 and the root of mean square errors (RMSE) are given in Table 5. The statistics in Table 5 and Fig.2 all sufficiently indicate that the fitting makes great effect.

Table 5 Nonlinear noise estimation of daily return time series of various stock indexes.

	SSE Composite Index	Hang Seng Index	S&P 500	FTSE 100	Nikkei 225
σ	0.0079	0.0101	0.0069	0.0066	0.0090
NSR	37.8%	70.8%	73.9%	69.1%	63.4%
R^2	0.9931	0.9994	0.9988	0.9994	0.9986
RMSE	0.00699	0.00437	0.00645	0.00555	0.00594

Results Analysis

Traditional financial models are all based on the hypothesis that stock return is normally distributed (i.e., acts as stationary distribution with characteristic parameter $\alpha = 2$, and its variance is limited and stationary), in which variance of return is applied to measure risk. A large amount of empirical research indicates, however, stock return significantly deviate from normal distribution and is characterized by leptokurtic, fat tail and state continuing. The above practical characteristics of stock return can be well depicted by stationary distribution with characteristic parameter $\alpha \in (1,2)$. As population variance in this case is uncertain or infinity, taking sample variance as the measurement of risk is of no significance.

Nonlinear dynamics and chaos and fractal theory which have rapidly developed in recent years provide a new perspective for us to study modern financial risk management and portfolio selection. Complex financial dynamic system is essentially characterized by intrinsic randomness of deterministic system and the fluctuation of the deterministic part is completely predicted in a short period (so we may believe this part of fluctuation to be of no risk in a short period), so as the essential financial risk measurement, the natural choice is the standard deviation of nonlinear noise of financial asset return time series σ. Not only

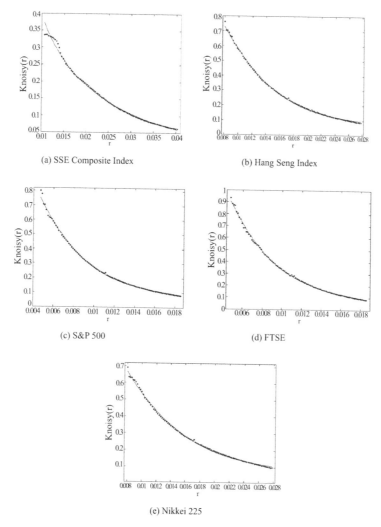

(a) SSE Composite Index

(b) Hang Seng Index

(c) S&P 500

(d) FTSE

(e) Nikkei 225

Figure 2 Fitting plots of coarse-grained entropy of daily time series of various stock indexes.

can we apply σ to guard against and control financial risk, but also we can use it to optimize portfolio selection (see Urbanowicz and Holyst (2004)). By observing the estimation values of σ, we can find that the risks of eastern stock markets including Chinese Mainland, Hong Kong and Japan are higher than that of western stock markets including US and UK. The main reason is that western stock markets are more mature than eastern markets and the later are influenced to a large extent by the free input or output of large amount of capital and regular changes in trading rules. Moreover, we can find that the risks of Hong Kong and Japan stock markets are higher than that of Chinese Mainland stock market. Strong "policy market" characteristic, limitation of margin of rise or fall, relatively weak international linkage and small size of QFII may account for the above finding. As we know, the trial of direct investment in Hong Kong stock market for individual investors of Chinese Mainland will be on the way, and in spite of the great significance of this policy, how to efficiently guard against risk is the primary challenge faced by investors.

Besides σ discussed above, NSR is another important index and it indicates the extent of operation efficiency of stock market. Higher NSR is (i.e., the proportion of intrinsic random fluctuation in the whole fluctuation is higher), market is more mature and operates more efficiently and is closer to EMH. By observing the values of NSR, we can find that the other four stock markets are more and more mature than the stock market of Chinese Mainland. The statistical results which respectively indicate nonnormality, leptokurtic, fat tail, nonlinearity, determinism and mutation also significantly support above finding. This arises from two aspects. On one hand, efficient securities legislation system, credit system and multi-dimensional supervision and management system are not completely established, and some illegal behaviors such as inside trading, manipulating price and issuing false information usually occur. On the other hand, a rational investment culture which core idea is "value investment in the long run" and a healthy and harmonious ecological environment in securities market are not formed and the size of institutional investors is relatively small, so there exist strong effects of "positive feedback" and "herding behavior", and with the accumulation of these effects in a long period financial bubble will be prone to arise, and more heavily financial crisis will probably break out.

Conclusion

With the step-by-step opening of China Stock Market and gradual strengthening of international linkage, how to efficiently measure and manage risk, evaluate and improve market operation efficiency is an important project in present financial research. According to nonlinear dynamics and chaos and fractal theory, we apply phase space reconstruction technique and coarse-grained entropy method to estimate the nonlinear noise levels in stock markets of Chinese Mainland, Hong Kong, US, UK and Japan, and we emphasize on discussing the standard deviation of nonlinear noise σ and noise-to-signal ratio NSR which are two important indexes about risk measurement and efficiency evaluation, and further we make a comprehensive comparison analysis on the risk and operation efficiency of stock markets of above countries or areas. The following research work will be applying these two indexes to the simulation of stock market based on agent to study the influence of microindividuals on the macromarket.

References

1. Banerjee, A. (1992). A simple model of herd behavior. *Quarterly Journal of Economics*, 107, 797–817.
2. Boldrin, M., &Montrucchio, L. (1986). On the indeterminism of capital accumulation paths. *Journal of Economic Theory*, 40, 26–39.
3. Cawley, R., & Hsu, G. H. (1992). Local-geometric projection method for noise reduction in chaotic maps and flows. *Physical Review A*, 46, 3057–3082.
4. De Bondt, W.F.M., &Thaler, R.H. (1985). Does the stock market overreact?. *Journal of Finance*, 40,793–805.

5. De Long, J.B., Shleifer, A., Summers, L.H. &Waldmann, R.J. (1990a). Noise trader risk in financial markets. *The Journal of Political Economy*, 98, 703–738.
6. De Long, J.B., Shleifer, A., Summers, L.H. &Waldmann, R.J. (1990b).
7. ositive feedback investment strategies and destabilizing rational speculation. *Journal of Finance*, 45, 379–395.
8. Diks, C. (1996). Estimating invariants of noisy attractors. *Physical Review E*, 53, R4263–R4266.
9. Eckmann, J.P., Kamphorst, S.O. &Ruelle, D. (1983). Recurrence plots of dynamical systems. *Europhysics Letters*, 4, 973–977.
10. Grassberger, P. &Procaccia, I. (1983).Characterization of strange attractors. *Physical Review Letters*, 50, 346–349.
11. Haihua Wu, & Daoye Li. (2001). Nonlinear dynamics analysis on stock market-in case of Shanghai stock market. *Journal of Qingdao University*,16, 1–7.
12. Longbing Xu, & Rong Lu. (1999) Exploring the nonlinearity of Chinese stock market based on R/S analysis. *Forecasting*,18,59–62.
13. Oltmans, H. &Verheijen, P.J.T. (1997). The influence of noise on power law scaling functions and an algorithm for dimension estimations. *Physical Review E*, 56, 1160–1170.
14. Schreiber, T. (1993). Determination of the noise level of a chaotic time series. *Physical Review E*, 48, R13–R16.
15. Urbanowicz, K. &Holyst, J.A. (2003) .Noise-level estimation of time series using coarse-grained entropy. *Physical Review E*, 67, 046218.
16. Urbanowicz, K. & Holyst, J.A. (2004). Investment strategy due to the minimization of portfolio noise level by observations of coarse-grained entropy. *Physica A*, 344, 284–288.
17. Xusong Xu, & Yanbin Chen. (2001). An empirical study on the nonlinearity of Shanghai and Shenzhen stock market. *The Journal of Quantitative & Technical Economics*,18, 110–113.

Section V

Models and Applications of Chaos Theory in Electronics

1. **Simple Chaotic Oscillator: From Mathematical Model to Practical Experiment**   317
    J. Petržela, Z. Kolka and *S. Hanus*

2. **A Robust Chaos-Based True Random Number Generator Embedded in Reconfigurable Switched-Capacitor Hardware**   328
    M. Drutarovský and *P. Galajda*

3. **Conservative Chaos Generators with CCII+ Based on Mathematical Model of Nonlinear Oscillator**   345
    J. Petržela and *J. Slezák*

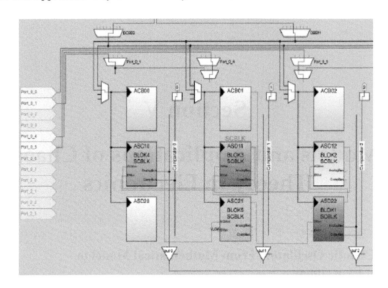

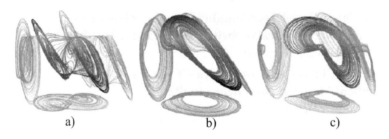

a)   b)   c)

Simple Chaotic Oscillator: From Mathematical Model to Practical Experiment[†]

Jiří Petržela,[1,a] Zdeněk Kolka[1] and Stanislav Hanus[1]

[1]Dept. of Radio Electronics, Brno University of Technology, Purkyňova 118, 612 00 Brno, Czech Republic.
[a]Email: petrzelj@feec.vutbr.cz.

ABSTRACT

This paper shows the circuitry implementation and practical verification of the autonomous nonlinear oscillator. Since it is described by a single third-order differential equation, its state variables can be considered as the position, velocity and acceleration and thus have direct connection to a real physical system. Moreover, for some specific configurations of internal system parameters, it can exhibit a period doubling bifurcation leading to chaos. Two different structures of the nonlinear element were verified by a comparison of numerically integrated trajectory with the oscilloscope screenshots.

Keywords: Nonlinear oscillator, chaos, Lyapunov exponents, circuit realization, measurement.

Introduction

It is well known that many physical systems with some time evolution can be described by a set of generally nonlinear differential equations. It seems that an electronic circuit is an easy way how to construct a dynamical system which represents a given mathematical model. That is why we often use this approach for the purpose of studying dynamical motion, which can be simple (fixed points, limit cycles) as well as very complicated (chaos). We can also see some universality behind the construction of chaotic oscillators, because we are not interested in the concrete physical interpretation of the individual state variables. For example, chaos was recently reported in the scientific fields such as chemistry, mechanics, economy, biology, etc. For lumped electronic circuit, the necessary condition is a finite number of the state variables and, of course, a complexity of the synthesized network grows in accordance with growing dimension of the dynamical system. The

---

[†]Reused with permission from: J. Petržela, Z. Kolka, S. Hanus, Simple Chaotic Oscillator: From Mathematical Model to Practical Experiment, *Radioengineering*, 15 (1), 2006.

very first discovered chaotic oscillator was deeply described in [1], and is so far the only dynamical system where the presence of chaos was confirmed numerically, experimentally as well as mathematically. These equations, taking name after its discoverer as the Chua's equations, belong to extensive group of dynamical systems covered by a following state space representation in compact matrix form

$$\dot{\mathbf{x}} = \mathbf{A}\,\mathbf{x} + \mathbf{b}\,h\left(\mathbf{w}^T\mathbf{x}\right), \tag{1a}$$

where $\mathbf{X} \in \Re^3$ is a vector of the state variables, $\mathbf{A}$ is a square matrix and $\mathbf{b}$, $\mathbf{w}$ are column vectors. The saturation-type nonlinearity of the form

$$h\left(\mathbf{w}^T\mathbf{x}\right) = 0.5\left(\left|\mathbf{w}^T\mathbf{x}+1\right| - \left|\mathbf{w}^T\mathbf{x}-1\right|\right), \tag{1b}$$

separates the state space by two parallel boundary planes $U_{\pm 1}$ into the three affine regions making a brief analysis possible and quite simple. Due to the symmetrical vector field with respect to the origin, the dynamical behavior in both outer regions $D_{\pm 1}$ of the state space is the same and determined by the roots of the characteristic polynomial

$$\det\left(s\,\mathbf{E} - \mathbf{A}\right) = \left(s - \upsilon_1\right)\left(s - \upsilon_2\right)\left(s - \upsilon_3\right) = s^3 - q_1 s^2 + q_2 s - q_3 = 0, \tag{2a}$$

and similarly for single inner region D_0

$$\det\left[\,s\,\mathbf{E} - \left(\mathbf{A} + \mathbf{b}\,\mathbf{w}^T\right)\right] = \left(s - \mu_1\right)\left(s - \mu_2\right)\left(s - \mu_3\right) =,$$
$$= s^3 - p_1 s^2 + p_2 s - p_3 = 0 \tag{2b}$$

where $\mathbf{E}$ is the unity matrix. In the parameter space of some interest we can assume one fixed point per state space region, namely

$$\left(\mathbf{A} + \mathbf{b}\,\mathbf{w}^T\right)\overline{\mathbf{x}}_{inner} = \mathbf{0} \rightarrow \overline{\mathbf{x}}_{inner} = \left(0 \quad 0 \quad 0\right)^T, \tag{3a}$$

and

$$\overline{\mathbf{x}}_{outer} = \pm\mathbf{A}^{-1}\,\mathbf{b}, \quad \det\left(\mathbf{A}\right) \neq 0 \tag{3b}$$

The main property of the chaotic solution is in extreme sensitive to the changes of the initial conditions. It implies that we can not obtain closed-form analytic solution, so our analysis is restricted to the numerical integration.

There is allways some uncertainty in the initial state so that any predictions about future behavior are no longer available. Moreover, the chaotic attractor is dense having a fractional topological dimension. No-intersection theorem means that for chaos there must be at least three state variables.

Mathematical Model

Note that the dynamical system expressed as (1) is linear in each region of the state space, with the behavior given uniquely by eigenvalues, roots of the characteristic polynomial. To model any type of motion of such a class of dynamical systems, the parameter space must be at least six-dimensional. Now, we try to reduce this amount as much as possible, preserving a global qualitative behavior. To specify the parameter window where the chaos becomes possible, we should first recall the relations between eigenvalues and their equivalent numbers p_i's and q_i's. Substituing $s = \alpha_1 / 3 + \Lambda$ into (2) leads to a reduced cubic polynomial of the form

$$\Lambda^3 + \frac{3\alpha_2 - \alpha_1^2}{9}\Lambda + \eta = 0 , \tag{4}$$

where

$$\eta = \frac{\alpha_1 \alpha_2}{6} - \frac{\alpha_3}{2} - \frac{\alpha_1^3}{27} . \tag{5}$$

Let denote

$$\Delta = \frac{\alpha_3^2}{4} - \frac{\alpha_1^2 \alpha_2^2}{108} + \frac{\alpha_1^3 \alpha_3}{27} - \frac{\alpha_1 \alpha_2 \alpha_3}{6} + \frac{\alpha_2^3}{27} . \tag{6a}$$

and

$$\psi_{pos,neg} = \sqrt[3]{-\eta + \sqrt{\Delta}} \pm \sqrt[3]{-\eta - \sqrt{\Delta}} . \tag{6b}$$

For $\Delta > 0$ we have one real

$$\lambda_3 = \psi_{pos} + \alpha_1 / 3 , \tag{7}$$

and a pair of complex conjugated numbers $\sigma \pm \omega i$ with the following real and imaginary part

$$\sigma = -\psi_{pos} / 2 + \alpha_1 / 3 , \; \omega = \sqrt{3}\psi_{neg} / 2 . \tag{8}$$

In these terms $\alpha_i = p_i$ or $\alpha_i = q_i$ depending on the region of the state space. The relations between eigenvalues and their equivalent numbers are given by the formulas

$$2\sigma + \lambda_3 = \alpha_1, \lambda_3 \left(\sigma^2 + \omega^2 \right) = \alpha_3, \sigma^2 + \omega^2 + 2\sigma \lambda_3 = \alpha_2. \tag{9}$$

In the case of double-scroll attractor, state space geometry is $D_{\pm 1}$: $\mathfrak{R}_u^2 \oplus \mathfrak{R}_s^1$ and D_0: $\mathfrak{R}_u^1 \oplus \mathfrak{R}_s^2$ while for producing the so-called double-hook attractor it should be $D_{\pm 1}$: $\mathfrak{R}_u^2 \oplus \mathfrak{R}_s^1$ together with D_0: $\mathfrak{R}_u^1 \oplus \mathfrak{R}_s^1 \oplus \mathfrak{R}_s^1$. In spite of the obvious restrictions of the eigenvalues resulting from the terms (9), the inner region can behave like an overdamped circuit.

To date, no double-hook attractor produced by a nonlinear oscillator was reported. Searching for such a configuration of system parameters leading to some structure of chaotic attractor [6] is the topic of our future study.

The process of reduction of internal system parameters is based on the equivalence of some pair q_i and p_i with the same order. Since the parameter α_3 directly corresponds to λ_3 we have immediately

$$\lambda_3^{outer} < 0 \rightarrow q_3 < 0, \lambda_3^{inner} > 0 \rightarrow p_3 > 0, \tag{10}$$

leading to the condition $q_3 \neq p_3$. If this is satisfied, there is just one equilibrium point per region as necessary for generation of chaotic attractors. Starting with the reference dynamical system of class C presented in [2] and taking into account $q_1 = p_1$ and $q_2 = p_2$ we can write the differential equations describing new dynamical system as

$$dx/dt = y, dy/dt = z, dz/dt = q_1 z - q_2 y + g(x), \tag{11}$$

where the piecewise-linear (PWL) function

$$g(x) = q_3 x + (p_3 - q_3) h(x). \tag{12}$$

The fixed points located inside regions $D_{\pm 1}$ migrate along the x axis with the ratio

$$\overline{\mathbf{x}}_{outer} = (\pm p_3 / q_3 \mp 1 \quad 0 \quad 0)^T. \tag{13}$$

Note that these fixed points can eventually enter D_0 region. It is evident that $q_3 \neq 0$. Similarly, $p_3 \neq 0$ otherwise there is an equilibrium line in the inner state space segment. Due to the dissipativity of the system, one must have $q_1 < 0$. Next, let write a basic conditions using a signum function, i.e., $sign(\sigma_{inner}) \neq sign(\sigma_{outer})$ and $\sigma_{inner} < 0$. Then, terms (9) lead to

$$q_2 < \sigma^2 + \omega^2, \; q_1 > \lambda_3^{outer}. \tag{14}$$

Note that all conditions above are in accordance with a formation of hyperbolic equilibria, where $\sigma \neq 0$. To preserve $\sigma_{outer} > 0$, i.e., the saddle-focus of index 2 equilibria in $D_{\pm 1}$ we have $\psi_{pos} < \min(-q_1/3, 2q_1/3)$. The process of double-scroll attractor evolution is briefly discussed by means of Fig. 4. It is evident that both boundary planes $U_{\pm 1}$ can not be coplanar with any eigenplane denoted as eP or eigenvector marked as eV. Nevertheless, to acquire a better overview on the global dynamics, one can make a linear transformation of coordinates. For example, if the state matrix is in a real Jordan form, eV is orthogonal to eP. In spite of this, making a rigorous mathematical proof of chaos is a nontrivial task. Here, we simply suppose that the so-called Shilnikov's theorems [7] are fulfilled, its detailed form will be given elsewhere. In further text, we adopt the following constant set of the parameters

$$q_1 = -0.6, q_2 = 0.846. \tag{15}$$

The same numbers also appear in [3]. Now, let see what eigenvalues hold for the partial cases of the Fig. 4. For the first picture it is $\sigma_{outer} = -0.05$, $\lambda_3^{outer} = -0.502$, $\sigma_{inner} = -0.572$, $\lambda_3^{inner} = 0.545$ and for the last picture the corresponding eigenvalues are $\sigma_{outer} = 0.171$, $\lambda_3^{outer} = 0.942$, $\sigma_{inner} = -0.572$, $\lambda_3^{inner} = 0.545$.

Let define the volume element in the tangent space of the vector field. We can classify the type of the attractor by using the concept of one-dimensional Lyapunov exponent (LE), which measures the average rate of the convergence or divergence of two neighborhood trajectories. There are just three LE and, in accordance with the natural definition of chaos, one LE must be positive and one equals to zero. Because of the dissipation, the last one must be negative with the largest absolute value, as the element is shrinking with the time progression. Fig. 1 and Fig. 2 illustrate how the largest one-dimensional LE vary with the two natural bifurcation parameters p_3 and q_3 (with respect to circuit). Each of these graphs consist of 10201 point, computed for final time t_{max}=400. A standard Gram-Smith orthogonalization were performed during the estimation after each Δt=1 step. The significance of colors is shown in Fig. 2. From this point of view, light blue marks a trivial fixed point and inside the white areas entire solution suddenly becomes unbounded. Selected state trajectories of some interest were also numerically simulated by using MathCAD and build-in 4th order Runge-Kutta iteration method with time interval $t \in (0, 600)$, iteration step Δ=0.06 and initial conditions $X_0 = (0.5 \quad 0 \quad 0)^T$. The results are shown in Fig. 3, with color scale growing along the vertical axis.

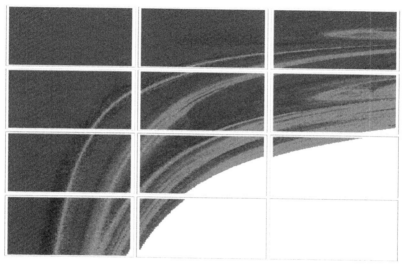

Figure 1 Largest LE as a function of two internal system parameters: horizontal axis $p_3 \in (0, 3)$ and vertical axis $q_3 \in (-1, -5)$.

Circuitry Implementation

The algebraic simplicity of the differential equations does not guarantee also its easy physical implementation. In spite of this, the new dynamical system (11) has one big advantage if compared to a classical first ODE equivalent of Chua's equation [4]. It is much more easy to practically realize (11) with four independent parameters than a system with six parameters. The straightforward circuit synthesis is usually based on basic building blocks such as inverting integrator, differential amplifiers and PWL transfer function. Such realization leads to large number of circuit elements. Here, we are going to use a completely different approach. It is evident that we can recast (11) into the single differential equation of the third-order. Thus, we can realize this dynamical system by parallel connection of the third-order admittance function together with a nonlinear resistor having PWL or polynomial AV curve. For admittance function shown in Fig. 5 we can derive the expression

$$Y(s) = C_1 C_2 C_3 R_1 R_2 s^3 + C_1 C_2 (R_1 + R_2)s^2 + (C_1 + C_2)s.$$

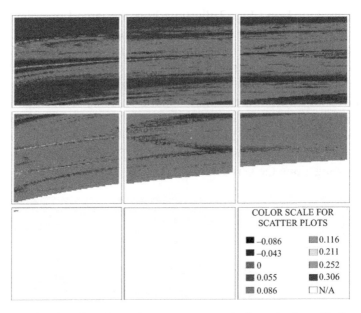

Figure 2 Largest LE as a function of two internal system parameter: horizontal axis $p_3 \in (3, 6)$ and vertical axis $q_3 \in (-1, -3)$.

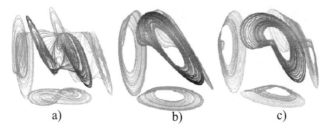

Figure 3 Perspective views on different state space attractors: a) double-scroll b) funnel c) single-scroll.

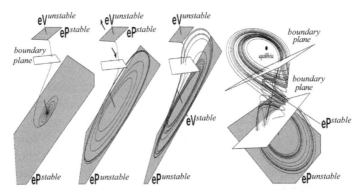

Figure 4 Evolution of the double-scroll attractor with $p_3 = 0.8$ and: a) $q_3 = -0.4$, b) $q_3 = -0.6$, c) $q_3 = -0.8$, d) $q_3 = -1.1$.

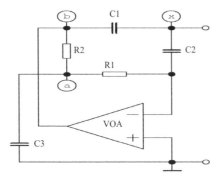

Figure 5 Third-order admittance function Y(s).

To obtain a common-valued circuit elements, a current and time rescaling have been performed leading to following listing of linear capacitors and resistors

$$C_1=C_2=5\text{nF},\ C_3=16\text{nF},\ R_1=10\text{k}\Omega,\ R_2=20\text{k}\Omega \tag{17}$$

and operational amplifier AD713. For experimental verification, we suppose that (17) does not change. Two different configurations of the active nonlinear resistor have been used, i.e., PWL resistor shown in Fig. 6 and polynomial resistor with the structure given by Fig. 7. In the first case, 1N4007 diodes and current-feedback amplifier AD844 has been used to form a basic negative slope of the final AV curve. The voltages of two independent external dc sources should be chosen carefully low to preserve CFOA working in the linear regime.

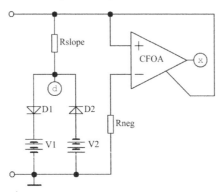

Figure 6 Three-segment PWL resistor.

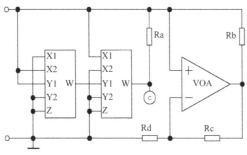

Figure 7 Nonlinear resistor with polynomial AV characteristics.

Let consider some level of idealization resulting into the equations

$$U^+ = U^-, I^- = I^C, I^+ = 0, U^C = U^O. \tag{18}$$

The input admittance of the PWL resistor in the region D_0 then becomes

$$Y_{input} = -1/R_{neg}, \tag{19a}$$

while in both outer segments of the vector field it is

$$Y_{input} = 1/(R_{slope} + R_d) - 1/R_{neg}, \tag{19b}$$

where R_d is a differential resistance of diode in the forward direction. We can derive the fundamental condition for the existence of three fixed points in the form $R_{slope} + R_d < R_{neg}$ which is equivalent to the intersection of the PWL AV characteristics of the resistor with the load line identical with horizontal axis $i \rightarrow 0$. For the same impedance norm used above, we can observe a double-scroll attractor if

$$R_{neg} = 10k\Omega, R_{slope} = 2k\Omega, \tag{20}$$

Note that the voltage output of CFOA is buffered and directly represents the state variable $x(t)$. We can approximate the PWL curve by a third-order polynomial function. For this purpose, two four-quadrant analog multipliers AD633 can be used, both with transfer function $W = K(X1 - X2)(Y1 - Y2) + Z$, where $K = 0.1$. It is not hard to derive the analytic expression

$$i = f(u) = \frac{K^2}{R_a} u^3 + \left(\frac{1}{R_a} - \frac{R_c}{R_d R_b} \right) u, \tag{21}$$

which generates a double-scroll attractor when

$$R_a = R_b = 2.2k\Omega, R_c = 12k\Omega, R_d = 10k\Omega. \tag{22}$$

Nonlinear oscillator works in hybrid voltage/current mode. Figure 8 is an example of the time dependance of the chaotic waveform, namely $b(t)$ and $d(t)$. In the end, some oscillos-cope HP54603B screenshots of selected chaotic attractors are presented in Fig. 9, making use of the XY view mode. Similarly, Fig.10 shows a qualitatively equivalent chaotic structure for a polynomial nonlinearity. It is worth nothing that also Pspice circuit simulator give us the same results for our oscillator with both types of the nonlinearity. Moreover, several other types of PWL AV characteristics can produce a chaotic waveform such as signum, sigmoid, goniometrical or exponential functions. For the first case, one can make use of a single comparator. The remaining functions are much more difficult to implement.

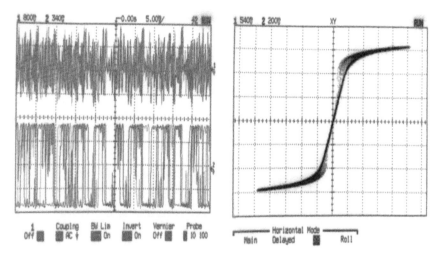

Figure 8 Chaotic waveform in time domain, on-diode voltage.

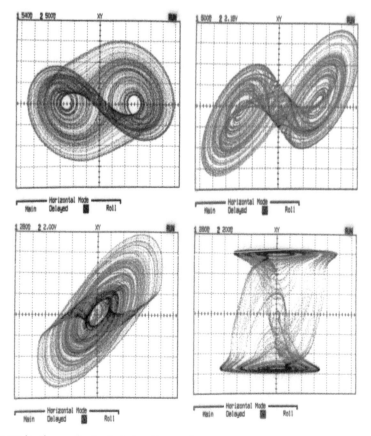

Figure 9 Plane projections of selected chaotic steady states.

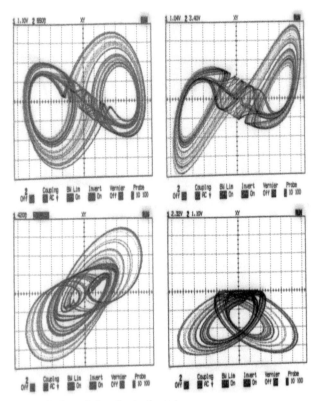

Figure 10 Plane projections of selected chaotic steady states.

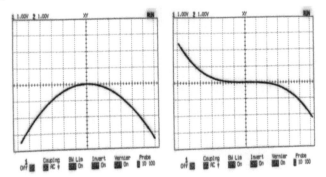

Figure 11 Voltages at the output of both multipliers.

Conclusion

Although the nonlinear oscillator belongs to the dynamical systems of class C, it can not be realized by using a classical circuit synthesis, since it leads to the unrealistic values of circuit components [5]. The physical realization is very simple, removing the necessity of using cumbersome inductors with weakly defined inductance. Precise tuning of the circuit parameters allows us to observe a different structure of chaotic attractors, corresponding transient events, periodic signals having a fundamental frequency as well as multiple-periodic harmonic waveforms. Practical measurements prove a very good final agreement with theoretical expectations. Thus, this circuit is well suited for laboratory experiments and educational purposed. During these measurements, we also verify that the chaotic signal has a continuous and broad-band frequency spectrum. This property can be used in many applications.

Acknowledgements

The research described in this paper was financially supported by the projects GAČR no. 102/03/H105, GAČR no. 102/04/0469 and MSM 021630513.

References

1. CHUA, L.O., KOMURO, M., MATSUMOTO, T. The double scroll family. *IEEE Trans. on CAS I*, 1986, vol. 33, no. 11, p. 1073–1117.
2. POSPÍŠIL, J., KOLKA, Z., HORSKÁ, J., BRZOBOHATÝ, J. New reference state model of the third-order piecewise-linear dynamical system. *Radioengineering*, 2000, vol. 9, no. 3, p. 1–4.
3. SPROTT, J.C., LINZ, S.J. Algebraically simple chaotic flows. *International Journal of Chaos Theory and Applications*, 2000, vol. 5, no. 2, p. 1–20.
4. POSPÍŠIL, J., KOLKA, Z., HORSKÁ, J., BRZOBOHATÝ, J. Simplest ODE equivalents of Chua's equations. *International Journal of Bifurcation and Chaos*, 2000, vol. 10, no. 1, p. 1–23.
5. CHUA, L.O., LIN, G.N. Canonical realization of Chua's circuit family. *IEEE Trans. on CAS I*, 1990, vol. 37, no. 7, p. 885–902.
6. SILVA, CH.P., CHUA, L. O. The overdamped double-scroll family. *International Journal of Circuit Theory and Applications*, 1988, vol. 16, p. 233–302.
7. SILVA, CH.P. Shilnikov's theorem-A tutorial. *IEEE Trans. on CAS I*, 1993, vol. 40, no. 10, p. 675–682.

A Robust Chaos-Based True Random Number Generator Embedded in Reconfigurable Switched-Capacitor Hardware[†]

Miloš Drutarovský[1,a] and *Pavol Galajda*[1,b]

[1]Dept. of Electronics and Multimedia Communications, Technical University of Košice, Park Komenského 13, 04120 Košice, Slovak Republic.
[a]Email: milos.drutarovsky@tuke.sk.
[b]Email: pavol.galajda@tuke.sk.

ABSTRACT

This paper presents a new chaos-based True Random Number Generator (TRNG) with decreased voltage supply sensitivity. Contrary to the traditionally used sources of randomness it uses well-defined deterministic switched-capacitor circuit that exhibits chaos. The whole design is embedded into commercially available mixed-signal Cypress PSoC reconfigurable device without any external components. Proposed design is optimized for reduction of influence of supply voltage to the quality of generated random bit stream. The influence of circuit non-idealities is significantly reduced by the proposed XOR corrector and optimized circuit topology. The ultimate output bit rate of proposed TRNG is 60 kbit/s and quality of generated bit-streams is confirmed by passing standard FIPS and correlation statistical tests performed in the full range of PSoC device supply voltages.

Keywords: Cryptography, chaos, Markov chains, TRNG, PSoC mixed-signal hardware, FIPS tests.

Introduction

With the rapid development of computer networks and wireless communications, information security becomes one of the major problems for effective use of information

---

[†]Reused with permission from: Miloš Drutarovský, Pavol Galajda, A Robust Chaos-Based True Random Number Generator Embedded in Reconfigurable Switched-Capacitor Hardware, *Radioengineering*, Vol. 16, No. 3, 2007.

technology. These problems can be solved by using cryptography. In many practical applications (e.g., large sensor networks) cryptographic primitives have to be implemented in a very cost effective way by using standard hardware components. In a cryptographic application where ultimate security is necessary, a True Random Number Generator (TRNG) is required. TRNGs are widely used for example as confidential key generators for symmetric key crypto-systems (e.g., AES) and public-key ones (e.g., RSA, ECC). In some algorithms (e.g., digital signatures) or protocols (e.g., zero-knowledge), random numbers are intrinsic to the computation [1].

Classical TRNG uses some random physical phenomenon [2]. Currently the most frequently used phenomenon in embedded TRNGs is direct noise amplification of a resistor [3] and jitter noise of digital clock signals [4]. Embedded TRNGs frequently use some external devices or rather heuristic source of randomness (e.g., metastability [5]). Although these limitations can be overcome by a design of proper custom circuits, randomness extraction is still a big challenge in the designs based on off the shelf devices as FPGA [6], [7] or general microprocessors [8].

Chaotic circuits [9] represent an efficient alternative to the classical TRNGs. Contrary to traditionally used sources of randomness they use a well-defined analog deterministic circuit that exhibits chaos. Chaotic systems are characterized by a "sensitive dependence on initial conditions", i.e., a small perturbation eventually causes a large change in the state of the system. However, the slightest uncertainty about the initial state (which is unavoidable in all analogue implementations of chaos systems) leads to a very large uncertainty after very short time. With such initial uncertainties, the system's behavior can be predicted only for a short time period. Additionally as it has been recently shown [10], if the state variable of chaos system is not available to the observer, and the chaos-based system map is well designed, the output of the system cannot be predicted at all. Such implementation is a source of infinite entropy that is absolutely required for good TRNG.

Many random number generators based on analog and deterministic chaotic phenomena have been proposed, see e.g., [11], [12] and references in [10]. Some of them have been simulated only. Others have not been sufficiently optimized for cryptographic applications as they provide certain bias or other deviations. The design [14] describes practical implementation of recently proposed deterministic chaos circuit based on Markov map [9], [13]. In contrary with Field Programmable Analog Array (FPAA) implementation [13], chaos based TRNG implementation [14] uses mixed-signal Cypress PSoC reconfigurable hardware [15], [16]. It was shown that after suitable modification of originally proposed Markov map [10] it can be easily embedded in the selected PSoC hardware [14], [17]. Although this TRNG provides much lower output speeds than pure FPAA implementation [13], used mixed-signal PSoC hardware includes also embedded microcontroller. Such hardware can be in principle used for other cryptographic tasks (e.g., in sensor networks). Moreover, special attention was devoted to the identification of analog circuit non-idealities. The original design [17] is optimized for reduction of their influence and is fully functional for supply voltage $V_{CC} = 5V \pm 10\%$. This paper describes a new robust implementation of TRNG in PSoC device that is functional in full range of supply voltages $V_{CC} = 3.1V \div 5.5V$.

The paper is organized as follows. A brief overview of the theory of Markov chaotic sources is given in Section 2. In Section 3, a new robust method of mapping chaos based TRNG into PSoC devices is presented. The experimental TRNG hardware used to test the proposed method is described in Section 4. In Section 5, statistical evaluations of internal and output TRNG signals are made. Finally, concluding remarks are presented in Section 6.

An Overview of the Theory of Markov Chaotic Sources

An ideal Random Number Generator (RNG) is a discrete memory-less information source that generates equiprobable symbols and its state chain is shown in Fig.1a.

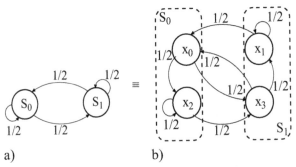

Figure 1 State chains a) of an ideal RNG, b) Markov chain of proposed RNG implementation.

The system is in the state S_0 when an output of the RNG has been "0" and in the state S_1 when an output has been "1". A RNG output has always the same probability $1/2$ to get the system in either state, S_0 or S_1. It is known that these sources can be build up from Piece-Wise Affine (PWA) Markov chaotic maps [10]. These maps are one dimensional, discrete-time ones, in which the state variable $x(n)$ is computed as

$$x(n+1) = M[x(n)] \tag{1}$$

where $M : [-1,1] \rightarrow [-1,1]$, and $x(0)$ is the initial condition. We focus on the recently proposed PWA chaotic map [9], [13]

$$M(x) = (2x+1) \bmod 2 - 1 \tag{2}$$

that is plotted in Fig.2. The map $M[(x)]$ can be expressed as

$$x(n+1) = \begin{cases} 2x(n) - 2 & \text{for } x > 1/2 \\ 2x(n) & \text{for } -1/2 < x < 1/2 \\ 2x(n) + 2 & \text{for } x < -1/2 \end{cases} \tag{3}$$

The main advantage of this chaotic map is its increased robustness in the case of an analogue implementation [9], [10]. The dynamics of equation (1) can be represented by the Markov chain and its evolution can be studied through a square matrix (often referred to as the kneading matrix) defined for the map (3) in Fig. 2 [9], [10].

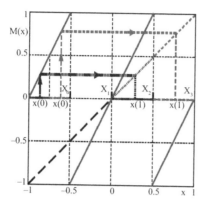

Figure 2 Proposed piece-wise chaotic map based on the eq. (3) and its kneading matrix.

This iterative process (1), (2) can be interpreted as a Markov state chain with 4 states (Fig.1b). The state machine is in its discrete state x_i when the chaotic system has its continuous state variable x in the partition interval X_i. The weights assigned to the graph arrows represent the probabilities by which the state machine changes from a one state to another. The chain in Fig.1b is not suitable for a direct realization of the ideal RNG since it is sequential (it has a memory). However, it is possible to easily build a rigorously independent binary sequence from $x(n)$. In fact, it is sufficient to aggregate the Markov states into two macro-states S_0 and S_1 shown with dotted lines in Fig.1b [17]. Note, that this aggregation is different from that introduced in [10] in order to allow simpler implementation in PSoC devices.

PSoC based TRNG

An Overview of PSoC Architecture

The Cypress PSoC™ device family [15] consists of a mixed-signal array with an on-chip CPU. A PSoC device includes configurable Analog Blocks (AB) [16] and Digital Blocks (DB), as well as programmable interconnections as is shown in Fig.3.

The basic features of the PSoC device are as follows:

- powerful Harvard CPU architecture,
- advanced peripherals (PSoC Blocks):
    4 rail-to-rail continuous analog PSoC blocks
    8 Switched Capacitor (SC) analog blocks
    8 digital PSoC blocks
- programmable references voltage $V_{ref} = 1.3\text{V}$ (set by internal bandgap reference) or $V_{ref} = V_{cc}/2$,

- internal 24 MHz oscillator (allows to build a real single chip application),
- precision, programmable clocking (provides two phase clocks- Φ_1, Φ_2 for SC),
- flexible on-chip memory,
- programmable pin configurations.

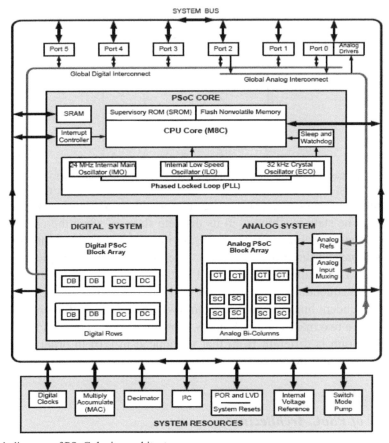

Figure 3 Block diagram of PSoC device architecture.

The PSoC architecture allows implementation of many different functions merely by altering the internal circuit's switches. The on-chip CPU controls the functionality of the digital and analog blocks and it can dynamically change the parameters (e.g., gain of SC block) and topology of these blocks.

Four SC Blocks TRNG Architecture

As shown in [10], the chaotic map in Fig. 2 can be implemented by a pipeline-ADC with 1.5-bit/stage architecture. In [17] we introduced different implementation of the chaotic map (3) by using four SC (4-SC) analog PSoC blocks. The equation (3) can be rewritten according to Table 1 where $b(n+1)$ is a binary output of the TRNG at time step $(n+1)$.

Table 1 TRNG chaotic map equations optimized for PSoC 4-SC blocks.

1st step	$x'(n) \quad = \begin{cases} x(n) - V_{ref} \text{ for } x(n) > 0 \\ x(n) + V_{ref} \text{ for } x(n) < 0 \end{cases}$
2nd step	$x''(n) \quad = \begin{cases} x'(n) - V_{ref} \text{ for } x'(n) > 0 \\ x'(n) + V_{ref} \text{ for } x'(n) < 0 \end{cases}$
3rd step	$x(n+1) = 2x''(n) \quad \text{for all } x(n)$ $b(n+1) \quad = \begin{cases} \log 1 \text{ for } x(n+1) > 0 \\ \log 0 \text{ for } x(n+1) < 0 \end{cases}$

The steps given by equations in Table 1 can be directly mapped into four PSoC SC blocks shown in Fig. 4. Addition/subtraction of the reference voltage and comparison operations from steps one and two are mapped to SC1 and SC3, respectively. Multiplication by 2.0 from step three is realized in SC3 and SC4. Comparison from step three is implemented in SC3. We have chosen a common SC implementation operating on two-phase clocks (Φ_1 and Φ_2 driven by the on-chip oscillator). During the first phase, block SC1 adds the reference voltage $\pm V_{ref}$ (actual polarity is controlled by on-chip CPU) to the input voltage V_{in} according to the polarity of the analogue voltage at the output of the block SC3. If this voltage is negative, the reference voltage $V_{ref+} = +V_{ref} = 1.3V$ will be added in the next clock phase to the input voltage, otherwise the voltage $V_{ref-} = -V_{ref} = -1.3V$ must be added. During the second clock phase, the block SC1 tests the polarity of the output voltage for the block SC3. The operation of the block SC3 is similar to that of the SC1.

Blocks SC2 and SC4 realize the Sample and Hold function between blocks SC1 and SC3. This is needed for a correct operation of SC blocks connected as a "ring oscillator". Additionally, blocks SC3 and SC4 define the gain of the circuit, which has to be as close as possible to the ideal gain equal to 2.0. In our case the gain is limited by discrete values of capacitors C_F and C_A to the value

$$(27 \times 19)/(16 \times 16) = 513/256 \approx 2.004 \tag{4}$$

It can be shown that the structure in Fig. 4 processes two independent bit streams, $b_1(n)$ and $b_2(n)$ in a pipeline processing structure. The bit streams are read out sequentially from this structure and they are overlapped according to the equations

$$b_1(n) = b(2n), \quad n = 0,1,\ldots$$
$$b_2(n) = b(2n+1), \quad n = 0,1,\ldots \tag{5}$$

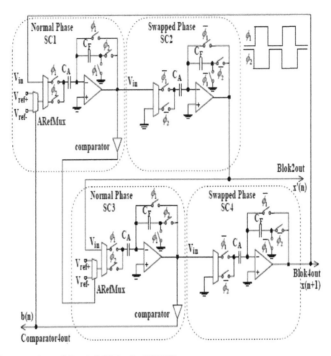

Figure 4 PSoC implementation of the 4-SC blocks TRNG.

Properties of 4-SC TRNG:
- it uses voltage range from $AGND$ to $\pm 2V_{ref}$ ($x(n)$ is actually between 0V and 5V, with low distortion in power supply boundary voltages),
- there exist two independent bitstreams (can be used for effective postprocessing, as was shown in [17]),
- it requires relatively simple control that must be performed by processor core (only one analog comparator interrupt is used),
- gain is split and can by set more precisely to the optimal value (equal to 2),
- it uses no external devices.

Disadvantage of 4-SC TRNG:
- proposed design requires supply voltage equal to 5V. This drawback can be resolved by the new design with 5-SC blocks TRNG implementation, as described in the next section,
- sensitivity to the V_{CC} can be removed only by using an external V_{ref}.

New Robust PSoC TRNG Architecture

Robust PSoC TRNG is similar to that realized by our 4-SC blocks TRNG with the difference that SC1 block in 4-SC TRNG is replaced by a pair of SC11 and SC12 ones, as is shown in Fig.5. Main practical disadvantage of 4-SC TRNG implementation is the above mentioned requirement of supply voltage $V_{cc} = 5\text{V}$. This is a consequence of mapping equations given in Table 1 that are optimized for use with the internal $V_{ref} = 1.3\text{V}$. This drawback can be resolved by the design referred as 5-SC blocks TRNG implementation, as described in this section.

The 5-SC TRNG implementation realizes the equation (3) in a way different than in the case with 4-SC blocks. The steps are given according to Table 2, where $b(n+1)$ is a binary output of the TRNG at time step $(n+1)$. The steps given by equations in Table 2 can be directly mapped into five PSoC SC blocks as is shown in Fig. 5 and its hardware implementation is depicted in Fig. 6.

The user module placement of 5-SC TRNG in PSoC reconfigurable hardware is shown in Fig. 7. The equations in the 1st and the 2nd steps generate modified map depicted in Fig. 8. Although this is a mirrored version of the original map (3), it has no influence on TRNG function. The voltage range of values $x(n)$ are from *AGND* to $\pm V_{ref}$, where $V_{ref} = V_{CC}/2$, whereby 5-SC TRNG has become independent on the supply voltage V_{CC}. The detailed operation of individual steps realized by SC blocks is similar as those described for 4-SC TRNG.

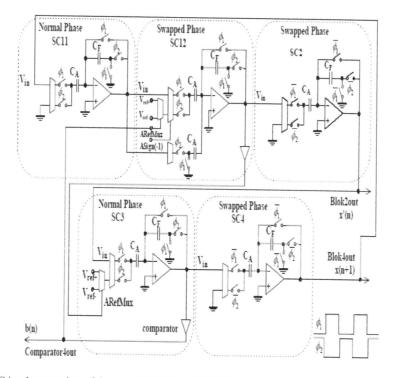

Figure 5 PSoC implementation of the new 5-SC blocks TRNG.

Table 2 TRNG chaotic map equations optimized for PSoC 5-SC blocks.

1st step	$x'(n)$	$= \begin{cases} V_{ref} - 2x(n) & \text{for } x(n) > 0 \\ -V_{ref} - 2x(n) & \text{for } x(n) < 0 \end{cases}$
2nd step	$x(n+1)$	$= \begin{cases} x'(n) - V_{ref} & \text{for } x'(n) > 0 \\ x'(n) + V_{ref} & \text{for } x'(n) < 0 \end{cases}$
3rd step	$b(n+1)$	$= \begin{cases} \log 1 & \text{for } x(n+1) > 0 \\ \log 0 & \text{for } x(n+1) < 0 \end{cases}$

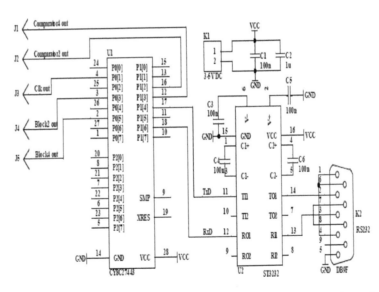

Figure 6 Schematic diagram of the TRNG testing hardware. Outputs $J1, J2,\ldots, J5$ are for testing.

It can be shown that the switched-capacitor structure in Fig.5 processes two independent bit streams, $b_1(n)$ and $b_2(n)$ in a pipeline processing structure analogous to 4-SC TRNG. This is a direct consequence of 2-phase clocks- Φ_1, Φ_2 used for SC. The bit streams $b_1(n)$ and $b_2(n)$ are read out sequentially from this structure and they are overlapped according to the equation (5).

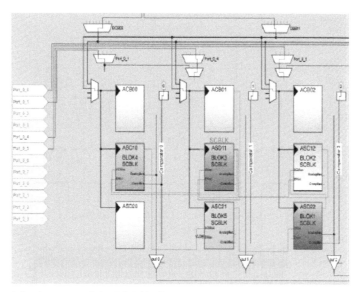

Figure 7 User module placement of the 5-SC blocks TRNG implemented in PSoC reconfigurable hardware.

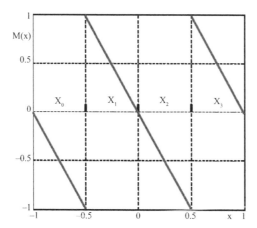

Figure 8 The modified piece-wise chaotic map of 5-SC TRNG implementation.

Experimental TRNG Hardware

The experimental TRNG bit streams were acquired from a PSoC based TRNG implementation depicted in Fig.6 [14]. In the Fig. 9, particular testing signals acquired from oscilloscope are shown, as well. These signals are connected to the output pins just for testing purposes. During normal TRNG operations their outputs are disabled. Presented results have been obtained using 28-pin CY8C27443 based PSoC module used in PSoC Cypress contest [14]. All tested bit streams analyzed in the next section were generated with 60 kbit/s.

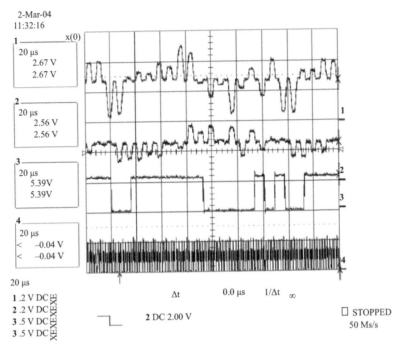

Figure 9 Waveforms 1, 2, 3, and 4 represent $x'(n)$ (main analog output), $x(n+1)$ (sub analog output), $b(n)$ (main compare output), and SC clock (clk source), respectively.

Testing of Generated TRNG Data

Any practical TRNG implementation behaves as an information source with a memory and generates nonequiprobable bits. This is caused by circuit nonidealities (e.g., SC circuit offsets, gain errors, temperature and supply voltage variations, etc.) and generated bits exhibit a certain redundancy. In the following section we will use some well-defined basic statistical tests [1], to measure the deviation of proposed TRNG from ideal RNG.

Correlation Tests

When two random variables b_1 and b_2 are statistically independent, their correlation is zero (note that the opposite statement is not generally valid). Correlations are always between -1 and 1 and are defined as

$$\text{corr}(b_1, b_2) = \frac{E\left[(b_1 - E[b_1])(b_2 - E[b_2])\right]}{\sqrt{\text{var}(b_1)\,\text{var}(b_2)}} \tag{6}$$

$E[b]$ denotes the expected value or mean value of b, that is, the average value of a large number of repeated trials. In the case of random bits, $E[b] = \Pr(b=1)$ where Pr denotes probability. The expression

$$\operatorname{var}(b) = E[(b - E[b])^2]$$ (7)

denotes the variance of b.

From previous analysis we know that the proposed TRNG generates two interleaved data streams, $b_1(n)$ and $b_2(n)$ that should be completely independent (and hence uncorrelated). Cross correlation values of 1,000,000 bit long sequences $b_1(n)$ and $b_2(n)$ are shown in Fig.10. They are within expected random fluctuations

$$\approx \pm 3/\sqrt{1,000,000} = \pm 0.003$$ (8)

expected for ideal RNG [1] and 1,000,000 bit records. This result confirms that deviation from independency of b_1 and b_2 cannot be detected for 1,000,000 bit records.

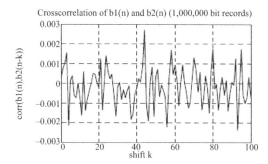

Figure10 Crosscorrelation of 1,000,000 bit records $b_1(n)$ and $b_2(n)$ for different shift values k.

Autocorrelation values for sequence $b_1(n)$ are shown in Fig.11 (the values for $b_2(n)$ have the same character and are not included due to space limitation). It is clearly visible that autocorrelation values for small values ($k = 1, 2, \ldots 7$) are higher than random fluctuations given by limit (7) for 1,000,000 bit records. These deviations are caused, e.g., by the circuit saturation, V_{ref} asymmetry and a gain error of SC circuits as it was shown in [17]. Although these deviations are clearly visible, they are relatively small and can be further decreased by suitable additional post-processing.

Postprocessing of raw TRNG data

Redundancy in an information source can be caused by two sources, namely difference in the probabilities of the two binary symbols, and memory of an information source. The simplest redundancy reduction technique that affects both sources of randomness is XOR correction [19]. In XOR correction, non overlapped bits from the original binary sequence are grouped in a block of p bits and summed up modulo 2 to produce one output bit per block. Example of XOR corrector for $p = 2$ is shown in Fig. 12. It must be stressed that XOR correction can improve statistical properties of input bitstream only if the input bits are statistically independent. It was shown in [19] that statistical properties of two weakly correlated and slightly biased sequences can be improved, as well.

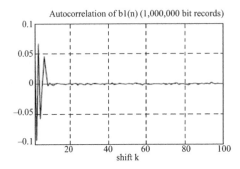

Figure 11 Autocorrelation of 1,000,000 bit record $b_1(n)$ for different shift values k.

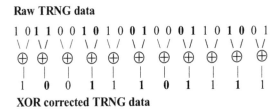

Figure 12 Example of XOR correction of TRNG output for $p = 2$ that is used for reduction of the output bitstream bias.

XOR corrector can be easily implemented in proposed TRNG as there exist two independent bitstreams $b_1(n)$ and $b_2(n)$. Each of these bitstream is only slightly autocorrelated, as is shown in Fig .11. Bias of $b_1(n)$ and $b_2(n)$ is also very small and XOR corrector with $p = 2$ improves the statistical properties of corrected TRNG bitstream. Fig. 13 shows autocorrelation of XOR corrected sequence. It is clearly visible that autocorrelation properties of XOR corrected sequence are significantly better and consistent with results [19], [17].

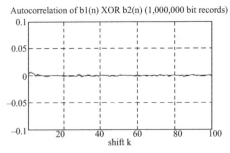

Figure 13 Autocorrelation of XOR decimated 1,000,000 bit record $b_1(n)$ XOR $b_2(n)$ for different shift values k.

FIPS Statistical Tests

Testing of the quality of random data is a very important issue, especially in cryptography. There are several ready to use test packages and recommendations. We have chosen FIPS140-2 tests that analyze 20,000 bit records and define thresholds to assess TRNG randomness. Note that FIPS140-2 tests are really only a very basic statistical tests [1] and they cannot reveal all possible TRNG deviations. If more deep analysis is required it is possible to use, e.g., the NIST statistical test suite [17] that can analyze much longer bit records. Longer records are necessary for detection of very small statistical deviations. FIPS 140-2 consists of standard and easily understood statistical tests (if one of the tests fails, the generator fails the test) [18]:

Monobit test

The number of 1's (T_1) in the tested bit stream should satisfy $9726 < T_1 < 10274$.

Poker test

Divide the 20,000 bit stream into 5,000 contiguous 4 bit segments. Count and store the number of occurrences of the 16 possible 4 bit values. Denote $f(i)$ as the number of each 4-bit value where $0 \le i \le 15$. Evaluate the following:

$$T_2 = \frac{16}{5000}\left(\sum_{i=0}^{15} f^2(i)\right) - 5000 \tag{9}$$

The test is passed if $2.16 < T_2 < 46.17$.

Run test

A run is defined as a maximal sequence of consecutive bits of either all ones or all zeros that is part of the 20,000 bit sample stream. The incidences of runs (for both consecutive zeros G_i and consecutive ones B_i) of all lengths (≥ 1) in the sample stream should be counted and stored.

The test is passed if the runs that occur (of lengths 1 through 6) are each within the corresponding interval specified in [18]. This must hold for both the zeros and ones (i.e., all 12 counts must lie in specified interval). For the purposes of this test, runs greater than 6 are considered to be of length 6.

Long runs test

A long run is defined to be a run of length 26 or more (of either zeros or ones). On the sample of 20,000 bits, the test is passed if there are no long runs. We had split 1,000,000 bit XOR corrected TRNG records (for 4-SC TRNG@5V and 5-SC TGNG@5V) into 100 non- overlapped 20,000 bit records and tested by all FIPS 140-2 tests. All XOR corrected records passed it without problems for both TRNG structures. Typical outputs of FIPS tests for one 20,000 bit record are given in Table 3. The results of FIPS140-2 tests for 5-SC and 4-SC TRNGs performed for different supply voltages V_{CC} are shown in Table 4 and Table 5.

Table 3 Results of FIPS tests for one 20,000 bit record (B_i , G_i is the number of consecutive zeros and ones, respectively, of length $i = 1, 2, 3, 4, 5, 6$ in the tested bitstream).

Monobit Test	Poker Test	Runs Test		Long Run Test
		B_1=2587	G_1=2569	
		B_2=1286	G_2=1183	
T_1=10,107	T_2=22.1312	B_3=588	G_3=662	No long run
		B_4=290	G_4=311	
		B_5=155	G_5=177	
		B_{6+}=147	G_{6+}=151	
Passed	Passed	Passed		Passed

Table 4 FIPS140-2 test results for 5-SC TRNG (F-FAIL, P-PASS, *-tested hardware is not working for this supply voltage).

Voltage Tests	3.0* [V]	3.1 [V]	3.2 [V]	3.5 [V]	4.0 [V]	4.5 [V]	5.0 [V]	5.5 [V]
Monobit	F	P	P	P	P	P	P	P
Poker	F	P	P	P	P	P	P	P
Run	F	P	P	P	P	P	P	P
Log runs	F	P	P	P	P	P	P	P

Table 5 FIPS140 test results for 4-SC TRNG (F-FAIL, P-PASS, *-tested hardware is not working for this supply voltage).

Voltage Tests	3.0* [V]	3.5 [V]	4.0 [V]	4.5 [V]	4.55 [V]	4.6 [V]	5.0 [V]	5.5 [V]
Monobit	F	P	F	F	P	P	P	P
Poker	F	F	F	F	P	P	P	P
Run	F	F	F	F	P	P	P	P
Log runs	F	P	P	P	P	P	P	P

Conclusions

In this paper we have described and evaluated new robust chaos-based TRNG embedded in a switched-capacitor reconfigurable hardware. The circuit topology was optimized in order to extend the range of supply voltages. Experimental results confirmed that new 5-SC

TRNG works under significantly larger supply voltage range than previously proposed 4-SC TRNG. The proposed TRNG provides FIPS compliant quality of random data at up to 60 Kbit/s data rates.

Acknowledgements

This work has been done in the frame of the Slovak scientific projects VEGA 1/4054/07, VEGA 1/4088/07 of Slovak Ministry of Education and under COST 297 grant. This is an extended and improved version of the paper published in the Proceedings of 17th International Conference Radioelektronika 2007, April 24- 25, 2007, Brno, Czech Republic.

References

1. MENEZES, J.A., OORSCHOT, P.C., VANSTONE, S. A. *Handbook of Applied Cryptography.* New York: CRC Press, 1997.
2. EASTLAKE, D., CROCKER, S., SCHILLER, J. Randomness recommendations for security. *Request for Comments 1750*, December 1994, www.ietf.org/rfc/rfc1750.txt.
3. PETRIE, C.S., CONNELLY, J.A. A noise-based IC random number generator for applications in cryptography. *IEEE Trans. Circuits and Systems I*, 2000, vol. 47, no. 5, pp. 615–621.
4. BUCCI, M., et al. A high-speed oscillator-based truly random number source for cryptographic applications on a smart card IC. *IEEE Transactions on Computers*, 2003, vol.52, no.4, pp.403–409.
5. EPSTEIN, M., HARS, L., KRASINSKI, R., ROSNER, M., ZHENG, H. Design and implementation of a True Random Number Generator based on digital circuit artifacts. In *C.D. Walter, C. K. Koc, Ch. Paar(Eds.): CHES 2003, LNCS 2779*, Springer, Berlin, 2003, pp.152–165.
6. FISCHER, V., DRUTAROVSKY, M. True Random Number Generator embedded in reconfigurable hardware. In *B.S. Kaliski, Jr., C.K. Koc, C. Paar (Eds.): Cryptographic Hardware and Embedded Systems, 4th International Workshop, CHES 2002. Redwood Shores (California, USA), August 13–15, 2003, LNCS 2523*, Springer, Berlin, 2003, pp. 415–430.
7. DRUTAROVSKY, M., FISCHER, V., SIMKA, M., CELLE, F. A Simple PLL-based True Random Number Generator for embedded digital systems. *Computing and Informatics*, 2004, vol. 23, no.5–6, pp. 501–516.
8. Open Random Bit Generator, available at: http://mywebpages.comcast.net/orb/index.html.
9. KENNEDY, M.P., ROVATTI, R., SETTI, G. (Eds.) *Chaotic Electronics in Telecommunications.* Boca Raton: CRC International Press, 2000.
10. CALLEGARI, S., ROVATTI, R., SETTI, G. Embeddable ADC Based True Random Number Generator for cryptographic applications exploiting nonlinear signal processing and chaos. *IEEE Trans. on Signal Processing*, 2005, vol. 53, no. 2, pp. 793–805.
11. BERNSTEIN, G. M., LIEBERMAN, M. A. Secure random number generation using chaotic circuits. *IEEE Transactions on Circuits and Systems*, 1990, vol.37, no.9, pp.1157–164.
12. STOJANOVSKI, T., PIHL, J., KONCAREV, L. Chaos-based Random Number Generators-Part II: Practical realization. *IEEE Transactions on Circuit and Systems-I: Fundamental Theory and Applications*, 2001, vol. 48, no.3, pp. 382–385.
13. CALLEGARI, S., ROVATTI, R., SETTI, G. First direct implementation of a true random source on programmable hardware, *Int. J. Circ. Theor. Appl.*, 2005; 33:1–16.
14. DRUTAROVSKY, M., BACA, M., GALAJDA, P. Chaos Based True Random Number Generator, *design entry to the International PSoC Design Contest, Cypress MicroSystems Inc.*, February 2004.
15. PSoC Mixed Signal Array Final Data Sheet, Document No. 38– 12012 Rev.C, August 28, 2003, pp. 1–332, available at: http://www.cypressmicro.com.

16. ESS, D. V. Understanding switched capacitor analog blocks. *Cypress Microsystems Application Note AN2041*, 2002, pp. 1–16.

17. DRUTAROVSKY, M., GALAJDA, P. Chaos-based True Random Number Generator embedded in a reconfigurable hardware. *Journal of Electrical Engineering*, 2006, vol. 57, no.4, pp. 218–225.

18. NIST FIPS PUB 140-2: Security Requirements for Cryptographic Modules (2001), available at: csrc. nist.gov/publications/fips/fips140- 2/fips1402.pdf.

19. DAVIES, R.B. Exclusive OR (XOR) and hardware random number generators. February 28, 2002, pp.1–11, available at: http://www.robertnz.net/pdf/xor2.pdf.

Conservative Chaos Generators with CCII+ Based on Mathematical Model of Nonlinear Oscillator[†]

Jiří Petržela[1,a] and *Josef Slezák*[1]

Dept. of Radio Electronics, Brno University of Technology, Purkyňova 118, 612 00 Brno, Czech Republic.
[a] Email: petrzelj@feec.vutbr.cz.

ABSTRACT

In this detailed paper, several novel oscillator's configurations which consist only of five positive second generation current conveyors (CCII+) are presented and experimentally verified. Each network is able to generate the conservative chaotic attractors with the certain degree of the structural stability. It represents a class of the autonomous deterministic dynamical systems with two-segment piecewise linear (PWL) vector fields suitable also for the theoretical analysis. Route to chaos can be traced and observed by a simple change of the external dc voltage. Advantages and other possible improvements are briefly discussed in the text.

Keywords: Analog oscillator, attractor, conservative dynamics, chaos, state space.

Introduction

A huge pile of the journal articles solving problems with chaos and other types of complex behavior have been published from its early discovery about fourty years ago. The reliable parts of these publications are about methods how to create chaotic waveform by an oscillator. With the latest progress in the analog integrated circuit design the new electronic devices are developed and thus used inside various applications including chaos generators. It is also the case of CCII+ which is modern functional block with three ports capable to work at high frequencies. For any circuit structure and purpose, having certain knowledge about the basic properties of the generated waveforms in time and frequency domain is essential. Understanding a complex dynamics is useful also from the theoretical standpoint. Chaos can be considered as the universal phenomenon due to the normalization

---

[†]Reused with permission from: Jiří Petržela, Josef Slezák, Conservative chaos generators with CCII+ based on mathematical model of nonlinear oscillator, *Radioengineering*, 17(3), 2008.

of system parameters and arbitrary interpretation of state variables. That is the reason why chaos has been experimentally confirmed and reported in many distinct scientific fields, for example in power electronics, digital circuits, ecology, chemical reactions, population growth, optics, etc.

The overall informations about conservative dynamics can be found in [1]. Several conservative systems taken over from the technical practice are also provided in this publication. The very simple example of driven dynamical system is mentioned in [2] where PWL type nonlinearity and periodic driving force is assumed. The integrator based conservative chaotic oscillator with two modifications is presented and verified in [3]. Both presented systems model dynamics of the Nosé-Hoover's thermostat. This paper is organized as follows. Chapter two gives the necessary mathematical background together with the numerical integration of the studied systems. Third chapter is focused on the individual circuits. They are synthesized in order to obtain different structures with qualitatively equivalent conservative dynamics. The next chapter is aimed on the experimental verification of final oscillators.

Numerical Analysis

Let us consider a mathematical model of the nonlinear oscillator with two-segment PWL vector field

$$a_3\ddot{x} + a_2\ddot{x} + a_1\dot{x} + a_0 x = b_0 \max(x - \wp, 0) + \Im,$$

(1)

where dots denote the time derivatives, $\wp$ is the breakpoint and $\Im$ represents offset of the nonlinear function. Note that we can assume $a_3 = 1$ without a loss of generality. Straightforward analysis of this differential equation following the rules of linear algebra leads to the three possibilities for the number of fixed points. There are no equilibria if $\Im > \wp$ and one fixed point located on the boundary plane occurs when $\Im = \wp$. We get two equilibria per region only in the particular case $\Im < \wp$ and they are in positions $\mathbf{X}_0 = [\Im/a_0 \quad 0 \quad 0]^T$ and $\mathbf{X}_0 = [(b_0\wp - \Im)/(b_0 - a_0) \quad 0 \quad 0]^T$ respectively. Only the latter case is important from the viewpoint of the conservative chaos generation. It is not hard to obtain the characteristic equation for the left segment of the vector field in the form

$$x < \wp \quad \Rightarrow \lambda^3 + a_2\lambda^2 + a_1\lambda + a_0 = 0,$$

(2)

and similarly for the right segment

$$x > \wp \quad \Rightarrow \lambda^3 + a_2\lambda^2 + a_1\lambda + a_0 - b_0 = 0.$$

(3)

It is well known that the eigenvalues are roots of these polynomials and determine the local behavior of the system in the neighborhood of the fixed points. The eigenvalues can be computed symbolically by means of the Cardan formulas. This kind of approach can give us an idea when eigenvalues are complex conjugated forming a composition of the eigenplane and eigenvector or if the geometry is more likely a combination of three eigenvectors. Also the question about the stability indexes can be easily answered. In spite of this, the existence of chaos will remain uncovered. Due to this, it is time to leave the symbols and work with numbers. The basic requirement for no dissipation tell us that $a_2=0$. Other values are given in the publication [4]

$$a_1 = 2.7, a_0 = 1, b_0 = 2, \tag{4}$$

or can be obtained using the stochastic optimization way described in [5]:

$$a_1 = 8.1, \; a_0 = 5.5, \; b_0 = 8.7 \tag{5}$$

In both cases the parameters of the nonlinear function were the same, in detail $\Im=-1$ and $\wp=0$. The set of parameters (4) leads to the following real and a pair of complex conjugated eigenvalues

$$x < \wp \;\Rightarrow\; \lambda_{1,2} = 0.177 \pm 1.672j \;\; \lambda_3 = -0.354$$
$$x > \wp \;\Rightarrow\; \lambda_{1,2} = -0.177 \pm 1.672j \;\; \lambda_3 = 0.354 \tag{6}$$

The same geometry of the vector field hold for the second case (5) as expected, i.e.

$$x < \wp \;\Rightarrow\; \lambda_{1,2} = 0.323 \pm 2.9j \;\; \lambda_3 = -0.646$$
$$x > \wp \;\Rightarrow\; \lambda_{1,2} = -0.194 \pm 2.866j \;\; \lambda_3 = 0.388 \tag{7}$$

The geometry of the vector field is

$$x < \wp \;\Rightarrow\; \mathfrak{R}_u^2 \oplus \mathfrak{R}_s^1 \; x > \wp \;\Rightarrow\; \mathfrak{R}_u^1 \oplus \mathfrak{R}_s^2. \tag{8}$$

The numerical integration using the fourth-order Runge-Kutta method with final time $t_{max}=400$, time step $t_{step}=0.04$ and zero initial condition is shown in Fig. 1 for the first set of parameters (4) with the state variables ranges $x \in (-1,1)$, $y \in (-2,2)$, $z \in (-3,3)$. Analogically, Fig. 2 corresponds to the attractor when (5) is set up. In this case, the axes are fixed at $x \in (-3,3)$, $y \in (-5,5)$, $z \in (-10,10)$. Note that such ranges of the individual state variables are narrow enough for the practical implementation. This is indeed important because of current and/or voltage saturation of the active devices. The numerical analysis involving the computation of the spectrum of Lyapunov exponents as well as the dimensions of the state space attractors reveals that the chaotic regions are significantly surrounded by the regions with unbounded solution. The system parameters $\Im$ and $\wp$ do not affect the eigenvalues so it can be hardly considered as a natural bifurcation parameters. Also a thin basin of attraction

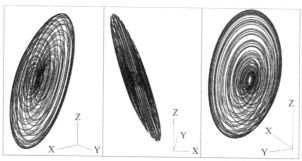

Figure 1 3D state space trajectories for first set of parameters.

can cause serious problems, especially in case of experimental verification. Chaos is often destructed simply by switching on/off the power supply.

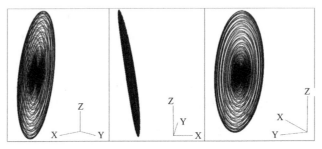

Figure 2 3D state space trajectories for second set of parameters.

Circuit Realization

Before speaking about particular circuit realizations, the universal active element CCII should be introduced by means of the hybrid three-port matrix

$$V_x = V_y, \ I_y = 0, \ I_z = \pm I_x. \tag{9}$$

The orientations of all currents in this paper follow standard conventions, i.e., inwards the three-port. Several integrated circuits can be used as CCII where the sign in equalities (9) determines the type of the conveyor. The negative variant can be constructed by means of EL2082C while positive CCII is commercially available under the notation AD844.

The best way how to create an oscillator for modeling the dynamical behavior of the differential equation (1) is a parallel connection of some third-order admittance function with the shifted approximation of the AV characteristic of the ideal diode. Each active block (CCII+) in the oscillator is supposed to be in the linear regime of operation for all possible values of node voltages or branch currents. This proposition has the consequence while choosing impedance normalization factor.

Nonlinear Function Synthesis

The constant current source (CCS) includes one active block X5, resistor R_x and external dc voltage V_x providing the output current $L_{out} = -I_x = -V_x / R_x$. It is obvious that the direction of the output current can be reversed and the value of resistor R_x can be computed such that voltage V_x can be taken as a positive or negative supply voltage. CCS can be connected in parallel with a diode-resistor composite. Using such conception, only a single CCII+ is needed. The nonlinear device (diode) is strictly passive and can be replaced with other element with similar AV curve. Such modification will not make the qualitative changes in the global behavior of the circuit. The final circuit can be simpler, taking advantage of the active admittance network with reversed signs of all coefficients in (11). It turns out after applying the reasonable scaling factors that the real values of the circuit components (including R_y) make an effect of the forward-biased diode differential resistance negligible. Unfortunately, numerical analysis as well as the laboratory experiments reveal the problems with other non-ideal diode property, breakpoint. This voltage is typically located around V_p=0.5V (1N4007, BAT42). To overcome this obstacle, another CCII+ marked as X4 has been used as a voltage controlled current switch [6]. In this case the third-order admittance is passive.

Individual Configurations

Each final circuit is composed of a few basic building blocks which work simultaneously but its function can be understood separately. The main components are shown in Fig. 3 and ideally behave like a lossy grounded frequency dependent negative resistors (FDNR) with the possibility of the current inversion (this holds only for the first circuit). The associated admittance functions are

$$\hat{Y}(s)= I_{in}(s)/V_{in}(s)= \pm\left[C_{i1}C_{i2}R_i s^2 + (C_{i1}+C_{i2})s\right],\tag{10}$$

where $V_{in}(s)$ and $I_{in}(s)$ are Laplace transform of the input voltage and input current respectively.

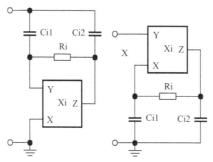

Figure 3 Two reciprocal realizations of the FDNR using CCII+.

The first type of the final oscillator arises from (10) where one linear capacitor is replaced by another FDNR. Evidently there are a lot of possible combinations as shown in Fig. 5. Let assume that switching on/off the resistor R_y must change the sign of the absolute coefficient of the characteristic polynomials as well as the real eigenvalue. For input voltage higher than zero the connection X4 and D behaves like a negative impedance converter. Due to this G_y=-1/R_y and the admittance $Y(s)$ must consume the energy from the rest of the circuit. Taking into account the goal restriction a_2=0 we immediately get

$$Y(s)= \tilde{Y}(s)+ \hat{Y}(s)\approx \tilde{a}_3 s^3 + \tilde{a}_2 s^2 + \tilde{a}_1 s + \tilde{a}_0 - \hat{a}_2 s^2 - \hat{a}_1 s = a_3 s^3 + a_1 s + a_0.\tag{11}$$

It means that another energy providing FDNR should be connected in parallel.

The second type of a third-order admittance function originated from the modified connection of a two CCII+ used for the filter design and with the general impedances of all branches. The final equation for the input admittance is then simplified by substituing resistors, capacitors or even inductors and its serial or parallel interconnections as shown in Fig. 6.

The third type of the oscillator is based on the fundamental admittance cell shown in Fig. 4. Input admittance of such a cell is

$$\hat{Y}(s) = -\left[Y_1(s) + Y_3(s) + Y_1(s)Y_3(s)/Y_2(s) \right]. \tag{12}$$

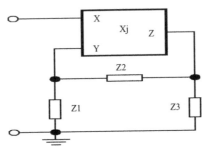

Figure 4 CCII+ based elemental admittance cell.

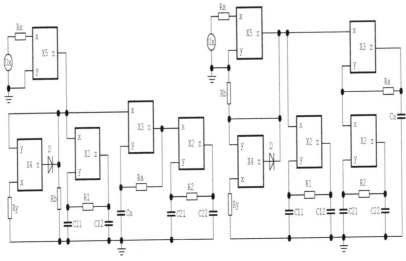

Figure 5 Type I conservative chaotic oscillators.

As this equation suggests, any of the individual admittance can be frequency dependent as well as any of the used general impedance can be substitued by more complicated network leading to the desired linear part of the vector field. For example, the final structures illustrated in Fig. 7 can play this role. Only a single FDNR is used comparing to the first type of the oscillator.

Note that each circuit works in the hybrid voltage-current mode. State variables of the original mathematical model are measurable via their linear combinations.

In accordance to common form (11) the coefficients of the admittance function for the circuits in Fig. 5 are

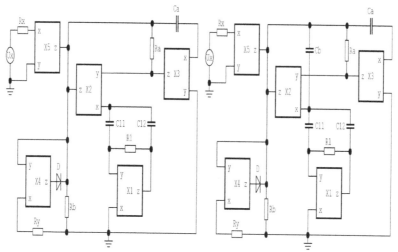

Figure 6 Type II conservative chaotic oscillators.

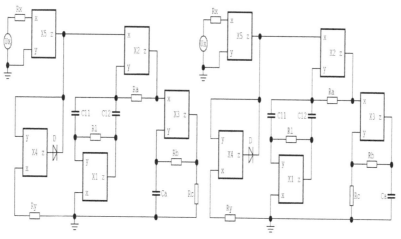

Figure 7 Type III conservative chaotic oscillators.

$$a_3 = c_{21}c_{22}c_a r_2 r_a \, , a_2 = c_a r_a (c_{21} + c_{22}) + c_{21}c_{22}r_2 - c_{11}c_{12}r_1 \, , a_1 = c_{21} + c_{22} - c_a - c_{11} - c_{22}, b_0 = 1/r_y,$$
$$a_0 = 1/r_b \, .$$

(13)

In these terms the normalized values of capacitors and resistors were used. Searching for concrete numbers is the problem from the area solving the system of the nonlinear algebraic equations. Such a task can cause the serious problems with finding the real positive values only. This is another advantage of the proposed structures of oscillators. Although the relative large amount of the circuit elements seems to be redundant it leads to the good overall results (quick convergence with zero error). It is not hard to learn that the satisfying performace also last for the second type of the circuit.

Namely for the left circuit in Fig. 6 we get

$$a_3 = c_{11}c_{12}c_a r_1 r_a, b_0 = 1/r_y, a_0 = 1/r_b, a_2 = c_a r_a (c_{11} + c_{12}) - c_{11}c_{12}r_1, \quad a_1 = 2c_a - c_{11} - c_{12}, \tag{14}$$

and for the right circuit in Fig. 6 the coefficients are

$$\begin{cases} a_3 = c_{11}c_{12}c_a r_1 r \\ b_0 = 1/r_y \\ a_0 = 1/r_b \\ a_2 = c_a r_a (c_{11} + c_{12} + 2c_b) - c_{11}c_{12}r_1 \\ a_1 = 2c_a - c_{11} - c_{12} \end{cases} \tag{15}$$

Note that slight modification of the circuit (adding a single capacitor) results into a change of the single equation. This feature can ensure convergence in many particular cases. The circuits shown in Fig. 7 differ only in the admittances of two branches. In this case this interchange does not affect the input admittance, which is in the form

$$a_3 = c_{11}c_{12}c_a r_1 r_a (r_b / r_c + 1), a_0 = 1/r_c, b_0 = 1/r_y, \quad a_2 = c_a r_a (c_{11} + c_{12})(r_b / r_c + 1) + c_{11}c_{12}r_1 (r_a / r_c - 1),$$

$$a_1 = (c_{11} + c_{12})(r_a / r_c - 1) + c_a (r_b / r_c + 1). \tag{16}$$

Experimental Results

AV characteristic of the nonlinear two-port can be visualized by a current-sensing resistor $100\,\Omega$ connected in series with a triangular wave generator. The result is shown in Fig. 8, where both input and output signals are centered. The concrete values of circuit components for first type of configuration and (5) are $C_{11}=C_{12}=C_{21}=10\text{nF}, C_{22}=47\text{nF}, C_a=10\text{nF}, R_1=81\text{k}\,\Omega, R_2=15.5\text{k}\,\Omega, R_a=1.4\text{k}\,\Omega, R_b=10\text{k}\,\Omega$ and $R=5\text{k}\,\Omega$. For the same configuration and (6) some values change $C_{22}=100\text{nF}, R_1=177\text{k}\,\Omega, R_2=17\text{k}\,\Omega, R_a=600\,\Omega, R_b=1.8\text{k}\,\Omega$ and $R_y=1.2\text{k}\,\Omega$.

For the second type of oscillator with (5) the list of components is $C_{11}=C_{12}=10\text{nF}, C_a=24\text{nF}, R_1=14\text{k}\,\Omega, R_a=3\text{k}\,\Omega, R_b=10\text{k}\,\Omega$ and $R=5\text{k}\,\Omega$. Analogically, for the set (6) following passive elements change $C_a=50\text{nF}, R_a=1.4\text{k}\,\Omega, R_b=1.8\text{k}\,\Omega$ and $R_y=1.2\text{k}\,\Omega$. Finally, the third type of oscillator with set (5) consists of $C_{11}=6\text{nF}, C_{12}=23\text{nF}, C_a=25\text{nF}, R_1=17.5\text{k}\,\Omega, R_a=800\,\Omega, R_b=11\text{k}\,\Omega, R_c=10\text{k}\,\Omega, R_y=5\text{k}\,\Omega$ and for (6) we get $C_{11}=9\text{nF}, C_{12}=17\text{nF}, C_a=9\text{nF}, R_1=14.7\text{k}\,\Omega, R_a=400\,\Omega, R_b=19\text{k}\,\Omega, R_c=1.8\text{k}\,\Omega, R_y=1.2\text{k}\,\Omega$. Definitely, other sets of values are also possible for the different initial guess (Mathcad optimizer were used).

The digital oscilloscope HP54603B screenshots from a number of laboratory experiments are shown by means of Fig. 8 (AV characteristic of the nonlinear element and chaotic waveforms) and Fig. 9 (plane projections). Also the initial conditions close to zero should be introduced into the circuit before making each particular measurement.

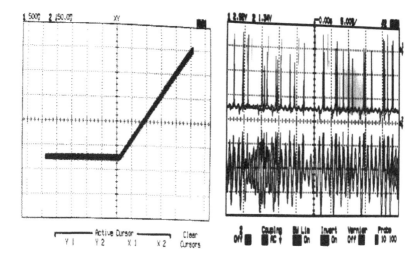

Figure 8 Nonlinear AV characteristic and chaotic waveforms.

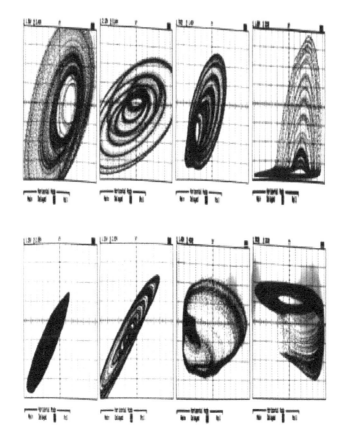

Figure 9 Plane projection of chaotic attractors using XY display.

Conclusion

In this article several novel inductorless structures of the simple oscillators with associated volume conserving dynamics have been suggested and experimentally verified. The good agreement between theory and practice can be considered since the orbits closely related to conservative ones have been found. It is worth nothing that Pspice circuit simulator gives the same results. Moreover, single external dc source can be used to turn the oscillator inside a single equilibrium regime or even into the equilibriumless one. The integrated circuit AD844 provides an extra node which acts as the output voltage follower. These buffered outputs allow us to observe other combinations of state variables without affecting the proper function of the oscillator.

The higher order admittances presented in this paper can be directly used for the generation of the multi-spiral chaotic attractors. On the other hand, mathematical model of the studied systems is not suited for 2D or 3D grid attractors. More details about such oscillators can be found in some recent publications [7]. The symmetry of the vector field and the possibility of mapping absolute value onto the AV characteristic of the diode allow synthesized oscillators to model any shape of the state space attractors associated with two segment PWL vector field. The authors believe that much simpler oscillators with the same dynamics and employing only three CCII± can be discovered. This is the topic for further study.

Acknowledgements

Research described in this paper was financially supported by the research program MSM 0021630513.

References

1. SPROTT, J.C. *Chaos and Time Series Analysis*. Oxford University Press, 2003.
2. GOTTLIEB, H.P.W., SPROTT, J.C., Simplest driven conservative chaotic oscillator. *Physics Letters A*, 2001, vol. 291, p. 385–388.
3. PETRŽELA, J. Wideband signal generator based on chaos. In *Proceedings of the 12th Electrotechnical and Computer Conference ERK 2004*. Portorož (Slovinsko), 2004, p. 61–64.
4. SPROTT, J.C., LINZ, S.J. Algebraically simple chaotic flows. *International Journal of Chaos Theory and Applications*, 2000, vol. 5, no. 2, p. 1–20.
5. PETRŽELA, J., HANUS, S. On the optimization of the specific class of chaotic oscillators. In *Proceedings of the 13th Electrotech-nical and Computer Conference ERK 2005*. Portorož (Slovinsko), 2005, p. 111–114.
6. LIU, S.I., WU, D.S., TSAO, H.W., WU, J., TSAY, J.H. Nonlinear circuit applications with current conveyors. In *IEE Proceedings-G*, 1993, vol. 140, no. 1, p. 1–6.
7. LU, J., CHEN, G. Generating multi-scroll chaotic attractors: theories, methods and applications. *International Journal of Bifurcation and Chaos*, 2006, vol. 16, no. 4, p. 775–858.

Section VI

Models and Applications of Chaos Theory in Human Sciences

1. **Ancient and Current Chaos Theories**   357
   G. Gündüz

2. **Complex Freedom**   376
   D. Pećnjak

3. **An Understanding of Language Development Models—Pidginization from the Perspective of Chaos Theory**   384
   G. Zhao

4. **Chaos and Natural Language Processing**   393
   M. Crisan

5. **Discrete Phase-Locked Loop Systems and Spreadsheets**   405
   S. Abramovich, E. Kudryashova, G.A. Leonov and *S. Sugden*

6. **Modeling Complex Spatial Dynamics of Two-Population Interaction in Urbanization Process**   429
   Y. Chen and *F. Xu*

7. **Characterizing Growth and Form of Fractal Cities with Allometric Scaling Exponents**   449
   Y. Chen

8. **Estimating the Distribution of Dynamic Invariants: Illustrated with an Application to Human Photo-Plethysmographic Time Series**   473
   M. Small

9. **Asymmetry, Symmetry and Beauty**   487
   H. Sabelli, A. Lawandow and *A.R. Kopra*

10. **Nonlinear Dynamics in Psychology**   520
    S.J. Guastello

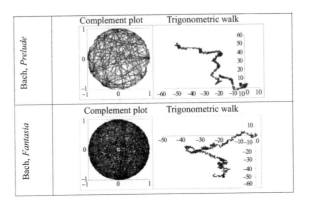

Ancient and Current Chaos Theories[†]

Güngör Gündüz

Department of Chemical Engineering, Middle East Technical University, Ankara, Turkey.
Email: ggunduz@metu.edu.

ABSTRACT

Chaos theories developed in the last three decades have made very important contributions to our understanding of dynamical systems and natural phenomena. The meaning of chaos in the current theories and in the past is somewhat different from each other. In this work, the properties of dynamical systems and the evolution of chaotic systems were discussed in terms of the views of ancient philosophers. The meaning of chaos in Anaximenes' philosophy and its role in the Ancient natural philosophy has been discussed in relation to other natural philosophers such as of Anaximander, Parmenides, Heraclitus, Empedocles, Leucippus (i.e., atomists), and Aristotle. In addition, the fundamental concepts of statistical mechanics and the current chaos theories were discussed in relation to the views in Ancient natural philosophy. The roots of the scientific concepts such as randomness, autocatalysis, nonlinear growth, information, pattern, etc. in the Ancient natural philosophy were investigated.

Keywords: Chaos, randomness, dynamical systems, natural philosophy, Ancient philosophy, information, biological evolution, nonlinear.

Introduction

In the last three decades, the chaos theories have born, grown, matured, and revolutionized our understanding of natural phenomena. Classical physics, statistical physics, electromagnetism, quantum theory, and relativity also had great revolutionary impacts in explaining natural phenomena, but none of these had so rapid influence on the fields other than basic physics. In the past, it usually took some decades for the application of physical laws in applied sciences and engineering after their discovery. However, chaos theories soon found applications in almost all branches of technical, medical, and social fields, and also in arts.

[†]Reused with permission from: Güngör Gündüz, Ancient and Current Chaos Theories, *Interdisciplinary Description of Complex Systems* 4(1), 1–18, 2006.

The term chaos is first seen in the Sumerian epic of Gilgamesh which is the oldest script strongly touching the fundamentals of human psychology and human's understanding of the earthly and heavenly events. In one of the paragraphs of the epic, the Earth God complains to other Gods about humans, and says 'humans have gotten so overcrowded, and they have run into dearth, starvation, and chaos, and they do not respect me; something has to be done'. Gods then decided to send water flood to extinct them. The flood of Noah in religious scripts is based on this story. In terms of chaos current theories, it is true that anything that multiplies can cause overcrowding and thus chaos, then the components becomes pretty much free from the general rules (the rules of the system or of Gods) [1]. It is possible to control the chaos, and to get out from it by external intervention of physical forces, i.e., Gods' intervention in Gilgamesh epic.

The chaos concept has been a fundamental metaphor for both natural and social events, and cosmogony in ancient societies. According to a Mesopotamian cosmogony, the conjunction of male Apsu which represents freshwater and of female Tiamat which represents seawater gives birth to Goddess Mummu who represents the chaotic fog and clouds. Mummu then gave birth to primitive ocean and water. First Gods came out from the primitive water, and one of them then became a creative God and created earth, stars, and all living things [2]. The Sumerians used to believe that the earth and stars were floating soil on water [3]. Similarly, the Egyptians also used to believe in that the universe was essentially water, and all stars were surrounded by water. There is also a phrase in Cor'an stating that the essence of all things is water.

The Egyptians attributed a kind of philosophical meaning to the word chaos and thought that it is the primordial state before genesis, and it is the medium for the coexistence of form and structure. It is in fact a kind of reservoir in which all kinds of field forces and forms dissolve in infinitesimal time. In Genesis, it is understood as a coarse but homogeneous structure with feasibility [4]. In Hesiod's Theogony, chaos meant an empty space or matter not yet formed. It was believed that order (i.e., cosmos) came out from chaos, which was also the beginning of time.

Rational Thinking

The predominating elements of Middle East cosmology before rational thinking were: (i) unlimited and infinity, (ii) the basic element of all things, and (iii) chaos and order. In this respect, water was believed to represent all the properties described as such; and the first natural philosopher, Thales of Miletus, considered water as the element of everything [5]. He also thought that the earth rests on water as in the Sumerian cosmogony. Thales considered water to be an element, and also as something that all other things can be reduced into [6]. He used to visualize water together with a force that revives or moves it. In other words, the mover and the moving are not separated and they coexist in the same object [7].

Water as the sole material principle, has been objected by Anaximander, the second philosopher of Miletus. He introduced the concept of 'apeiron, the unlimited or infinity' and proposed that it is both a principle (arche) and an element. Apeiron is a kind of reservoir where all varieties are not yet differentiated; they are entangled in the form of a composite state. Apeiron is like a tank of all qualities, and it serves to conserve all beings.

He rejected the idea of a single element (like water) to be the founding principle of all existing. He claimed that the single element wouldn't allow the appearance of others. According to him, any 'elemental stuff' can change into one or more of the other elemental stuffs, and every 'coming into being' is due to the change of a pre-existent. He, in fact, is the founder of a dynamic universe model by claiming the continuous generation of new things.

Anaximander thought that the contradictions like earth (dry), water (wet), fire (hot), and air (cold) pre-existed before apeiron, and they were at fight (i.e., competition) with each other. He thus pointed the importance of four-stuff long before Empedocles.

Anaximander also thought that the need for earth's stability like water bears in some problems, and one needs something stronger than an analogy and deeper than a cushion of water [5]. He in fact thought the earth to be at rest at mid-space. He also claimed a kind of primitive biological evolution theory.

The third philosopher of Miletus, Anaximenes, did not respect much the uncertainty in the apeiron concept, and he returned to more concrete basic elements. He was also disturbed with the idea that the earth suspends in space without any support. He thought that something light and mobile, and thus readily available everywhere should surrender it. Anaximenes considered air to be the essential material of everything. The Anaximander's principle of unlimited was indeterminate and metaphysical, and Anaximenes considered air to be the 'unlimited'. The word 'gas' was derived from the word 'chaos' that meant 'air'. The Anaximenes' chaos does not only represent a material, but it also represents a principle.

The carrying medium for all stuff in the universe was later changed into something celestial and indestructible by Aristotle, and he called it 'aither', which may correspond to his 'primeval matter' [4]. Aither was assumed to be the carrying medium needed for the propagation of electromagnetic waves by Maxwell in 19th century. Einstein discarded it by his special theory of relativity. However, the developments in quantum electromagnetism and in quantum field theories introduce a kind of space which is not absolute empty but owns something, and it interacts with material medium, i.e., particles. In fact, Dirac proposed a vacuum full with oscillators.

In Anaximenes' cosmology, some sort of motion produces variations in the density of 'air', and hence, the basic stuffs of the universe are generated. This view has some parallelism with the production of mass from quantum fluctuations. In quantum electrodynamics vacuum fluctuations result in pair production of virtual electrons and positrons, which immediately annihilate; however, the interaction of a real electron with the virtual particles (i.e., vacuum fluctuations) increases its mass at small but measurable quantity. Anaximenes' primordial state and our quantum vacuum are both full, they both have some kind of contents, and they are not empty.

According to Anaximenes, the motion (i.e., force) that causes the changes in the density of air is also in air. He also thinks as Thales that the moving and the mover coexist in the same object. This monism realizes the universe as an animate object of which deriving force is inside the body. For nonliving objects the force needed for motion is supplied from outside. In fact in most physical theories, classical and quantum mechanics, electromagnetism, and relativity, an external force is needed to move the object. However, in general relativity and in Bohm's quantum world the action and the potential coexist in a unified form as in Anaximenes's state of chaos. In Aristotle's philosophy a change happens in the chronological order 'from potentiality to actuality'. In Bohm's world the potentiality and actuality form a composite state and the things interact through 'implicate order'. The relationship between the potential and the actual is the relationship between implicate and explicate order. The general relativity equations inherit a dynamic property that implies a dynamic and expanding universe. Einstein had introduced a cosmological constant for a static universe, but, he then removed it after Hubble's discovery of expanding universe.

As mentioned above, any supporting medium for the universe needs itself also another supporting medium according to Anaximander. Anaximenes overcame this difficulty by air (or chaos) hypothesis; it is so light that it does not need anything else and it is self-sufficient. That is, 'the cause and the source of chaos is itself'. This philosophy has made important influence on the proceeding philosophers, especially on Aristotle. His aither concept based on the ultra light property of Anaximenes' air, and he removed its materialistic properties and attributed metaphysical properties to it. Aristotle attributed 'to be self sufficient and to be the cause of itself' only to the 'first mover'. Science and religion both used this postulate; the former said that the universe is self-sufficient while the latter said God is self-sufficient. This postulate could not be yet changed or improved further throughout the history of philosophy.

Anaximenes considers the chaos (or gas) as the most disordered state. We do not know if disorder (i.e., randomness) prevails in quantum vacuum, but it prevails in the world of chemical atoms. The exchange of properties takes place only through the principles of conservation of momentum and energy. Boltzmann called this random exchange dynamics as 'molecular chaos', and the word chaos has been a scientific term since then [8].

In a system where there are no attractions between the components (i.e., molecules) the changes could be described only by kinetic terms. In fact, Anaximenes explained the changes in nature in terms of 'hot and cold', and 'densification and rarefaction' mechanisms. In modern sciences especially in statistical physics, 'hot and cold' refers to the change of kinetic energy of molecules. In Maxwellian distribution of speeds, the shift of the speed distribution to the right or to left makes the system hotter or colder. It is not clear if Anaximenes supposed that the change in density is sufficient to produce all the existing stuff. In fact, what is meant by density is somehow blurred. It may mean the number density or the mass density per unit volume. However both can be correlated with each other. If the mass density is of nuclear origin then the mass density mostly changes with the number of protons of which changes go by number. The change of the number of neutrons and the binding mass are also functions of the number of protons in the nucleus. Therefore 'densification and rarefaction' essentially corresponds to changes in number, which in turn changes the chemical potential. It seems that in Anaximenes' view 'densification and rarefaction' corresponds to change in number. In this respect, 'the continuous creation and destruction' principle of Aristotle differs from the 'densification and rarefaction' principle of Anaximenes.

Although the change in nature through chemical changes is implicitly seen in Anaximenes, it is well established by Aristotle by his 'continuous generation and destruction' principle. His thoughts are based on the philosophy of atomists though Aristotle stood against them.

The isomeric changes in organic molecules keep the number constant but change the chemical potential, which is associated with the shape or 'form' besides the atomic mass content. In modern sciences, 'densification and rarefaction' can imply both the kinetic changes due translational energies and also the structural changes due to 'vibration & rotation & electronic excitation' modes of the atomic motion. These modes of motion are bound to the shape of the object, but it was too early for the philosophers of the Miletus to talk about the evolution of a preferred shape in natural events. As believed, they probably could explain the wood stuff but not trees and plants. In this sense, it is not clear how evolutionary processes take place in Anaximenes' universe. Neither quantum fluctuations nor the axioms of most physical theories say anything about the evolutionary processes. The evolution is inherently associated with irreversibility, and it can be tackled with the entropy concept of thermodynamics.

Parmenides pinned down an important corner stone in ancient natural philosophy. His philosophy emphasizes on 'being', and he rejects the dynamical 'change'. According to him 'all' in the universe is unified and the change is an illusion. Parmenides' philosophy can be summarized as,

i) Nothing perishes; nothing comes from nothing.
ii) 'Change' is the loss of one quality and the gain of another.
iii) 'Quality' and 'object' are indistinguishable.

These assertions make to explain the 'change' in nature impossible. Both 'ii' and 'iii' fall in contradiction with 'i' [6]. The first assertion has stood as the keystone of all natural philosophies throughout the history. The attempt to change the second assertion led Anaxagoras to develop his 'information' (or 'sperma') concept, and led Aristotle to develop his 'potential' concept, which was actually first introduced by Platon.

The attempt to change the third argument yielded two new developments introduced by Empedocles and by atomists. Empedocles being a member of the Parmenides School emphasized on four-stuff view (i.e., earth (solid), water (liquid), air (gas), and fire (energy)), and claimed that all changes in nature can be interpreted in terms of four-stuff, which, cannot change into other things; and so, their number is fixed. Four-stuff are elements and unlimited. Historically, the four-stuff developed from water of Thales, gas of Anaximenes, earth of Xenophanes and Parmenides, and fire of Heraclitus. All changes occur by the combination of four-stuff at different ratios. In order to account for the continuous combination and dissociation in nature Empedocles introduced 'love' and 'strife'. This was a revolutionary concept in natural philosophy, and today we know in the world of science that attraction (e.g., love) and repulsion (e.g., strife) are the two fundamental classifications of all forces in nature.

The indivisible particle (or 'a-tomos') concept introduced by atomist philosophers was a kind of antithesis to the philosophy of the Parmenides School (i.e., Eleatic School). According to atomists (i.e., Leucippus, Democritus, Epicurus, and Lucretius) the infinite numbers of atoms each with its own identity (i.e., chemical potential) collide and react freely and randomly, and the change of any quality (or property) depends on the types of atoms reacted. In the views of atomists 'atoms are infinite in number each having a different shape, they randomly collide with each other until they find the best fit of shape, and then they combine'.

The principles proposed by atomists essentially form the fundamental bases of the dynamics of chemical atoms in our age. Chemical atoms randomly collide, and exchange momentum and energy. The random collisions may be elastic or reactive, the latter yields molecules. The elastic collisions redistribute the momentum and energy carried by each atom, and a system when kept isolated for sufficiently long time goes to equilibrium where all observable properties remain constant in time. The approach to equilibrium is one-directional, or irreversible. However, this fact bears some problems. The total mass and total energy are both invariable in an isolated system; but all other observable properties are apt to undergo deviations from thermostatic equilibrium values. In other words, they undergo 'fluctuations' around equilibrium values. These fluctuations are usually neglected since the relative frequency of the occurrence (or the probability) of a fluctuation decreases very sharply with its size and duration. Large deviations from equilibrium are exceedingly rare, and small fluctuations occur more frequently. The observable fluctuations manifest themselves only under very favorable conditions, such as the Brownian movement of suspended particles in a liquid, or the opalescence of liquids near the critical point. About a hundred years ago Einstein, Smoluchowski, and other physicists demonstrated that these phenomena can be explained by the idea of fluctuations. So fluctuation is also a physical fact as well as a philosophical concept [9,10].

Randomness

When the collisions are elastic, the speed distribution of molecules is given by the Maxwellian distribution [8]. Randomness has been objected by many thinkers who claim that the universe must work as a perfect clock, and not by a mechanism relying on chance. However, randomness must be interpreted as 'nonpreference'. That is, the objects collide with each other without any preference. This implies that there is no divine preference and intervention in natural processes. In this respect, the Milesian School of philosophy (i.e., Thales, Axaximander, and Anaximenes) did a pioneering work by setting up the most fundamental grounds for scientific thinking. The physics of stochastic processes relies on Anaximenes' nature.

Boltzmann showed that the equilibrium solution of his transport equation yields the Maxwellian distribution. That means, if the restrictions are removed from a system, its components obey the Maxwellian distribution, otherwise they obey a non-Maxwellian distribution. In other words, all distributions in nature tend to be Maxwellian in time, which, somehow stands like a background distribution.

If the total energy (i.e., temperature) of a system increases, the distribution curve shifts to the right and gets broadened. In other words, the differences between the components increase under the effect of increased energy. In ancient natural philosophy 'being' is realized by 'fire'; in other words, the things come into 'being' under the effect of fire. Energy increases the possibilities, that is, it creates new states that the components of the system can get into. As the energy is decreased the Maxwellian curve becomes steeper and shifts to the left; it finally becomes a pulse function or an arrow on the y-axis (e.g., number or population axis) when the temperature was decreased to zero degrees Kelvin. At this temperature all components are at perfect order and they all have same zero energy. The perfect equality could be achieved only at complete nothing, i.e., at zero degrees Kelvin.

Whenever the system is given little energy, every member of the system gains different amounts of energies and become different from each other. All changes become possible with the exchange of fire as proposed by Heraclitus. Although Heraclitus' fire is not identical to energy of our time, fire serves as energy in some respects. The addition of fire increases differentiation and hence the differences between the properties of different components. In other words, energy increases the driving force (or the contradiction) between the components of a system. It is no wonder that Heraclitus introduced both 'fire' and 'contradictions (duality or dialectics)' as the fundamental entities of nature.

The interaction of molecules or species in a random system takes place entirely through probabilistic interactions. According to Popper it is difficult to have a satisfactory theory of probability free from contradictions, and numerical probability denotes how frequent an event takes place [11]. Probability theory serves as the best tool to make predictions about random (or Maxwellian) systems. In order to predict how much something happens depends entirely on the energy distribution of species. So energy (or fire) does not only provide the needed threshold, but also arranges how many are involved and what happens. In fact, in Heraclitus' world, things are not 'made of', but 'made from' fire.

The shift of the Maxwellian curve to the right with temperature is due to the increase of 'complexions' according to Boltzmann. The randomness can increase both in the coordinate and the velocity spaces so-called the phase space of a system. A peculiar property of Maxwellian distribution is that, it is an asymmetric curve and starts from zero and goes to infinity; that is, there is no upper limit, and the number of complexions can go to infinity. The increase of the energy of a system creates new possibilities or states that the system can go. Boltzmann showed by his famous H-theorem that once the molecules left their places, it becomes difficult for them to go back to their original positions. The increase in complexions is an increase in new possibilities for a system, so the number of choices, and thus the degrees of freedom or randomness increases. Boltzmann identified this property with the increase of entropy. In classical thermodynamics entropy is defined as the less availability of energy, while Boltzmann defined entropy as the increase of randomness. On the same line, Shannon showed in early 1950s that entropy is the decrease of information of the system available to us. In fact both the Boltzmann's entropy equation and the Shannon's information equations have the same mathematical structure with opposite signs.

The right hand side of the Maxwellian distribution represents the superior property and those having these properties are always low in number. However, it is this part of the system that provides larger possibility for a change in nature. As an example, let this part to represent the number of fast runners (needed for Darwinian natural selection in animals), or very beautiful species (needed for Darwinian sexual selection), or very clever or rich people (needed for economic development), or very honest people (needed for social stability and cooperation), etc. They are low in number but have large capability to affect others. Although the distributions of these properties do not perfectly fit the Maxwellian distribution we can use the Maxwellian (or non-Maxwellian) distribution as a tool to make interpretations. These properties are transmitted to others through different means; economic wealth is transmitted through the exchange of goods, while genetic properties are transmitted through cross breeding. The superior properties, which are 'actuality' at present, make the 'potentiality' of future in the words of Aristotle. In terms of genetic science, a property of grand parents may not show up in sons or daughters who carry them as potential values, but may appear in grandchildren.

The Maxwellian distribution does not tell us why biological evolution takes place, but it tells us the simple mechanism of natural selection. As an example let the speed distribution at a lower temperature represent the speed distribution of predators and that at high temperature represent that of the preys. That is, the speed distribution of preys is shifted beyond the velocity distribution of predators. It is clear that the low speed preys have no chance to survive, nor the low speed predators. Only the preys and the predators on the right part of the curves (i.e., high speed species) can survive. Natural selection pushes the species in this example in such a direction that the offspring have the potentiality to be fast runners. Depending on the ecological conditions the biological species are always pushed to gain new properties (i.e., higher speed, better vision, better hearing, higher intelligence, higher birth rate, etc.). The 'change at present' is an attempt to have 'future equilibrium'. The equilibrium is always transient and it is apt to change always causing irreversible changes in the biological structures of living organisms.

If everything is exchanged randomly, how can the objects, the patterns, and order come out from randomness? The view of Anaximenes that everything comes from air cannot easily explain the reason and the mechanism of generation of order (or cosmos). Because a perfectly disordered system cannot go to an ordered state; this is also what Boltzmann says. One thing we can think of is that the disordered system may have tiny order at the very micro level so that each tiny ordered form can behave as the 'sperma' of Anaxagoras; they grow and form the objects. Recently Hong questioned microscopic irreversibility, that is, a possibility of the generation of form in an irreversible way in relation to biosystems [12]. The existence of sperma in completely disordered state is usually difficult to accept. Anaxagoras criticized the second assertion of Parmenides for accepting 'change' as the loss of one quality and the gain of another. He said there is no smallest part of anything, and the things can be infinitely divisible. Everything was mixed at the beginning and it is so now.

According to Anaxagoras the change that we observe in the objects is due to change in the proportions of infinitely small constituents of the objects. According to atomists any object can be a collection of different atoms but only some specific atoms can make up an object. Anaxagoras' philosophy bases on the existence of everything in a single object, and some of them are dominant in their proportions. In this respect Anaxagoras' philosophy forms a kind of foreground for the 'potential' concept of Aristotle. What Anaxagoras emphasized is that the smallest constituents can persistently stay together making a 'seed' (or sperma) which behaves like the nucleus of objects. Anaxagoras' sperma probably corresponds for instance to unit cell in crystals, DNA in biological cells, etc. Sperma owns specific information that tells to the system how or in what way to grow. The sperma regulates the proportions of the constituents in a growing object. In the atomists' philosophy each atom has a specific property; they may be round, cornered, sharp, etc. All these, in fact, correspond to chemical potential in atoms. In Pythagorean philosophy the abstract geometric shapes with proportioned edges are the constituents of matter.

The interpretation of entropy in terms of information by Shannon has much deeper philosophical importance than 'degradation of energy'. The increase of entropy denoting the increase of randomness and thus the loss of information about the system can also be used not only for quantitative changes, but also for qualitative changes. The loss of beauty of a flower in time, or the decrease of the moral attitudes of a society due to poor economy and corruption all can be treated as an increase in entropy in the corresponding parameter state. The Boltzmann entropy equation and Shannon's information equation thus serve as mathematical relations at the interface of physical and nonphysical worlds, because, both the physical and the nonphysical phenomena have their own appropriate 'configuration'. For instance, an ordinary speech can be made either more or less effective by rewording it without changing the number of words in it. Its influence on people can be in a way either to improve the ties between people or to worsen it. The entropy decreases in the former while it increases in the second.

According to Boltzmann, the Maxwellian distribution represents the maximum entropy (i.e., disordered) state and order does not come out from it. In maximum entropy state the system has fluctuations at micro level but they do not cause creation of any sperma, because whatever created is subsequently destructed. This is due to the principle of 'microscopic reversibility', that is, any molecular process and its reverse takes place with the same frequency.

Autocatalysis

The current chaos theories started after the pioneering work of Lorenz [13]. The nonlinear thermodynamics provided an important ground for chaos theories and also for the theories on pattern formation. The creation of pattern in chemically reacting systems was first handled by Turing [14] by using a set of differential eqautions, and then studied in detail by Prigogine [15]. The work of Turing is of utmost importance to understand the pattern formation that he called it 'morphogenesis'. The Lotka-Volterra problem, the Belousov-Zhabotinsky reactions, and the Bruselator problem of Prigogine were the model problems of nonlinear thermodynamics, and they were all autocatalytic (i.e. self-multiplying) reactions. That is, in all these problems the product catalyzes itself and changes other things (i.e., reactants) into product.

The Lotka-Volterra problem discusses the change of populations in wild life. It is a simple but powerful model for ecological systems. As an example we may consider the relation between grass (G), rabbit (R), and fox (F). The reaction between them can be given by,

$$G + R \rightarrow R + R$$
$$R + F \rightarrow F + F$$
$$F \rightarrow E.$$

Overall:

$$G \rightarrow E \tag{1}$$

Rabbit feeds on the grass (e.g., $G + R$), and after a while, it gives birth and thus multiplies in number (e.g., $R + R$). Fox eats the rabbit (e.g., $R + F$), and it also multiplies (e.g., $F + F$). Foxes then die and go extinct (e.g., E). The dead fox decays, on which grass grows, and it multiplies, too.

The competition between species and the conversion of others into a specific species or state has grounds in the philosophy of Empedocles. The four-stuff philosophy of Empedocles was a way out from the very static philosophy of Parmenides to explain the change in nature [6]. According to him the four-stuff (i.e., earth, water, air, and fire) cannot be converted completely into each other, because each of the four-stuff is an element, and there is always an eternal 'competition' between them. In this respect he opened a door to atomists who said each atom is unique by itself and cannot be changed into anything else. The philosophers defending four-stuff attributed autocatalytic property to each of the four-stuff. They said the universe cannot be made from one single element (i.e., entity); otherwise everything would finally be converted into it.

The change in nature takes place by mutual interaction of four-stuff, and each tries to increase its amount. Ice immersed into water cools it and tries to freeze it, but water warms up ice and tries to convert it into water. In this sense, all natural phenomena are autocatalytic, and each 'being' tries to convert others into itself, as rabbit converts grass into rabbit, fox converts rabbit into fox, etc.

The events are also somehow autocatalytic in Bohm's world. According to Bohm the events are constantly generated by the whole and swollen back by it. Bohm's 'whole' resembles Anaximender's apeiron, and it is also somehow autocatalytic, because each event reproduces another event that looks like itself. Each whole aims to reproduce its subwholes [16].

If we consider grass, rabbit, or fox with their unique configurations (or structure) each tries to convert the other into itself. In biological world natural selection appears to be a competition phenomenon between configurations (i.e., species). Since each configuration (i.e., species) owns specific information, natural selection results in reshaping of a specific configuration by selecting those of which some zones of its DNA is more (or less) expressed over others. The species, which have the chance to survive for sufficiently long time, have appropriate sets of biological reactions driven by their DNA to stabilize themselves against their environments. In single cells, the natural mutations cause the change of configurations in the offspring. Therefore the ensemble partition function of translation, vibration, rotation, and excitation energies of DNA becomes the determining basis for natural or spontaneous mutations. The chemical environment and the mutagens accelerate mutations. However, the spontaneous mutations do not seem to be sufficient in more complex species. The high rank species have opposite sexes, which tremendously accelerate the reconfiguration of DNA creating the birth of quite different new offspring (or configurations, differently structured DNA s, information, Anaxagoras spermas, or whatever).

Atomists substituted four-stuff by atoms and they assigned different shape or a kind of hook to each atom as mentioned earlier. Different tiny shape or hook corresponds to the chemical potential of an atom in today's language. Every chemical potential is a result of the distribution of electrons on the shells of atoms. Atomists were not well understood in Antiquity, and most other philosophers defended four-stuff theory.

Schlögl did an interesting work enlightening the relation between phase change (i.e., change of four-stuff) and chemical kinetics (i.e., atomic reactions) [17]. He considered autocatalytic reactions similar to (1).

One of his systems is,

$$A + 2X \xrightarrow{\leftarrow} 3X$$

$$B + X \xrightarrow{\leftarrow} C$$

Overall:

$$A + 2X \xrightarrow{\leftarrow} C \tag{2}$$

Here two X molecules are involved in an autocatalytic reaction and they change A into another X. One of X then combines with B and gives C. The rate equations for all steps can be written and solved for the change of the concentrations of the reactants. The simple chemical kinetics approach to express C in terms of X at steady state conditions yields an equation, which is, mathematically identical to the equations of phase change (i.e., Van der Waals and virial) of classical thermodynamics [17]. In fact the changes of earth, water, and air into each other is nothing but the changes of states in thermodynamics. The four-stuff philosophers in fact visualized the changes in nature to take place through phase changes. These state changes are called first order phase transition. We also have second order and λ-phase transitions, and they deal with order-disorder transitions. Another system studied by Schlögl is,

$$A + X \xrightarrow{\leftarrow} 2X$$

$$B + X \xrightarrow{\leftarrow} C$$

Overall:

$$A + B \xrightarrow{\leftarrow} C \tag{3}$$

At steady state the relation between C and X yields a mathematical equation, which is of the form of second order phase transition. In second order phase transition the appearance of the object does not change, but the inner order or structure changes. A magnet has an ordered pattern at atomic level, but it is destroyed above a critical temperature, and it becomes demagnetized. The very striking discovery of Schlögl was that the explanation of change in nature based on four-stuff could be explained by atomic dynamics (i.e., chemical reactions). It is interesting that in this explanation the molecules have to be autocatalytic, that is, they compete for themselves as each of the four-stuff. The autocatalytic or self-multiplying systems were shown to exhibit so-called sigmoid growth in time. It was first observed by Verlhurst for the growth of microbial organisms; it is interesting that some empires also display same growth pattern [18]. The rate of such growths can be mathematically described by a parabolic equation so-called logistic equation. Its iteration gives an idea about how a self-multiplying system goes into chaos. The original parabolic shape becomes two overlapping parabola in the second iteration, and it gives more complex structures on further iterations. In the sixth iteration a complex structure with different hyper symmetries is obtained with no resemblance to the original parabola [19]. Figures 3.1–3.10 in [19] give beautiful description of the iterations of logistic map. Four important observations can be made on the chaotic route of the system:

1. in every step of iteration the original parabolic shape repeats itself in decreasing sizes and with some deformations of the parabolic shape. However its number increases in the overall domain but spread out to varying positions on the curve (self-similarity and memory),
2. the decrease of the dimensions of the parabolic shape in the proceeding iterations obeys some mathematical proportioning (fractal dimension and patterns),
3. the curved structure is lost in the final shape and it changes very abruptly with the loss of differentiability (unpredictability),
4. new geometrical structures and higher symmetries formed indicate higher complexities (evolution of order and complexity).

Self Similarity and Memory

The first of the above indicates that in chaotic growth fragmentation takes place, and one or more of the parameters grow in number and/or in magnitude, but the system tends to loose its original shape. Every fragment carries some of the properties of its precursors, and also gains new structures (or information sites). The system reproduces new pattern of its kind at smaller sizes. The self-similarity is a unique property of chaotic systems, and it can be clearly seen especially if the function studied has complex terms as in Mandelbrot transformations, which also yield Julia sets [20–22]. Chaotic growth inherits self-similarity [20–30]. Whatever descends from the precursor serves as memory to the new system. As chaos increases, the total sizes of memory regions decrease and the system runs into difficulty in memorizing its past. Every chaotic or nonlinearly growing process has its own history. Some of the original information is lost in each step of change while new information (or structural changes) is gained at the expense of loss of former information. The system has a kind of mechanism which weakens the memory due to the continuous generation-destruction mechanism of Aristotle; the more number of steps of new generations the more difficult to get the original shape on going backward. The living organisms also show the same behavior. They have both common and different genes on DNA. In the developmental stages (i.e., iteration steps) some are expressed and some not. For instance, tail-forming genes are expressed in the fetus of baboon, but they are not expressed in chimpanzee. Tail information centre is lost in chimpanzee. The system gains new information in the course of evolution. The new information can result from two sources, one is due to the new additions of atoms or species, and the second is due to the change of the overall shape of the object. The second one is due to the configurational change of the system. A system may have different configurational structures without changing its material content but changing only the order of the alignment of its atoms or molecules. The difference between two same-material structures can be characterized by their configurational entropies. The Anaxagoras' sperma depends not only on atoms but also on their alignment in space. The sperma unlike the atoms undergoes changes in evolutionary processes. Non-evolving sperma may function as atoms, which stay unchanged according to Ancient atomists. Anaxagoras' sperma is unlimited in number and present in primordial mass. Aristotle's interpretation of unlimited sperma is unlimited principles. In our scientific world they should be interpreted as information centres whatever the structure of the object is. The information can be piecewise as atoms or structurewise as ensembles or configurations. In chaotic or nonlinear dynamic growth new information is created in the form of micro structures, and they all depend on the chemical potentials of atoms (shapes or hooks), energy (fire), and the way of organization (i.e., principles).

In chaos, the sperma may correspond to microstructures formed in the system, or in the chaotic attractor. However, Anaxagoras proposed his sperma theory to find a way out from the Parmenides philosophy that rejects change. Anaxagoras thought that everything contains tiny bits of everything, and the cluster of some similar bits can dominate and thus form a sperma, which then informs the system about what shape or property to undergo. Anaxagoras' sperma resembles micro structures in chaos or strange attractor. In this respect strange attractor is full of information, and can yield different outcomes depending on which microstates dominate inside. Anaximander had proposed a historically important biological evolution theory which bases on the combination of best fitting parts. It also represents a kind of change from chaos to order.

Fractal Dimension and Patterns

Chaos theories introduced a revolutionary concept to explain the shapes of objects or patterns in nature. The change from one phase into another becomes possible under the effect of fire that exists at different proportions according to Heraclitus. In chaotic growth the energy supplied does not only affect the dynamics, it also changes the geometric dimensions at every step of iteration. The proportionality (or scaling) occurring in the change of shapes or structures can be mathematically expressed by a power law. The scaling power for the nonlinear growth was called 'fractal dimension' by Mandelbrot [20]. Most objects in nature own a scaling dimension [20–24, 31–35].

The change of dimensions at certain proportions results in generation and destruction of similar geometric objects at varying dimensions. In fact, the solution of the mathematical equations of Bénard cells, which occur in viscous liquids heated from the bottom shows that some geometries are repeatedly produced in the system. Nonlinearly growing systems draw our attention to geometric shapes which are the elements of the Pythagorian School. In fact, as mentioned earlier, Aristotle claimed that atoms correspond to the geometric entities of the Pythagorian School.

Not all energy given to a system causes immediate change. The system absorbs the energy given for a while, and after reaching a critical amount it becomes unstable and splits itself into two, that is, it bifurcates. Every bifurcation can be viewed as a kind of phase change operation. A chaotic state can be arrived after a series of subsequent bifurcations. This introduces a kind of history to the final pattern accomplished. Therefore every final pattern generated in chaotic growth represents a 'happening' not only a 'being' or 'event' [36]. In the history of natural philosophy only Aristotle had a sound proposal about the importance of the shapes of objects. He considered shape (i.e., form) as a priori entity which when filled by matter makes an object. It cannot be separated from matter; it exists when matter exists.

In biological world, pattern forms as a result of information supplied by DNA. In nonlinearly growing objects every step changes the total information content of the former step, and the evolution of a form (shape, or morph) comes out from the evolving information content of the system. Therefore it is time dependent and bases only on finally achieved material (i.e., atomic) content, and also the way of combination of atoms.

Chaotic systems are in continuous exchange with the environment meanwhile they maintain their self-similarity, and this establishes a new framework to guide research in many fields [37]. As the similarity is concerned Popper points out that 'something repeats itself' is based on the judgment of similarity. Similarity concept heavily depends on our experience so we can never demonstrate that the world of phenomena is inherently repetitious [11]. In nonlinear growth dynamics the repetitions are not static repetitions as in the growth of a crystal where the information obtained from the unit cell always remains same; rather, the units change their shape and their inner contents resulting in a change also in their information content. That is, the present structure is not sustainable in the next step, and finally complex systems evolve spontaneously from simple precursors interacting haphazardly or weakly. The self-organization involves self-similarity, and according to Mandelbrot's fractal principle it extends everywhere in the universe [38]. The fractal dimension is not the property of points, but of how they are arranged with respect to each other; therefore it is inherently nonlocal [39]. So it influences every point around.

Unlike the patterns like shores, leaves etc. the chaotic attractors are true fractals [40]. In phase space the trajectories exponentially diverge from each other, and they wrap on each other imparting fractal geometry to the attractor as time goes to infinity. As a result, fine structures form at all scales, and complexity is created [23–35,37,41]. The Lyapunov exponents give the divergence of trajectories; and the system goes into chaos if we can assign a Lyapunov exponent to the system. The chaotic attractor represents the coherent coexistence of diverging points [42]. The border of the attractor is the most dynamic region where, the turbulence takes place. The system either forms patterns or goes into randomness, or perhaps extinction as in Mandelbrot pattern [21,38]. No matter how turbulent the system is, it remains chaotic if a correlation exists between the components of the system. When the Hamiltonian is non-separable due to nonlinear terms, the system is apt to chaos [41,43].

Chaos theories introduced an important concept into the physics world. In all physical theories there was no place for numbers; in fact, number theory had never been a concern of physicists. Since the chaotic attractor involves too many cycles, it is of concern to find out how all these varying oscillations can coexist in the system. It was found out by Lorenz that some frequency ratios expressed in terms of prime or semi prime numbers have higher durability [13]. In fact, in circle map, mode locking occurs when the ratio of two fundamental frequencies of toroidal motion is rational [29]. It corresponds to a relatively more stable state. When the system parameters are expressed by irrational numbers, then, the system is directed to chaos. In fact the fractal dimension is also an irrational number. In other words, chaotic dynamics uses numbers, and especially prime numbers. The prime numbers can be considered to be the atoms of ordinary numbers. According to Pythagorean School the truth can be tracked only by mathematical reasoning, and numbers can express the attributes of all things. The fractal dimension is an irrational number and never ends, it is a kind of unlimited in the sense of Pythagorian School; and it is a property of our nature.

Another interesting outcome of the role of numbers in chaotic dynamics is in understanding of aesthetics. The stability achieved through mode locking introduces a hierarchy of rational numbers, which are established according to Farey tree construction [44]. Farey tree can be constructed by using continued fractions. The irrational numbers can be also expressed in terms on continued fractions, which end at infinity. The least convergent continued fraction can be constructed by using '1'. This continued fraction leads to what is known as 'golden ratio', which has been used in numerous artifacts, architecture, sculptures, pictures, etc. throughout the human civilization. Golden ratio has been a kind of measure of aesthetics [45–49]. The chaos theories imply that the philosophy of aesthetics should be based on the importance of numbers in the dynamics of nonlinearly growing systems. The conservative and the stretching forces balance each other to generate such patterns, which create the feeling of aesthetics in our minds. Among all those patterns the one, which is the least stable and apt to easiest change (i.e., golden ratio configuration), is perceived as the most aesthetic configuration. It is interesting that the golden ratio shows up also in the growth of empires, which are also dynamical systems where stability during growth should be of primary issue [19].

Unpredictability

A chaotic system loses its continuity with abrupt changes, and it cannot be described by smooth functions. Therefore integrability is lost, and one cannot predict the future of the system. Classical chaotic systems are deterministic but not predictable. Therefore one cannot easily go backward in coordinate space to obtain the initial pattern from the last pattern. Renormalizability techniques are useful but one cannot get the full recovery [50]. The recovery can be improved by defining several points around the starting origin [39]. Therefore, determinism is lost at this point. In fact an infinitesimal change in the initial conditions may amplify some growing parameters and take the system into chaos, a property so-called 'sensitive dependence on initial conditions'. In addition, if the initial conditions are expressed by irrational numbers such as π and $2^{1/2}$ there is no way to measure it exactly. Therefore predictability is lost in chaotic systems. According to chaos theories most nonlinear dynamical systems cannot be reduced to the superpositions of elementary functions [51]. So the analysis of the system for forecasting usually becomes impossible.

In Laplacian thought the determinism and the predictability are the same thing; they are image of each other. Thus there exists a parallelism between the Laplacian thought and the religious destiny or fate. There can be no real possibilities in such a world; because every happening has to be necessarily so. The truths, which are necessary, do not have to be logically necessary [52].

The separation of determinism and predictability from each other can be on the grounds that determinism is an ontological concept while predictability is an epistemic concept [53]. This separation has made important contribution to the concept of free will. It is believed that unpredictability can be a source of human cognitive powers [54]. There is a conflict between free will and determinism. Heisenberg's uncertainty principle shed a light on free will, however Schrödinger claimed that free will is an illusion. There is almost no place for free will also in Bohm's quantum world. However, some philosophers talk in favour of indeterminism [55].

Complexity

The generation of new patterns, higher order symmetries, or cosmos in terms of ancient philosophy is a unique property of chaotic dynamics. Chaos can be viewed as the science of pattern formation. The critical question here is why chaos can lead to order. According to Landsberg the disorder of a system can be defined as the entropy divided by the system's maximal entropy. If the rate of increase in the number of micro states due to increase of entropy is less than the rate of increase of maximal entropy, then the disorder decreases [56,57]. The basin of strange attractor is full of microstates, which behave like the sperma of Anaxagoras. In other words the strange attractor is full of information, which can lead to different new formations. The variety of information increases with the extent of chaos, that is, the more chaotic the system is the more variety of micro states; and thus the higher chance for new patterns. The number of existing microstates determines the pathways of new evolving systems from chaos. The complexity of pathways increases the diversity of new formations; however, the rate of evolution decreases as it has higher degree of branching. The decrease in the number of microstates and thus the decrease in the complexity of pathways can accelerate the evolution of certain species. For instance, small mammals first showed up by the end of third geological period. The sharp change by the end of third period and the extinction of almost 95 % of all species including dinosaurs accelerated the evolution of mammals in the fourth geological period.

The genetic information of a living species denotes the maximum information content, but not its organizational information. Only some of them are used in the structural and metabolic organizations. The loss of some of the current information can lead to the increase of the accumulating potential information, which then leads to mutations for new species. The increase in the maximal information content provides a ground for the increase of the potential information, and thus the possibilities for new species. The amphibians like frog have much longer DNA compared to most other species. Such huge potential information was actually needed to have special organization to adopt a new life on earth rather than in water. It helped also to have a variety of new species in the course of evolution such as reptiles, birds, etc.

The structural and metabolic organization in a living species is all controlled by active sites of DNA. The active sites can be considered as active microstructures of DNA. Anaximander had proposed a primitive evolution theory 2500 years ago, and claimed life originated in the seas. He said new parts form under solar radiation from cracking of shells of forms, which in turn join to give the living species. Only best fitting parts can join and give the species. This primitive theory more or less shows a parallelism with the current evolution theory assuming the active sites of DNA refer to the parts of Anaximander.

In chaotic systems hyper symmetries are generated while fragmentation takes place (see Figs. 3.7 and 3.10 in [19]). In other words the increase of complexity is a natural consequence of nonlinearly growing systems. Forster considers a possibility of a kind of conservation law that says simplicity achieved at the higher level is at the expense of complexity at the lower level [58]. This approach is interesting in the sense that it somehow inspires a kind of metaphysical conservation law. This thought may have roots in Parmenides' philosophy.

The understanding of complexity will help also to understand the living organisms [59]. The property of the whole cannot be reduced to the properties of components but related to the interaction between them as Aristotle said. In the complexity theory, the self-organizing systems get involved in new interactions and form new connections between the components.

The rate of the change of entropy of a system that nonlinearly grows is a measure of how it undergoes complexity [41]. In fact the rate of change of Shannon entropy gives Kolmogorov entropy. The Kolmogorov entropy is zero for steady systems, and positive for chaotic systems. So the Kolmogorov entropy can be used to understand how complexity develops. The high Kolmogorov entropy means high rate of change of the internal structure, and of the information content; therefore, the faster development of complexity. According to Gatlin the Shannon entropy denotes capacity to carry the real 'potential information' [60]. Therefore the Kolmogorov entropy is a measure of the gain of new information and thus of complexity. In other words, it indicates the rate of Aristotelian change 'from potentiality to actuality'.

References

1. Köklügiller, A.: *Gılgamesh*. In Turkish. Can Yayınları, 1984.
2. Sayılı, A.: Mathematics, Astronomy, and Medicine in Egyptians and Mesopotamians. 3rd edition. Türk Tarih Kurumu Basımevi, 1991.
3. Kramer, S.N.: *History Starts at Sumer*. Kabalcı Yayınevi, 1992.
4. Šesták, J. and Mackenzie, R.C.: *The Fire/Heat Concept and Its Journey from Prehistoric Time into the Third Millennium*. Journal of Thermal Analysis and Calorimetry 64(1), 129–147, 2001.
5. Barnes, J.: *The Presocratic Philosophers*. Routledge, 1993.
6. Denkel, A.: Democritos/Aristoteles, Natural Philosophy in Ancient Age. Kalamış Yayıncılık, 1987.
7. Cogniot, G.: *Ancient Age Materialism*. Sarmal Yayınevi, 1992.
8. Holt, E.H. and Haskell, R.E.: *Foundations of Plasma Dynamics*. The Macmillan Co., 1965.
9. Landau, L.D. and Lifshitz, E.M.: *Statistical Physics*. Pergamon Press, 1969.
10. Yourgrau, W.; Merwe, A. and Raw, G.: *Treatise on Irreversible and Statistical Thermophysics*. The Macmillan Company, 1966.
11. Popper, K.R.: *The Logic of Scientific Discovery*. In Turkish. Yapı Kredi Yayınları, 1989.
12. Hong,F.T.: *Towards physical dynamic tolerance: an approach to resolve the conflict between free will and physical determinism*. Biosystems 68(2–3), 85–105, 2003.
13. Lorenz, E.N.: *Deterministic Nonperiodic Flow*. Journal of the Atmospheric Sciences 20(2), 130–141, 1963.
14. Turing, A.M.: *The Chemical Basis of Morphogenesis*. Philosophical Transactions of the Royal Society of London. Series B, Biological Sciences 237, 37–72, 1952.
15. Glansdorff, P. and Prigogine, I.: Thermodynamic Theory of Structure, Stability, and Fluctuations. Wiley-Interscience, 1971.
16. Bohm, D.: Wholeness and the Implicate Order. Routledge, 1995.
17. Schlögl, F.: Chemical reaction models for non-equilibrium phase transitions. Zeitschrift für Physik 253, 147–161, 1972.
18. Gündüz G.: *The nonlinear and scaled growth of the Ottoman and Roman Empires*. Journal of Mathematical Sociology 26(3), 167–187, 2002.
19. Rasband, S.N.: *Chaotic Dynamics of Nonlinear Systems*. John Wiley and Sons, Ch. 3, 1990.
20. Mandelbrot, B.B.: *The Fractal Geometry of Nature*. W.H. Freeman and Comp., 1983.
21. Peitgen, H.O. and Richter P.H.: *The Beauty of Fractals*. Springer-Verlag, 1986.
22. Peitgen, H.O.; Jürgens, H. and Saupe, D.: *Chaos and Fractals*. Springer-Verlag, 1992.
23. Kaye, B.H.: *A Random Walk Through Fractal Dimensions*. VCH Verlagsgesellschaft mbH, Ch. 1, 1989.
24. Feder, J.: *Fractals*. Plenum Press, 1988.
25. Gardner, M.: *Mathematical Games*. Scientific American 235, 124–133, December 1976.
26. Blumenthal, L.M. and Menger, K.: *Studies in Geometry*. W.H. Freeman and Comp., 1970.
27. Ben-Jacob, E. and Garik, P.: *Ordered shapes in nonequilibrium growth*. Physica D 38(1–3), 16–28, 1989.
28. Barnsley, M.F.: *Fractals Everywhere*. 2nd edition. Academic Press Professional, 1993.
29. Jensen, M.H.; Bak, P. and Bohr, T.: Transition to chaos by interaction of resonances in dissipative systems. I. Circle maps. Physical Review A 30(4), 1960–1969, 1984, http://prola.aps.org/abstract/PRA/v30/i4/p1960_1.
30. Hénon, M.: *A two-dimensional mapping with a strange attractor*. Communications in Mathematical Physics 50(1), 69–77, 1976.
31. Feigenbaum, M.J.: *Quantitative universality for a class of nonlinear transformations*. Journal of Statistical Physics 19(1), 25–52, 1978.
32. Feigenbaum, M.J.: *The universal metric properties of nonlinear transformations*. Journal of Statistical Physics 21(6), 669–706, 1979.

33. Metropolis, N.; Stein, M.L. and Stein P.R.: *On finite limit sets for transformations on the unit interval.* Journal of Combinatorial Theory—Series A 15, 25–44, 1973.
34. Hu, B. and Mao, J.M.: *Period doubling: Universality and critical-point order.* Physical Review A 25(6), 3259–3261, 1982, http://prola.aps.org/abstract/PRA/v25/i6/p3259_1.
35. Hu, B. and Mao, J.-M.: *Fractal dimension and degeneracy of the critical point for iterated maps.* Journal of Physics A: Mathematical and General 20(17), 1809–1818, 1987.
36. Prigogine, I.: *From Being to Becoming.* W.H. Freeman and Company, 1980.
37. Coppa, D.F.: *Chaos theory suggests a new paradigm for nursing science.* Journal of Advanced Nursing 18(6), 985–991, 1993.
38. Jones, D. and Culliney, J.: Confucian Order at the Edge of Chaos: The Science of Complexity and Ancient Wisdom.Zygon 33(3), 395–404, 1998.
39. Collier, J.D.: *Holism in the New Physics.* Descant 79/80, 135–154, 1993, http://www.ukzn.ac.za/undphil/collier/papers/holism.pdf.
40. Koperski, J.: *Has Chaos Been Explained?* British Journal for the Philosophy of Science 52(4), 683–700, 2001, http:///bjps.oxfordjournals.org/cgi/reprint/52/4/683.
41. Kronz, F.M.: *Nonseparability and Quantum Chaos.* Philosophy of Science 65(1), 50–75, 1998.
42. Gleick, J.: *Chaos: Making a New Science.* Penguin, 1988.
43. Arrowsmith, D.K. and Place, C.M.: *An Introduction to Dynamical Systems.* Cambridge University Press, Ch. 6, 1990.
44. Cvitanovic, P.; Soderberg, B. and Shraiman, B.I.: *Scaling Laws for Mode Locking in Circle Maps.* Physica Scripta 32, 263–270, 1985.
45. Ghyka, M.: The Geometry of Art and Life. Dover, 1978.
46. Huntley, H.E.: *The Divine Proportion.* Dover, 1970.
47. Bergil, M.S.: *Golden Ratio.* In Turkish. Arkeoloji Ve Sanat Yayınları, 1993.
48. Gündüz, G. and Gündüz, M.: *Randomness, Chaos, and Pattern Formation.* In Turkish. METU Press, 2002.
49. Livio, M.: The Golden Ratio: The Story of Phi, the World's Most Astonishing Number. Broadway Books, 2002,
50. Ref.19, Ch. 7,
51. Leiber, T.: Deterministic Chaos and Computational Complexity: the Case of Methodological Complexity Reductions. Journal for General Philosophy of Science 30(1), 87–101, 1999.
52. Arsenijević, M.: *Determinism, Indeterminism and the Flow of Time.* Erkenntnis 56(2), 123–150, 2002.
53. Bishop, R.C.: *On Separating Predictability and Determinism.* Erkenntnis 58(2), 169–188, 2003.
54. Garson, J.W.: *Chaos and free will.* Philosophical Psychology 8(4), 365–375, 1995.
55. Popper, K.R.: The Open Universe: An Argument for Indeterminism. Routledge, 1982.
56. Landsberg, P.T.: *Can entropy and "order" increase together?* Physics Letters A 102(4), 171–173, 1984.
57. Gündüz, G. and Gündüz, U.: *The mathematical analysis of the structure of some songs.* Physica A 357(3–4), 565–592, 2005.
58. Forster, M.R.: How do Simple Rules 'Fit to Reality' in a Complex World? Minds and Machines 9(4), 543–564, 1999.
59. Gilbert, S.F. and Sarkar, S.: *Embracing complexity: Organicism for the 21st Century.* Developmental Dynamics 219(1), 1–9, 2000.
60. Gatlin, L.L.: *Information Theory and the Living System.* Columbia University Press, 1972.

Chapter 2

Complex Freedom [†]

Davor Pećnjak

Institute of Philosophy, Zagreb, Croatia and University of Zagreb, Centre for Croatian Studies, Department of Philosophy, Zagreb, Croatia. Email: davor@ifzg.hr.

ABSTRACT

We have a very strong intuition and a very strong feeling that we, as human beings, generally have freedom of the will and freedom of the action. It seems that in most situations we can do this or that; namely, we can do action A or we can refrain from doing action A under the same conditions. The view which argues that this is not an illusion and that we have genuine freedom is the libertarian view. I would like to examine could that view be plausible under scientific understanding of the world. It seems that physical sciences strongly support determinism. Chaos theory and indeterminism in quantum mechanics could not save freedom because chaos is a deterministic theory and indeterminate events in quantum mechanics happen by pure chance. Pure chance is not something we want as freedom. But, perhaps, we can have freedom reconciled (although maybe in a restricted form) if actions or decisions can be described by equations which allow more than one solution and if these solutions can be interpreted as refering to different contents of the will or to different actions.

Keywords: free will, determinism, libertarianism, chaos theory, complexity.

Introduction

In this article, I would like to examine some points in the philosophical debate of free will and what points from complexity or chaos theory can be, perhaps, interesting for the debate. I will simplify things a bit, but I certainly try not to make an oversimplification!

Freedom and libertarianism

We have a very strong intuition and a very strong feeling that we, as human beings, generally have freedom of the will and freedom of the action. It seems that in most situations we can do this or that; namely, we can do action A or we can refrain from doing action A. It seems that

---

[†]Reused with permission from: Davor Pećnjak, Complex Freedom, *Interdisciplinary Description of Complex Systems* 7(1), 14–21, 2009.

it is the case that it is so even in situations in which there is no reason to refrain from doing *A*—for example in escaping from fire—still it seems that "ontological" situation is thus that nothing inside or outside us (as agents) is such that would not allow us to refrain from doing *A* in that circumstances. Of course, there may be, and there are, situations in which some agents are completely determined what they will do next or what will happen. There may be cases of complete determination even if our world is such that it contains real freedom for the will and action—for example, full psychological determination (psychopathological cases or acting under alcohol or drugs, etc.) or environmental situation in which there is no possibility to "exercise freedom" because we instinctively do what we do or environmental situation in which "physical forces" are so overwhelming that agent's will and agent's acting is without significance and without impact (even on agent himself).

But leaving these possible situations aside, I would like to speculate about some prerequistes that must be present for us to have freedom of the will and freedom of the action in most ordinary circumstances if all these can be scientifically described. I do not endorse compatibilism, namely the view that freedom and determinism are compatible. Moreover, I think that compatibilism is untenable. (But thorough arguing for untenability of compatibilism is not the subject of present article and is not needed for present purposes.) Only viable way for genuine freedom is the libertarian view (for an excellent overview of libertarianism and its variants see Clarke [3]). Those who endorse libertarianism are incompatibilists regarding freedom and determinism—freedom cannot be reconciled with determinism. Libertarians argue exactly for what is said at the beginning: not only that we have strong intuition and feeling of freedom—becuse that can be an illusion—but that real "metaphysical" situation is that, in most cases, under exactly the same circumstances we can do action *A* or we can do action *not-A* (refrain from doing action *A*). We can use the language of possible worlds and say that if in this actual world a certian agent *X* does *A* at time *t*, then there is a possible world with the same laws of nature and overall history up to time *t* as actual world, but in which, at *t*, agent *X* does *not-A* (refrains from doing *A*). (To make things shorter at this very point, I consider deliberating, agent's deciding and agent's willing as types of action as well, see for example Pink [12]) How could that be possible, especially under the scientific description of nature and agents, which seems hostile to libertarianism? It seems that modern science, especially physical sciences, strongly supports determinism and deterministic view. The cases of (genuine) indeterminism are seemingly rare and tied to some special situations; and there is widespread doubt that examples of indeterminism (in physical sciences, as it will be clear a bit later taking an example) are also not of help for libertarian construction of freedom.

Incompatibility of Free will and Determinism

First of all, let me sketch the essence of an argument for incompatibility of free will and determinism (see for example, van Inwagen [13,14], Lamb [9]).

Universe is governed by the laws of nature. They are as they are and we cannot change them—no one has the power to make laws of nature different from what they are (except God, of course). Likewise, no one has the power over the past. It is not possible that someone acts now in a way that would make the past different from what in fact it was. No one has the power to act that something which is a fact of the past would not have been a fact about a past [5, p.9]. The doctrine of determinism says the following: for any state or event X in the universe there is a set of previous states and events that, together with the laws of nature, inevitably entail state or event X. This kind of argument is called "The Consequence Argument" as well. It is standardly interpreted causally: Laws of nature and the set of previous states and events causally necessitate X. First clear explication of such a thought had been given by Laplace. So, by Laplaceian determinism today we simply mean that when we choose certain state of the universe at any instant of time together with the laws of nature, what happend before and what will happen after that instant of time is uniquely determined. Now it follows, according to these explications, that if freedom requires the possibility to do otherwise (than what is done in fact), namely the possibility of doing A and refraining from doing A, it is incompatible with determinism.

But, to repeat the question already posed—how could that be, because it seems that modern science strongly, if not completely, supports determinism. The cases of indeterminism are seemingly rare and tied to some special situations. But certainly even some of these indeterminisms could not be of much help for a libertarian.

Determinism and Indeterminism

There may be some doubts whether Laplacian determinism firmly holds accross (the whole) of modern physics and it is still unresolved matter according to Earman [4]. He analyses some very interesting examples from physics—both classical and quantum—and shows where there are cases which involve indeterminism; he also try to show how some of them, perhaps, could be reinterpreted in a deterministic outset, though then other problems arise for those reinterpretations. These cases of indeterminism are special cases though maybe they could provide some framework how to think about how to help a libertarian view.

Of course, at the level of quantum mechnics, there are some cases of chance events which are really just pure chance (for example, the decay of a neutron in a free state). But that would not much help a libertarian: because at the level of decision making, at the level of will and the level of action and behaviour, pure chance or randomness in this sense is also something over which an agent does not have a control and influence just like an agent does not have a control and influence over the laws of nature and past states of the universe in the deterministic outset. Also, we do not frequently observe purely random behaviour. When we do, then in most cases it is the behaviour of a mentally ill person. So, randomness in action would not be a mark of freedom, but it is a mark of mental illnes.

The classical general relativistic physics [4, pp. 34–40] also admits indeterminism in an interpretation. In a nutshell, regarding the inital value problem for source-free Einstein gravitational field equations, Earman [4, pp. 35–36] says that "specifying the metric field and its normal derivative on some space-like slice Σ does not suffice to determine ... the values of the field at points of four-dimensional manifold to the future or the past of Σ. Indeed, specifying Lorentz signature metric on Σ and the entire causal past of Σ does not suffice to determine Lorentz signature metric at points to the future." It means that we can have completely the same past of the metric field and the same causality in it, but that, from some point, the future is not the same in an evolution of the manifold.

For another example, not tied to Earman, here I shall mention the work of the so-called Bruxelles-Austin group led by Prigogine on far-from-equilibrium systems. They look at the complex systems as a whole and take a new approach to describe them. Fundamental to their description and explanation is *distribution*. So, the *structure of the distribution* of complexes of particles from which a system is build is something elementary important and not the classical description and explanation of single particle with its trajectory, momentum and direction. As, for example, says Robert Bishop [1, p.121]-these kind of theories and explanations which are concerned primarily on distribution functions open possibilities for genuine indeterminacy, namely, that macroscopic far-from equilibrium systems are irreducibly indeterministic. If so, that would mean that some indeterminism is inherent in complex macroscopic systems.

Chaos is of No Help for Libertarian Construction of Freedom

I think that it is pretty much obvious that chaos theory could not help libertarians in explaining their notion of freedom. Chaos theory is in fact a deterministic theory [6,10,16] and could nicely fit into Laplaceian vision of determinism. However, there are several very interesting properties of ingredients of chaos theory. Among others, the theory incorporates many non-linear equations and there is so-called sensitive dependence. A system is sensitivly dependent on initial conditions if very slight, indeed, very tiny, difference in initial conditions leads to great differences in later development. "In fact, in some dynamical systems it is normal for two almost identical states to be followed, after a sufficient time lapse, by two states bearing no more resemblance than two states chosen at random from long sequence" writes Lorenz [10, p.8]. So, a system of non-linear equations can produce huge diffrences between initially almost identical dynamical systems (for a concrete example, see Wolf [17]). This means that the priciple which says that from similar conditions and similar causes we should arrive to similar effects is no longer universally valid. The other property which follows is that chaotic dynamical systems are very complicated systems: though some of them can be governed even by simple equations, their appearance is very complicated. Because of that complicated appearance, they may even look random. But they are *not* random: they *just look* random and they are *just* very complex and complicated.

One thing that must not mislead us is that at the practical level, there could be many cases of poor predictability or predictability could be completely impossible. This is due only to sensitive dependence in chaotic or complex systems. It is not a mark of freedom. When investigating and measuring real systems, we are bounded how precise we can measure important values. So if we can, for example, be precise in measurement to fifth decimal, but two similar systems show sensitive dependence only to the sixth or further decimal, we would be in no position to predict what will happen to those systems and how much would they differ, perhaps even after just a few steps. But this situation arises only due to our limitations or the limitations of our instruments. It is not that, in reality, the systems in question are not completely deterministic systems.

So, for genuine freedom, it seems that we must steer between randomness and complete determination. Let's see what could be done.

Freedom and Non-solvable Equations

In an article with the title "Free Will Remains a Mystery", van Inwagen [15] argues, among other things, precisely for that-free will is a mystery! Namely, he is an incompatibilist regarding free will and determinism, but also he thinks that free will is something that exists and that we have it. In his words: "But if free will is incompatible with determinism, we are faced with a mystery, for free will undeniably exists, and it also seems to be incompatible with *in*determisism" [15,p.158]. Robert Kane [7, p.12] comments: "Van Inwagen believes that noone to date has been able to give an intelligible account of incompatibilist freedom; and he has doubts about the possibility of doing so. Yet because he also thinks the Consequence Argument is undeniably sound, he argues that we must continue to believe in an undetermined will even if we do not know how to give an intelligible account of it." If such a mysterian view is right, namely that we cannot explain how we have (if we have) free will, how can that view be reflected upon, regarding scientific view of the world?

First of all, maybe we have fundamental freedom of will and of action and indeed it is not explicable in any theory that can be available to human beings. Only God can know how it is possible, but we, with our cognitive capabilities are, to borrow the phrase from Colin McGinn [11] used in another context, "cognitively closed" for such an explanation. Simply, as chameleon is cognitively closed for a physical theory of colours and light that we, humans, have, perhaps we cannot come to formulate and understand what lies in the foundations of free will and free action and to explain them. On the other hand, perhaps we shall be able to formulate a very precise and complex(ity) theory of our deliberating and acting on the results of that deliberating. Such a theory, if it will be a mathematically formulated physical theory, will perhaps contain all, some or at least one of the equations which will be in such a form that they have no solution. Let me cite Edward Lorenz [10, p. 13], though from other context, in support of this speculation: "Very often, when the flow is defined by a set of differential equations, we lack suitable means for solving them —some differential equations are intrinsically unsolvable. In this event, even though the difference equations of the associated mapping must exist as relationships we cannot find out what they look like. For some real-world systems we even lack the knowledge needed to formulate the differential equations; can we honestly expect to write any equations that realistically describe surging waves, with all their bubbles and spray, being driven by a gusty wind against a rocky shore?"

We can interpret this in two ways: in an epistemic sense and in a metaphysical sense. Epistemic interpretation would suggest that we are limited in the possibility of knowing how something happens—but that what happens happens in a (complex) determined way. So that is not a rescue for a libertarian account of freedom and free will. The other interpretation, a metaphysical one, seems to be more promising. This would suggest that there is not a determinate process inherent in reality that would be computationally solvable. The lack of suitable means of solving the equations may perhaps mean that there is no inherent process in reality which happens in a completely determined way (but not perhaps completely randomly).

Freedom and Rationality

I would like to say something about the connection of rationality and freedom as well. Here, I would not go into assesing what people *actually do and how they actually behave* —we know, of course, that people are too much irrational in practical everyday life—but I would like to examine what rationality would require how to choose and how to behave.

Rationality also can be an obstacle for freedom. However complex may be our intertwining of our preferences and however complex we must think about them, for most situations in which we can find ourselves, there is one and only one solution which is the best solution for that situation. Rationality would dictate that we take the solution which is the best and take a course of action which it prescribes. There may be situations in which more than one action would be equally rational for us to take, so it would not matter which one of that equal actions is undertaken (from the viewpoint of rationality). But, in most cases, there is only one solution available as the best solution. So, according to rationality, there would not be genuine choosing—only one course of action is possible as the most rational and, if we want to maximize our rationality and be completely rational beings there will be no freedom for us. But, of course, anyone who would like to argue for libertarianism, would not like to lose rationality. Libertarian would like to have a situation that we (can) act rationally, but that we do so *freely*. So, rationality (or maximization of rationality) somehow has to be reconciled with freedom. I have no offhand solution for this problem (but Thomas Pink [12, pp. 44–54] offers a plausible solution) but I would like to say the following: Perhaps we should distinguish abstract theory of rationality on one side and how that rationality is realized in human beings as, for example, a complex interaction of components of a system of beliefs, desires, preferences and representations of the situations in which subject finds himself. This system is mentally and physically realized as dispositions and/or states and processes in the brain.

The physical description (if something as that would be ever available) should be then in some form which does not yield a unique solution which would be a definite determination of the undertaking of the most rational action of the agent. That description should allow for diferent possible outcomes in this situation. In other words, the physical situation of the agent should be so that it allows for different actions and not only the most rational. So, we shall perhaps have descriptions and explanations at two levels: at more abstract level—the rational (intentional) level there would be only one solution which would be the most rational for the agent and at the more basic level-let's provisionally say physical level-we shall have a situation of the realization of that rationality as part of a complete (physical) situation of the agent which does not uniqely determine and causally necessitate undertaking the most rational even though an agent does the most rational action.

Conclusion

So, bearing in mind what is said above, where should freedom be between determinism and chance? I'll try to sketch just a general frame and I admit that there are many *ifs* in my conclusion!

First, perhaps, we should narrow the scope of possible actions under the same set of circumstances. It would mean that not everything is possible to will and to do under the same set of (antecedent) circumstances. But it would also mean that not only one inevitable action is possible but a certain range of them. So, both at the level of rational (intentional) explanation and "physical" explanation we should have descriptions which do not (causally) necessitate. Rational reasons provide what is best or most rational for an agent to do but they do not *causally* necessitate that agent would inevitably do what it prescribes; and rational explanation in virtue of these reasons *does* explain why agent does according to it if agent really takes that course of action which is prescribed by what is most rational to do. But it does not explain agent's actions as inevitable and comlpetely determined by previous states and laws of nature. So, perhaps "physical" situation of an agent should also be such that it does not necessitate the outcome what agent will do.

There are equations or systems of (differential) equations which have multiple solutions (more than one solution). If we could interpret these different (numerical) solutions that they refer to different contents of the will or to different actions, then it could mean that different actions are compatible with the same situation which obtains before taking a certain action. So, an agent would be in situation with open possibilities, though it could be a restricted range of possibilities. But, in that range there would be a genuine openess (which action to take and whichever action is then taken, it would not be one that inevitably followed). But that what is chosen and which action is undertaken would not be random on the other hand, because it would be compatible with some intentional (broad) rational explanation, even in the cases where the action undertaken is not the most rational, and it would be compatible with specified previous states and a physical description of a situation; and whatever else is in that range what equations allow, is, by that very fact, compatible with previous states and a physical description of a situation.

Of course, details of such an approach, if possible at all, are yet to be worked out, but it seems that it provides a general framework how libertarian and scientific worldviews could be reconciled.

Acknowledgments

I would like to thank Hinko Wolf for improving my understanding of non-linearity and chaos, Josip Stepanić, Tomislav Janović, and all the participants to DECOS 2008 Conference. This article is part of a project "Question of Free Will and the Problem of Consciousness" which is supported by the Ministry of Science, Education and Sport of the Republic of Croatia.

References

1. Bishop, R.: *Chaos, Indeterminism and Free Will.* in [8, pp.111–124].
2. Campbell, J.K.; O'Rourke, M. and Shier, D. (eds.): *Freedom and Determinism.* MIT Press, Cambridge, Massachusets, 2004.
3. Clarke, R.: *Libertarian Accounts of Free Will.* Oxford University Press, Oxford, 2003.
4. Earman, J.: *Determinism: What We Have Learned and What We Still Don't Know.* in [2, pp. 21–46].
5. Fischer, J.M.: *The Metaphysics of Free Will.* Blackwell, Oxford, 1994.
6. Gleick, J.: *Kaos: rađanje nove znanosti,* Izvori, Zagreb, 1996, (in Croatian, translated by Milica Lukšić).
7. Kane, R.: *Introduction: The Contours of Contemporary Free Will Debates.* in [8, pp. 3–41].
8. Kane, R. (ed.): *The Oxford Handbook of Free Will.* Oxford University Press, Oxford, 2002.
9. Lamb, J.: *On a Proof of Incompatibilism.* Philosophical Review, 86(1), 20–35, 1977.
10. Lorenz, E.: *The Essence of Chaos.* University of Washington Press, Seattle, 1993.
11. McGinn, C.: *The Mysterious Flame.* Basic Books, New York, 1999.
12. Pink, T.: *Free Will.* Oxford University Press, Oxford, 2004.
13. van Inwagen, P.: *The Incompatibility of Free Will and Determinism.* Philosophical Studies, 27,185–199, 1975.
14. van Inwagen, P.: *An Essay on Free Will.* Oxford University Press, Oxford, 1983,
15. van Inwagen, P.: *Free Will Remains a Mystery.* in [8, pp.158–177].
16. Williams, G.P.: *Chaos Theory Tamed.* Joseph Henry Press, Washington D.C., 1997.
17. Wolf, H.: *Forced Steady-state Vibrations of Systems with Clearances,* Strojarstvo, 42(3–4), 109–118, 2000.

An Understanding of Language Development Models—Pidginization from the Perspective of Chaos Theory[†]

Guodong Zhao

Foreign Languages Department of Inner Mongolian Finance and Economics College. Hailaer Street 47. Huhhot, Zip code: 010051 China. Email: zhaoliu1116@163.com.

ABSTRACT

With the accelerated globalization, domestic and international communications become more frequent than ever before. As the major media of international communication, languages contact with each other more actively by day. And in the active contact any language would gradually develop and change. Pidgin language is a unique linguistic phenomenon resulted from the frequent language contact. By taking pidgin as an example, this article attempts to shed some light on the language development models in the frequent contact and how one language reacts to its environment, which includes a foreign language, a new culture and new information. With "Chaos Theory" brought to the horizon, this paper figures out three ways, namely, to progress, to decay, and to remain untouched, for any language to choose from when it is in contact with a new environment.

Keywords: Pidgin, Language Contact, Environment, Complexity.

Introduction

In 1863, August Schleicher formulated his theory of the Family Tree of Language, and thus he started the famous Evolutionists view of human language. This was a big stride in the field of linguistic study. But Family tree theory was not reliable in interpreting the way language develops, because firstly, Family tree theory does not allow convergence of

[†]Reused with permission from: Guodong Zhao, An Understanding of Language Development Models-Pidginization from the Perspective of Chaos Theory, English Language Teaching, Vol. 3, No. 2; June 2010. www.ccsenet.org/elt.

different languages; however language convergence is an unquestioned social phenomenon today; secondly it seems that language bifurcated at a certain time in history suddenly, while in reality language change is a very common happening people can witness almost everyday; thirdly, this theory puts emphasis on the linear development of language rather than the interrelationship between different languages for example English, Latin and French, but a lot of evidence, for example loanwords and borrowings, show that language change is resulted to a certain extent from language contact; at last, Family Tree Theory idealized the uniform speech communities, but in fact everyone is speaking a dialect.

Later the structuralists put forward the structuralist view of language by treating language as closed structure of words and rules, independent from its environment. This idea is not sound enough either, for any language is a part of the whole system of society and language often reacts to the stimulus from the complexity of the system. The founder of modern linguist Saussure wanted to establish the autonomy of linguistics, freeing it from reliance on other factors. But language is so involving and so closely related to the environment that any language could not undergo a slight change without relying on some force from outside. Chaos Theory is a set of theories and methods developed in different disciplines to understand and research complex, dynamic systems and non-linear development of these systems. Being open and able to self-organize are the features of the theory. Chaos, with reference to chaos theory, refers to an apparent lack of order in a system that nevertheless obeys particular laws or rules.

Pidgin language is a contact language with unique linguistic features. Decades ago, no light was shed on it for there seems to be no means by which to explain it fully. From the perspective of Chaos Theory, pidgin can be well explained as a third way a language reacts to the environment. To understand pidgin language in this angle, it is advisable to start with the analysis of language development.

Language Development Model

The Family Tree Model

The evolutionist's study of linguistic development in history concerns mainly about the Indo-European languages as is shown in the Fig. 1. Languages like Chinese are excluded from this model. It is easy to imagine that English and German have a lot in common in that they are both derived from the same parent language. A lot of English words have their counterparts in German, and the counterparts display similarities to the English words. For example:

German	English	German	English
Artikel	article	lernen	learn
Ball	ball	studieren	study
Bier	beer	kommen	come
Familie	family	hören	hear
Literatur	literature	bilden	build

These phenomena show that the similarities between "sister languages" are due to the same origin. But by examining the vocabulary of French and English, we can still find almost the same circumstance. Moreover, we can find some words in Chinese which demonstrate the influences of English, such as "*Safa*", "*tanke*", "*kafei*", "*kekoukele*", "*youmo*", "*balei*", "*qiaopai*", "*VCD*", "*GDP*", which is also the case in English, for instance, "kang", "taiji", "qigong", "tea", "China", and these words are from Chinese.

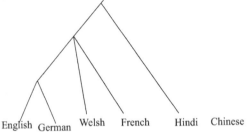

Figure 1

The findings above illustrate that language not only develops vertically, but horizontally. In reality, we still can find that many cases like loanwords, borrowings, transliterations and blending do exist in every language, no matter how far away these languages are in origin from each other, which is contradicted to the Theory of Family Tree formulated by August Schleicher. The Family Tree Theory put emphasis on the linear relationship of languages, and it does not allow language convergence as is shown in Fig. 2, even the languages like D and E, which may be quite close to each other in geographical relation, could never come into terms with each other. But even the most irrelevant two languages are now starting to infiltrate into each other one way or another. Language is no longer isolated from its environment due to the fast pace of globalization; Languages contact becomes inevitable.

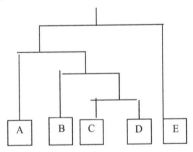

Figure 2

The evidence above reveals that the Theory of Family Tree of Languages has its limitations. It cannot objectively illustrate the actual change of language by taking every aspects of the environment into consideration. Language frequently reacts to its environment through absorbing new information and adapting to new situation. A good case in point is the English language. The Old English adopted a lot of Latin words in its interrelation with Latin after the Romans conquer. Later, the old English again borrowed a lot of words from French after the Norman Conquest. The simile language change occurred so far as many other languages are concerned. So, it can be concluded that language exists in a system of complexity, and language frequently react to the influences of this system in various ways.

The Structuralism's View on Language

In the early 20th century, the Swiss linguist Ferdinand de Saussure, by developing the comparative methods, put forward the structural model, which started the modern linguistics and was a great revolution in the field. According to this theory, language is a structure made up of words organized according to certain rules. Language development is a linear development independent of the environment. Linguistics is an isolated discipline, which has nothing to do with other fields. This theory cannot explain the actual model of language because it depicts language in a 2-dimension way, which contains only words and the grammar of language, without taking the external aspects into consideration.

In reality, language cannot survive without social and environmental soil. Foreign language learners would easily find the cultural and social traces in the vocabulary of the language. For example, English language contains a lot of words and idioms related to the sea, which could be explained as England is a nation on islands; the language of this nation is affected by its location; Arab, on the contrary, is a nation on desert, so, in Arabian language, there must be a lot of words describing desert and camels, which have been proved true. See Fig. 3. If this model tells the truth of language, a language learner could master any language when he or she has the knowledge of the grammar and vocabulary of that language.

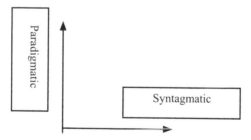

Figure 3

This may be true to the native language learners. When it comes to the second language learning, we find that a person with a large vocabulary and a good command of the grammar of the second language dose not necessarily use the language well, which is often the biggest obstacle for any second language learner (SLL). It is easy to be accepted that language is three-dimensional. Basically, the whole universe is three dimensional, and language is one part of the universe, we would simply draw such a conclusion that language is a 3-dimension body with its own internal organization and the external relations to the universe, as is shown in Table 4. In a whole, language is not a linear structure of words and rules, but a system of complexity, where each element has its own organization and influences each other. This model is what the Chaos Theory describes as the "complexity".

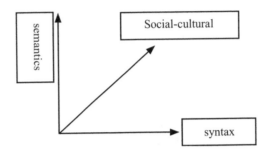

Figure 4

Modern Linguistics and Chaos Theory

Chaos Theory is set of theories and methods developed in different disciplines to understand and research complex, dynamic systems and non-linear development of these systems. Complex systems are more than the sum of their components, are open towards their environment and have the ability of self-organization. With the development of Chaos Theory our knowledge about complex system, their actions and reactions and their internal structure will increase. In many disciplines Chaos theory and its discoveries are successfully applied. According to the Chaos Theory, language is a contributor of a complexity. Language development is not an isolated, independent and linear process, but demonstrates the nature of the complexity. The new model of language development is not a tree-like model, but a river delta web with different branches interrelated with each other, which can be shown in Table 5 below. So language change is a much-involved process, in which both the historical development of the language and the horizontal relationship between this language and other cultures and languages contribute a lot to the process. The study of language change cannot be carried out without taking these two aspects into consideration. And accordingly the synchronic study of any language should involve not only the study of the characteristics of one language, but also the relations between this language and other languages. Language contact, one very common linguistic phenomenon, which was once ignored by some linguists, now gradually draws the attention of many researchers. See Table 5.

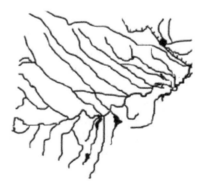

Figure 5

Language Contact and Pidgins

Two Attractors: Strive for New Information and Cling to Tradition

With language contact as a prerequisite of this research based on the Chaos Theory, two important attractors, namely "strive for new information" and "cling to tradition" are introduced into broad light. Being exposed to the environment, a language has mainly two ways to go—progress or decay. It may generate into a higher level of development by adjusting to the new environment, or it may become obsolete and backward because it can hardly keep pace with the time. But there is still a third way to go. It may remain untouched by the new environment but reacts to the environment by taking a mid-way measure.

Due to frequent social contact, a language may inevitably be exposed to a lot of new information. If the language is quite open enough, and it is ready to deal with the new information in an active manner, a lot of new words and expressions will get into this language step by step and finally become a part of the vocabulary of this language. In this case, the force of "strive for new information" of this language is strong, and this language will generate into a higher level of development. This is the first and the most active way in handling language contact. With a large number of new words being absorbed into the language, the morphology and phonology of this language will undergo certain changes, and the new vocabulary from another language will also be modified so that it may easily accepted by the language communities in the target language. A very good case in point is the Old English (OE for short). The OE had lost a lot of its own words and absorbed few thousands words mainly from French, Latin and the languages of other low countries. In this way, the OE gradually generated into Middle English with many loanwords and borrowings, but it was still English. So, from the formation of the vocabulary of English in history we can find the clear clue to the horizontal relationship between English and its environment.

However, if a language is not open enough, or not ready to absorb the new information, it will react to the influences of the environment very passively. At this time, the second way is taken. Facing large amount of new information, this language may directly reflect the input with regular feedback and cling to the tradition. Thus, the vocabulary of the language would be out-of-date for it cannot meet the need of social development. But in order to keep pace with the time, many members of this language community would turn to speaking foreign languages in order to communicate with the people beyond their nation. In the long run, with fewer people go on speaking their own language, this language will gradually lose its front in the fierce competition if no measures are taken to save their language from degenerating. And in reality some languages of the minority groups develop in this direction. Fortunately, actually no language today is absolutely closed or completely open. Being closed or open is a matter of degree instead of a matter of quality.

Pidgins, A Mid Way to Go

Quite often, the above two attractors would form balance. If one national language community (for instance English people) come into terms with one native language community (for instance Chinese) for economic interests or political necessity, but they are reluctant to learn or speak the native language because they feel the superiority of their own language, however, the native language speakers have no interest in learning English because they believe they are least benefited from the communication with this English speaking community, but they are forced to learn to some extent, a balanced situation will emerge. In this case, English speakers will make concessions on their language so as to break the dilemma. They would simplify their language so that the native speakers can easily accept it. The native speakers, on the other hand will accept the simplified language by adopting some phonological, morphological and syntactical patterns of their own language in the new language. Finally a language of blending—pidgin language comes into being. And this is the mid-way taken between the above two ways.

Pidgins, A Special Way in Handling New Information

The Forming of Pidgin

The words pidgin is derived from the Chinese inaccurate pronunciation of the English word "business". Pidgin is a special linguistic phenomenon in many parts of the world other than China. The forming of pidgins in different places displays different characteristics. It is very difficult to define the forming of the special language. But by examining many pidgins in history, Holm presents such an explanation: "A reduced language that results from extended contact between groups of people with no language in common; it evolves when they need some verbal communication, perhaps for trade, but no group learns the native language of the other language for social reasons that may include lack of trust or of close contact."

The Chinese linguist Ye-Baokui states that pidgins refer to the languages form by mixing different natural language elements in the areas of frequent contact. Ralph Fasold puts it this way: "Roughly a pidgin language is generally understood to be a simplified language with a vocabulary that comes mostly from another language".

From the definitions presented by the linguists above, a conclusion could be drawn that pidgin is a contact language. It is special in that it concerns the contact of two or more completely different languages. These languages bifurcated, if the did, at a very early time in history. According to the Evolutionists' view, convergence is almost impossible, not mentioning two completely different languages far away from each other in the family tree. But actually, pidgin gives us a good example of language contact, especially the contact between two obviously different languages. This situation illustrates that language exists in a system of complexity and reacts to this system actively.

Characteristics of Pidgins

The uniqueness of the pidgins lies in how they cope with the new information. Most pidgins were formed from the colonization. "To be specific, pidgins are typically formed in trading and plantation situations in which speakers of one language need to communicate with speakers of another language in a limited way. Under either circumstance, the speakers of both languages are not equal economically and socially. Furthermore, the two languages have distinctive cultural backgrounds. So each party is not willing to accept other's language. To either speaker, the other language is new information. It is necessary to create a new variety out of two or more existing languages, and this process is pidginization, or just as William A. Foley put it "a cultural contact induced change". From the perspective of sound, for the convenience of the economically inferior language speakers, pidgins are likely to be fewer and less complicated in their possible arrangement than those of corresponding standard languages of the economically superior groups. For example, after the Opium War in China, owing to frequent contact with foreigners, a English-based Chinese pidgin was developed in Shanghai. Pidgin words of this kind like "*kangbaidu*" (comprador), "*lasika*" (last car, referring to the last one), "*heluosanmu*" (all same), "*wentaoluo*" (one dollar), "*kangbaiyin*" (combine), "*namowen*" (number one) are comparatively easy to pronounce, but the meanings are mostly English.

The morphological and syntactical structures of the words are also simplified. A pidgin is lack of inflectional endings in nouns, verbs, pronouns and adjectives. The complicated syntactical structures like the clauses are also out of sight in pidgins. All these are concessions made by the socio-economically dominant language. Ye-Fei sheng etc. take Tok Pisin as an example to show the morphological characteristics of pidgins in the book An Outline of Linguistics. Without the inflectional endings, a pidgin language is not as flexible as the dominant language, so outflanking is a common method in description. For example, "grassbelong face" refers to mustache; "jump inside" means surprised; and "inside tell him" means thinking.

The characteristics of pidgins discussed above demonstrate the mid-way measure taken by one language in coping with new information. The unique way a language deals with new information shows that language contact is possible and universal regardless of the distances between the language communities or between the languages themselves.

Where to Go, Pidgin?

Pidgin is a language with no stable speakers. It is an agreement reached by two or more parties for special reason. The future of pidgin languages is unpredictable. If the two or more parties don't come into terms with each other, the language is put aside. If a pidgin can get a large population who speak the pidgin as their mother tongue, the pidgin will become a Creole. This is a diachronic phenomenon that involves a process from quatitive change to qualitive change.

Conclusion

Pidgin is a unique language phenomenon different from the ordinary language varieties. The special feature of pidgin—unpredictability, is a manifestation of Chaos Theory in application. The analysis of the form and the characteristics of pidgin shed some light on the basic model of language development. Language contact is inevitable, but how to react to the influence coming from the environment will determine the future of a language. With such awareness, a language community will consciously take corresponding measures to cope with the system of complexity so that their language could not only survive, but also keep pace with the time.

References

1. Albert C. and Baugh, Thomas Cable. (2001). A History of English Language. Beijing: Foreign Language Teaching and Research Press.
2. J.M. Aitchison. (1981). Language Change: Progress or Decay? London.
3. William A. Foley. (2001). Anthropological Linguistics: An Introduction. Blackwell Publishers Ltd,.
4. Ronald Wardhaugh. (2000). An Introduction to Sociolinguistics. Foreign Language Teaching and Research Press.
5. Ralph Fasold. (2000). The Sociolinguistics of Language. Foreign Language Teaching and Research Press.
6. R. A. Hudson, Sociolinguistics. 2nd edition. (2000). Foreign Language Teaching and Research Press.
7. Diane Larsen-Freeman and Michael H. Long. (2000). An Introduction to SLA Research. Foreign Language Teaching and Research Press.
8. Holm. (1988). Pidgins and Creoles. Cambridge University Press.
9. Gregory Rae, Chaos Theory: A Brief Introduction. [Online] Available: http://www.imho.com/grae/chaos/chaos.html (May 22, 2006).

Chapter 4

Chaos and Natural Language Processing

M. Crisan

Department of Computer and Software Engineering, Polytechnic University of Timisoara, V. Parvan 2, 300223 Timisoara, RO. Email: marius.crisan@cs.upt.ro.

ABSTRACT

The paper discusses the possibility for a new approach in developing models for the dynamics of natural language based on the deterministic chaotic behavior of dynamical systems. The simulation results proved that at the word level the meaning can be modeled dynamically in a unitary way according to the contribution of the component phonemes.

Keywords: natural language modeling, dynamic systems, chaos, semantics.

Introduction

Attempts to explain language and identify a model comprehensive enough to account for language generation and understanding are rooted in ancient times. In modern approaches, scientists are interested in developing computational techniques for both speech (or character/word sequences) recognition and synthesis. An important class of methods of language processing is based on probabilistic models [1], another one uses neural networks (in particular self-organizing maps) [2]–[4], but the main challenge for any approach is dealing with the nonlinear character of language phenomenon. The main progress achieved so far pertains in principal with a concise articulation of the constraints that language obeys rather than to provide a working model of language performance. However, the recent evidences of neural sciences and the development in the domain of dynamical systems may offer promising perspectives in modeling such nonlinear processes as language processing from the chaos theory point of view. Therefore, starting from the fact that natural language phenomenon can be viewed as a dynamical system, the purpose of this work is to investigate the possibility of modeling the linguistic components by chaotic attractors.

Meaning and Dynamical Systems

We start by presenting some results obtained in previous works [5–6]. By defining meaning as something that must have a finite description, we introduced the concept of undivided meaning-whole (UMW). This is structured information which exists internally at the knowledge-base level. Although UMW is a unitary information structure, it can be described rationally in terms of cognitive semantic units. These semantic units form the generating principle of producing the sequence of uttered words.

When an agent wants to communicate, it starts from the UMW which exists internally in its knowledge base (KB). A sentence (utterance) is significant or meaningful if it can generate knowledge in an ideal receiver (reader or hearer). This knowledge is a result of a reaction mechanism triggered by the series of words in the sentence. An ideal receiver has to possess the "capacity" to extract meaning from a sentence. This capacity is what qualifies an ideal receiver, and can be described by the cognition of four cognitive properties assigned to a sentence by the transmitter: (1) semantic competency, (2) expectancy, (3) contiguity in space and time, and (4) transmitter's intention. These cognitive properties are the requirements for defining a grammatical and meaning-bearing sentence. When words are uttered producing different sounds in sequence, only apparently there is differentiation. Ultimately, the sound sequence is perceived as a unity or UMW and only then the word meaning, which has also to be present in the receiver's mind, is identified.

The above described capacity of the receiver to extract meaning from series of words is dependent on another assumption. The meaning of the whole word/sentence has to be inherently present in the KB of the receiver. In this way, it can be explained how it is possible to grasp the meaning even before the whole word or sentence has been uttered. The sounds which differ from one another because of difference in pronouncement cause the cognition of the one changeless UMW without determining any change in it. Sometimes, reasoning may have to be applied to the components of the sentence in order to clear the cognition and make possible the perception of the meaning-whole. It appears that the unitary word-meaning is an object of each agent's own cognitive perception. When a word, such as "tree" is pronounced or read there is the unitary perception or simultaneous cognition of trunk, branches, leaves, fruits, etc. in the receiver's mind. Communication (verbal or written) between peoples is only possible because of the existence of the UMW which is potentially perceivable by everyone, and is only revealed by words' sounds or symbols.

The concept of UMW is consistent with a more general view, suggested by Bohm in [7], regarding the possibilities for wholeness in the quantum theory to have an objective significance. This is in contrast with the classical view which must treat a whole as merely a convenient way of thinking about what is considered to be in reality nothing but a collection of independent parts in a mechanical kind of interaction. If wholeness and non-locality is an underlying reality then all the other natural phenomena must, one way or another, be consistent with such a model. We suggest that natural language generation and understanding is a phenomenon that might be modeled in such a way. UMW is like "active information" in Bohm's language, and is the activity of form, rather than of substance. As Bohm puts it suggestively [7], "...when we read a printed page, we do not assimilate the substance of the paper, but only the forms of the letters, and it is these forms which give rise to an information content in the reader which is manifested actively in his or her subsequent activities." Similar so called mind-like quality of matter reveals itself strongly at the quantum level. The form of the wave function manifests itself in the movements of the particles. From here, a new possibility of modeling the mind as a dynamical system may be considered.

In line with Kantian thought, in [8] we find a similar insight, as above, regarding the linguistic apprehension. This is the interplay of two factors of different levels: (1) the empirical manifold of the separate letters or words and (2) the *a priori* synthesis of the manifold which imparts a unity to those elements which would otherwise have remained a mere manifold.

According to the above statements it appears motivated to use the concept of manifold for modeling the mind as the seat of language generation and understanding. Manifolds are defined as topological spaces possessing families of local coordinate systems that are related to each other by coordinate transformations pertaining to a specific class. They may be seen also as the multidimensional analogue of a curved surface. This property seems suitable to represent both the natural language constraints and semantic content of linguistic objects.

Usually, a dynamical system is a smooth action of the real numbers or the integers on a manifold. The manifold is the state space or phase space of the system. Having a continuous function, F, the evolution of a variable x can then be given by the equation:

$$x_{t+1} = F(x_t). \tag{1}$$

The same system can behave either predictably or chaotically, depending on small changes in a single term of the equations that describe the system. Equation (1) can also be viewed as a difference equation ($x_{t+1} - x_t = F(x_t) - x_t$). It can generate iterated maps. An important property of dynamical systems is that even very simple systems, described by simple equations, can have chaotic solutions. This doesn't mean that chaotic processes are random. They follow rules, but even the simple rules can produce amazing complexity. In this regard, another important concept in our approach is that of an attractor. An attractor is a region of state space invariant under the dynamics, towards which neighboring states in a given basin of attraction asymptotically approach in the course of dynamic evolution. The basin of attraction defines the set of points in the space of system variables such that initial conditions chosen in this set dynamically evolve to a particular attractor. It is important to note that a dynamical system may have multiple attractors that may coexist, each with its own basin of attraction [9]. This type of behavior appears suitable for modeling self-organizing processes, and is considered to be a condition for a realistic representation of natural processes.

One example of a self-organizing model is the topological feature map, proposed by Kohonen [10] [11], for the projection of high dimensional pattern data into a low-dimensional feature space. The process of ordering an initial random map is named self-organization. The result is the topological ordering of pattern projections, or in other words the self-organizing map (SOM). Each input dimension is called a feature and is represented by an N-dimensional vector. Each node in the SOM is assigned an N-dimensional vector and is connected to every input dimension. The components or weights of this vector are adjusted following an unsupervised learning process. First, it is found the winning node, i.e., the node whose weight vector shows the best match with the input vector in the N-dimensional space. Next, all weight vectors in the neighborhood in the direction given by the input vector are adjusted. This process requires numerous iterations until it converges, i.e., all the adjustments approach zero. It begins with a large neighborhood and then gradually reduces it to a very small neighborhood. Consequently, the feature maps achieve both ordering and convergence properties, and offer the advantages, of reducing dimensions and displaying similarities. However, SOM solutions (and neural networks in general) are yet in the need for improvement. For instance, in [12], an important problem for SOM is discussed. In order to obtain a realistic speech projection, the problem is to find a hypercubical SOM lattice where the sequences of projected speech feature vectors form continuous trajectories. In another work [13], both SOM and a supervised multilayer perceptron were used for bird sounds recognition. The conclusion was that although the tested algorithms proved to be quite robust recognition methods for a limited set of birds, the proposed method cannot beat a human expert listener. Considering these limitations, a new approach in the field may be necessary. The insufficiently explored domain of dynamical systems and chaos theory may offer promising perspectives in modeling natural processes, and natural language might be one of them.

Word Modeling

In quantum experiments, when particles interact, it appears they were all connected by indivisible links into a single whole. The same behavior is manifested by the chaotic solutions in an attractor, as we will see in this section. In spite of the apparent random behavior of both phenomena, there is an ordered pattern given by the form of the quantum wave (or potential) in the former case, and by the equations of the dynamic system in the latter.

Let's consider the simplest case of the quadratic iterated map described by the equation:

$$x_{t+1} = a_1 + a_2 x_t + a_3 x_t^2 \tag{2}$$

Although so simple, it is nonlinearly stable and can manifest chaotic solutions. The initial conditions may be drawn to a special type of attractor called a chaotic attractor. This may appear as a complicated geometrical object which gives the form of the dynamic behavior.

In nonlinear dynamics the problem is to predict if a given flow will pass through a given region of state space in finite time. One practical way to decide if the nonlinear system is stable is to actually simulate the dynamics of the equation. The primary method in the field of nonlinear dynamic systems is simply varying the coefficients of the nonlinear terms in a nonlinear equation and examining the behavior of the solutions. The initial values of the components of the model vector, $m_i(t)$, were selected at random in a process of finding a chaotic attractor. Strange attractors are bounded regions of phase space corresponding to positive Lyapunov exponents. We found easily more than 100 chaotic attractors. In Table 1 we presented a list of several coefficients along with the Lyapunov exponent for which the attractors were found by random search. The initial condition x_0 was selected in the range 0.01–1 and lies within the basin in many cases. The Lyapunov exponent is computed in an iterated process according to the following equation [14]:

$$LE = \Sigma \log_2 |a_2 + 2a_3 x_n| /N \qquad (3)$$

Table 1 The coefficients values and the lyapunov exponent for 10 attractors of (2).

Attractor	a_1	a_2	a_3	LE
1	1.2	−0.9	−0.9	0.3106
2	1.1	−1	−0.6	6.6073
3	1.1	−0.7	−0.9	0.1538
4	0.8	−1.1	−1	0.2805
5	0.7	−1.2	−0.8	0.2001
6	−0.4	−1.2	1.2	0.3144
7	−0.7	−1.1	1.2	0.3033
8	−0.8	−1.1	0.7	6.9382
9	−0.8	−0.9	1.1	0.2214
10	−1.2	−0.9	0.8	0.2793

The sum is computed from the value of $n=1$ to the value of $n=N$, where N is some large number. The exponent LE gives the rate of exponential divergence from perturbed initial conditions. If the value is positive (for instance, greater than 0.005) then there is sensitivity to initial conditions and a chaotic attractor can manifest. If the solution is chaotic, the successive iterates get farther apart, and the difference usually increases exponentially. The larger the LE, the greater is the rate of exponential divergence, and the wider the corresponding separatrix of the chaotic region. If LE is negative, the solutions approach one another. If LE is 0 then the attractors are regular. They act as limit cycles, in which trajectories circle around a limiting trajectory which they asymptotically approach, but never reach.

It's interesting to analyze into more details the behavior of a chaotic attractor. The idea in the case of the self-organizing maps is to project the N-dimensional data into something that be better understood visually. A similar idea we follow in constructing iterated maps. It is convenient to plot the values in the iterated process versus their nth previous iterate for a more suggestive aspect. In Fig. 1 it is presented the iterated map for the attractor No. 9, for $n=4$. A remarkable property of the chaotic solutions, as noted above in connection with quantum physics, is the "ballet-like" behavior as iterations progress. Each new dot on the map, which represents the solution x_{n+1}, appears in a random position but orderly following the chaotic attractor's form.

In Fig. 2 it is shown the same attractor only after a few iterates (80). It can be seen the sparse distribution of dots but along with the ordered path. This type of behavior is similar with the quantum phenomena, as in the case of the distribution of photons along

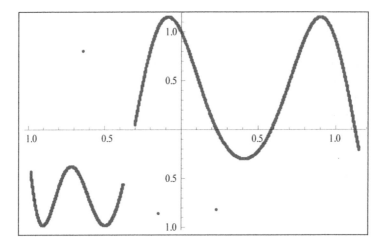

Figure1 Quadratic iterated map of (2).

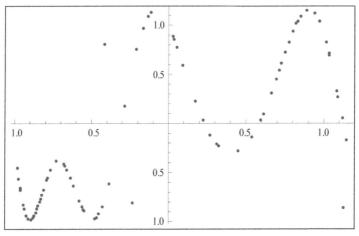

Figure 2 Quadratic iterated map of (2) after 80 iterates. Note the sparse distribution of dots along the regular pattern of the strange attractor.

the interference pattern lines in the two-slit interference experiment, when the photons are emitted in series one after the other. Interesting enough, this is also akin to the quality of the perception act (understanding word meaning). It's an established observation fact that a word meaning is at first perceived vaguely and then more and more clearly. Thus, through the process of repeated perception or iterations finally the meaning is revealed. Therefore, we may suggest that meaning can be mathematically modeled as a basin of attraction.

Another interesting property is the symmetry between a_1 and a_3 and the corresponding iterated map. Considering again the chaotic attractor $a_1=-0.8$, $a_2=-0.9$, $a_3=1.1$, a symmetric behavior can be obtain for the values $a_1=0.8$, $a_2=-0.9$, $a_3=-1.1$ as in Fig. 3.

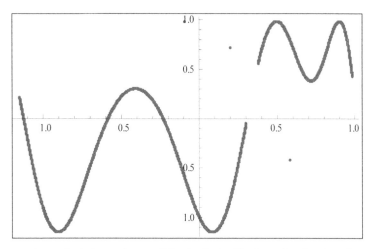

Figure 3 The symmetric quadratic iterated map of Fig. 1, obtained by inverting the sign of a_1 and a_3.

There is a large possibility to obtain other attractors by tuning the values of the coefficients. The shape of the attractor changes smoothly with small variations of the coefficients. Even if the interval of variation is rather small, important changes in the shape of the map can be obtained. In Fig. 4, the dynamic behavior of (2) can be observed for a_1 = –1.3, a_2 = –0.65, and a_3 = 0.8. If –1.38 ≥ a_1 ≥ –0.94 (a_2 = –0.9 and a_3 = 0.8) the value of *LE* is negative and fixed point patterns manifest. Trajectories approach a limit cycle for a_1 = –1.3, a_2 = –1, and a_3 = 0.9615. If a_3 > 0.9615 solutions grow unbounded.

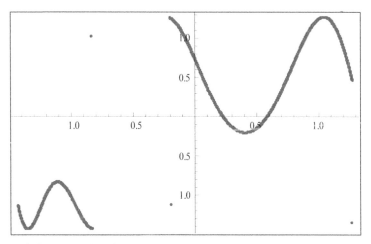

Figure 4 The dynamic behavior of (2) for a_1 = –1.3, a_2 = –0.65, and a_3 = 0.8.

An important change in shape can be obtained for a_1 = –1.3, a_2 = –1.06, and a_3 = 0.8, with the value of LE = 0.3648, as shown in Fig. 5.

The above analysis revealed the fact that chaotic attractors offer dynamic properties that can map in a continuous manner the feature vectors according to some input patterns.

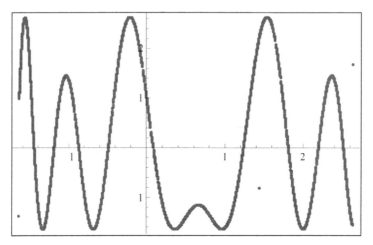

Figure 5 A different behavior of (2), obtained for $a_1 = -1.3$, $a_2 = -1.06$, and $a_3 = 0.8$.

In the process of language communication, the dynamics of each phoneme, as it is uttered, has a contribution to the dynamics of the entire word. The goal is to construct a unified word feature that may account for the word meaning or UMW, by encapsulating the phonemes' dynamics into a unitary description of a chaotic attractor.

For a generic word w, composed by a series of m phonemes $p_1 p_2 \ldots p_m$, the word feature vector is $W = [P_1, P_2, \ldots P_m]$, where P_i, $i=1$, m, are the quadratic maps (2) corresponding to each phoneme. In order to encapsulate the phonemes' dynamics into a resulting attractor at the word level, we explored two possibilities: (i) to map the phonemes' attractors as coefficients of a higher-order polynomial type equation, and (ii) to linearly superpose the phonemes' attractors.

In the first approach, the quadratic maps P_i, $i = 1$, m, form the coefficients of the following polynomial type equation,

$$w_{t+1} = k_1 + k_2(P_{1t}w_t + P_{2t}w_t^2 + \ldots + P_{mt}w_t^n), \tag{4}$$

where k_1 and k_2 are scale parameters. Eq. (4) describes the chaotic behavior at the word level.

Each valid word of length m will determine a corresponding attractor with a unique dynamic behavior. Small variations in the input will be tolerated and recognized with the same meaning, but other illegal combinations will be rejected. For words with higher length, similar higher-order iterated maps can be used.

A second possibility is to use a linear superposition of P_i, $i= 1$, m, of the following form:

$$w_{t+1} = z_1 P_{1t} + z_2 P_{2t} + \ldots z_{m-1} P_{m-1t} + P_{mt}, \tag{5}$$

where $z_1, \ldots, z_{m-1}$ are subunitary superposition parameters. These parameters account for the progressive accumulation of the individual phoneme dynamics into the word meaning as the phonemes are uttered in sequence.

In order to exemplify our approach, let's consider the phonemes /a/, /e/, /d/, and /r/ as they may form the English words *dear* and *dare*. We disregard here the pronunciation issue and consider only the phonemes as phonetic letters. The corresponding feature vectors $A = [a_1, a_2, a_3]$, $E = [e_1, e_2, e_3]$, $D = [d_1, d_2, d_3]$ and $R = [r_1, r_2, r_3]$, are mapped by the following equations:

$$a_{t+1} = a_1 + a_2 a_t + a_3 a_t^2, \tag{6}$$

$$e_{t+1} = e_1 + e_2 e_t + e_3 e_t^2, \tag{7}$$

$$d_{t+1} = d_1 + d_2 d_t + d_3 d_t^2, \tag{8}$$

$$r_{t+1} = r_1 + r_2 r_t + r_3 r_t^2, \tag{9}$$

where a_t, e_t, d_t, and r_t are the dynamic variables. The four trajectories (6)–(9) are presented in Fig. 6 for the following feature vectors: $A = [-0.9, -1.6, 0.6]$, $E = [-1, -1, 0.7]$, $D = [0.5, -1.4, -0.6]$ and $R = [0.8, -1.1, -1]$.

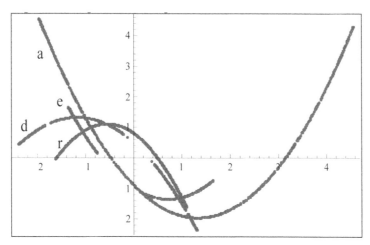

Figure 6 Chaotic attractors for phonemes /a/, /e/, /d/, and /r/.

The resulting attractor for the word *dear* is constructed according to

$$w_{dear(t+1)} = k_1 + k_2(d_t w_t + e_t w_t^2 + a_t w_t^3 + r_t w_t^4), \tag{10}$$

where k_1 and k_2 are scale coefficients, and is represented in Fig. 7.
Similarly, the resulting attractor for the word *dare* appears in Fig. 8, according to

$$w_{darc(t+1)} = k_1 + k_2(d_t w_t + a_t w_t^2 + r_t w_t^3 + e_t w_t^4). \tag{11}$$

There is a clear difference between the dynamics of (10) and (11), although common trajectory patterns can be indentified in both chaotic attractors. This is according to our expectations since both words are composed of the same phonemes. Concomitantly, the meanings encapsulated by the respective words dynamics are clearly different.

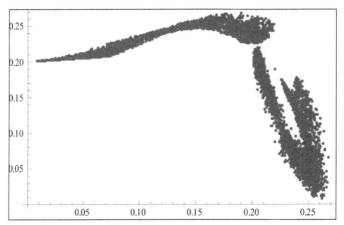

Figure 7 Chaotic attractor for *dear* according to (10).

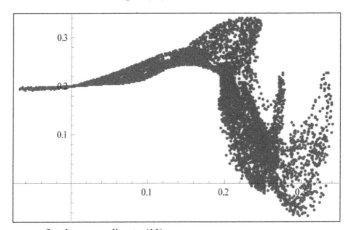

Figure 8 Chaotic attractor for *dare* according to (11).

In the second approach involving linear superposition, according to (5), the dynamics of the word *dear* is modeled as follow:

$$w_{\text{dear}(t+1)} = z_1 d_t + z_2 e_t + z_3 a_t + r_t, \tag{12}$$

where $z_1 < z_2 < z_3 < 1$. The resulting chaotic behavior is shown in Fig. 9. It is interesting to compare the dynamics of (12) with that of (10), for the same word *dear*. They are different because the process of phonemes' encapsulation is linear in (12) and nonlinear in (10). However, the dynamic contribution of the component phonemes is suggestively captured in both cases.

A similar linear superposition can be used for the word *dare* in the following form:

$$w_{\text{dare}(t+1)} = z_1 d_t + z_2 a_t + z_3 r_t + e_t, \tag{13}$$

keeping the same superposition parameters as in (12). The dynamics of (13) appears in Fig. 10. When comparing to (12), a clear global difference can be noticed concomitantly with the identification of common trajectories patterns.

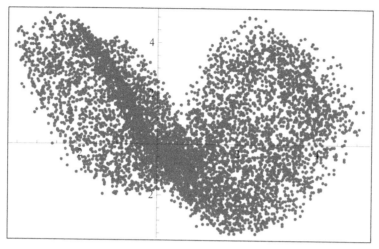

Figure 9 Chaotic attractor for *dear* according to (12).

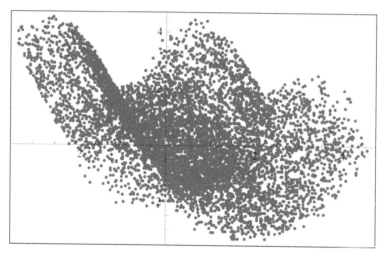

Figure 10 Chaotic attractor for *dare* according to (13).

The simulation results for both linear and nonlinear superposition of phonemes' dynamics have proved the validity of the dynamic approach in modeling meaning and semantics. The model can also account the synthetic interplay between the separate linguistic components and the ultimate unitary manifestation of meaning.

Conclusions

Our purpose was to study the possibility of using dynamical systems in modeling the dynamics of natural language. Of particular interest in any model of meaning, and semantics in general, is to account the interplay process of the empirical manifold of individual phonemes or separate words and the unitary characteristic of meaning as a whole. We started from the premise of UMW and the observation facts of language apprehension and noted a similitude with the chaotic behavior of dynamical systems. The attractor behavior as studied for the quadratic iterated map seems to be robust enough to accept feature vectors for phonemes that may compose any word of length m in the dictionary. The separate dynamics of the phonemes participate in the manifestation of a unique dynamic behavior of the entire word that may represent the UMW. Two approaches have been proposed for modeling the formation of word's dynamics based on individual phonemes, and the simulation results proved to be satisfactory in both cases.

References

1. D. Jurafsky, J.H. Martin, *Speech and Language Processing: An Introduction to Natural Language Processing, Computational Linguistics, and Speech Recognition*, Prentice-Hall, 2000.
2. T. Kohonen, "Self-organizing maps of symbol strings", Report A42, Helsinki University of Technology, Laboratory of Computer and Information Science, Espoo, Finland, 1996.
3. T. Kohonen, P. Somervuo, "Self-organizing maps of symbol strings with application to speech recognition", in Proc. of Workshop on Self-Organizing Maps (WSOM'97), pp. 2–7, Espoo, Finland, 1997.
4. T. Honkela., "Self-Organizing Maps in Natural Language Processing", Espoo, Finland, 1997.
5. M. Crisan, "Meaning as Cognition," *Proceedings of the I International Conference on Multidisciplinary Information Sciences and Technologies-InSciT2006*, Merida, Spain, 2006, pp. 369–373.
6. M. Crisan, "Chaos and Natural Language Processing," *Acta Polytechnica Hungarica*, Vol. 4, No.3, 2007, pp. 61–74.
7. D. Bohm, "A new theory of the relationship of mind and matter," *Philosophical Psychology*, Vol. 3, No. 2, 1990, pp. 271–286.
8. H.G. Coward, *The Sphota Theory of Language*, Motilal Banarsidass Delhi, 3rd ed. 1997.
9. G. Pulin and X. Jianxue, "On the multiple-attractor coexisting system with parameter uncertainties using generalized cell mapping method," *Journal Applied Mathematics and Mechanics*, Vol. 19, No. 12 December, 1998, pp. 1179–1187.
10. T. Kohonen, "Self-organized formation of topologically correct feature maps," *Biological Cybernetics*, 43: 59–69, 1982.
11. T. Kohonen, *Self-Organizing Maps*, Springer, Berlin (3rd extended ed. 2001), 1997.
12. P. Somervuo, "Speech Dimensionality Analysis on Hypercubical Self-Organizing Maps," *Neural Processing Letters*, Vol. 17-2, April 2003, pp. 125–136.
13. A. Selin, J. Turunen, and J. T. Tanttu, "Wavelets in Recognition of Bird Sounds," *EURASIP Journal on Advances in Signal Processing*, Vol. 2007, Article ID 51806, 9 pages, doi:10.1155/2007/51806.
14. J. C. Sprott, *Chaos and Time-Series Analysis*, Oxford University Press, 2003.

Chapter 5

Discrete Phase-Locked Loop Systems and Spreadsheets[†]

Sergei Abramovich,[1] Elena Kudryashova,[2,a] Gennady A. Leonov[2,b]
and Steve Sugden[3]

[1]State University of New York at Potsdam, USA. Email: abramovs@potsdam.edu.
[2]Saint Petersburg State University, Russia.
[a]Email: elena@centrobalt.spb.ru.
[b]Email: leonov@math.spbu.ru.
[3]Bond University, Australia. Email: ssugden@staff.bond.edu.au.

ABSTRACT

This paper demonstrates the use of a spreadsheet in exploring non-linear difference equations that describe digital control systems used in radio engineering, communication and computer architecture. These systems, being the focus of intensive studies of mathematicians and engineers over the last 40 years, may exhibit extremely complicated behavior interpreted in contemporary terms as transition from global asymptotic stability to chaos through period-doubling bifurcations. The authors argue that embedding advanced mathematical ideas in the technological tool enables one to introduce fundamentals of discrete control systems in tertiary curricula without learners having to deal with complex machinery that rigorous mathematical methods of investigation require. In particular, in the appropriately designed spreadsheet environment, one can effectively visualize a qualitative difference in the behavior of systems with different types of non-linear characteristic.

Keywords: phase locked loops, clock skew, difference equations, spreadsheet modeling, orbit diagrams, period-doubling bifurcations, tertiary education.

Introduction

The use of an electronic spreadsheet to support the exploration of topics in exact sciences and engineering through computing is well documented. A monograph edited by Filby

[†]Reused with permission from: S. Abramovich, E. Kudryashova, G.A. Leonov, and S.J. Sugden, Discrete Phase-Locked Loop Systems and Spreadsheets, *Spreadsheets in Education*, 2(1), 23–49, 2005.

[16] includes well-chosen examples from a number of areas of contemporary science and engineering, including biology, chemistry, geology and electronics. Kreith and Chakerian [21] focus on the use of a spreadsheet in modeling iterative mathematical structures ranging from Fibonacci numbers to fractals. Neuwirth and Arganbright [36], using a bicycle metaphor as a reference to software simple in structure yet powerful in educational applications, presented a spreadsheet as a user-friendly and self-sufficient exploratory tool enabling mathematical modeling of various concepts studied across the secondary and tertiary mathematics curriculum. Silva [43] used a spreadsheet as a modeling and simulation tool for studying electrical circuits. El-Hajj with co-authors [13,14] used a spreadsheet as a tool for the simulation and modeling of various systems of automatic control, both linear and nonlinear, in order to avoid the use of complicated mathematical methods in the context of engineering education. Levin and Talis [30] used a spreadsheet in the context of teaching academically challenging topics in digital design associated with concurrent error detection.

The material of this paper stems from a number of educational applications of the software at the tertiary level, including its use in discrete mathematics course at Bond University, in secondary mathematics teacher education at SUNY Potsdam, and in the framework of collaboration between Intel Corporation and the Faculty of Mathematics and Mechanics of Saint Petersburg State University on a project "Control in the Distributed Systems of Clock Pulse Generators in Multiprocessor Clusters." One of the topics included in the project is the research and development of mathematical theory of discrete phase locked loops for array processors commonly used in radio engineering, communication, and computer architecture [24]. Such digital control systems showed their high efficiency in eliminating a clock skew—an undesirable phenomenon arising in parallel computing. To clarify, consider Fig. 1 in which clock C sends pulses via conduit L to parallel processors P_k. The work of parallel algorithms requires that processors involved perform certain operations simultaneously. However, each clock pulse travels a different distance in time to drive each processor. As a result, a time-discordance in the activation of the processors arises. This phenomenon is called *clock skew*. Digital phase-locked loops have gained widespread recognition and preference over their analog counterparts because of their ability to deal with this phenomenon effectively. From a mathematical perspective, this gives rise to a problem associated with the analysis of global stability of non-linear difference equations that serve as mathematical models of the phase-locked loops [25]; that is, the analysis can be formulated in terms of parameters for such systems.

It should be noted that the study of difference equations (alternatively, iterative maps) has a long and fascinating history. Leonardo Pisano (a 13th century Italian mathematician better known by his nickname Fibonacci) was probably the first to study the relation (a second order difference equation)

$$x(t+2) = x(t+1) + x(t), t \in Z \tag{1}$$

satisfying the initial conditions $x(0)=x(1)=1$ (hereafter Z denotes the set of whole numbers) in connection with his investigations of the growth of the population of rabbits breeding in ideal circumstances.

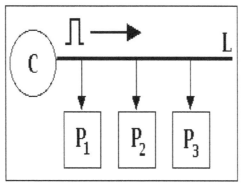

Figure 1 Travel of clock pulses.

In 1843, a French mathematician Jacques Binet showed that the solution to equation (1) satisfying the given initial conditions has the following closed representation

$$x(t) = \frac{1}{\sqrt{5}}\left[\left(\frac{1+\sqrt{5}}{2}\right)^{t+1} - \left(\frac{1-\sqrt{5}}{2}\right)^{t+1}\right], t \in Z \qquad (2)$$

which was known, however, more than a century earlier to Euler and Daniel Bernoulli [44] as well as to De Moivre [20]. Despite its rather complex form (resembling a solution to a linear differential equation of the second order), sequence (2) generates integers only, nowadays commonly known as Fibonacci numbers.

In 1845, a Belgian mathematician Pierre Verhulst introduced the following non-linear difference equation of the first order

$$x(t+1) = rx(t)(1-x(t)), t \in Z, r > 0 \qquad (3)$$

as a mathematical model of population dynamics within a closed environment that takes into account internal competition [38]. Since then, equation (3) which can be generalized to the form

$$x(t+1) = f(x(t)), t \in Z \qquad (4)$$

has been studied intensively. However, only in the second part of the 20th century, due to the efforts of many mathematicians, the limiting structure of solutions to equation (4) has come to be recognized as extremely complex [41,31,42,34,35,15]. In particular, in equation (3), period-doubling bifurcations were discovered. Surprisingly, whereas solutions to a linear multidimensional discrete equation $x(t+1) = Ax(t), x \in R^n, t \in Z$ and its continuous analogue $dx/dt = Ax(t), x \in R^n, t \in N$ (here A is a constant $n \times n$-matrix), to a large extent, as formula (2) may suggest, possess similar behavior, solutions to equation (4) and its continuous one-dimensional analogue $dx/dt = f(x(t)), x,t \in R$, bear qualitatively different structure.

Another example of a non-linear difference equation important in applications is the following equation of the first order

$$x(t+1)= x(t)-\alpha \sin x(t)+\gamma, \quad t \in N \tag{5}$$

where α and γ are non-negative parameters. Over the last 40 years, many authors conducted rigorous studies of equation (5) both as a pure mathematical object [6,19] and as a mathematical model of a phase-locked loop [37,17,32,33]. Osborne's [37] pioneering use of exact methods, such as the Contraction Mapping Theorem, applicable to the direct study of non-linear effects in digital phase-locked loops described by equation (5) with $\gamma = 0$, enabled for the discovery of inadequacy of the methods of linearization as means for understanding their complex behavior. However, even the exact methods used by Osborne, while revealing what then appeared as multiple cycle slipping followed by a divergent behavior of iterations, did not allow for a precise interpretation of non-linear effects discovered as transition from global asymptotical stability to chaos through period-doubling bifurcations [27]. In particular, in the latter study, the values of parameter α in equation (5) with $\gamma = 0$ that correspond to period-doubling bifurcations have been found by using a combination of analytical and computational methods. In the present paper, a spreadsheet is used as a self-sufficient mathematical/pedagogical tool allowing for both finding these values and further refining them.

Global Asymptotical Stability as a Desirable Regime in Phase-locked Loops

Consider a difference equation

$$x(t + 1) = x(t) - \alpha \sin(x(t)), \quad t \in Z, \alpha > 0, \tag{6}$$

which serves as a mathematical model of the simplest discrete phase-locked loop without filters with a sinusoidal characteristic of its phase detector and zero frequency misalignment. Equation (6), being a recurrence relation, has a unique solution for any initial value $x(0)$. The purpose of control in digital phase-locked loops is the elimination of a clock skew for almost all initial values $x(0)$ so that $x(t)$ tends to a stationary solution of equation (6). By equating $x(t+1) = x(t)$, one can conclude that equation (6) have two stationary solutions, $x(t)= 2\pi j$ and $x(t)=(2j+1)\pi$. As demonstrated in [27], the former solution is asymptotically stable and the latter solution is Lyapunov unstable. Therefore, whereas the unstable solution can not be realized physically, the elimination of clock skew in the phase-locked loop described by equation (6) can be guaranteed if the relations $\lim_{t \to +\infty} x(t) = 2 j\pi$ and $\lim_{t \to +\infty} (x(t+1) - x(t)) = 0$ hold true for almost all orbits $x(0)$. In turn, such property of solutions ensures a global asymptotical stability of equation (6).

As mentioned in [26], the global asymptotical stability of equation (6), in mathematical terms is equivalent to the existence of a function defined on its solutions and satisfying several conditions. One such condition requires the function's derivative to be negative. It can be shown [27] that the fulfillment of this condition implies the inequality $\alpha < 2$. Thus equation (6) is globally asymptotically stable for any $\alpha \in (0,2)$.

The global asymptotical stability of equation (6) as a physical phenomenon can be demonstrated in a variety of forms within a spreadsheet by modeling equation (6) as a recurrence relation both numerically and graphically. First, this modeling may include the construction of a bifurcation diagram which shows the limiting behavior of iterations as parameter α varies along the interval $(0,2)$ and then extends beyond it. Second, based on the results of modeling through the construction of this diagram, one can narrow the region of parameter α and explore equation (6) as a recurrence relation. In what follows, a spreadsheet will be presented as a user-friendly tool in modeling equation (6) for different values of parameter α. In such a way, as several authors have argued regarding the effect of technology on curriculum [18,9,3], embedding advanced mathematical ideas in a technology tool like a spreadsheet enables for the investigation of problems without learners having to deal with complex machinery of non-linear control theory that mathematically rigorous methods of investigation require.

Spreadsheet-based Approach to Mapping a Segment into Itself

The problem of mapping a segment into itself that will be discussed in this section in connection with the behavior of orbits defined by equation (6) is a special case of a more general problem of mapping the circle into itself. In the specific case of phase-locked loops, this more general problem was studied in [39,40,7]. Whereas a spreadsheet's computational capability allows for the representation of problems involving complex variables by dealing separately with real and imaginary parts [5,10], in what follows the discussion will be limited to a spreadsheet-based demonstration of what it means for the segment to be mapped into itself.

In exploring the behavior of orbits generated by equation (6), one has to be concerned with only those $x(t)$ that do not leave the segment $\left[-\pi,\pi\right]$ as t grows larger. This gives rise to the following question: What is the range of parameter α that keeps orbits $x(t)$, whatever their qualitative behavior is, being bounded by π in the absolute value? With this in mind, consider the function $g(x) = x - \alpha \sin x$, for $\alpha > 2$. The use of a spreadsheet in combination with the basic techniques of differential calculus makes it possible to find those values of α for which the function $g(x)$ maps the segment $\left[-\pi,\pi\right]$ into itself. To this end, one can begin with constructing an interactive graphing environment allowing for the simultaneous variation of a (slider-controlled) parameter and the corresponding graph of a function that depends on this parameter. By using a slider that controls the variation of α (Fig. 2, cell B2), one can first locate $\alpha = 4.6$ (Fig. 3) as an approximate value of the parameter such that when $\alpha > 4.6$ the inequality $|g(x)| \leq \pi$ cannot be satisfied for all $x \in [-\pi,\pi]$ (Fig. 4).

In order to find a more accurate value of α than the one found through graphing, note that $g(-x) = -g(x)$ and $g'(x) = 1 - \alpha \sin x$. Therefore, the equation $g'(x) = 0$ has two roots: $\pm \arccos(\frac{1}{\alpha})$, in the interval $(-\pi,\pi)$. The graph pictured in Fig. 3 suggests that $g\left[-\arccos\dfrac{1}{\alpha}\right] = -g\left[\arccos\dfrac{1}{\alpha}\right] = \pi$; thus the value of α should be chosen to satisfy the equation $-\arccos(\frac{1}{\alpha}) + \alpha \sin[\arccos(\frac{1}{\alpha})] = \pi$, whence

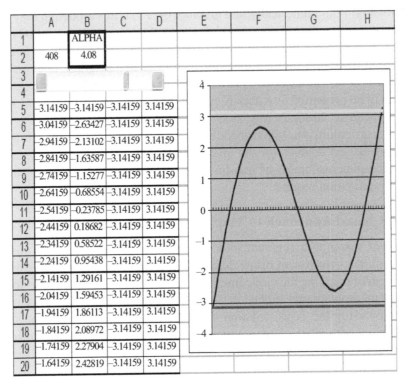

	A	B	C	D
1		ALPHA		
2	408	4.08		
3				
4				
5	-3.14159	-3.14159	-3.14159	3.14159
6	-3.04159	-2.63427	-3.14159	3.14159
7	-2.94159	-2.13102	-3.14159	3.14159
8	-2.84159	-1.63587	-3.14159	3.14159
9	-2.74159	-1.15277	-3.14159	3.14159
10	-2.64159	-0.68554	-3.14159	3.14159
11	-2.54159	-0.23785	-3.14159	3.14159
12	-2.44159	0.18682	-3.14159	3.14159
13	-2.34159	0.58522	-3.14159	3.14159
14	-2.24159	0.95438	-3.14159	3.14159
15	-2.14159	1.29161	-3.14159	3.14159
16	-2.04159	1.59453	-3.14159	3.14159
17	-1.94159	1.86113	-3.14159	3.14159
18	-1.84159	2.08972	-3.14159	3.14159
19	-1.74159	2.27904	-3.14159	3.14159
20	-1.64159	2.42819	-3.14159	3.14159

Figure 2 $|g(x)| < \pi, x \in (-\pi, \pi), \alpha = 4.08.$

$$\sqrt{\alpha^2 - 1} = \pi + \arccos(\frac{1}{\alpha}) \tag{7}$$

Equation (7) can be solved numerically within a spreadsheet using the method of iterations as described in [29]. To this end, equation (7) can be rewritten in the form

$$\alpha = \sqrt{1 + (\pi + \arccos\frac{1}{\alpha})^2}, \tag{8}$$

enabling the construction of the following recurrence

$$\alpha_{n+1} = \sqrt{1 + (\pi + \arccos\frac{1}{\alpha_n})^2} \tag{9}$$

which can be iterated by using the spreadsheet's remarkable feature of recurrent counting.

The result of iterating recurrence relation (9) with $\alpha_0 = 4.5$ is shown in Fig. 5. Note that if after several iterations, the equality $\alpha_{n+1} = \alpha_n$ is satisfied within a specified precision, then α_n is the root of equation (8) with the same precision. The very precision of iterations can be controlled by a slider. To this end, one can use the spreadsheet function TRUNC (number, num_digits) which, when applied to iterations of sequence (9) displayed

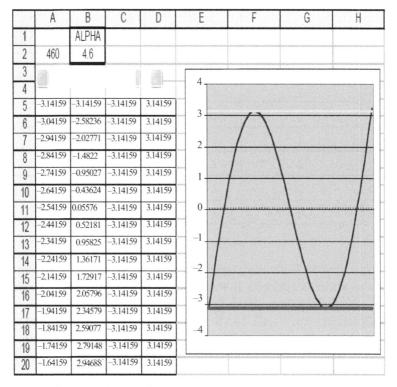

Figure 3 $\max\limits_{x\in(-\pi,0]} g(x) = \pi$; $\min\limits_{x\in[0,\pi)} g(x) = -\pi$.

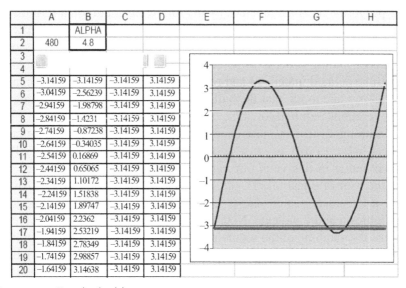

Figure 4 For $\alpha > 4.6$ $\exists x_0$, $|g(x_0)| > \pi$.

		A	B	C
1	number of iterations	alpha		precision
2	0	4.5		7
3	1	4.5983475		
4	2	4.603103		
5	3	4.6033277		
6	4	4.6033383		
7	5	4.6033388		
8	6	4.6033388		

Figure 5 Finding the root of equation (7) through iterating sequence (9).

in column B, truncates each iteration (number) with precision (num_digits) displayed in cell D2 to which a slider is attached. In other words, by using a slider one can control the precision of a root's finding through the formula:

=IF(OR(B2=B1,B2=""),"",TRUNC(SQRT(1+(PI()+ACOS(1/B2))^2),D$2))

defined in cell B3 and replicated down column B. As Figure 5 indicates, one can see that the value $\alpha_1 = 4.6033388$ (cell B8) is the root of equation (7) found by he method of iterations with the precision of seven digits after the decimal point.

Moreover, the environment makes it possible to terminate the iteration process once two consecutive iterations coincide within a specified accuracy (precision). To this end, the last (action-type) formula can be replaced by the following conditional formula

=IF(OR(B2=B1,B2=""),"",TRUNC(SQRT(1+(PI()+ACOS(1/B2))^2),D$2))

In turn, the so refined formula can be connected to computing the number of iterations required for a root finding process to be completed. To this end, the formula =IF(B3=" "," ",1+A2) can be defined in cell A3 and replicated down column A. This computational improvement allows one to see that the higher the precision of an iterated root, the bigger the number of iterations required for achieving that precision.

Finally, assuming $2 < \alpha \leq \alpha^*$ and choosing $x(0) \in [-\pi,\pi]$, one can use equation (6) to conclude that $x(1) = g(x(0)) \in [-\pi,\pi]$. Similarly, the inclusion $x(2) = g(x(1)) \in [-\pi,\pi]$ holds true. Continuing in this vein, one can come to the conclusion that for $2 < \alpha \leq \alpha^*$ where α^* is the root of equation (7), the sequence $x(t) \in [-\pi,\pi]$ $\forall t \in Z$. Also, one can see that when α becomes greater than α^*, the sequence $x(t)$ leaves the segment $[-\pi,\pi]$ as graphs in Figures 6 and 7 indicate. Note that the described approach provides the exact estimate (α^*) of the mapping of the segment $[-\pi,\pi]$ into itself and can be applied to other nonlinearities, like $g(x) = x - \alpha\ sign(\sin x)$ (see equation (10) below) for which the estimate of parameter α is different, of course. A slight modification of the environment pictured in Fig. 2 would result in π being that estimate.

To conclude, note that, as a technical improvement of the visual component of the environment described in this section, one can format the charts of Figures 6 and 7 allowing

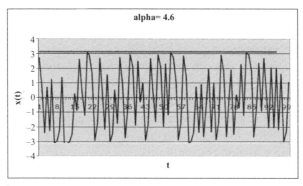

Figure 7 $\alpha = 4.6 : |x(t)| > \pi, t \in N, t > N.$

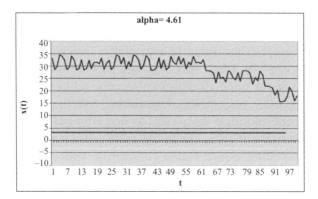

Figure 6 $\alpha = 4.6 : |x(t)| \leq \pi, t \in N.$

for the interactive appearance of the value of parameter α in a chart's title that corresponds to a graph of $g(x)$ being displayed. This can be done, for example, by clicking at the chart area, setting the transparency at 100% in the Colors and Lines dialogue box, and entering the formula =B2 in cell I7 (Fig. 2). It is also possible to clear the box Gridlines in the View window of the Preferences in Excel menu so that to make gridlines invisible in the transparent chart area.

The Bifurcation Diagram as a Tool for Refining Bifurcation Parameters

The bifurcation diagram (sometimes called the *orbit diagram* or *Feigenbaum plot*) of a dynamic system represents a set of attracting orbits that follow a transient behavior of an orbit. A simple example known as "the *3x+1* problem" [23] can be used to clarify the point. Let $x(t+1)=3x(t)+1$ if $x(t)$ is odd, and $x(t+1)=x(t)/2$ if $x(t)$ is even, $t \in Z$. The sequence 5,16,8,4,2,1,4,2,1, 4,2,1,…is the orbit of the point $x(0)=5$ under this iteration. That kind of sequence, attracted by the 3-cycle (4,2,1), is sometimes called eventually periodic sequence, the first three numbers of which—5, 16, 8 characterize its transient behavior. Here, the initial value $x(0)$ may be considered as an integer parameter, on which the iteration depends. The attractors represent the set of values (called a cycle), which the iteration converges to; in other words, it shows what happens with iterations of $x(0)$ after they have settled down. In that way, a bifurcation diagram can be used to demonstrate the limiting behavior of iterations.

In the case of equation (6), one can use a spreadsheet to construct a bifurcation diagram which relates any value of parameter α to the corresponding set of attracting orbits. This diagram can then be zoomed-in (e.g., by changing the maximum and minimum values of the α-axis scale using the *Format Axis* feature) to locate approximate values of parameter α for which different bifurcations of orbits occur. Once such values are known, one can refine them by iterating equation (6) within the appropriate range of the parameter using spreadsheet techniques described in the next section.

In order to explain the bifurcation diagrams of Figures 8, 9 and 10, note that a cycle is considered to be globally stable if the change in initial data does not change the cycle's qualitative behavior. However, when the limiting cycle that attracts an orbit depends on initial value, such a cycle is considered to be locally stable. As shown in Figure 8, the diagram does not depend on the value of $x(0)$ until α reaches π -at that point both the upper and lower branches split into two branches each of which corresponds to either $x(0) > 0$ or $x(0) < 0$. Figures 9 and 10 show two separate bifurcation diagrams constructed for $x(0) = 0.5$ and $x(0) = -0.5$, respectively. One can see how the symmetry of cycles of the mapping of the segment into itself disappears as one attempts to construct a bifurcation diagram for a specific value of $x(0)$ - the initial values ± 0.5 are used to demonstrate this phenomenon. It is interesting to note that in comparison with well known orbit diagrams for quadratic maps (e.g., equation (3); for their spreadsheet-based representation, see [11], [21], [36]), the bifurcation diagram for equation (6) exhibits a qualitatively different type of orbital behavior; namely, it shows the emergence of two locally stable non-symmetrical 2-cycles from one globally stable symmetrical 2-cycle as the parameter α approaches π. It appears that such qualitative difference is due to the difference in the number of points of extremum of the two systems' non-linear characteristics [38].

A computational part of the spreadsheet that generates the bifurcation diagram is shown in Fig. 11. Following are spreadsheet formulas used to construct the environment and defined, respectively, in cells F2, E2, E3, D2, B3 and C3: =IF(B2=1,OFFSET(B2,L$1-1,3),F1-C2*SIN(F1));=IF(P1=1,O1,-O1);=IF(B3=1,$E$2,E2-C2*SIN(E2));=F2;=IF (B2= L$1,1,B2+1);=IF(B3=1,C2+M$1,C2).

In addition, cells A2 and B2 are entered with the numbers 1 and 1.5, respectively. The first of the above listed spreadsheet formulas include a spreadsheet function OFFSET consisting of three parts (arguments): a cell, a vertical offset, and a horizontal offset. For example, the formula =OFFSET(A1, 3, 5) returns the value of cell F4 which is located three rows down and five columns to the right of cell A1. Note that the OFFSET function is a generalization of another spreadsheet function, INDEX, which use is described in [2]. Indeed, whereas the latter function makes it possible to locate the value of a designated cell within a fixed range (array) only, the former function makes it possible to work with evolving arrays. In particular, the use of the OFFSET function in the construction of bifurcation diagrams makes it possible to control the number of iterations required for a transient process to be completed.

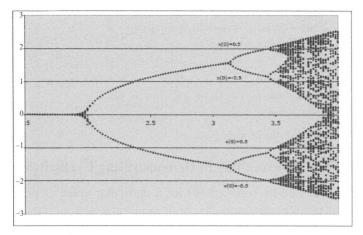

Figure 8 Bifurcation diagram for equation (6).

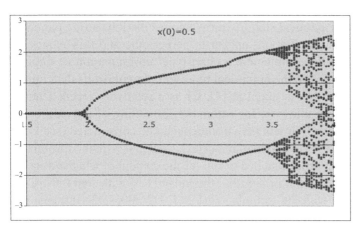

Figure 9 Bifurcation diagrams for *x(0)=0.5*.

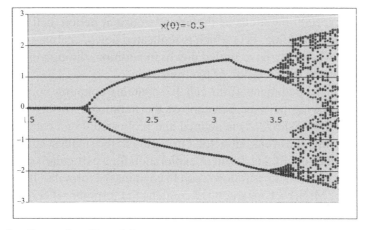

Figure 10 Bifurcation diagram for $x(0) = -0.5$.

	A	B	C	D	E	F	G	H	I	J	K	L	M	N	O	P
1			its	parameter	attractors	iterations	attractors		its	attractors	iterations	attractors	40	0.02	50	0.5	0
2		1	1	1.5	7.72E-13	-0.5	7.7E-13		1	-8E-13	0.5	-8E-13				initial value	change sign of
3		2	2	1.5	-3.9E-13	0.21914	-4E-13		2	3.9E-13	-0.21914	3.9E-13				(abs)	initial value
4		3	3	1.5	1.93E-13	-0.1069	1.9E-13		3	-2E-13	0.10694	-2E-13	number of increment				
5		4	4	1.5	-9.7E-14	0.05317	-1E-13		4	9.7E-14	-0.05317	9.7E-14	iterations				
6		5	5	1.5	4.83E-14	-0.0265	4.8E-14		5	-5E-14	0.02655	-5E-14					

Figure 11 Numerical part of the bifurcation diagram.

The use of Conditional Formatting in Demonstrating Cyclic Behavior of Orbits

Behavior of orbit generated by equation (6) for a particular value of parameter a can be demonstrated in alternative environments, both numerical and graphical. As mentioned above, by using the option of zooming-in at a bifurcation diagram which shows the limiting behavior of the orbits as the function of α, a specific value of the parameter for which such representations are sought can be selected. For example, Fig. 12 and 13 shows graphically how the global asymptotical stability turns into oscillations as the parameter α leaves the interval (0,2). These oscillations appear to be periodic and this suggests that conditional formatting (CF) could complement a chart-type representation of the cyclic behavior of orbits by being applied to the numerical representation of orbits that, otherwise, is difficult to make sense visually. As noted in [4], CF is a relatively recent feature of the modern graphical spreadsheet allowing for automatic formatting of any cell based on its current value. To this end, the following environment can be constructed (see Fig. 14).

First, cell B2—the value of parameter α – is entered with the formula =A2/100 and it is controlled through a slider attached to cell A2 the content of which varies in the range [100, 500] enabling the variation of a within the range [1,5] with step 0.01. Cell C2 – the value of $x(0)$—is entered with the formula =IF(E2=1, D2/100, -D2/100) enabling the change of sign and absolute value of the initial value through the sliders attached to cells D2 and E2 respectively.

Second, column B, beginning from cell B6, includes orbits $x(t)$ with $x(0)$ duplicated in cell B6. Setting 1000 iterations as the orbits' transient period, cell D7 is entered with the formula =TRUNC(B1005, I$1) which is replicated down, enabling one to consider the iterations of $x(0)$ when they have settled down. In addition, through the slider attached to cell I1 one can control the precision of the truncation.

Third, column E, beginning from cell E7 contains counting numbers that serve as mediators in identifying periodic behavior of the iterations. To this end, in cell F8 the formula=IF(D$7=D8,E8-E$7," ") is defined and replicated down column F, enabling the spreadsheet to identify the orbits' attractor. Furthermore, the smallest number in this column represents the period sought. The result of exploration for a particular value of parameter α is displayed in the range F5:I5 as follows: cell I5 contains the formula =IF(SUM(F7:F73)=0, "INSTABILITY", MIN(F7:F73)); cell H5 contains the formula =IF(I5="INSTABILITY", " ", "Period=").

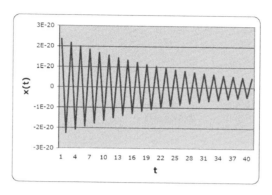

Figure 12 Approaching the first bifurcation: asymptotic stability with $\alpha = 1.96$.

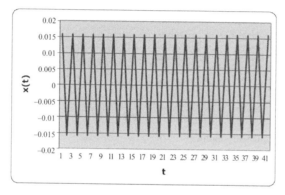

Figure 13 The first bifurcation: oscillations with $\alpha = 2.00$.

Finally, CF of iterations can be utilized by highlighting the range D7:D73, opening the dialogue box of CF and defining two conditions—cell value is equal to =$D$7, and cell value is equal to =$D$8—depending on which a cell in this range may be colored (formatted) in one of two chosen colors. Concurrently, the behavior of iterations can be shown graphically using the chart-wizard feature of a spreadsheet. As Figure 14 illustrates, for $\alpha = 2.1$ the orbits behave as a two-cycle and all cell containing the settled down iterations (column D) have been formatted to demonstrate this type of behavior.

Also, by changing the values of a and $x(0)$, global asymptotical stability of cycles can be demonstrated in the CF setting. However, as parameter a approaches one of its bifurcation values, the corresponding cycle looses its stability-a phenomenon represented through a non-regular behavior of iterations and the disappearance of regular coloring patterns. As parameter a changes further, a new globally asymptotically stable cycle emerges with period being the double of a previous one. Thus, by changing the value of parameter a one can observe the phenomenon of period-doubling bifurcations using a combination of spreadsheet graphing and CF applied to numerical iterations (Figs. 15–18). In particular, one can observe (Figs. 19, 20) in an alternative environment how in the neighborhood of $\alpha = \pi$ a globally stable 2-cycle is replaced by two locally stable 2-cycles—a characteristic property of orbits generated by equation (6).

	A	B	C	D	E	F	G	H	I
change data	alpha	x(0)	change x(0)	change sign of x(0)			Truncate to	7	digits
1	3.141592654	-0.07							
2									
3									
4					ALPHA=	3.1416		INSTABILITY	
5	-0.07								
6	0.14973193536675		1.5848265000						
7	-0.31890909347347		-1.5564568000						
8	0.66607711065742		1.5848127000						
9	-1.27513300531516		-1.5564712000						
10	1.73014304602607		1.5847990000						
11	-1.3716424846873		-1.5564856000						
12	1.70785193342589		1.5847853000						
13	-1.40428066478432		-1.5564999000						
14	1.69385833186984		1.5847716000						
15	-1.423978480648		-1.5565141000						
16	1.68381713845756		1.5847580000						
17	-1.43773197750040		-1.5565283000						
18	1.67608897989823		1.5847445000						
19	-1.448105056		-1.5565425000						
20	1.669871815		1.5847309000						
21	-1.456314586		-1.5565566000						
22	1.664713578		1.5847174000						
23	-1.463034127		-1.5565707000						
24	1.660334999		1.5847040000						

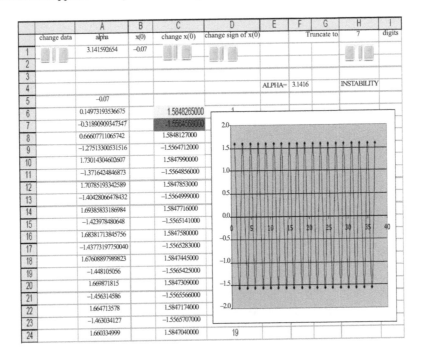

Figure 14 CF enhances spreadsheet graphics.

Amplitude of a Cycle as the Function of Parameter

As mentioned by Durkin [12] using computer experimentation in exploring dynamic systems with a complex behavior of orbits often results in experimental findings that stimulate pure mathematical investigations seeking to determine whether those findings are accurate. Leonov and Seledzhi [27] used a combination of computational experiments and formal mathematical explorations to reveal hidden behavior of cycles generated by equation (6). In educational settings, a spreadsheet can be used as a medium for experimental findings that can motivate students' mathematical explorations. In other words, through appropriately designed computational experiments, a spreadsheet can become an agent of mathematical activities for students [2]. Furthermore, as will be shown below, once a mathematical model is constructed, it can be explored within a spreadsheet, thus providing an alternative representation of the concepts involved. One such a spreadsheet-based experiment can deal with the study of the dependence of the amplitude of a cycle on the parameter α.

The term *amplitude* is used here to refer to the maximum offset of a function from its baseline level which is equal to zero in the case of the cycles studied in this paper. Therefore, one can use the spreadsheet function =MAX(LARGE(range,1), ABS(SMALL(range,1))) for the determination of the amplitude of a cycle for each value of parameter α Here, the first argument of the functions LARGE and SMALL refers the range of cells that includes the values of $x(t)$ for sufficiently large values of t; the second argument of the two functions refers to the largest and the smallest values within the corresponding range.

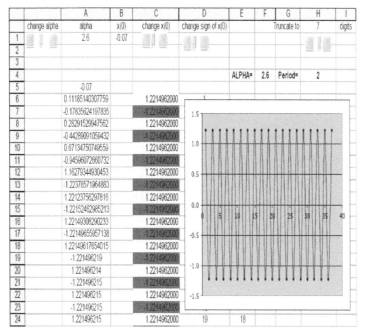

	A	B	C	D	E	F	G	H	I
change alpha	alpha	x(0)	change x(0)	change sign of x(0)			Truncate to	7	digits
1	2.6	-0.07							
2									
3									
4					ALPHA=	2.6	Period=	2	
5	-0.07								
6	0.11185140307759		1.2214962000						
7	-0.17835624197835		-1.2214962000						
8	0.28291529947562		1.2214962000						
9	-0.44289091059432		-1.2214962000						
10	0.67134750749559		1.2214962000						
11	-0.94596072660732		-1.2214962000						
12	1.16279344930453		1.2214962000						
13	-1.22378571964863		-1.2214962000						
14	1.22123756297816		1.2214962000						
15	-1.22152462985213		-1.2214962000						
16	1.22149308290233		1.2214962000						
17	-1.22149655957138		-1.2214962000						
18	1.22149617654015		1.2214962000						
19	-1.221496219		-1.2214962000						
20	1.221496214		1.2214962000						
21	-1.221496215		-1.2214962000						
22	1.221496215		1.2214962000						
23	-1.221496215		-1.2214962000						
24	1.221496215		1.2214962000						

Figure 15 CF shows a 2-cycle.

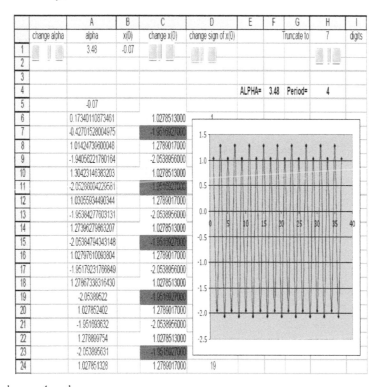

	A	B	C	D	E	F	G	H	I
change alpha	alpha	x(0)	change x(0)	change sign of x(0)			Truncate to	7	digits
1	3.48	-0.07							
2									
3									
4					ALPHA=	3.48	Period=	4	
5	-0.07								
6	0.17340110873461		1.0278513000						
7	-0.42701528004975		-1.9516927000						
8	1.01424739600048		1.2789017000						
9	-1.94056221780164		-2.0538956000						
10	1.30423146383203		1.0278513000						
11	-2.05266004229581		-1.9516927000						
12	1.03055934490344		1.2789017000						
13	-1.95384277603131		-2.0538956000						
14	1.27396279863207		1.0278513000						
15	-2.05384794343148		-1.9516927000						
16	1.02797610093804		1.2789017000						
17	-1.95179231766849		-2.0538956000						
18	1.27867338316430		1.0278513000						
19	-2.05389522		-1.9516927000						
20	1.027852402		1.2789017000						
21	-1.951693632		-2.0538956000						
22	1.278899754		1.0278513000						
23	-2.053895631		-1.9516927000						
24	1.027851328		1.2789017000						

Figure 16 CF shows a 4-cycle.

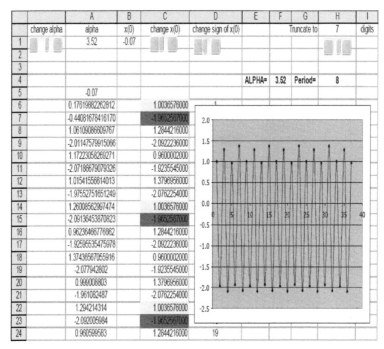

Figure 17 CF shows an 8-cycle.

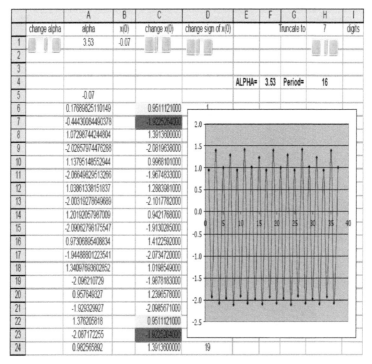

Figure 18 CF shows a 16-cycle.

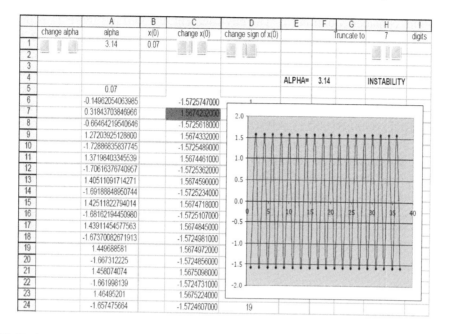

Figure 19 CF shows the loss of global stability for $\alpha = \pi$: $x(0) = 0.07$.

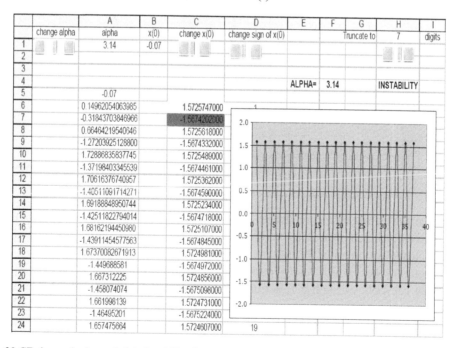

Figure 20 CF shows the loss of global stability for $\alpha = \pi$: $x(0) = -0.07$.

To begin, consider equation (6). The environment pictured in Figure 21 is designed to construct the graph of $A(\alpha)$ and it includes three distinct parts. First, in column B, beginning from cell B7, the sequence $x(t)$ is generated through the formula =B6-B\$2*SIN(B6), where cell B6 contains the initial value $x(0)$ (controlled by a slider attached to cell C2) and cell B2 contains a slider-controlled (via cell A2) value of parameter *a*. For each value of *a*, the amplitude of a corresponding cycle is calculated in cell F3 through the formula =MAX(LARGE(B900:B1000,1), ABS(SMALL(B900:B1000,1))) which refers to a range of cycle's values that includes sufficiently many iterations.

Second, the numbers in column D, beginning from cell D6, represent values of *a* increased by the factor of 100. Column E, beginning from cell E7, contain values of the function $A(\alpha)$; that is, the amplitudes of the corresponding cycles.

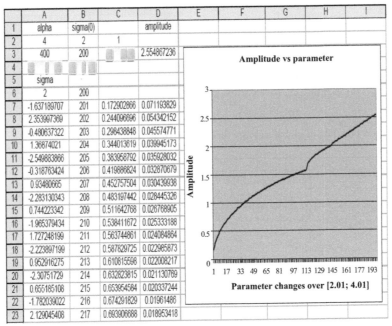

	A	B	C	D	E	F	G	H	I
1	alpha	sigma(0)		amplitude					
2	4	2	1						
3	400	200		2.554867236					
4									
5	sigma								
6	2	200							
7	-1.637189707	201	0.172902866	0.071193829					
8	2.353997369	202	0.244096696	0.054342152					
9	-0.480637322	203	0.298438848	0.045574771					
10	1.36874021	204	0.344013619	0.039945173					
11	-2.549883866	205	0.383958792	0.035928032					
12	-0.318763424	206	0.419686824	0.032870679					
13	0.93480665	207	0.452757504	0.030439938					
14	-2.283130343	208	0.483197442	0.028445326					
15	0.744223342	209	0.511642768	0.028768905					
16	-1.965379434	210	0.538411672	0.025333188					
17	1.727248199	211	0.563744861	0.024084864					
18	-2.223897199	212	0.587629725	0.022965873					
19	0.952916275	213	0.610815598	0.022008217					
20	-2.30751729	214	0.632823815	0.021130769					
21	0.655185108	215	0.653954584	0.020337244					
22	-1.782039022	216	0.674291829	0.01961486					
23	2.129045408	217	0.693906688	0.018953418					

Chart: Amplitude vs parameter. Y-axis: Amplitude (0 to 3). X-axis: 1, 17, 33, 49, 65, 81, 97, 113, 129, 145, 161, 177, 193. Parameter changes over [2.01; 4.01]

Figure 21 $A(\alpha)$ in equation (6) increases monotonically with the parameter α.

A powerful feature of a spreadsheet is its computational and operational functionality of interactive dynamic construction of the table representation of a function defined neither through a closed formula nor through recursion [1].

The function $A(\alpha)$ which relates parameter α to the amplitude *A* of the corresponding cycle, is an example of such a function. The basic computational idea that enables its table representation is the use of a circular reference in a spreadsheet formula (that is, a reference to a cell in which the formula is defined). Indeed, the use of a circular reference in the spreadsheet formula =IF($A$2=200," ",IF($A$2=D6,F$3,E6)) defined in cell E6 and replicated down column E makes it possible to keep the values of already computed amplitudes unchanged as parameter *a* varies in cell B2 controlled by a slider attached to cell A2. This can be done through a simple relationship between the cells defining the formula =A2/100 in cell B2.

Finally, the third part of this environment includes a Chart Wizard that represents the graph of the function $A(\alpha)$ constructed over the range [2.01; 4.01]. This graph shows that the amplitude of an orbit $x(t)$ monotonically increases on this interval, uniformly about $x(0) \in (-\pi, \pi)$. The only peculiar behavior of $A(\alpha)$ that can be discovered through spreadsheet graphing is the existence of the point in which the graph appears to lose its smoothness as α changes from $\alpha = 3.14$ to $\alpha = 3.15$ (Fig. 21). Alternatively, this behavior can be demonstrated by graphing the difference $\Delta A = A(\alpha + \Delta\alpha) - A(\alpha)$.

A much more interesting behavior of the function $A(\alpha)$, that may stimulate significant mathematical work on the part of students, can be observed in the case of the following equation with a discontinuous non-linearity

$$x(t+1) = x(t) - \alpha \, sign(\sin t) \tag{10}$$

typically used in floating phase-locked loops that harness both the effects of sampling and quantizing [25,28,8]. To this end, the environment described in this section can be retrofitted by entering cell B7 with the formula =B6-\$B\$2*SIGN(SIN(B6)) and replicating it down column B thus allowing one to explore the behavior of the amplitude $A(\alpha)$ of the cycles generated by equation (10). In doing so, one can discover that the behavior of the function $A(\alpha)$ strongly depends on the value of $x(0)$. For example, Figs. 22 and 23 show the graphs of $A(\alpha)$ in the case of $x(0)=1.1$ and $x(0)=1.5$, respectively. This kind of behavior changes, however, as $x(0)$ becomes greater than 2. For example, when $x(0)=2.1$ and $x(0)=2.6$, the corresponding graphs of $A(\alpha)$ have the form shown in Figs. 24 and 25 respectively.

Computational experiments which enable the dynamic construction of the function $A(\alpha)$ for different positive values of $x(0)$ suggest two different formulas depending on whether $x(0) < \alpha_{min}$ or $x(0) > \alpha_{min}$.

If $0 < x(0) < \alpha_{min}$, then

$$A(\alpha) = \begin{cases} x(0), & if \ \alpha < 2x(0) \\ \alpha - x(0), & if \ \alpha > 2x(0) \end{cases}.$$

If $x(0) > \alpha_{min}$, then

$$A(\alpha) = \begin{cases} 2\alpha - x(0), & if \ \alpha < x(0) \\ 0, & if \ \alpha = x(0) \\ x(0), & if \ \alpha > x(0) \end{cases}$$

Similarly, the case of $x(0) < 0$ can be considered.

An interesting fact can be discovered through a computational experiment involving a spreadsheet function OFFSET is that there exists $\alpha = \alpha^*$ such that $A(\alpha, x(0)) = x(0)$ for $\alpha > \alpha^*$. In order to locate such value of α^*, a means of selecting a value from a designated

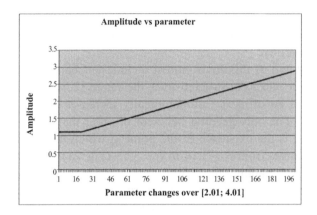

Figure 22 $x(0)=1.1$.

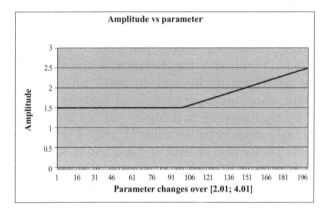

Figure 23 $x(0) = 1.5$.

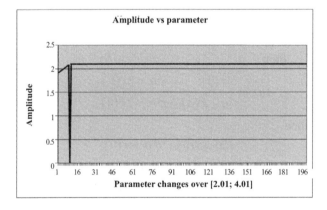

Figure 24 $x(0)=2.1$.

cell is required. The spreadsheet pictured in Fig. 26 shows zero in cell C2 as the return of the formula =OFFSET(D6, C6, 1), where C6 relates to the value of parameter a for which the amplitude equals zero. Such value of a can be located through changing the value of

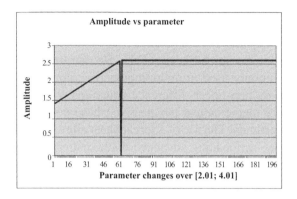

Figure 25 $x(0)=2.6$.

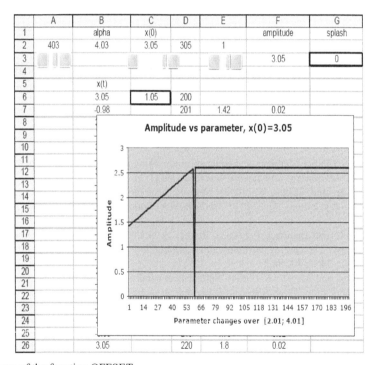

	A	B	C	D	E	F	G
1		alpha	x(0)			amplitude	splash
2	403	4.03	3.05	305	1		
3						3.05	0
4							
5		x(t)					
6		3.05	1.05	200			
7		-0.98		201	1.42	0.02	
8							
...							
26		3.05		220	1.8	0.02	

Figure 26 The use of the function OFFSET.

$x(0)$ using the slider located above cell C6. This can give rise to a mathematical activity aimed at the formal demonstration of the results of a computational experiment. Indeed, as mentioned in [27], the solution to equation (10) with initial data $x(0) \in (0,\alpha)$ can be represented either as $x(t)=x(0)$ or $x(t)=x(0)-\alpha$ as $x(t)=x(0)-\alpha$, depending on whether t is an even or odd number. This shows that eventually, after the orbits of equation (10) have settled down, they exhibit oscillating behavior with period two, resembling oscillations that occur in two-dimensional conservative continuous dynamical systems [26].

Conclusion

This paper has demonstrated how the combination of a number of computational techniques made possible by the use of a spreadsheet enables for educational applications to digital phase-locked loops commonly used in radio engineering, communication, and computer architecture. Described by non-linear difference equations which dynamics is extremely complicated from a pure mathematical perspective, these control systems can be found across various tertiary curricula. The use of a spreadsheet allows for an alternative approach to the study of this topic—embedding advanced mathematical ideas in the software frees learners from the necessity of dealing with complex mathematical machinery of non-linear control systems that rigorous methods of investigation require. In particular, this alternative approach has been used to introduce this topic to students studied discrete mathematics at Bond University.

Acknowledgement

The authors gratefully acknowledge the support of the Ministry of Education of Russian Federation, grant number 1216P07, under the program "Development of scientific potential of institutes of higher education".

References

1. Abramovich, S. (2000). Mathematical concepts as emerging tools in computing applications. *Journal of Computers in Mathematics and Science Teaching*, **19**(1): 21–46.
2. Abramovich, S., and Strock, T. (2002). Measurement model for division as a tool in computing applications. *International Journal of Mathematical Education in Science and Technology*, **33**(2): 171–185.
3. Abramovich, S., and Brouwer, P. (2003). Revealing hidden mathematics curriculum to pre-teachers using technology: The case of partitions. *International Journal of Mathematical Education in Science and Technology*, **34**(1): 81–94.
4. Abramovich, S., and Sugden, S. (2004). Spreadsheet conditional formatting: An untapped resource for mathematics education. *Spreadsheets in Education*, **1**(2): 85–105. (Available: http://www.sie.bond.edu.au).
5. Arganbright, D.E. (1993). Practical Handbook of Spreadsheet Curves and Geometric Constructions. Boca Raton, FL: CRC Press.
6. Arnold, V.I. (1983). Geometrical Methods in the Theory of Ordinary Differential Equations. New York: Springer.
7. Belykh, V.N., and Lebedeva, L.V. (1983). Investigation of a particular mapping of a circle. *Journal of Applied Mathematics and Mechanics*, **46**(5): 771–776.
8. Belykh, V.N., and Maksakov, V.P. (1979). Difference equations and dynamics of a first-order digital system of phase synchronization. *Radiotekhnika i Elektronika*, **24**(5): 958–964. (In Russian).
9. Blum, W., and Niss, M. (1991). Applied mathematical problem solving, modelling, applications, and links to other subjects—state, trends and issues in mathematics instruction. *Educational Studies in Mathematics*, **22** (1): 37–68.
10. Levie, R. (2004). *Advanced Excel for Scientific Data Analysis*. New York: Oxford University Press.
11. Durkin, M.B., and Nevils, B.C. (1994). Using Spreadsheets to See Chaos. *Journal of Computers in Mathematics and Science Teaching*, **13**(3): 321–338.

12. Durkin, M.B. (1991). The accuracy of computer algorithms in dynamical systems. *International Journal of Bifurcation and Chaos*, **1**(3): 625–639.
13. El-Hajj, A., Karaki, S., Al-Husseini, M., and Kabalan, K.Y. (2004). Spreadsheet solution of systems of nonlinear differential Equations. *Spreadsheets in Education*, **1**(3): 195–207. (Available: http://www.sie. bond.edu.au).
14. El-Hajj, A., Karaki, S., Kabalan, K.Y. (2004). Nonlinear control systems simulation using spreadsheets. *Spreadsheets in Education*, **1**(2): 67–84. (Available: http://www.sie.bond.edu.au).
15. Feigenbaum, M.J. (1978). Quantitative universality for a class of nonlinear transformations. *Journal of Statistical Physics*, **19**: 25–52.
16. Filby, G. (1998). *Spreadsheets in Science and Engineering*. Berlin: Springer-Verlag.
17. Gupta, S.C. (1975). Phase-locked loops. *Proceedings of the IEEE*, **63**(2): 291–306.
18. Howson, G. (1985). The impact of computers on mathematics education. *The Journal of Mathematical Behavior*, **4**: 295–303.
19. Jakobson, M.V. (1971). On smooth mappings of the circle into itself. *Mathematics USSR Sbornik*, **14**(2): 161–185.
20. Knuth, D.E. (1977). *The Art of Computer Programming*, volume 1: Fundamental Algorithms. Reading, MA: Addison Wesley.
21. Kreith, K., and Chakerian, D. (1999). *Iterative Algebra and Dynamic Modeling*. New York: Springer.
22. Kung, S.Y. (1998). *VLSI Array Processors*. Englewood Cliffs, N.J.: Prentice Hall.
23. Lagaris, J.C. (1985). The 3x+1 problem and its generalizations. *American Mathematical Monthly*, **92**(1): 3–23.
24. Lapsley, P., Bier, J., Shoham, A., and Lee, E.A. (1997). *DSP Processor Fundamentals: Architectures and Features*. New York: IEEE Press.
25. Leonov, G.A. (2001). *Mathematical Problems of Control Theory: An Introduction*. Singapore: World Scientific.
26. Leonov, G.A., and Seledzhi, S.M. (2002). *Phase-Locked Loops in VLSI* (in Russian). Saint Petersburg: Nevsky Dialect.
27. Leonov, G., and Seledzhi, S. (2005). Stability and bifurcations of phase-locked loops for digital signal processors. *International Journal of Bifurcation and Chaos*, **15**(4): 1347–1360.
28. Leonov, G.A., and Seledzhi, S.M. (2005). As astatic phase-locked system for digital signal processors: circuit design and stability. *Automation and Remote Control*, **66**(3): 348–355.
29. Levin, I., and Abramovich, S. (1992). Solving equations within spreadsheet. *Journal of Computers in Mathematics and Science Teaching*, **11**(3/4): 337–345.
30. Levin, I., and Talis, V. (2004). Using spreadsheets for teaching principles of on-line checking of logic circuits. *Spreadsheets in Education*, **1**(3): 131–141. (Available: http://www.sie.bond.edu.au).
31. Li, T.Y., and Yorke, J.A. (1975). Period three implies chaos. *American Mathematical Monthly*, **82**(10): 985–992.
32. Lindsey W. C., and Chie C. M. (1981). A survey of digital phase-locked loops. *Proceedings of the IEEE*, **69**(4): 671–685.
33. Lindsey, W. C. (1972). *Synchronization Systems in Communication and Control*. Englewood Cliffs, N. J.: Prentice-Hall.
34. May, R. M. (1976). Simple mathematical models with very complicated dynamics. *Journal of Theoretical Biology*, **51**: 511–524.
35. Metropolis, N., Stein, M., and Stein, P. (1973). On finite limit sets for transformations of the unit interval. *Journal of Combinatorial Theory*, **15**: 25–44.
36. Neuwirth, E., and Arganbright, D. (2004). *The Active Modeler: Mathematical Modeling with Microsoft®Excel*. Toronto: Brooks/Cole.
37. Osborne, H.C. (1980). Stability analysis of an Nth power digital phase-locked loop—Part 1: First-order DPLL. *IEEE Transactions on Communications*, **28**(8): 1343–1354.
38. Schuster, H.G. (1984). *Deterministic Chaos*. Weinheim: Physik-Verlag.

39. Shakhtarin, B.I. (1977). On the oscillation stability criterion of phased AFC. *Radio Engineering and Electronic Physics* (English translation of *Radiotekhnika i Elektronika*), **22**(3): 101–107.

40. Shakhtarin, B.I., and Arkhangel'skiy, V.A. (1977). Dynamic characteristics of discrete automatic phase-lock systems. *Radio Engineering and Electronic Physics* (English translation of *Radiotekhnika i Elektronika*), **22**(5): 62–69.

41. Sharkovsky, A.N. (1995). Coexistence of cycles of a continuous map of the line into itself. *International Journal of Bifurcation and Chaos in Applied Sciences and Engineering*, **5**(5): 1263–1273.

42. Sharkovsky, A.N., Kolyada, S.F., Sivak, A.G., and Fedorenko, V.V. (1997). Dynamics of one-dimensional maps. *Mathematics and its Applications*, 407. Dordrecht: Kluwer.

43. Silva, A.A. (1994). Simulating electrical circuits with an electronic spreadsheet. *Computers and Education*, **22**(4): 345–353.

44. Weisstein, E.W. (1999). *CRC Concise Encyclopedia of Mathematics*. Boca Raton, FL: Chapman & Hall/ CRC.

Modeling Complex Spatial Dynamics of Two-Population Interaction in Urbanization Process[†]

Yanguang Chen[1] and *Feng Xu[2]*

[1]Department of Geography, College of Urban and Environmental Sciences, Peking University, Beijing 100871, China. Email: chenyg@pku.edu.cn.
[2]China Center for Town Reform and Development, Beijing 100045, China. Email: fengxu.pku@gmail.com.

ABSTRACT

This paper is mainly devoted to laying an empirical foundation for further research on complex spatial dynamics of two-population interaction. Based on the US population census data, a rural and urban population interaction model is developed. Subsequently a logistic equation on percentage urban is derived from the urbanization model so that spatial interaction can be mathematically connected with logistic growth. The numerical experiment by using the discretized urban-rural population interaction model of urbanization shows a period-doubling bifurcation and chaotic behavior, which is identical in patterns to those from the simple mathematical models of logistic growth in ecology. This suggests that the complicated dynamics of logistic growth may come from some kind of the nonlinear interaction. The results from this study help to understand urbanization, rural-urban population interaction, chaotic dynamics, and spatial complexity of geographical systems.

Keywords: Allometric scaling, Bifurcation, Chaos, Complex dynamics, Logistic growth, Two-population interaction, Urbanization.

Introduction

The study of the logistic equation as viewed from ecology indicates that a simple deterministic system can present periodic oscillation and chaotic behavior along with the model parameter change (May 1976). However, why the simple model contains complex dynamics still remains ambiguous. Urban study can provide us with facilities for exploring the springhead of complicated dynamics of logistic process. In the urbanization process,

[†]Reused with permission from: Yanguang Chen, Feng Xu, Modeling Complex Spatial Dynamics of Two-Population Interaction in Urbanization Process, *Journal of Geography and Geology*, Vol. 2, No. 1; September 2010.

the level of urbanization follows the sigmoid curve and can be described with the logistic function (Karmeshu 1988; United Nations 1980; United Nations 1993). Moreover, the urban system and ecological system show comparability in several aspects (Dendrinos 1992; Dendrinos and Mullally 1985), which implies that the process of urbanization might have period-doubling bifurcation or chaotic dynamics. In theory, urban system and the process of urbanization can generate complex behaviors such as chaos (e.g., Dendrinos 1996; Dendrinos and El Naschie 1994; Nijkamp 1990; Nijkamp and Reggiani 1998; Van der Leeuw and McGlade 1997; Wong and Fotheringham 1990).

Many studies of chaotic cities are relative to spatial interaction and logistic growth. On the other hand, a great number of simulation analyses and empirical researches show that urban system bears the fractal structure (e.g., Batty and Longley 1994; Chen and Zhou 2003; Frankhauser 1994; White and Engelen 1994). Fractal structure and chaotic behavior coexist in lots of systems. Fractal property of urban systems suggests complex dynamics of urban evolution. What concerns us is not only the bifurcation and chaos in the sheer numerical simulation experiments but also the ones that can be captured from the observation data. One of the viewpoints is that fractal actually appears at the edge of chaos and the coexistence phenomenon of fractal and chaos does not imply the certain correlation between them (Bak 1996). Perhaps this is true, but we still intend to investigate it from the standpoint of urban systems and urbanization dynamics in order to reveal the relation between the chaotic behavior and fractal structure of nonlinear systems.

Now chaotic cities and fractal cities have become important branch ranges of self-organized cities (Portugali 2000). *Fractal cities* mean the cities with self-similarity or scaling invariance, while *chaotic cities* suggest the cities with spatial regularity behind random behaviors. The studies of fractal cities and systems of cities are supported by a great number of observations (e.g., Batty and Longley 1994; Chen and Zhou 2004; Chen and Zhou 2006; Frankhauser 1994; White et al. 1997). However, most applications of chaos theory in the social sciences lack empirical content (Nijkamp and Reggiani 1992). This situation has changed little for more than ten years. In fact, cities and networks of cities are typical complex systems suitable for exploring complicated dynamics (Allen 1997; Wilson 2000). The key lies in how to associate theory with practice and reality. The principal aim of this paper is at two aspects. One is to lay an empirical foundation for researching chaotic cities, and the other is to prepare for revealing the essence of complicated behaviors of simple models and the relation between chaotic cities and fractal cities.

The following parts of this paper are structured as follows. In section 2, we build a nonlinear dynamical model about the urban-rural interaction based on the population census data of the United States of America (USA), and then derive the logistic equation of urbanization level from the model. In section 3, with the aid of the US census data, we demonstrate the feasibility and rationality of the model based on statistical analysis, logistic analysis and numerical simulation analysis. This part is used to consolidate the empirical foundation of the model. In section 4, we implement numerical simulation experiment with the model of urbanization dynamics, testifying whether or not such a model presents all the behavior characters of the logistic equation, including periodic oscillation and chaotic behavior. Finally, in Section 5, the discussion is concluded by making some remarks on the significance of the complicated dynamics research from the aspect of the two-population interaction in urban geography.

Choice and Transform of Mathematical Models

Urban-rural Interaction Models

A variety of mathematical models has been made to describe the spatial dynamics of the urban-rural population migration. Among these models, two are attention-getting. One is the Keyfitz-Rogers linear model (Keyfitz 1980; Rogers 1968), and the other, the United Nations nonlinear model (United Nations 1980; Karmeshu 1988). The United Nations adopted a pair of nonlinear equations to characterize the urbanization dynamics

$$\begin{cases} \dfrac{dr(t)}{dt} = ar(t) + \varphi u(t) - b\dfrac{r(t)u(t)}{r(t)+u(t)} \\ \dfrac{du(t)}{dt} = cu(t) + \psi r(t) + d\dfrac{r(t)u(t)}{r(t)+u(t)} \end{cases}, \tag{1}$$

where $r(t)$ and $u(t)$ denotes the rural and urban population in time t respectively, a, b, c, d, φ and ψ are parameters. If parameters $\varphi = \psi = 0$, we can derive the logistic model of urbanization level from the UN model. For many years, the United Nations has been using the logistic function to forecast the level of urbanization of each country in the world (United Nations 1993; United Nations 2004).

However, empirical studies and statistical analyses show that the urbanization dynamics of many countries such as America, China, and India can be effectively described neither by the Keyfitz-Rogers model nor by the United Nations model. In short, rural population couldn't migrate into urban regions and *vice versa* without spatial interaction between urban and rural population. In other words, population migration and exchange between urban and rural regions depends only on urban-rural population interaction. Consequently, two items of the United Nations model are actually excrescent and equation (1) should be simplified to such a form

$$\begin{cases} \dfrac{dr(t)}{dt} = ar(t) - b\dfrac{r(t)u(t)}{r(t)+u(t)} \\ \dfrac{du(t)}{dt} = cu(t) + d\dfrac{r(t)u(t)}{r(t)+u(t)} \end{cases}. \tag{2}$$

According to equation (2), the rural population can not spontaneously flow into the cities and *vice versa*. The exchange of urban and rural population relies mainly on the urban-rural interaction. The *urban-rural interaction* bears an analogy with the predator-prey interaction in ecology (Dendrinos and Mullally 1985). The size of urban population is influenced by rural population size and in turn reacts on it. So both the growth rate of urban population and that of rural population depend to a great extent on the coupling or cross correlation between the urban and rural population. For a close region, it is theoretically expected $b=d$. As will be shown later, the US model of urbanization dynamics might be simpler than equation (2). That is $c=0$ in reality.

Derivation of the Logistic Model

In order to research into the above model, we need to examine it from two ways: one is the logical analysis, and the other empirical analysis. The logical analysis involves at least two aspects. First, whether or not the level of urbanization derived from the above model is similar to the logistic increase, and whether or not the total population in a region is limited. Second, whether or not the result of the numerical simulation is coincident with that of the mathematical deduction.

First, we derive the well-known logistic model on the level of urbanization, i.e., *urbanization ratio*. The *level of urbanization* is defined as the proportion or share of urban population in relation to the total population in a region (United Nations 2004). Thus, we have

$$L(t) = \frac{u(t)}{P(t)} = \frac{u(t)}{r(t)+u(t)} = \frac{V(t)}{1+V(t)}, \tag{3}$$

where $L(t)$ refers to the level of urbanization, $P(t)=r(t)+u(t)$ to the total population, and $V(t)=u(t)/r(t)$ to the *urban-rural ratio* of population. Differentiating $L(t)$ with respect to t gives

$$\frac{dL(t)}{dt} = \frac{du(t)/dt}{r(t)+u(t)} - \frac{u(t)}{[r(t)+u(t)]^2}\left[\frac{dr(t)}{dt}+\frac{du(t)}{dt}\right]. \tag{4}$$

Substituting equation (2) into equation (4) yields

$$\frac{dL(t)}{dt} = \frac{cu(t)}{r(t)+u(t)} + \frac{dr(t)u(t)}{[r(t)+u(t)]^2} - \frac{u(t)}{[r(t)+u(t)]^2}\left[ar(t)+cu(t)+(d-b)\frac{r(t)u(t)}{r(t)+u(t)}\right] \tag{5}$$

For simplicity, taking a region as a close system, then we have $b=d$. In terms of the definition of urbanization level, equation (5) can be transformed into the following form

$$\frac{dL(t)}{dt} = cL(t)[1 - L(t)] + (d-a)\frac{r(t)}{u(t)}L(t)^2. \tag{6}$$

According as equation (3), we have an urban-rural ratio

$$V(t) = \frac{u(t)}{r(t)} = \frac{L(t)}{1-L(t)}. \tag{7}$$

This implies $1/V(t)=r(t)/u(t)=1/L(t)-1$. Therefore, equation (6) can be transformed into a logistic equation

$$\frac{dL(t)}{dt} = cL(t)\big[1 - L(t)\big] + (b - a)\frac{1}{V(t)}L(t)^2 = (b + c - a)L(t)\big[1 - L(t)\big].$$

(8)

Thus, we have constructed a mathematical relation between models for two interacting population and the logistic growth. Let $k=b+c-a=c+d-a$ represent the *intrinsic rate of growth*. Then equation (8) can be simplified as the usual form

$$\frac{dL(t)}{dt} = kL(t)\big[1 - L(t)\big].$$

(9)

Solving equation (9) yields the well-known expression of the logistic curve

$$L(t) = \frac{1}{1 + (1/L_0 - 1)e^{-kt}},$$

(10)

where L_0 represents the initial value of $L(t)$. That is, when $t=0$, we have $L(t)=L_0$.

A key criterion to judge the urbanization model is the rationality of the increase curve of the total population. Taking derivative of population $P(t)$ with respect to time t yields

$$\frac{dP(t)}{dt} = \frac{dr(t)}{dt} + \frac{du(t)}{dt}.$$

(11)

Substituting equation (2) into equation (11) gives

$$\frac{dP(t)}{dt} = ar(t) + cu(t).$$

(12)

Obviously, from equation (12) we can get two inconsistent equations as follows

$$\frac{dP(t)}{dt} = cP(t) + (a - c)r(t),$$

(13)

$$\frac{dP(t)}{dt} = aP(t) + (c - a)u(t).$$

(14)

According to equation (13), when $a > c$, the total population grows more quickly; while according to equation (14), when $a > c$, the total regional population grows slower. These two equations collide with each other. The inconsistency can be eliminated by two conditions: $a=c$ or $c=0$. If $a=c$ as given, then the total population will grow infinitely in the exponential way predicted by Malthus (1798/1996); On the other, if $c=0$, the total population will stop growing when it increases to certain extent. Under the latter circumstance, according as equation (13), since the rural population $r(t) \to 0$, the growth rate of the total population $P(t)$ will gradually decrease to 0; According as equation (14), because the whole population will be completely urbanized, i.e., $u(t) \to p(t)$, the growth rate of the total population will tend toward 0 ultimately. In the real world, we do have $c=0$, as will be illustrated in the following empirical analysis.

It is easy to see that b or d is a very significant parameter in equation (2). On the one hand, it controls the developing trend and quantity of the total population; on the other hand, it affects the original rate of growth k value of the logistic equation on level of urbanization. As we know, parameter k dominates the behavior characters of the dynamical system. When $k > 2.57$, the logistic map coming from the discretization of equation (9) will present very complicated behaviors (May, 1976). So what is the case in reality? In the next section, we will validate the above models in virtue of the US observation data. Then we perform numerical simulation experiment to unfold some intrinsic regularity of the urbanization dynamics.

Empirical Foundation of Two-population Interaction Model

Data and method

The main purpose of this study, as indicated above, is to lay an empirical foundation for further research on complex spatial dynamics of urban-rural interaction. So it is necessary to make relevant statistical analysis of the dynamical equations. There are two central variables in the study of spatial dynamics of urban development: population and wealth (Dendrinos 1992). According to our theme, we only choose the first variable, population, to test the models. Generally speaking, the population measure falls roughly into four categories: rural population $r(t)$, urban population $u(t)$, total population $P(t) = r(t)+u(t)$, and level of urbanization, $L(t) = u(t)/P(t)$.

The American data comes from the population censuses whose interval is about 10 years. Although the website of American population census offers 22 times of census data from 1790 to 2000, we only use the data from 1790 to 1960 (Table 1). The reason is that the US changed the definition of cities in 1950, and the new definition came into effect in 1970. From then on, the American urban population was measured with the new standard. As a result, the statistic caliber of the population data from 1970 to 2000 might be different from those before 1970 although they approximately join with each other (Fig. 1).

The data displayed in table 1 are fitted to the discretization expressions of the United Nations model and the Lotka-Volterra-type model respectively (r.e. Dendrinos and Mullally 1985; Lotka 1956; Volterra 1931). Since the Keyfitz-Rogers model and the American urbanization model are both special cases of the United Nations model, there is no need to try Keyfitz-Rogers model particularly. The parameters of models are made by the least squares computation, which can make the key parameters, slopes, fall into the most reasonable range.

After estimating the model parameters, we should make tests in two ways. One is the well-known statistical test, and the other is the logical test, which is often ignored in practice. If the model fails to pass the statistical test, it has problems such as incomplete or redundant variables, or inaccurate parameter values; if the model cannot pass the logical test, it has structure problem so that it cannot explain the phenomena at present and predict the developing trend in future. Statistical tests can be made in definite procedure, while the logical test needs to be done with the help of mathematical transformation and numerical simulation experiment.

Table 1 The US rural and urban population and the related data (1790–1960).

Time (year) [t]	Interval (years) [Δt]	Rural population [$r(t)$]	Urban population [$u(t)$]	$\dfrac{r(t)u(t)}{r(t)+u(t)}$	Rural rate of growth [$\Delta r(t)$]	Urban rate of growth [$\Delta u(t)$]
1790	10	3727559	201655	191305.67	125855.30	12071.60
1800	10	4986112	322371	302794.21	172831.00	20308.80
1810	10	6714422	525459	487322.03	223077.60	16779.60
1820	9.8125	8945198	693255	643391.97	284153.58	44228.48
1830	10	11733455	1127247	1028443.23	348484.30	71780.80
1840	10	15218298	1845055	1645549.78	439908.20	172944.10
1850	10	19617380	3574496	3023569.39	560942.30	264202.20
1860	10	25226803	6216518	4987478.10	342920.70	368584.30
1870	10	28656010	9902361	7359287.97	740346.40	422737.40
1880	10	36059474	14129735	10151800.00	481402.70	797653.00
1890	10	40873501	22106265	14346837.12	512383.50	810856.70
1900	9.7917	45997336	30214832	18235956.49	425582.20	1210127.90
1910	9.7917	50164495	42064001	22879255.97	163788.26	1244862.74
1920	10.25	51768255	54253282	26490822.68	221831.22	1454372.39
1930	10	54042025	69160599	30336844.29	341720.60	554473.90
1940	10	57459231	74705338	32478532.68	373837.30	1542285.60
1950	10	61197604	90128194	36448706.03	506197.80	2293539.90
1960	10	66259582	113063593	41776788.81		

Source: http://www.census.gov/population.

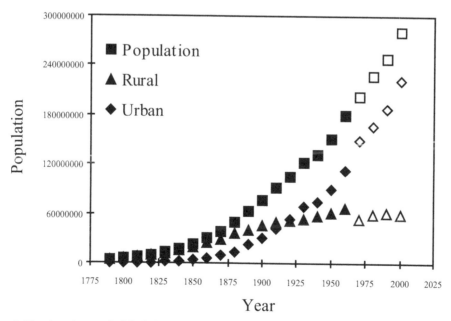

Figure 1 The changing trend of the US urban, rural and total population (1790–2000). (Notes: The solid points are data from 1790 to 1960; the hollow points are data from 1970 to 2000. The definition of the city after 1960 is different from before, but the two calibers generally fit with each other.)

Parameters Estimation and Model Selection

In order to make statistical analysis, we must discretize the United Nations model so that it transform from differential equations into difference expressions, i.e., a 2-dimension map. Then the analysis of continuous dynamics changes to that of discrete dynamics. If $\Delta t = 10$ as taken, then $dx/dt \propto \Delta x/\Delta t$. Let $r(t)$, $u(t)$ and $r(t)^*u(t)/[r(t)+u(t)]$ be independent variables, and $\Delta u(t)/\Delta t$ and $\Delta r(t)/\Delta t$ be dependent variables. A multivariate stepwise regression analysis based on the least squares computation gives the following model

$$\begin{cases} \dfrac{\Delta r(t)}{\Delta t} = 0.02584r(t) - 0.03615\dfrac{r(t)u(t)}{r(t)+u(t)} \\ \dfrac{\Delta u(t)}{\Delta t} = 0.05044\dfrac{r(t)u(t)}{r(t)+u(t)} \end{cases}. \tag{15}$$

This is a pair of difference equations of which all kinds of statistics including F statistic, P value (or t statistic), variance inflation factor (VIF) value and Durbin-Watson (DW) value can pass the tests at the significance level of $\alpha = 0.01$ (Appendix 1). In this model, $c=0$. Although we should have $b=d$ in theory, they are not equal in the empirical results. There might be two reasons for this. One is that the US is not a truly closed system because of mass foreign migration; the other is that the natural growth of the urban population is dependent on the urban-rural interaction. The second reason might be more important. But on the whole, the equations as a special case of the United Nations model can better describe the American urban and rural population migration process in the recent 200 years.

In light of equation (10), the level of urbanization should follow the logistic curve. It is easy to calculate the urbanization ratio using the data in Table 1. A least squares computation involving the percentage urban data gives the following results

$$L(t) = \frac{1}{1 + 20.41573e^{-0.02238t}}. \tag{16}$$

The goodness of fit is about $R^2=0.9839$. For convenience, we set t=year-1790 (Fig. 2). Thus we have $k \approx 0.02238$ as the estimated value of the intrinsic growth rate. On the other hand, we could estimate the original rate of growth k value by equation (15): one is k_1=b-a≈0.03615-0.02584=0.01031, and the other is k_2=d-a≈0.05044-0.02584=0.02460. The intrinsic growth rate should come into between k_1=0.01031 and k_2=0.03615 and indeed it does. The parameter values estimated from the dynamical system model, equation (15), are similar to that from the logistic model, equation (16). There are some differences between different estimated results due mainly to three factors. The first is non-closed region, the second imprecise data, and the third the computation error resulting from transformation from continuous equation to discrete expression.

For comparison and selection, we also fit the American rural and urban data to the discretization of the predator-prey interaction model. Let $r(t)$, $u(t)$ and $r(t)^*u(t)$ be independent variables and $\Delta u(t)/\Delta t$ or $\Delta r(t)/\Delta t$ dependent variables. The multivariable stepwise regression based on least squares computation gives an abnormal result, which cannot be accepted. If we loosen the requirements, then the American urbanization process could be expressed with the Keyfitz model. However, this mathematical expression has two vital shortcomings, which defies us to accept the Keyfitz model for the US urbanization. In short, neither the linear Keyfitz-Rogers model nor the usual non-linear Lotka-Volterra model is as good as the United Nations model in terms of logic sense and statistic effect (Appendix 2).

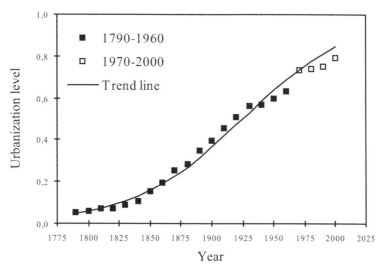

Figure 2 Logistic process of the US level of urbanization (1790–2000). (Note: The solid points are the data from 1790 to 1960, and the hollow points the data from 1970 to 2000)..

We can generate the data of American urban, rural and total population and the urbanization level by using discrete dynamics model, and then draw a comparison between the simulation value and observed data. Figures 3 and 4 respectively show the simulation results based on equations (15). It is easy to see that the change of the urban and total population approximately follow the path of the S-curve, while the rural population first increases, then decreases, and finally turns itself into the urban population completely (Fig. 3). Moreover, the level of urbanization increases in the logistic way (Fig. 4). The changing trend of the numerical simulation results displayed in figures 3 and 4 is roughly coincident with the actual observation data (Figs. 1 and 2). Although it is unpractical that the saturation value of the urbanization level is 100%, the characters of evolvement of the urban and rural population reflected by the discrete dynamical model, i.e., equation (15), comply with logic rules well. The total population converges, and the change of the percentage urban conforms to the logistic curve.

To sum up, the American model of urban-rural population interaction can be expressed by equation (2) but the parameter $c = 0$. This is the experimental foundation of theoretical analysis of discrete urbanization dynamics. So far, we have finished the building work of the model of urbanization based on the population observation in the real world. In the following section, we will discuss the complicated behaviors of the above model of urbanization dynamics in the possible world in theory.

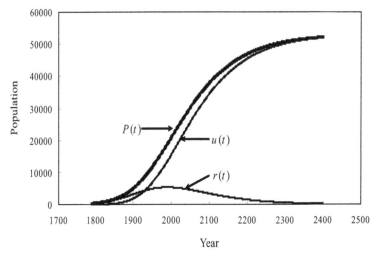

Figure 3 Numerical simulation curve of rural, urban, and total population in the American urbanization process. (Notes: The numerical simulation results are based on the discrete dynamical equations of urbanization, equation (15), the unit of population is taken as 10,000 persons).

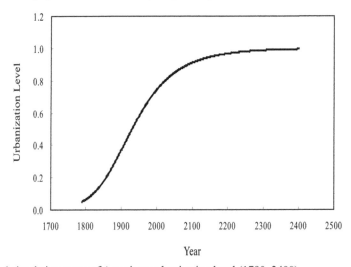

Figure 4 Numerical simulation curve of American urbanization level (1790–2400).

Complex Behaviors of Urbanization Dynamics

One of the purposes of this work is to prepare for revealing the essence of complicated behaviors of simple models. The discrete equations of two-population interaction between urban and rural systems can exhibit all the complex dynamics arising from the logistic map, including period-doubling bifurcation and chaos. Chaos theory is a field on the random behavior and latent order of certain dynamical systems, which are highly sensitive to initial conditions (Malanson et al. 1990). Chaos is often defined as intrinsic unpredictability of deterministic systems. In other words, a difference equation is regarded as a chaotic system if the solution to the equation is sensitively dependent on its initial conditions.

The discrete urban-rural interaction model can show richer details of complicated behaviors than what the logistic map does, and especially, it can offer a new way of looking at complex dynamics of simple mathematical models. According to equation (15), the parameter $c=0$, thus equation (8) can be reduced to

$$\frac{dL(t)}{dt} = (b-a)L(t)[1-L(t)] = kL(t)[1-L(t)], \tag{17}$$

where the intrinsic rate of growth is $k=b-a$. The discretization of equation (17) is a finite-difference equation

$$L_{t+1} = (1+k)L_t - kL_t^2, \tag{18}$$

Defining a new variable $x_t = kL_t/(1+k)$, we can turn equation (18) into the familiar parabola, i.e., a 1-dimension map $x_{t+1} = (1+k)x_t(1-x_t)$.

As we know, according to May (1976), the quadratic map can present periodic oscillation and even more complicated chaotic behaviors under certain conditions. Since equation (17) is derived from equation (2), the behavior characters of equation (18) should be able to be produced by the discretization of equation (2). For testing this hypothesis, we can perform some numerical simulation experiment by using equation (2), which can be discretized as a 2-dimension map

$$\begin{cases} r(t+1) = (1+a)r(t) - b\dfrac{r(t)u(t)}{r(t)+u(t)} \\ u(t+1) = (1+c)u(t) + d\dfrac{r(t)u(t)}{r(t)+u(t)} \end{cases}. \tag{19}$$

The conversion between differential equation and difference will result in some subtle change of parameter values. But for simplicity, we don't modify the parameter symbols after converting equation (2) into equation (19). The numerical solutions of equation (19) shows that when the difference between b and a increases (please notice $k=b-a$), the growth curve of urbanization level L_t indeed changes from simplicity to complexity, from S shape to periodic oscillation and even to chaos. In short, all the behaviors of logistic map revealed by May (1976) can be exhibited by the discrete two-population interaction model (Fig. 5).

As a matter of conciseness, we may as well set $b=d$ based on the theoretical hypothesis. According to the estimated results of the US urbanization model, let $a=0.025$ and $c=0$. In addition, the US census data in 1790 are taken as the original urban and rural population values. Then, we increase the value of b and d continually. The numerical simulation result shows that when $b=d<1.31$, the urbanization level presents the S-shaped curve growth, i.e., a fixed state curve; when $b=d>2.025$ ($k=b-a>2$), the dynamical system comes into 2-period oscillation state; when $b=d>2.475$ ($k=b-a>2.45$), the system takes on 4-period oscillation state; then the system will fall into 8, 16, and 2^n-period state as the values of b and d

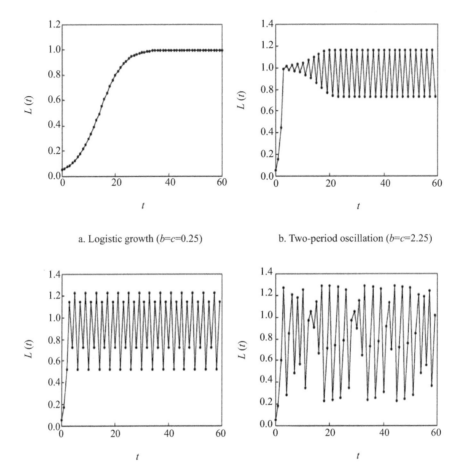

a. Logistic growth (*b*=*c*=0.25)   b. Two-period oscillation (*b*=*c*=2.25)

Figure 5 Four types of changes of urbanization level by urban-rural interaction model: from fixed state to chaos. (Notes: The parameter values are taken as *a*=0.025, *c*=0. In order to correspond to the logistic model, we make *b*=*d*. The initial urban and rural population values are based on the US census in 1790, i.e., $r(0)=r_0=3,727,559$, $u(0)=u_0=201,655$. It is easy to see that the numerical simulation results from the two-population interaction model are identical in curves to those from the logistic model by May in 1976).

increase (*n* is a positive integer); when *b*=*d*>2.6 (*k*=*b*-*a*>2.575), the system will perform random period or chaotic state. The growing limit of parameters is *b*=*d*=3.03. Compared with the work of May (1976), the period-doubling bifurcation route to chaos of the discrete urban-rural interaction model is identical in patterns to that of the logistic map. Of course, there might be subtle difference sometimes. Why the 2-dimension map exhibits the same complex dynamics with that arising from the 1-dimension map? Maybe the two-population interaction model poses a new question about the essence of chaos (Fig. 6).

Further, if we ignore the connection between the urban-rural interaction model and the logistic equation by permitting *b*≠*d*, then the behavior features of the dynamical system will become much richer. When we fix *a*, *c*, and *d*, the system will exhibit periodic oscillation or even chaos; however, when we fix *b*, the behavior characters of the system

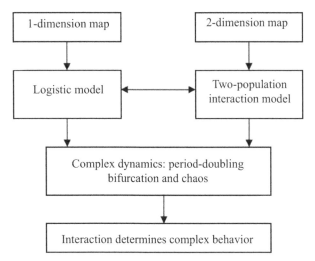

Figure 6 A 1-dimension map and a 2-dimension map reach the same goal by different routes.

do not change along with the changes of the other parameters. It is obvious that the key parameter that determines the system behavior is b, or strictly speaking, is the difference between a and b. In detail, for instance, let's consider a=0.025, c=0, and d=0.05 according to the aforementioned empirical analysis. The curve of the urbanization level changes along with b is in the same way with the result based on b=d, but the critical values of the period-doubling bifurcation route to chaos increases.

Under the circumstances, when the system comes into the chaotic state, it still presents periodic oscillation. However, the period is not only a multiple of 2 any more, but a random integer. For example, when b=3.2, system will enter into period 5 state (Fig. 7). More experimental results show that system will present period 3 or period 6 in the chaotic state. This illustrates the well-known Sharkovsky's theorem, and remind us of Li and

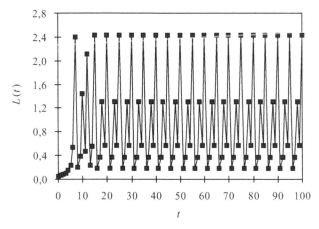

Figure 7 Period 5 oscillation of urbanization level: a special chaotic state. (Note: The parameter values are such as a=0.025, b=3.2, c=0, d=0.05).

Yorke (1975)'s discovery. The proposition "period three implies chaos" can be expressed equivalently as "period any number beyond 2-multiple implies chaos".

Based on the above simulation analysis, we can reach conclusions as follows. First, the key element of the urbanization process lies in rural population rather than urban population. According as the dynamical model of urbanization for America, the non-linear term of urban-rural interaction connects the city on one end and the village on the other. It is the difference of parameters *a* and *b* that dominates the behavior features of the urbanization dynamics. This implies that it is the rural region and urban-rural interaction that determine the progress of urbanization. Secondly, only when *b=d*, there is strict mathematical relation between the urban-rural interaction model and the logistic equation. On the one hand, the logistic model parameter derived by mathematics is *k=b-a*, whose value controls the behavior characters of the logistic map; on the other, numerical simulation shows that it is the value of (*b-a*) that determines the behavior of the urban-rural interaction model. Evidently, the precondition of connecting the urban-rural interaction model with the logistic equation is *b=d*. Third, the periodic oscillation and chaotic behavior of urban-rural interaction maybe only belong to the results of the sheer theoretical analysis. As we know, since the number of the urban and rural population can not be negative in reality, namely $r(t) \geq 0$ and $u(t) \geq 0$, it is certain that $L(t) \leq 1$ in terms of equation (3). However, if the dynamic system exhibits periodic oscillation or even chaotic behavior, the simulation value of the rural population will be negative, namely $r(t) < 0$. Thus the numerical simulation shows the abnormal phenomenon of urbanization ratio $L(t) > 1$ (see Figs. 7). Actually, this case is impossible. So period-doubling bifurcation and chaos of urbanization seem not to happen in the real world, it only appears in the imaginary world as a theoretical product (Chen, 2009).

For the sake of understanding the essence of complex dynamics, let's implement a simple mathematical transformation. Removing the nonlinear terms indicating interaction in equation (2) yields

$$\begin{cases} \dfrac{dr(t)}{dt} = ar(t) \\ \dfrac{du(t)}{dt} = cu(t) \end{cases} \tag{20}$$

The difference between the growth rates of the urban and the rural population is as follows

$$c - a = \frac{du(t)}{u(t)dt} - \frac{dr(t)}{r(t)dt} = \frac{d}{dt}\left[\ln\frac{u(t)}{r(t)}\right]. \tag{21}$$

Taking equation (7) into consideration, we have

$$c - a = \frac{d}{dt}\left[\ln V(t)\right] = \frac{dL(t)}{dt}\left[\frac{1}{L(t)} + \frac{1}{1 - L(t)}\right]. \tag{22}$$

Thus equation (22) can be transformed into a logistic equation

$$\frac{\mathrm{d}L(t)}{\mathrm{d}t} = (c - a)L(t)\big[1 - L(t)\big].$$ (23)

Although equation (20) also leads to logistic equation, the dynamical patterns of the linear differential equations are very simple. The numerical simulation based on the discretization of equation (20) shows no periodic oscillation, say nothing of chaos. This suggests that complicated dynamics such as period-doubling bifurcation and chaos coming from the logistic map is in fact rooted in interaction associated with nonlinearity.

An allometric scaling relation between urban population and rural population can be derived from equation (20) such as

$$u(t) = \eta r(t)^{b/a} = \eta r(t)^{D_u/D_r} = \eta r(t)^{\sigma},$$ (24)

in which $\eta = u_0 r_0^{-b/a}$ is a proportionality coefficient, and here r_0 and u_0 denote the initial values of rural and urban population respectively. Apparently, the allometric scaling exponent σ is given by

$$\sigma = \frac{b}{a} = \frac{D_u}{D_r},$$ (25)

where D_u refers to the fractal dimension of urban population, and D_r to the fractal dimension of rural population. This suggests that the logistic equation and the related transformation may be the mathematical link between the chaos and fractals of urban evolvement.

Another interesting discovery is that the logistic equation comes between the exponential growth models indicating simplicity, equation (20), and the two-population interaction model indicating complexity, equation (2). This reminds us that the logistic model maybe implies a mathematical transform between simple expressions and complex dynamics. As space is limited, it is impossible to make all these questions clear here and the pending questions will be discussed in future reports.

Conclusions

Urbanization is a complex process of spatial dynamics with two-population interaction. The mathematical model based on the US observation data can be proposed to describe the nonlinear evolvement. From the urban-rural interaction model, we can strictly derive the logistic equation about level of urbanization. As logistic growth exists widely in the nature and human society, the two-population interaction model may reflect a kind of ubiquitous dynamical systems. Therefore, some research conclusions can be generalized to other fields, including ecology, economics, and geology. Furthermore, since the logistic model is related to chaos, this implies that the process of urbanization has the potential possibility to bear periodical oscillation and chaotic behavior. Accordingly, the urban-rural nonlinear interaction models could help us better understand the urban system by chaos theory, and meanwhile comprehend more the nature of chaos by urban evolution. The main conclusions of this article are summarized as follows.

First, the urban-rural population interaction model can exhibit periodical oscillation and chaotic behavior in theory. The period-doubling bifurcation diagram route to chaos is identical in patterns to that from the logistic model in ecology. This provides us a new perspective to understand the complicated dynamics of simple systems. The numerical simulations show that the urban-rural population interaction model can exhibit 2-period oscillation, 4-period oscillation, ..., 2^n-period oscillation, and finally chaos, along with the change of parameter values. The changing regularity reminds us of the logistic map. Further research suggests that the discrete two-population interaction model can tell us more about the complex systems than what the logistic model does.

Secondly, complicated dynamics such as period-doubling bifurcation and chaos result from interaction instead of sheer logistic processes. The logistic model on the level of urbanization can be derived not only from urban-rural population interaction models, but also from the allometric equations of rural and urban population growth. However, the behavior patterns of the exponential growth models are very simple. In other words, periodical oscillation and chaotic behaviors can never be generated by means of the exponential growth equations. This implies that no complicated dynamics appears without interaction between rural and urban population.

Thirdly, the logistic equation may possibly form a mathematics transform relation between simplicity and complexity, which provide us a new way to look at complexity. As indicated above, the logistic model can be derived from the urban-rural interaction models or from a pair of exponential equations. This suggests that the logistic equation may act as the mathematical transform from the nonlinear interaction models to the simple linear equations. Such a relation may become a bridge connecting simplicity and complexity. In particular, if such transform can be testified to have universal property, it can be developed into a fire-new logistic transform method. If so, the transform will probably associate the unanalysable nonlinear equation with simple rules and take the opportunity to solve some long-standing unanalysable nonlinear problems.

References

1. Allen, P.M. (1997). *Cities and regions as self-organizing systems: models of complexity.* Amsterdam: Gordon and Breach Science Pub.
2. Bak, P. (1996). *How nature works: the science of self-organized sriticality.* New York: Springer-Verlag.
3. Batty, M., & Longley, P.A. (1994). *Fractal cities: a geometry of form and function.* London: Academic Press.
4. Chen, Y.G. (2009). Urban chaos and perplexing dynamics of urbanization. *Letters in Spatial and Resource Sciences*, 2(2), 85–95.
5. Chen, Y.G., & Zhou, Y.X. (2003). The rank-size rule and fractal hierarchies of cities: mathematical models and empirical analyses. *Environment and Planning* B: *Planning and Design*, 30(6), 799–818.
6. Chen, Y.G., & Zhou, Y.X. (2004). Multi-fractal measures of city-size distributions based on the three-parameter Zipf model. *Chaos, Solitons & Fractals*, 22(4), 793–805.

7. Chen, Y.G., & Zhou, Y.X. (2006). Reinterpreting central place networks using ideas from fractals and self-organized criticality. *Environment and Planning B: Planning and Design*, 33(3), 345–364.
8. Dendrinos, D.S. (1992). *The dynamics of cities: ecological determinism, dualism and chaos.* London and New York: Routledge, Chapman and Hall.
9. Dendrinos, D.S. (1996). Cites as spatial chaotic attractors. In L.D. Kiel, & E. Elliott (Eds.), *Chaos theory in the social sciences: foundations and applications.* Ann Arbor, MI: The University of Michigan Press, pp 237–268.
10. Dendrinos, D.S., & El Naschie, M.S. (Eds.) (1994). Nonlinear dynamics in urban and transportation anaylysis. *Chaos, Solitons & Fractals* [Special Issue], 4, 497–617.
11. Dendrinos, D.S., & Mullally, H. (1985). *Urban evolution: studies in the mathematical ecology of cities.* New York: Oxford University Press.
12. Frankhauser, P. (1994). *La fractalité des structures urbaines.* Paris: Economica.
13. Karmeshu. (1988). Demographic models of urbanization. *Environment and Planning B: Planning and Design*, 15(1), 47–54.
14. Keyfitz, N. (1980). Do cities grow by natural increase or migration? *Geographical Analysis*, 12, 142–156.
15. Li, T-Y, & Yorke, J.A. (1975). Period three implies chaos. *American Mathematical Monthly*, 82, 985–992.
16. Lotka, A.J. (1956). *Elements of mathematical biology.* New York: Dover Publications.
17. May, R.M. (1976). Simple mathematical models with very complicated dynamics. *Nature*, 261, 459–467.
18. Malanson, G.P., Bulter, D., & Walsh, S.J. (1990). Chaos theory in physical geography. *Physical Geography*, 11(4): 293–30.
19. Malthus, T.R. (1996). *An essay on the principle of population.* London: Routledge (reprinted).
20. Nijkamp, P. (1990). Theory of chaos: relevance for analyzing spatial process. In M.M. Fisher, P. Nijkamp, & Y. Papageorgiou (Eds), *Spatial choices and processes.* Amsterdam: North-Holland, pp 49–79.
21. Nijkamp, P., & Reggiani, A. (1992). Impacts of multiple-period lags in dynamic logit model. *Geographical Analysis*, 24(2), 159–173.
22. Nijkamp, P., & Reggiani, A. (1998). *The economics of complex spatial systems.* Amsterdam: Elsevier.
23. Portugali J. (2000). *Self-organization and the city.* Berlin: Springer-Verlag.
24. Rogers, A. (1968). *Matrix analysis of interregional population growth and distribution.* Berkeley, CA: University of California Press.
25. United Nations. (1980). *Patterns of urban and rural population growth.* New York: U.N. Department of International Economic and Social Affairs, Population Division.
26. United Nations. (1993). *World urbanization prospects.* New York: U.N. Department of Economic and Social Information, Population Division.
27. United Nations. (2004). *World urbanization prospects: the 2003 revision.* New York: U.N. Department of Economic and Social Affairs, Population Division.
28. Van der Leeuw, S.E., & McGlade, J. (1997). Structural change and bifurcation in urban evolution: a non-linear dynamical perspective. In: S.E. Van der Leeuw, & J. McGlade (Eds.), *Time, process and structured transformation in archaeology.* London: Routledge, pp 331–372.
29. Volterra, V. (1931). Variations and fluctuations of the number of individuals in animal species living together. In R.N. Chapman (Ed.), *Animal ecology.* New York: McGraw-Hill, pp 409–448.
30. White, R., & Engelen, G. (1994). Urban systems dynamics and cellular automata: fractal structures between order and chaos. *Chaos, Solitons & Fractals*, 4(4), 563–583.
31. White, R., Engelen, G., & Uljee, I. (1997). The use of constrained cellular automata for high-resolution modeling of urban-land dynamics. *Environment and Planning B: Planning and Design*, 24, 323–343.
32. Wilson, A.G. (2000). *Complex spatial systems: the modelling foundations of urban and regional analysis.* Singapore: Pearson Education.
33. Wong, D, Fotheringham, A.S. (1990). Urban systems as examples of bounded chaos: exploring the relationship between fractal dimension, rank-size, and rural to urban migration. *Geografiska Annaler B*, 72, 89–99.

Appendices

Regression Analysis Results of the US Urbanization Model

The regression analysis results of the US model of urbanization based on the least squares computation are tabulated as follows (Table A1, Table A2). The contents include ANVOA summary, estimated values of model's coefficients and related statistics.

Table A1 ANVOA summary of the US urbanization model (case 1).

Dependent variable	Model	Independent variable	R^2	Std. Error of the Estimate	Durbin-Watson	F	P-value (Sig.)
$\Delta r(t)$	1	$r(t)$	0.682	233020.670	0.527	34.383	2.399E-05
	2	$r(t), r(t)u(t)/P(t)$	0.873	152286.817	1.472	51.482	1.917E-07
	3	$r(t), r(t)u(t)/P(t), u(t)$	0.946	102505.113	3.116	82.122	3.986E-09
$\Delta u(t)$	1	$r(t)u(t)/P(t)$	0.905	295964.057	1.756	152.865	1.333E-09

Table A2 Coefficients and related statistics of the US urbanization model (case 1).

Dependent variable	Model	Independent variable	Regression Coefficients	Std. Error	t	P-value (Sig.)	VIF
$\Delta r(t)$	1	$r(t)$	0.00908	0.00155	5.864	2.399E-05	1
	2	$r(t)$	0.02584	0.00368	7.026	4.100E-06	13.201
		$r(t)u(t)/P(t)$	−0.03615	0.00763	−4.739	2.635E-04	13.201
	3	$r(t)$	0.04128	0.00431	9.572	1.602E-07	40.048
		$r(t)u(t)/P(t)$	−0.14059	0.02444	−5.753	5.000E-05	299.145
		$u(t)$	0.03419	0.00782	4.371	6.393E-04	145.326
$\Delta u(t)$	1	$r(t)u(t)/P(t)$	0.05044	0.00408	12.364	1.333E-09	1

When we take the rate of growth of the rural population as dependent variable, a least squares computation yields the following alternative model

$$\frac{\Delta r(t)}{\Delta t} = 0.04128r(t) - 0.14059\frac{r(t)u(t)}{r(t)+u(t)} + 0.03419u(t).$$

If this model is employed to describe the growth of the rural population, the saturation value of the urbanization ratio will be less than 1, which tallies with the actual situation better. However, this model gives rise to two problems. First, the model cannot avoid multicollinearity, which could be detected from the VIF value in Table A2. Second, based on the model, the total population will not converge but increase infinitely.

Regression Analysis Results Based on the Lotka-Volterra Model

The multivariable stepwise regression based on least squares computation gives the following expressions

$$\begin{cases} \dfrac{\Delta r(t)}{\Delta t} = 0.03166r(t) - 0.07763u(t) + 0.0000000010113r(t)u(t) \\ \dfrac{\Delta u(t)}{\Delta t} = 0.01567r(t) + 0.00000000016266r(t)u(t) \end{cases}$$

The first equation has serious problems. Firstly, the estimated values of the parameters cannot pass logical test. The coefficient of the linear term, $u(t)$ should be positive, but here it is negative. The physical meaning of the negative coefficient is inexplainable. What is more, the coefficient of the non-linear term, i.e., the cross term, $r(t)u(t)$, should be negative, indicating that the urban-rural interaction can transform the rural population into the urban population, but it is positive here. This conflicts with the symbol of the non-linear term of the second equation, in which the coefficient value of the cross term is positive, too.

Secondly, some values fail to pass the statistic test either. VIF value is far more than 10, which implies that there exists serious multi-collinearity between the three independent variables. If we eliminate the non-linear term, then new problems will rise. The test of serial correlation of residual errors can not be acceptable (DW=1.082), and the symbol problem of the linear term $u(t)$ remains unresolved. If we further remove the linear term $u(t)$, then the DW value will decrease to 0.527, which is more unacceptable (Table A3, Table A4). This suggests that the variables are insufficient, or the serial correlation is serious, either of which is against the basic rules of regression modeling. As for the second equation, nothing seems wrong statistically, but it cannot be understood in logic. According to this equation, the rural population will automatically flow into cities without the urban-rural interaction and the urban population will increase without any relation to itself.

Table A3 ANVOA summary of the US urbanization model (case 2).

Dependent variable	Model	Independent variable	R^2	Std. Error of the Estimate	Durbin-Watson	F	P-value (Sig.)
$\Delta r(t)$	1	$r(t)$	0.682	233020.670	0.527	34.383	2.399E-05
	2	$r(t)$, $u(t)$	0.819	181628.502	1.082	33.965	2.694E-06
	3	$r(t)$, $u(t)$, $r(t)u(t)$	0.889	147333.597	2.225	37.343	6.232E-07
$\Delta u(t)$	1	$r(t)$	0.887	322617.571	1.402	126.115	5.344E-09
	2	$r(t)$, $r(t)u(t)$	0.919	282511.849	1.913	85.165	6.473E-09
	3	$r(t)$, $u(t)$	0.914	290705.035	1.897	80.015	9.939E-09

If we loosen the requirements, then the American urbanization process could be expressed with the Keyfitz model such as

$$\begin{cases} \dfrac{\Delta r(t)}{\Delta t} = 0.00908r(t) \\ \dfrac{\Delta u(t)}{\Delta t} = 0.01433r(t) + 0.01011u(t) \end{cases}$$

Table A4 Coefficients and related statistics of the US urbanization model (case 2).

Dependent variable	Model	Independent variable	Regression Coefficients	Std. Error	t	P-value (Sig.)	VIF
$\Delta r(t)$	1	$r(t)$	0.00908	0.00155	5.864	2.399E-05	1
	2	$r(t)$	0.01854	0.00306	7.026	2.168E-05	6.413
		$u(t)$	-0.00980	0.00291	-4.739	4.237E-03	6.413
	3	$r(t)$	0.03166	0.00507	9.572	2.152E-05	26.806
		$u(t)$	-0.07763	0.02299	-5.753	4.520E-03	607.776
		$r(t)u(t)$	1.011E-09	3.410E-10	4.371	1.022E-02	419.272
$\Delta u(t)$	1	$r(t)$	0.02409	0.00214	11.230	5.344E-09	1
	2	$r(t)$	0.01567	0.00395	3.967	1.241E-03	4.424
		$r(t)u(t)$	1.62657E-10	6.716E-11	2.422	2.858E-02	4.424
	3	$r(t)$	0.01433	0.00489	2.928	1.038E-02	6.413
		$u(t)$	0.01011	0.00466	2.169	4.654E-02	6.413

This mathematical expression has two vital shortcomings. The first is the logical problem. According as the model, urban population, rural population and the total population will increase exponentially without any limit, which is against our common sense. The second is the statistic problem. That is, the second equation can not pass the DW-test (Table A3, Table A4).

Characterizing Growth and Form of Fractal Cities with Allometric Scaling Exponents[†]

Yanguang Chen

Department of Geography, College of Urban and Environmental Sciences, Peking University, Beijing 100871, China. Email: chenyg@pku.edu.cn.

ABSTRACT

Fractal growth is a kind of allometric growth, and the allometric scaling exponents can be employed to describe growing fractal phenomena such as cities. The spatial features of the regular fractals can be characterized by fractal dimension. However, for the real systems with statistical fractality, it is incomplete to measure the structure of scaling invariance only by fractal dimension. Sometimes, we need to know the ratio of different dimensions rather than the fractal dimensions themselves. A fractal-dimension ratio can make an allometric scaling exponent (ASE). As compared with fractal dimension, ASEs have three advantages. First, the values of ASEs are easy to be estimated in practice; second, ASEs can reflect the dynamical characters of system's evolution; third, the analysis of ASEs can be made through prefractal structure with limited scale. Therefore, the ASEs based on fractal dimensions are more functional than fractal dimensions for real fractal systems. In this paper, the definition and calculation method of ASEs are illustrated by starting from mathematical fractals, and then, China's cities are taken as examples to show how to apply ASEs to depicting growth and form of fractal cities.

Keywords: urban form; urban growth; urban hierarchy; urban structure; fractal; fractal dimension; allometric growth; scaling exponent.

Introduction

Dimension is a measurement of space, and measurement is the basic link between mathematical models and empirical research. So dimension is a necessary measurement for spatial analysis. Studying geographical spatial phenomena of scaling invariance such as

[†]Reused with permission from: Yanguang Chen, "Characterizing Growth and Form of Fractal Cities with Allometric Scaling Exponents," *Discrete Dynamics in Nature and Society*, vol. 2010, Article ID 194715, 22 pages, 2010. doi:10.1155/2010/194715.

cities and systems of cities has highlighted the value of fractal dimension (Batty 2005; Batty and Longley 1994; Frankhauser 1994; Haag 1994; Mandelbrot 1983; White and Engelen 1994). However, there are two problems in practical work. On the one hand, sometimes it is difficult for us to determine the numerical value of fractal dimension for some realistic systems, but it is fairly easy to calculate the ratio of different fractal parameters; On the other hand, in many cases, it is enough to reveal the a system's information by the fractal-dimension ratios and it is unnecessary to compute fractal dimension further (Chen and Jiang 2009). The ratio of different dimensions of a fractal can constitute an allometric coefficient under certain conditions. As a parameter of scale-free systems, the allometric coefficient is in fact a scaling exponent, and the fractal-dimension ratio can be called allometric scaling exponent (ASE).

The use of ASEs for the simple regular fractals in the mathematical world is not very noticeable. But for the quasi-fractals or random fractals in the real world, the function of ASEs should be viewed with special respect. A city can be regarded as an evolutive fractal. The land use patterns, spatial form and internal structure of a city can be modeled and simulated with ideas from fractal geometry (Batty 1991; Batty and Longley 1994; Benguigui et al. 2000; Benguigui et al. 2006; De Keersmaecker et al. 2003; Thomas et al. 2007; Thomas et al. 2008). There are many kinds of fractal dimensions which can be employed to characterize urban form and structure. The fractal parameters in the most common use include grid dimension, radial dimension and boundary dimension. Based on the box-counting method, we can estimate the values of boundary dimension and grid dimension at the same time and calculate their ratio to get ASE. The scaling exponent can reflect the geographical information that we need. However, it's not convenient to evaluate the fractal dimensions of the complicated systems. In practice, we can directly estimate the ASE values by skipping the calculation of fractal dimensions, and thus gain our ends of spatial analysis.

This work is devoted to discussing how to characterize growth and form of fractal cities with ASEs. The remaining part of this paper is structured as follows. Section 2 presents the basic classification of allometric relations, and brings to light the mathematical relationships between allometric scaling and fractal dimension. Section 3 gives the definition of ASE based on regular fractals, including self-similar and self-affine fractals. The concept of allometry is extended becomingly. Section 4 shows the function and use of ASE in urban studies through examples, and generalize this scaling exponent from form to function. Finally, the article is concluded with a brief summary.

Allometric Growth and Fractals

The Law of Allometric Growth

The law of allometric growth originated from biological sciences (Gayon 2000; Lee 1989), and allometric analysis is introduced to social science by Naroll and Bertalanffy (1956). From then on, the formulation gradually becomes a law of the urban geography, describing the relationship observed to be invariable between urban area and population for all cases in which the specified conditions are met (Batty and Longley, 1994; Lee 1989; Nordbeck 1971; Tobler 1969). In biology, the allometric rule was defined formally as follows (Beckmann 1958, page 247): "The rate of relative growth of an organ is a constant fraction of the rate of relative growth of the total organism." Actually it can also be restated in a broad sense as will be stated next (Chen and Jiang 2009): "The rate of relative growth of a part of a system is a constant fraction of the rate of relative growth of the whole or another part of the system." Where urban systems are concerned, allometric growth can be divided into two types: longitudinal allometry and cross allometry. The cross allometry is also called transversal allometry, which can be divided into two types, too. One is based on the rank-size distribution of cities, and the other based on hierarchy of cities or cascade structure of urban systems. The relationships among the three types of allometric growth is as follows:

$$
\text{Allometric growth} \begin{cases} \text{Longitudinal allometry based on time } (t) \\ \text{Cross allometry} \begin{cases} \text{Cross-sectional allometry based on rank } (k) \\ \text{Hierarchical allometry based on class } (m) \end{cases} \end{cases} \tag{1}
$$

The allometric equation usually takes the form of power law. Let x and y as functions of the variable v represent two measures of a system. If the relation between x and y follows the allometric scaling law, then we have

$$y(v) \propto x(v)^b, \tag{2}$$

where v refers to the variable such as time (t), rank (k), and class (m), and b to the scaling exponent. If we research the process of urban evolvement in different years, we'll get two time series, $x(t)$ and $y(t)$, which represent a city's population, area or perimeter (boundary length), etc., in the tth year. Thus we have an equation of longitudinal allometry:

$$y(t) \propto x(t)^b. \tag{3}$$

If we study n cities in a region, we should put these cities in order by ranking them according to population size. Let $v=k=1,2,\ldots,n$ be city rank, we will get an equation of cross-sectional allometry in the following form

$$y(k) \propto x(k)^b. \tag{4}$$

Further, we can put these cities into a hierarchy with cascade structure in terms of city size (Chen and Zhou 2004). Let $v=m=1,2,\ldots,M$ be city class, and M represents the bottom class of the hierarchy. Then we have a hierarchical allometric equation such as

$$y(m) \propto x(m)^b. \tag{5}$$

The three kinds of allometric relations can be applied to urban studies. If x refers to the urban population and y to the corresponding urban area, the above equations reflect the well-known allometric relations between the urban area and population (Batty and Longley 1994; Lee 1989; Longley et al. 1991); If x represents urban area, and y denotes the corresponding urban boundary length, then the above equations suggest the allometric scaling relations between the urban area and perimeter. The concept of allometric growth reminds us of the scaling relation between urban area and population at first. As a matter of fact, the geometrical relation between urban area and perimeter is a more typical allometric relation. If an urban area is compared to the volume of an animal, then the corresponding urban perimeter can be compared to the animal's epidermal area, and the urban population within the boundary to the animal's body weight.

The allometric relations between urban area and population are more familiar to geographers, but this paper attaches more importance to the scaling relationship between urban area and perimeter. A problem arises about defining urban boundary. Since urban form can be treated as a fractal body like Fournier dusts (Thomas et al. 2008), where is an unambiguous urban boundary? Since an urban boundary can be treated as a fractal line (Batty and Longley 1994), how can we figure out the length of the boundary? The key is to utilize the resolution of remote sensing (RS) image and scaling range of statistical self-similarity. For given resolution of a RS image or digital map, we can find a continuous close curve around a city figure, and this curve can be defined as urban boundary. Although a city can be theoretically treated as a fractal, the cities in the real world are actually prefractals rather than proper fractals. There is usually a scaling range on the log-log plot of estimating fractal dimension. The lower limit of the scale-free range suggests an urban boundary, by which we can calculate a length of the prefractal geographical line. A lot of methods and approaches have been developed for defining urban boundaries and estimating the fractal dimension of the boundaries (Longley and Batty 1989a; Longley and Batty 1989b). It is not difficult to constitute the allometric scaling relation between urban perimeter and area.

Fractal Properties of Allometric Growth

Scaling and dimensional analysis actually proceeded from physics, starting with Newton. Nowadays, allometric analysis, scaling analysis and dimensional analysis have reached the same goal in theoretical exploration by different routes of development (He and Liu, 2009; West, 2004). Allometric scaling relation is essentially a kind of geometrical measure relation. According to the principle of dimensional consistency in mathematics, one measure x is in proportion to another measure y only when the two measures, x and y, have the same dimension as one another (Lee, 1989). Otherwise, the two measures should be transformed into another two measures with identical dimension. Let the dimension of x be D_x and that of y be D_y. Then we have

$$y(v)^{1/D_y} \propto x(v)^{1/D_x}, \tag{6}$$

or

$$y(v) \propto x(v)^{D_y/D_x}. \tag{7}$$

Comparing equations (1) with equation (6) shows

$$b = \frac{D_y}{D_x}. \tag{8}$$

This suggests that the scaling exponent, b, is just the ratio of one dimension, D_y, to the other, D_x.

Allometric growth, including longitudinal allometry and cross allometry, can be divided into five types in terms of b value. This kind of classification of allometry will help us describe growing fractals efficiently. When $b > 1$, i.e., $D_y > D_x$, positive allometry results, and this implies that y increases at a faster rate than x. When $0 < b < 1$, i.e., $D_y < D_x$, negative allometry results, and this suggests that y increases at a slower rate. When $b=1$, i.e., $D_y = D_x$, isometry results and this implies that the two variables, x and y, increase in linear proportions with respect to one another (Lee 1989). When $b<0$, inverse allometry results. When $b=0$, no allometry or isometry results, and this value corresponds to constant proportionality. Isometry is a special case of allometry, and it is originally defined as follows: the growth rates in different parts of a growing organism are the same. For allometric scaling, a small change in a system can lead to an enormous and disproportionate increase in the size of subsystems. In contrast, isometric scaling means that growth does not lead to any change in geometry of a system.

In the next section we'll reveal the allometric scaling relations in fractal bodies. To give a better explanation to allometric scaling of fractals, we should draw a mathematical analogy between self-similar structure and urban cascade structure. For an urban hierarchy with cascade structure, we can use three exponential functions based on hierarchical structure to characterize it. These three exponential functions can be changed into a set of power functions (Chen and Zhou 2006; Chen and Zhou 2008). For a fractal, we can take a set of power laws to describe it, and the set of power laws can also be transformed into a set of exponential laws based on the steps of fractal generation (Chen and Zhou 2004). In practice, the exponential function for continuous distribution can be substituted with geometric series in the discrete framework.

The relation between fractal geometry and Euclidean geometry is of "duality". For a Euclidean geometrical body, the dimension is known (0, 1, 2, or 3), but its direct measures such as length, area and volume are unknown without calculation. On the contrary, for a fractal geometrical body, the direct measures, including length, area and volume, are always known, but its dimension required to be calculated. For example, the dimension of a rectangle is $d=2$, which can be known by common sense. However, the area of the rectangle cannot be known before the side lengths are measured. In contrast, the area of the Sierpinski gasket is zero and its inner boundary length is infinite, which can be known before implementing computation. However, the dimension of the gasket could not be known without any measuring and calculation. For the regular fractals in mathematics, the concept of length, area, and volume begin to make little sense. It is dimension that reflects the quality and quantity of fractal bodies. Of course, the dual relation is valid only for the proper fractals. As for the prefractals or the statistical fractals, the conventional measure is still serviceable. Precisely because of this, we can investigate the allometric scaling relation of cities by means of length, area, and population size among others.

However, there are no any real fractals in the real world. The so-called fractals in natural and human systems are of scaling invariance only within certain scale range. They are actually prefractals or quasifractals in the statistical sense. Therefore, we need the concepts of length, area, or even volume to build the allometric models although a fractal body has no length (e.g., it is 0 for Cantor set, or infinite for the Koch curve) and area (it equals 0) in theory. A solution to this problem is to rely on the scale concept: we examine fractals within certain scaling range. For a fractal hierarchy with infinite classes ($m=1,2,3,...$), we can investigate the finite classes, say, the first M classes ($m=1,2,3,..., M$). Here m denotes the order of class, and M is a positive integer indicative of the last class. In this instance, it is more suitable to use ASE rather than fractal dimension to characterize the real-world fractal phenomena. The reason is that the fractal dimension calculation requires the radius of covering balls or the side length of boxes to tend towards infinitesimal in theory, but there exists no infinitesimal length scale in reality. The allometric scaling is out of this restriction--an allometric analysis can be made within a certain scaling range.

Allometric Scaling of Fractals

Allometric Scaling of Self-similar Fractals

A fractal usually suggests the allometric scaling relations in the broad sense, and this can be illustrated with several classical fractal patterns. Figure 1 shows a growing fractal, which was constructed twenty years ago (Jullien and Botet 1987; Vicsek 1989). In urban studies, this kind of growing fractal is sometimes employed to model or even simulate urban growth by introducing chance factors (Batty and Longley 1994; Longley et al. 1991; White and Engelen 1993). Indeed, there is an analogy between the way a fractal develops and the way a city grows. The fractal of isotropic growth can be made through two ways: one is ceaseless cumulative process, which represents stepwise expansion of urban population (Fig. 1a); the other is endless subdivision process, which represents gradual aggregation of regional population (Fig. 1b). Finally, these two fractal evolutions reach the same goal by different routes. The similarity dimension of this fractal is

$$D_a = -\frac{\ln(N_{m+1}/N_m)}{\ln(s_{m+1}/s_m)} = \frac{\ln 5}{\ln 3} \approx 1.465, \tag{9}$$

where $N_m = 5^{m-1}$, $s_m = 3^{1-m}$, D_a denotes the fractal dimension, m is the order indicating operation step, N_m refers to the number of self-similar parts of length scale s_m in the mth step needed to cover the whole structure. For this regular fractal without overlapped parts, similarity dimension equals Hausdorff dimension and box dimension.

The boundary of the growing fractal is a kind of the quadratic Koch curve (Fig. 1c), and its dimension is

$$D_l = -\frac{\ln(N_{m+1}/N_m)}{\ln(s_{m+1}/s_m)} = \frac{\ln 5}{\ln 3} \approx 1.465, \tag{10}$$

in which D_l denotes the dimension of fractal boundary, the remaining notation is the same as in the foregoing formula. We can see that the dimension of the growing fractal form equals that of the growing fractal curve in this special case (Vicsek 1989).

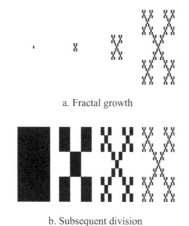

a. Fractal growth

b. Subsequent division

c. Fracal line of boundary

Figure 1 A growing fractal and its self-similar boundary line (The first four steps, by Vicsek 1989; Longley et al. 1991, and Frankhauser 1998).

For the growing fractal of infinite accumulative process displayed in Figure 1a, the longitudinal allometry is the same as the cross allometry. We can use a geometrical series to describe the number, area, and perimeter of the self-similar copies of step m in the form

$$f_m = f_1 5^{m-1},\qquad(11)$$

where m denotes the step number, f_m represents the number of self-similar parts ($f_m = N_m$), or the area ($f_m = A_m$) or the perimeter ($f_m = L_m$) of the fractal copies in the mth step, and $f_1 = 1$ is a parameter. Then, the allometric scaling relation between area and perimeter is

$$L_m \propto A_m^b = A_m^{D_l / D_a}.\qquad(12)$$

The ratio of boundary dimension D_l to form dimension D_a is 1, i.e., $b = D_l/D_a = 1$, which indicates an isometric scaling relation. The parameter b refers to the scaling exponent. Actually it is very easy to describe the fractal growth in Figure 1a using an allometric relation. The longitudinal allometry is the same as the cross allometry, and the scaling exponent equals unity.

For the fractal growth of continuous subdivision process illustrated by Figure 1b, the case is different. First of all, let us examine the longitudinal allometry. We needs three equations to describe the number, area, and perimeter of self-similar copies in the mth step, and the results are as follows

$$N_m = N_1 5^{m-1}, \tag{13}$$

$$A_m = A_1 (\frac{5}{9})^{m-1}, \tag{14}$$

$$L_m = L_1 (\frac{5}{3})^{m-1}, \tag{15}$$

Obviously, if m approaches to infinity, both the number of fractal copies and the length of fractal boundary will diverge, while the area of fractal body will tend to zero.

Where the growing process is concerned, the scaling relations between fractal-part number and boundary length can be derived by taking the logarithm of equations (10) and (12) and by eliminating m-1, and the result is

$$N_m \propto L_m^a = L_m^{\ln(5)/\ln(5/3)}. \tag{16}$$

The scaling exponent $a=\ln(5)/\ln(5/3) \approx 3.151$. The scaling of fractal-part number vs boundary length is a positive allometry. Conversely, the scaling of fractal-boundary length vs fractal-part number is a negative allometry. The allometric scaling relation between the fractal boundary length and fractal-part area is in the following form

$$L_m \propto A_m^b = A_m^{\ln(5/3)/\ln(5/9)}. \tag{17}$$

The scaling exponent $b=\ln(5/3)/\ln(5/9) \approx -0.869$. The scaling phenomenon with an exponent $b<0$ is so-called inverse allometric growth.

Next, we should investigate the cross allometry. In fact, the cross allometry based on Figure 1b is the same as that based on Figure 1a. Suppose that the area of each fractal copies is one unit. Then the area and boundary length of fractal copies in the mth step of Figure 1b can be formulated as

$$A_m = A_1 5^{m-1}, \tag{18}$$

$$L_m = L_1 5^{m-1}. \tag{19}$$

Thus we have

$$L_m \propto A_m^{\ln(5)/\ln(5)} = A_m^{[\ln(5)/\ln(3)]/[\ln(5)/\ln(3)]} = A_m^{D_l/D_a}. \tag{20}$$

This suggests that the cross allometry based on Figure 1b is an isometry because the scaling exponent is $b=D_l/D_a=1$. Indeed, equations (15) to (17) hold for Fig. 1a, but for the inner mapping procedure of Fig. 1b, the surface of the elements change at each step when going on with iteration.

Another well-known example of fractals is the Sierpinski gasket which is displayed in Fig. 2a (Mandelbrot 1983), and its interior border curve is also a fractal line, which is shown in Fig. 2b. If we introduce a chance factor indicating randomicity, then the interior boundary can be used to simulate random walk curve. The fractal dimension of the Sierpinski gasket is

a. Sierpinski gasket

b. Boundary of Sierpinski gasket

Figure 2 The Sierpinski gasket and its interior boundary lines (The first four steps).

$$D_a = -\frac{\ln(N_{m+1}/N_m)}{\ln(s_{m+1}/s_m)} = \frac{\ln 3}{\ln 2} \approx 1.585, \tag{21}$$

where $N_m = 3^{m-1}$, $s_m = 2^{1-m}$. If we use the box-counting method to evaluate the fractal dimension of its interior boundary, then we have minimum box number such as

$$N_m = 3N_{m-1} + 2^{m-1}, \tag{22}$$

where $m = 1, 2, 3, \ldots$ and $N_0 = 0$. By recurrence, we get

$$N_m = \sum_{j=0}^{m-1}(3^{m-1-j}2^j) = 3^{m-1}\sum_{j=0}^{m-1}[(\frac{2}{3})^j] \xrightarrow{m \to \infty} 3^{m-1}\frac{1}{1-2/3} = 3^m. \tag{23}$$

where $j = 1, 2, \ldots m-1$. This suggests that when m becomes large enough, N_m will approaches 3^m. So the box dimension of the interior boundary is

$$D_l = -\frac{\ln N_m}{\ln s_m} = \frac{m \ln 3}{(m-1)\ln 2} \xrightarrow{m \to \infty} \frac{\ln 3}{\ln 2} \approx 1.585. \tag{24}$$

It is easy to understand the cross allometry of the Sierpinski gasket. For simplicity, let us consider the result of step m by ignoring its growing process. The pattern is in fact a prefractal. Assuming that the area of the smallest fractal part in the geometrical figure is one unit, we can get an allometric scaling relation between the fractal-part area and the interior boundary length as

$$L_m \propto A_m^{\ln(3)/\ln(3)} = A_m^{[\ln(3)/\ln(2)]/[\ln(3)/\ln(2)]} = A_m^{D_l/D_a}. \tag{25}$$

This suggests that, for the cross allometric relation, the scaling exponent is $b = D_l/D_a = 1$, indicating an isometric relation.

The longitudinal allometry of the gasket is some different. In the mth step, the number, area and interior boundary of the self-similar copies can be formulated as follows

$$N_m = N_1 3^{m-1},$$

$$N_m = N_1 3^{m-1}$$

$$A_m = A_1 (\frac{3}{4})^{m-1} \qquad (26)$$

$$L_m = L_1 (\frac{5}{2})^{m-1}$$

So the allometric scaling relation between fractal-copy number and interior boundary length can be expressed as

$$N_m \propto L_m^a = L_m^{\ln(3)/\ln(5/2)}, \qquad (27)$$

Thus we have a scaling exponent $a=\ln(3)/\ln(5/2) \approx 1.199$. This is a kind of positive allometry. The relations between the interior boundary length and the fractal-copy area can be written as

$$L_m \propto A_m^b = A_m^{\ln(3/2)/\ln(3/4)}. \qquad (28)$$

Therefore, $b=\ln(5/2)/\ln(3/4) \approx -3.185$, and this is a kind of inverse allometry.

Generally speaking, we can describe a fractal in two ways based on allometric relations: one is longitudinal allometry, and the other is cross allometry. For the longitudinal allometry indicative of growing process, the measures of number, area and length can be expressed with three exponential equations

$$N_m = N_1 r_n^{m-1},$$

$$A_m = A_1 r_a^{1-m}, \qquad (29)$$

$$L_m = L_1 r_l^{m-1},$$

in which the parameter $r_n=N_{m+1}/N_m$, $r_a=A_m/A_{m+1}$, and $r_l=L_{m+1}/L_m$ represent number ratio, area ratio and length ratio, respectively. The proportionality coefficient N_1, A_1, and L_1 are constant (commonly we have $N_1=1$, $A_1=1$, and $L_1=1$). Then we derive three power laws or negative power laws from equations (26), (27), and (28) (Chen and Zhou, 2008). In practice, we can make use of any two of the following three power laws

$$N_m = (N_1 A_1^{\ln r_n / \ln r_a}) A_m^{-\ln r_n / \ln r_a} \propto A_m^{-a},$$

$$L_m = (L_1 A_1^{\ln r_l / \ln r_a}) A_m^{-\ln r_l / \ln r_a} \propto A_m^{-b}, \qquad (30)$$

$$N_m = (N_1 L_1^{-\ln r_n / \ln r_l}) L_m^{\ln r_n / \ln r_l} \propto L_m^c.$$

The parameters

$$a = \frac{\ln r_n}{\ln r_a}, \ b = \frac{\ln r_l}{\ln r_a}, \ c = \frac{\ln r_n}{\ln r_l} \qquad (31)$$

represent different ASEs.

For the cross allometry indicating hierarchical structure, we can derive an area-length scaling relation. If m value is limited and the area of the basic fractal parts is one unit, then the area measure (A_m) and the number measure (N_m) can be mathematically regarded as equivalent to one another. Thus the area or number and length will be formulated as

$$A_m = A_1 r_a^{m-1},$$

$$L_m = L_1 r_l^{m-1}. \tag{32}$$

Based on equations (32) and (33), an allometric scaling relation between the area and the length is

$$L_m = (L_1 A_1^{-\ln r_l / \ln r_a}) A_m^{\ln r_l / \ln r_a} \propto A_m^{D_l / D_a}. \tag{33}$$

Here the scaling exponent is defined by $b = \ln r_l / \ln r_a = D_l / D_a$.

For the same scaling relation between area and length, there exist a difference between the longitudinal allometry and the cross allometry. Sometimes the longitudinal allometry is an inverse allometry, while the cross allometry is always a positive allometry or a negative allometry. The allometric scaling relations are very useful for us to characterize the random fractals. In fractal theory, if m is large enough, the length L_m approaches infinity while the area A_m tends to zero. However, in allometric analysis, we usually take $m=1, 2, \ldots, M$, and M is a finite number. In this instance, both area A_m and length L_m are limited value. This suggests that a fractal concept is defined under an extreme condition, while the concept of allometric growth is defined within certain scale range.

In geographical studies, the scaling exponents have been associated with allometry and dimension in the context of fractal properties of cities for many years. A comparison of relations between urban area and border length, between urban population and radius, between urban border length and radius, and so forth, were discussed by Longley et al. (1992). In the novel paper on form following function, Batty and Kim (1992) once discussed several relationships among scaling law, allometry, and urban form. Frankhauser (1994) tackled the relations between urban perimeter and area as well as with surface classes. Imre and Bogaert (2004) presented that the scaling exponent of the urban area-perimeter relation is just ratio of the fractal dimension of urban boundary to that of urban form. More recently, Benguigui et al. (2006) and Thomas et al. (2008) considered also the relation between perimeter and area for cities by relating the scaling exponents with fractal dimension. The innovative point of this paper rests with that all these relations will be examined within the theoretical framework of hierarchy of cities.

Allometric Scaling of Self-affine Fractals

The self-affine fractals can also be described with allometric scaling equations. If the way of isotropic growth is turned into the way of anisotropic growth, the self-similar growing fractal in Figure 1a will become a self-affine growing fractal displayed in Fig. 3 (Jullien and Botet 1987; Vicsek 1989). To characterize this kind of self-affine fractals, we need two equations

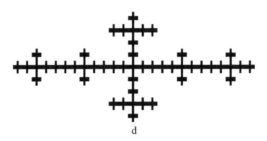

Figure 3 A sketch map of the self-affine growing fractal (the first four steps, by Vicsek 1989).

$$N(S) = S_h^{D_h},$$

$$N(S) = S_v^{D_v}, \tag{34}$$

where N is the number of self-affine copies in a certain step, S_h refers to the length scale of horizontal direction, S_v to the length scale of vertical direction, D_h denotes the fractal dimension based on horizontal measurement, and D_v represents the fractal dimension based on vertical measurement. Obviously, $N=7^{m-1}$, $S_h=5^{m-1}$, $S_v=3^{m-1}$, in which $m=1, 2, 3,\dots$. Thus we have two fractal dimension values

$$D_h = \frac{\ln N(S)}{\ln S_h} = \frac{\ln 7}{\ln 5} = 1.209, D_v = \frac{\ln N(S)}{\ln S_v} = \frac{\ln 7}{\ln 3} = 1.771. \tag{35}$$

From equations (35) and (36), an allometric scaling relation can be derived as follows

$$S_h = S_v^{D_v/D_h} = S_v^{\ln(5)/\ln(3)} = S_v^{\alpha}. \tag{36}$$

The scaling exponent of the two growing directions is $\alpha=D_v/D_h=1.465$ (Chen and Lin 2009). It is evident that, for the growing fractals, isotropy indicates isometry (Fig. 1a), while anisotropy suggests allometry of different directions (Fig. 3).

The allometric scaling relations between fractal-part area and boundary length can be characterized from the perspective of fractal growth. The area (A_m) and length (L_m) in the mth step can be formulated as

$$A_m = A_1 7^{m-1}, \tag{37}$$

$$L_m = L_1 7^{m-1}. \tag{38}$$

For simplicity, only one side of boundary length is considered in equation (39). The result implies that the scaling relation between area and length is an isometry, i.e., $L_m \propto A_m$. But in the real world, the case is more complicated, and the area-length scaling may suggest an allometric growth.

The relation between area and perimeter of the self-affine fractal can be expressed by a linear equation. Based on the fractal structure in Figure 3, the perimeter P_m can be formulated as

$$P_m = 2L_m + 2.$$ (39)

Thus we have

$$P_m = (\frac{2L_1}{A_1})A_m + 2.$$ (40)

The area-perimeter relation of the fractal in question is an isometry. This suggests that linear relation sometimes represent a kind of isometry.

Empirical Analysis

The Scaling Relation between Urban Area and Boundary

The scaling relation between urban area and perimeter bears an analogy with the allometric relation of fractal growth. Let us take China's cities as an example to illuminate this question. The processed data came from the study by Wang et al. (2005), who estimated the fractal dimension of the boundaries of 31 megacities in China—a magecity is the city with a population of more than 1,000,000. The approach to evaluating fractal dimension is grid coverage method, which is similar to the box-counting method. In fact, the original data of Wang et al. (2005) are from the Institute of Geographical Sciences and Natural Resources Research (IGSNRR), Chinese Academy of Science (CAS), China. By means of remote-sensing data and Geographical Information System (GIS), IGSNRR built the national resources and environmental database based on the maps with a scale of 1:100,000. The database mainly includes the land use data in 2000, which is inputted in 2000, and data in 1990, which is inputted in 2001. By using the results from the study by Wang et al. (2005), we can explain the allometric scaling relations between urban area and perimeter.

In the digital map, we can use a set of grid consisting of uniform squares to cover the urban figure. Changing the side length of squares of the grid s, the number of the squares occupied by urban boundary, $L(s)$, and the number of the squares occupied by urban figure, $N(s)$, will vary correspondingly. The urban area $A(s)$ can be defined by square number, $N(s)$, i.e., $A(s)=N(s)$, and the boundary length is represented by the number, $L(s)$. If the relation between urban area and perimeter follows the allometric scaling law, then, according to equation (34), we have

$$\ln L(s) = C + \frac{D_l}{D_a} \ln A(s) = C + b \ln A(s),$$ (41)

where C is a constant, D_a refers to the fractal dimensions of urban form, D_l to the fractal dimensions of urban boundary. Thus, the ASE value is given by $b=D_l/D_a$. Wang et al. (2005)

changed the side length of squares of the grid for 19 times, and the minimum area of the square is 200×200 m^2. The scaling exponents of the 31 megacities in China are evaluated by using the advanced language of GIS software (Table 1). Assuming the dimension of urban form $D_a=2$, Wang et al. (2005) estimated the fractal dimension of urban boundary and the result is $D_1=D_a b=2b$ (see "boundary dimension 1" in Table 1). The information of spatio-temporal evolution of China's cities from 1990 to 2000 can be revealed through ASE values.

Table 1 The fractal dimension values of urban boundaries and ASEs of the area-perimeter relations of 31 megacities in China in 1990 and 2000.

City	1990			2000		
	Scaling exponent (b)	Boundary dimension 1 $(D_1)^*$	Boundary dimension 2 $(D_1)^{**}$	Scaling exponent (b)	Boundary dimension 1 $(D_1)^*$	Boundary dimension 2 $(D_1)^{**}$
Anshan	0.735	1.469	1.250	0.690	1.380	1.174
Beijing	0.751	1.502	1.277	0.722	1.444	1.228
Changchun	0.702	1.404	1.194	0.701	1.401	1.192
Changsha	0.766	1.532	1.303	0.763	1.526	1.298
Chengdu	0.838	1.676	1.425	0.837	1.674	1.424
Chongqing	0.753	1.505	1.281	0.723	1.446	1.230
Dalian	0.745	1.489	1.267	0.737	1.474	1.254
Fushun	0.706	1.411	1.201	0.683	1.366	1.162
Guangzhou	0.702	1.403	1.194	0.772	1.544	1.313
Guiyang	0.874	1.748	1.487	0.871	1.742	1.482
Hangzhou	0.800	1.599	1.361	0.783	1.565	1.332
Harbin	0.685	1.369	1.165	0.654	1.307	1.112
Jilin	0.712	1.424	1.211	0.716	1.432	1.218
Jinan	0.717	1.433	1.220	0.732	1.463	1.245
Kunming	0.794	1.588	1.351	0.736	1.472	1.252
Lanzhou	0.741	1.482	1.260	0.736	1.471	1.252
Nanchang	0.727	1.454	1.237	0.751	1.502	1.277
Nanjing	0.785	1.569	1.335	0.747	1.494	1.271
Qingdao	0.689	1.377	1.172	0.653	1.305	1.111
Qiqihaer	0.678	1.355	1.153	0.670	1.340	1.140
Shanghai	0.741	1.481	1.260	0.711	1.422	1.209
Shenyang	0.650	1.300	1.106	0.639	1.278	1.087
Shijiazhuang	0.786	1.571	1.337	0.733	1.466	1.247
Taiyuan	0.777	1.554	1.322	0.769	1.538	1.308
Tangshan	0.750	1.500	1.276	0.728	1.456	1.238
Tianjin	0.688	1.376	1.170	0.678	1.356	1.153
Urumchi	0.724	1.447	1.232	0.721	1.441	1.226
Wuhan	0.738	1.475	1.255	0.747	1.494	1.271
Xian	0.731	1.461	1.243	0.683	1.366	1.162
Zhengzhou	0.753	1.506	1.281	0.713	1.426	1.213
Zibo	0.763	1.525	1.298	0.747	1.493	1.271
Anshan	0.735	1.469	1.250	0.690	1.380	1.174

Notes: * The scaling exponent and the first kind of fractal dimension values (boundary dimension 1) were estimated and provided by Dr. Xinsheng Wang, who assumed that the dimension of urban form is $D_a=2$ (see Wang et al. 2005). ** The second kind of fractal dimension values (boundary dimension 2) was estimated through the scaling exponent values by this paper's author, who assumed that the dimension of urban form is $D_a=1.701$.

The true dimension of urban form, however, is expected to be less than 2, namely $D_a < 2$. Theoretically, a city boundary have an analogy with the triadic Koch curve, which has a fractal dimension $D_l \approx 1.262$. Empirical analyses and computer simulation have shown that the fractal dimension of urban boundary is close to the Koch curve's dimension on the average (Batty and Longley 1988; Longley and Batty 1989b). On the other hand, urban growth and form can be modeled by diffusion-limited aggregation (DLA) model and dielectric breakdown model (DBM), and the average fractal dimension D_a is close to 1.701 (Batty 1991a; Batty and Longley 1994). The DBM-simulation process has been well illustrated by Batty (1991b). By using the value $D_a = 1.701$, we can re-evaluate the boundary dimension of the 31 cities in China (see "boundary dimension 2" in Table 1). Therefore, the scaling exponent of urban area-perimeter relation is expected to be $b \approx 1.262/1.701 \approx 0.742$. Actually, the average value of ASEs of the 31 megacities is about 0.742 in 1990, and around 0.727 in 2000. In the sense of statistical average, the results approach to the theoretical expectation on the whole.

A question may be put as follows: what are the real values of the boundary dimension and form dimension of China's cities? We cannot know them by the data from Wang et al. (2005). For example, the boundary dimension of Beijing city in 2000 may be about 1.444, may be about 1.228, may be other numerical values. We cannot make sure. However, we know the approximate value of ASE, that is, $b \approx 0.722$ (Table 1). According to equation (7), the higher value of ASE suggests higher boundary dimension, or lower form dimension. Where geographical space is concerned, the ASE values of Northern China's cities are higher than those of Southern China's cities in the mass. The landform of Northern China is mainly plain, while Southern China is principally of mountainous terrain. Generally speaking, the form dimensions of the cities in plain are higher than those of cities in mountainous region. For boundary dimension, the opposite is true. The cities with the highest ASE values is in northeastern plain of China, such as Shenyang and Harbin, while the cities with the lowest ASE values is in southwestern mountainous region, such as Guiyang and Chengdu. Where geographical evolution is concerned, as a whole, the ASE values in 2000 are lower than those in 1990. As we know, the boundary dimension of a city become lower and lower over time (Longley and Batty 1989b). For the form dimension, the opposite is true (Benguigui et al. 2000). This suggests that the ASE values of urban boundary and form tend to descending with the lapse of time. The function and use of a spatial measure can be thrown out by comparison and relation. ASE is just the result of comparing or relating one fractal dimension with the other fractal dimension.

The Scaling Relations between Urban Area and Population

The allometric scaling analysis of urban form can be generalized to the relation between urban area and population. The allometric model of urban area-population scaling can be used to predict regional population growth (Lo and Welch 1977). Let us take China's system of cities as example to illustrate the pattern of allometric relation. We will take the national capital, Beijing, as example to make a longitudinal allometric analysis, and employ the system of cities in China as another example to make a cross allometric analysis. The original data came from the Ministry of Housing and Urban-Rural Development of China.

The two basic measures, urban population and built-up area, are used as variables. For Chinese, the "built-up area" is also called as "surface area of built district". In fact, the term "urban area" is a concept of administrative sense in China. There is no certain relationship between urban area and urban landscape; therefore the concept of urban area cannot reflect urban form effectively. The built-up area in Chinese is similar to the concept of urbanized area in the western world. Perhaps the former is smaller than the latter. In short, it is built-up area rather urban area that is suitable for us to make an allometric analysis for China's cities. From now on, "urban area" will be used to mean "built-up area" in the context.

First of all, we carry out a longitudinal allometric analysis of Beijing's growth. The data are urban population and built-up area from 1991 to 2005. Through the log-log plot we can find that city area and population from 1991 to 2004 satisfies the allometric scaling relation in the mass (Fig. 4). However, the data point of 2005 is an exceptional value by reason of demography (Appendix). So far there has been no strict definition for cities, and both urban population and urban area are varied frequently in China. The alteration is sometimes caused by administrative factors rather than urban growth itself. From 1991 to 2000, the land use of Beijing is grimly restricted with urban policy. From 2001 to 2003, urban region of the city was enlarged suddenly by governmental behavior instead of urban natural growth. Since 2004, urban land use was restricted again so that Beijing's area is not proportional to its population size. If the abnormal variation is on the small side, it could not influence the appearance of the statistical law of urban evolvement. However, if the change is too large, the scaling relation will be broken. The abovementioned outliers will appear if administrative factors disturb urban development violently.

Table 2 The urban area and population of Beijing from 1991 to 2005.

Year	Original data		Logarithmic value		Standardization	
	Urban Population P_t (10,000)	Built-up Area A_t (km²)	$\ln P_t$	$\ln A_t$	Sdz($\ln P_t$)	Sdz($\ln A_t$)
1991	629.6	397.4	6.445	5.985	-0.772	-1.028
1992	634.7	429.4	6.453	6.062	-0.742	-0.843
1993	640.8	454.1	6.463	6.118	-0.707	-0.709
1994	649.6	467.0	6.476	6.146	-0.657	-0.642
1995	656.1	476.8	6.486	6.167	-0.620	-0.592
1996	662.8	476.8	6.496	6.167	-0.582	-0.592
1997	670.0	488.1	6.507	6.191	-0.542	-0.536
1998	675.3	488.3	6.515	6.191	-0.513	-0.535
1999	682.4	488.3	6.526	6.191	-0.474	-0.535
2000	690.9	490.1	6.538	6.195	-0.429	-0.526
2001	861.4	747.8	6.759	6.617	0.387	0.485
2002	949.7	1043.5	6.856	6.950	0.748	1.282
2003	962.7	1180.1	6.870	7.073	0.799	1.577
2004	1187.0	1182.3	7.079	7.075	1.573	1.581
2005	1538.0	1200.0	7.338	7.090	**2.532**	1.617
Average	806.1	667.3	6.654	6.415	0.000	0.000
Stdev	260.3	313.4	0.270	0.418	1.000	1.000

Original data source: The Ministry of Housing and Urban-Rural Development of China. *1991-2005 Statistic Annals of China's Urban Construction.*

Excluding the data of 2005 from our consideration according to the double standard error, we can make a regression analysis by using the data from 1991 to 2004. On the whole, the process of urban growth conforms to the allometric scaling law to some extent (Figure 4). A least squares calculation of 14 years data yields the following scaling model of longitudinal allometry

$$A_t = 0.002P_t^{1.885},\qquad(42)$$

where t denotes year (t=1991,1992,…, 2004), A_t and P_t are built-up area and urban population of year t, respectively. The scaling exponent is estimated as about $b \approx 1.885$, and the goodness of fit is $R^2 \approx 0.948$. Compared with the urban population, the city area expanded in the mode of positive allometry. That is to say, if the populations of Beijing increase a unit, the city area will increase more than a unit. In other words, the urban area grows quicker than the urban population. This is an allometric mode of wasteful land use, and the scaling exponent suggests that the urban expansion of Beijing should be restricted by taking strong economic measures.

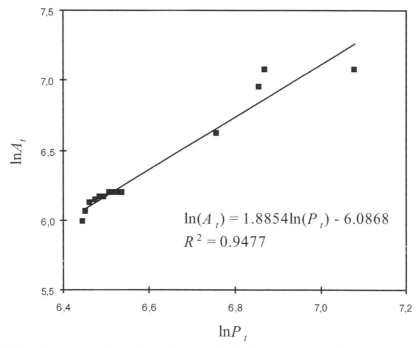

$$\ln(A_t) = 1.8854\ln(P_t) - 6.0868$$
$$R^2 = 0.9477$$

Figure 4 The allometric scaling relation between the built-up area and urban population of Beijing (1991–2004).

(**Note:** The unit of area is square kilometer, and the unit of population is ten thousands. The data point in 2005 is taken as an outlier. Because that the scale is small, the data points on double logarithmic plot are distributed concentratively so that it is not clear for us to read it. Therefore, the axes of the plot are of ordinary scale.)

It will not be surprising if some readers doubt the result from the Beijing case. The effect of fitting the Beijing's data set to equation (2) is not very satisfying. The application of the scaling law to Beijing is mainly based on an apriori idea rather than some statistical evidences. The apriori idea is that urban systems should follow but sometimes offends the law of allometric growth (Chen 2008). In fact, human systems are different from physical systems to a degree. The empirical laws on physical systems are of spatio-temporally translational symmetry, while the empirical laws on human systems are not of translational symmetry in both space and time. For physical systems, a counterexample or exceptional case is enough to overthrow an empirical law. However, for human systems, a few counterexamples or exceptional cases are not enough to overrule any empirical law which is supported by many observational data. Just because of this, August Lösch, the well-known German economist, once said that if a mathematical model supported by many cases does not agree with reality, it may be reality rather than the mathematical model that is wrong[*] (Chen 2008). It is hard to clarify this notion in a few lines of words. Many empirical evidences from other cities lend support to the scaling relation between urban area and population (e.g., Batty and Longley 1994; Chen and Lin 2009; Lee 1989; Lo and Welch 1977). If a city, say, Beijing, fails to follow this law, it is the city instead of the scaling law that is wrong. If so, the urban man-land relation should be improved according to the allometric scaling law. The meaning of Beijing case lies in three aspects. First, it gives an approximate scaling exponent for our understanding Beijing's growth. Second, it reveals the problem of Beijing's development to be resolved in the future. Third, it suggests that we should develop an urban theory by the ideas from fractals and allometric growth for urban planning and spatial optimization of cities.

Next, let us make a cross sectional allometric analysis based on the rank-size relationships. For simplicity, only the allometric pattern in 2005 is shown (Fig. 5). This year, there were 660 cities which were approved officially in China. We rank the population size of these cities from the largest to the smallest, and put urban area in order coinciding with population size. As a result, we have an allometric scaling relation such as

$$A_k = 1.917 P_k^{0.817}, \tag{43}$$

where k denotes city rank by population ($k=1, 2,\ldots, 660$), A_k refers to the area of the city of rank k, and P_k to the city population of rank k. The scaling exponent is estimated as $b \approx 0.817$, and the goodness of fitting is $R^2 \approx 0.842$. According to our rule of sorting order, we have an inequality $P_k \geq P_{k+1}$ to a certainty. However, another inequation, $A_k \geq A_{k+1}$, will not necessarily come into existence. The reason is that a city has more population does not imply that it has larger built-up area.

Finally, we can make an allometric analysis based on hierarchy of cities with cascade structure. Putting the 660 cities in order by population size, we can classify them in a top-down way in terms of the 2^n principle of cities: the first class has one city—the city of rank 1, the second class has two cities—the cities of ranks 2 and 3, the third class has four cities—the cities of ranks 4, 5, 6, and 7, and so on. Evidently, the mth class will have

---

[*]This opinion refers to a letter from Professor Michael Woldenberg at State University of New York (2004).

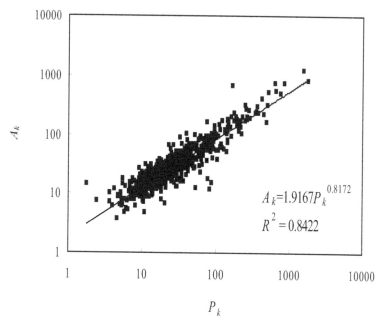

Figure 5 The allometric scaling relation between built-up area and urban population of China's cities (2005).

2^{m-1} cities (Chen and Zhou, 2004). Thus, the 660 cities can be divided into ten classes, and the city number in the last class is expected to be $N_{10}=2^9=512$ according to the theoretical rule. However, the cities in the last class are less developed, and we have only 149 cities $(660-(2^0+2^1+2^2+\ldots+2^8)=149)$. Moreover, the sizes of the cities in the tenth class are smaller than that is expected theoretically. Therefore, the bottom class is a lame-duck class (Davis 1978). A least squares computation gives the following allometric scaling model

$$A_m = 1.786P_m^{0.856}, \qquad (44)$$

in which, m denotes the order of class ($m=1, 2,\ldots, 10$), A_m refers to the average area of the cities of order m, and P_m to the average population size of the cities in the mth class. The scaling exponent is estimated as $b\approx0.856\approx6/7$, and the goodness of fitting is $R^2\approx0.987$ (Fig. 6).

What is the expected value of the scaling exponent of urban area and population relation? Through spectral analysis, we can learn that the dimension of the urban population is $D_p=2$. In the other hand, simulation experiment analyses showed that the expected fractal dimensions of the urban form is $D_a\approx1.701$ (Batty 1991). Therefore, ASE is expected to be $b=D_a/D_p\approx1.701/2\approx0.85$. ASE of China's cities in 2005 is close to this value.

The allometric scaling pattern based on the rank-size distribution is equivalent in theory to the allometric scaling relation based on hierarchical structure. The former differs from the latter to some extent in empirical analysis, but the difference between the two is not very significant. One gives the scaling exponent $b\approx0.817$, and the other yields $b\approx0.856$. As far as our example is concerned, the scaling exponent based on hierarchical system is closer to the theoretical expectation. After all, the two scaling models belong to the negative allometry ($b<1$). This suggests that, by and large, the land use of China's cities is comparatively reasonable.

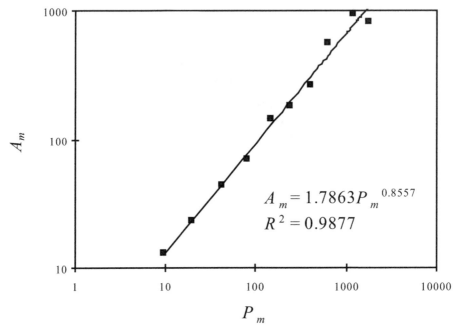

Figure 6 The allometric scaling relation between the average built-up area and urban population of China's hierarchy of cities (2005).

Further, by using the least squares calculation, we can easily fit the data to the scaling relation between the urban number and population as well as the scaling relation between the urban number and area. Of course, the lame duck class should be removed as an outlier from the regression analysis because the number of cities in this class is too few to support the scaling relation. Based on the hierarchy of cities without the tenth class, the modeling results are as follows

$$N_m = 14784.254 P_m^{-1.262} ,$$

$$N_m = 28133.543 A_m^{-1.435} . \tag{45}$$

The coefficients of determination are $R^2 \approx 0.995$ and $R^2 \approx 0.975$, respectively. This implies that the fractal dimension of hierarchy of cities is $D \approx 1.262$ by population measure or $D \approx 1.435$ by area measure.

In light of the nature of allometric growth, we can classify the geographical space into three types. The allometric patterns reflected by Figs. 1, 2, and 3 as well as Table 1 correspond to the real space (R-space); Figure 4 illustrates a longitudinal allometry which corresponds to the phase space (P-space); The cross allometry reflected by Figs. 5 and 6 corresponds to the order space (O-space). The real space is the conventional concept of geographical space, while the phase space and the order space belong to the generalized space, a kind of abstract space (Chen and Jiang 2009). In the 2-dimension real space, the fractal dimension, especially the box-counting dimension, of urban form is generally less than 2; in the generalized space, however, the fractal dimension of urban form cannot be confined by the dimension of the Euclidean space in which the urban form exist.

In addition, there are some inherent relationship among fractal structure, allometric relation and self-organized networks. In the process of measuring the fractal dimension of the form displayed in Figs. 1, 2, and 3, the linear size s_m can be replaced by length scale d_m, which denotes the distance between two centers of immediate fractal copies of order m. Thus we have a network-based definition of fractal dimension $D=-\ln(N_{m+1}/N_m)/\ln(d_m/d_{m+1})$ (Chen and Zhou 2008). In fact, fractal geometry, allometric concepts, and network science are being slowly integrated into a new theory, which can be employed to explain the evolvement and development of cities (Batty 2008). As space is limited, many questions are pending further discussion in future studies.

Conclusion

A fractal growth is actually an allometric growth, the process of allometric growth is always involved with a number of fractal dimension relations. For a simple regular fractal, say, the Koch curve, one fractal dimension is enough to characterize its geometrical property. However, for a random fractal, especially, for a prefractal phenomenon, e.g., a city, it is not sufficient to characterize its form and structure with only one fractional dimension. We should employ a set of fractal parameters including various fractal dimensions, the ratios of fractal dimensions, among others, to describe the complicated systems of scaling invariance. The ratio of one fractal dimension to the other related fractal dimension can constitute an ASE discussed above. Now, the main conclusions in this article can be drawn and summarized as follows.

First, if the form of the growing phenomena such as cities is self-similar, the boundaries of the phenomena will be of self-similarity also. The geometric relationship between the boundary length and the whole form is always an allometric scaling relation. The allometric relation can be described from two angles of view. One is the longitudinal allometry, the other is the cross allometry. The former reflects the progressively evolutive process of the fractal growing from an initiator, while the latter reflects the hierarchy with cascade structure corresponding to the growing process. The method of allometric scaling analysis can be applied not only to the isotropic growing phenomena indicating self-similar fractals, but also to the anisotropic growths indicative of self-affine fractals. For the self-affine patterns, there exists an allometric scaling relation between the parts in different directions.

Second, the mathematical description of the allometric growth rests with two aspects: one is various fractal dimensions, and the other is ASEs. The fractal characterization is a static method, laying emphasis on the best result by assuming the linear size of fractal elements to approach to zero. Theoretically, if the linear size of fractal measure becomes infinitesimal, the fractal dimension value will approach to a real constant. By contrast with fractal dimension, ASE lays stress on an evolutive process or a kind of spatial relations. The linear size of yardmeasure for estimating ASEs not necessarily tend towards infinitesimal. As long as the scaling range is long enough, the result will be satisfying. This is significant for urban studies because the self-similarity of cities is valid only within certain scale limits. By means of allometric analysis we can reveals the regularity and complexity of urban evolvement and structure efficiently.

Third, the fractal studies can be generalized from real space to the abstract space in terms of allometric growth. All the fractals that can be directly exhibited by maps or pictures are fractal in actual space. However, there are lots of fractals which cannot be immediately represented by graphics. The scaling invariance of this kind of fractals can be indirectly revealed with mathematical transformation and log-log plots. Majority of these fractals often come from the generalized space. Urban form and boundaries belong to the real space, but the scaling relation between urban area and population belong to the generalized space. It is difficult to evaluate the dimensions of the fractals in the abstract space, but it is easy to estimate the ratio of different fractal dimensions based on the generalized space. In many cases, what we want to know is just the fractal dimension ratios rather than the fractal dimensions themselves. Through allometric analyses we can directly calculate the ratio of fractal dimensions and thus obtain ASEs, thereby we can research the structure and functions of fractal systems.

Appendix: How to Reveal the Outliers of an Allometric Scaling Relation

The city of Beijing is taken as an example to show how to reveal the outliers of an allometric scaling relation. The allometric growth is in fact based on exponential growth theoretically. Suppose that both urban area and population increase exponentially with the passage of time. Taking the logarithm of urban population P_t and built-up area A_t, we get the logarithmic values of the two measures, $\ln P_t$ and $\ln A_t$. Then we can standardize the logarithmic measures by z-score, and the results are represented by $\text{Sdz}(\ln P_t)$ and $\text{Sdz}(\ln A_t)$ in Table A. If one or two of the standardized data are greater than double standard error, 2, the values can be regarded as outliers on the statistical significance of $\alpha=0.05$. Clearly, the data point in 2005 is an exceptional value.

Acknowledgement

This research was sponsored by the National Natural Science Foundation of China (Grant No. 40771061) and Beijing Natural Science Foundation (Grant No. 8093033). Many thanks to two anonymous referees who provided interesting suggestions. The errors and omissions which remain are all mine.

References

1. M. Batty, Cities and Complexity: Understanding Cities with Cellular Automata, MIT Press, Cambridge, Mass, USA, 2005.
2. M. Batty and P. A. Longley, *Fractal Cities: A Geometry of Form and Function*, Academic Press, London, UK, 1994.
3. P. Frankhauser, *La Fractalité des Structures Urbaines*, Economica, Paris, France, 1994.

4. G. Haag, "The rank-size distribution of settlements as a dynamic multifractal phenomenon," *Chaos, Solitons and Fractals*, vol. 4, no. 4, pp. 519–534, 1994.

5. B.B. Mandelbrot, *The Fractal Geometry of Nature*, W.H. Freeman and Company, San Francisco, Calif, USA, 1982.

6. R. White and G. Engelen, "Urban systems dynamics and cellular automata: fractal structures between order and chaos," *Chaos, Solitons and Fractals*, vol. 4, no. 4, pp. 563–583, 1994.

7. Y. Chen and S. Jiang, "An analytical process of the spatio-temporal evolution of urban systems based on allometric and fractal ideas," *Chaos, Solitons and Fractals*, vol. 39, no. 1, pp. 49–64, 2009.

8. M. Batty, "Generating urban forms from diffusive growth," *Environment & Planning A*, vol. 23, no. 4, pp. 511–544, 1991.

9. M. Batty, "Cities as fractals: simulating growth and form," in *Fractals and Chaos*, A.J. Crilly, R.A. Earnshaw, and H. Jones, Eds., pp. 43–69, Springer, New York, NY, USA, 1991.

10. L. Benguigui, D. Czamanski, M. Marinov, and Y. Portugali, "When and where is a city fractal?" *Environment and Planning B*, vol. 27, no. 4, pp. 507–519, 2000.

11. L. Benguigui, E. Blumenfeld-Lieberthal, and D. Czamanski, "The dynamics of the Tel Aviv morphology," *Environment and Planning B*, vol. 33, no. 2, pp. 269–284, 2006.

12. M.-L. De Keersmaecker, P. Frankhauser, and I. Thomas, "Using fractal dimensions for characterizing intra-urban diversity: the example of Brussels," *Geographical Analysis*, vol. 35, no. 4, pp. 310–328, 2003.

13. I. Thomas, P. Frankhauser, and M.-L. De Keersmaecker, "Fractal dimension versus density of built-up surfaces in the periphery of Brussels," *Papers in Regional Science*, vol. 86, no. 2, pp. 287–308, 2007.

14. I. Thomas, P. Frankhauser, and C. Biernacki, "The morphology of built-up landscapes in Wallonia (Belgium): a classification using fractal indices," *Landscape and Urban Planning*, vol. 84, no. 2, pp. 99–115, 2008.

15. J. Gayon, "History of the concept of allometry," *American Zoologist*, vol. 40, no. 5, pp. 748–758, 2000.

16. Y. Lee, "An allometric analysis of the US urban system: 1960–80," *Environment & Planning A*, vol. 21, no. 4, pp. 463–476, 1989.

17. R.S. Naroll and L. von Bertalanffy, "The principle of allometry in biology and social sciences," *General Systems Yearbook*, vol. 1, Part II, pp. 76–89, 1956.

18. R.S. Naroll and L. von Bertalanffy, "The principle of allometry in biology and social sciences," *Ekistics*, vol. 36, no. 215, pp. 244–252, 1973.

19. S. Nordbeck, "Urban allometric growth," *Annaler B*, vol. 53, no. 1, pp. 54–67, 1971.

20. W.R. Tobler, "Satellite confirmation of settlement size coefficients," *Area*, vol. 1, no. 3, pp. 30–34, 1969.

21. M. J. Beckmann, "City hierarchies and distribution of city sizes," *Economic Development and Cultural Change*, vol. 6, no. 3, pp. 243–248, 1958.

22. Y. Chen and Y. Zhou, "Multi-fractal measures of city-size distributions based on the three-parameter Zipf model," *Chaos, Solitons and Fractals*, vol. 22, no. 4, pp. 793–805, 2004.

23. P.A. Longley, M. Batty, and J. Shepherd, "The size, shape and dimension of urban settlements," *Transactions of the Institute of British Geographers*, vol. 16, no. 1, pp. 75–94, 1991.

24. P.A. Longley and M. Batty, "On the fractal measurement of geographical boundaries," *Geographical Analysis*, vol. 21, no. 1, pp. 47–67, 1989.

25. P. A. Longley and M. Batty, "Fractal measurement and line generalization," *Computers and Geosciences*, vol. 15, no. 2, pp. 167–183, 1989.

26. J.-H. He and J.-F. Liu, "Allometric scaling laws in biology and physics," *Chaos, Solitons and Fractals*, vol. 41, no. 4, pp. 1836–1838, 2009.

27. B.J. West, "Comments on the renormalization group, scaling and measures of complexity," *Chaos, Solitons and Fractals*, vol. 20, no. 1, pp. 33–44, 2004.

28. Y. Chen and Y. Zhou, "Reinterpreting central place networks using ideas from fractals and self-organized criticality," *Environment and Planning B*, vol. 33, no. 3, pp. 345–364, 2006.

29. Y. Chen and Y. Zhou, "Scaling laws and indications of self-organized criticality in urban systems," *Chaos, Solitons and Fractals*, vol. 35, no. 1, pp. 85–98, 2008.

30. R. Jullien and R. Botet, *Aggregation and Fractal Aggregates*, World Scientific, Teaneck, NJ, USA, 1987.

31. T. Vicsek, *Fractal Growth Phenomena*, World Scientific, Singapore, 1989.

32. R. White and G. Engelen, "Cellular automata and fractal urban form: a cellular modelling approach to the evolution of urban land-use patterns," *Environment & Planning A*, vol. 25, no. 8, pp. 1175–1199, 1993.

33. P. Frankhauser, "The fractal approach: a new tool for the spatial analysis of urban agglomerations," *Population*, vol. 10, no. 1, pp. 205–240, 1998.

34. P. Longley, M. Batty, J. Shepherd, and G. Sadler, "Do green belts change the shape of urban areas? A preliminary analysis of the settlement geography of South East England," *Regional Studies*, vol. 26, no. 5, pp. 437–452, 1992.

35. M. Batty and K. S. Kim, "Form follows function: reformulating urban population density functions," *Urban Studies*, vol. 29, no. 7, pp. 1043–1069, 1992.

36. A. R. Imre and J. Bogaert, "The fractal dimension as a measure of the quality of habitats," *Acta Biotheoretica*, vol. 52, no. 1, pp. 41–56, 2004.

37. Y. Chen and J. Lin, "Modeling the self-affine structure and optimization conditions of city systems using the idea from fractals," *Chaos, Solitons and Fractals*, vol. 41, no. 2, pp. 615–629, 2009.

38. X. Wang, J. Liu, D. Zhuang, and L. Wang, "Spatial-temporal changes of urban spatial morphology in China," *Acta Geographica Sinica*, vol. 60, no. 3, pp. 392–400, 2005.

39. M. Batty and P. A. Longley, "The morphology of urban land use," *Environment & Planning B*, vol. 15, no. 4, pp. 461–488, 1988.

40. C. P. Lo and R. Welch, "Chinese urban population estimates," *Annals of the Association of American Geographers*, vol. 67, no. 2, pp. 246–253, 1977.

41. Y.G. Chen, *Fractal Urban Systems: Scaling, Symmetry, and Spatial Complexity*, Scientific Press, Beijing, China, 2008.

42. K. Davis, "World urbanization: 1950–1970," in *Systems of Cities*, I. S. Bourne and J. W. Simons, Eds., pp. 92–100, Oxford University Press, New York, NY, USA, 1978.

43. M. Batty, "The size, scale, and shape of cities," *Science*, vol. 319, no. 5864, pp. 769–771, 2008.

Estimating the Distribution of Dynamic Invariants: Illustrated with an Application to Human Photo-Plethysmographic Time Series[†]

Michael Small

Department of Electronic and Information Engineering, Hong Kong Polytechnic University, Hung Hom, Kowloon, Hong Kong. Email: ensmall@polyu.edu.hk

ABSTRACT

Dynamic invariants are often estimated from experimental time series with the aim of differentiating between different physical states in the underlying system. The most popular schemes for estimating dynamic invariants are capable of estimating confidence intervals, owever, such confidence intervals do not reflect variability in the underlying dynamics. We propose a surrogate based method to estimate the expected distribution of values under the null hypothesis that the underlying deterministic dynamics are stationary. We demonstrate the application of this method by considering four recordings of human pulse waveforms in differing physiological states and show that correlation dimension and entropy are insufficient to differentiate between these states. In contrast, algorithmic complexity can clearly differentiate between all four rhythms.

Background

Various dynamic invariants are often estimated from time series in a wide variety of scientific disciplines. It has long been known that these estimates (and in particular correlation dimension estimates) alone are not sufficient to differentiate between chaos and noise. Most notably, the method of surrogate data [1] was introduced in an attempt to reduce the rate of false positives during the hunt for physical examples of chaotic dynamics. Although it is not possible to find conclusive evidence of chaos through estimation of dynamic invariants, surrogate methods are often used to generate a distribution of statistic values (i.e., the estimates of the dynamic invariant) under the hypothesis of linear noise.

[†]Reused from: M. Small, Estimating the distribution of dynamic invariants: illustrated with an application to human photo-plethysmographic time series, *Nonlinear Biomedical Physics* 2007, 1:8.

In the most general form, the standard surrogate methods can generate the distribution of statistic values under the null hypothesis of a static monotonic nonlinear transformation of linearly filtered noise. In this communication, we introduce a significant generalisation of a recent surrogate generation algorithm [2,3]. The *pseudo-periodic surrogate* (PPS) algorithm allows one to generate data consistent with the null hypothesis of a noise driven periodic orbit—provided the data exhibits pseudo-periodic dynamics.

Previously, this algorithm has been applied to differentiate between a noisy limit cycle, and deterministic chaos. By modifying this algorithm and applying it to noisy time series data, we are able to generate surrogate time series that are independent trajectories of the same deterministic system, measured via the same imperfect observation function. That is, we assume that there is a deterministic dynamical system subject to additive independent and identically distributed (i.i.d.) observational noise. This ensemble of *attractor trajectory surrogates* (ATS) can then be used to estimate the distribution of statistic values for estimates of any statistic derived from these time series.

The statistics of greatest interest to us are dynamic invariants of the underlying attractor, and in particular correlation dimension and entropy estimates provided by the *Gaussian kernel algorithm* (GKA) [4,5]. Our choice of the GKA is entirely arbitrary, but based on our familiarity with this particular algorithm. True estimation of dynamic invariants from noisy data is a process fraught with difficulty, in this paper we are only concerned with estimating the distribution of estimates. To emphasise this point further we repeat out analysis with another quantity, Lempel- Ziv complexity [6], which *does not* constitute a dynamics invariant. Nonetheless, our algorithm provides a reliable estimate of the distribution of statistic values for this statistic as well.

An important application for the ATS technique is to determine whether dynamic invariants estimated from distinct time series are significantly different. The question this technique can address is whether (for example) a correlation dimension of 2.3 measured during normal electrocardiogram activity is really distinct from the correlation dimension of 2.4 measured during an episode of ventricular tachycardia [7,8]. Estimates of dynamic invariants (including the GKA [4,5]) often do come with confidence intervals. But these confidence intervals are only based on uncertainty in the least-mean-square fit, not the underlying dynamics. Conversely, it is standard practice to obtain a large number of representative time series for each (supposedly distinct) physical state, and compare the distribution of statistic values derived from these. But, this approach is not always feasible: in [7,8] for example, the problem is not merely that these physiological states are difficult and dangerous to replicate, but that inter-patient variability makes doing so infeasible. In the remainder of this paper we describe the new ATS algorithm and demonstrate that it can be used to estimate the distribution of dynamic invariant estimates from a single time series of a known dynamical system (we demonstrate this with the Hénon map and the chaotic Rössler system). We then apply this same method to four recordings of human pulse waveforms, measured via photoplethysmography [9,10]. Each of the four recordings corresponds to a distinct physiological state. We compute correlation dimension and entropy using the GKA method and show that the expected distribution of correlation dimension and entropy estimates are insufficient to differentiate between these four physiological states. In contrast, we show that algorithmic complexity can clearly differentiate between all four rhythms.

In Section 2 we describe the algorithm we employ in this paper, and in Section 2.2 we demonstrate that, for suitable parameter values, this technique will preserve the deterministic dynamics of the underlying system. In Section 3 we present some numerical case studies and in Section 4 we finally present our conclusions.

Attractor Trajectory Surrogates

In the first part of this section we will review the PPS algorithm presented in [2] and describe the novel features of the ATS approach. In section 2.2 we examine the foundation of this technique's ability to preserve the underlying deterministic dynamics.

The Algorithm

In what follows we assume that the measured scalar time series x_t represents discretely sampled measurements of a deterministic dynamical system (possibly continuous) under the influence of observational noise. In other words, the dynamics are determined by a smooth manifold M and deterministic evolution operator $\varphi: M \to M$. The output of the evolution of an initial condition $m_0 \in M$ under φ (i.e., $m_t = \varphi^t(m_0)$) are observed via the differentiable function $h: M \to \mathbf{R}$. Unfortunately, experimental measurement is not perfect and the observed time series (x_t) is subject to observational noise, hence, $x_t = h(\varphi^t(m_0)) + \varepsilon_t$ where $\varepsilon_t \approx N$ is drawn from some stationary noise distribution. For the case of dynamic noise, the situation is complicated further as the evolution of m_t is governed by $m_{t+1} = \varphi(m_t) + \xi_t$ where ξ_t is stochastic.

The ATS algorithm may now be described as follows. Embed the observed scalar time series (x_t) to obtain a vector time series (z_t) $z_t \in \mathbf{R}^d$, of N observations. The choice of embedding is arbitrary, but has been adequately discussed in the literature (there are numerous works in this field, [11] for example, provides references to several of them). We assume that the embedding is such that there exists a continuously differentiable map $\Xi: M \to \mathbf{R}^d$ between the underlying manifold M and the embedding space $\mathbf{R}^d$ such that both Ξ and $D\Xi$ are one-to-one. Under these conditions, the dynamics of (φ, M) and the evolution of $z_t = \Xi(m_t) \in \mathbf{R}^d$ are considered to be equivalent.

From the embedded time series, the surrogate is obtained as follows. Choose an initial condition, $w_1 \in (z_j: j = 1, 2, \ldots, N)$. Then, at each step i, choose the successor to w_i with probability

$$P\left(w_{i+1} = z_{j+1}\right) \propto \exp \frac{-\left\| w_i - w_j \right\|}{\rho} \tag{1}$$

where the *noise radius* ρ is an as-yet unspecified constant. That is, the successor of w_i, w_{i+1} is chosen to be the point z_{j+1} with probability proportional to $\exp \dfrac{-\left\| w_i - w_j \right\|}{\rho}$, where z_j is the antecedent of z_{j+1}. In other words, the successor to w_i is the successor of a randomly chosen neighbour of w_j. Equation (1) may then be written as $P\left(w_{i+1} = z_{j+1}\right) = \dfrac{P_{i,j}}{\sum\limits_{k=1}^{N} P_{i,k}}$, where

$P_{i,k} = \exp\dfrac{-\left\|w_i - w_k\right\|}{\rho}$ (and similarly $p_{i,k} = \exp\dfrac{-\left\|w_i - w_j\right\|}{\rho}$). Finally, from the vector time series $\{w_i\}$ the ATS $\{s_i\}$ is obtained by projecting w_i onto $[1\ 0\ 0\ 0\cup 0] \in \mathbf{R}^d$ (the first coordinate). Hence

$$s_t = w_i \cdot [1\ 0\ 0\ 0\cup 0] \qquad (2)$$

In [2,3] this algorithm was shown to be capable of differentiating between deterministic chaos and a noisy periodic orbit. In the context of the current communication we assume that $\{x_t\}$ is contaminated by additive (but possibly dynamic) noise and we choose the noise radius ρ such that the observed noise is replaced by an independent realisation of the same noise process. Furthermore, we assume that the deterministic dynamics are preserved by suitable choice of embedding parameters. Under these two assumptions, (z_t) and (w_t) have the same invariant density and (x_t) and (s_t) are therefore (noisy) realisation of the same dynamical system with (for suitable choice of ρ) the same noise distribution. We illustrate this more precisely in the following section.

Invariance

As in [2,3] the problem remains the correct choice of ρ. This is the major difference between the ATS described here and the PPS of [2,3]. However, since the null hypothesis we wish to address is different from (and more general than) that of the PPS, choice of ρ for the ATS is less restrictive. For $t=T$ given, one can compute $P\left(w_{i+1} \neq z_{j+1} \wedge \left\|w_i - z_j\right\| = 0 : t = T\right)$ directly from the data by applying (1) to the embedded time series (we use the symbol $\wedge$ here in the usual manner to denote logical conjunction). Assuming the process is ergodic (that is, ergodic with respect to the standard measure-this assumption is sufficient rather than necessary) one can then sum

$$P\left(w_{i+1} \neq z_{j+1} \wedge \left\|w_i - z_j\right\| = 0\right) = \frac{1}{N}\sum_{T=1}^{N} P\left(w_{i+1} \neq z_{j+1} \wedge \left\|w_i - z_j\right\| = 0 : j = T\right) \qquad (3)$$

to get the probability of a temporal discontinuity in the surrogate at any time instant. By temporal discontinuity we mean that $w_i = z_j$ but $w_{i+1} \neq z_{j+1}$. That is, a point where the surrogate trajectory does not exactly follow the data. There is a one-to-one correspondence between a value $p = P\left(w_{i+1} \neq z_{j+1} \wedge \left\|w_i - z_j\right\| = 0\right)$ and ρ, and we choose to implement (1) for a particular value of p (i.e., a particular transition probability) rather than a specific noise level ρ. In what follows we find that studying intermediate values of p ($p \approx 0.1$) is sufficient. For $p \in [0.1, 0.8]$ the qualitative behaviour over the corresponding narrow range of ρ is uniform. We choose to illustrate with $p = 0.1$, but the results for other choices are similar. Of course, for $p \to 1$ or $p \to 0$ the algorithm will not work well.

Now, suppose that the embedding parameters τ and de have been selected correctly and the noise in the data is not too large, then the transformation $x_t \mapsto z_t$ dictated by these parameters is an embedding. That is, the operator $\Xi: M \to \mathbf{R}^d$ with $\Xi(m_t) = z_t$ (in the absence of noise) and its derivative $D\Xi$ are both one-to-one. Hence, the dynamic evolution of can be represented by

$$z_{j+1} = \Phi(z_j) + e_j \tag{4}$$

where $\Phi(\cdot)$ is diffeomorphic to the true evolution operator (i.e., $\Phi = \Xi \circ \varphi \circ \Xi^{-1}$ where φ: $M \to M$ is the underlying evolution operator, defined earlier) and e_j are uncorrelated noise vectors (corresponding to the terms ε_t and possibly ξ_t described earlier). Now we consider the process of constructing a surrogate. Let denote the surrogate vector time series of length N. Clearly, setting $w_1 = z_k$ for some randomly chosen k is simply some new initial condition. Now, $w_{i+1} = z_{j+1}$ where j is chosen randomly from a distribution such that $\|w_i - z_j\|$ is small. Let $\varepsilon_i = z_j - w_i$ corresponding to the small (random) perturbation introduced by selection according to (1), then

$$w_{i+1} = \Phi(w_i + \varepsilon_i) + e_j. \tag{5}$$

Note that, ε_j is the perturbation introduced in taking z_j's successor to be the successor of w_i (it is a *dynamic noise* term, and it is a perturbation introduced by the ATS method). Conversely, e_j is the dynamic error in applying Φ (this term is inherent to the data, and to our model of the data). By taking n-th iterates of (4) and (5) we see that the two noise terms e_j and ε_{i+1} will combine. In other words, from (5) we get

$$w_{i+2} = \Phi(\Phi(w_i + \varepsilon_i) + e_j + \varepsilon_{i+1}) + e_{j+1}, \tag{6}$$

and so on. Suppose that $e_j \approx D$ where D is some noise distribution. Then, for the surrogates (s_t) to be a new realisation of the system that generated (x_t) we require that $e_j + \varepsilon_{i+1} \approx D$. But this is equivalent to the condition that $z_j - w_i \approx kD$ for sufficiently small k. Hence, the critical issue is the choice of ρ such that these two noise terms are drawn from the same distribution and that therefore the surrogate dynamic (5) is equivalent to (4). This requires sufficient data, ergodicity, and ρ small enough. Note that, as ρ becomes smaller and the surrogate data become more like realisations of the same system, we also see less randomisation. This is a natural and unavoidable tradeoff.

Results

The following subsections present the application of this method for data generated from the Hénon map (section 3.1), the Rössler system (section 3.2) and experimental measurements of human pulse pressure waves (section 3.3).

The Hénon Map

One potential difficulty of this method is that the stretching- and-folding characteristic of Smale horseshoe type chaos could easily destroy the dynamics of (5) and therefore produce surrogate trajectories that short-cut across the attractor. Although we can see from equations (5) and (6) that for sufficiently small perturbations this will never be the case, we would like to test this possibility in practise. For this purpose we apply the method described in the previous section to the extremely well studied Hénon map: one of the archetypes of Smale horseshoe chaos. Figure 1 illustrates typical ATS calculations for this data set. Using short (1000 point) sections of the Hénon system, with the addition of observational noise (the Figures show 1% and 10% noise levels), we computed typical ATS data for different values of transition probability p.

We find that in almost all cases (see Fig. 1) the results for the ATS data agree qualitatively with the data. Comparison of estimated dynamic invariant (results omitted) confirms this. In all cases, for moderate range of p (i.e., p neither approaching 0 nor 1) and moderate observational noise, we find data and surrogate agree closely. When this same computation was repeated for *dynamic* noise, we found data and surrogates to be similarly indistinguishable (see Fig. 2): except for the case of large p and small noise (in this case, 1% dynamic noise and $p = 0.8$). Note that, for the Hénon map larger values of dynamic noise will actually force the system into an unstable régime.

The Rössler System

We now demonstrate the applicability of this method for a more realistic example: noisy time series data simulated from the Rössler differential equations (during "broadband" chaos). We integrated (one thousand points with a time step of 0.2) the Rössler equations both with and without multidimensional dynamic noise at 5% of the standard deviation of the data. As far as possible, we generated realisations of the Rössler system that superficially resemble the physiological data of 3.3.

The purpose of this is to provide a more realistic test of our method. We then studied the x-component after the addition of 5% observational noise. We selected embedding parameters using the standard methods (yielding $d_e=3$ and $\tau = 8$) and then computed ATS surrogates for various exchange probabilities $p = 0.05, 0.1, 0.15,\ldots, 0.95$. For the data set and each ensemble of surrogates we then estimated correlation dimension D, entropy K and noise level S using the GKA algorithm [4,5]. The GKA embedding used embedding dimension $m=2,3,\ldots,10$ and embedding lag of 1. It is important to note that, a correlation dimension estimate is not the same thing as the actual correlation dimension. In particular, this algorithm estimates correlation dimension and noise level simultaneously (as well as entropy). A lower correlation dimension (associated with the presumed determinism in the system) is accompanied by an increase in the estimated noise level. That is, the estimated dimension can be lower because the algorithm is attributing more of the variation in the data to noise, and therefore estimating a higher noise level (and hence, in some case, the correlation dimension falls below 1). Similarly, the fact that the entropy is negative in the first case is associated with the system noise. Nonetheless, we are using these numbers only as measures, that is, as test statistics. Figure 3 depicts the results when the GKA is applied with embedding dimension $m=4$ and the exchange probability is $p=0.1$. Other values of m gave equivalent results, as did various values of p in the range [0.1, 0.8].

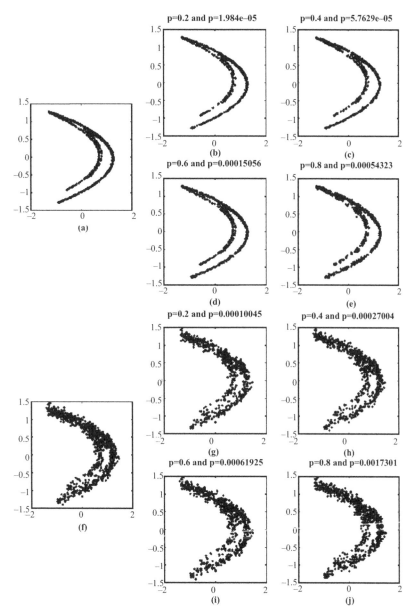

Figure 1 Sample reconstructed attractors for data and surrogates of the Hénon map. Panels (a) and (f) are embedded time series data from the *x*-component of the Hénon system with the addition of 1% and 10% observational noise (respectively). The remaining panels are representative ATS time series. Panels (b), (c), (d) and (e) are surrogates for panel (a), and Panels (g), (h), (i) and (j) are for panel (f). Each surrogate is computed with a different level of transition probability *P*. In panels (b) and (g), *p*=0.2; in panels (c) and (h), *p*=0.4; in panels (d) and (i), *p*=0.6; and, in panels (e) and (j), *p*=0.8. In each case the attractors reconstructed from the surrogates have the same qualitative features as that of the data—with the possible exception of panel (e). The likely reason for this noted exception is the relatively high transition probability (*p*=0.8) and the relatively low noise level (1%). Of course, for smaller values of *p* (i.e., *p*=0.1) the similarity is even more striking.

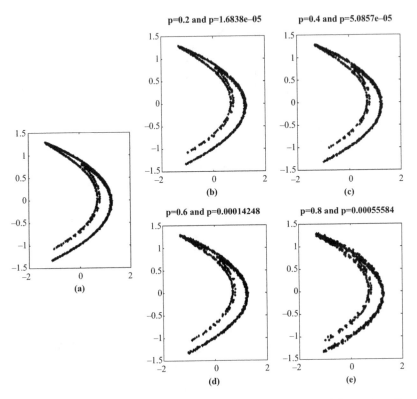

Figure 2 Sample reconstructed attractors for data andsurrogates of the Hénon map. Panel (a) is an embedded time series data from the *x*-component of the Hénon system with the addition of 1% *dynamic* noise. The remaining panels are representative ATS time series. Each surrogate is computed with a different level of transition probability *P*. In panel (b) *p*=0.2; in panel (c) *p*=0.4; in panel (d) *p*=0.6; and, in panels (e) *p*=0.8. In each case the attractors reconstructed from the surrogates have the same qualitative features as that of the data.

For such moderate *p* we found that the estimate of noise *S* from the GKA algorithm coincided for data and surrogates, but this was often not the case for more extreme values of *p*.

This estimate of signal noise content is therefore a strong test of the accuracy of the dynamics reproduced by the ATS time series. One expects this to be the case as noise level is precisely the parameter upon which the ATS method depends. Furthermore to confirm the spread of the data we also estimated *D*, *K*, and *S* for 20 further realisations of the same Rössler system (with different initial conditions). In each case, as expected, the range of these values lies within the range predicted by the ATS scheme. We do see, for example, in Fig. 3(c) that the range of noise level exhibited by the true Rössler system is not as expansive as that for the surrogates (to some extent, we can also observe the same problem with entropy in Fig. 3(e)). This is due to the fact that the ATS method can be made to introduce more randomisation than absolutely necessary.

Rossler with observational noise

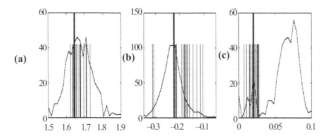

Rossler with observational and dynamic noise

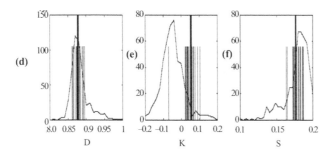

Figure 3 Distribution of statistics D, K and S for short and noisy realisations of the Rössler system. The histogram shows the distribution of statistic estimates (D, K and S) for 500 ATS time series generated from a 1000 point realisation of the Rössler system. The tall vertical line on each plot is the comparable value for the data and the shorter vertical lines indicate 20 independent realisations of the same process. The top row of figures depicts results for the Rössler system with observational noise only, the bottom row of figures has both observational and dynamic noise. Panels (a) and (d) show correlation dimension estimates, (b) and (e) are entropy, and (c) and (f) are noise level.

By tuning down the randomisation we (obviously) will converge to the true data. By increasing the randomisation we cover an ever widening range, which will always include the true value. For large randomisation, and for statistics that are most sensitive to noise (in this case K and S) there may also be some bias—the observed difference in the means. Although it is desirable that both distributions coincide exactly, it is re-assuring (and sufficient) that the ATS distribution contains the true distribution.

Photo-plethysmographic Recordings

We now consider the application of this method to photoplethysmographic recordings of human pulse dynamics over a short time period (about 16.3 seconds). We have access to only a limited amount of data representative of each of four different dynamic regimes. In any case, we would expect the system dynamics to change if measured over a significantly longer time frame. The data collection and processing with the methods of nonlinear time series analysis are described in [9,10]. Previously, we have studied nonlinear determinism in cardiac dynamics measured with electrocardiogram (ECG) [7,8]. Although we do not consider ECG data here, this data would be another useful system to examine with these methods.

Actually, the problem with ECG data is that we have too much data and it is therefore difficult to fairly select a "representative" small number of short time series. However, we intend to examine this data more carefully in the future. However, we do note in passing that both PPG and ECG are measures of cardiac activity and are therefore potentially equivalent [12,13]. The four data sets we examine in this communication are depicted in Fig. 4. For each data set we repeated the analysis described for the Rössler time series. Results for GKA embedding dimension $m=6$ and $p=0.1$ are depicted in Fig. 5.

As with the Rössler system, variation of the parameters m and p did not significantly change the results. We find that in every case (except for extreme values of p) the distribution of D, K and S estimated from the ATS data using the GKA included the true value. Most significantly, this indicates that the range of values of p is appropriate. Moreover, these

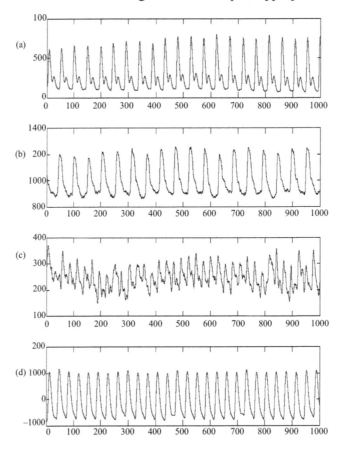

Figure 4 Human pulse waveform recorded withphoto-plethysmography. Four recordings of human pulse waveform (61Hz) in four different physiological conditions. The four time series correspond to: (a) normal, (b) quasi-stable, (c) unstable, and (d) post-operative (stable).

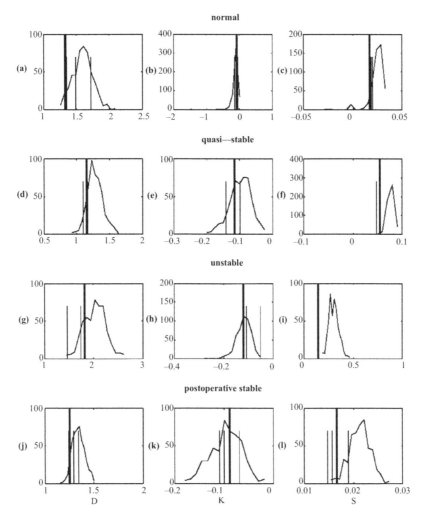

Figure 5 Distribution of statistics D, K and S for human pulse waveforms. The histogram shows the distribution of statistic estimates (D, K and S) for 500 ATS time series generated from each of the four time series depicted in figure 4. The taller vertical line on each plot is the comparable value for the shorter vertical lines are for the (limited) subsequent data recorded from each patient. In each case only two or three subsequent contiguous but non-overlapping time series were available. The figures are: (a) correlation dimension (D), (b) entropy (K), and (c) noise (S) for the normal rhythm; (d) D, (e) K, and (f) S for the quasi-stable rhythm; (g) D, (h) K, and (i) S for the unstable rhythm; and (j) D, (k) K, and (l) S for the post-operative stable rhythm.

results are consistent with the hypotheses that the noise is effectively additive and can be modelled with this simple scheme, and that the underlying deterministic dynamics can be approximated with a local constant modelling scheme.

We also estimated the statistics D, K and S for additional available data (subsequent, contiguous, but non-overlapping) from each of the four rhythms. This small amount of data afforded us two or three additional estimates of each statistic for each rhythm. For the unstable and quasistable rhythm we observed good agreement. For the stable (normal and post-operative) rhythms, this is not the case. On examination of the data we find that this result is to be expected. Both the stable rhythms undergo a change in amplitude and baseline subsequent to the end of the original 16 second recording, this non-stationarity is reflected in the results. This same non-stationarity has also been observed independently in Bhattacharya and co-workers [9,10].

We now return to the question that the ATS test was designed to address: can we differentiate between these four rhythms based on the GKA? Figure 6 provides the answer. In Fig. 6 we see the estimated distribution of statistic values (*D*, *K* and *S*) for each of the four rhythms shown in Fig. 4. Clearly (and not surprisingly), the correlation dimension estimate and noise level of the unstable rhythm is significantly different from the other three rhythms. Our analysis indicates that, contrary to what one may expect from individual measurements, the stable or "quasi-stable" rhythms cannot be properly distinguished based on these nonlinear statistics derived from the GKA.

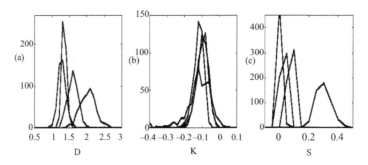

Figure 6 Discriminating power of the statistics *D*, *K* and *S* for human pulse waveforms. The distribution (a binned histogram) of statistic values estimated via the ATS method (as described in Fig. 5) for each of the four distinct physiological waveforms is shown. The four rhythms correspond to those in Figure 4. These figures show that correlation dimension alone is sufficient to differentiate between three of these four physiological states: on the left, "post-operative" and "quasi-stable" are indistinguishable, the correlation dimension for "normal" is bigger, and "unstable" is larger again We see that these three statistics are insufficient to differentiate between the "quasi-stable" and "post-operative" states, moreover, there is considerable overlap with the "normal" group.

Moreover, we find that entropy estimated with the GKA algorithm *K* is of no use in differentiating between any of these four rhythms. Although it is not the purpose of this paper to provide a discriminating statistic for this data, it would be nice to do so. Therefore, in Fig. 7 we repeat the calculation of surrogates and statistic distribution for the same data, but using algorithmic complexity (see [11] and the references therein) with binary, ternary, and quaternary encodings with equal likelihood for each symbol. Using this scheme it can be seen from Fig. 7 that it is possible to distinguish, with a high level of certainty between three of these rhythms. Distinguishing between all four is also possible, with a small likelihood of error (see Fig. 7(a)).

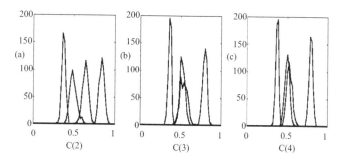

Figure 7 Discriminating power of complexity for human pulse waveforms. The distribution (a binned histogram) of statistic values estimated via the ATS method (as described in Fig. 5) for each of the four distinct physiological waveforms is shown. The four rhythms correspond to those in figure 4. These figures show that complexity with 2, 3, and 4, symbols (plots (a), (b), and (c), respectively) is sufficient to differentiate between at least three of these four physiological states. The lowest complexity corresponds to "post-operative" state, the next highest to "quasi-stable" followed by "healthy" and finally "unstable". As in figure 6 there is considerable overlap between the "normal" and "quasi-stable" samples. However, for complexity with a binary partition (panel (a)) the four rhythms do appear to be distinct.

Conclusion

The results of this analysis are in general agreement with those presented in [9,10]. Independent linear surrogate analysis [1] has confirmed that each of these four rhythms is inconsistent with a monotonic nonlinear transformation of linearly filtered noise (these calculations are routine, and not presented in this paper). The only significant difference is that the correlation dimension estimates we present here are significantly lower than those in [9,10]. This is due to the different correlation dimension algorithm. Unlike the algorithm employed in [9,10], the GKA separates the data into purely deterministic and stochastic components, and hence estimates both D and S. The correlation dimension estimated in [9,10] is the combined effect of both components of the GKA. Although we have considered the specific application to human pulse dynamics, the algorithm we have proposed may be applied to a wide variety of problems. We have shown that provided time delay embedding parameters can be estimated adequately, and an appropriate value of the exchange probability is chosen, the ATS algorithm generates independent trajectories from the same dynamical system. When applied to data from the Rössler system we confirm this result, and we demonstrate its application to experimental data.

When the ATS algorithm is applied to generate independent realisation of a hypothesis test, one is able to construct a test for non-stationarity. If two data sets do not fit the same distribution of ATS data then they can not be said to be from the same deterministic dynamical system. Unfortunately, the converse is not always true and the power of the test depends on the choice of statistic. The utility of this technique as a test for stationarity remains uncertain.

Acknowledgements

This research was fully supported by a grant from the Research Grants Council of Hong Kong (Project No. PolyU 5269/06E). The author wish to thank J. Bhattacharya for supplying the photo-plethysmographic time series.

References

1. Theiler J, Eubank S, Longtin A, Galdrikian B, Farmer JD: **Testing for nonlinearity in time series: The method of surrogate data.** *Physica D* 1992, **58**: 77–94.
2. Small M, Yu D, Harrison RG: **Surrogate test for pseudo-periodic time series data.** *Physical Review Letters* 2001, **87**: 188101.
3. Small M, Tse C: **Applying the method of surrogate data to cyclic time series.** *Physica D* 2002, **164**:187–201.
4. Diks C: **Estimating invariants of noisy attractors.** *Physical Review E* 1996, **53**: R4263–R4266.
5. Yu D, Small M, Harrison RG, Diks C: **Efficient implementation of the Gaussian kernel algorithm in estimating invariants and noise level fromnoisy time series data.** *Physical Review E* 2000, **61**: 3750–3756.
6. Lempel A, Ziv J: **On the complexity of finite sequencess.** *IEEE Trans Inform Theory* 1976, **22**: 75–81.
7. Small M, Yu D, Simonotto J, Harrison RG, Grubb N, Fox K: **Uncovering non-linear structure in human ECG recordings.** *Chaos, Solitons and Fractals* 2001, **13**: 1755–1762.
8. Small M, Yu D, Grubb N, Simonotto J, Fox K, Harrison RG: **Automatic identification and recording of cardiac arrhythmia.** *Computers in Cardiology* 2000, **27**: 355–358.
9. Bhattacharya J, Kanjilal P: **Assessing determinism of photoplethysmographic signal.** *IEEE Transactions on Systems, Man and Cybernetics A* 1999, **29**: 406–410.
10. Bhattacharya J, Kanjilal P, Muralidhar V: **Analysis and characterization of photo-plethysmographic signal.** *IEEE Transactions on Biomedical Engineering* 2001, **48**: 5–11.
11. Small M: *Applied Nonlinear Time Series Analysis: Applications in Physics, Physiology and Finance, Volume 52 of Nonlinear Science Series A* Singapore: World Scientific; 2005.
12. Zhao Y, Small M: **Equivalence between 'feeling the pulse' on the human wrist and the pulse pressure wave at fingertip.** *International Journal of Neural Systems* 2005, **15**: 277–286.
13. Zhao Y, Small M: **Evidence consistent with deterministic chaos in human cardiac data: surrogate and nonlinear dynamical modeling.** *International Journal of Bifurcation and Chaos* 2007. To appear.

Chapter 9

Asymmetry, Symmetry and Beauty[†]

Hector Sabelli,[1,a] Atoor Lawandow[1,b] and Abbe R. Kopra[1]

[1]Chicago Center for Creative Development, 2800 N. Lakeview, Chicago, Illinois 60614, USA.
[a]Email: Hector_Sabelli@rush.edu.
[b]Email: alawandow@uchicago.edu.

ABSTRACT

Asymmetry and symmetry coexist in natural and human processes. The vital role of symmetry in art has been well demonstrated. This article highlights the complementary role of asymmetry. Further we show that the interaction of asymmetric action (recursion) and symmetric opposition (sinusoidal waves) are instrumental in generating creative features (relatively low entropy, temporal complexity, novelty (less recurrence in the data than in randomized copies and complex frequency composition). These features define Bios, a pattern found in musical compositions and in poetry, except for recurrence instead of novelty. Bios is a common pattern in many natural and human processes (quantum processes, the expansion of the universe, gravitational waves, cosmic microwave background radiation, DNA, physiological processes, animal and human populations, and economic time series). The reduction in entropy is significant, as it reveals creativity and contradicts the standard claim of unavoidable decay towards disorder. Artistic creations capture fundamental features of the world.

Keywords: poetry; music; bios; chaos; creativity.

Introduction

Here we study music and poetry with mathematical methods that analyze symmetry and creativity. The relation between music and mathematics has been recognized since Pythagoras. The concept of symmetry has found many applications in science and art, but the term is given several different interpretations. Thermodynamics defines the spontaneous increase in entropy as an increase in symmetry, and, according to nineteenth century thermodynamics, natural processes tend to decay. In contrast, physics interprets

[†]Reused with permission from: Hector Sabelli , Atoor Lawandow and Abbe R. Kopra, Asymmetry, Symmetry and Beauty, *Symmetry* 2010, *2*, 1591–1624; doi:10.3390/sym2031591.

symmetry as regularity, and esthetics highlights symmetry as a defining feature of beauty and art. Evolutionary theory states that natural processes are creative; human and artistic creativity thus continue natural creativity.

The term symmetry originally meant a sense of harmonious, aesthetically pleasing proportionality that embodies beauty. Soon the term symmetry acquired a more precise meaning of mathematical balance such as when one shape becomes exactly like another if you flip ("mirror" symmetry), slide, or turn it. In time, symmetry came to be defined when this internal identity could be demonstrated according to the rules of a formal system. A mathematical object is symmetric with respect to a given mathematical operation, if, when applied to the object, this operation preserves some property of the object. The set of operations that preserve a given property of the object form a group. Symmetry has been transformed in physics to mean invariance under any kind of transformation. According to Noether's theorem, a continuous mathematical symmetry corresponds to a conserved quantity.

Symmetry has come to mean any form of conservation; conversely, creation is associated with change and asymmetry. There is an ideological implication in interpreting symmetry as conservation. In this sense, it is said that most if not all laws of nature originate in symmetries [1]. Asymmetries such as the golden proportion are described as "symmetries". This overlooks the fundamental role of asymmetry in physical and esthetic processes.

Symmetry does not exclude its opposite, asymmetry, but it is paired with it, as all opposites are. Indeed, for reasons of symmetry, symmetry itself must be paired with asymmetry. The concept of asymmetry has evolved from the mere lack of symmetry to the asymmetry of time in physics and biology, the discovery of asymmetry in biological molecules by Pasteur, and the demonstration of symmetry breaking as a core process in physical and biological evolution, from Pierre Curie to the 2008 Nobel prize winners Nambu, Kobayashi, and Maskawa. Parity symmetry is violated by phenomena that involve the weak force, and there also are violations of charge symmetry and of CP symmetry.

Asymmetry is central in natural and human processes. Asymmetry is also central to beauty and art, as illustrated by the golden proportion 1:1.618, utilized in classic architecture, renaissance and contemporary painting, books, and postcards. Asymmetry and symmetry coexist as fundamental forms among mathematical and psychological structures. Bourbaki, a seminal group of mathematicians, described the architecture of mathematics as based on three foundations, lattices, groups and topology [2]. Lattices, groups and topology are mathematical abstractions of order, symmetry, and spatial form. The leading Swiss psychologist Jean Piaget [3] was astonished to realize that the fundamental cognitive structures he had discovered in the study of mental development corresponded to Bourbaki's three 'mother structures' of mathematics.

Bourbaki's mother forms-linear asymmetry, two-dimensional and bipolar symmetry, and vertical hierarchy as a sequence of steps-were prefigured in one of the earliest artistic archetypes: the first pyramid constructed by the prime minister and physician Imhotep (author of the first medical (and scientific) textbook, identified with Asclepius by the Greeks and the Romans), the Ziggurats built by the Sumerians, Elamites and Assyrians, and the step pyramids built by the Mayans, Mexicans and Incas (Fig. 1).

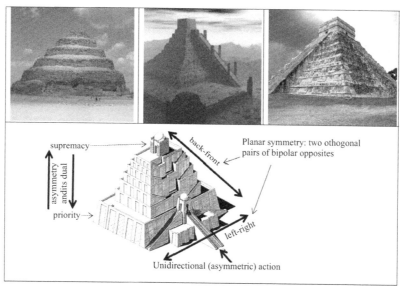

Figure 1 Step pyramid archetype. Top: early pyramids in Egypt, Middle East, and Mexico. Bottom: Diagram illustrating how it captures asymmetry, symmetry of opposites, and transformation from simple and larger to a smaller top.

The Bourbaki forms also correspond to macroscopic features of the three dimensions of the Central Nervous System [4] that is the most complex system known: the asymmetry of sensory to motor function in its dorsal ventral axis, symmetry in its right-left axis, and the creation of complex forms from the folding of simpler ones in the vertical axis. Thus Sabelli [4–6] proposed that asymmetry, symmetry and continuous topological transformation from simple to complex (evolution) are the three fundamental patterns of processes, and that asymmetry and symmetry represent a primordial opposition that accounts for creativity in nature as well as in art.

Asymmetry is evident in non-stationary and non-periodic processes, both stochastic (random walks) and causal (Bios). Together, asymmetry, creativity and nonrandom causation define Bios [5], a fractal pattern generated mathematically by recursions that involve both asymmetry and symmetry (Fig. 2). We first identified Bios as a distinct pattern when studying heartbeat series; this led us to develop analytic methods that demonstrate causality and creativity in empirical processes. Bios is commonly found in natural and human processes: quantum processes [7], the expansion of galaxies and quasars in three surveys [8,9,10] gravitational waves [11], cosmic microwave background radiation [10], DNA [5], physiological processes [5,12–17], animal populations [18] and human populations [5], economic time series [19–21] and the prime numbers [22]. Levy et al. [23] have already demonstrated Bios in several musical scores; we have expanded this work in this and other publications [24]. In other words, asymmetry is natural, a core property of complex and creative processes. Causality and creativity define Bios and differentiate it from chaos and from random walks. Bios can be generated mathematically by several recursions that combine the temporal asymmetry of action and the symmetry of bipolar feedback [5,9]. The

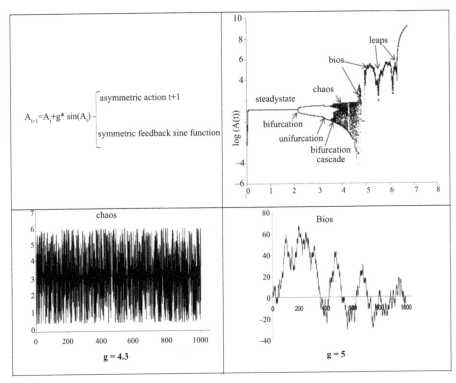

Figure 2 Process equation. The sequence of patterns in steps of increasing complexity in the time series generated by the equation as g increases (logarithmic scale).

sinusoidal function provides a mathematical model for the philosophical tenet advanced by Heraclitus, Empedocles, Lao-tzu, Hegel, Darwin, and Salk that the interaction of opposites creates novel and complex processes [4]. This connects Bios with both creativity and a well-supported hypothesis regarding its generation.

To study artistic creations and archetypes in relation to natural processes is meaningful. Artistic archetypes portray fundamental patterns of nature, as highlighted by Pythagoras who found in music the mathematics of the universe. In this paper we demonstrate asymmetry, symmetry and creativity in music and poetry using mathematical techniques that measure these features in time series.

Methods

The data presented in the figures above are time series and therefore have no physical dimensions. We analyzed the sequence of notes in musical scores and the sequence of letters in literary texts in English and Spanish. The data were compared with mathematical models of non-periodic variations: random series, random walks, and stationary Chaos and non-stationary Bios generated by the process equation [25] (Fig. 2) for *g = 4.3* and *5*.

We measure the degree of asymmetry these musical and poetic texts and its changes in time by dividing the series into epochs and computing the middle, mean and median of each epoch. To measure symmetry and opposition we use complement plots [26] (graphs of the sine versus cosine of each term of the series) and trigonometric walks [27] (cumulative sum of the sine versus cosine of each term of the series). Complement plots omit the linearity of time but conserve temporal order (in the lines connecting successive points of the series). As sine and cosine are orthogonal to each other, this creates a 2-D plot. Sine and cosine are out of phase; one of these opposites is larger at each time and the data are plotted in a 3-D helix rather than a 2-D circle. Complement plots thus assume the priority of time, the asymmetry of opposites and the alternation of their relative size. The helix is a classic portrait of dialectic oscillation of opposites.

We analyze these musical and poetic texts with recurrence methods using the Bios Data Analyzer [28]. Given a series $A_1, A_2, ..., A_N$, we construct the sequences (Euclidean vectors) y_i of N successive members of a time series $[y_i=(A_i, A_{i+1}, A_{i+2}, ..., A_{i+N})]$ starting with each data point A_i. This is referred to as embedding in dimension N. *Recurrence isometries* are calculated by comparing the Euclidean norms of these vectors, and if the difference between them is less than a chosen cutoff radius (1%), a recurrence is plotted and counted. We have made calculations with many cutoff radii, anging from 0.1 to 50 % and the results are similar. The quantification of recurrences at low and high embeddings allows one to consider both simple and complex patterns. *Recurrence plots* graph isometries as a function of time. Recurrences are colored according to their distance from red to violet. Both types of recurrence generate the same pattern in recurrence plots, but the number of isometries increases and the number of similarity recurrences decreases with the length of the vector. *Isometry* is the number of isometric recurrences as a percentage of the total number of possible recurrences in the sample ($N.N/2$). The number of isometries (as % of the total number of possible recurrences) is computed for the original data and for a copy of the data randomized by shuffling. *Novelty* is defined as the increase in recurrence isometry produced by shuffling the data [29,30], and is quantified as the ratio of isometries after shuffling over isometries in the original data. Novelty is a key measure to distinguish creative from non-creative processes. Novelty is demonstrable for recurrence isometry but not for similarity recurrence. Embedding plots present the value of novelty computed with $2, 3, ..., 100$ embeddings.

In addition to analyzing the time series, we also analyzed the time series of difference between consecutive terms. Periodic series as well as causally generated non-periodic series (chaos, bios) show morphological and numerically quantifiable pattern in the series of differences, while random series and random walks do not show pattern (randomness) in the series of differences between consecutive terms (except at very high embeddings). The standard deviation $S.D = \sqrt{\sum \dfrac{(x_i - mean)^2}{N}}$ of the series was recorded as a function of time. Series converging to an attractor display a decrease in $S.D.$ with increasingly larger samples. In contrast, diversification is the increase in variance of a time series with an increase of the size of the sample (global diversification) or with increasing the embedding (local diversification): the $S.D$ is computed for sets ("embeddings") of $2, 3, ..., 100$ consecutive terms of the time series, starting with each term in the series. The values obtained for each embedding are averaged for the entire series, and these averages are plotted as a function of the number of embeddings.

The power spectrum of these series was calculated using a program developed by Patel (Patel and Sabelli 2003).

We measured entropy *H* with the de Moivre equation adopted by Boltzmann, and by Shannon to measure entropy, using a range of bins ($N = 2, 4, 8,\ldots$), and plot the value of entropy as a function of the logarithm of the number of bins (Fig. 3). The values for entropy H a linear function of the logarithm of the number of bins:

$$H = s + d.\log 2n$$

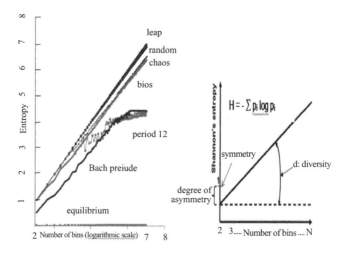

Figure 3 Entropy-bin plot. Plot of informational entropy as defined by Shannon. It shows the calculation of the degree of symmetry (entropy at 2 bins = 1 for perfect symmetry and less than 1 for asymmetry) and diversity (slope). Many natural and human processes are highly asymmetric. Biotic series and random walks have a small degree of asymmetry, while chaos, random, and periodic series are symmetric.

where *n* is the number of bins used to calculate entropy [31]. The slope d (for diversity) of the entropybin regression line measures diversity of the data (Fig. 2); the slope is *0* for numerical series with two equally probable values, regardless of their temporal arrangement (random or periodic), and increases with the number of different values in the data from period 2 to period 3,4,…, and chaos. Linearly increasing series, random series, statistical noise (random walks, pink noise), chaotic series, biotic series, and many empirical time series have a near 1 slope.

The intercept s (for symmetry *s*) of the entropy-bin regression line at 2 bins (the least number of bins required for the calculation of entropy) is 1 for all symmetric distributions (random, sinusoidal, chaotic); asymmetry *y* ($s < 1$)is present in mathematically generated biotic series, in random walks and in empirical data (physical, biological, economic).

Results

Asymmetry and Symmetry

Histograms (Fig. 4) and Pareto histograms (Fig. 5) demonstrate asymmetry in the distribution of notes and letters in musical and literary texts. Statistical measures quantify asymmetry and demonstrate non-stationarity (Fig. 6 and Table 1).

Two dimensional plots (Fig. 7) do not show simple patterns as in chaos or bios. Trigonometric plots (Fig. 8) show rounded complement plots (not partial as in chaotic attractors), significant asymmetry. Table 1 presents statistical and trigonometric measures of the asymmetry of opposites.

Entropy

Table 2 shows that maximal entropy occurs for ordered series such as a sequence of linearly increasing numbers, not for random data, as often assumed. Musical and literary texts have relatively low entropy. Figure 9 shows that this low entropy results from low symmetry rather than from decreased diversity. Both of these results are also observed in biotic series generated mathematically as well as in natural processes.

Power Spectrum Analysis

The power spectrum analysis of poems and music (Table 3) shows a broad spectrum. The slope of musical compositions is low, in the range of chaos, well below -1 (pink noise) and −1.7 (Bios generated with the process equation). There is great diversity in the slope of poems.

Biotic Patterns in Music and Poetry

Isometry quantification (Fig. 10) demonstrates less recurrence (novelty) in many musical pieces, Bios, and random walks than in randomized copies, and more recurrences in many literary texts and in stationary chaos than in randomized copies.

Recurrence portraits of musical and literary (Fig. 11) series show changes in pattern with time (*temporal complexity*), as observed in mathematical Bios and in random walks; in contrast, the recurrence plots of stationary random series or chaotic attractors are uniform.

Recurrence plots and quantification of the series of differences between consecutive terms (Fig. 12) in these data also show temporal pattern in music, chaos and Bios, indicating *non-random causation*, in contrast to random walks.

Figure 13 shows temporal variation in variance, with overall increase in S.D. with embedding (local diversification) but not with time (global diversification).

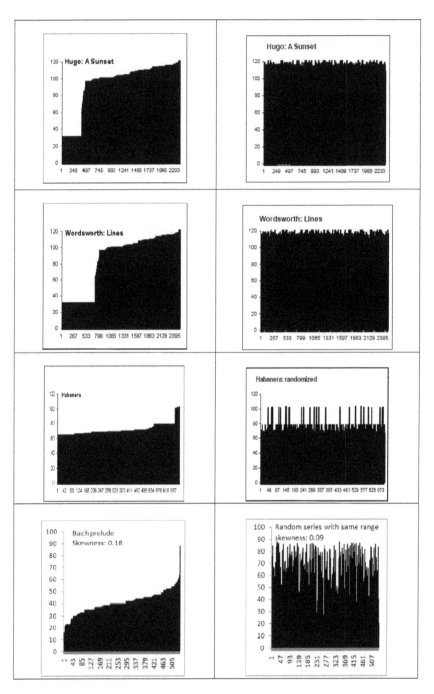

Figure 4 Histogram showing asymmetry in the statistical distribution of letters in poetry (Hugo's *A Sunset* and Wordsworth's *Lines Written as a School Exercise at Hawkshead*) and musical notes (Georges Bizet's "Habanera" from *Carmen* and Bach's *Prelude*). Units: the number of occurences are plotted in the y-axis.

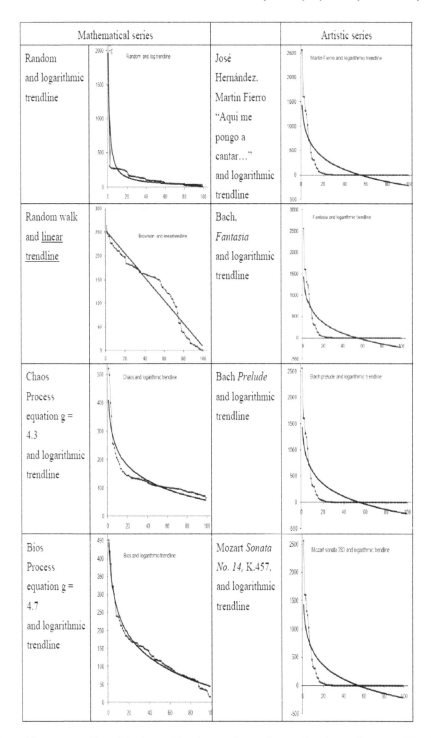

Figure 5 Pareto histogram and logarithmic trendline (except for random walk, where a linear trendline fits).

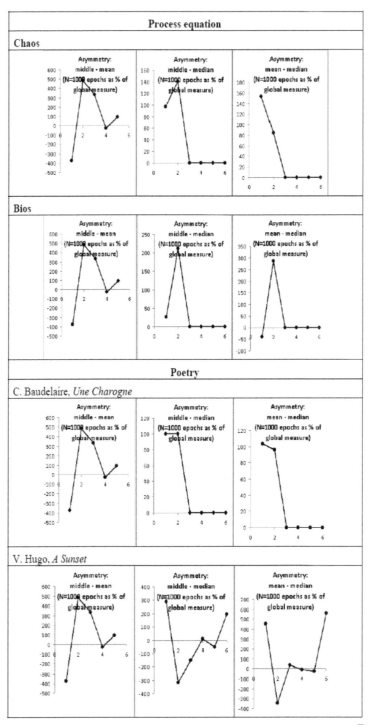

Figure 6 contd....

Figure 6 contd....

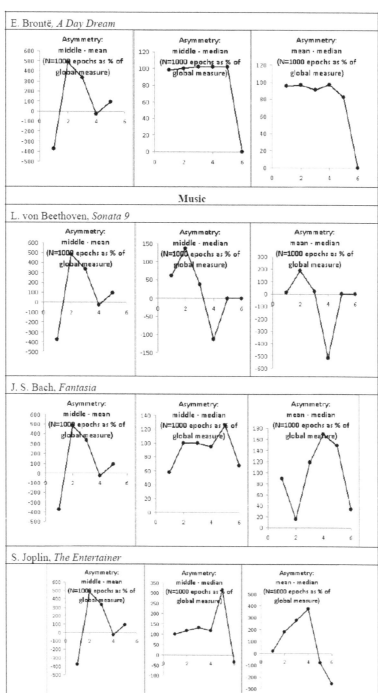

Figure 6 Quantification of asymmetry in music and poetry.

Table 1 Asymmetry of opposites.

Series	Max	Min	Middle	Mean	Median	Mean median	Middle median	Middle mean	Cosine/sine average
Random	9.97	0.02	5.00	4.91	4.94	−0.03	0.039	0.065	4.91
Random Walk	5.97	−5.00	5.49	0.33	0.16	0.16	5.32	5.159	0.33
Chaos	5.79	0.49	2.65	3.16	3.10	0.06	−0.45	−0.50	3.16
Bios	17	−43.87	30.43	−14.02	−13.02	−1.00	43.45	44.45	−14.02
Prime Numbers	7919	2	3959	3683	3576	106.91	382.5	275.59	−0.004
Music									
Bach: *Prelude*	63	16	23.5	40.52	40	0.522	−16.5	−17.02	40.52
Bach: *Famasia*	59	11	24	39.2	41	−1.80	−17	−15.20	39.20
Freeman: *Affair in San Miguel*	59	8	25.5	36.33	42	−5.67	−16.5	−10.83	36.33
Duke: *April in Paris*	59	3	28	37.47	36.5	0.97	−8.5	−9.47	37.47
Poetry									
Whitman:*A Sight in Camp*	121	32	44.5	92.3	104	−11.70	−59.5	−47.80	92.30
Rumi: *Descent*	122	32	45	90.59	104	−13.41	−59	−45.59	90.59
Neruda:*Puedo escribir*	122	32	45	92.16	105	−12.84	−60	−47.16	92.16
Fierro: *Aqui me pongo a cantar*	243	32	105.5	96.12	105	−8.88	0.5	9.38	96.12
Zorrilla: *Don Juan Tenorio*	233	32	100.5	87.1	101	−13.90	−0.5	13.4	87.10

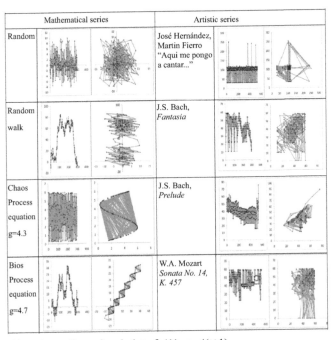

Figure 7 Time series $A(t)$, and two dimensional plot of $A(t)$ vs. $A(t+1)$.

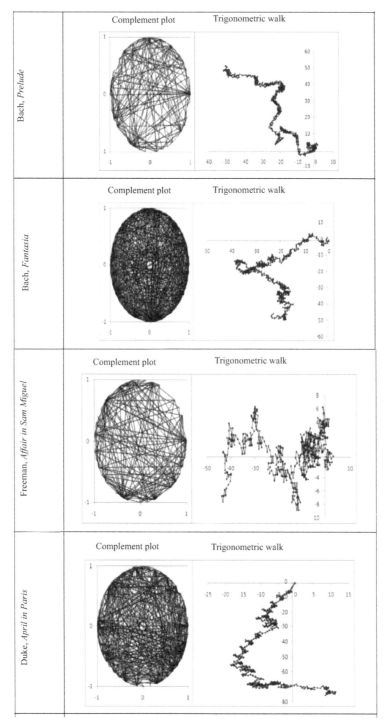

Figure 8 contd....

Figure 8 contd....

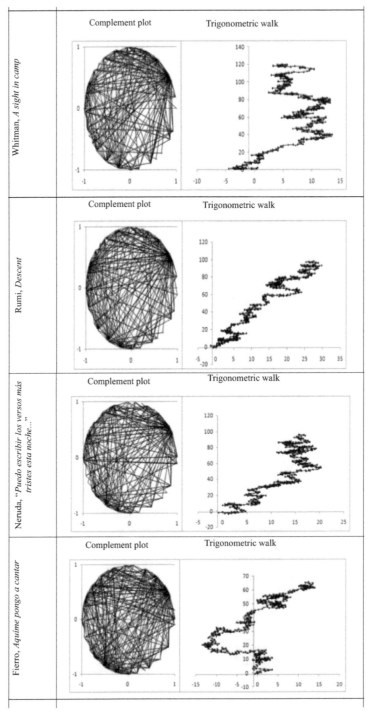

Figure 8 contd....

Figure 8 contd....

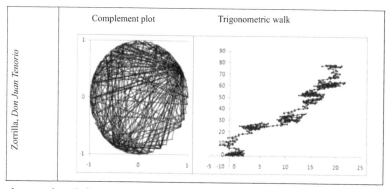

Figure 8 Complement plots (left) and trigonometric walks (right) of musical compositions and of poems.

Table 2 Low entropy of music and literary compositions.

Series	Entropy at 16 bins
Music	
Debussy.*Etudes*	3.20296
Di Capua. *O Sole Mio*	2.15
Beethoven. *Fur Elise*	2.07
Chopin. *Ballade* No. 4 in F	3.54696
Denver. *Country Roads*	3.638
Bach. *Fantasia*	3.65822
Poetry	
Hugo. *A sunset*	1.75256
Pushkin. *Eugene Onegin* book I & II	1.8449
Goethe. *To Luna*	2.99606
Mandelstam. *Tristia*	2.94824
Blake. *On Another's Sorrow*	2.62664
Whitman. *A sight in camp*	2.67581
Mathematical series	
Random	3.57727
Random Walk	3.54639
Linearly increasing numbers	4
Leap (process equation. $g = 2*n$)	4
Sine wave	3.78594
Devil staircase	3.17955
Lorenz choas	3.83145
Rossler choas	3.94242
Process equation choas $A_{t+1} = A_t + 4.3 * \sin(A_t)$	3.76694
Bios:Process equation $A_{t+1} = A_t + 4.65 * \sin(A_t)$	3.62817

Table 2 contd....

Table 2 contd....

Bois: Sum of 4 sine waves	3.75031
Bois: Sum of 3 sine waves	3.93651
Natural and social processes	
Distance between galaxies	3.96239
Schrodinger's equation	3.64343
El Nino	3.66817
Air Temperatures	3.23883
Clear Water River	2.48224
Earthquakes	2.37551
MSN Temperature	3.67427
Heartbeat intervals	3.50399
Electroencephalogram	3.19064
Human DNA	3.67782
Human RND	3.96081
France population	3.51594
Daily bond yield DAAA	3.62622

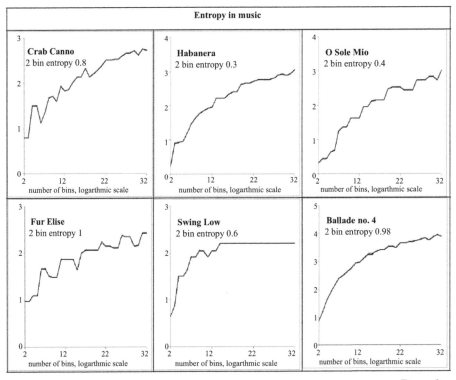

Figure 9 contd....

Figure 9 contd....

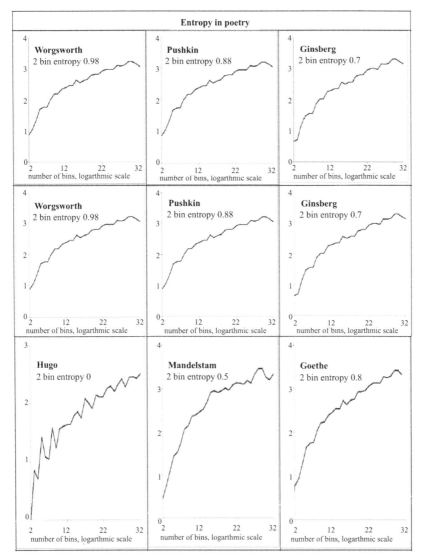

Figure 9 Entropy of musical compositions and of literary texts as a function of the number of bins. The musical compositions analyzed include: Eduardo di Capua's *O Sole Mio*, Georges Bizet's Habanera from *Carmen*, Bach's *Crab Canon*, Mozart's *Fur Elise*, the American Spiritual *Swing Low Sweet Chariot*, and Chopin's *Ballade no. 4*. The poems are: Wordsworth's *Lines Written as a School Exercise at Hawkshead*, Pushkin's *Eugene Onegin*, Ginsberg's *Howl*, Hugo's *A Sunset*, Mandelstam's *Tristia*, and Goethe's *To Luna*.

Table 3 Power spectrum analysis.

Model time series	Slope-standard error	Model time series	Slope
Random	0	Random walk	0
Process Chaos	0.06	Logistic chaos	–0.28
Bios	–1.76	Sun of sine waves	–3.64+0.02
Pink noise	–1.00	Browman noise	–2.11
Physical processes	Slope+standard error	Human process	Slope+standard error
Quantum	–4.08+0.11	Heartbeat intervals	–1.4943+0.03
Gravitational waves	–0.27+0.04	Economic	–0.1396+.03
Music	Slope+standard error	Poetry	Slope+standard error
Piano Sonata	–0.33+0.06	Ginsberg: *Howl*	0.4087+0.04
April in Paris	–0.49+0.06	Wordsworth	–0.1182+0.04
Fantasia	–0.29+0.06	Hugo	0.3746+0.04
Chopin Ballad	–0.151+0.06	Mandelstam: *Iristia*	0.065+0.12

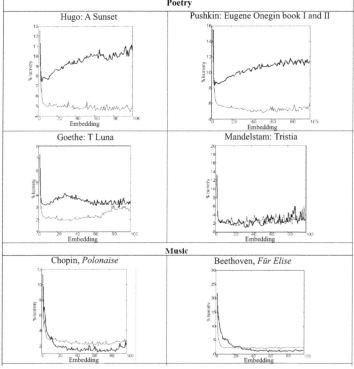

Figure 10 contd....

Figure 10 ***contd....***

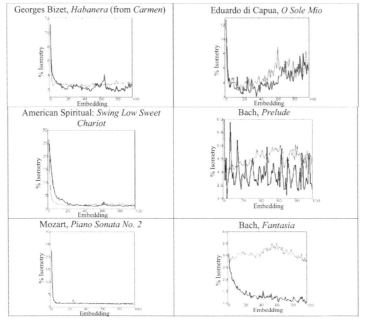

Figure 10 Quantification of isometry as a function of the number of embeddings in musical compositions and literary texts as compared with their randomized copy (blue line).

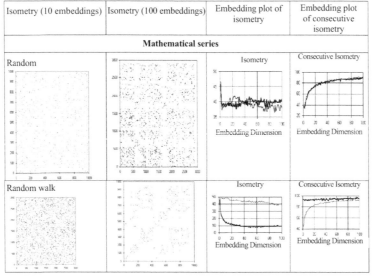

Figure 11 contd....

Figure 11 contd....

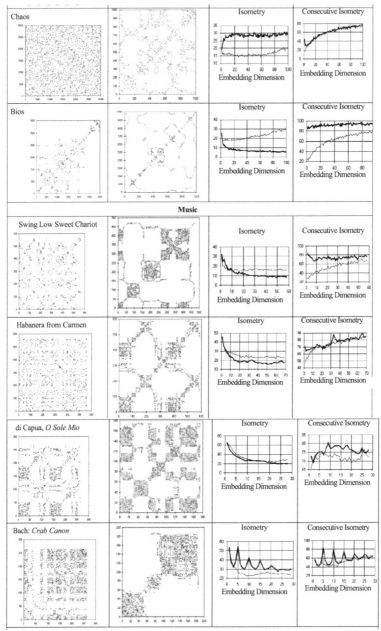

Figure 11 contd....

Figure 11 contd....

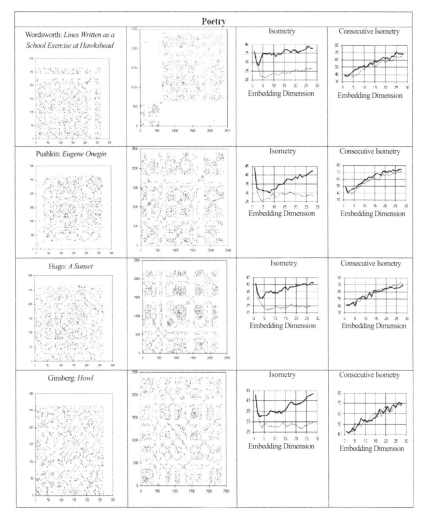

Figure 11 Recurrence plots of musical compositions show temporal complexity. The quantification of isometries shows novelty and the quantification of consecutive isometries supports non-random causation.

Discussion

These results indicate that asymmetry, symmetry and creativity are evident in literary and musical series. Asymmetry is indicated by histograms (Fig. 4), Pareto histograms (Fig. 5) and the difference between middle, median and mean (Fig. 6). Symmetry is indicated by complement plots (Fig. 8).

Poems and musical scores show low entropy; low entropy indicates order and creativity just as high entropy indicates disorder and decay. Simple mathematical analyses of literary and musical series also reveal *local diversification, temporal complexity*, and *complex frequency composition* (power spectrum). Musical compositions also show *novelty*.

It is significant that these simple analyses demonstrate these features in artistic productions, as this validates their interpretation as indicators of creativity in natural and socioeconomic processes. This also suggests that natural and mental processes share deep similarities.

These hallmarks of creativity may be detected in series generated randomly, but they may also be generated causally. Causal (non-random) creativity defines Bios. We found empirically that fundamental physical, biological and human processes are causal and creative. Recursions such as the process equation $A_{t+1}=A_t kt.sin(A_t)$ and the diversifying equation $A_{t+1}=A_t+sin(ktA_t)$ generate a sequence of patterns as time t increases: equilibrium, periodicities, chaos, bios and leaps (Fig. 2).

These equations involve *asymmetric action* (recursion in time), *bipolar opposition* (sine function), and *continuity* (conserved term A_t). These three elements correspond to the defining properties of the three mother structures of mathematics (Bourbaki): lattice asymmetry, group opposition, and topological continuous transformation of spatial form.

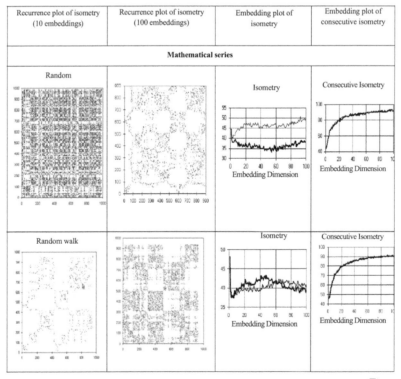

Figure 12 contd....

Figure 12 contd....

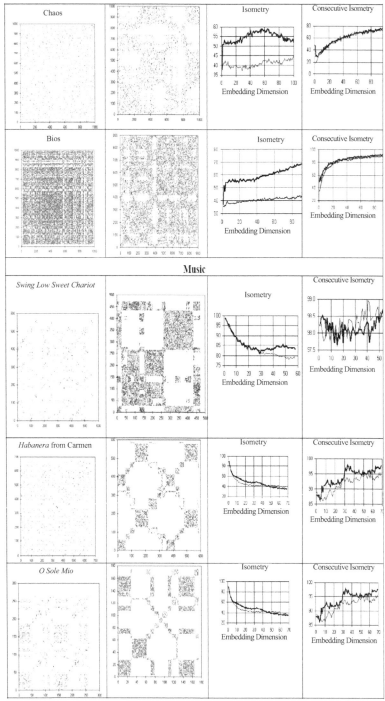

Figure 12 contd....

Figure 12 contd....

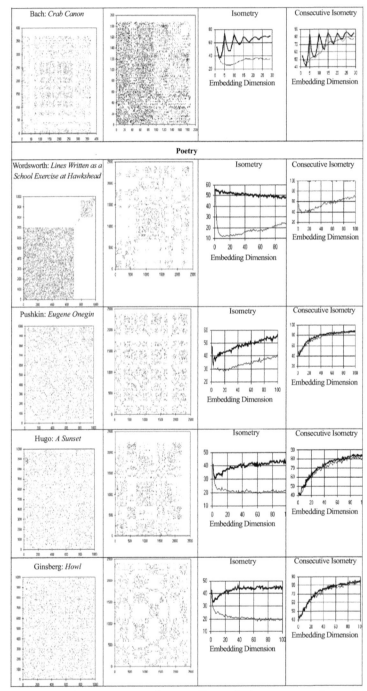

Figure 12 Series of differences between consecutive terms Pattern in the recurrence plots and the quantification of isometry and of consecutive isometries in the show non-random causation.

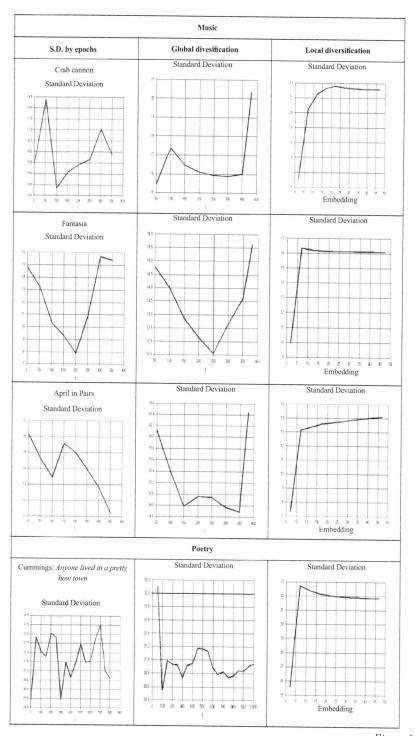

Figure 13 contd....

Figure 13 contd....

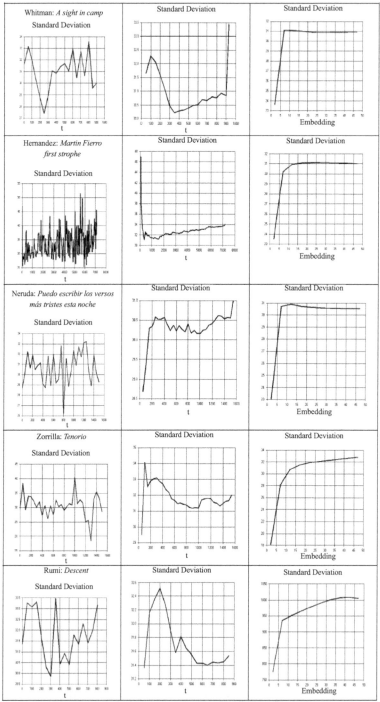

Figure 13 contd....

Figure 13 contd....

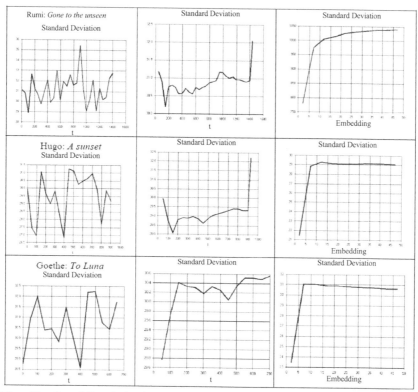

Figure 13 Diversification. *S.D* variation by epochs, global and local diversification.

These three elementary forms are also found by Piaget in fundamental cognitive operations, in the three dimensions of the Central Nervous System [4] and in many other physical, biological and human processes as illustrated by Table 4. Some of these may be mere analogies (as between the wing of a bird and the wing of an airplane) but we as hypotheses to be tested that many of them are homologies (as between the leg of a horse and the wing of a bird). We thus conceived the idea that these three forms—unidirectional flow of energy in asymmetric time (action), bipolar opposition such as the bidirectional communication of information by bidimensional and bipolar electromagnetic radiation, and tripolar, continuous transformation of tridimensional matter—constitute basic principles of nature [4]. This is Process theory (Table 4).

These equations involve *asymmetric action* (recursion in time), *bipolar opposition* (sine function), and *continuity* (conserved term A_i). These three elements correspond to the defining properties of the three mother structures of mathematics (Bourbaki): lattice asymmetry, group opposition, and topological continuous transformation of spatial form. These three elementary forms are also found by Piaget in fundamental cognitive operations, in the three dimensions of the Central Nervous System [4] and in many other physical, biological and human processes as illustrated by Table 4.

Table 4 Asymmetry and symmetry.

	Asymmetry	Dyadic asymmetry and symmetry	Triadic asymmetry and symmetry
Integers	1	2	3
Example of numerical form	Uni-directional time	Bi-polar and bi-dimensional electro-magnetic change	Tri-polar nuclear forces
Numerical archetypes	Divine proportion φ =1.618	Perfect circular symmetry:Π=3.1415	e=2.71828
Form	Asymmetry	Opposition asymmetric and symmetric	Triadicity
Bourbaki's structures	Order: lattices	Group opposition and other algebras	Structure 3D geometry
Physics	Action: energy flow in time	Communication(two-valued information opposites)	Tri-dimensional matter
Physical forces	Gravitation (attraction)	Electromagnetic (attraction and repulsion)	Strong nuclear force creates structure Weak nuclear force creates asymmetry
Light	Linear propagation	Sine wave	Color Sptial radiation
Schrodinger equation	Engery total E and potenial U	sine and cos.if E>U sine and cosh if E>U	Momentum = mass* velocity
Einstein equation	E	c^2	m
Chemistry	Asymmetric molecules and biomolecules (Pasteur)	Covalent and	Asymmetric carbons
Biology	Metabolism	Anabolism and catabolism Sexuality (Linnaeus)	Organisms. Mother. father and child
Evolutionary theory	Life,survival,evolution	Competition and (Darwin) and mutual aid (Kropotkin)	Multicellularity, symbiosis, mutual aid sociality
Central Nervous system	Asymmetric dorsal-sensory to ventral motor axis	Quasi-symmetric right-left axis	Vertical axis from simple to complex
Physiology,Medicine	Homeostasis:equilibrium	Sympathetic and parasympathetic.Periodicity	Anatomy
Social groups	Age, generations	Sexes.Parent child Master slave.	Upper, middle and lower classes
Social aims organization	Health of persons and enviroment	Sexes.Class cooperation (socialism) and competition (smith) and struggle (Marx)	Executive. Legislative and Judicial powers
Economics	Consumption	Distribution,trade	Production
Psychology	Flux (James) Action(Moreno)	Conhlct (Freud) Fight or flight (Cannon). Role reversal(Moreno)	Id.ego.superego (Freud) Creativity. (Moreno).Confict behaviors and emotions[4]

Table 4 contd....

Table 4 contd....

Cognition[3]	Mental operations regarding order	Mental operations regarding classes	Mental operations regarding space
Education[34]	Pragmatic	Emotional	Cognitive and artistic
Methodology	Priority of the simple and objective[6,33]	Analyze synergy and conflict of opposites	Supremacy of the complex or the subjective [6,33]
Research techniques	Time series	Phase plane of opposites[35], trigonometric analysis	N-dimensional recurrence
Dynamics	Asymmetric factor of catastrophes	Bifurcation. Bifurcating factor of catastrophes	Form and transformation Complex patterns: fractal chaos,bois
Process philosophy	Process(not isolated events or stability)	Dialectic:coexisting interacting, opposites	Matenal embodiment Creative synthesis of opposites and triads Supremacy of complex
Logic	Implication	Mutual exclusion (Aristotle Boole) and implication	Triadic categories (Hegel.Pierce) and operations
Biotic logic	Evolving concepts	(Sebelli) of opposites	
Process equation	RecursionA(t+1)	Sin(A(t))	Sequence of pattems:convergence periodicity, chaos, bios
Music	Rhythm, tempo	Melody	Harmony

Some of these may be mere analogies (as between the wing of a bird and the wing of an airplane) but we as hypotheses to be tested that many of them are homologies (as between the leg of a horse and the wing of a bird). We thus conceived the idea that these three forms-unidirectional flow of energy in asymmetric time (action), bipolar opposition such as the bidirectional communication of information by bidimensional and bipolar electromagnetic radiation, and tripolar, continuous transformation of tridimensional matter-- constitute basic principles of nature [4]. This is Process theory (Table 4).

The study of literary and musical creations suggests that this theory extends to art. Bourbaki's three forms may thus provide a mathematical foundation for a theory of creative evolution and of artistic creation.

The small numbers appear to capture some important aspects of Bourbaki's mother structures of mathematics: lattice asymmetry is unidirectional, group opposition is bidirectional, and space-like topological forms are tridimensional. We can identify these numerical forms in the three fundamental physical forces, **uni**directional gravitation, **bi**polar and **bi**dimensional electromagnetic force, and **tri**polar nuclear forces that unite 3 quarks to form protons and neutrons and thereby construct **tri**dimensional atomic nuclei. More generally, action (energy * time) is asymmetric, information requires the distinction between two opposite entities, and matter is tridimensional.

Asymmetry and symmetry are the defining features of lattices and groups, two of the three fundamental mathematical structures described by Bourbaki. Lattice abstracts asymmetry and thereby one-dimensional order, which is physically embodied in the sequence of time, but every lattice has a dual ($a < b$ implies $b > a$). Logically and methodologically, we describe the *priority of the simple* and the *supremacy of the complex* as complementary, opposite asymmetries embodied in actual processes, a concept developed by considering the vertical organization of the Central Nervous System in the context of integrating biological, social and psychological aspects of medical diagnosis and treatment [33], and also applicable to social issues [34] and to scientific research [5].

The third fundamental structure is modeled by topology, the study of space-like forms, their formation and transformation. Material structures are tridimensional; correspondingly, topological knots are described by three 'colors'. The human body itself shows temporal asymmetry in the dorsalventral axis, rough right-left symmetry, and the asymmetry of complexity in its vertical dimension. These three patterns also exist in the multiple levels of complexity that form the "vertical dimension" of nature from physical to mental processes.

A hierarchy of complexity is also evident in the sequence of patterns generated by recursions such as the logistic, the process and the diversifying equations. Indeed, symmetry and entropy decrease as the process equation enters the biotic phase.

Entropy is also relatively low in natural processes with a biotic pattern [35,36]. This may account for the development of complexity in physical and biological processes, so far justified by the local development of pockets of low entropy due to the selective intake of low entropy materials [37] and the selective excretion of high entropy waste products [37,38].

In statistics, entropy measures the degree of symmetry [39–48], while information is the degree of asymmetry [31]. According to classic thermodynamics, processes spontaneously flow towards higher entropy, i.e., from asymmetry to symmetry, and thereby towards equilibrium, and rest; actually physical and biological evolution demonstrates that processes often spontaneously flow from symmetry to asymmetry. Evolution increases information, and thereby generates complexity and beauty.

Symmetry is beautiful because it renders stability, asserts Shu-Kun Lin [49]; nevertheless, he adds, symmetry is in principle ugly because it is associated with information loss. From a different esthetic perspective, asymmetry is beautiful because it overcomes stability and drives evolution and the generation of complexity. Symmetry is beautiful because it embodies order rather than decay, disorder and rest. High entropy implies complexity, not necessarily disorder. But also asymmetry contributes to order and beauty; in fact, the mathematical definition of order is lattice asymmetry, and the asymmetric divine or golden proportion $\varphi = 1.618...$ has been a cannon of beauty in architecture and painting since Phidias. While the geometric symmetry of a face often conveys beauty, in reality human faces always show a degree of mathematical asymmetry, and their asymmetry contributes to esthetic pleasure.

Mathematical modeling demonstrates that Bios is generated by recursions that involve both asymmetry and symmetry. Opposites play a major role in creative evolution, artistic creativity, and beauty, rather than annulling each other in featureless thermodynamic equilibrium.

Asymmetry and symmetry are evident contributors to beauty, which is a major component of our emotional life and of art. Both art and life also involve creative processes that transform simple forms to generate complexity. Indeed, beauty requires complexity, because it may exist in simple as well as complex patterns, but it is perceived only by relatively complex organisms (e.g., mate selection, bees and birds attraction to flowers, etc.). Here we find in art Bourbaki's three generative forms, the asymmetry that defines lattices, the symmetry of opposites that defines groups, and the topological transformation of simple forms into complex ones.

Bourbaki's concepts provide a way to formulate the notion of causal, non-random creativity in an exact manner. Causal creativity represents an alternative to the theological concept of creation of matter from the void, classic determinism in which causality is not creative, and the current models of random variability in theories of the early universe, quantum mechanics, biological evolution by random mutation, and stochastic markets that offer no way to understand or influence processes. In contrast to portraits of change and creativity as "noise", the observation of Bios from quantum mechanics to music supports Pythagoras and Kepler's perspective in which music is a model for the mathematics of the universe.

In summary, both natural and artistic creative processes display asymmetry, symmetry, and creative transformation, indicating a common, possibly fractal, process (Bios) extending from quantum entities to mental activity.

Acknowledgements

We are thankful to the Society for the Advancement of Clinical Philosophy and to Mrs. Maria McCormick for their support.

References

1. Anderson, P.W. More Is Different. *Science* **1972**, *177*, 393–396.
2. Bourbaki, N. *Éléments de mathématique*; Actualités Scientifiques et Industrielles: Paris, France, 1952.
3. Beth, E.W.; Piaget, J. *Epistémologie mathématique et psychologie*; Essai sur les relations entre la logique formelle et la pensée réelle: Paris, France, 1961.
4. Sabelli, H. *Union of Opposites: A Comprehensive Theory of Natural and Human Processes*; Brunswick Publishing: Lawrenceville, VA, USA, 1989.
5. Sabelli, H. *Bios: A Study of Creation*; World Scientific: Singapore, 2005.
6. Sabelli, H.; Carlson-Sabelli, L. As simple as one, two, three. Arithmetic: a simple, powerful, natural and dynamic logic. In *Sustainable Peace in the World System and the Next Evolution of Human Consciousness*, Proceeding of International Systems Society 40th meeting, Hall, M.L.W., Ed.; Louisville, KY, USA, 1996; pp. 543–554.
7. Sabelli, H.; Kovacevic, L. Quantum Bios and Biotic Complexity in the Distribution of Galaxies. *Complexity* **2006**, *11*, 14–25.

8. Sabelli, H.; Kovacevic, L. Biotic Expansion of the Universe. *International Conference on Advances in Internet, Processing, Systems, and Interdisciplinary Research*, Sveti Stefan, Montenegro, 2003. Electronic Publication IPSI-2003.

9. Thomas, G.; Sabelli, H.; Kauffman, L.; Kovacevic, L. Biotic patterns in Schrödinger's equation and the evolution of the universe. *InterJournal* **2006**, 1787.

10. Sabelli, H.; Thomas, J.; Kovacevic, L; Horan, D. Biotic Dynamics of Galactic Distribution, Gravitational Waves, and Quantum Processes. A Causal Theory of Cosmological Evolution. *Black Holes and Galaxy Formation*; Wachter A.D., Propst, R.J., Eds; Nova Science Publishers: Hauppauge, NY, USA, 2009.

11. Sabelli, H. Complex Biotic Patterns in LIGO Recordings point to the creativity of gravitational interactions. *Complexity* **2010**, *15*, 12–24.

12. Carlson-Sabelli, L.; Sabelli, H.; Zbilut, J.; Patel, M.; Messer, J.; Walthall, K.; Tom, C.; Fink, P.; Sugerman, A.; Zdanovics, O. How the heart informs about the brain. A process analysis of the electrocardiogram. *Cybernetics and Systems 94*; Trappl, R., Ed.; World Scientific: Singapore, 1994.

13. Carlson-Sabelli L.; Sabelli, H.; Patel, M.; Messer, J.; Zbilut, J.; Sugerman, A.; Walthall, K.; Tom, C.; Zdanovics, O. Electropsychocardiography. Illustrating the Application of Process Methods and Chaos Theory to the Comprehensive Evaluation of Coronary Patients. *Complex. Chaos Nurs.* **1995**, *2*, 16–24.

14. Carlson-Sabelli, L.; Sabelli, H.; Messer, J.; Patel, M.; Sugerman, A.; Kauffman, L.; Walthall, K. Process method: Part I. An empirical measure of novelty differentiates creative organization from static order and chaos. In *Systems thinking, globalization of knowledge, and communitarian ethics,* Proceeding of International Systems Society, Rhee, Y.P., Bailey, K.D., Eds.; Kwanak Press: Seoul, Korea, 1997; pp. 1072–1090.

15. Sabelli, H.; Carlson-Sabelli, L.; Patel, M.; Zbilut, J.; Messer, J.; Walthall, K. Psychocardiological portraits: A clinical application of process theory. In *Chaos theory in Psychology*; Abraham, F.D., Gilgen, A.R., Eds.; Greenwood Publishing Group: Westport, CT, USA, 1995; pp. 107–125.

16. Sabelli, H.; Carlson-Sabelli, L.; Patel, M.; Sugerman, A. Dynamics and psychodynamics: Process Foundations of Psychology. *J. Mind Behav.* **1997**, *18*, 305–334.

17. Sabelli, H.; Messer, J.; Kovacevic, L.; Walthall, K. The biotic pattern of heartbeat intervals. *Int. J.Cardiol.***2010** (in press).

18. Sabelli, H.; Kovacevic, L. Biotic Complexity of Population Dynamics. *Complexity* **2008**, *13*, 47–55.

19. Patel, M.; Sabelli, H. Autocorrelation and Frequency Analysis Differentiate Cardiac And Economic Bios From 1/F Noise. *Kybernetes* **2003**, *32*, 692–702.

20. Sabelli, H. Bios, creative organization in economic, biological, and meteorological data.*International Conference on Advances in Internet, Processing, Systems, and Interdisciplinary Research*, Sveti Stefan, Montenegro, 2003. Electronic Publication IPSI-2003.

21. Sabelli, H.; Sugerman, A.; Kauffman, L.; Kovacevic, L.; Carlson-Sabelli, L.; Patel, M.; Messer, J.; Konecki, J.; Walthall, K.; Kane, K. Biotic Patterns in Biological, Economic and Physical Processes. *J. Appl. Syst. Stud.* **2004**, *5*, 14–26.

22. Sabelli, H. The Biotic Pattern of Prime Numbers. *Cybern. Syst. J.* **2008** (in press).

23. Levy, A.; Alden, D.; Levy, C. Biotic patterns in music. *Society for Chaos Theory in Psychology and Life Sciences Meeting, SCTPLS2006*; Johns Hopkins University, Baltimore, MD, USA, 4–6 August 2006.

24. Sabelli, H. Music, Poetry, Painting, and Bipolar Illness. *Nonlinear. Dynam. Psychol. Life Sci.* 2010 (in press).

25. Kauffman, L.; Sabelli, H. The Process equation. *Cybern. Syst.* **1998**, *29*, 345–362.

26. Sabelli, H. Complement plots: Analyzing opposites reveals Mandala-like patterns in human heartbeats. *Int. J. Gen. Syst.* **2000**, *29*, 799–830.

27. Dekking, M.; Mendès-France, M. Uniform Distribution Modulo One *Journal für die reine und angewandte Mathematik* **1981**, *239*, 149–153.

28. Sabelli, H.; Sugerman, A.; Kovacevic, L.; Kauffman, L.; Carlson-Sabelli, L.; Patel, M.; Konecki, J. Bios Data Analyzer. *Nonlinear. Dynam. Psychol. Life Sci.* **2005**, *9*, 505–538.

29. Sabelli, H. Novelty, a Measure of Creative Organization in Natural and Mathematical Time Series. *Nonlinear. Dynam. Psychol. Life Sci.* **2001**, *5*, 89–113.

30. Sabelli, H.; Abouzeid, A. Definition and Empirical Characterization of Creative Processes. *Nonlinear. Dynam. Psychol. Life Sci.* **2003**, *7*, 35–47.

31. Sabelli, H.; Patel, M.; Sugerman, A.; Kovacevic, L.; Kauffman, L. Process Entropy, a Multidimensional Measure of Diversity and Symmetry. http://creativebios.net/webjass/10Entropy.pdf.

32. Sabelli, H.; Sugerman, A.; Carlson-Sabelli, L.; Patel, M.; Kauffman, L. Embedding Plots: A Tool to Measure Simplicity, Complexity and Creativity. *J. Appl. Syst. Stud.* **2004**, *5*, 159–201.

33. Sabelli, H.; Carlson-Sabelli, L. Biological Priority and Psychological Supremacy, a New Integrative Paradigm Derived from Process Theory. *Am. J. Psychiatry* **1989**, *146*, 1541–1551.

34. Torre, C. Chaos, Triadic Theory of Psychological Competence in the Academic Setting. In *Chaos Theory in Psychology*; Gilgen, A., Abraham, F., Eds.; Praeger/Greenwood Publishing: Westport, CT, USA, 1995; pp. 279–294.

35. Carlson-Sabelli, L.; Sabelli H. Phase plane of opposites: A Method to study change in complex processes, and its application to sociodynamics and psychotherapy. *Social Dynam.* **1992**, *3*, 1–6.

36. Sabelli, H. Biothermodynamics. *Open Cybern. Syst. J.* **2009** (in press).

37. Schrödinger, E. *What is Life? The Physical Aspect of the Living Cell*; The Macmillan Company: New York, NY, USA, 1945.

38. Prigogine, I. *From Being to Becoming: The New Science of Connectedness*; Doubleday: New York, NY, USA, 1987.

39. Prigogine, I. *The End of Certainty*; The Free Press: New York, NY, USA, 1997.

40. Cohen J.; Steward, I. *The Collapse of Chaos*; Penguin: New York, NY, USA, 1994.

41. Leyton, M. *Symmetry, Causality, Mind*; MIT Press: Cambridge, UK, 1992.

42. Petitjean, M. Order, entropy and symmetry: An awkward relation? *Symmetry Cult. Sci.* **2005**, *16*, 5–6.

43. Jakulin, A. Symmetry and information theory. *Symmetry Cult. Sci.* **2005**, *16*, 7–26.

44. Matsuno, K. Symmetry and Information: Symmetry as an Emergent Property of Information. *Symmetry Cult. Sci.* **2005**, *16*, 27–36.

45. Smith, A. A hierarchical perspective. *Symmetry Cult. Sci.* **2005**, *16*, 37–46.

46. Salthe, N. Asymmetry and self-organization. *Symmetry Cult. Sci.* **2005**, *16*, 71–90.

47. Darvas, G. Order, entropy and symmetry. *Symmetry Cult. Sci.* **2005**, *16*, 91–108.

48. Lin, S.K. The Nature of the Chemical Process. 1. Symmetry Evolution-Revised Information Theory, Similarity Principle and Ugly Symmetry. *Int. J. Mol. Sci.* **2001**, *2*, 10–39.

Chapter 10

Nonlinear Dynamics in Psychology[†]

Stephen J. Guastello

Dept. Psychology, Marquette University, P.O. Box 1881, Milwaukee, WI 53201-1881 USA.
Email: stephen.guastello@marquette.edu.

ABSTRACT

This article provides a survey of the applications of nonlinear dynamical systems theory to substantive problems encountered in the full scope of psychological science. Applications are organized into three topical areas-cognitive science, social and organizational psychology, and personality and clinical psychology. Both theoretical and empirical studies are considered with an emphasis on works that capture the broadest scope of issue that are of substantive interest to psychological theory. A budding literature on the implications of NDS principles in professional practice is reported also.

The purpose of this article is to provide a survey of the recent developments in the application of nonlinear dynamical systems (NDS) theory to theoretical and practical problems encountered in the domain of psychology. For the benefit of non-psychologists, it is important to note that the scope of psychology is expansive. Introductory textbooks are typically organized around the following themes: brain physiology and behavior, psycho-physics, sensation, perception, learning, memory, cognition, intelligence and mental measurement, development, social psychology, motivation and emotion, personality of normal range people, abnormal psychology, psychotherapy and counseling, and industrial-organizational psychology. At the other end of the professional spectrum, the largest professional organization for psychologists, the American psychological Association, contains more than 50 topical interest groups in addition to its general membership core.

Only about half of academic psychologists can be found in university psychology department (Brookhart 2000). Psychologists are often found in departments or schools of education, communication, medicine or health sciences, business, law, and occasionally engineering. It is against that backdrop that interdisciplinary studies of NDS have emerged. Although the literature is already large and growing rapidly,

[†]Reused with permission from: Stephen J. Guastello, "Nonlinear dynamics in psychology," *Discrete Dynamics in Nature and Society*, vol. 6, no. 1, pp. 11–29, 2001. doi:10.1155/S1026022601000024.

the density of coverage in any particular topic area named above is thin. For that reason it would be unwise to engage a critical reviews of topic areas that are changing quickly, and it would be pretentious to propose to include the pedantic details of the literature as it is known now. Rather, I engage a more modest objective of providing a survey of that literature with an emphasis on the books and articles within the NDS-psychology area that have the broadest scope.

The Broad Organization of NDS in Psychology

The early years of NDS in psychology can be traced back to catastrophe theory (Thom 1975), its widespread conceptual applications (Zeeman 1977), a special issue of Behavioral Science (Cobb and Ragade 1978), and a parade of articles that subsequently appeared in that journal. After a period of waning in the later 1980s, the broader history of NDS, which dates back to works by Poincaré, Cantor and Lyapunov in the 19th century, collided with psychological theory in an extensive way. Psychologist F. Abraham, mathematician R. Abraham, and graphic artist Shaw (1990) teamed up to produce a landmark well-reasoned speculation on the possible applications of NDS to psychology in most of its content domains. The roster of dynamics now included attractors of varying types, bifurcations in several forms, chaos, fractals, and self-organization.

The Society for Chaos Theory in Psychology and Life Sciences was founded in 1991 in San Francisco. The scope of its interest in dynamics has expanded to include neural nets, evolutionary computation, cellular automata, and other related forms of dynamics. The scope of its substantive interests extends from microbiology to macroeconomics, although its productivity in psychology is perhaps its most prominent feature. Its earliest contributions to the literature consists of three edited collections (Abraham and Gilgen 1995; Robertson and Combs 1995; Sulis and Combs 1996) and the research journal Nonlinear Dynamics, Psychology, and Life Sciences, the inaugural issue of which appeared in January 1997. (See Guastello 1997, for the editorial purview.) Two later edited collections, which were orchestrated by the Society's European counterparts, are also historically significant as well (Guindani and Salvadori 1998; Tschacher and Dauwalder 1999).

Each of the foregoing publications unpacks many interesting authors with vibrant lines of research. The prominent themes among them are considered next, but they are not limited to Society publications. The path of least resistance suggests that the speci®c works in psychology should be organized into three themes: cognitive science, social organizations, and clinical studies. I will forego a section on neuroscience, largely because of the continuity of that topic with non-psychological studies. Works by Freeman (1995); Basar (1998, 1999); Waters (1999); Ferro Milone, Minelli

and Turicchia (1998) and Minelli and Turicchia (1999) should be considered significant contributions in that area, however.

Cognitive Science Applications

Current thinking in NDS theory is that consciousness is an integrated process consisting of psycho-physics and sensation processes, perception, cognition, learning, memory, and action. Although it has been convenient to think of these processes as separate entities, the separations are somewhat contrived. Similarly, real-world stimuli largely consist of continuous flows and random shocks.

Figure 1 displays the set of relationships that are thought to exist among cognitive process as seen by an outside observer (researcher). Starting at the upper right, an incoming flow of stimuli is first encountered by the human processes of sensation and psychophysical transduction. Perception pro-cesses organize the incoming stimuli into recognizable wholes through combinations of learned regimes and innate capabilities. Cognition involves a wide range of processes by which the recognized patterns are compared, associated with information already in memory, transformed in simple or complicated ways, and organized into responses.

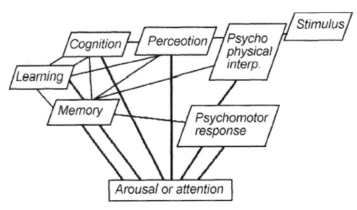

Figure 1 Integrative network of cognitive processes.

Learning involves one or more processes by which the individual organism acquires knowledge, skills, abilities, and adaptive responses. Memory pertains to how what is learned is organized and stored, and without which learning would be impossible.

The psychomotor response section of the model signifies how the response is produced by the individual (or not produced).

The heavy black connections between the boxes in Fig. 1 indicate information flows that have been acknowledged in conventional thinking as portrayed in introductory textbooks. The light blue connections acknowledge additional patterns of information flow that characterize the complex systems nature of human cognition. For instance, cognition-action patterns are now studied as "autonomous agents" (Tschacher and Dauwalder 1999). The heavy red connections indicate a connection to a general driving force of arousal or attention. Attention processes limit the extent, and probably the speed, by which a combination of the foregoing cognitive processes will be engaged.

Each function depicted in Fig. 1 has lent itself to empirical study of nonlinear dynamical processes (Clayton and Frey 1996, 1997; Davids, Button and Bennett 1999; Goertzel 1993a, 1993b, 1994, 1997; Gregson 1992, 1995, 1998, 1999; Guastello 1995; Kahonen 1989; Kelso 1995; Port and van Gelder 1995; Porter and Hogue 1998; Pressing 1999; Ward and West 1998). There have been substantial theoretical contributions as well

(Combs 1995; Gilgen 1995; Hardy 1998; MacCormac and Stamenov 1996; Sulis 1997; Tschacher and Dauwalder 1999).

Psychophysics

The landmark contribution from the 19th century, due to Weber and Fechner, was to find the nonlinear relationships that exist between the strength of physical stimuli, such as brightness of lights, loudness of sounds, and so forth. Those relationships carried with them concepts of the absolute threshold and difference threshold, which initially were conceptualized as numbers that generalized across humans.

The landmark contribution by the middle 20th century, due to Stevens, was the formation of signal detection theory. The threshold concepts were now regarded as stochastic entities, and physical stimuli were detected relative to noise. To be brief, it became possible to separate the sensory acuity of the individual from experimental response set, and from the properties of the stimuli themselves. The study of simple decision errors gave way to the foundation of decision theory, which today involves decisions of a much more complex nature.

The NDS contributions to psychophysics are largely the work of Gregson (1992, 1995, 1998, 1999). They address additional questions not considered previously such as what happens when: stimuli contain multiple parameters as in the taste of wine, stimuli have complex decay functions, and stimuli are generated. The mathematical model that is central to the new nonlinear psychophysics is the gamma recursion formula:

$$Y_{j+1} = a\left(Y_j - 1\right)\left(Y_j - ie\right)\left(Y_j + ie\right) \tag{1}$$

(Gregson 1992, p. 20) where Y is the strength of a response at j points in time, a is a control parameter representing signal strength, e is a situational control parameter, and i is the imaginary number $(-1)^{1/2}$. The gamma function is a cubic variant of the logistic map, and expands into real and imaginary response components. It expands further into multi-parameter signal and response functions.

Perception

Psychologists have historically paid great attention to optical illusions with the thinking that illusions betray underlying invariant processes of how the mind, and by inference the brain, functions. NDS approaches to the study of perceptual illusions date back to Stewart and Peregoy (1983) and Tàeed, Ta'eed and Wright (1988), who used a cusp catastrophe model for the process:

$$df\left(y\right)/ dy = y^3 - by - a \tag{2}$$

where y is the response on the continuum between selecting one perceptual image over the other, b is the bifurcation parameter representing the total amount of detail in the ambiguous pictorial stimuli, and a is the asymmetry parameter representing a bias in the pictorial details toward one perceived image over another. Guastello (1995) also suggested that the cusp model for ambiguous stimuli transfer well to the study of more complex decisions, such as those made by juries in legal proceedings. A later wave of NDS research in this area targets multi-stable response patterns more generally, including the classic ambiguous stimuli used in the experiments named above (Haken 1999; Kelso 1995; Porter and Hogue 1998). The favored mathematical model is the Ginzberg-Landau model of phase transitions:

$$df(y)/dy = y^4/4 - y^2/2 - ay \tag{3}$$

A quick look, however, indicates that Eq. (3) is a potential function for the cusp catastrophe model where the bifurcation effect is held constant. This function has been useful, nonetheless, for modeling the organization of motor response patterns and speech configurations in addition to strictly perceptual phenomena.

Learning

Learning theory in psychology has undergone numerous developments in psychology in the past century. The oldest regime, due to Thorndike, first identified the learning curve, which is a plot of the number of correct responses given by an organism as a function of the number of learning trials. Reinforcement (e.g., feedback, reward, punishment, etc.) given to the learning organism strengthens the probability of the desired response. In the classic experiment, a cat was placed in a puzzle box, from the cat had to figure out how to escape. The cat began to emit a variety of hissing, scratching, and other irrelevant responses until it discovered the pedal on the door of the box that opened the cage door. With subsequent trials the cat emits fewer irrelevant responses and eventually goes right to the pedal. The explanation for the underlying psychological events was predicated on associations being formed in the "mind" of the cat.

The second regime, due to Pavlov, explored the role of conditioned reflexes, and how natural reflex mechanisms can be shaped into complex behaviors. The explanation for the underlying psycho-logical events (observed mostly from dogs in the early years) was again predicated on associations being formed. The third regime, however, due to Skinner, emphasized the role of reinforcement shaping behavior, where behavior started with ad lib, or operant behaviors. Skinner took a hard core position of not invoking any associationist explanations, or even assuming that a structure of "mind" existed.

The fourth regime grew up alongside and slightly before Pavlov's. Gestalt psychologists, who were best known for their contributions to perception, provided evidence based on primate behavior that learning can occur in one trial. In cases where it does, it is known as insight. Insight is a well-known critical event in the creative thinking process. The Gestalt psychologists were also known for the dictum that "The whole is greater than the sum of its parts", which, by some stretches of the imagination, foreshadowed the post-Newtonian, system-wise thinking that characterizes most NDS applications today.

The fourth regime grew up alongside Skinner's approach. Cognitive learning theory, due to Tolman, was based on experiments that showed that the rat knew where the cheese was located in the maze, and in what relative probability. Here learning produced a cognitive map, which denotes spatial representations, along with what we would call expectancy.

A substantial amount of cognitive motivation theory (decision-makers gravitate toward options that are associated with the odds of the best results) is based on this line of thinking. Germane to NDS, cognitive motivation theory is essentially the psychological explanation behind why game theory explains a wide range of human phenomena.

The fifth major movement was statistical learning theory, due to Estes, which addressed the issue of whether partial learning can actually occur. According to this view, learning on each trial is partial. The learned elements eventually add up, often quickly, to the final learning objective. The sixth was social learning theory, due to Bandura, which was based on the observation that learning can occur through vicarious learning, i.e., by watching the behavior of others and their outcomes. The most recent regime is implicit learning theory (Seger 1994), which focuses attention on things that are learned while the learner is trying more deliberately to learn something else.

The NDS interpretation of learning reflects a few different approaches. One centers around neural networks and learning programs. For the most part, this frame of reference does not proffer central equations. If anything, the nonlinear structures are hidden from the observer-research by the computer program executing the neural network simulation, but see Vickers and Lee (1998, 2000) for the psychological implications of what neural networks have to offer.

Other approaches reflect a return to associationist thinking. Ideas, behaviors, and the stimuli that trigger them are thought to self-organize into wholes (Gilgen 1995; Haken, 1999; Kampis, 1991). If the self-organization process is complete, as in the case of a feasible task, the common learning curve would result, which would in turn be characterized as a fixed point attractor, as in Eq. (4):

$$Z_2 = e^{-az_1},$$
(4)

Where z is behavior and a is the Lyapunov exponent which is constrained to negative values (Guastello, 1995; Guastello and Guastello 1998). We can have a variable, b, in the model that sharpens or flattens the infection in the learning curve, as in Eq. (5):

$$Z_2 = bz_1 e^{-az_1},$$
(5)

In the in event that learning is incomplete, an asymptote does not form, and the exponent in Eq. (4) or (5) is positive. In other words we have chaos before self-organization (Guastello and Guastello 1998). Figure 2 shows ensembles of learning curves for an experiment involving a difficult task which the learners frequently did not master (Guastello, Bock, Caldwell and Bond, 2005). In the left panel, learners were allowed to talk, which would have facilitated learning. In the right panel, talking was not allowed.

The results shown in Fig. 2 were actually obtained from a group learning task, which is revisited below under the heading of work group coordination. Implicit learning was involved in that particular task.

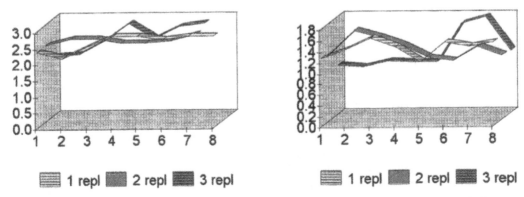

Figure 2 Results for a coordination learning task where groups of learners were allowed to talk (left) or not (right), and in which one, two, or three members of the learning group (repl) were replaced during the course of the experiment (from Guastello, Bock, Caldwell and Bond, 2005, p.195). Reprinted with permission of the Society for Chaos Theory in Psychology & Life Sciences.

Memory

There is reason to believe that memory is a distributed process that involves many groupings of neurons that are relatively small, and that the temporal patterns of neuron ®ring contain a substantial amount of information about memory storage processing (Kahonen 1989). The temporal dynamics of memory experiments can be analyzed for two broad classes of information. Inter-trial analyses would indicate whether and how the response to one experimental trial would impact on the subsequent responses. While inter-trial response times have been traditionally regarded as random or probabilistic processes, dynamical analyses indicates that is not the case, and that modicum of structure and long-term patterning is evident (Clayton and Frey 1996, 1997). On the other hand, the meaning of those patterns is yet to be revealed. Intra-trial analyses of memory experiment data would provide information on the cue encoding, retrieval, and decision processes.

Figure 3 depicts the results from a memory experiment by Clayton and Frey (1997) in which the researchers examined power spectra from time of response times from a set of memory tasks. The three tasks varied in memory load; the task labelled X-0 required the least load, and SD2 required the greatest. Three spectra were obtained, which corresponded to the three memory loads. Similar FFT results were obtained for each series, however. An important conclusion from the experiment was that the noise in the time series of responses did not correspond to a plethora of uncontrolled variables, as mental measurement theorists have historically assumed. Rather, there was no need to assume that a complex high-dimensional system was involved in the generation of "mental noise".

Cognition

Theories of cognition are closely aligned with theories of intelligence and the measurement of intelligence. It has always been controversial as to what "intelligence" contains, and how much of it is the result of heredity, and how much is the result of the environment, including the educational environment. Nonetheless, the objective is to understand what

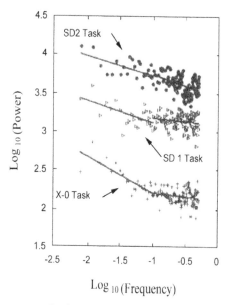

Figure 3 Power spectra for response series in memory tasks with low, medium, and high levels of memory demand (from Clayton and Ferry, 1997, p. 177). Reprinted with permission of the Sociaty for Chaos Theory in Psychology & Life Sciences.

is innate ability, and to separate that (or to determine its inseparability from) particular content domains.

One of the earliest view of intelligence, due to Spearman in the early 1930s, was that there was only one component of intelligence, which was general intelligence, or g. A competing view, due to Thurstone, was that intelligence consisted of several independent factors, and should be measured as such. Further work by Guilford in the 1950s indicated that just about everything measured by commonly-used intelligence tests pertained to convergent intelligence; creative thinking involved divergent intelligence which was typically missed in measurement batteries. The objective of convergent thinking is to find the one correct answer, or the best answer of the options available. Examples include: numerical computation; memory for ideas, sights, sounds, and events; word choices and interpretation; interpretation of a set of logical propositions; or the conversion of one set of symbols into another set. The goal of divergent thinking is to come up with many possible solutions, answers, or options. Often the more unique or aesthetically pleasing responses are given preference. Examples include producing many possible adjectives that could be used to describe a given object, organizing and reorganizing a set of objects into categories, producing novel uses for common objects, drawing analogies between one relationship and another, and suggesting possible consequences of an unusual but significant event. The current landing point is the triarchic theory of intelligence, due to Sternberg, which holds that intelligence contains three broad components: convergent thinking, divergent thinking, and the ability to learn from one's experience.

Applications of NDS to cognition have taken a few distinctive approaches. Goertzel (1993a,1993b,1994,1997) developed computer simulations of cognition with the objective of integrating as many components of the psyche as possible. Here we find systems and subsystems of logic with feedback loops that sustain or alter the logical sets. Perhaps his most poignant finding (1994) is the existence of "chaotic logic" by which irrational belief systems are held in place by strategically organized loops that prevent the illogical system from unraveling. The external validity of the models to actual human thought processes has not yet received experimental attention.

Hardy (1998) studied semantic lattices in verbal cognition. A semantic lattice is a network of mental elements found in memory. The network is organized through associations between a characteristic of an idea and other ideas that share the same characteristic. A semantic lattice, furthermore, is thought to be organized into modules, which Hardy (1998) calls semantic constellations. For instance, if we were to think of two people, e.g., John and Mary, we can immediately connect to everything we know about John, his job, preferences in clothing and food, his car, and so forth. We can do the same with Mary, the relationship between John and Mary, and their relationships with everyone else in their social network. According to this line of evolving theory, the percolation and growth of a lattice is based on attractor strength, bifurcations, and self-organization. Although much of the supporting argument for the theory comes from experimental pre-NDS psychology, the empirical verification of the literal dynamics has not yet been observed.

Other productive lines of thought in NDS involve autonomous agents and collective intelligence. Autonomous agents are self-organized configurations of perception, cognition, and action; the special ingredient here is that the search for stimuli and the actions taken are intentional (Guillot and Meyer, 2000; Tschacher and Dauwalder 1999). The formation of agents, as well as the percolation of semantic lattices, can be better appreciated with a brief discussion of the processes of assimilation and accommodation discussed below in the section on human development.

Collective intelligence may be considered a social process as well as cognitive process insofar as ants and bees are social insects. Here we consider the cognitive substance of the phenomenon, as did Sulis (1997, 1999). The important attribute of the phenomenon is that a colony of ants will engage in coordinated activities such as hunting for food in repeated forays to different locations, maintenance of the nest, and reassigning labor to tasks as needed. Ants do so without any one ant having any apparent knowledge or cognitive picture of the entire plan. In other words, the local interaction among ants gives rise to a self-organized collective behavior.

Human Development

Developmental psychology is concerned with the changes in human cognition, social behavior, and physical development across the life span. The three domains are related, of course. Theories of development have often involved a stage process which comprise a hierarchical series of abilities and behaviors. The three basic principles of stage theories are: (a) Each stage is qualitatively different from each other stage. (b) Each stage in the series is progressively more complex than the previous stage. (c) Each person goes through an invariant sequence of stages. There has been some lively debate over the years regarding whether stages are really as discrete as their proponents claim, whether stages of social and cognitive development are as synchronized as they might appear to be, whether discrete stages really exist in some aspects of development. Furthermore, stage theories tend to focus on the innate unfolding of cognitive, social, or behavioral schemata, and the relative contribution of the environment needs to be considered as well. The latter counterpoint has become known as the nature versus nurture controversy. NDS studies in human development have to date centered primarily on the cognitive and motor development of young children. Much of the coordination of limbs, as in learning to walk, is a self-organizing process that begins at the local level of limbs rather than beginning from a central plan that was somehow encoded in the brain. The self-organization process is similar that invoked by Kelso (1995) in his work on multi-stable responses.

Importantly, the sequence of motor development stages displays substantial individual differences among children (Metzger 1997; Thelen and Smith 1994; Turvey and Carello 1995).

Metzger (1997) mentioned that progress in cognitive development studies is limited by the availability of a set of cognitive measurements applicable to development. Nonetheless there has been some recent progress in the area of cognitive development. In one of the better-accepted stage theories, due to Piaget, the developing child interprets the world and makes responses using a process of assimilation or accommodation. Assimilation is the first response of the child to interpret and respond using existing behavioral and cognitive sequences (schemata). When the existing repertoire is not longer adequate, the child responds by accommodating the new situation by developing one or more new schemata. Feldman, Csikszentmihalyi and Gardner (1994) observed that the process of assimilation, accommodation, and eventual cognitive growth is essentially a self-organization process. Partridge (2000) observed further that the accommodation and assimilation process is a fundamental phenomenon of complex adaptive systems. He observed, furthermore, that the operating individual characteristic is temperament, which is a child's proclivity to engage in a new situation or approach another person, to avoid, or to delay before approaching. The approach-avoidance process is thought to be the primary driver in environmental interaction, which in turn facilitates or retards the assimilation-accommodation process. The transition between stages is a gradual increase of a new response and a graduate phasing out of an old response. The two gradients of behavior change, however cross at a critical point, producing discontinuous responses similar to those found in catastrophe models.

The approach-avoidance dynamics lends itself to dynamical modeling. That is considered in the broader context of the motivation models presented in the social and organizational psychology section of this article.

Social and Organizational Psychology

This group of topics includes social cognition, motivation, interpersonal attraction, attitudes, and several other topics of occupational relevance. Creativity, which has an important cognitive component, is included here largely because of the group dynamics that are also part of the process.

Motivation

Psychological theories of motivation have taken many forms over the years. Hunger and thirst predispose animals to behave as desired in learning experiments. If the rat knows where the cheese is, we can leap quickly to cognitive theories of motivation, due to Vroom, whereby the decision maker chooses behavior options that will produce the desired expected reward levels. There is also a theory of equity, due to Adams, in which the agent takes action to restore or maintain equity with other agents. Another important theme that pervades many social and organizational theories of motivation is the distinction between intrinsic and extrinsic motivation. Extrinsic motivation and extrinsic reward describe situations where the agent receives reward from an outside source. It contrasts with intrinsic motivation, where the agent receives reward, usually intangible, from the activity itself. Examples of intrinsic motivation would include the motives for achievement, affiliation, and power.

Physiological motivation consists of only one form, which is arousal. Arousal originates in the reticular formation of the brain, transfers to the thalamus, and transfers again to the cortical areas where it is interpreted. The same essential process applies to emotion as well; NDS studies of emotion are included in the personality and clinical sections of this article.

The butterfly catastrophe model of motivation in organizations draws together many of the previously-known dynamics affecting personnel selection and training, motivation, and work performance, absenteeism, and turnover (Guastello 1981, 1987, 1995; Fig. 4). The principles of several motivational theories are represented in the model. The butterfly catastrophe model consists of three stable states of performance and four control parameters. The four control parameters are ability (asymmetry), extrinsic motivation (bifurcation), intrinsic motivation (swallowtail), and a management climate that tolerates individual differences and encourages intrinsic motivation to dominate over extrinsic motivation (butterfly). Although all parts of the model, including the butterfly structure itself, have been empirically verified, it should be noted that some practical applications of this model may involve only subsets of the butterfly dynamics.

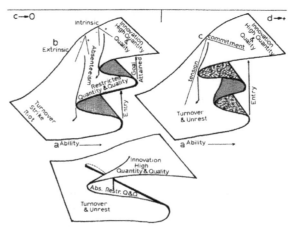

Figure 4 Butterfly catastrophe model of motivation in organizations (from Guastello, 1987). Reprinted with permission of the American Psychological Association.

The gradients on the butterfly responses surface that run between the stability points and the point of degenerate singularity are interpretable as approach and avoidance gradients in motivation theory. If one were to define a more complex dynamical field, such as having many jobs to choose from, the dynamics become progressively more complicated (Abraham 1995; Guastello, Johnson and Rieke 1999).

Social Cognition

Vallacher and Nowak (1994, 1997) and their contributors prepared a foundation for studying phenomena known to social psychologists. Attitudes, social judgments, and social relationships change over time, but conventional psychology has been, at best, clumsy about representing the temporal character of social events. Correlational dimensions that vary across experimental conditions, analyses of attractors and stability, catastrophes, and cellular automata are all promising tools in this regard.

As an example, Latane and Nowak (1994) examined the diffusion of an attitude change attempt in a hypothetical population. The cellular automata results (Fig. 5) reflect the effect of physical proximity of one cell, representing 400 people, to another. The results of the diffusion process were consistent with mathematical predictions from earlier social theory.

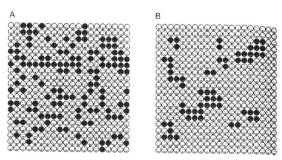

Figure 5 Spatial distribution of attitudes before (A) and after (B) social influence. Faces reflect one of two attitudinal positions (from Latane and Nowak, 1994, p. 244).

Interpersonal Attraction

People are attracted to each other for all sorts of reasons; an extensive summary of reasons is beyond the scope of this article. Nonetheless, once two people find that they have something in common, the forces of approach-avoidance and cooperation-competition then ensue. In the particular case of romantic relationships, the time evolution of the relationship can be volatile or not, depending on the strength of the attraction and propensity toward avoidance.

In Fig. 6 Rinaldi and Gregnani (1998) model the phase portrait of the mutual attraction for robust and fragile couples. For robust couples, there are simple rather than multiple saddle points.

When enough people join the social group, a network of relationships forms. The evolution and stability of networks depend on the intimacy, intensiveness, and exclusiveness, of the relationships. Alisch, Azizighanbari and Bargfeldt (1997) have been working on simulations of children's friendship networks based on those three variables using 3-D phase portraits.

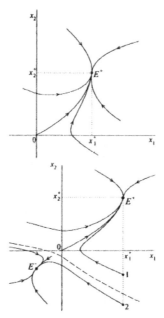

Figure 6 Dynamical fields for robust (upper) and fragile (lower) couples for varying initial conditions (from Rinaldi and Gragnani 1998, p. 291). Reprinted with permission of the Sociaty for Chaos Theory in Psychology & Life Sciences.

Creative Problem Solving

Creativity is recognized as a complex phenomenon involving divergent thinking skills, some personality traits that are commonly associated with creative individuals across many professions, an environment rich in substantive and interpersonal resources, and cognitive style. Cognitive style is a combination of personality and cognition; it refers to how people might use their talents rather than the quantity of such talents. According to an early version of the "chance-configuration" concept (Simonton 1988), creative products are the result of a random idea generation process. Greater quantities of ideas are generated from enriched personal and professional environments. Idea elements recombine into configurations as part of the idea generation process. When the creative thinker latches on to a new configuration and explores it as a possible solution to a problem, a form of self-organization of the idea elements takes place.

In the context of nonlinear dynamics, however, the generation and recombination of idea elements is chaotic rather than random at all. The self-organization of idea elements is largely a response to a chaotic system state. The idea elements, meanwhile, are generated by determined, human systems, whether individually or in groups. The individuals filter out some ideas and attract others depending on their goals for problem solving. They also organize idea elements according to their own unique mental organization and experience; some of these mental organizations are shared with other people, and other problem solvers in the group, and other mental organizations are not so shared. The process of idea generation retraces the paths that the individuals have mentally created already among idea elements, prior to any one particular problem-solving event (Guastello 1995, 1998a).

The mushroom (parabolic umbilic; Fig. 7) catastrophe was found to explain the dynamics of creative problem solving in groups who were working together in real time in an experimental situation (Guastello 1995). The response surface represents two interacting clusters of social interaction patterns. General Participation included information giving, asking questions, and statements of agreement with other people's ideas; it was found to be a bi-stable variable. Especially Creative Participation included statements that initiated courses of action for the group, elaboration of ideas, and rectifying intellectual conflicts; it displayed one stable state with instability at the high contribution end of the scale.

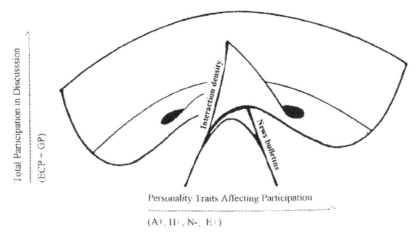

Figure 7 Mushroom catastrophe model for creative production in problem solving groups. ECP=Especially Creative Participation; GP=.General Participation; A+=.warmth and participativeness; H+=.social boldness; N-=unpretentiousness; E+=.assertiveness. "News bulletins" are experimentally applied shocks that were specific to the problem the group was trying to solve (from Guastello, 1995, p. 362). Reprinted with permission of the Sociaty for Chaos Theory in Psychology & Life Sciences.

Two of the four system control parameters, both of which were asymmetry variables, were occupied by personality traits. One cluster of traits distinguished high-production participants from low-production participants on the factor for general contributions. Assertiveness distinguished those who most often gave especially creative responses from others. The two bifurcation control parameters were overall group activity level, which captured a social dynamic, and the effect of particular experimental stimuli, which captured an environmental contribution. The news bulletins were introduced periodically as part of the game; they contained unexpected changes in the problem situation that should provoke an adaptive response from the players.

The mushroom structure itself was verified through a polynomial regression technique. In this case, a nonlinear regression technique was also used for estimating a Lyapunov exponent, which was positive and translated into a dimensionality of 5.46. This high dimensionality, which is also fractal, was an important observation because, according to the emerging theory, chaos leads to self-organization, and as creative self-organized systems engender more instability, it would follow that creative problem solving groups are systems operating at the edge of chaos.

Recent studies have also explored whether computer-facilitated communication can enhance the group's overall level of production compared to the production of a collection of non-interacting individuals, so long as the group is large enough to produce a critical mass of ideas. Computer media can facilitate chaotic levels of idea production. In this situation, "chaotic" refers to bursts of high and low idea production over time, on the part of either individuals or groups. Additionally, greater amounts of variability in production on the part of an individual are associated with greater quantities of ideas that are produced by other group members in between two successive inputs from a particular person. These dynamics conform to the logistic map structure (Guastello 1995).

At the group level of analysis, greater productivity is associated with a relatively complex problem task, where the task can be broken down into subtopics. At that time the group members can work on any subtopic in any order they choose, go back and forth among the subtopics, and so on. In the actual groups studied (Guastello 1998a), the number of active topics increased and decreased in a periodic fashion. The level of output by the group was chaotic overall, but it also showed periodic rises and drops in activation level in accordance with the change in the number of active topics. Thus the result, in the thinking of synergetics, is a coupled dynamic consisting of a periodic driver

$$A_2 = 0.75 A_1 e^{-0.36A1} + 0.33; \tag{6}$$

and a chaotic slave

$$Z_2 = e^{(0.25z1)} + 0.43A - 0.26C - 0.34. \tag{7}$$

In Eqs. 4 and 5, z_i represents group production levels can be observed depending on the topic that the group is working on (C); and, A is the number of active discussion threads during the time interval of z_i; time was measured in 4-day periods. The exponent in Eq. (6) was negative, and the exponent in Eq. (5) was positive.

Leadership Emergence

The rugged landscape model of self-organization has been received increased attention lately as an explanation for organizational phenomena, particularly where strategic management is involved (McKelvey 1999). The rugged landscape model of self-organization explains how leaders emerge from a leaderless group, and the possible ways in which their emergence could take form (Guastello 1998b; Zaror and Guastello 2000). The group activity selected for study involved a complex creative problem solving task. Once presented with the task and an hour (of experimental time) to complete it, numerous verbal interactions transpire among group members. These local interactions are thought to culminate in the eventual self-organization of the group such that the role of a general leader emerges along with several other, more specific roles. This form of differentiation had been reported without the non-NDS concepts in personality studies of emerging leaders decades beforehand.

The formation of roles would constitute fitness peaks, which denote relative fitness, local stability, and clusters of similar subspecies with regard to shared adaptive traits. The probability density function that is associated with the swallowtail catastrophe model (Eq.

(8)) describes the distribution of people into unstable and locally stable social roles. The swallowtail catastrophe structure contains a response surface of discontinuous events, or qualitatively different outcomes, such that there are two stable states, with a minor anti-mode between them, an unstable state, and a major anti-mode separating the unstable state from the two stable ones.

$$pdf(z) = \xi \exp\left[-\theta_1 z^5 + \theta_2 z^4 - \theta_3 cz^3 - \theta_4 bz^2 - \theta_5 az\right],\tag{8}$$

In Eq. (8), z is the extent to which members of the group endorse a particular group member as the leader; a, b, and c are control parameters; ζ is a constant that maintains unit density; and θ_i are nonlinear regression weights. Research continued in the direction of identifying the control variables in different types of work situations.

McClure (1998) investigated similar ideas, but with both task and therapy groups in mind. He emphasized the importance of the content of conversations for inducing desired outcomes for the group.

Work group Coordination

One important problem in group productivity is that the quality of the outcome for a group is not consistently better than the outcome or product from the most talented individual in the group. Psychologists have found partial remedies and explanations for this phenomenon by defining the task as a group task, allocating rewards to the group rather than to individuals, and defining responsibilities to prevent social loading, and enhancing group cohesion. The differential productivity problem still persists, nonetheless, and the internal dynamics of group workers is the current point focus of NDS research (Guastello et al. 2005; Guastello and Guastello 1998).

Coordination occurs when group members make the same or compatible responses at the right time for optimal production. In game theory, there are several distinct forms of coordination; the type described here is based on the four-way stop intersection. If the drivers correctly perceive the turn-taking system adopted by the preceding drivers and follow the sequence, then all cars pass through the intersection in a minimum amount of time with the lowest odds of a collision. In a real-life intersection, any of several possible rule systems could be adopted by the drivers, and each driver approaching the intersection needs to observe the strategy that is actually in effect, then make the correct move. If a car tries to go through the intersection out of turn, then an accident could occur, or at the very least, other players would need to revert to ad lib turn-taking to untangle the confusion at the intersection.

The process of group coordination involves the development of nonverbal communication links among the participants. These links evolve with repeated practice with each other. The evolution of the links is essentially a self-organization process. Furthermore, the basic process of coordination is non-hierarchical, meaning that a leader, who usually contributes task structuring activities of some sort, is not required. This state of affairs is not unlike the flocking of birds, herds of beasts, or schools of fish, which operate without leaders.

The results of experiments to date show that if the experimental task is not excessively difficult, the group will display a coordination learning curve. The coordination acquired during one task session will transfer to the learning and performance curve of a second task. If the task is too difficult, self-organization will not be complete, and the time series of coordination data will be chaotic. Verbalization enhances performance to some extent. A coordinated group can withstand changes in personnel up to a point before coordination breaks down.

General Perspective on Management

Some of the organizational literature focuses attention on the broad concept of the organization as a whole entity and the implications for management. One line of thinking emphases the contrasts between systems thinking and the nonlinear dynamics of change against the mechanistic and static view of conventional management (Kiel 1994; Thietart and Forgues 1995; Anderson, Meyer, Eisenhardt, Carley and Pettigrew 1999). Another considers how strategic management is best accomplished for a complex adaptive system (Dooley 1997; McKelvey 1999; Stacey 1992). A third line of thinking emphasizes practical ways for inducing entropy for organizational change and means by which managers can develop intuitive understandings of nonlinear organizational dynamics (Eoyang 1997; Goldstein 1994; Wheatley 1992; Zimmerman, Lindberg and Plsek 1998).

Personality and Clinical Psychology

The area of personality and clinical psychology is concerned with theories of the normal range personality, the etiology and treatment of psycho-pathology, and related topics. Some of the prominent theories in psychology have devoted more attention to the pathological side of human nature, while others have had the opposite emphasis, which is to maximize mental health and the human condition. The contributions of NDS have addressed concepts that were initially framed by a variety of different psychological schools of thought. The following summary is based on that "natural" organization of topics.

Classical Psychoanalysis

The earliest and most comprehensive theory of human personality, pathology, and the treatment thereof came from Sigmund Freud's psychoanalysis at the turn of the 20th Century. The key aspects of the theory were propositions concerning: instincts and the primacy of sexuality and destruction in motivation, levels of conscious awareness, psychic energy, psychic structures and the defense mechanisms of the ego, a theory of psychosexual development, and the psychoanalytic method for diagnosis and treatment. Although Freudian psychoanalysis enjoyed widespread popularity as an operating theory of the psyche and as a treatment method, the theory eventually succumbed to deficits in its ability to explain psychological phenomena adequately. It remains valuable, however, for its historical impact on psychologists who either followed or disagreed with its general principles or specific ideas. Of interest to NDS is that Freudian psycho-analysis invoked a concept of psychic energy and principles pertaining to energy movement around the psyche. Energy was assumed to limited in quantity, and moving around a closed psychic system according to rules of thermodynamics. To date, no one has ever found, let alone been able to measure, a construct of psychic energy. The thermodynamics principles were consigned long ago to the realm of Quaint Metaphor. Dynamicists are now revisiting Freudian concepts for their dynamical components, and how they might be recycled into contemporary theory (Goldstein 1995, 1997).

Others have focused attention on the conversation and redirection of thought that takes place in psychotherapeutic sessions (Stein 1999). Of further interest, the psychological dynamics of motivation and behavior change follow the same mathematical rules as phase transitions for physical materials.

Psychoanalytic theory took a historically interesting turn with the work of Carl Jung. Jung was initially one of Freud's protégés but inevitably had a few ideas of his own: a theory of psychological types based on four variables, the fundamental notion of psychic energy as an unlimited quantity in an open or dissipative system, the assumption that a life force was central to psychic existence instead of sexuality, the concepts of archetypes, and the existence of a racial memory in the unconsciousness. Archetypes are symbolic images that represent some important themes that are found in major mythological systems throughout the world. Archetypes were thought to transfer from generation to generation genetically through the structure of racial memory, and the collective unconscious. For instance, one might observe the visual and conceptual similarities between the cupids in Roman art of the first century B.C.E. and the angels depicted in Renaissance art, and consider how they might qualify as an archetype. One might further ruminate on the similarity between symbols and mystical concepts found in European, Egyptian, Hindu, and Chinese mythologies, and hence the credence of a collective unconscious common to all human being.

Today, scientific psychology does not take the concept of racial memory and its genetic transmission seriously, although the collective unconscious is a bit more palatable. Jung's theory of psychic energy is no more established than Freud's theory of energy. Archetypes might be regarded as special cases of cognitive prototypes. As symbolic objects they remain interesting, but today they are regarded as memes-deeply meaningful idea elements that float around the culture through artifacts, written works, and social structures that maintain interest in their content (e.g., libraries, educational systems, government funding).

To his credit, Jung found a bridge between cultural experience and the individual psyche. The circulation of memes is interesting to contemporary NDS theorists in clinical psychology (Butz 1998; Chamberlain and Butz 1998) as well as those concerned with broader issues of creativity and culture (Combs and Holland 1996; Csanyi 1989; Feldman et al. 1994). Others theorists have ruminated on the relationships between fundamental mathematical ideas since the Renaissance and the idea elements captured by many of the principal Jungian archetypes (Robertson 1995).

Humanistic Psychology

Humanistic psychology is usually associated with the work of Abraham Maslow, Rollo May and Carl Rogers, which gained prominence in the 1950s and 1960s. Those theorists diverged from their predecessors by aiming at a concept of optimal human functioning and optimal human experience instead of just an escape from psychic trauma and pathological conditions. Their work produced a concept of the self, self-esteem, and self-actualization. Although a concept of self was known in previous times, research was finally generated on the subject, some of which was picked up by experimental social psychologists.

Of importance to this review was the humanists' emphasis on the whole person. Mental health and psychic development require an integrated personal understanding of all the aspects of one's being on the part of the client and therapist. In recent times, after a decade or two of distraction from the humanistic principles, the emphasis on the whole person has evolved into a concept of the person as a complex system in the NDS sense (Butz 1998; Chamberlain and Butz 1998; Orsucci 1998; Massimini and Delle Fave 2000; Schiepek 1999).

The traditional philosophical counterpoint between free will and determinism takes a new turn with the discovery and understanding of mathematical chaos; the chaos concept must have an impact on the way people are taught to interpret events for themselves, or on behalf of their clients (Mosca 1995). There is a certain stability to human personality in the form of psychological traits (a long story in contemporary psychology by itself), and at the same time a flexibility in self-concept that is dependent on us having multiple social roles (another long story). Add in the cultural contributions and the stages of development, and it is compelling to regard the psycho-with the discovery and understanding of mathematical chaos; the chaos concept must have an impact on the way people are taught to interpret events for themselves, or on behalf of their clients (Mosca 1995). There is a certain stability to human personality in the form of psychological traits (a long story in contemporary psychology by itself), and at the same time a flexibility in self-concept that is dependent on us having multiple social roles (another long story). Add in the cultural contributions and the stages of development, and it is compelling to regard the psycho- logical structure of the self as a complex adaptive system (Marks-Tarlow 1999).

Psychopathologies and Treatment

NDS theorists have turned some attention and empirical study to the severe psychopathologies such as schizophrenia (Hornero, Alonso, N. Jimeno, A. Jimeno and L_opez 1999) bipolar disorder (Abraham et al. 1990; Scott 1985; Thomasson, Pezard, Allilaire, Renault and Martinerie, in press), and psychopathic deviance (Lange 1999). Others have considered the stability (Clair 1998; Guastello 1982; Byrne, Mazanov and Gregson, in press) and symptom complexity (Tschacher, Scheier and Grawe 1998) of outpatient treatments.

In the final category, entire families may present themselves for therapy. The family is now conceptualized as a complex system, and much can be learned by studying the temporal patterns of interactions among family members (Butz, Chamberlain and McCown 1996; Koopmans 1998; Pincus, in press).

Summary

The word "conclusions" is not a good choice in this context, but a few summary remarks are warranted based on the history and inventory of nonlinear dynamics in psychology presented here. The days of "Here's Chaos!" have been gone for quite some time. The value of NDS concepts and methods for explaining phenomena is being actively explored and tested. All the main areas of psychology have been visited by NDS theory to some extent. A promising development is that NDS can address phenomena that were observed or conceptualized in past theory, but for which a set of modeling concepts and methods was not yet available. What results is an integration of NDS and basic principles in each percolation of psychological theory.

In the cognitive area, it is possible that NDS applications will be instrumental in uniting several aspects of cognition into a comprehensive theory or model of mental processes. In the social and organizational area, the possible outcomes are more diffuse, but there is at least a core of thinking that purports to describe how two people can become connected, then more people become connect to form a networks, cohesive social groups, and groups with differentiated internal structures. In the clinical applications, there is an emphasis on understanding the whole person, and that such an understanding is critical to development of mentally healthy situations.

There are also implications for professional practice in organizations or clinical treatment. The state of practical application is expected to evolve in concert with the theory and methodologies.

References

1. Abraham, F.D. (1995) Dynamics, bifurcation, self-organization, chaos, mind, conflict, insensitivity to initial conditions, time, unification, diversity, free will, and social responsibility. In: Robertson, R. and Combs, A. (Eds.), Chaos theory in psychology and life sciences (pp. 155–173). Mahwah, NJ: Lawrence Erlbaum Associates.
2. Abraham, F.D., Abraham, R.H. and Shaw, C.D. (1990) A visual introduction to dynamical systems theory for psychology. Santa Cruz: Ariel.
3. Abraham, F. D. and Gilgen, A. R. (Eds.) (1995) Chaos theory in psychology. Westport, CT: Praeger.
4. Alisch, L.-M., Azizighanbari, S. and Bargfeldt, M. (1997) Dynamics of children's friendships. In: Eve, R.A., Horsfall, S. and Lee, M.E. (Eds.), Chaos, complexity, and sociology (pp. 163–181). Thousand Oaks, CA: Sage.
5. Anderson, P., Meyer, A., Eisenhardt, K., Carley, K. and Pettigrew, A. (1999) Introduction to the special issue: Application of complexity theory to organization science. Organization Science, 10, 233–236.
6. Basar, E. (1998) Brain oscillations I. New York: Springer-Verlag.
7. Basar, E. (1999) Brain oscillations II. New York: Springer-Verlag.
8. Brookhart, S. (2000) Psychology on campus: Degrees of separation. APS Observer, 13(2), 1, 6–9, 26.
9. Butz, M.R. (1998) Chaos and complexity: Implications for psychological theory and practice. Philadelphia, PA: Taylor & Francis.
10. Butz, M.R., Chamberlain, L. L. and McCown, W. G. (1996) Strange attractors: Chaos, complexity, and the art of family therapy. New York: Wiley.
11. Byrne, D.G., Mazanov, J. and Gregson, R.A.M. (in press). A cusp catastrophe analysis of changes to adolescent smoking behavior in response to smoking prevention programs. Nonlinear Dynamics, Psychology, and Life Sciences, 5.

12. Chamberlain, L.L. and Butz, M.R. (Eds.) (1998) Clinical chaos: A therapist's guide to nonlinear dynamics and therapeutic change. Philadelphia, PA: Brunner/Mazel.

13. Clair, S. (1998) A cusp catastrophe model for adolescent alcohol use: An empirical test. Nonlinear Dynamics, Psychology, and Life Sciences, 2, 217–242.

14. Clayton, K. and Frey, B.B. (1996) Inter- and intra-trial dynamics in memory and choice. In: Sulis, W. and Combs, A. (Eds.), Nonlinear dynamics in human behavior (pp. 90–106). Singapore: World Scientific.

15. Clayton, K. and Frey, B.B. (1997) Studies in mental "noise". Nonlinear Dynamics, Psychology, and Life Sciences, 1, 174–180.

16. Cobb, L. and Ragade, R.K. (Eds.) (1978) Applications of catastrophe theory in the behavioral and life sciences. Behavioral Science, 23, 291–419.

17. Combs, A. (1995) Psychology, chaos, and the process nature of consciousness. In: Abraham, F. D. and Gilgen, A.R. (Eds.), Chaos theory in psychology (pp. 129–137). Westport, CT: Praeger.

18. Combs, A. and Holland, M. (1996) Synchronicity: Science, myth, and the trickster. New York: Marlowe & Co. Csanyi, V. (1989) Evolutionary systems and society: A general theory of life, mind, and culture. Durham, NC: Duke University Press.

19. Davids, K., Button, C. and Bennett, S. (1999) Modeling human motor systems in nonlinear dynamics: Intentionality and discrete movement behaviors. Nonlinear Dynamics, Psychology, and Life Sciences, 3, 3–30.

20. Dooley, K.J. (1997) A complex adaptive system model of organization change. Nonlinear Dynamics, Psychology, and Life Sciences, 1, 69–97.

21. Eoyang, G.H. (1997) Coping with chaos: Seven simple tools. Cheyenne, WY: Lagumo.

22. Feldman, D.H., Csikszentmihalyi, M. and Gardner, H. (1994) Changing the world: A framework for the study of creativity. Westport, CT: Praeger.

23. Ferro Milone, F., Minelli, T.A. and Turicchia, L. (1998) Neuron synchronization and human EEG phenomenology. Nonlinear Dynamics, Psychology, and Life Sciences, 2, 21–34.

24. Freeman, W. (1995) Society of brains. Mahwah, NJ: Lawrence Erlbaum Associates.

25. Gilgen, A.R. (1995) A search for bifurcations in the psychological domain. In: Abraham, F. D. and Gilgen, A.R. (Eds.), Chaos theory in psychology (pp. 139–144) Westport, CT: Praeger.

26. Goertzel, B. (1993a) The evolving mind. New York: Gordon & Breach.

27. Goertzel, B. (1993b) The structure of intelligence: New mathematical model of mind. New York: Springer-Verlag.

28. Goertzel, B. (1994) Chaotic logic: Language, thought and reality from the perspective of complex systems science. New York: Plenum.

29. Goertzel, B. (1997) From complexity to creativity: Explorations in evolutionary, autopoetic, and cognitive dynamics. New York: Plenum.

30. Goldstein, J. (1994) The unshackled organization. Portland, OR: Productivity Press.

31. Goldstein, J. (1995) Unbalancing psychoanalytic theory: Moving beyond the equilibrium model of Freud's thought. In: Robertson, R. and Combs, A. (Eds.). Chaos theory in psychology and life sciences (pp. 239–252) Mahwah, NJ: Erlbaum.

32. Goldstein, J. (1997) Embracing the random in the self-organizing psyche. Nonlinear Dynamics, Psychology, and Life Sciences, 1, 181–202.

33. Gregson, R.A.M. (1992) n-Dimensional nonlinear psychophysics. Hillsdale NJ: Erlbaum.

34. Gregson, R. A.M. (1995) Cascades and fields in nonlinear psychophysics. Singapore: World Scientific.

35. Gregson, R.A.M. (1998) E€ects of random noise and internal delay in nonlinear psychophysics. Nonlinear Dynamics, Psychology, and Life Sciences, 2, 73–94.

36. Gregson, R.A.M. (1999) Confidence judgments for discrimination in nonlinear psychophysics. Nonlinear Dynamics, Psychology, and Life Sciences, 3, 31–48.

37. Guastello, S.J. (1981) Catastrophe modeling of equity in organizations. Behavioral Science, 26, 63–74.

38. Guastello, S.J. (1982) Moderator regression and the cusp catastrophe: Application of two-stage personnel selection, training, therapy, and policy evaluation. Behavioral Science, 27, 259–272.

39. Guastello, S.J. (1987) A butterfly catastrophe model of motivation in organizations: Academic performance. Journal of Applied Psychology, 72, 165–182.
40. Guastello, S.J. (1995) Chaos catastrophe and human affairs. Mahwah, NJ: Erlbaum.
41. Guastello, S.J. (1998-a) Creative problem solving groups at the edge of chaos. Journal of Creative Behavior, 32, 38–57.
42. Guastello, S.J. (1998-b) Self-organization and leadership emergence. Nonlinear Dynamics, Psychology, and Life Sciences, 2, 303–316.
43. Guastello, S.J., Bock, B.R. Caldwell, P. & Bond, R. W., Jr. (2005) Origins of group coordination: Nonlinear dynamics and the role of verbalization. Nonlinear Dynamics, Psychology, and Life Sciences, 9, 175–208.
44. Guastello, S.J. and Guastello, D.D. (1998) Origins of coordination and team effectiveness: A perspective from game theory and nonlinear dynamics. Journal of Applied Psychology, 83, 423–437.
45. Guastello, S.J., Johnson, E.A. and Rieke, M.L. (1999) Nonlinear dynamics of motivational flow. Nonlinear Dynamics, Psychology, and Life Sciences, 3, 259–274.
46. Guillot, A. and Meyer, J.-A. (in press). Chaotic dynamics underlying action selection in mice. Nonlinear Dynamics, Psychology, and Life Sciences, 4.
47. Guindani, F.M. and Salvadori, G. (Eds.) (1998) Chaos, fractals, and models. Pavia, Italy: Italian University Press.
48. Haken, H. (1999) Synergetics and some applications to psychology. In: W. Tschacher and J.-P. Dauwalder (Eds.), Dynamics, synergetics, and autonomous agents (pp. 3–12). Singapore: World Scientific.
49. Hardy, C. (1998) Networks of meaning. Westport CT: Praeger.
50. Hornero, R., Alonso, A., Jimeno, N., Jimeno, A. and L_opez, M. (1999) Estimation of correlation dimension to evaluate cognitive performance in schizophrenia patents using a new computer technique. Nonlinear Dynamics, Psychology, and Life Sciences, 3, 49–64.
51. Kampis, G. (1991) Self-modifying systems in biology and cognitive science. Oxford, UK: Pergamon.
52. Kelso, J.A.S. (1995) Dynamic patterns: Self-organization of brain and behavior. Cambridge, MA: MIT Press.
53. Kiel, L.D. (1994) Managing chaos and complexity in government. San Francisco: Jossey-Bass.
54. Kohonen, T. (1989) Self-organization and associative memory. (3rd edn.). NY: Springer-Verlag.
55. Koopmans, M. (1998) Chaos theory and the problem of change in family systems. Nonlinear Dynamics, Psychology, and Life Sciences, 2, 133–148.
56. Lange, R. (1999) A cusp catastrophe approach to the prediction of temporal patterns in the kill dates of individual serial murderers. Nonlinear Dynamics, Psychology, and Life Sciences, 3, 143–160.
57. Latan_e, B. and Nowak, A. (1994) Attitudes as catastrophes: From dimensions to categories with increasing involvement. In: Vallacher, R. R., and Nowak, A. (Eds.), Dynamical systems in social psychology (pp. 219–249) New York: Academic Press.
58. MacCormac, E. and Stamenov, M.I. (1996) Fractals of brain, fractals of mind: In search of a symmetry bond. Philadelphia: Benjamins.
59. Marks-Tarlow, T. (1999) The self as a dynamical system. Nonlinear Dynamics, Psychology, and Life Sciences, 3, 311–346.
60. Massimini, F. and Delle Fave, A. (2000) Individual development in a bio-cultural perspective. American Psychologist, 55, 24–33.
61. McClure, B. (1998) Putting a new spin on groups. Mahwah, NJ: Lawrence Erlbaum Associates. McKelvey, B. (1999) Avoiding complexity catastrophe in coevolutionary pockets: Strategies for rugged landscapes. Organization Science, 10, 294–321.
62. Metzger, M.A. (1997) Applications of nonlinear dynamical systems theory in developmental psychology: Motor and cognitive development. Nonlinear Dynamics, Psychology, and Life Sciences, 1, 55–68.
63. Minelli, T.A. and Turicchia, L. (1999) Progressive coherence patterns for electroencephalographic phenomenology. Nonlinear Dynamics, Psychology, and Life Sciences, 3, 129–142.

64. Mosca, F. (1995) Freedom in chaos theory: A case for choice in a universe without a bottom line. In: Abraham, F.D., and Gilgen, A.R. (Eds.). Chaos theory in psychology (pp. 181–191). Westport, CT: Praeger.

65. Orsucci, F. (1998) The complex matters of the mind. Singapore: World Scientific.

66. Partidge, T. (in press). Temperament development modeled as a nonlinear complex adaptive system. Nonlinear Dynamics, Psychology, and Life Sciences, 4.

67. Pincus, D. (in press). A framework and methodology for the study of nonlinear self-organizing family systems. Nonlinear Dynamics, Psychology, and Life Sciences, 5.

68. Port, R.E. and van Gelder, T. (Eds.) (1995) Mind as motion. Cambridge, MA: MIT Press.

69. Porter, R.J. and Hogue, D.M. (1998) Nonlinear dynamical systems in speech perception and production. Nonlinear Dynamics, Psychology, and Life Sciences, 2, 95–132.

70. Pressing, J. (1999) Referential dynamics of cognition and action. Psychological Review, 106(4), 714–747.

71. Rinaldi, S. and Gragnani, A. (1998) Love dynamics between secure individuals: A modeling approach. Nonlinear Dynamics, Psychology, and Life Sciences, 2, 283–301.

72. Robertson, R. (1995) Jungian archetypes: Jung, Gödel, and the history of archetypes. York Beach, ME: Samuel Weiser.

73. Robertson, R. and Combs, A. (Eds.) (1995) Chaos theory in psychology and life sciences. Mahwah, NJ: Erlbaum.

74. Schiepek, G. (1999) Die grundlagen der systemischen therapie: Theorie, praxis, forschung. Gottingen: Vandenhoeck and Ruprecht.

75. Scott, D.W. (1985) Catastrophe theory applications in clinical psychology: A review. Current Directions in Psychological Research and Reviews, 4, 69–86.

76. Seger, C.A. (1994) Implicit learning. Psychological Bulletin, 115, 163–196.

77. Simonton, D.K. (1988) Creativity, leadership, and change. In: Sternberg, R.J. (Ed.), The nature of creativity: Contemporary psychological perspectives (pp. 386–426). Cambridge, MA: MIT Press.

78. Stein, A.H. (1999) Whose thoughts are they, anyway? Dimensionally exploding Bion's "double-headed arrow" into coadapting, transitional space. Nonlinear Dynamics, Psychology, and Life Sciences, 3, 65–92.

79. Stewart, I.N. and Peregoy, P.L. (1983) Catastrophe theory modeling in psychology. Psychological Bulletin, 94, 336–362.

80. Sulis, W. (1997) Fundamental concepts of collective intelligence. Nonlinear Dynamics, Psychology, and Life Sciences, 1, 35–54.

81. Sulis, W. (1999) A formal theory of collective intelligence. In: Tschacher W. and Dauwalder J.-P. (Eds.), Dynamics, synergetics, and autonomous agents (pp. 224–237) Singapore: World Scientific.

82. Sulis, W. and Combs, A. (Eds.) (1996) Nonlinear dynamics in human behavior. Singapore: World Scientific.

83. Ta'eed, L.K., Ta'eed, O. and Wright, J.E. (1988) Determinants involved in the perception of the Necker cubes: An application of catastrophe theory. Behavioral Science, 33, 97–115.

84. Thelen, E. and Smith, L. (1994) A dynamic systems approach to the development of cognition and action. Cambridge, MA: MIT Press.

85. Thietart, R.A. and Forgues, B. (1995) Chaos theory and organizations. Organization Science, 6, 19–31.

86. Thom, R. (1975) Structural stability and morphegenesis. New York: Benjamin-Addison-Wesley.

87. Thomasson, N., Pezard, L., Allilaire, J.-F., Renault, B. and Martinerie, J. (in press). Nonlinear EEG changes associated with clinical improvement in depressed patients. Nonlinear Dynamics, Psychology, and Life Sciences, 4.

88. Tschacher, W. and Dauwalder, J.-P. (1999) (Eds.), Dynamics, synergetics, and autonomous agents. Singapore: World Scientific.

89. Tschacher, W., Scheier, C. and Grawe, K. (1998) Order and pattern formation in psychotherapy. Nonlinear Dynamics, Psychology, and Life Sciences, 2, 195–216.

90. Turvey, M.T. and Carello, C. (1995) Some dynamical themes in cognition and action. In: Port, R. E., and van Gelder T. (Eds.), Mind as motion (pp. 373–401). Cambridge, MA: MIT Press.

91. Vallacher, R.R. and Nowak, A. (Eds.) (1994) Dynamical systems in social psychology. New York: Academic Press.

92. Vallacher, R. R. and Nowak, A. (1997) The emergence of dynamical social psychology. Psychological Inquiry, 8, 73–99.

93. Vickers, D. and Lee, M.D. (1998) Dynamic models of simple judgments: I. Properties of a self-regulating accumulator model. Nonlinear Dynamics, Psychology, and Life Sciences, 2, 169–194.

94. Vickers, D. and Lee, M.D. (2000) Dynamic models of simple judgments: II. Properties of a self-organizing PAGAN (parallel, adaptive, generalized accumulator network) model for multi-choice tasks. Nonlinear Dynamics, Psychology, and Life Sciences, 4, 1–32.

95. Ward, L.M. and West, R.L. (1998) Modeling human chaotic behavior: Nonlinear forecasting analysis of logistic iteration. Nonlinear Dynamics, Psychology, and Life Sciences, 2, 261–282.

96. Watters, P.A. (1999) Psychophysiology, cortical arousal, and dynamical complexity (DCx). Nonlinear Dynamics, Psychology, and Life Sciences, 3, 211–234.

97. Wheatley, M.J. (1992) Leadership and the new science. San Francisco: Berrett-Koehler.

98. Zaror, G. and Guastello, S.J. (2000) Self-organization and leadership emergence: A cross-cultural replication. Nonlinear Dynamics, Psychology, and Life Sciences, 4, 113–119.

99. Zeeman, E.C. (1977) Catastrophe theory: Selected papers 1972–1977. Reading, MA: Addison-Wesley.

100. Zimmerman, B., Lindberg, C. and Plsek, P. (1998) Edgeware: Insights from complexity science for health care leaders. Irving, TX: VRH, Inc.

Section VII

Models and Applications of Chaos Theory in Mechanical Sciences

1. **Chemical Reactivity Dynamics and Quantum Chaos in Highly Excited Hydrogen Atoms in an External Field: A Quantum Potential Approach**   547
   P.K. Chattaraj and *B. Maiti*

2. **Regular and Chaotic Motion of a Bush-Shaft System with Tribological Processes**   569
   J. Awrejcewicz and *Y. Pyryev*

3. **Relaxed Plasma Equilibria and Entropy-Related Plasma Self-Organization Principles**   582
   R.L. Dewar, M.J. Hole, M. MeGann, R. Mills and *S.R. Hodson*

4. **Generalized Complexity and Classical-Quantum Transition**   595
   A.M. Kowalski, A. Plastino and *M. Casas*

5. **A New Mechanical Model for Particle Transport by Surface Waves and Applications**   609
   M. Ragulskis, E. Sakyte, J.M. Seoane and *M.A.F. Sanjuán*

6. **Chaotic Behavior of the Biharmonic Dynamics System**   627
   V.S. Aslanov

7. **Modeling, Chaotic Behavior and Control of Dissipation Properties of Hysteretic Systems**   645
   J. Awrejcewicz and *L. Dzyubak*

8. **Nonlinear Dynamics and Chaos of Microcantilever-Based TM-AFMs with Squeeze Film Damping Effects**   667
   W.M. Zhang, G. Meng, J.B. Zhou and *J.Y. Chen*

9. **Mathematical Identification of Homogenisation Processes in Argon Stirred Ladle**   689
   K. Michalek and *K. Gryc*

10. **Analysis of a Nonlinear Aeroelastic System with Parametric**   697
    Uncertainties Using Polynomial Chaos Expansion
    A. Desai and *S. Sarkar*

11. **Chaos Synchronization Criteria and Costs of Sinusoidally**   717
    Coupled Horizontal Patform Systems
    J. Cai, X. Wu and *S. Chen*

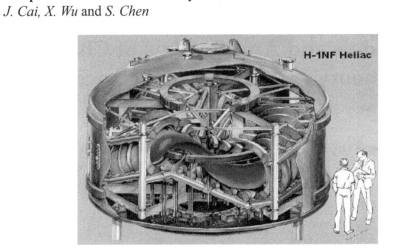

Chemical Reactivity Dynamics and Quantum Chaos in Highly Excited Hydrogen Atoms in an External Field: A Quantum Potential Approach[†]

P.K. Chattaraj[1,a] and *B. Maiti*[1]

[1]Department of Chemistry, Indian Institute of Technology, Kharagpur, 721302, India.
[a]Email: pkc@chem.iitkgp.ernet.in.

ABSTRACT

Dynamical behavior of chemical reactivity indices like electronegativity, hardness, polarizability, electrophilicity and nucleophilicity indices is studied within a quantum fluid density functional framework for the interactions of a hydrogen atom in its ground electronic state ($n=1$) and an excited electronic state ($n=20$) with monochromatic and bichromatic laser pulses. Time dependent analogues of various electronic structure principles like the principles of electronegativity equalization, maximum hardness, minimum polarizability and maximum entropy have been found to be operative. Insights into the variation of intensities of the generated higher order harmonics on the color of the external laser field are obtained. The quantum signature of chaos in hydrogen atom has been studied using a quantum theory of motion and quantum fluid dynamics. A hydrogen atom in the electronic ground state ($n=1$) and in an excited electronic state ($n=20$) behaves differently when placed in external oscillating monochromatic and bichromatic electric fields. Temporal evolutions of Shannon entropy, quantum Lyapunov exponent and Kolmogorov – Sinai entropy defined in terms of the distance between two initially close Bohmian trajectories for these two cases show marked differences. It appears that a larger uncertainty product and a smaller hardness value signal a chaotic behavior.

Keywords: Chemical reactivity, Electronegativity, Hardness, Hydrogen atom, Quantum theory of motion (QTM), Quantum chaos, Quantum fluid dynamics (QFD), Chaotic dynamics.

[†]Reused with permission from: P. K. Chattaraj and B. Maiti, Chemical Reactivity Dynamics and Quantum Chaos in Highly Excited Hydrogen Atoms in an External Field: A Quantum Potential Approach, *International Journal of Molecular Sciences*, 2002, *3*, 338–359.

Introduction

The chaotic ionization of hydrogen atoms [1–3] in highly excited states by microwave fields has become an important area of research for both experimentalists [1–7] and theoreticians [4]. In 1974 Bayfield and Koch [8] first studied the chaotic ionization of hydrogen atoms which has been considered to be very important in atomic theory [1,2,4,5,9–28]. Sanders and Jensen [4] have studied the chaotic ionization of hydrogen and helium using classical mechanics [4]. When the hydrogen atom is promoted to a highly excited state it gets ionized in case the field intensity is above some threshold value and the ionization probability depends on the field intensity [4,6,7]. Standard diagnostics used for the present study include electronegativity (χ), hardness (η), polarizability (α), phase-volume (Vps), electrophilicity index (ω), nucleophilicity index ($\dfrac{1}{\omega}$), Shannon entropy (S), quantum Lyapunov exponent (Λ) and Kolmogorov-Sinai entropy (H) defined in terms of the distance between two initially close Bohmian trajectories. In this paper we have generated the higher-order harmonics [3,29,30]. The response of the atom when it interacts with the external field vis–á–vis the variation of its reactivity is an important area of research. Electornegativity (χ) [31] and hardness (η) [32] are two cardinal indices of chemical reactivity. Pauling [33] introduced the concept of electronegativity as the power of an atom in a molecule to attract electrons to itself. The concept of hardness was given by Pearson [34] in his hard-soft acid-base (HSAB) principle which states that, "hard likes hard and soft likes soft". These popular qualitative chemical reactivity concepts have been quantified in density functional theory (DFT) [35]. Another important hardness-related principle is the maximum hardness principle (MHP) [36,37], which states that, " there seems to be a rule of nature that molecules arrange themselves so as to be as hard as possible". The quantitative definitions for electronegativity [38] and hardness [39] for an N-electron system with total energy E can respectively be given as

$$\chi = -\mu = -\left(\frac{\partial E}{\partial N}\right)_{v(F)} \tag{1}$$

and

$$\eta = \frac{1}{2}\left(\frac{\partial^2 E}{\partial N^2}\right)_{v(F)} = \frac{1}{2}\left(\frac{\partial \mu}{\partial N}\right)_{v(F)}. \tag{2}$$

In eqs.(1) and (2) μ and $v\left(\vec{r}\right)$ are chemical potential (Lagrange multiplier associated with the normalization constraint of DFT [34,36]) and external potential respectively. An equivalent expression [40,41] for hardness is

$$\eta = \frac{1}{N}\iint \eta\left(\vec{r},\vec{r}'\right) f\left(\vec{r}'\right) \rho\left(\vec{r}\right) d\vec{r}\, d\vec{r}' \tag{3}$$

where $f\left(\overrightarrow{r}\right)$ is the Fukui function [40] and $\eta\left(\overrightarrow{r},\overrightarrow{r}'\right)$ is the hardness kernel given by [40]

$$\eta = \frac{1}{N}\iint \eta\left(\overrightarrow{r},\overrightarrow{r}'\right) f\left(\overrightarrow{r}'\right) \rho\left(\overrightarrow{r}\right) d\overrightarrow{r}\, d\overrightarrow{r}'$$

$$\eta\left(\overrightarrow{r},\overrightarrow{r}'\right) = \frac{1}{2}\frac{\delta^2 F[\rho]}{\delta\rho\left(\overrightarrow{r}\right)\delta\rho\left(\overrightarrow{r}'\right)} \tag{4}$$

where $F(\rho)$ is the Hohenberg-Kohn universal functional of DFT [35].

The complete characterization of an N-particle system acted on by an external potential $v\left(\overrightarrow{r}\right)$ requires only N and $v\left(\overrightarrow{r}\right)$. The response of the system subjected to a change in N at fixed $v\left(\overrightarrow{r}\right)$ is given by χ and η while the linear response function [34] measures the response of the system when $v\left(\overrightarrow{r}\right)$ is varied at constant N. If the system is kept under the influence of the weak electric field, polarizability (α) takes care of the corresponding response. During molecule formation the electronegativities of the pertinent atoms get equalized [42,43]. A stable configuration or a favorable process is generally associated with maximum hardness [36,37], minimum polarizability [44–47] and maximum entropy [48] values. The conditions for maximum hardness and entropy and minimum polarizability complement the usual minimum energy criterion for stability.

Recently Parr et al. [49] have defined the electrophilicity index (W) as

$$W = \frac{\mu^2}{2\eta} \tag{5}$$

We also study the behavior of (*1/W*), a valid candidate for the nucleophilicity index. Note that the quantity (*1-W*) will also serve the purpose of a nucleophilicity index. It has also been shown recently [50] that the uncertainty product or the phase space volume (Vps) is a measure of quantum fluctuations and hence has bearing in the studies of quantum domain behavior of classically chaotic systems.

It has been already demonstrated [51] that in case we focus our attention to a specific atom/molecule taking part in a chemical reaction the whole procedure can be simulated by the interaction of an atom/molecule with an external field of the strength of the order of the "chemical reaction field". A molecular reaction dynamics can be envisaged [44] by monitoring the time evolution of the electronegativity of a specific atom from its isolated atom value to the equalized molecular electronegativity value as well as by studying the dynamic profiles of hardness and entropy and how they get maximized and that of the minimization of polarizability during the course of the chemical reaction. In the present work we study the interaction of a hydrogen atom in its ground electronic state and an excited electronic state with laser fields of different colors. The effect of the frequency of the external laser field on the overall reactivity of the atom in its various electronic states vis–á–vis the validity of the associated electronic structure principles in a dynamical context as well as the intensities of the generated higher order harmonics [52] would be understood in this study.

Dynamics of these reactivity parameters (η and α) have been studied [44,46,53] in the contexts of various time dependent processes. Whether η and α can provide some insight into the quantum domain behavior of a classically chaotic system is yet to be analyzed. Hydrogen atoms and molecules in an oscillating electric field have been considered to be "veritable gold mines for exploring the quantum aspects of chaos" [54]. Depending on the frequency and the field intensity, hydrogen [54,55] atoms in the presence of an external field have been shown to exhibit regular/chaotic dynamics. Both quantum fluid dynamics (QFD) [56,57] and quantum theory of motion (QTM) [58,59] have provided quantum signatures of chaos in hydrogen atoms. In QFD [56] the overall motion of the system under consideration is mapped onto that of a "probability fluid" having density $\rho\left(\vec{r},t\right)$ and current density $j\left(\vec{r},t\right)$ under the influence of the external classical potential augmented by a quantum potential [55–59] and $\rho\left(\vec{r},t\right)$ and $\chi\left(\vec{r},t\right)$ ($j=\rho\nabla\chi$) are respectively obtained [55–59] from the amplitude and the phase of the wave function. In QTM [58], the wave motion is governed by the solution to the time dependent Schrödinger equation (TDSE) and the particle motion is followed by solving the pertinent Newton's equation of motion with forces originating from both classical and quantum potentials. Important insight into the chaotic dynamics has been obtained [57] through ρ vs $-\chi$ plots which can be considered to be "canonically conjugate". In QTM it is obtained [59] in terms of the distance between two initially close Bohmian trajectories and the associated Kolmogorov—Sinai entropy.

In the present paper we monitor the possible regular/chaotic dynamics through the time evolution of various reactivity indices of a hydrogen atom in the ground and highly excited electronic states in the presence of one-color and two-color laser pulses. The theoretical background of the present work is provided in section II. Section III presents the numerical details, and the results and discussions are given in section IV. Finally, section V contains some concluding remarks.

Theoretical Background

Classical interpretation of quantum mechanics is as old as the quantum mechanics itself. In the Madelung representation [55] the time-dependent Schrödinger equation for a single particle moving under potential $V\left(\vec{r}\right)$ (in au), viz.

$$\left[-\frac{1}{2}\nabla^2 + V\left(\vec{r}\right)\right]\psi\left(\vec{r},t\right) = i\frac{\partial\psi\left(\vec{r},t\right)}{\partial t}, i=\sqrt{-1} \tag{6}$$

is transformed into two fluid dynamical equations. Substituting the following polar form of the wave function

$$\psi\left(\vec{r},t\right) = \rho^{1/2}\left(\vec{r},t\right)\exp\left(i\chi\left(\vec{r},t\right)\right) \tag{7}$$

in eq.(6) and separating the real and the imaginary parts, one obtains an equation of continuity

$$\frac{\partial\rho}{\partial t} + \nabla.j = 0 \tag{8a}$$

and an Euler-type equation of motion

$$\frac{\partial v}{\partial t} + \left(v.\nabla\right)v = -\nabla\left(V + V_{qu}\right). \tag{8b}$$

In eqs (8) the charge density, $\rho\left(\vec{r},t\right)$ and current density, $j\left(\vec{r},t\right)$ is

$$j\left(\vec{r},t\right) = \rho\left(\vec{r},t\right)v\left(\vec{r},t\right) \tag{9a}$$

where the velocity $v\left(\vec{r},t\right)$ can be defined in terms of the phase of the wave function as

$$v(\vec{r},t) = \left(\dot{\vec{r}}\right) = \nabla\chi(\vec{r},t) \tag{9b}$$

The quantity V_{qu} appearing in eq.(8b) is called the quantum potential or Bohm potential of hidden variable theory [60] and defined as

$$V_{qu} = -\frac{1}{2}\frac{\nabla^2\rho^{1/2}}{\rho^{1/2}} \tag{9c}$$

Therefore, in this quantum fluid dynamics [55] the overall motion of the system under consideration can be thought of as a motion of a "probability fluid" having density $\rho\left(\vec{r},t\right)$ and velocity $v\left(\vec{r},t\right)$ under the influence of the external classical potential augmented by a quantum potential, Vqu. For the ground state of a many-particle system, $\rho\left(\vec{r},t\right)$ contains all information [35]. In a time-dependent situation also the time-dependent density functional theory [52] asserts that any physical observable can be expressed as a functional of $\rho\left(\vec{r},t\right)$ and $j\left(\vec{r},t\right)$ and thus allows us to formulate the dynamics in terms of "classical-like" 3D quantities. Although Madelung transformation in terms of $\rho\left(\vec{r},t\right)$ and $j\left(\vec{r},t\right)$ is not straightforward in a many particle situation, we can make use of the time dependent density functional theory in constructing two fluid-dynamical equations in 3D-space. The formalism is termed as quantum fluid density functional theory [61] which has been applied in understanding ion-atom collisions [61-63], atom-field interactions [64,65] and electronegativity [51,66], hardness [66-68] and entropy dynamics [68] in a chemical reaction. Quantum potential plays a crucial role in the quantum theory of motion [58] as well. In this representation of quantum mechanics developed by de Broglie [69] and Bohm [70], the overall motion of the system is understood in terms of the motion of a particle experiencing forces originating from the classical and quantum potentials. The Newton's equation of motion for this particle guided by a wave (represented by $\psi\left(\vec{r},t\right)$, a solution to eq.(6)) can be written as

$$\left(\frac{\partial}{\partial t}+\dot{\vec{r}}.\nabla\right)\left(\ddot{\vec{r}}\right)=-\nabla\left(V+V_{qu}\right)\Big|_{\vec{r}=\vec{r}(t)} \tag{9d}$$

At a particular instant the solution to the time dependent Schrödinger equation (6) fixes the velocity of the particle (cf. eq.9b) and, hence, for a given initial position the particle motion can be studied through the solution $\vec{r}(t)$ to the eq.(9b).

Theories based on quantum potential idea have been applied in solving various physico-chemical problems [58,71–83]. Because of the presence of nonlinearity and also the "classical language", these theories have been found [57–59,81–85] to be helpful in understanding the quantum domain behavior of classically chaotic systems which is described as quantum chaology by Berry [86]. The quantum theory of motion, however, allows one to study the quantum chaos in a system without any resort to its classical domain dynamics [58]. The time-dependent Schrödinger equation (in a u.) for the present problem is

$$\left[-\frac{1}{2}\nabla^2+V\left(\vec{r}\right)\right]\psi\left(\vec{r},t\right)=i\frac{\partial\psi\left(\vec{r},t\right)}{\partial t} \tag{10a}$$

where the potential $V\left(\vec{r},t\right)$ is given by

$$V\left(\vec{r},t\right)=-\frac{1}{r}+v_{ext}\left(\vec{r},t\right). \tag{10b}$$

In eq.(10b) the external potential for the monochromatic and bichromatic laser pulses may be written as

$$v_{ext}\left(\vec{r},t\right)=\varepsilon_1 z, \textit{ for monochromatic pulse} \tag{10c}$$

$$=\varepsilon_2 z, \textit{ for bichromatic pulse} \tag{10d}$$

where

$$\varepsilon_1=\varepsilon\,\cos\!\left(\omega_0 t\right) \tag{10e}$$

and

$$\varepsilon_2 = 0.5\varepsilon\left[\cos(\omega_0 t) + \cos(\omega_1 t)\right] \tag{10f}$$

To have slow oscillations during and after the source being switched on, ε is written in terms of the maximum amplitude ε_0 and the switch-on time t' as

$$\varepsilon = \varepsilon_0 t / t' \qquad for\ 0 \le t \le t' \tag{10g}$$

$$= \varepsilon_0 \qquad\qquad otherwise. \tag{10h}$$

It may be noted that for a many-electron problem one may either solve the associated TDSE or the corresponding generalized nonlinear Schrödinger equation within a quantum fluid density functional framework [46,53,55,63–66,69,89], the latter being three dimensional even in the case of a many-electron system. To construct the hardness kernel (eq.(4)), we need the Hohenberg-Kohn universal functional $F[\rho]$. For a many-electron system $F[\rho]$ may be taken as [53]

$$F[\rho] = \frac{1}{2}\int\rho\left(\vec{r},t\right)\left|\nabla\chi\left(\vec{r},t\right)\right|^2 d\vec{r} + T[\rho] + \frac{1}{2}\int\int\frac{\rho\left(\vec{r},t\right)\rho\left(\vec{r}',t\right)}{\left|\vec{r}-\vec{r}'\right|}d\vec{r}\,d\vec{r}' + E_{xc}[\rho] \tag{11a}$$

where the first term is the macroscopic kinetic energy, the last term is the exchange-correlation energy, and $T[\rho]$ is the intrinsic kinetic energy given by [53]

$$T[\rho] = T_0[\rho] + T_w[\rho] - a(N)\lambda\int\frac{\rho^{4/3}/\vec{r}}{1+\frac{\vec{r}\rho^{1/3}}{0.043}}d\vec{r} \tag{11b}$$

where $T_0[\rho]$ is the Thomas-Fermi functional [88], $T_w[\rho]$ is the Weizsäcker functional [88], λ is a constant [53], $a(N)$ is an N-dependent parameter [53].

For obtaining the global hardness η (eq. 3) we also require the Fukui function $f\left(\vec{r}\right)$. We employ the following local formula for $f\left(\vec{r}\right)$,

$$f\left(\vec{r}\right) = \frac{s\left(\vec{r}\right)}{\int s\left(\vec{r}\right)d\vec{r}} \tag{11c}$$

where the local softness $s\left(\vec{r}\right)$ is given as follows as prescribed by Fuentealba [89]

$$s\left(\vec{r}\right) = \frac{\delta\left(\vec{r}-\vec{r}'\right)}{2\eta\left(\vec{r},\vec{r}'\right)}. \tag{11d}$$

For calculating $\eta\left(\overrightarrow{r},\overrightarrow{r}'\right)$ of the above equation the following local form for $F[\rho]$ is used [53]:

$$F[\rho] = T^{local}[\rho] + V_{ee}^{local}[\rho] \tag{11e}$$

where the local kinetic energy [90] and the electron-electron repulsion energy [91] may be taken as [53]

$$T^{local}[\rho] = T_0[\rho] + \frac{3}{4\pi}\left(3\pi^2\right)^{1/2}\int\frac{\rho^{4/3}/\overrightarrow{r}}{1+\frac{\overrightarrow{r}\rho^{1/3}}{0.043}}d\overrightarrow{r} \tag{11f}$$

and

$$V_{ee}^{local}[\rho] = 0.7937\left(N-1\right)^{2/3}\int\rho^{4/3}d\overrightarrow{r}. \tag{11g}$$

Note that the above treatment is applicable to many-electron systems and all electron-electron interaction terms would be absent in the case of a hydrogen atom.

To follow the polarizability dynamics the dynamic polarizability is defined as [44,53]

$$\alpha(t) = \left|\mathbf{D}_{ind}^z(t)\right|/\Im_z(t) \tag{12a}$$

where $\mathbf{D}_{ind}^z(t)$ is the electronic part of the induced dipole moment given as

$$\mathbf{D}_{ind}^z(t) = \int z\rho\left(\overrightarrow{r},t\right)d\overrightarrow{r} \tag{12b}$$

and $\Im_z(t)$ is the z-component of the external field.

The phase space volume or the uncertainty product Vps has been shown [92] to be an important diagnostic of the quantum signature of classical chaos [92] as related to the compactness of the electron cloud [93]. For the present problem it may be defined as

$$\mathbf{V}_{ps} = \left\{\left\langle\left(\mathbf{p}_{\tilde{\rho}}-\left\langle\mathbf{p}_{\tilde{\rho}}\right\rangle\right)^2\right\rangle\left\langle\left(\mathbf{p}_z-\left\langle\mathbf{p}_z\right\rangle\right)^2\right\rangle\left\langle\left(\tilde{\rho}-\left\langle\tilde{\rho}\right\rangle\right)^2\right\rangle\left\langle\left(z-\left\langle z\right\rangle\right)^2\right\rangle\right\}^{1/2}. \tag{13}$$

A sharp increase in $\mathbf{V}_{ps}(t)$ implies a chaotic motion [92] since it is a measure of the associated quantum fluctuations [92].

To generate the harmonic spectrum the induced dipole moment, $\mathbf{D}_{ind}^z(t)$ is Fourier transformed to obtained $d(\omega)$. It has been shown [94] that the absolute square of the Fourier transform, $|d(\omega)|^2$ is roughly proportional to the experimental harmonic distribution.

The Shannon entropy is given by

$$S = k\int\rho\ln(\rho)\,d\overrightarrow{r}, \tag{14}$$

where k is the Boltzmann constant.

We can generate the "quantum trajectory" of a particle for a given initial position from equation (9b). Now, we are in a position to analyze the sensitive dependence on initial condition, a characteristic of a chaotic system. Equation (9b) is solved with two dif ferent initial positions of the particle, $(\tilde{\rho}, z)$ and $(\tilde{\rho} + d\tilde{\rho}, z + dz), d\tilde{\rho} = 0, dz = 0.01)$. Initial momentum of the particle is taken as zero in all cases. We study the time evolution of phase space distance (D) for the corresponding quantum trajectories defined as [56,59,82,83]

$$D(t) = \left[\left(\tilde{\rho}_1(t) - \tilde{\rho}_2(t) \right)^2 + \left(z_1(t) - z_2(t) \right)^2 + \left(p_{\tilde{\rho}_1}(t) - p_{\tilde{\rho}_2}(t) \right)^2 + \left(p_{z_1}(t) - p_{z_2}(t) \right)^2 \right]^{1/2}, \quad (15a)$$

where $(\tilde{\rho}, p_{\tilde{\rho}}, z, p_z)$ refers to a point in phase space. We also calculate the associated Kolmogorov-Sinai entropy as defined [82,83] below

$$H = \sum_{\Lambda_+ > 0} \Lambda_+, \quad (15b)$$

where the Lyapunov exponent is given by [82,83]

$$\Lambda = \lim_{D(0) \to 0, t \to \alpha} \frac{1}{t} \ln \left[D(t) / D(0) \right] \quad (15c)$$

According to the Hamilton-Jacobi formulation of quantum mechanics, a positive *KS* entropy is associated with a chaotic quantum dynamics [59,87].

Numerical Solution

The TDSE (eq.10a) is solved numerically in cylindrical polar coordinates $(\tilde{\rho}, \tilde{\phi}, z)$ as an initial boundary value problem using an alternating direction implicit method [95]. The solution procedure begins with the ψ_{1s} and ψ_{20s} analytical wave functions of the hydrogen atom. Since the electron density varies rapidly near the nucleus and relatively slowly elsewhere, we transform the variables as follows

$$y = \tilde{\rho}\phi \quad (16a)$$

and

$$\tilde{\rho} = x^2. \quad (16b)$$

Eq.(10a) takes the following form in the transformed variables once an analytical integration is carried out over $0 \leq \tilde{\phi} \leq 2\pi$,

$$\left\{ \left(\frac{3}{4x^3} \right) \frac{\partial y}{\partial x} - \left(\frac{1}{4x^2} \right) \frac{\partial^2 y}{\partial x^2} - \frac{\partial^2 y}{\partial z^2} \right\} - \left(\frac{1}{x^4} - 2v_{eff} \right) y = 2i \frac{\partial y}{\partial t}. \quad (17)$$

The resulting tridiagonal matrix equation is solved using a Thomas algorithm. The mesh sizes adopted here are $\Delta x = \Delta z = 0.4 au$ and $\Delta t = 0.01 au$, ensuring the stability of the forward-time-central-space type numerical scheme adopted here.

The initial and boundary conditions associated with this problem are

$y(x, z)$ *is known for* $\forall\ x, z\ at\ t = 0$ \quad (18a)

$$y(0,z)=0=y(\infty,z)\forall\ z,t \tag{18b}$$

$$y(x,\pm\infty)=0\ \forall\ x,t. \tag{18c}$$

The numerical scheme is stable [96] due to the presence of $i=\sqrt{1}$. As a further check of the numerical accuracy, we have verified the conservation of norm and energy (in zero field cases). The wave function is moved forward to the end of the simulation and then taken back to its initial position by reversing the time direction, where the original profile is reproduced well within the tolerance limit of the present calculation. We have also solved eq. (9b) using a second order Runge-Kutta method to generate the "quantum trajectories" of a given initial position. The field parameters are in atomic units unless otherwise specified.

Results and Discussions

The time evolution of different reactivity parameters are depicted in Figs. 1–12. All quantities are in atomic units. Unless otherwise specified, in all figures a and b refer to the ground state ($n=1$) and excited state ($n=20$) of the hydrogen atom, respectively, and a red colored solid line and blue colored solid line respectively signify monochromatic and bichromatic pulses.

Figure 1 presents the time dependence of the external field with different frequencies and the same amplitude.

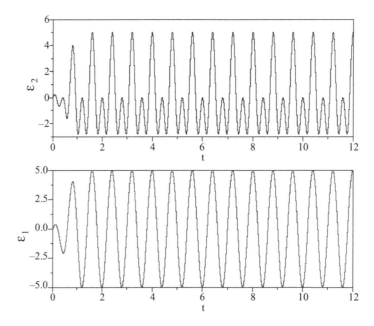

Figure 1 Time evolution of the external electric field: ε_1 (—) monochromatic pulse, ε_2 (—) bichromatic pulse. Field parameters: $\varepsilon_0=5.0$; $\omega_0=2.5\pi$, $\omega_1=2\omega_0$.

Temporal evolution of the chemical potential is depicted in Fig. 2. It exhibits characteristic oscillations. The oscillations in μ is not in phase with the external field. It is important to note that the amplitude of μ -oscillations becomes very large for both the electronic states and both monochromatic and bichromatic pulses.

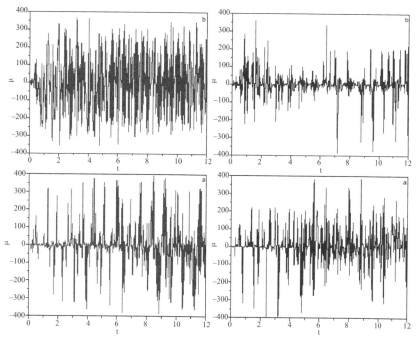

Figure 2 Time evolution of chemical potential (μ) when a hydrogen atom is subjected to external electric fields: a-Ground state; b-Excited state. (−) Monochromatic pulse, (−) bichromatic pulse. Field parameters: $\varepsilon_0 = 5.0$, $\omega_0 = 2.5\pi$, $\omega_1 = 2\omega_0$.

Chemical hardness (η) is presented in Fig. 3. For both one and two—color cases η is much larger for $n=1$ state than that of the $n=20$ state for the whole time range. This may be considered to be a dynamical variant of the MHP. Hardness oscillates in time in all the cases. However, the oscillation is neither in phase nor out of phase with respect to the oscillations in the external one-and two-color fields. It is expected because of the fact that as soon as the laser is switched on, there starts a tug-of-war between the atomic nucleus and the external field to govern the electron-density distribution.

The nucleus tries to make the density distribution spherically symmetric owing to the central nature of the nuclear coulomb field while the cylindrical symmetry of the applied electric field tries to create an oscillating dipole that emits radiation including higher harmonics. Overall density oscillation becomes nonlinear due to the interplay of two different types of effects. Hardness for the $n=1$ state decreases (for both one-and two-color situations) and attains a more or less steady value at the end of the simulation, which is still large in comparison to the corresponding value for the $n=20$ state. For both one-and two-color situations, η values relative to the corresponding values in absence of the field (not shown) are much larger for the *n=1* state. It appears that a relatively smaller η value signals a possible chaotic dynamics.

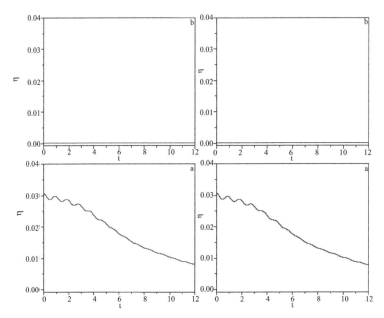

Figure 3 Time evolution of hardness (η) when a hydrogen atom is subjected to external electric fields:a-Ground state; b-Excited state. (–) Monochromatic pulse, (–) bichromatic pulse. Field parameters: $\varepsilon_0 = 5.0$, $\omega_0 = 2.5\pi$, $\omega_1 = 2\omega_0$.

Polarizability values as they evolve in the course of time are presented in Fig. 4. It oscillates with a frequency that is double that of the external field. The extrema in the external field corresponding to the minima in α and the latter blows up when the field is zero. Here also if we compare the respective minimum α values (α_{min}) for the two electronic states, α_{min} for the ground state is smaller than that of the excited state which is conspicuous for the bichromatic pulse. This is in conformity with minimum polarizability principle (MPP). The MPP reveals itself in a time-dependent situation.

Figure 5 depicts the dynamics of the uncertainty product (phase volume). As in the cases of μ and η, Vps also oscillates neither in phase nor out of phase with the external field. The magnitude of *Vps* retains its initial (*t=0*) small value for the *n=1* state whereas for the *n=20* state it increases quickly to a very large value. Since Vps measures the quantum fluctuations, a chaotic trajectory is generally associated with large *Vps* values [92]. "... large increases in *Vps* can be expected to accompany a chaotic trajectory. Conversely, small to moderate increases in *Vps* can be evidence that given quantum mechanical trajectory should be regarded a nonchaotic [92a]". In general, the electrons are "tightly bound" and hence the distribution is "less diffuse" for the *n=1* state and "loosely bound" for the *n=20* state and the system is expected to be harder and less polarizable for the ground state [32,34,46,53,88,93]. Again, the electron density being more compact in the ground state,

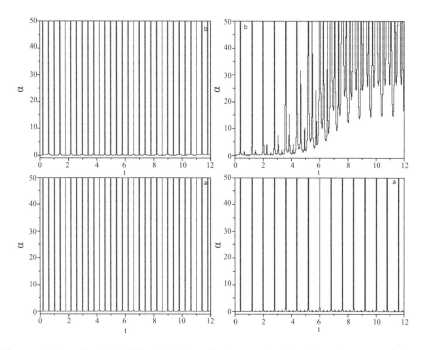

Figure 4 Time evolution of polarizability (α) when a hydrogen atom is subjected to external electric fields: a-Ground state; b-Excited state. (–) Monochromatic pulse, (–) bichromatic pulse. Field parameters: $\varepsilon_0 = 5.0$, $\omega_0 = 2.5\pi$, $\omega_1 = 2\omega_0$.

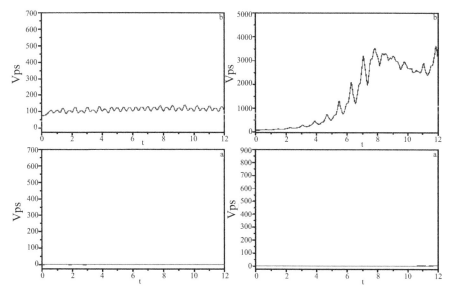

Figure 5 Time evolution of phase volume (*Vps*) when a hydrogen atom is subjected to external electric fields: a-Ground state; b-Excited state. (–) Monochromatic pulse, (–) bichromatic pulse. Field parameters: $\varepsilon_0 = 5.0$, $\omega_0 = 2.5\pi$, $\omega_1 = 2\omega_0$.

the corresponding uncertainty product is expected [93] to be small. Once the external field is switched on, the ground state density would be distributed over a larger volume and consequently there would be a decrease in η and increase in α and *Vps* of the system. Since a smaller η value is accompanied with a large *Vps* value and vice versa and Vps is known [92] to bear the signature of the classical chaos in the corresponding quantum domain behavior, hardness can as well be considered to be a diagnostic of the chaotic dynamics in a quantum system.

Figures 6 and 7 depict respectively the dynamical profiles of electrophilicity and nucleophilicity indices respectively. Both W and $1/W$ show oscillations characteristic of the resultant field of two competing ones for both the electronic states and for the one-and two-color pulses.

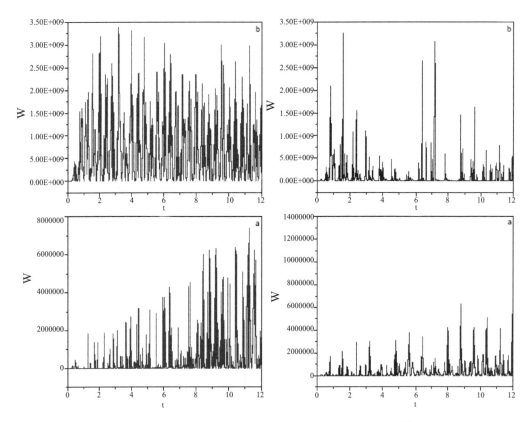

Figure 6 Time evolution of electrophilicity index (W) when a hydrogen atom is subjected to external electric fields: a-Ground state; b-Excited state. (–) Monochromatic pulse, (–) bichromatic pulse. Field parameters: $\varepsilon_0 = 5.0$, $\omega_0 = 2.5\pi$, $\omega_1 = 2\omega_0$.

The harmonic spectra are presented in Fig. 8. The overall domain of the spectra and their envelopes look like those reported by Erhard and Gross [52]. We found that the harmonics generated by the monochromatic and bichromatic pulses look similar and those generated from the former is less intense than those resulted from the latter [52].

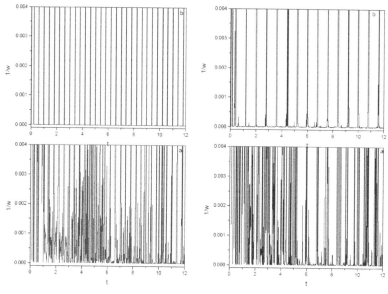

Figure 7 Time evolution of nucleophilicity index ($1/W$) when a hydrogen atom is subjected to external electric fields: a-Ground state; b-Excited state. (—) Monochromatic pulse, (—) bichromatic pulse. Field parameters: $\varepsilon_0 = 5.0$, $\omega_0 = 2.5\pi$, $\omega_1 = 2\omega_0$.

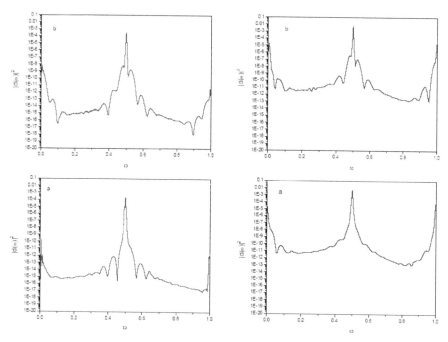

Figure 8 $|d\omega|^2$ vs ω plot when a hydrogen atom is subjected to external electric fields: a-Ground state; b-Excited state. (—) Monochromatic pulse, (—) bichromatic pulse. Field parameters: $\varepsilon_0 = 5.0$, $\omega_0 = 2.5\pi$, $\omega_1 = 2\omega_0$.

Figure 9 depicts the phase ($p_{\tilde{\rho}}, \tilde{\rho}$ and p_z vs z) of cases a and b for monochromatic laser pulse, The fraction of the total phase space visited by the Bohmian trajectories is much more for the excited state. These plots reflect that the case a is for regular motion whereas the case b is for chaotic motion.

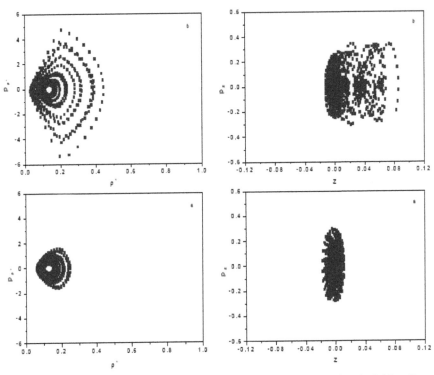

Figure 9 Phase space trajectories when a hydrogen atom is subjected to external electric field: a-Ground state; b-Excited state. (....) Monochromatic pulse. Field parameters: $\varepsilon_0 = 5.0$, $\omega_0 = 2.5\pi$, $\omega_1 = 2\omega_0$.

Figure 10 depicts the phase ($p_{\tilde{\rho}}, \tilde{\rho}$ and p_z vs z) plots case a and b for bichromatic laser pulse. These plots also reflect that the case a is for regular motion whereas the case b is for chaotic motion.

Figure 11 depicts the Kolmogorov-Sinai (KS) entropy for both ground ($n=1$) and excited ($n=20$) states for monochromatic and bichromatic laser pulses. For both monochromatic and bichromatic laser pulses the KS entropy (H) retains its initial very small value for $n=1$. For $n=20$ case H remains small initially and then increases rapidly to a high positive value. The small H value in the former case vis-á-vis the very large H value in the latter provides unmistakable signature of chaos in the highly excited state of the hydrogen atom in presence of an external electric field.

Shannon entropy has been shown in Fig. 12. In the figure a and b refer to the ground and $n=20$ states of the hydrogen atom respectively. It increases in the ground state and decreases in the excited state for both the laser pulses, a possible signature of the maximum

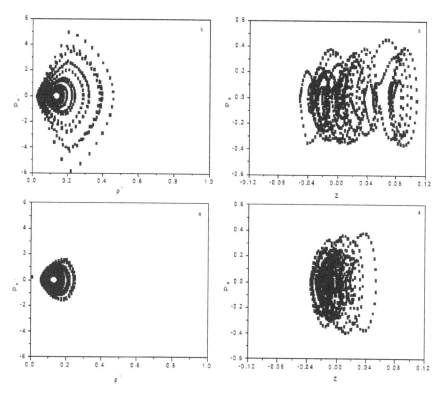

Figure 10 Phase space trajectories when a hydrogen atom is subjected to external electric field: a-Ground state; b-Excited state. (....) Bichromatic pulse. Field parameters: $\varepsilon_0 = 5.0$, $\omega_0 = 2.5\pi$, $\omega_1 = 2\omega_0$.

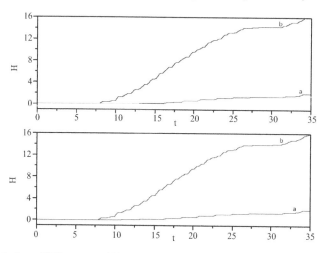

Figure 11 Time evolution of KS entropy (H) when a hydrogen atom is subjected to external electric fields: a-Ground state; b-Excited state. (—) Monochromatic pulse, (—) bichromatic pulse. Field parameters: $\varepsilon_0 = 5.0$, $\omega_0 = 2.5\pi$, $\omega_1 = 2\omega_0$.

entropy principle vis-à-vis chaotic ionization from the highly excited state. It is important to note that the calculations have been carried out up to 3500 au with no change in the qualitative trends. Plots are truncated at a much smaller time steps for easy visualization.

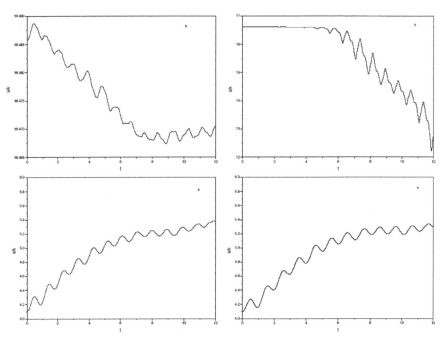

Figure 12 Time evolution of S/k, where S is the Shannon entropy and k is the Boltzmann constant when a hydrogen atom is subjected to external electric fields: a-Ground state; b-Excited state. (–) Monochromatic pulse, (–) bichromatic pulse. Field parameters: $\varepsilon_0 = 5.0$, $\omega_0 = 2.5\pi$, $\omega_1 = 2\omega_0$.

Concluding Remarks

Quantum potential based theories are adopted to study the reactivity dynamics and chaos of a hydrogen atom in its ground and excited electronic states interacting with z-polarized laser pulses of different colors. Dynamical variants of the principles of electronegativity equalization, maximum hardness, minimum polarizability and maximum entropy manifest themselves. A tug-of-war between the spherically symmetric nuclear coulomb field and cylindrically symmetric external electric field to govern the electron density distribution is delineated through the dynamical profiles of various reactivity indices like electronegativity, hardness, polarzability, electrophilicity, nucleophilicity and phase volume for the external field and in different electronic states. Harmonic spectra of the higher order harmonics included in the radiation emitted by the resulting oscillating dipole have been analyzed. Temporal evolution of Bohmian trajectory, KS entropy and Shannon entropy has easily differentiated the regular and chaotic behavior of hydrogen atom respectively in ground and excited states in presence of an oscillating electric field. For both the laser pulses the increase in the uncertainty product for the excited state is very large, which implies a possible chaotic dynamics. A large hardness value, on the other hand, is expected to characterize a regular behavior.

Acknowledgment

We thank CSIR, New Delhi for financial assistance and Mr. U. Sarkar for help in computation.

References and Notes

1. Jensen, R.V.; Susskind, S.M.; Sanders, M.M. *Phys. Rep.* **1991**, *201*, 1–56.
2. Koch, P. M.; van Leeuwen, K.A.H. *Phys. Rep.* **1995**, *255*, 289–403.
3. *Atoms in Intense Laser Fields;* Gavrila, M. Ed.; Academic Press: Boston, 1992.
4. Sanders, M.M.; Jensen, R.V.; *Am. J. Phys.* **1996**, *64*, 21–31; **1996**, *64*, 1013.
5. Mariani, D.R.; *The Ionisation of Hydrogen and Helium Atoms by Static and Microwave ionization of highly excited hydrogen atoms: Experimental and theory, in the physics of phase space,* edited by Y. S. Kim and W.W. Zachary; Springer, New York, 1987, pp. 106–113.
6. Mariani, D.R.; van de Water, W.; Koch, P.M.; Bergeman, T. *Phys. Rev. Lett.* **1983**, *50*, 1261.
7. van de Water, W.; Yoakum, S.; van Leeuwen, K. A. H.; Sauer, B. E.; Moorman, L.; Galvez, E. J.; Mariani, D. R.; Koch, P.M. *Phys. Rev. A.* **1990**, *42*, 573.
8. Bayfield, J.E.; Koch, P.M. *Phys. Rev. Lett.* **1974**, *33*, 258–261.
9. Koch, P.M.; van Leeuwen, K.A.H.; Rath, O.; Richards, D.; Jensen, R.V. *Microwave ionization of highly excited hydrogen atoms: Experiment and theory, in the physics of phase space,* edited by Y. S. Kim and W. W. Zachary; Springer, New York, 1987, pp. 06–113.
10. van Leeuwen, K.A.H.; Oppen, G.V.; Renwick, S.; Bowlin, J.B.; Koch, P.M.; Jensen, R.V.; Rath, O.; Richards, D,; Leopold, J.G. *Phys. Rev. Lett.* **1985**, *55*, 2231–2234.
11. Galvez, E.J.; Sauer, B.B.; Moorman, L.; Koch, P. M.; Richards, D. *Phys. Rev. Lett.* **1988**, *61*, 2011–2014.
12. Leopold, J.G.; Richards, D.J. *Phys. B At. Mol. Phys.* **1989**, *24*, 1209–1240.
13. Sanders, M.M. *Chaotic Ionisation of One and Two Electron Atom. Ph. D. thesis,* Yale University, New Haven, CT, 1991.
14. Bayfield, J.E. *Am. Sci.* **1983**, *71*, 375–383.
15. Leopold, J.G.; Percival, I.C. *J. Phys. B. At. Mol. Phys.* **1979**, *12*, 709–721.
16. Rath, O.; Richards, D. *J. Phys. B* (submitted).
17. Born, M. *The Mechanics of the Atom; Frederick Ungar,* New York, 1960.
18. Leopold, J.G. and Richards, D. *J. Phys. B. At. Mol. Phys.* **1986**, *19*, 1125.
19. Jensen, R.V. *Phys. Rev. Lett.* **1982**, *49*, 1365–1368.
20. Jensen, R.V. *Phys. Rev. A.* **1982**, *30*, 386–397.
21. Schwieters, C.D.; Delos, J.B. *Phys. Rev. A* **1995**, *51*, 1030–1041.
22. Shepelyansky, D.L. *Chaotic Behavior in Quantum System: Theory and Application; Plenum, New York,* 1985, pp. 187–197.
23. Lichtenberg, A.J.; Liebergman, M.A. *Regular and Stochastic Motion,* Springer, New York, 1983.
24. Casati, G.; Chirikov, B. V.; Shepelyansky, D.L. *Phys. Rev. Lett.* **1984**, *53*, 2525–2528.
25. Casati, G.; Guarneri, L.; Shepelyansky, D.L. *IEEE J. Quantum Electron.* **1988**, *QE–24*, 1420–1444.
26. Jensen, R.V. *Phys. Scr.* **1987**, *35*, 668–673.
27. Landau, L.D.; Lifshitz, E.M. *Mechanics, Course of Theoretical Physics*; Pergamon: New York, 1976; Vol. 1.; 3rd ed.
28. Goldstein, H. *Classical Mechanics. Addison—Wesley, Reading, MA,* **1980**, 2nd ed.
29. Mcphersion, A.; Gibson, G.; Jara, H.; Johann, U.; Luk, T.S.; McIntyre, I.A.; Boyer, K.; Rhodes, C.K. *J. Opt. Soc. Am.* **1987**, *B4*, 595.

30. *Super-Intense Laser-Atom Physics;* Piraux, B.; L'Huillier, A., Rzazewski, K. Eds.; NATO ASI Series B316; Plenum Press: New York, 1993.

31. *Electronegativity: Struct. Bonding;* Sen, K.D.; Jorgenson, C.K. Eds.; Springer–Verlag: Berlin, 1987, Vol. 66.

32. *Chemical Hardness: Struct. Bonding;* Sen, K.D; Mingos, D.M.P.; Eds.; Springer–Verlag: Berlin, 1987; Vol. 66.

33. Pauling, L. *The Nature of the Chemical Bond,* 3rd ed.; Cornell University Press: Ithaca, NY, 1960.

34. Pearson, R.G. *Coord. Chem. Rev.* **1990,** *100,* 403; *Hard and Soft Acids and Bases;* Dowden, Hutchinson and Ross: Stroudsberg, PA, 1973.

35. Hohenberg, P.; Kohn, W. *Phys. Rev. B* **1964,** *136,* 864; Kohn, W.; Sham, L.J. *Phys. Rev. A* **1965,** *140,* 1133; Parr, R.G.; Yang, W. *Density Functional Theory of Atoms and Molecules;* Oxford University Press; Oxford, 1989; *Annu. Rev. Phys. Chem.* **1995,** *46,* 701; Chattaraj, P.K. *J. Indian. Chem. Soc.* **1992,** *69,* 173; Kohn, W.; Becke, A.D.; Parr, R.G. *J. Phys. Chem.* **1996,** *100,* 12974.

36. Pearson, R. G. *J. Chem. Educ.* **1987,** *64,* 561; *Acc. Chem. Res.* **1993,** *26,* 250; *J. Chem. Educ.* **1999,** *76,* 267.

37. Parr, R.G.; Chattaraj, P.K. *J. Am. Chem. Soc.* **1991,** *113,* 1854; Chattaraj, P.K.; Liu, G.H.; Parr, R.G. *Chem. Phys. Lett.* **1995,** *237,* 171; Pearson, R. G. *Chemtracts Inorg. Chem.* **1991,** *3,* 317; Liu, S.; Parr, R. G. *J. Chem. Phys.* **1997,** *106,* 5578; Chattaraj, P. K. *Proc. Indian Natl. Sci. Acad. Part A* **1996,** *62,* 513; Pearson, R. G. *Chemical Hardness: Application from Molecules to Solid;* Wiley–VCH Verlag GMBH: Weinheim, 1997; Ayers, P. W.; Parr, R. G. *J. Am. Chem. Soc.* **2000,** *122,* 2010.

38. Parr, R.G.; Donnelly, D.A.; Levy, M.; Palke, W.E. *J. Chem. Phys.* **1978,** *68,* 3801.

39. Parr, R.G.; Pearson, R.G. *J. Am. Chem. Soc.* **1983,** *105,* 7512.

40. Berkowitz, M,; Ghosh, S. K.; Parr, R. G. *J. Am. Chem. Soc.* **1985,** *107,* 6811; Ghosh, S. K.; Berkowitz, M.*J. Chem. Phys.* **1985,** *83,* 2976.

41. Parr, R.G.; Yang, W. *J. Am. Chem. Soc.* **1984,** *106,* 4049.

42. Sanderson, R.T. *Science* **1951,** *114,* 670; *Science* **1955,** *121,* 207; *J. Chem. Educ.* **1954,** *31,* 238; Chattaraj, P.K.; Maiti, B.; *J. Chem. Educ.* **2001,** *78,* 811–813.

43. Politzer, P.; Weinstein, H. *J. Chem. Phys.* **1979,** *70,* 3680; Parr, R. G.; Bartolotti, L.J. *J. Am. Chem. Soc.* **1982,** *104,* 3081; Nalewajski, R.F.*J. Phys. Chem.* **1985,** *89,* 2831; Mortier, W.J.; Ghosh, S.K.; Shankar, S. *J. Am. Chem. Soc.* **1986,** *108,* 4315.

44. Chattaraj, P.K.; Sengupta, S.*J. Phys. Chem.* **1996,** *100,* 16126.

45. Ghanty, T.K.; Ghosh, S.K.*J. Phys. Chem.* **1996,** *100,* 12296.

46. Chattaraj, P.K.; Sengupta, S.*J. Phys. Chem. A* **1997,** *101,* 7893.

47. Chattaraj, P.K.; Poddar, A.*J. Phys. Chem. A* **1998,** *102,* 9944; **1999,** *103,* 1274; Chattaraj, P.K.; Fuentealba, P; Gomez, B.; Contreas, R. *J. Am. Chem. Soc.* **2000,** *122,* 348; Fuentealba P.; Simon– Manso, Y.; Chattaraj, P.K. *J. Phys. Chem. A* **2000,** *104,* 3185; Chattaraj, P.K.; Fuentealba, P.; Jaque, P.; Toro–Labbe, A. *J. Phys. Chem. A* **1999,** *103,* 9307; Chattaraj, P. K.; Maiti, B. *J. Phys. Chem. A* **2001,** *105,* 169–183.

48. Jaynes, E.T. *In Statistical Physics;* Ford, K.W., Ed.; Brandeis Lectures, Vol–3; Benjamin: New York, 1963. Levine, R.D.; Bernstein, R.B. In *Dynamics of Molecular Collisions; Miller, W.H., Ed.; Plenum Press: New York, 1976;* Gadre, S. R.; Bendale, R. D. *Curr. Sci.* **1985,** *54,* 970.

49. Parr, R.G.; Szentpaly, L. v.; Liu, S. *J. Am. Chem. Soc.* **1999,** *121,* 1922.

50. Chattaraj, P.K.; Sengupta, S.; *J. Phys. Chem. A* **1999,** *103,* 6122.

51. Chattaraj, P.K.; Nath, S. *Int. J. Quantum Chem.* **1994,** *49,* 705.

52. Runge, E.; Gross, E.K.U. *Phys. Rev. Lett.* **1984,** *52,*997; Dhara, A.K.; Ghosh, S.K. *Phys. Rev. A* **1987,** *35,* 442; Erhard, S.; Gross, E.K.U. *In. Multiphoton Processes* 1996; Lambropoulos, P.; Walther, H. Eds.; IOP Publishing: London, 1997; pp. 37–45 and reference therein.

53. Chattaraj, P.K.; Sengupta, S.; Poddar, A. *Int. J. Quantum Chem.* **1998,** *69,* 279; *In Nonlinear Dynamics and Computational Physics;* Sheorey, V. B. Ed.; Narosa: New Delhi, 1999; pp. 45 – 53.

54. Lakshmanan, M.; Ganeshan, K. *Curr. Sci.* **1995,** *68,* 38.

55. Casati, G.; Chirikov, B.V.; Guarneri, I; Shepelyansky, D.L. *Phys. Rep.* **1987,** *154,* 77; Hasegawa, H.; Robnik, M.; Wunner, G. *Prog. Theo. Phys. Suppl.* **1989,** *98,* 198; Ganeshan, K.; Lakshmanan, M. *Phys.*

Rev. A **1990**, *42,* 3940. Howard, J.E.; Farelly, D. *Phys. Lett. A* **1993**, *178,* 62; Delande, D.; Gay, J.C. *Phys. Rev. Lett.* **1987**, *59,* 1809; Delande, D. *Chaos and Quantum Physics;* Elsevier: Amsterdam, 1991; Friedrich, H.; Wintgen, D. *Phys. Rep.* **1989**, *183,* 37; Holle, A.; Weibusch, G.; Main, J.; Hager, B.; Rottke, H.; Welge, K. H. *Phys. Rev. Lett.* **1986**, *56,* 2594; Holle, A; Marini Rottke, H.; Welge, K.H. *Phys. Rev. Lett.* **1986**, *56,* 2594; Holle, A.; Marini, J.; Weibusch, G.; Rottke, H.; Welge, K.H. *Phys. Rev. Lett.* **1988**, *61,* 161.

56. Madelung, E.*Z. Phys.* **1926**, *40,* 322; Ghosh, S.K.; Deb, B. M. *Phys. Rep.* **1982**, *92,* 1–44.
57. Chattaraj, P.K.; Sengupta, S. *Phys. Lett. A* **1993**, *181,* 225; *Ind. J. Pure Appl. Phys.* **1996**, *34,* 518; Chattaraj, P.K. *Ind. J. Pure Appl. Phys.* **1994**, *32,* 101.
58. Holland, P.R. *The Quantum Theory of Motion;* Cambridge University Press: Cambridge, U.K., 1993.
59. Sengupta, S.; Chattaraj, P. K. *Phys. Lett. A* **1996**, *215,* 119; Chattaraj, P. K.; Sengupta, S. *Curr. Sci.* **1996**, *71,* 134; Chattaraj, P. K.; Sengupta, S. *J. Phys. Chem. A* **1999**, *103,* 6122–6126.
60. Belinfante, F. J. *A Survey of Hidden Variable Theories,* Pergamon Press, New York, 1973.
61. Deb, B.M.; Chattaraj, P.K. *Phys. Rev. A* **1989**, *39,* 1696–1713.
62. Deb, B.M.; Chattaraj, P.K. *Chem. Phys. Lett.* **1988**, *148,* 550–556.
63. Deb, B.M.; Chattaraj, P.K.; Mishra, S. *Phys. Rev. A* **1991**, *43,* 1248–1257.
64. Chattaraj, P. K. *Int. J. Quant. Chem.* **1992**, *41,* 845–859.
65. Deb, B.M.; Dey, B.K. *Int. J. Quant. Chem.* **1995**, *56.*
66. Nath, S.; Chattaraj, P.K. *Pramana,* **1995**, *45,* 65–73.
67. Chattaraj, P.K.; Nath, S. *Chem. Phys. Lett.* **1994**, *217,* 342–348.
68. Chattaraj, P.K.; Nath, S. *Proc. Indian. Acad. Sci. (Ch. Sci.)* **1994**, *106,* 229–249.
69. de Broglie, L. *Nonlinear Wave Mechanics: A Causal Interpretation,* Elsevier, Amsterdam, 1993.
70. Bohm, D. *Phys. Rev.* **1952**, *85,* 166–179, 180–193.
71. Gutzwiller, M. C. *Chaos in Classical and Quantum Mechanics,* Springer, Berlin, 1990.
72. Eckhardt, B. *Phys. Rep.* **1988**, *103,* 205–297.
73. Jensen, R.V. *Nature,* **1992**, *355,* 311–317; **1995**, *373,* 16 and references therein.
74. Dewdney, C.; Hiley, B. *J. Found. Phys.* 1982, *12,* 27–48.
75. Takabayasi, T. Prog. *Theor. Phys.* **1952**, *8,* 143–182; **1953**, *9,* 187–222; **1955**, *14,* 283–302.
76. Kan, K.K.; Griffin, J. *Phys. Rev. C* **1977**, *15,* 1126–1157.
77. Weiner, J.H; Partom, Y. *Phys. Rev.* **1969**, *187,* 1134–1147; Weiner J. H.; Askar, A. *J. Chem. Phys.* **1972**, *54,* 3534–3541.
78. McCullough, E.A.; Wyatt, R.E. *J. Chem. Phys.* **1971**, *54,* 3534–3541.
79. Hirschfelder, J. O.; Christoph, A. C.; Palke, W. E. *J. Chem. Phys.* **1974**, *61,* 5435–5455; Hirschfelder, J. O.; Tang, K.T.*J. Chem. Phys.* **1976**, *64,* 760–785; **1976**, *65,* 470–486.
80. Skodie, R.T.; Rohrs, H.W.; VanBuskirk, J. *Phys. Rev. A* **1989**, *40,* 2894 – 2916.
81. Parmenter, R.H.; Valentine, R.W. *Phys. Lett. A* **1995**, *201,* 1–8.
82. Schwengelbeck, U.; Faisal, F.H.M. *Phys. Lett. A* **1995**, *199,* 281–286.
83. Faisal, F.H.M.; Schwengelbeck, U. *Phys. Lett. A* **1995**, *207,* 31–36.
84. Misner, C.W.; Thorne, K.S.; Wheeler, J. A. *Gravitation*; W.H. Freeman and Company: San Francisco, 1973.
85. McDonald, S.W.; Kaufman, A. N. *Phys. Rev. Lett.* **1979**, *42,* 1189–1191.
86. Berry, M. V. *Proc. R. Soc. A, London,* **1987**, *413,* 183–198.
87. Chattaraj, P.K. in *Symmetries and Singularity Structures: Intregability and Chaos in Nonlinear Dynamical Systems:* Lakshmanan, M; Daniel, M, Eds.; Springer–Verlag; Berlin, 1990; pp. 172–182.
88. Parr, R.G.; Yang,W. *Density Functional Theory of Atoms and Molecules;* Oxford University Press: Oxford, U.K. 1989.
89. Fuentealba, P.*J. Chem. Phys.* **1995**, *103,* 6571.
90. Ghosh, S.K.; Deb, B. M. *J. Phys. B* **1994**, *27,* 381.
91. Parr, R.G. *J. Phys. Chem.* **1988**, *92,* 3060.

92. (a) Feit, M.D.; Fleck, J.A., Jr. *J. Chem. Phys.* **1984,** *80,*2578; (b) Choudhury, S.; Gangopadhayay, G.; Ray, D.S. *Ind. J. Phys. B* **1995,** *69,* 507; (c) Graham, R.; Hohnerbach, M. *Phys. Rev. A* **1991,** *43,* 3966; (d) *Idem. Phys. Rev. Lett.* **1990,** *64,* 637.

93. Pearson, R G. *Chemical Hardness: Applications from Molecules to Solids;* Wiley – VCH Verlag GMBH; Weinheim, 1997; pp. 116–119.

94. L'Huillier, A.; Lompre, L.A.; Mainfray, G.; Manus, C. *In Atoms in Intense Laser Fields;* Gavrila, M. Ed.; Academic Press; Boston, 1992, pp. 139.

95. Ames, W.F. *Numerical Methods for Partial Differential Equations:* Academic: New York, 1977, pp. 252.

96. Chattaraj, P.K.; Rao. K.S.; Deb, B. M. *J. Comput. Phys.* 1987, *72,* 504.

Regular and Chaotic Motion of a Bush-Shaft System with Tribological Processes[†]

Jan Awrejcewicz[1] and Yuriy Pyryev[2]

[1]Department of Automatics and Biomechanics (K-16), Technical University of Lodz, 1/15 Stefanowskiego Street, 90-924 Lodz, Poland. Email: awrejcew@p.lodz.pl.
[2]Institute of Mechanics and Printing, Warsaw University of Technology, 2 Konwiktorska Street, 00-217 Warsaw, Poland. Email: y.pyryev@wip.pw.edu.pl.

ABSTRACT

The methods of both analysis and modeling of contact bush-shaft systems exhibiting heat generation and wear due to friction are presented [3–5]. From the mathematical point of view, the considered problem is reduced to the analysis of ordinary differential equations governing the change of velocities of the contacting bodies, and to the integral Volterra-type equation governing contact pressure behavior. In the case where tribological processes are neglected, thresholds of chaos are detected using bifurcation diagrams and Lyapunov exponents identification tools. In addition, analytical Mel'nikov's method is applied to predict chaos. It is shown, among the others, that tribological processes play a stabilizing role. The following theoretical background has been used in the analysis: perturbation methods, Mel'nikov's techniques [7,8], Laplace transformations, the theory of integral equations, and various variants of numerical analysis.

Introduction

It should be emphasized that in bibliography devoted to this research, either tribological processes occurring on the contact surfaces are not accounted [1], or inertial effects are neglected [6]. In other words, both mentioned processes are treated separately. In this work, both elements of complex contact behavior are simultaneously included into consideration, which allow for a proper modeling of the real contact system dynamics.

A classical problem concerning the vibration of a friction pair consisting of a rotating shaft and bush fixed to a frame by mass-less springs (a simple model of typical braking

---

[†]Reused with permission from: Jan Awrejcewicz and Yuriy Pyryev, "Regular and chaotic motion of a bush-shaft system with tribological processes," *Mathematical Problems in Engineering*, vol. 2006, Article ID 86594, 13 pages, 2006. doi:10.1155/MPE/2006/86594.

Analytical and numerical analyses are carried out in a wide range through the investigation of various types of nonlinearities, dampings, and excitations applied to the analyzed system. A Duffing-type elastic nonlinearity, a nonlinear density of the frictional energy stream, a nonlinear friction dependence versus velocity, and a nonlinear contact temperature characteristic, as well as nonlinear character of wear are accounted, among the others.

It is clear that from the engineering point of view it is important to understand and control the dynamics occurring in kinematic pairs of the contacting bodies, it is expected also to obtain recipes for an optimal choice of frictional materials as well as other parameters required for realization of long-term and reliable work of various elements of machines and mechanisms. Therefore, it is highly required that progress in mathematical modeling of processes that appears in contacting systems yields finally the results close to those observed in the real systems.

In [4], critical values of the parameters responsible for chaos occurrence are found using Mel'nikov's approach. Originality of the research lies in the following: (i) Mel'nikov's function is constructed for the case of our analyzed dissipative system; (ii) the obtained analytical results are confirmed by extended numerical studies with the use of the Lyapunov exponents, Poincarè maps and bifurcation diagrams; and (iii) analysis of the contact characteristics is carried out.

Section 2 is devoted to a mathematical modeling of the problem of vibrations of a friction pair consisting of a rigid body (a bush) connected with a basing by means of springs and dampers and a rotating thermoelastic shaft. Frictional heat generation, wear of a bush, and thermal expansion of a cylinder (shaft) are taken into account. Eventually, the analyzed problem is expressed as the system of nonlinear differential equation and an integral equation describing the angular velocities of a bush and contact pressure.

Calculation of the Lyapunov exponents are presented in Section 3. The model of vibrations of a rigid body (a bush) placed on a cylinder (shaft) rotating at variable speed is analyzed in Section 4, without taking tribological processes into account. Mel'nikov's method is applied in the analysis of chaotic phenomena of a bush for external excitations. In Section 5, we show how important role various tribological processes play and, in particular, heat

generation due to friction and wear. Conclusions of our study are presented in Section 6.

Mathematical Modeling of the Analyzed System

Consider thermoelastic contact of a solid isotropic circular shaft (cylinder) of radius R_1 with a cylindrical tube-like rigid bush of external radius R_2, which is fitted to the cylinder according to the expression $U_* h_U(t)$ $(h_U(t) \to 1, t \to \infty)$. The internal bush radius is: $R_1 - U_*$ ($U_*/R_1 \ll 1$) (Fig.1.1). The bush is linked with the housing by springs and a damper with viscous coefficient c.

We assume that the bush is a perfect rigid body, and that radial springs have the stiffness coefficient k_1, whereas tangent springs are characterized by nonlinear stiffness k_2 and k_3 of Duffing type. In addition, the bush is subjected to a damping force action in tangent direction. The cylinder rotates with such angular velocity $\Omega(t) = t_*^{-1} \Omega_1(t)$, that the centrifugal forces may be neglected. We assume that the angular speed of the shaft rotation changes in accordance with $\omega_1 = \omega_k + \xi_k \sin\omega't$. We assume that between the bush pad or the so-called Pronny's brake) has been investigated in [1]. In [9], the so-called thermoelastic contact between a rotating cylinder and a fixed noninertial pad has been studied. Next, a more complicated axially symmetric problem of chaotic self-excited vibrations (caused by friction) and wear of the rotating cylinder and the bush (fixed to the frame by springs and viscous damping elements) is investigated.

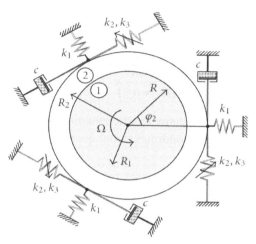

*Figure 1 The analyzed system.

and shaft dry friction appears defined by the function $F_t(V_r)$, where V_r is the relative velocity between the two given bodies $V_r = \Omega R_1 - \varphi_2' R_1.B_2$ denotes the mass moment of inertia. We assume also that in accordance with the Amontos assumption, the friction force is $F_t = f(V_r)N(t)(f(V_r)$ is the kinetic friction coefficient).

The friction force F_t yields heat generated by friction on the contact surface $R=R_1$, and wear U^w of the bush occurs. Observe that the frictional work is transformed to heat energy. Let the shaft temperature, denoted by $T_1(r,t)$, be initially equal to T_0. It is further assumed that the bush transfers heat ideally, and that between both the shaft and bush Newton's heat exchange occurs and the bush has constant temperature T_0. Vibrations of the bush being in thermoelastic contact with the rotating shaft are governed by the following nondimensional equation [4]:

$$\varphi''(\tau) + 2h\kappa\varphi'(\tau) - \varphi(\tau) + b\varphi^3(\tau) = \varepsilon F(\omega_1 - \varphi')p(\tau), \quad 0 < \tau < \infty, \tag{2.1}$$

with the initial condition $\varphi(0) = \varphi^0, \varphi'(0) = \omega^0$, where the nondimensional contact pressure is defined through solutions to the equation [3, 5]

$$p(\tau) = h_U(\tau) - u^w(\tau) + 2\gamma\tilde{\omega}\int_0^\tau G_p'(\tau - \xi)F(\omega_1 - \phi')p(\xi)(\omega_1 - \phi')d\xi, \quad 0 < \tau < \tau_c. \tag{2.2}$$

---

*Figure 1.1 in original source

The bush wear $u^w(\tau)$ and the shaft temperature $\theta(r,\tau)$ are defined through the following equations [3]:

$$u^w(\tau) = k^w \int_0^\tau |\omega_1 - \phi'| p(\xi) d\xi, \quad 0 < \tau < \tau_c, \tag{2.3}$$

$$\theta(r,\tau) = \gamma\tilde{\omega} \int_0^\tau G_\theta'(\tau,\tau-\xi) F(\omega_1 - \phi') p(\xi)(\omega_1 - \phi') d\xi, \tag{2.4}$$

where

$$\{G_p(\tau), G_\theta(1,\tau)\} = \frac{\{0.5,1\}}{Bi\tilde{\omega}} - \sum_{m=1}^\infty \frac{\{2Bi, 2\mu_m^2\}}{\mu_m^2 \tilde{\omega}(Bi^2 + \mu_m^2)} e^{-\mu_m^2 \tilde{\omega}\tau}, \tag{2.5}$$

μ_m (m=1,2,3,...) are the roots of characteristic equation $BiJ_0(\mu) - \mu J_1(\mu) = 0$.

In (2.1)–(2.5), the following nondimensional quantities are introduced

$$\tau = \frac{t}{t_n}, \quad r = \frac{R}{R_1}, \quad p = \frac{P}{P_*},$$

$$\theta = \frac{T_1 - T_0}{T_*}, \quad \varphi(\tau) = \varphi_2(t_*\tau), \quad u^w = \frac{U^w}{U_*}, \quad \varepsilon = \frac{P_* t_*^2 2\pi R_1^2}{B_2},$$

$$h = \frac{cR_2^2}{2t_* B_2}, \quad k^w = \frac{P_* k^w R_1^2}{U_*}, \quad \gamma = \frac{(1-\eta)E_1\alpha_1 R_1^2}{\lambda_1(1-2\gamma)t_*},$$

$$Bi = \frac{\alpha_T R_1}{\lambda_1}, \quad \tau_c = \frac{t_c}{t_*}, \quad \tilde{\omega} = \frac{t_* a_1}{R_1^2}, \tag{2.6}$$

$$\omega_0 = \omega' t_*, \quad h_U = h_U(t_*\tau), \quad F(\omega_1 - \varphi') = f(V_*(\omega_1 - \varphi')),$$

$$b = \left(k_3 R_2^4 - \left(\frac{2}{3}\right)k_2 R_2^2 + (l_1 + R_2)R_2\left(\left(\frac{l_0}{l_1}\right)\left(1 + 3\left(\frac{R_2}{l_1}\right) + 3\left(\frac{R_2}{l_1}\right)^2\right) - 1\right)\frac{k_1}{6}\right)\left(\frac{t_*^2}{B_2}\right)$$

where

$$V_* = \frac{R_1}{t_*}, \quad t^* = \sqrt{\frac{B_2}{k_* R_2^2}}, \quad k_* = k_1\left(\frac{l_0}{l_1} - 1\right)\left(\frac{l_1}{R_2} + 1\right) - k_2,$$

$$T_* = \frac{U_*}{\alpha_1(1+v_1)R_1}, \quad P_* = \frac{\alpha_1 E_1 T_*}{(1-2v_1)}, \tag{2.7}$$

and l_0 is the no stretched spring length, l_1 is the length of the compressed spring for $\varphi_2 = 0$, ($k_* >$ 0), E_1 is the elasticity modulus, v_1 is the Poisson coefficient, α_1 is the coefficient of thermal expansion of the shaft, α_T is the heat transfer coefficient, a_1 is the thermal diffusivity, λ_1 is the heat transfer coefficient, $\varphi_2(t)$ is the angle of bush rotation, K^w is the wear coefficient, η denotes the part of heat energy associated with wear $\eta \in [0,1]$, t_c is the time of contact ($0 < t < t_c$, $P(t) > 0$).

Note that the stated problem is modeled by both nonlinear differential equation (2.1) and integral equation (2.2) governing rotational velocity $\varphi'(\tau)$ and contact pressure $p(\tau)$. Temperature and wear are defined by (2.4) and (2.3), respectively.

Calculation of Lyapunov Exponents

A particular case of our problem is further studied ($\gamma = 0$, $k^w = 0$, $p(\tau) \to 1$). The dependence of kinematics friction on relative velocity is approximated by the function $F(y) = F_0 \operatorname{sgn}(y) - \alpha y + \beta y^3$. Since the latter is non smooth due to the presence of the $\operatorname{sgn}(y)$ function in the kinematic friction, the methods commonly used to compute the exponents require smoothness of the vector fields as a necessary condition. Non smooth systems yield only approximations for the Lyapunov exponents, which can be considered valid as long as we do not bother too much with the vicinity of the non smoothness points [2]. The function $\operatorname{sgn}(y)$ is approximated by the following one [3]:

$$\operatorname{sgn}_{\varepsilon_0}(y) = \begin{cases} \operatorname{sgn}(y), & |y| > \varepsilon_0 \\ (2 - |y|/\varepsilon_0)(y/\varepsilon_0), & |y| < \varepsilon_0 \end{cases} \tag{3.1}$$

Note that while computing Lyapunov exponents, besides the following equations:

$$x' = y, \quad y' = x - bx^3 + \varepsilon\left[F_0 \operatorname{sgn}_{\varepsilon_0}(v_r) - \alpha v_r + \beta v_r^3\right] - \varepsilon h_1 y, \quad z' = \omega_0, \tag{3.2}$$

also three additional systems of equations ($n=1,2,3$) with respect to perturbations are solved:

$$\tilde{x}'^{(n)} = \tilde{y}^{(n)}, \quad \tilde{y}'^{(n)} = \tilde{x}^{(n)} - 3bx^2\tilde{x}^{(n)} + \varepsilon\left[F_0\delta_{\varepsilon_0}(v_r) - \alpha + 3\beta v_r^2\right]\tilde{v}_r^{(n)} - \varepsilon h_1\tilde{y}^{(n)}, \quad \tilde{z}'^{(n)} = 0, \tag{3.3}$$

where
$$x = \varphi(\tau), y = \varphi'(\tau), z = \omega_0\tau, v_r = \omega_k + \xi_k \sin z - y, \quad h_1 = 2h/\varepsilon, \quad \tilde{v}_r^{(n)} = \xi_k\tilde{z}^{(n)}\cos z - \tilde{y}^{(n)},$$

$$\delta_{\varepsilon_0}(y) = \begin{cases} 0, & |y| > \varepsilon_0 \\ \dfrac{2}{\varepsilon_0}\left(1 - \dfrac{|y|}{\varepsilon_0}\right), & |y| < \varepsilon_0. \end{cases} \tag{3.4}$$

Twelve equations of system (3.2) and (3.3) are solved using the fourth-order Runge-Kutta method and Gram-Schmidt reorthonormalization procedure.

Let $\tilde{X}_0^0, \tilde{Y}_0^0, \tilde{Z}_0^0$ be initial values of perturbation vectors which are orthonormal. After time T, an orbit $X(\tau)$ reaches the point X_1 with the associated perturbations $\tilde{X}_1, \tilde{Y}_1, \tilde{Z}_1$. Then, the so-called *Gram-Schmidt reorthonormalization procedure* is carried out and the following new initial set of conditions is formulated:

$$\tilde{X}_1^0 = \tilde{X}_1 / \left\| \tilde{X}_1 \right\|, \quad \tilde{Y}_1^0 = \tilde{Y}_1' / \left\| \tilde{Y}_1' \right\|, \quad \tilde{Y}_1' = \tilde{Y}_1 - \left(\tilde{Y}_1, \tilde{X}_1^0 \right) \tilde{X}_1^0$$

$$\tilde{Z}_1^0 = \tilde{Z}_1' / \left\| \tilde{Z}_1' \right\|, \quad \tilde{Z}_1' = \tilde{Z}_1 - \left(\tilde{Z}_1, \tilde{X}_1^0 \right) \tilde{X}_1^0 - \left(\tilde{Z}_1, \tilde{Y}_1^0 \right) \tilde{Y}_1^0. \tag{3.5}$$

Next, after time interval T, a new set of perturbation vectors $\tilde{X}_2, \tilde{Y}_2, \tilde{Z}_2$ is defined, which is also reorthonormalized due to the Gram-Schmidt procedure (3.5). This algorithm is repeated M times. Note that $\left(\tilde{X}_1^0, \tilde{Y}_1^0 \right) = 0$, $\left(\tilde{X}_1^0, \tilde{Z}_1^0 \right) = 0$, $\left(\tilde{Y}_1^0, \tilde{Z}_1^0 \right) = 0$, and if $X = (x, y, z)$, $Y = (x_1, y_1, z_1)$, then $\|X\| = \sqrt{x^2 + y^2 + z^2}$, and the scalar product $(X, Y) = xx_1 + yy_1 + zz_1$.

Finally, a spectrum of three Lyapunov exponents is computed via formulas

$$\lambda_1 = \frac{1}{MT} \sum_{i=1}^{M} \ln \left\| \tilde{X}_i \right\|, \quad \lambda_2 = \frac{1}{MT} \sum_{i=1}^{M} \ln \left\| \tilde{Y}_i' \right\|, \quad \lambda_3 = \frac{1}{MT} \sum_{i=1}^{M} \ln \left\| \tilde{Z}_i' \right\|, \tag{3.6}$$

where the occurring vectors are taken before the normalization procedure.

Our numerical computations are carried out for the particular case ($\gamma = 0$, $k^w = 0$). The following nondimensional parameters are taken: $F_0 = \alpha = \beta = 0.3$, $\omega_0 = 2$, $\omega_k = 0.4$, $b = 1$, $\varepsilon = 0.1$. Numerical analysis is carried out for the bifurcation diagram with respect to x versus ζ_k, for $\zeta_k \in (0, 12)$ and $\zeta_k \in (3.5, 4)$. The obtained results are shown in Figs. 3.1(a) and 3.1(b) for $h_1 = 0$, in Fig. 3.2(a) for $h_1 = 0.5$, and in Fig. 3.2(b) for $h_1 = 1$. The Lyapunov exponents in time interval $\tau \in (1200, 1514)$ ($\tilde{X}_0^0 = (1,0,0)$, $\tilde{Y}_0^0 = (0,1,0)$, $\tilde{Z}_0^0 = (0,0,1)$, $T = 0.005$, $M = 80000$, $\varepsilon_0 = 0.01$) are computed due to formulas (3.6) for the same values of parameters. In Figs. 3.1(c), 3.1(d), 3.2(c), 3.2(d), dependencies of Lyapunov exponents on the control parameter ζ_k are reported. A study of both Lyapunov exponents and bifurcation diagrams implies that chaos begins for (i) $\zeta_k = 3.78$, for $h_1 = 0$; (ii) for $\zeta_k = 3.8$, for $h_1 = 0.5$; (iii) for $\zeta_k = 4.25$, for $h_1 = 1$ (note that the largest Lyapunov exponent λ_1 is positive). An increase of the parameter h_1 responsible for damping yields an increase of the amplitude of the bush, where chaos is born.

Note that since our system (3.2) is autonomous, one of the Lyapunov exponents is always zero.

Mel'nikov's Method

In order to estimate analytically the critical parameters responsible for chaos occurrence Mel'nikov's technique [8] is often used. In this case the Mel'nikov's function is (see [4, 8])

$$M(\tau_0) = -\int_{-\infty}^{\infty} y_0(t) \left[F_0 \, \text{sgn}(\omega_r) - \alpha \omega_r + \beta \omega_r^3 - h_1 y_0(t) \right] dt = I(\tau_0) + J(\tau_0) \tag{4.1}$$

where

$$\omega_r(t) = \omega_k + \zeta_k \sin \left(\omega_0 (t + \tau_0) \right) - y_0(t), \quad y_0(\tau) = -\sqrt{2/b} \, \sinh(\tau) / \cosh^2(\tau),$$

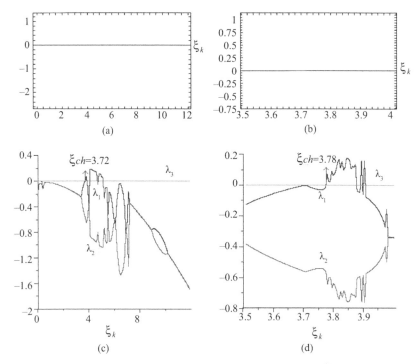

(a)

(b)

(c)

(d)

*Figure 2 Bifurcation diagrams (a), (b) and Lyapunov exponents (c), (d) using ξ_k as control parameter, $h_1 = 0$, $\gamma = 0$, $k^w = 0$: (a), (c) $\xi_k \in (0,12)$; (b), (d) $\xi_k \in (3.5,4)$.

$$J(\tau_0) = 2C + 2\zeta_k \sqrt{A^2 + B^2} \sin(\omega_0\tau_0 + \varphi_0)$$

$$+6\beta\zeta_k^2 \left(I_{220}\cos^2\omega_0\tau_0 + I_{202}\sin^2(\omega_0\tau_0) - 2\omega_*I_{111}\sin(\omega_0\tau_0)\cos(\omega_0\tau_0)\right)$$

$$+2\beta\zeta_k^3 \left(-I_{130}\cos^3\omega_0\tau_0 - 3I_{112}\sin^2(\omega_0\tau_0)\cos(\omega_0\tau_0)\right),$$

$$A = \left(\alpha - 3\beta\omega_k^2\right)I_{110} - 3\beta I_{310}, \quad B = 6\beta\omega_k I_{201},$$

$$C = \beta I_{400} - \left(\alpha - h_1 - 3\beta\omega_k^2\right)I_{200}, \quad \varphi_0 = \arctan\left(\frac{A}{B}\right). \tag{4.2}$$

In (4.1), the term $I(\tau_0)$ is defined by the formula

$$I(\tau_0) = -F_0 \int_{-\infty}^{\infty} y_0(t)\,\mathrm{sgn}(\omega_r)\,dt = 2F_0\sqrt{\frac{2}{b}}\sum_m \frac{\mathrm{sgn}(\omega'_r(t_m))}{\cosh t_m}, \tag{4.3}$$

where t_m are the roots of the equation

$$\omega_r(t_m) = \omega_k + \zeta_k \sin(\omega_0(t_m + \tau_0)) - y_0(t_m) = 0,$$

$$\omega'_r(t) = -\zeta_k \cos(\omega_0(t + \tau_0)) - x_0(t) + bx_0^3(t) \tag{4.4}$$

---

*Figure 3.1 in original source

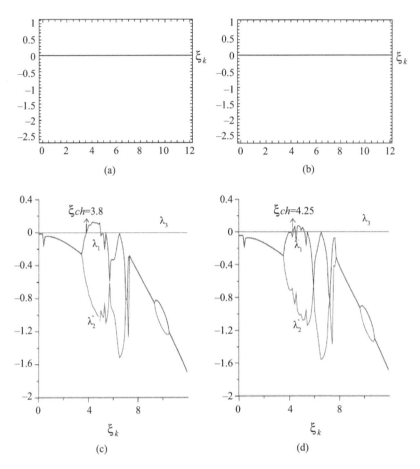

*Figure 3 Bifurcation diagrams (a), (b) and Lyapunov exponents (c), (d) using ξ_k as control parameter, $\gamma = 0$, $k_z=0$, $b=1$, $k^w = 0$: (a), (c) $h_1=0.5$; (b), (d) $h_1=1$.

If the Mel'nikov's function (4.1) changes sign, then chaos may occur. In order to apply combined Mel'nikov's and numerical methods, a perturbation of the Hamiltonian system, where the function sgn(y) occurs, has been approximated by a continuous perturbation with an application of a small parameter. The multivalued relation sgn(y) is approximated by the function $\text{sgn}_{\varepsilon_0} (y)$ defined by (3.1), where the regularization parameter ε_0 is a "small" positive real number. The differential equation (inclusion) (2.1) is then approximated by (3.2).

In the so-called first improvement of Mel'nikov's function $M(\tau_0)$ (see the expression standing by ε) for $0 < \varepsilon < 1$ in the expression representing a distance between stable and unstable manifolds of the critical saddle point, a transition of the parameter ε_0 to zero ($\varepsilon_0 \rightarrow 0$) can be realized. In order to be sure of neglecting the so-called second improvement of Mel'nikov's function standing by ε^2 [7] (the under integral function includes the differential of the approximated perturbation), the following condition should be satisfied

---

*Figure 3.2 in original source

$\varepsilon / \varepsilon_0 < 1$. Then, if the mentioned condition is satisfied, only the first improvement of Mel'nikov's function can be applied to estimate the distance between stable and unstable manifolds of the critical point.

In Fig. 4.1, Mel'nikov's function $M(\tau_0)$ for different values of parameter ζ_k before and after sign change of $M(\tau_0)$ is reported. One may be convinced that both analytical and numerical predictions of chaos coincide.

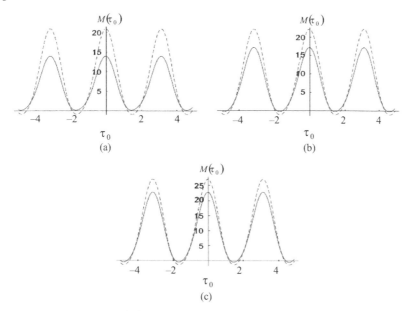

Figure 4 Mel'nikov's function $M(\tau_0)$ versus parameter τ_0: (a) for $\zeta_k = 3.2$ (solid curves) and for $\zeta_k = 3.81$ (broken curves), $h_1 = 0$; (b) for $\zeta_k = 3.5$ (solid curves) and for $\zeta_k = 3.9$ (broken curves), $h_1 = 0.5$; (c) for $\zeta_k = 3.9$ (solid curves) and for $\zeta_k = 4.2$ (broken curves), $h_1 = 1$.

Numerical Analysis

In a general case, numerical analysis is carried out of a steel-made shaft ($\alpha_2 = 14.10^{-6°} C^{-1}$, $\lambda_1 = 21W / (m.^0 C^{-1})$, $v_1 = 0.3$, $a_1 = 5.9 \cdot 10^{-6}$ m²/s, $E_1 = 19 \cdot 10^{10}$ Pa). Observe that no accounting of tribological processes ($h_1 = 0.5$, $\zeta_k = 3.9$, $\gamma = 0$, $k^w = 0$) yields chaotic dynamics (Figure 5.1, curve 2). For $h_1 = 0.5$, $\zeta_k = 3.5$, $\gamma = 0$, $k^w = 0$, regular motion takes place (Fig. 5.1, curve 1). An account of thermal shaft extension ($\gamma = 1.87$) removes chaotic behavior of our system (Fig. 5.1, curves 3 and 4). For $\zeta_k = 3.5$, a subharmonic motion with frequency $\omega_0 / 2$ is obtained (Figure 5.1, curve 3), whereas for $\zeta_k = 3.9$ periodic motion is exhibited (Fig. 5.1, curve 4).

Owing to an account of wear ($k^w = 0.01$) and neglecting shaft thermal extension ($\gamma = 0$), contact pressure tends to zero, whereas cylinder wear approaches $U_*(p(\tau) \to 0, u^w(\tau) \to 1)$. The nondimensional bush wear is presented in Fig. 5.2, curve 1. In addition, in Fig. 5.2, curves 1 and 2 represent time histories of the nondimensional contact pressure.

---

*Figure 4.1 in original source

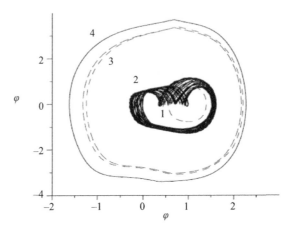

***Figure 5** Phase plane of bush motion for h_1=0.5, k^w 0: curve 1: $\zeta_k = 3.5$, $\gamma = 0$, curve 2: $\zeta_k = 3.9$, $\gamma = 0$, curve 3: $\zeta_k = 3.5$, $\gamma = 1.87$, curve 4: $\zeta_k = 3.9$, $\gamma = 1.87$.

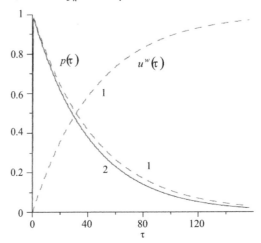

*Figure 6 Dimensionless contact pressure $p(\tau)$ and wear $u^w(\tau)$ versus dimensionless time τ: curve 1: $\zeta_k = 3.5$, $\gamma = 0$, $k^w = 0.01$, h_1=0.5, curve 2: $\zeta_k = 3.9$, $\gamma = 0$, $k^w = 0.01$, h_1=0.5.

A simultaneous account of shaft extension and bush wear yields a finite time of contact between both bodies. For instance, for h_1=0.5, $\zeta_k = 3.9$, $\gamma = 1.87$, k^w=0.01, contact pressure versus time is exhibited by curve 4 in Fig.5.3. The nondimensional time contact interval is $\tau_c = 72$. For $\zeta_k = 3.5$ time contact is $\tau_c = 65.8$. In Fig. 5.4, curves 3 and 4 represent the dependence of nondimensional wear on the nondimensional time in a general case. Curve 3 corresponds to h_1=0.5, $\zeta_k = 3.5$, $\gamma = 1.87$, k^w=0.01, whereas curve 4 is associated with the following parameters: h_1=0.5, $\zeta_k = 3.9$, $\gamma = 1.87$, k^w=0.01. Owing to heat shaft extension, the wear of bush is increased thirty times (see curves 4 and 2 in Fig. 5.4).

---

*Figure 5.1 in original source
**Figure 5.2 in original source

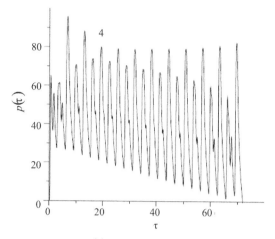

*Figure 7 Dimensionless contact pressure $p(\tau)$ versus dimensionless time τ : curve 4 : $\zeta_k = 3.9$, $\gamma = 1.87$, $k^w = 0.01$, $h_1 = 0.5$.

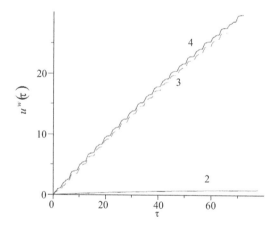

**Figure 8 Time history of dimensionless wear $u^w(\tau)$, $h_1 = 0.5$: curve 2: $\zeta_k = 3.9$, $\gamma = 0$, $k^w = 0.01$, curve 3: $\xi_k = 3.5$, $\gamma = 1.87$, $k^w = 0.01$, curve 4: $\zeta_k = 3.9$, $\gamma = 1.87$, $k^w = 0.01$.

Conclusions

This paper extends the analysis carried out in [4]. Contrary to the previous results, a novel mechanism of contact between the bush and shaft is proposed, a viscous damping is added, and an influence of tribological factors on both regular and chaotic dynamics is analyzed. The analytical formula of the Mel'nikov's function of the investigated system has been first formulated, and then numerical analysis of nonlinear phenomena is carried out.

The influence of tribological processes on dynamic behavior of the analyzed system in the vicinity of chaos has been illustrated and discussed. An account of bush wear and

---

*Figure 5.3 in original source
**Figure 5.4 in original source

neglecting of shaft thermal expansion implies that the contact pressure tends to zero, the bush wear approaches the values of the shaft compressing, and bush vibrations are damped. On the other hand, taking into account the shaft thermal extension and neglecting of bush wear results in chaos disappearance and the occurrence of a regular motion. In a general case (both shaft thermal extension and bush wear are taken into account), time interval of the contact of two bodies is bounded. In the lack of contact, the bush stops due to an extensive wear process.

Appendix

Here we report the expressions of the following functions:

$$I_{200} = \frac{2}{3b}, \quad I_{400} = \frac{8}{35b^2}, \quad I_{201} = \frac{\pi\omega_0\left(2-\omega_0^2\right)}{6b\sinh\left(\pi\omega_0/2\right)},$$

$$I_{110} = -\frac{\pi\omega_0}{\sqrt{2b}\cosh\left(\pi\omega_0/2\right)}, \quad I_{112} = \frac{\pi\omega_0\cosh\left(\pi\omega_0/2\right)}{\sqrt{2b}\left(1-2\cosh\left(\pi\omega_0/2\right)\right)},$$

$$I_{111} = -\frac{\pi\omega_0}{\sqrt{2b}\cosh\left(\pi\omega_0\right)}, \quad I_{220} = \frac{\pi\omega_0\left(2\omega_0^2-1\right)+\sinh\left(\pi\omega_0\right)}{3b\sinh\left(\pi\omega_0\right)},$$

$$I_{202} = \frac{\pi\omega_0\left(1-2\omega_0^2\right)+\sinh\left(\pi\omega_0\right)}{3b\sinh\left(\pi\omega_0\right)},$$

$$I_{310} = \frac{\omega_0\left(11+10\omega_0^2-\omega_0^4\right)}{120b\sqrt{2b}}\left\{ \begin{array}{l} \psi\left(\dfrac{1-i\omega_0}{4}\right)-\psi\left(\dfrac{3-i\omega_0}{4}\right)+\psi\left(\dfrac{1+i\omega_0}{4}\right) \\ -\psi\left(\dfrac{3+i\omega_0}{4}\right) \end{array} \right\},$$

$$I_{130} = -\frac{3\pi\omega_0}{8\sqrt{2b}}\left\{ \begin{array}{l} \cot\left(\dfrac{\pi\left(1-i\omega_0\right)}{4}\right)+\cot\left(\dfrac{3\pi\left(1-i\omega_0\right)}{4}\right)-\cot\left(\dfrac{\pi\left(3-i\omega_0\right)}{4}\right) \\ -\cot\left(\dfrac{\pi\left(1-3i\omega_0\right)}{4}\right) \end{array} \right\},$$

$$\psi\left(z\right)=\frac{\Gamma'(z)}{\Gamma(z)},$$

(A.1)

Acknowledgment

This work has been partially supported by the Grant of the State Committee for Scientific Research of Poland (KBN) No. 4 TO7A 031 28.

References

1. A.A. Andronov, A.A. Vitt, and S.E. Khaikin, *Theory of Oscillators*, Pergamon Press, Oxford, 1966.
2. J. Awrejcewicz and C.-H. Lamarque, *Bifurcation and Chaos in Nonsmooth Mechanical Systems*, World Scientific Series on Nonlinear Science. Series A: Monographs and Treatises, vol. 45,World Scientific, New Jersey, 2003.
3. J. Awrejcewicz and Yu. Pyryev, *Thermoelastic contact of a rotating shaft with a rigid bush in conditions of bush wear and stick-slip movements*, International Journal of Engineering Science **40** (2002), no. 10, 1113–1130.
4. J. Awrejcewicz and Yu. Pyryev, *Influence of tribological processes on a chaotic motion of a bush in a cylinder-bush system*, Meccanica **38** (2003), no. 6, 749–761.
5. J. Awrejcewicz and Yu. Pyryev, *Contact phenomena in braking and acceleration of bush-shaft system*, Journal of Thermal Stresses **27** (2004), no. 5, 433–454.
6. J.R. Barber, *Thermoelasticity and contact*, Journal of Thermal Stresses **22** (1999), no. 4-5, 513–525.
7. S. Lenci and G. Rega, *Higher-order Melnikov functions for single-DOF mechanical oscillators: theoretical treatment and applications*, Mathematical Problems in Engineering **2004** (2004), no. 2, 145–168.
8. V.K. Mel'nikov, *On the stability of the center for time-periodic perturbations*, Transactions of the Moscow Mathematical Society **12** (1963), 1–56.
9. Yu. Pyryev and D.V. Grylitskiy, *Transient problem of frictional contact for the cylinder with heat generation and wear*, Journal of Applied Mechanics and Technical Physics **37** (1996), no. 6, 99–104.

Relaxed Plasma Equilibria and Entropy-Related Plasma Self-Organization Principles[†]

Robert L. Dewar,[1] *Matthew J.Hole,*[1] *Mathew MeGann,*[1] *Ruth Mills*[1]
and *Stuart R. Hodson*[2]

[1]Plasma and fluids theory Group, Research School of physics & Engineering, the Australian National University, Canberra ACT 0200, Australia.
[a]Email: robert.dewar@anu.edu.au.
[2]Princetion Plasma physics Laboratory, PO Box 451, Princeton, N.J. 08543 , USA ,
Email: shudson@pppl.gov.

ABSTRACT

The concept of plasma relaxation as a constrained energy minimization is reviewed. Recent work by the authors on generalization this approach to partially relaxed three-dimensional plasma systems in a way consistent with chaos theory is discussed, with a view to clarifying the thermodynamic aspects of the variational approach used. Other entropy-related approaches to finding long-time steady states of turbulent or chaotic plasma systems are also briefly reviewed.

Keyword: Plasma, Fusion, Relaxation Chaos, Turbulence.

Introduction

Magnetically confined fusion plasmas are thermodynamically nonequilibrium systems, where particles and energy are injected (or generated by fusion reaction) deep in the plasma, providing heat which flows towards the much colder edge region . This creates a kind of heat engine that drives both turbulent flows and more laminar zonal flows, somewhat analogous to the way solar energy deposition near the equator drives the dynamics of planetary atmospheres and oceans (see Fig. 1).

[†]Reused with permission from: Robert L. Dewar, Matthew J.Hole, Mathew MeGann, Ruth Mills, and Stuart R.Hodson, Relaxed plasma equilibria and entropy-related plasma self-organization principles, *Entropy* 2008, *10,* 621- 634; DOI: 10.3390/e10040621.

Confinement of strongly heated plasmas against turbulent diffusion across the magnetic field has been found, surprisingly, to improve in some circumstances due to the spontaneous formation of transport barriers related to strongly sheared zonal flows [1] driven by [2.3] turbulence arising from instabilities.

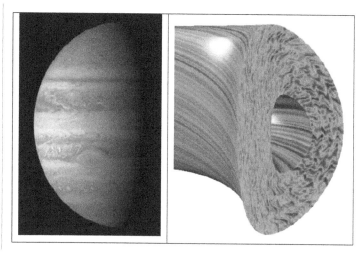

Figure 1 Analogue systems exhibiting self-organization in their quasi-two-dimensional turbulent dynamic: Left panel shows Cassini Jupiter Portrait (NASA .imagePIA04866), showing the strongly zonal nature of the solar-energy-driven turbulent atmospheric dynamics. Right panel (courtesy Jeff Candy http://fusion.gat.com/theory/Gyro) shows a simulation of the analogous turbulent dynamics in a tokomak plasma , driven by heat primarily coming from the central part of the plasma (not shown).)N.B the zonal direction in this case in the short way around the torus.

that tap large free energy provided by the heating and fueling of the plasma. The best-known example is the Low to high (L–H) confinement transition, where the transport barrier forms at the edge of the plasma , but internal have been found as well. Similar sheared-zonal-flow transport barriers also occur in the atmosphere [1]. For instance at the edges of the equatorial jet and the polar vortices.

The type of plasma turbulence referred to above is driven by temperature and density gradients, causing low-frequency plasma instabilities of the *drift wove* class (analogous in geophysics to Rossby waves [4,5]). These modes have little effect on the magnetic field but degrade confinement by eddy motions transporting plasma *across* the magnetic field lines.

Another type of instabilities, of the *tarring mode* class, driven by electric currents in the plasma, gives rise to electromagnetic turbulence, These modes cause magnetic reconnection, changing the topology of magnetic field lines. This effect is also potentially deleterious to plasma confinement because transport *along* magnetic field lines is very rapid. Toroidal magnetic confinement systems are designed with the intent that magnetic field lines stay on topologically toroidal surface (*invariant tori* in the language of Hamiltonian nonlinear dynamics [6,7]),but field-line tearing can destroy such surface and give rise to chaotic that allow anomalous transport along the magnetic field lines.

Because of the complexity of these phenomena, modelling the long-time behaviour of a fusion plasma *ab initio* is very difficult and various quasi-thermodynamic variational approaches [8] have been proposed to predict the steady state to which a plasma will relax given some global constraints .

In Sec.2 we review the variational principle first introduced in astrophysics by Woltjer and developed physically in the fusion context by Taylor and other authors . We then, in Sec.3, in dicate how this approach in being extended to three-dimensional magnetic confinement systems, spelling out the (very elementary) thermodynamics involved in more detail than elsewhere in the literature. In Sec.4 we mention very briefly other approaches that may have application in plasma physics, and point the way future research in Sec.5.

The Plasma Relation Concept

Although plasmas are definitely not in global thermal equilibrium, we assume that most degrees of freedom relax quickly. Thus, after an initial transient, the system reaches a statistical quasi-equilibrium characterized by only a few parameter [9], which evolve slowly due, *inter alia*, to the smallness of flows of matter and energy between the plasma and the outside world over the short relaxation timescale .Relaxation theory describes states of a system on an intermediate timescale, long compared with relaxation times, but short compared with heating and confinement times. Thus we take the plasma to be closed and thermally isolated and freeze the slow parameters, imposing them as constraints.

We shall seek a *static* equilibrium-we assume the plasma relaxes to a state with no mass flow. Also, we model the plasma as a single magntohydrodnamic (MHD) fluid, a crude but surprisingly good approximation for the purpose of constructing background equilibrium solutions on top of which more sophisticated physics can be modeled. Finally, we use only constraints that are conserved in ideal MHD (*ideal* here meaning a single–fluid, in viscid, electrically perfectly conducting, perfect gas model). The general variational theory of ideal–MHD equilibria was enunciated by Kruskal and Kulsrud [10], basing their theory on the minimization of the total plasma energy, electromagnetic plus kinetic:

$$W = \int_P \left(\frac{B^2}{2\mu_0} + \frac{P}{\gamma - 1} \right) d\tau \tag{1}$$

Subject to the full set of ideal- MHD constraints. Here the plasma volume P is assumed to be bounded by a perfectly conducting wall, $d\tau$ denotes a volume element, B is the magnetic field strength (SI units- μ_0 is the permeability of free space) P is the plasma pressure [11], and $\gamma \equiv c_p / c_v$ is the ratio of specific heats [so the internal energy of the plasma is $U = \int p d\tau (\gamma - 1)$] We likewise base our variational principal on the minimization of W [12], and, because we use constraints that are ideal- MHD invariants, our equilibria are automatically a subset of those treated by Kulsrud. Approach is analogous to the Energy-Casimir method [13, p. 511],often called Arnold's method, and is illustrated schematically in Fig. 2. It is the conceptual basis for our generalization of relaxation theory discussed in the next section.

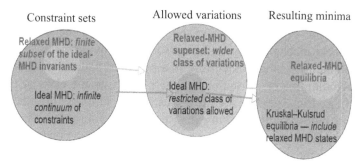

Figure 2 Constraint sets, spaces of allowed variations, and equilibrium states: Illustrating how broadening the space of allowed variation narrows the class of equilibria.

Woltjer [14] observed that the "magnetic helicity"

$$K = \int_P A.B \, d\tau, \tag{2}$$

Where A is the vector potential such that $B = \nabla \times A$, is an ideal-MHD invariant and used this as the only constraint, giving a constant-pressure equilibrium with a force-free Beltrami field.

$$\nabla \times B = \mu B \tag{3}$$

As Euler-Lagrange equation (the constant μ being a Lagrange multiplier). Taylor [15,16], argued that the helicity K is the "most conserved" invariant for a plasma in which turbulence causes field-line reconnection and showed that Beltrami solutions modeled the results from the UK Zeta experiment well. The Woltjer-Taylor relaxation approach has been generalized to two-fluid Hall MHD by Yoshida et al. [17,18] using an additional helicity constraint involving both the magnetic field and the fluid vorticity. Also, Ito and Yoshida [19] developed a statistical mechanical form of relaxation theory using the Shannon or Renyi entropy, and Minardi [20] has derived the force-free relaxed state from an argument based on his magnetic entropy concept .

Nonuniform Relaxation

The work mentioned in the previous section assumed that the plasma relaxes uniformly throughout its volume, which is both undesirable form a confinement Point of view and unrealistic in all but the most turbulent experiments. To allow spatially nonuniform relaxation to be modeled, Bhattarcharjee and Derwas [21] expanded of the set of ideal-MHD invariants used as constraints in the minimization of W by taking moments of $A.B$ with ideal-MHD-invariant weight functions that were smooth functions of position .

To be more precise, in this early work the magnetic field, which can be described as a Hamiltonian dynamical system, was assumed to be integrable, so the plasma volume was foliated by invariant tori. Thus the Weight functions were taken to be smooth functions of the flux threading these tori. However the assumption of integrability is appropriate only for systems with a continuous symmetry (known as two-dimensional systems because of the existence of an ignorable coordinate). This is a reasonable assumption for tokamaks ,which are, neglecting the discreteness of the coils providing the toroidal magnetic field, axisymmetric. These machines rely on a large toroidal provide the poloidal magnetic field required for confinement, and this current is a potential source of reconnection-causing instabilities, including major disruptions of the plasma. In the class of machines known as stellarators (e.g., Fig. 3),external coils are used instead of the toroidal plasma current, producing Amore puiescent plasma but expense of axisymmetry. It is the development of a theory of MHD equilibria in stellarators, one which takes into account the problem of field-line chaos is, that is the main motivation for our current work on finding a generalization of variational relaxation theory to three-dimensional systems.

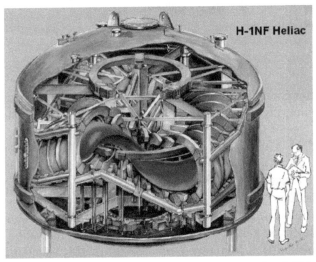

Figure 3 H-1 National Facility heliac stellarator at the Australian National University. The strongly nonaxisymmetric, helically deformed toroidal plasma is indicated in red.

A nonaxisymmetric system is generically not integrable- there will be islands and chaotic regions in the magnetic field of a stellarator. (By *chaotic* magnetic field region we simply mean a volume filed ergodically by a single field line.) Since transport along magnetic field lines. *Parallel transport*, is very rapid in a hot plasma [9] the temperature, density and pressure will rapidly become uniform in a chaotic region.

However, the Kolmogorov-Arnol'd-Moser (KAM) theorem (e.g. [6,p. 330] or [7,p. 174]) gives reason to believe that some invariant tori survive smooth perturbation away from integrability, provided their winding number(*rotational transform* in magnetic confinement jargon) is sufficiently irrational that they obey a Diophantine criterion relating to approximation by sequences of rationals. While the magnetic field can be described by a Hamiltonian, it cannot simply be written in the standard from assumed in proving the KAM theorem: $H_0 + \varepsilon H_1$, with H_0 an integrable field and H_1 a known perturbation. This is because plasma currents, as yet unknown, also change with geometric perturbation and they modify the Hamiltonian. Thus the KAM theorem, as normally understood, is not directly applicable to this Hamiltonian. However, generalizing previous work [22,23], we have studied a simpler Hamiltonian problem associated with force balance across a *fixed* KAM barrier that shows the rotational transforms on either side of such a surface must be strongly irrational, as in KAM theory.

By definition, magnetic field lines cannot cross an invariant torus, so such a torus will separate chaotic regions of the plasma and be impermeable to parallel transport ,allowing a pressure differential to exist between the regions. We proceed on the assumption that some invariant tori L_i do exist (Fig. 4), and, for simplicity, assume maximal chaos in the regions P_i between them, so that the pressure P_i in each such region is constant. We term such pressure-jump-sustaining interfaces, which can be thought of as impermeable idral-MHD membranes, *KAM barriers*. Thus we have recently proposed [24] that the generalization of the Woltjer-Taylor approach appropriate to three-dimensional geometry is the minimization of the total plasma energy

$$W = \sum_i \left(\int_{P_i} \frac{B^2}{2\mu_0} d\tau + \frac{p_i V_i}{\gamma - 1} \right) + \int_v \frac{B^2}{2\mu_0} d\tau \tag{4}$$

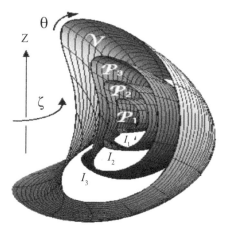

Figure 4 Nested annular toroidal relaxation regions P_i and vacuum region V separated by KAM transport barriers, L_i, as described in the text. Also shown are arbitrary poloidal and toroidal angles, θ and ζ, respectively which allow the toroidal interfaces L_i to be specified parametrically by $r = r_i(\theta, \zeta)$.

Subject to the helicity constraints K_i=const, where K_i is defined as in Eq. (2) with P replaced by P_i, and V_i is the volume of region P_i. The magnetic fluxes threading the P_i are conserved holonomically by restricting allowed (Eulerian) variation in A at the boundaries ∂P_i to be of the form

$$\delta A = \delta r_i \times B + \delta \alpha n_i + \nabla \delta \chi \quad on \, S_i \tag{5}$$

Where δr_i is the variation in the position vector $r = r_i(\theta, \zeta)$ (see Fig. 4) of a point on the boundary n_i is the unit normal, δa is an arbitrary function that allows nonideal variations, and $\delta\chi$ is an arbitrary, gauge term. Thus constraint leaves loop integrals of A as Lagrangian invariants so fluxes are conserved. As we allow shape variations in the barrier surfaces, in addition to helicity conservation we need to constrain the pressure variations. Since we are working on the intermediate timescale, short compared with heating and isentropic. For an ideal (perfect) gas the entropy S is given in terms of the pressure p and volume V by

$$S = S_0 + \frac{Nk}{\gamma - 1} \ln\left(\frac{pV^\gamma}{p_0 V_0^\gamma}\right) \tag{6}$$

Where S_0 and $p_0 V_0^\gamma$ are arbitrary reference values, N is the number of particles and k is Boltzmann's constant. Thus constancy of N and S implies the well-known pressure-volume relation.

$$pV^\gamma = p_0 V_0^\gamma \exp\left[(\gamma - 1)\frac{S - S_0}{Nk}\right] = const \tag{7}$$

We assume Eq. (7) applies to both the ion and electron gases, so the total pressure $p = p_i + p_e$ also obeys $pV^\gamma = const$, or, equivalently, $p^{\frac{1}{\gamma}} V = const$ (This constant is sometimes [10,25] called "mas," but Eq.(7) shows it is a nonlinear function of both mass, mN, *and* entropy, S, and therefore this terminology is best avoided.) Thus, introducing Lagrange multipliers μ_i and v_i for the helicity and pressure constraints respectively, our generalized relaxed-MHD equilibrium criterion is that extremizing the "free energy"

$$F \equiv \sum_i \left[\frac{1}{2} \int_{p_i} \left(\frac{B^2}{\mu_0} - \mu_i A.B\right) d\tau + \left(\frac{p_i}{\gamma - 1} - v_i p_i^{\frac{1}{\gamma}}\right) \int_{p_i} d\tau \right] + \int_v \frac{B^2}{2\mu_0} d\tau \tag{8}$$

with respect to the vector potential A, the pressures p_i and the barrier surfaces $r = r_i(\theta, \zeta)$ gives a ststic- equilibrium consistent with existence of magnetic-filed-line chaos between the KAM barriers. Because the constraints are a subset of the ideal- MHD constraints (Fig. 2), such equilibria will also be Kruskal- Kulsrud ideal- MHD equilibria.

The numerical implementation of this program is the proceeding towards a practical 3-D equilibrium code. Figure 5 shows a two-region solution for a test case where the innermost interface is a circularcross-section axisymmertric torus: $R = R_0 + r\cos\theta$, $Z = \sin\theta$, with $R_0 = 1.0$ and $r = 0.1$ while the outermost boundary is given by $R = R_0 + r(\theta, \phi)\cos\theta$, $Z = r(\theta, \phi)\sin\theta$ with $R_0 = 1.0$ and $r = 0.3 + d\cos(2\theta - \phi) + d\cos(3\theta - \phi)$, where d is an adjustable parmeter which introduces nonaxisymmetry and thus chaotic fields. (In the above we are assuming a cylindrical coordinate system R, Z, ϕ). The (strongly irrational) transforms and appropriate fluxes and pressure jump were prescribed, the Beltrami equation Eq.(3) was solved numerically in the two regions as in [24] and the position and shape of the KAM barrier surface was adjusted iteratively to satisfy the force balance jump condition $\left[\left[p + B^2/2\mu_0\right]\right] = const$ across it. (This relation can be derivd as an Euler-Lagrange equation from Eq.(8), $\left[\left[\ \right]\right]$ denoting the

Figure 5 Poincaré plots of magnetic field lines intersecting with a surface of section $\phi = const$ for the two-relaxation-region test case described in the text, the KAM barrier being shown as a red curve topologically equivalent to the inner and outer boundaries. In the case on the left, the symmetry breaking deformation parameter *d=0.01* wille on the right it is greater, *d=0.04* In the latter case the islands are clearly fatter and the chaotic regions around the island separatrices are larger.

jump across the barrier surface). A finite element method for solving the Beltrami equation, based on the variational principle, is being developed. Also the variational nature of the problem suggests the use of gradient- based optimization methods may be better than the iterative methods so far used, but, as the constraints do not automatically keep the rotational transforms fixed at the required irrational values [24], care will need be taken to control the rotational transforms at the boundaries during the minimization. The plasma will be stable not only to ideal- MHD instabilities but also to tearing and other non-ideal instabilities if the second variation of F is positive definite with respect to infinitesimal perturbations respecting the constraints. The stability problem has been studied in cylindrical geometry as a generalized eigenvalue problem by defining a Lagrangian $L = \delta^2 F - \lambda N$, with N a positive definite normalization. The stability condition is $\lambda \geq 0$ for all eigenvalues. Using normalization concentrated on the ideal- MHD barrier interfaces, the perturbed field in plasma regions is computed to be Beltrami ($\nabla \times B = \mu B$), with the same Lagrange multiplier μ as the equilibrium field. The interface equations between the relaxed regions produce an eigenvalue problem. In cylindrical geometry with axial periodicity, the displacement is Fourier decomposed, and displacements of the form $\exp i(m\theta + kZ)$ sought, where m is the poloidial mode number, and k the axial wave number. Hole et al. [26,27] have studied the stability of these configurations as a function of mode number and number of ideal barriers, and benchmarked these results to earlier single interface studies. Hole et al. also revealed a singular limit problem: the relaxed-MHD stability of a two-interface plasma differs, in the limit that the two interfaces merge, from that of the corresponding single-interface plasma.

The discrepancy has been resolved by Mills[28], who studied the stability of configurations in which the inter-barrier region was taken to be an ideal- MHD fluid rather than a relaxed region. In this case, the ideal stability of resonances in the inter-barrier region was handled explicitly, as opposed to the Woltjer-Taylor relaxed treatment, in which resonances do not explicitly feature. Plasmas with finite-width ideal- MHD barriers showed similar stability to the single-interface configuration in the limit as the barrier width went to zero. Mills concluded it is the different of treatment of resonances, which are implicit in Woltjer-Taylor relaxed plasmas, but explicit when computing ideal- MHD stability, that is responsible for reconciling the vanishing-interface-separation paradox. In more recent work, we have also shown that the tearing mode stability threshold of the plasmas is in agreement with found from the variational principle studied here. In ongoing work, we are also studying whether quantization in the toroidal direction leaves a stable residue of configurations in the parameter space. If so, these constrained minimum-energy states be related to internal transport barrier configurations, which are plasma configurations with good confinement properties that form that at sufficiently high heating power.

Other Entropy-Related Approaches

Maximum Entropy(MaxEnt) Principles in Plasmas and Fluids

By the second law of thermodynamics the entropy of a closed system increases monotonically, asymptoting towards a maximum as the approaches thermal equilibrium. Thus the application of equilibrium statistical mechanics theory can be viewed as a method for implementing the principle that systems tend towards a state of Maximum Entropy (MaxEnt). A review by Eyink and Sreenivasan [29] traces the use of the MaxEnt principle in turbulence theory back to Onsager's work (some un- published) in the 1940s. and cites some of the plasma and atmospheric physics literature where this approach has been used. [Onsager's equilibrium statistical mechanics is based on the Hamiltonian nature of inviscid vortex dynamics and is necessarily nondissipative. However it is very appropriate to the quasi-two-dimensional turbulence observed in strongly magnetized plasmas and planetary atmospheres (Fig. 1) where there is an inverse cascade to long wavelengths where viscous dissipation is weak]. Developments of the equilibrium statistical mechanics approach in the geophysical fluid dynamics context are further discussed in the article of Frederiksen and O'Kane [30] in this issue.

MaxEnt principles also occur in information theory as the least-informative estimate possible on the given information. This information theoretic entropy concept is used in Bayesian data analysis [31] and image processing, but Jaynes [32] also used it in physics to reinterpret statistical mechanics, with ramifications that are still being worked out today, including the Maximum Entropy Production(MEP) principle discussed below.

The traditional statistical mechanics approaches assume the system to be closed or in contact with a single heat bath, neither of which is appropriate to a real plasma/geophysical system where there are always fluxes of energy (and often matter) passing through the system due to heating near the centre of the plasma/planetary equator and cooling in the edge/polar regions. This problem can be overcome by the use of Jaynes [32] information-theory-based generalization of statistical mechanics with intensive variable(or parameter both) constraints, for thermodynamic systems, this equivalent to the use of canonical-like ensembles. This, to date, has been little used in fluid mechanics (and not at all in plasma physics), though it has been used to infer steady velocity distributions in internal turbulent flows. Such as hydraulic channels and pipes [33].

Maximum Entropy Production(MEP) Principles

In recent years the idea that a nonequilibrium system develops so as to maximize its entropy production (the MEP principle) has began to attract increasing attention as a potentially powerful way to predict how a complex open system will tend to evolve. Interest in the environmental sphere was originally sparked by the work of Paltridge [34] and the recent revival is partly due to the work of Roderick(C.) Dewar[35] using a general Jaynesian approach. Following the recent review of entropy production principles by Martyushev and Seleznev [36] we distinguish MEP in the nonequilibrium thermodynamics context from MEP in nonequilibrium statistical mechanics. In thermodynamics the entropy of the system (or subsystem) S is a state function like the internal Energy U, and these must satisfy the first and second laws of thermodynamics, $\delta Q = \delta W + dU$ and $TdS \geq \delta Q$, respectively, where T is the temperature, δQ is the net heat (in whatever form) entering the system (or subsystem) and δW the work done by the system or subsystem on the outside world (or other subsystems.) If the subsystem is a small volume element and local thermal equilibrium is assumed then equality can be assumed in the second law, but globally entropy increases due to heat flows described by generalized thermodynamic fluxes and forces. The authors of [36] base their discussion of nonequilibrium thermodynamics on a principle due to Ziegler and argue that this is sufficiently general that it covers both linear Onsager and Prigogine minimum entropy production principle, and linear and nonlinear maximum entropy production principles, reconciling then by their different interpretations.

Recently Yoshida and Mahajan [37] have constructed a nonlinear thermodynamic MEP "heat engine" model of transport barrier formation, in a plasma or fluid system between two heat baths at different temperature, exhibiting a critical temperature difference beyond which there is useful works δW available to drive flows (e.g., zonal flows) that reduce turbulent transport. The authors of [36] also discuss statistical mechanical formulations of a principle of MEP based on a principle of the most probable path in n-body phase space they trace back to work by Filyukov and Karpov in the late '60s. The modern approaches (e.g., [35]) also incorporate Jaynes' [32] ideas based on Bayesian statistics. However, [36] conclude that attempts to derive the MEP principle from microscopic first principles are so far unsatisfactory, as they require the introduction of additional hypotheses. They then go on to review the application of the MEP principle sciences.

Minimum Entropy Production Principles

In 1947 Prigogine proved a principle minimum entropy production and subsequently popularized his principle and applied it in physics chemistry and biology [36]. Although its use is limited to close-to-equilibrium systems, where thermodynamic forces and fluxes are linearly related, and it has its detractors on other grounds [38,39] a version [40] of minimum entropy production has had a small following in fusion plasma physics with claims of success in modeling some discharges [41]. However, because of its linearity it is not suited to modeling the emergence of nonlinear phenomena such as transport barriers.

Minimax Entropy Production Principles

Struchtrup and Weiss [42,43] introduce what they call a minimax principle in the context of extended thermodynamics, in which the global maximum of the local entropy production is minimized. There may also be transitional cases where entropy production is minimal with respect to some parameters and maximal with respect to others.

Conclusion

The multiregion-relaxation variational approach described in Sec. 3 holds strong promise of being the most satisfactory mathematical foundation on which to base the solution of the MHD toroidal equilibrium problem posed by Grad [44] over forty years ago. The numerical program we are developing will not only provide a practical tool for design and analysis of fusion experiments, but will allow numerical investigation of such fundamental issues as the critical point at which a KAM barrier ceases to be able sustain a pressure jump. One hopes this will stimulate further mathematical developments beyond KAM theory (see e.g., [45]) on the existence of KAM barriers and their breakup. While dissipationless MHD is a standard first-cut model in fusion plasma physics because of its (relative) mathematical tractability, it is clearly inadequate to describe much of the physics of the complex self-organizing system that is a hot, toroidally confined plasma. In particular, the lack of diffusive transport in the model allows the unphysically strong (infinite) gradients we have postulated to occur at a KAM barrier, and also more or less dictates our assumption of complete relaxation between the barriers. A first step away from this oversimplification has recently been taken by Hudson and Breslau [46], who used a simple anisotropic thermal diffusion to resolve the structure of the temperature profile in a chaotic magnetic field, revealing a much more complex structure than our current relaxation model can represent. Once dissipation is present a simple energy minimization variational principle is no longer appropriate, but it may still be possible to construct a variational relaxation model of plasma steady states by using the thermodynamic MEP principle with a phenomenological Ziegler entropy production function [36]. The Onsager MaxEnt approach has been partially explored in plasma physics, but its utility in climate modeling [30] suggests that it should be developed further. The use of Jaynesian MaxEnt approaches would appear to be an entirely open field in plasma physics, as is the use of statistical-mechanical MEP principles. Given the need for robust variational principles for predicting the overall behavior of fusion we plasmas, we expect entropy-based methods to be increasing importance in this field.

Acknowledgements

The first author wishes to thank the organizers of the AMSI/MASCOS Concepts of Entropy and their Applications Workshop, in particular Professor Philip Broadbridge, for the opportunity to present and discuss the topic of this paper in a stimulating environment. He acknowledges useful discussions with Drs Robert Niven and Jorgen Frederiksen. Some of the work presented was supported by U.S. Department of Energy Contact No. DE-AC02-76CH0373 and Grant No. DE-FG02-99ER54546 and the Australian Research

Council (ARC), Discovery Projects DP0452728 and DP0343765. The ARC Complex Open Systems Research Network, COSNet, grant RN0460006 also provided some support for the workshop.

References and Notes

1. Terry, P. W. Suppression of turbulence and transport by sheared flow. *Rev. Mod. Phys.* **2000**, 72, 109.
2. Diamond. P. H., Itoh, K., Itoh, S.-I., Hahm, T.S. Overview of zonal flow physics. In *E-Proceedings of the 20th IAEA Fusion Energy Conference, Vilamoura, Portugal 1-6 November 2004,* pages OV/2-1, Vienna, **2004**. IAEA, IAEA.
3. Dewar, R.L.: Abdullatif, R.F. Zonal flow generation by modulational instability. In Denier, J.P., Frederiksen, J. S.,editors, *Frontiers in Turbulence and Coherent structures: Proceedings of* the CSIRO/ COSNet Workshop on Turbulence , and Coherent *Structures, Canberra, Australia, 10–13 January 2006, volume 6 of World Scientific Lecture Notes in Complex Systems,* pages 145–430, Singapore, **2007**. World Scientific.
4. Hasegawa, A.; Maclennan, C. G.; Kodama, Y. Nonlinear behavior and turbulence spectra of drift waves and Rossby waves. *Phys. Fluids* **1979**, 22.2122.
5. Hasegawa, A.; Kodoma, Y. Erratum on "Nonlinear behavior and turbulence spectra of drift waves and Rossby waves" [Phys. Fluids **22**, 2122(1979)]. *Phys. Fluids* **1982**, 25, 2415.
6. Arrowsmith, D.K., Place, C. M. *An Introduction to Dynamical Systems.* Cambridge University Press, Cambridge, U.K., **1991**.
7. Lichtenberg, A.J.; Lieberman, M.A. *Regular and Chaotic Dynamics, 2nd ed.* Springer-Verlag, New York,**1992**.
8. Dnestrovskij, Y. N.; Dnestrovskij, A.Y.; Lysenko, S. E. Self- organization of plasma in tokamaks. *Plasma Phys. Rep.* **2005**, *31,529.*
9. Finn, J. M.; T. M. Antonsen. J. Turbulent relaxation of compressible plasmas with flow. *Physics of Fluids* **1983**, 26, 3540.
10. Kruskal, M.D.; Kulsrud, R. M. Equilibrium of a magnetically confined plasma in a toroid. *Phys. Fluids* **1958**, *1, 265.*
11. The pressure p in a fusion reactor is on the order of atmospheric, while the temperature is on the order of 10^8 K, so the particle density is on the order of a millionth of that in the atmosphere-insofar as local thermal equilibrium applies, the electrons and ions are ideal gases to a very good approximation.
12. Finn and Antonsen [9] observe that, in a fully relaxed system, extremizing energy at fixed entropy is equivalent to maximizing entropy at fixed energy. However, this is problematical in the multi-region relaxation model described in Sec.3 as there are separately conserved entropies in each region.
13. Morrison, P.J. Hamiltonian description of the ideal fluid. *Rev. Mod. Phys.* **1998**, *70, 467.*
14. Woltjer, L.A theorem on the force-free magnetic fields. *Proc. Nat. Acad. Sci. (U.S)* **1958**, *44, 489.*
15. Taylor, J.B. Relaxation and magnetic reconnection in plasmas. . Rev. Mod. *Phys.* **1986**, 58, 741.
16. Taylor, J.B. Relaxation of troroidal plasma and generation of reverse magnetic fields. Phys. Rev.Lett. **1974**, 33, 1139.
17. Yoshida, Z; Mahajan, S.M.; Ohsaki, S.; Iqbal, M.; Shatashvili, N. Beltrami fields in plasmas: High-confinement mode boundary layers and high beta equilibria . Phys. Plasmas **2001**, 8, 2125.
18. Yoshida, Z; Mahajan, S. M. Variational principles and self-organization in two- fluid plasmas. *Phys. Rev.* Lett. **2002**, *88, 085001.*
19. Ito, N.;Yoshida, Z. Statistical mechanics of magnetohydrodynamics. *Phys. Rev. E***1996**,*53, 5200.*
20. Minardi, E. The magnetic entropy concept. *J. plasma Phys.* **2005**, *71,53.*
21. Bhattacharjee, A.; Dewar, R.L. Energy principle with global invariants. *Phys. Fluids 1982, 25, 885.*
22. Berk, H.L,; Freidberg, J.P.; LIobet, X,; Morrison, P. J.; Tataronis, J. A. Existence and calculation of sharp boundary magnetohydrodynamic equilibrium in three-dimensional toroidal geometry. *Phys. Fluids* **1986**, *29, 3281.*

23. Kaiser, R.; Salat, A. Surface current equilibria from a geometric pont of view. Phys. Plasmas **1994**, 1, 281.

24. Hudson, S.R; Hole, M.J.; Dewar, R.L. Eigenvalue problems for Beltrami fields arising in a three-dimensional toroidal magnetohydrodynamic equilibrium, Phys. Plasmas 2007, 14, 052505.

25. Spis, G. O. Relaxed plamsa-vacuum systems with pressure. Phys. Plasmas **2003**, 10, 3030.

26. Hole, M.J.; Hudson, S.R.; Dewar, R.L Stepped pressure profile equilibria in cylindrical plasmas via partial Taylor relaxation. *J. Plasma Phys.* **2006**, *72,1167.*

27. Hole, M.J.; Hudson, S.R; Dewar, R.L Equilibria and stabilally relaxed Plasma-vacuum systems. *Nucl Fusion* 2007, *47,476.*

28. Mills, R. *3D plasma Equilibria* . Honours thesis, The Australian National University, Canberra ACT 0200, Australia, 2007.

29. Eyink, G.L.; Sreenivasan, K.R. Onsager and the theory of hydrodynamic turbulence. *Rev. Mod.Phys.* 2006. *78.87.*

30. Frederiksen, J.S.; O'Kane, T.J Entropy, closures and subgrid modeling. *Entropy* 2008, 10, submitted.

31. Sivia, D.S.; Skilling, J. *Data Analysis-a Bayesian Tutorial.* Oxford, University UK, 2nd edition, 2006.

32. Jaynes, E.T Information theory and statistical mechanics. *phys. rev.* 1957, *106, 620.*

33. Chiu, C-L.; Lin, G.F.; Lu, J.-M. Application of probability and entropy concepts in pipe-flow study. *J. Hydraulic. Eng.* (ASCE) 1993, *119,742.*

34. Palridge, G.W. A physical basis for a maximum of thermodynamic dissipation of the climate system. *Quart. J. Roy Met. Soc.* 2001, *127, 305.*

35. Dewar, R Information theory explanation of the fluctuation theorem maximum entropy production and self-organized criticality in non –equilibrium stationary states. *J. Phys. A: Math. Gen.* 2003, *36, 631.*

36. Martyushev, L.M.; Seleznev, V. D. maximum entropy production principle in physics, chemistry and biology, Physics.Reports **2006**, 426, 1.

37. Yoshida, Z.; Mahajan, S. M. "Maximum" entropy production in self- organized plasma boundary layer: a thermodynamic discussion about turbulent heat transport. Phys. Plasmas **2008**, 15, 032307.

38. Barbera, E. On the principle of minimal entropy production for navier-stokes-fourier fluids. Continuum Mechanics and Thermodynamics 1999, 11, 327.

39. Attard, P. Statistical mechanical theory for state systems. vi. variational principle. J. Chem. phys. 2006,125.214502.

40. Hameiri, E.; Bhattacharjee, A. Entropy production and plasma relaxation. Phys. Rev. A 1987, 35, 768.

41. Kucinski, M.Y. Minimum entropy production and mean field configuration in toroidal pinches. Plasma Phys. Control. Fusion 1998, 40,111.

42. Stuchtrup, H.; Weiss; W. Maximum of the local entropy production becomes minimal in stationary processes. Phys. Rev. Lett. 1998,80, 5048.

43. Struchtrup, H.; Weiss, W. Struchtrup and Weiss reply. Phys. Rev. Lett. 1998, 81, 5701.

44. Grad, H. Toroidal containment of a plasma. Phys. Fluids 1967, 10, 137.

45. Mackay, R.S. Greene's residue criterion. Nonlinearity 1992, 5, 161.

46. Hudson, S.R.; Breslau, J. Temperature contours and ghost surfaces for chaotic magnetic fields. Phys. Rev. Lett. 2008,100, 095001.

Generalized Complexity and Classical-Quantum Transition[†]

A.M. Kowalski,[1] Angelo Plastino[2] and Montserrat Casas[3]

[1]Comision de Investigaciones Cientificas (CIC), Argentina. Email: kowalski@fisica.unlp.edu.ar.
[2]La Plata Physics Institute (IFLP), Exact Sciences Fac., National University (UNLP), Argentina's National Research Council (CCT-CONICET), C.C. 727, (1900) La Plata, Argentina. Email: plastino@fisica.unlp.edu.ar.
[3]Departament de Física and IFISC, Universitat de les Illes Balears, 07122 Palma de Mallorca, Spain. Email: montse.casas@uib.es.

ABSTRACT

We investigate the classical limit of the dynamics of a semi classical system that represents the interaction between matter and a given field. On using as a quantifier the q-Complexity, we find that it describes appropriately the quantum-classical transition, detecting the most salient details of the changeover. Additionally the q-Complexity results a better quantifier of the problem than the q-entropy, in the sense that the q-range is enlarged, describing the q-Complexity, the most important characteristics of the transition for all q-value.

Keywords: Generalized entropy; Semiclassical theories; Quantum chaos; Statistical complexity.

Introduction

Quantifiers based on information theory, like entropic forms and statistical complexities (see as examples [1–4]) have proved to be quite useful in the characterization of the dynamics associated to time series, in the wake of the pioneering work of Kolmogorov and Sinai, who converted Shannon's information theory into a powerful tool for the study of dynamical systems [5,6]. In turn, information theory measures and probability spaces

---

[†]Reused with permission from: Kowalski, A.M.; Plastino, A.; Casas, M. Generalized Complexity and Classical-Quantum Transition. *Entropy* 2009, *11*, 111–123.

Ω are inextricably linked quantifiers. In the evaluating them, the determination of the probability distribution P associated to the dynamical system or time series under study is the basic ingredient. Many procedures have been proposed for the election of $P \in \Omega$. We can mention techniques based on symbolic dynamics [7]. Fourier analysis [8], and wavelet transform [9] (among others). The applicability of these approaches depends on the data-characteristics, i.e., stationarity, length of the series, parameter-variations, levels of noise-contamination, etc. The distinct treatments at hand "capture" the global aspects of the dynamics, but they are not equivalent in their ability to discern physical details. However, one should recognize that we are here referring to techniques defined in an ad-hoc fashion, not derived directly from the dynamical properties of the pertinent system themselves.

Statistical complexity. In [3], López-Ruiz, Mancini and Calbet (LMC) advanced a statistical complexity measure (SCM) based on the notion of "disequilibrium" as a quantifier of the degree of physical structure in a time series. Given a probability distribution associated with a system's state, the LMC measure is the product of a normalized entropy H times a distance to the uniform-equilibrium state Q. It vanishes for a totally random process and for a periodic one. Martín et al. [10] improved on this measure by modifying the distance-component (in the concomitant probability space). In Ref. [10], Q is built-up using Wootters' statistical distance while H is a normalized Shannon-entropy. Regrettably enough, the ensuing statistical complexity measure is neither an intensive nor an extensive quantity, although it does yield useful results. A reasonable complexity measure should be able to distinguish among different degrees of periodicity and it should vanish only for periodicity unity. In order to attain such goals it would seem desirable to give this statistical measure an intensive character. This was achieved in Ref. [4] obtaining a SCM that is (i) able to grasp essential details of the dynamics, (ii) an intensive quantity, and (iii) capable of discerning among different degrees of periodicity and chaos.

Deformed q-statistics. It is a well-known fact that physical systems that are characterized by either long-range interactions, long-term memories, or multi-fractal nature, are best described by a generalized statistical mechanics' formalism [11] that was proposed 20 years ago: the so-called q-statistics. More precisely, Tsallis [12] advanced in 1988 the idea of using in a thermodynamics' scenario an entropic form, the Harvda-Chavrat one, characterized by the entropic index $q \in R$ ($q = 1$ yields the orthodox Shannon measure):

$$S_q = \frac{1}{q-1} \sum_{i=1}^{N_S} \left[p_i - (p_i)^q \right], \tag{1}$$

where p_i are the probabilities associated with the associated N_s different system-configurations. The entropic index (or deformation parameter) q describes the deviations of Tsallis entropy from the standard Boltzmann-Gibbs-Shannon-one

$$S = -\sum_{i=1}^{N_S} p_i \ln(p_i). \tag{2}$$

It is well-known that the orthodox entropy works best in dealing with systems composed of either independent subsystems or interacting via short-range forces whose subsystems can access all the available phase space [11]. For systems exhibiting long-range correlations, memory, or fractal properties, Tsallis' entropy becomes the most appropriate mathematical form [13–16].

Quantum-classical frontier. The classical limit of quantum mechanics (CLQM) continues attracting the attention of many theorists and is the source of much exciting discussion (see, for instance, Refs. [17–18] and references therein). In particular, the investigation of a "quantum" chaotic motion is considered important in this limit. Recent literature provides us with many examples, although the adequate definition of the underlying phenomena is understood in diverse fashion according to the different authors (see Ref [19] and references therein).

It is reasonable to relay on q-statistics [36], so as to gather insights into the

Quantum—semiclassical—classical transition (CLQM). (3)

Why? Because we know that the classic to quantum route traverses high complexity regions of the appropriate phase space where chaos reigns, interrupted often by quasi-periodic windows [19–21]. In the semiclassical parcel of the associated trajectory one encounters also strong correlation between classical and quantum degrees of freedom [20–21].

In [36] we showed that a wavelet-evaluated q-entropy not only describes correctly the quantum-classical border but also that the associated deformation-parameter q itself characterizes the different regimes involved in the concomitant process, detecting the most salient fine details of the transition. The purpose of the present effort is to gather new insights into the q-statistics contribution to this problem by recourse to a new tool: the q-statistical complexity. Since in this work the pertinent q-quantifiers are computed using "wavelet techniques" (whose utility has been evidenced in ([20–21]), we provide a brief wavelet-résumé in the Appendix.

A Semi-classical Model and the CLQM

Quite a bit of quantum insight is to be gained from semiclassical perspectives. Several methodologies are available (WKB, Born-Oppenheimer approach, etc.). Here we consider two interacting systems: a classical and quantal ones. This can be done whenever the quantum effects of one of the two systems are negligible in comparison to those of the other one. Examples can be readily found. We can just mention Bloch-equations [23], two-level systems interacting with an electromagnetic field within a cavity, Jaynes-Cummings semiclassical model [24–27], collective nuclear motion [28], etc. We shall focus attention upon a special bipartite model [29–31] that has been found useful with reference to problems in such diverse fields as chaos, wave-function collapse, measurement processes, and cosmology [32]. In order to investigate the q-statistics' contribution to the CLQM problem by recourse to the q-statistical complexity (our goal here) we shall consider a trivial generalization of the semi-classical Hamiltonian that represents the zero-th mode contribution of a strong external field to the production of charged meson pairs [30–31]. It reads

$$\hat{H} = \frac{1}{2}\left(\frac{\hat{p}^2}{m_q} + \frac{P_A^2}{m_{cl}} + m_q\omega^2\hat{x}^2 \right), \tag{4}$$

where i) $\hat{x}$ and $\hat{p}$ are quantum operators, ii) A and P_A classical canonical conjugate variables and iii) $\omega^2 = \omega_q^2 + e^2 A^2$ is an interaction term that introduces nonlinearity, ω_q being a frequency. The quantities m_q and m_{cl} are masses, corresponding to the quantum and classical systems, respectively. As shown in Ref. [33], in dealing with (4) one faces an autonomous system of nonlinear coupled equations

$$\frac{d\langle\hat{x}^2\rangle}{dt} = \frac{\langle\hat{L}\rangle}{m_q}, \quad \frac{d\langle\hat{p}^2\rangle}{dt} = -m_q\omega^2\langle\hat{L}\rangle, \quad \frac{d\langle\hat{L}^2\rangle}{dt} = 2\left(\frac{\langle\hat{p}^2\rangle}{m_q} - m_q\omega^2\langle\hat{x}^2\rangle \right),$$

$$\frac{dA}{dt} = \frac{P_A}{m_{cl}}, \quad \frac{dP_A}{dt} = -e^2 m_q A\langle\hat{x}^2\rangle, \quad \hat{L} = \hat{x}\hat{p} + \hat{p}\hat{x}. \tag{5}$$

The system of Eqs. (5) follows immediately from Ehrenfest's relations [33]. To study the classical limit we need to also consider the classical counterpart of the Hamiltonian (4)

$$H = \frac{1}{2}\left[\frac{p^2}{m_q} + \frac{P_A^2}{m_{cl}} + m_q\left(\omega_q^2 + e^2 A^2\right)x^2 \right], \tag{6}$$

where all the variables are classical. Recourse to Hamilton's equations allows one to find the classical version of Eqs. (5) (see Ref. [33] for details). Let i) E stand for the total energy of the system and ii) I be an invariant of the motion described by the system (5), related to the Uncertainty Principle, that reads

$$I = \langle\hat{x}^2\rangle\langle\hat{p}^2\rangle - \frac{\langle\hat{L}^2\rangle}{4}. \tag{7}$$

It is easy to see that a classical computation of I yields $I = x^2 p^2 - \dfrac{L^2}{4} \equiv 0$. The classical limit is obtained by letting [33] the "relative energy"

$$E_r = \frac{|E|}{I^{\frac{1}{2}}\omega_q} \rightarrow \infty. \tag{8}$$

A measure of the degree of convergence between classical and quantum results in the limit of Eq. (8) is given by the norm N of the vector $\Delta u = u - u_{cl}$ [33]

$$N_{\Delta u} = |u - u_{cl}|, \tag{9}$$

where the three components vector $u = \left(\langle\hat{x}^2\rangle, \langle\hat{p}^2\rangle, \langle\hat{L}^2\rangle\right)$ is the "quantum" part of the solution of the system Eqs. (5) and $u_{cl} = \left(x^2, p^2, L\right)$ its classical partner. A detailed study of our present model was performed in Refs. [33-34]. We summarize here the main results of these references that are pertinent for our discussion. In plotting diverse dynamical quantities versus E_r (as it grows from unity to ∞), one finds *an abrupt change in the system's dynamics for special values of E_r to be denoted by E_r^{cl}*. From this value onwards, the pertinent dynamics starts converging to the classical one. It is thus possible to assert that E_r^{cl} provides us with an indicator of the presence of a quantum-classical "border". The zone

$$E_r < E_r^{cl} \tag{10}$$

corresponds to the semi-quantal regime investigated in Ref. [34]. This regime, in turn, is characterized by two different sub-zones [33]. i) One of them is an almost purely quantal one, in which the microscopic quantal oscillator is just slightly perturbed by the classical one, and ii) the other section exhibits a transitional nature (semi-quantal). The border between these two sub-zones can be well characterized by a "signal" value E_r^{P}. A significant feature of this point resides in the fact that, for $E_r \geq E_r^{P}$, chaos is always found. The relative number of chaotic orbits (with respect to the total number of orbits) grows with E_r and tends to unity for $E_r \to \infty$ [33–34]. Thus, as E_r grows from $E_r = 1$ (the "pure quantum instance") to $E_r \to \infty$ (the classical situation), a significant series of morphology-changes is detected, specially in the transition-zone ($E_r^{P} \leq E_r \leq E_r^{cl}$). The concomitant orbits exhibit features that are not easily describable in terms of Eq. (9), which is a global measure of the degree of convergence in amplitude (of the signal). What one needs instead is a statistical type of characterization, as that described in Refs. [20–21–35].

Previous q-entropy Results

In [36], we found that the normalized Tsallis wavelet entropy H_{S_q}, in the range $0 < q < 5$, correctly describes the "E_r-evolution", identifying the stages of the transition. As a second result we ascertained that within the subrange $0.1 < q < 0.4$, H_{S_q}, portrays the quantum sector, something that Shannon's measure is unable to do, concluding that it is the most appropriate entropy, and not the orthodox, $q = 1$ of Shannon's. Additionally, we discovered other transition-detectors in addition to the normalized Tsallis-entropy, specially its curvature when we plot it for that particular q-value, q_M, for which H_{S_q} has a minimum. q_M itself turned out to a good transition-indicator. These last results affirm that the deformation parameter q by itself can be regarded as the "looking glass" through which one can observe the quantum-classical transition. We are ready now to start presenting the new results of this contribution: the role played by the q-complexity in describing the route from the quantum regime to the classical one.

Present Results

Introducing the q-statistical Complexity

The Statistical Complexity can be viewed as a functional $C[P]$ that characterizes the probability distribution P associated to the time series generated by the dynamical system under study. It quantifies not only randomness but also the presence of correlational structures [3-4-10]. This quantity is of the form [22]

$$C[P] = Q[P, P_e] H_S[P], \tag{11}$$

where, to the probability distribution P, we associate the entropic measure $H_S[P] = S[P]/S_{max}$, with $S_{max} = S[Pe] (0 \leq H_S \leq 1)$. P_e is the uniform distribution and S is an entropy. We take here the disequilibrium Q to be defined in terms of the extensive Jensen divergence [4] by

$$Q = Q_J[P, P_e] = Q_0 \{S[(P + P_e)/2] - S[P]/2 - S[P_e]/2\}. \tag{12}$$

with Q_0 a normalization constant ($0 \leq Q_J \leq 1$). We denote the q-entropy (1) by S in (11) and (12). Our wavelet statistical complexity adopts then the following form

$$C_{q,J}[P] = Q_{q,J}[P, P_e] . H_{Sq}[P], \tag{13}$$

with

$$H_{S_q}[P] = S_q[P]/S_{q,max} = \frac{1}{1 - N_J^{1-q}} \sum_{j=-1}^{-N_J} (p_j - p_j{}^q), \tag{14}$$

a   normalized   wavelet   q-entropy   (NTWE)   (see   Appendix)   and

$Q_J[P, P_e] = Q_0 \{S_q[(P + P_e)/2] - S_q[P]/2 - S_q[P_e]/2\}$. The set $P = \{p_j\}$ is given by Eq. (19).

Numerical Results

By recourse to the wavelet statistical complexity $C_{q,J}$ (13), we will be able to characterize the details that pave the road towards the classical limit, accruing additional advantages over the q-entropy description. In obtaining our numerical results we choose $m_q = m_{cl} = \omega_q = e = 1$ for the system's parameters. As for the initial conditions for solving the system (5) we take $E = 0.6$, i.e., we fix E and then vary I so as to obtain our different E_r-values. Additionally, we have $\langle L \rangle(0) = L(0) = 0$ and $A(0) = 0$ (both in the quantum and the classical instances). $\langle x^2 \rangle(0)$ takes values in the interval $x^2(0) < \langle x^2 \rangle(0) \leq 0.502$, with $x^2(0) = 0.012$.

The first task is to evaluate the set $P = \{p_j\}$ with p_j given by (19) [Cf. (11) and (13)]. Our data points are the solutions of (5), from which we extract the values of $\langle x^2 \rangle$ and the (classical) values of x^2 at the time t (for a fixed E_r) (We have also performed these calculations extracting instead $\langle p^2 \rangle - p^2$ together with $\langle L \rangle - L$ and obtained entirely similar results to those reported below). We will deal with 2^{12} data-points, for each orbit. We define eight ($N_J = 8$) resolution levels $j = -1, 2, ..., -N_J$ for an appropriate wavelet analysis within the multi-resolution scheme of the Appendix . The p_j yield, at different scales, the energy probability distribution and in very many instances the NTWE has been found to constitute a suitable tool for detecting and characterizing specific phenomena.

We find, as first result, that $C_{q,J}$ correctly distinguishes the three zones or sections of our process, i.e., quantal, transitional, and classic, as delimited by, respectively, $E_r^P = 3.3282$ and $E_r^{cl} = 21.55264$, for all values of q, although the quality of the description steadily worsens for $q \to \infty$ (See Figs. 1, 2, 3, and 4, where we depict $C_{q,J}$ vs. E_r for different q-values). In Fig (1b) we have include the "Shannon case" C_J, i.e., the corresponding wavelet complexity evaluated with the Shannon entropy in (11). Notice the abrupt change of in the slope of the curve taking place at E_r^P, where a local minimum is detected for $q > 0.2$ (Fig. 1a). The transition zone is clearly demarcated between that point and E_r^{cl}. From here on $C_{q,J}$ tends to its classical value at the same time that the solutions of (5) begin to converge towards the classical ones. There are however some transition-details that are not well represented by $C_{q,J}$, for some q-values. We thus need to ascertain which is the appropriate q-range.

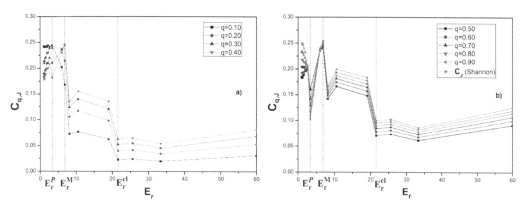

Figure 1 q-Statistical Complexity $C_{q,J}$ vs. E_r for $q \le 0.4$ (Fig. 1a) and $0.5 \le q < 1$ (Fig. 1b). Shannon's complexity are also displayed. Three zones are to be differentiated. They are delimited by special E_r-values, namely, $E_r^P = 3.3282$ and $E_r^{cl} = 21.55264$. Notice the local complexity maximum at $E_r^M = 6.8155$.

In general, the most noticeable $C_{q,J}$-modifications as q varies take place in the quantal zone, specially for $q < 1$ (Figs. 1a-b) and in the transition zone. In the quantum-classic route, an important milestone is found at $E_r = E_r^M$. This point can be detected, within the transition zone, at the value $\approx E_r = E_r^M = 6.8155$, for $0.2 < q \le \approx 17$, where a local complexity maximum (lcm) can be appreciated (Figs. 1, 2, 3 and 4) together with a great alteration in the system's dynamics (i.e., in the solutions of (5), which can be verified via Poincare's sections [34]). E_r^M divides into two sections the transitional region, one in

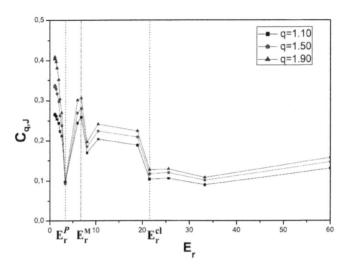

Figure 2 q-Statistical Complexity $C_{q,J}$ vs. E_r for $1 < q < 2$. The three zones and the point E_r^M of Fig. 1 are also seen here.

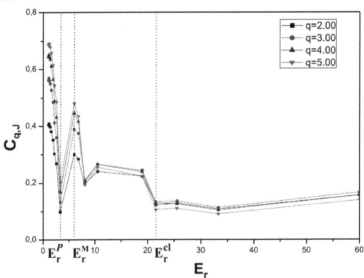

Figure 3 q-Statistical Complexity $C_{q,J}$ vs. E_r for $2 \leq q \leq 5$. No great changes are observed.

which the quantum-classical mixture characterizes a phase-space with more non-chaotic than chaotic curves and other, in which this feature is reversed [34]. The lcm becomes more pronounced as q grows up to $q=9$, and then becomes less and less noticeable, disappearing for $q > \approx 17$ (Fig. 4).

Notice also that for $0.7 \leq q < 1$ (Fig. 1b) and $1 < q < 2$ (Fig. 2), if $q \to 1$, the q-complexity behavior resembles more and more the Shannon-one of C_J. The above picture suffers no great changes for q, save for the above mentioned changes of the local maximum.

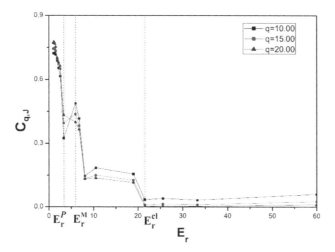

Figure 4 q-Statistical Complexity $C_{q,J}$ vs. E_r for $10 \leq q \leq 20$. The local maximum at $E_r{}^M$ disappears.

In view of these considerations, together with the fact that one obviously wishes for a H_{S_q}-minimum at $E_r{}^P$, we can assert that our q-quantifiers should be built up in q-range $0.2 < q \leq 17$. The q-influence on our transition-processes is clearly appreciated in Figs. 5, that plots $C_{q,J}$ vs. q for different values of our all important quantity E_r. The corresponding Shannon statistical complexity value (horizontal line) is included in all graphs for comparison's sake. Figs. 5a)-5b) correspond to the quantum sector, while Figs. 5c), 5d), 5e), and 5f) refer to the transitional one, and, finally, Figs. 5g)-5h) allude to the classical region. Although H_{S_q} possesses only one minimum as a function of q [36], $C_{q,J}$ instead may exhibit either a minimum and/or a maximum, plus one or more saddle-points. Consequently, $C_{q,J}$ intersects Shannon's curve C_J at least at one point, i.e., i) at $q=1$ and at one or more points, depending on E_r.

We find that distinct quantum-zone's graphs resemble each other. Ditto for the classical counterparts. Both kinds are clearly different objects, though. See Figs. 5a)-5b) and 5g)-5h), respectively. Moreover, plots corresponding to the neighborhood of $E_r{}^P$ point, where the transition region together with chaos begin exhibit another kind of morphology (Fig. 5c). For the transition zone two types of picture can be drawn, corresponding to $E_r \leq E_r{}^M$ (with a phase-space with more non-chaotic than chaotic curves) and $E_r > E_r{}^M$ (with a phase-space in which this aspect is reversed). The two subregions are displayed in Figs. 5d)-5e) (corresponding to $E_r = E_r{}^M$ (1st. sub-zone), and Fig. 5f) (2nd. subregion).

The later tends to resemble the aspect of Figs. 5g)-5h) (classical sector). Thus, $C_{q,J}$ as a function of q is perfectly able, by itself, i) of "detecting" important dynamical features like the "Signal Point" $E_r{}^P$ and ii) of distinguishing between the two transitional sub-regions, and iii) registering the similarities between the second of these two and the classical one.

Figure 5 q-Statistical Complexity $C_{q,J}$ for different E_r-values. Quantal (Figs. 5a-5b), transitional (Figs. 5c, 5d, 5e and 5f) and classic (5g-5h). The curves corresponding to the quantal zone resemble each other and exhibit a different aspect compared to those pertaining to the classical region.

Conclusions

We have studied in this communication, the classical-quantal frontier of the dynamics governed by a semi-classical Hamiltonian that represents the zero-th mode contribution of an strong external field to the production of charged meson pairs. This study was encompassed within the strictures of the so-called q-statistics and by recourse to a new tool: the *q*-Statistical Complexity (13) evaluated by performing a wavelet-band analysis.

The highlights of the road towards classicality are described by recourse to the relative energy E_r given by (8). As E_r grows from $E_r = 1$ (the "pure quantum instance") to $E_r \to \infty$ (the classical situation), a significant series of morphology-changes is detected for the solutions of the system of nonlinear coupled equations (5). The concomitant process takes place in three stages: quantal, transitional, and classic, delimited, respectively, by special values of E_r, namely, E_r^P and E_r^{cl}.

We encounter as a first result that $C_{q,J}$ distinguishes correctly for all value of q, the three sections of our process, i.e., quantal, transitional, and classic, as delimited by, respectively, E_r^P and E_r^{cl}. The description suffers a gradual deterioration process as $q \to \infty$, a rather important result in view of the fact that the q-entropy H_{S_q} is only able to distinguish our three regions in the range $0 < q < 5$. Such a fact makes the q-Statistical Complexity a much better quantifier than the q-Entropy for the description of a very involved process. As a second we determine an optimal q-range $O = [0.2 < q \le 17]$, much larger than the above quoted one for H_{S_q}. Within O our complexity-tool distinguishes a value $E_r = E_r^M$ within the transition zone (TZ) in which the complexity exhibits a local maximum. We can partition the TZ at $E_r = E_r^M$ into two subsections: one in which the quantum-classical mixture characterizes a phase-space with more non-chaotic than chaotic curves and another in which this aspect is reversed. Finally, we find that $C_{q,J}$, as a function of q is a good "detector" of transitional features (see Figs. 5): a) it identifies E_r^P, starting point of the transitional sector, where chaotic behavior begins to emerge and b) it identifies E_r^M, i.e., it distinguishes between the two subsections into which the transitional region divides itself. These last results reconfirm a previous one obtained for H_{S_q} in [36], namely, that the parameter q by itself can be regarded as the "looking glass" through which one can observe the quantum-classical transition.

Acknowledgments

AMK are supported by CIC of Argentina. The authors thank Prof. Maria Teresa Martin for her help in the computational aspects.

A. Normalized Tsallis Wavelet Entropy

Wavelet analysis is a suitable tool for detecting and characterizing specific phenomena in time and frequency planes. The *wavelet* is a smooth and quickly vanishing oscillating function with good localization in both frequency and time.

A wavelet family $\psi_{a,b}(t) = |a|^{-1/2} \psi\left(\dfrac{t-b}{a}\right)$ is the set of elementary functions generated by dilations and translations of a unique admissible *mother wavelet* $\psi(t)$. $a, b \in R, a \ne 0$ are the scale and translation parameters respectively, and t is the time. One have a unique analytic pattern and its replications at different scales and with variable time localization.

For special election of the mother wavelet function $\psi(t)$. and for the discrete set of parameters, $a_j = 2^{-j}$ and $b_{j,k} = 2^{-j}k$, with $j, k \in Z$ (the set of integers) the family

$$\psi_{j,k}(t) = 2^{j/2}\psi\left(2^j t - k\right), \quad j, k \in Z \tag{15}$$

constitutes an orthonormal basis of the Hilbert space $L^2(R)$ consisting of finite-energy signals.

The correlated decimated discrete wavelet transform provides a non-redundant representation of the signal X, and the values $\langle X, \psi_{j,k}\rangle$ constitute the coefficients in a wavelet series. These wavelet coefficients provide relevant information in a simple way and a direct estimation of local energies at the different scales. Moreover, the information can be organized in a hierarchical scheme of nested subspaces called multiresolution analysis in $L^2(R)$. In the present work, we employ orthogonal cubic spline functions as mother wavelets. Among several alternatives, cubic spline functions are symmetric and combine in a suitable proportion smoothness with numerical advantages.

In what follows, the signal is assumed to be given by the sampled values $x(n)$, $n = 1,..., N$. If the decomposition is carried out over all resolutions levels the wavelet expansion will read $(N_J = \log_2(N)$

$$X(t) = \sum_{j=-N_j}^{-1} \sum_k C_j(k)\psi_{j,k}(t) = \sum_{j=-N_j}^{-1} r_j(t), \tag{16}$$

where $C_j(k)$ are the wavelet coefficients and $r_j(t)$ is the detail signal at scale j.

Since the family $\{\psi_{j,k}(t)\}$ is an *orthonormal basis* for $L^2(R)$, the concept of energy is linked with the usual notions derived from Fourier's theory. The signal energy, at each resolution level $j = -1,..., -N_J$, will be the energy of the corresponding detail signal,

$$E_j = \left\|r_j\right\|^2 = \sum_k \left|C_j(k)\right|^2 \tag{17}$$

The total energy can be obtained in the fashion

$$E_{tot} = \left\|X\right\|^2 = \sum_{j=-N_j}^{-1}\sum_k \left|C_j(k)\right|^2 = \sum_{j=-N_j}^{-1} E_j. \tag{18}$$

Finally, we define the normalized p_j-values, which represent the relative wavelet energy

$$p_j = E_j / E_{tot} \tag{19}$$

for the resolution levels $j = -1,..., -N_J$. The p_j yield, at different scales, the probability distribution for the energy. Clearly, $\sum_j p_j = 1$ and the distribution $\{p_j\}$ can be considered as a time-scale density that constitutes a suitable tool for detecting and characterizing specific phenomena in both the time and the frequency planes.

The normalized Tsallis wavelet entropy (NTWE) is just the normalized Tsallis entropy associated to the probability distribution P,

$$H_{S_q}[P] = S_q[P] / S_{q,\max} = \frac{1}{1 - N_J^{1-q}} \sum_{j=-1}^{-N_J}\left(p_j - p_j^q\right), \tag{20}$$

where $S_{max} = \dfrac{1 - N_J^{1-q}}{q-1}$ is attained for the equiprobable distribution $P_e = \left\{ 1/N_j, ..., 1/N_j \right\}$.

The NTWE appears as a measure of the degree of order/disorder of the time series. It provides useful information about the underlying dynamical process associated with the series. Indeed, a very ordered process can be represented by a periodic mono-frequency signal (signal with a narrow band spectrum). A wavelet representation of such a signal will be resolved at one unique wavelet resolution level, i.e., all relative wavelet energies will be (almost) zero except at the wavelet resolution level which includes the representative series frequency. For this special level the relative wavelet energy will (in our chosen energy units) almost equal unity. As a consequence, the NTWE will acquire a very small, vanishing value. A signal generated by a totally random process or chaotic one can be taken as representative of a very disordered behavior. This kind of signal will have a wavelet representation with significant contributions coming from all frequency bands. Moreover, one could expect that all contributions will be of the same order. Consequently, the relative wavelet energy will be almost equal at all resolutions levels, and the NTWE will acquire its maximum possible value.

References and Notes

1. Shannon, C.E. A mathematical theory of communication. *Bell System Technol. J.* **1948**, *27*, 379–390.
2. Shiner, J.S.; Davison, M.; Landsberg, P.T. Simple measure for complexity. *Phys. Rev. E* **1999**, *59*, 1459–1464.
3. López-Ruiz, R.; Mancini, H.L.; Calbet, X. A statistical measure of complexity. *Phys. Lett. A* **1995**, *209*, 321–326.
4. Lamberti, P.W.; Martin, M.T.; Plastino, A.; Rosso, O.A. Instensive entropic non-triviality measure. *Physica A* **2004**, *334*, 119–131.
5. Kolmogorov, A.N.; A new metric invariant of transitive dynamic system and automorphysms in Lebesgue spaces. *Dokl. Akad. Nauk SSSR* **1958**, *119*, 861–864.
6. Sinai, Y.G.; On the concept of entropy of dynamical system. *Dokl. Akad. Nauk SSSR* **1959**, *124*, 768–771.
7. Mischaikow, K.; Mrozek, M.; Reiss, J.; Szymczak A. Construction of Symbolic Dynamics from Experimental Time Series. *Phys. Rev. Lett.* **1999**, *82*, 1144–1147.
8. Powell, G.E.; Percival, I.C. A spectral entropy method for distinguishing regular and irregular motion of hamiltonian systems. *J. Phys A: Math. Gen.* **1979**, *12*, 2053–2071.
9. Rosso, O.A.; Mairal, M.L. Characterization of time dynamical evolution of electroencephalographic records. *Physica A* **2002**, *312*, 469–504.
10. Mart´ın, M.T.; Plastino, A.; Rosso, O. A. Statistical complexity and disequilibrium. *Phys. Lett. A* **2003**, *311*, 126–132.
11. Hanel, R.; Thurner, S. Generalized Boltzmann Factors and the Maximum Entropy Principle: Entropies for Complex Systems. *Physica A* **2007**, *380*, 109–114.
12. Tsallis, C. Possible generalization of Boltzmann-Gibbs statistics. *J. Stat. Phys.* **1988**, *52*, 479–487.
13. Alemany, P.A.; Zanette, D.H. Fractal random walks from a variational formalism for Tsallis entropies. *Phys. Rev. E* **1994**, *49*, R956–R958.
14. Tsallis, C. Nonextensive thermostatistics and fractals. *Fractals* **1995**, *3*, 541–547.
15. Tsallis, C.; Generalized entropy-based criterion for consistent testing. *Phys. Rev. E* **1998**, *58*, 1442–1445.

16. Kalimeri, M.; Papadimitriou, C.; Balasis, G.; Eftaxias, K. Dynamical complexity detection in preseismic emissions using nonadditive Tsallis entropy. *Physica A* **2008**, *387*, 1161–1172.

17. Paz, J.P.; Zurek, W.H. Quantum Limit of Decoherence: Environment Induced Superselection of Energy Eigenstates. *Phys. Rev. Lett.* **1999**, *82*, 5181–5185.

18. Emerson J.; Ballentine, L.E. Quantum-classical correspondence for the equilibrium distributions of *Entropy* 2009, *11* 123 two interacting spins. *Phys. Rev. E* **2001**, *64*, 026217:1-026217:11.

19. Kowalski, A.M.; Martin, M.T.; Plastino, A.; Proto, A.N. Classical Limit and Chaotic Regime in a Semi-Quantum Hamiltonian. *Int. J. Bifurcation Chaos* **2003**, *13*, 2315–2325.

20. Kowalski, A.M.; Martin, M.T.; Plastino, A.; Proto, A.N.; Rosso, O.A.Wavelet statistical complexity analysis of the classical limit. *Phys. Lett. A* **2003**, *311*, 180–191.

21. Kowalski, A.M.; Martin, M.T.; Plastino, A.; Rosso, O.A. Entropic Non-Triviality, the Classical Limit, and Geometry-Dynamics Correlations. *Int. J. Mod. Phys. B* **2005**, *14*, 2273–2285.

22. Martin, M.T.; Plastino, A.; Rosso, O.A. Generalized statistical complexity measures: geometrical and analytical properties. *Physica A* **2006**, *369*, 439–462.

23. Bloch, F.; Nuclear Induction. *Phys. Rev.* **1946**, *70*, 460–474.

24. Meystre, P.; Sargent, M., III. *Elements of Quantum Optics*; Springer-Verlag, New York/Berlin, **1991**.

25. Bulgac, A.; Configurational quasidegeneracy and the liquid drop model. *Phys. Rev. C* **1989**, *40*, 1073–1076.

26. Milonni, P.W.; Shih, M.L.; Ackerhalt, J. R. Chaos in Laser-Matter Interactions; World Scientific Publishing: Singapore, 1987.

27. Kociuba, G.; Heckenberg, N. R. Controlling the complex Lorenz equations by modulation. *Phys. Rev. E* **2002**, *66*, 026205:1-026205:5.

28. Ring, P.; Schuck, P. *The Nuclear Many-Body Problem*; Springer-Verlag, New York/Berlin, **1980**.

29. Bonilla, L.L.; Guinea, F. Collapse of the wave packet and chaos in a model with classical and quantum degrees of freedom. *Phys. Rev. A* **1992**, *45*, 7718–7728.

30. Cooper, F.; Habib, S.; Kluger, Y.; Mottola, E. Nonequilibrium dynamics of symmetry breaking in $\dot{A}4$ theory. *Phys. Rev. D* 1997 *55*, 6471–6503.

31. Cooper, F.; Dawson, J.; Habib, S.; Ryne, R. D. Chaos in time-dependent variational approximations to quantum dynamics. *Phys. Rev. E* **1998**, 57, 1489–1498.

32. Chung, D.J.H.; Classical inflaton field induced creation of superheavy dark matter. *Phys. Rev. D* **2003**, *67*, 083514:1-083514:14.

33. Kowalski, A.M.; Martin, M.T.; Nu~nez, J.; Plastino, A.; Proto, A. N. A quantitative indicator for semi-quantum chaos. *Phys. Rev. A* **1998**, *58*, 2596–2599.

34. Kowalski, A.M.; Plastino, A.; Proto, A.N. Classical limits. *Phys. Lett. A* **2002**, *297*, 162–172.

35. Kowalski, A.M.; Martin, M.T.; Plastino, A.; Proto, A.N.; Rosso, O.A. Bandt-Pompe approach to the classical-quantum transition. *Physica D* **2007**, *233*, 21–31.

36. Kowalski, A.M.; Martin, M.T.; Plastino, A.; Zunino, L. Tsallis' deformation parameter q quantifies the classical-quantum transition. [arXiv:0812.4221v1], 2008; *Physica A*, 2009 (in Press).

A New Mechanical Model for Particle Transport by Surface Waves and Applications[†]

Minvydas Ragulskis,[1] *Edita Sakyte,*[1] *Jesús M. Seoane*[2]
and *Miguel A.F. Sanjuán*[2,a]

[1]Department of Mathematical Research in Systems, Kaunas University of Technology, Studentu 50-222, 51638 Kaunas, Lithuania.
[2]Nonlinear Dynamics, Chaos and Complex Systems Group, Department of Physics, University Rey Juan Carlos, Tulipan s/n, 28933 Mostoles, Madrid, Spain.
[a]Email: miguel.sanjuan@urjc.es.

ABSTRACT

We present a study of the behavior of a ball under the influence of gravity on a platform. A propagating surface wave travels on the surface of the platform while the platform remains motionless. This is a modification of the classical bouncing ball problem and describes the transport of particles by surface waves. Phase and velocity maps cannot be expressed in the explicit form due to implicit formulations, and no formal analytical analyses is possible. Numerical analysis shows that the transition to chaos is produced via a period doubling route which is a common property for classical bouncers. These numerical analysis have been carried out for the conservative and for the viscous cases and also for elastic and for inelastic collisions. The bouncing process can be sensitive to the initial conditions and can be useful for control techniques which can dramatically increase the effectiveness of particle transport in practical applications. Finally, we also consider the mechanical model of a particle sliding on a surface which is also important because it has important physical implications such as the transportation of thin films in biomedical applications, among others.

Introduction

A particle falling down, in a constant gravitational field, on a moving platform is called a bouncing ball problem, or a bouncer. This model was suggested more than thirty years ago

[†]Reused with permission from: Minvydas Ragulskis, Edita Sakyte, Jesús M. Seoane, and Miguel A.F. Sanjuán, "A New Mechanical Model for Particle Transport by Surface Waves and Applications," *Mathematical Problems in Engineering*, vol. 2009, Article ID 731358, 17 pages, 2009. doi:10.1155/2009/731358.

[1,2] as an alternative to the Fermi-Ulam model [3] of cosmic ray acceleration [4]. In the ensuing years many approaches to the bouncer model have been studied theoretically and experimentally [5,8]. It has been proved to be a useful system for experimentally exploring several new nonlinear effects [9,10]. Moreover, it has been implemented into a number of engineering applications [11,12].

The bouncer model can be briefly characterized by the following basic statements. (i) Maps derived for the bouncer model can be exactly iterated for any time function describing the moving platform [7,9] (though usually the platform is assumed to oscillate with a single frequency). (ii) The ball-platform collisions can be characterized by a coefficient of restitution α changing from $\alpha = 1$ for a perfectly elastic case to $\alpha = 0$ for a completely inelastic situation. (iii) The chaotic bouncer can be easily used to relate theoretical predictions to experimental results, [9,10] what makes it a paradigm model in nonlinear dynamics.

In this paper we assume that a particle is falling down in a constant gravitational field on a stationary platform. A propagating surface wave travels on the surface of the platform while the platform remains motionless. Such a model can be used to describe the transport of particles by propagating surface waves, which is an important problem with numerous applications. Powder transport by piezoelectrically excited ultrasonic surface waves [13], manipulation of bioparticles using traveling wave electrophoresis [14,15], and conveyance of submerged buoys in coastal waters [16] are just a few examples of problems involving the interaction between propagating waves and transported bouncing particles.

This paper is organized as follows. In Section 2 we present a complete description of our model, the bouncer system. Section 3 presents numerical simulations showing that the modified bouncer model possesses such an inherent chaotic dynamics. These results are carried out for both, the conservative and the dissipative cases. The case of a particle sliding on the surface is fully analyzed in Section 4. Conclusions and discussions of the main results of this paper are presented in Section 5.

Model Description

We consider the two-dimensional system shown in Fig. 1, where the surface of an elastic plate is represented by a solid line which coincides with the x-axis in the state of equilibrium. A point of the surface in the state of equilibrium $(x,0)$ is translated to coordinates (X,Y) when a wave process takes place. This translation is sensitive to time t and coordinate x:

$$X = x + \eta(x,t),$$
$$Y = \xi(x,t),$$

(2.1)

where the functions $\eta(x,t)$ and $\xi(x,t)$ determine deflections from the state of equilibrium. Explicitly, the longitudinal and transverse displacements of the medium at the surface of flat boundary with travelling Rayleigh wave can be expressed like [17]

$$u_x = \frac{\omega}{\chi} \sqrt{\frac{2\rho(1+v)}{E}} \left(1 - \frac{\sqrt{1-\chi^2}\sqrt{2(1-v)-\chi^2(1-2v)}}{(1-0.5\chi^2)\sqrt{2(1-v)}} \right) C \sin(\omega t \pm kx)$$

$$u_y = \frac{\omega}{\chi} \sqrt{\frac{2\rho(1+v)}{E}} \left(1 - \frac{\sqrt{2(1-v)-\chi^2(1-2v)}}{\sqrt{2(1-v)}} \right) 0.5 \, \chi^2 C \cos(\omega t \pm kx)$$

(2.2)

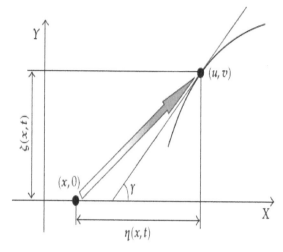

Figure 1 Schematic diagram illustrating the collision between the particle and the surface.

Where u_x and u_y are the longitudinal and transverse displacements, x the coordinate of the surface point of the medium before the wave process took place; C is a constant, k the wavenumber and ρ the density. χ can be found from the following algebraic equation:

$$\chi^6 - 8\chi^4 + 8\left(3 - \frac{1-2v}{1-v}\right)\chi^2 - 16\left(1 - \frac{1-2v}{2(1-v)}\right) = 0,$$

(2.3)

and the angular velocity ω can be found from the following transcendental equation:

$$\left(k^2 + \beta^2\right)^2 \cosh(\alpha a)\sinh(\alpha\beta) - 4k^2\alpha\beta \cosh(\beta a)\sinh(\alpha\alpha) = 0,$$

(2.4)

where $\alpha = \sqrt{\dfrac{v_l k^2 - \omega^2}{v_l^2}}$, $\beta = \sqrt{\dfrac{v_t k^2 - \omega^2}{v_t^2}}$, $v_l = \sqrt{\dfrac{(\lambda + 2\mu)}{\rho}}$ and $v_t = \sqrt{\dfrac{\mu}{\rho}}$, v the Poisson's

ratio and λ the first Lamé constant. It can be noted that the ratio between the amplitudes of transverse and longitudinal deformations depends on v. In usual elastic media it is quite normal that the transverse displacement is about 1.5 times larger than the longitudinal displacement [18]. The motion of a point in the medium is an ellipse. Also, the direction of the velocity of the particles at the peaks of the wave is opposite to the direction of wave propagation.

Rayleigh waves are dispersive due to a dependence of the wave's speed on its wavelength. Typical example is Rayleigh waves in the Earth where waves with a higher frequency travel more slowly than those with a lower frequency. Rayleigh waves thus often appear spread out on seismograms recorded at distant earthquake recording stations [13]. Surface acoustic waves (SAWs) generated by SAW devices on rough anisotropic materials also experience considerable dispersion [20]. On the other hand, film waves generated on a surface of a finite liquid bed [21] can be characterized by a single frequencywave component. Therefore we concentrate on a one single frequency steady-state Rayleigh wave propagation and disregard dispersion.

Whenever a traveling nondispersive Rayleigh surface wave occurs in a medium, it can be characterized by a retrograde elliptic motion of the particles of that medium:

$$\eta(x,t) = a \sin(\omega t - kx),$$
$$\xi(x,t) = b \cos(\omega t - kx),$$

(2.5)

where a and b are longitudinal and transverse amplitudes of the oscillations; ω is the angular frequency, and k is the wave number. Remind that in a usual elastic medium it is quite normal for the transverse displacement to be about 1.5 times larger than the longitudinal displacement [21].

The coordinates of the particle are denoted as (u,v). Assume that the particle is in contact with the surface at time moment t, then the following constrain takes place:

$$v = \xi(x,t),$$

(2.6)

where x is to be found from the following algebraic equality (where u and t are given and x is the unknown):

$$x + \eta(x,t) = u,$$

(2.7)

In other words, the instantaneous shape of the surface cannot be described by an explicit function. Nevertheless, the tangent to the surface at the point with abscise u can be expressed explicitly:

$$\tan\gamma = \frac{\partial\xi(x,t)/\partial x}{1 + \partial\eta(x,t)/\partial x},$$

(2.8)

where γ is the angle between the tangent and the x-axis. Instantaneous velocities (x- and y-components) of the point of the surface in contact with the particle can be expressed as $\partial\eta(x,t)/\partial t$ and $\partial\xi(x,t)/\partial t$ accordingly.

The governing equations of motion of a particle in a free flight mode are

$$mu'' + hu' = 0,$$
$$mv'' + hv' = -mg,$$

(2.9)

where top dots denote full derivative by time, m is the mass of the particle, h is the coefficient of viscous damping of the media above the surface, and g is the free fall acceleration. Initial conditions $u(t_0)=u_0$; $u'(t_0)=u'_0$; $v(t_0)=v_0$; $v'(t_0)=v'_0$ yield partial solutions:

$$u(t) = u_0 + u_0' \frac{m}{h}\left(1 - \exp\left(-\frac{h}{m}(t - t_0)\right)\right);$$

$$v(t) = v_0 + \left(v_0' + \frac{mg}{h}\right)\frac{m}{h}\left(1 - \exp\left(-\frac{h}{m}(t - t_0)\right)\right) - \frac{mg}{h}(t - t_0);$$

$$v'(t) = v_0' \exp\left(-\frac{h}{m}(t - t_0)\right) - \frac{mg}{h}\left(1 - \exp\left(-\frac{h}{m}(t - t_0)\right)\right).$$

(2.10)

The free flight stage continues until the particle collides with the surface. Unfortunately, it is impossible to determine the explicit time moment of the collision due to the fact that the instantaneous shape of the surface cannot be expressed by an explicit function. Instead, one has to use iterative numerical techniques in order to determine the exact moment of the bounce.

Localization of the root (the time moment of the collision) is performed using a time marching technique starting from the initial conditions until

$$v(t_0 + i.\Delta t) < \xi(x_i, t_0 + i.\Delta t)$$

(2.11)

where Δt is the time step; $i=1,2,...,r$; r is the step number for which (4.8) is satisfied for the first time, and x_i is the solution of (4.4) at fixed i:

$$x_i + \eta(x_i, t_0 + i.\Delta t) = u(x_i, t_0 + i.\Delta t);$$

(2.12)

and $u(t_0 + i.\Delta t)$, $v(t_0 + i.\Delta t)$ are determined by (4.7). Solution of (4.9) also requires an iterative numerical algorithm. When the root $\hat{t}$ is localized in the interval $t_0 + (r-1)\Delta t < \hat{t} < t_0 + r\Delta t$, one needs to fine down the value of $\hat{t}$ using an iterative computational algorithm. This iterative algorithm can be a most simple bisection method, though more sophisticated algorithms comprising the golden section rule or Newton's iterations for example can be used instead until the desirable accuracy is achieved. As the collision moment $\hat{t}$ is fined down in every iteration, the coordinate $\hat{x}$ corresponding to the collision point $\hat{u} : \hat{x} + \eta(\hat{x}, \hat{t}) = \hat{u}$ is also made more precise. Initially, $x_{r-1} < \hat{x} < x_r$; every iteration helps to reach a better accuracy.

Such iterative method of determination of the collision moment leads us to the important conclusion that phase and velocity maps cannot be expressed in an explicit form, and no formal analytical analysis is possible.

Nevertheless, the geometrical coordinates of the point of collision are $(\hat{u}, \xi(\hat{x}, \hat{t}))$ and can be reconstructed using computational techniques. Velocities of the particle just before the collision are $u'(\hat{t})$ and $v'(\hat{t})$. Similarly, instantaneous velocities of the surface in contact with the particle are $\eta_t'(\hat{x}, \hat{t})$ and $\xi_t'(\hat{x}, \hat{t})$.

Projections of the particle's velocities just before the collision to the normal and to the tangent to the surface at the contact point can be expressed in the following form:

$$\hat{P}_n' = -u'(\hat{t})\sin\gamma + v'(\hat{t})\cos\gamma,$$

$$\hat{P}_t' = u'(\hat{t})\cos\gamma + v'(\hat{t})\sin\gamma,$$

(2.13)

where the angle γ is determined from (4.5) at the point of collision.

Analogously, projections of velocities of the point of the surface in contact with the particle to the normal and to the tangent take the following form:

$$\hat{S}'_n = -\eta'_t(\hat{x},\hat{t})\sin\gamma + \zeta'_t(\hat{x},\hat{t})\cos\gamma,$$
$$\hat{P}'_t = \eta'_t(\hat{x},\hat{t})\cos\gamma + \zeta'_t(\hat{x},\hat{t})\sin\gamma, \tag{2.14}$$

Then, the velocities of the particle just after the collision (in the normal and tangent directions) are

$$P_n = (1+\alpha)S_n - \alpha\hat{P}_n,$$
$$P'_t = \beta S'_n - (1+\beta)\hat{P}'_t, \tag{2.15}$$

where α is the coefficient of restitution for the collision in the normal direction. This constant is a measure of the energy loss at each impact. For elastic collisions $\alpha = 1$, and $\alpha < 1$ for inelastic collisions. Coefficient β determines the friction between the particle and the surface at the moment of collision. There is no friction between the particle and the surface when $\beta = 0$. The utmost value $\beta = 1$ represents the situation when the projection of the particle's velocity (immediately after the impact) and the projection of the surface's point velocity to the tangent are equal.

The free flight stage starts over again immediately after the collision, and the initial conditions are

$$u(\hat{t}) = \hat{u},$$
$$u'(\hat{t}) = -\hat{P}'_n\sin\gamma + \hat{P}'_t\cos\gamma$$
$$v(\hat{t}) = \zeta(\hat{x},\hat{t}), \tag{2.16}$$
$$v'(\hat{t}) = P'_n\cos\gamma + P'_t\cos\gamma.$$

The presented model is a modification of the classic bouncer model which can be derived assuming $\zeta(x,t) = b\cos(\omega t)$ and $\eta(x,t) = 0$. In that case $u = x$, and the model becomes explicit.

Complex Dynamics in the Modified Bouncer Model

We will demonstrate that the modified bouncer model possesses such an inherent feature as chaotic dynamics. Moreover, we will show that the sensitivity to initial conditions can be exploited for the control of the process of conveyance. We will show these results for the conservative or nonviscous case ($h = 0$) and for the viscous case ($h \neq 0$).

Nonviscous Case

We take $h = 0$, for which the media above the surface is non-viscous, $\alpha = 1$ (elastic collisions) and $\beta = 0$ (no damping generated by sliding). The dynamics of a bouncing particle on a surface of a propagating wave is very sensitive to the initial conditions if the dynamics is Hamiltonian. Apparently, it is possible to find such a set of initial conditions which lead to regular and periodic dynamics. This is illustrated in Fig. 2 where collision heights $v(t)$ are plotted versus initial velocity $u'(0)$. 50 successive collisions are used for every discrete value of $u'(0)$ to produce this diagram.

The initial condition $u'(0) = \omega / k$ produces a period 1 motion at $\omega = 1$, $k=1$, $m=0.5$, $g=1$, $a = b/1.5$, $u(0) = \pi$, $v(0) = 0$, and $v'(0) = 0.5$. This is illustrated in Fig. 3.

We plot the trajectory of the particle in $3D$ for better visual interpretation. Collision moments are marked as black dots. At every moment of collision we also plot the instantaneous shape of the surface (one can note that the instantaneous shape of the surface is not harmonic). One can clearly see the difference in the complexity of the particle dynamics at $u'(0)=1$ (see Fig. 3) and $u'(0)=2$ (see Fig. 4).

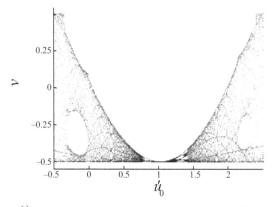

Figure 2 Collision heights $v(t)$ plotted versus initial velocity $u'(0)$ for $u'(0) = \omega / k$ at $\omega = 1$, $k=1$, $m=0.5$, $g=1$, $a = b/1.5$, $u(0) = \pi$, $v(0) = 0$, and $v'(0) = 0.5$.

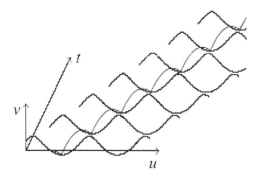

Figure 3 Period 1 trajectory of a bouncing particle for $\omega = 1$, $k=1$, $m=0.5$, $g=1$, $a = b/1.5$, $u(0) = \pi$, $v(0) = 0$, and $v'(0) = 0.5$.

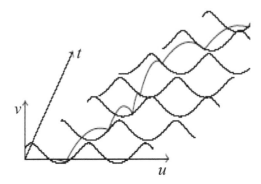

Figure 4 Chaotic trajectory of a bouncing particle for $\omega = 1$, $k=1$, $m=0.5$, $g=1$, $a= b/1.5$, $u(0)=\pi$, $u'(0)=2$, $v(0)=0$, and $v'(0)=0.5$.

Viscous Case

For this case, we assume that collisions are completely elastic ($\alpha = 1$), and there is no tangential friction between the particle and the surface ($\beta = 0$), but the media above the surface is viscous fixing the value of $h=0.1$ as in [22]. We use the reduced impact representation, where the height of the bouncing ball is sampled at each impact with the surface (impact sampling). Since the system is dissipative, we plot the bouncing process after the initial transients cease down _see Fig. 5(a). We skip 1500 successive bounces before starting to plot the collision heights $v(i)$ for every discrete value of ω.

Parameter ω is varied following the rule:
$\omega_i = 1 + (3/\ln 21).\ln(1 + 20i/1024)$; $i = 1,...,1024$ which helps to expand the cascade of period doubling bifurcations. The control parameter in our case is not the amplitude of the platform's oscillation but the velocity of the wave propagation; the collision height is used instead of collision velocity for a reduced impact representation. Moreover, the media above the surface of the plate is viscous. It appears that the transition to chaos via a period doubling route observed for a classical bouncer [23] is observed also for a particle bouncing on a surface of a stationary platform with a propagating wave traveling on its surface.

A phenomenological model could be used to exemplify the bifurcation diagram presented in Fig. 5(a). The logistic map [24] is probably the simplest model ever used to study the transition to chaos via a period doubling route. Simple computational experiments with appropriately chosen parameter values of the logistic map would illustrate the universality of the bifurcation diagram in Fig. 5(a).

An important parameter characterizing the effectiveness of the transport is the average longitudinal velocity of the particle $\bar{u}'$. We average it over a long period of time after the initial transients cease down. In order to calculate a nondimensional quantity we divide it from the velocity of the traveling wave $T_u = k\bar{u}'/\omega$. Thus, the average velocity of conveyance is equal to the velocity of the traveling wave if T_u is equal to 1 as shown in Fig. 5(b).

It is interesting to observe that the particle is transported with the average velocity of the traveling wave until the period 3 bouncing mode after a cascade of period doubling bifurcations (see Fig. 5(a)). The particle's average transportation velocity drops down only when the period 3 bouncing mode experiences its own cascade of period doubling bifurcations. External damping forces acting to the particle prevent its motion with the average wave's velocity in the direction of the wave propagation when this velocity becomes large enough _even though the collisions are elastic_. Also, the bouncing process is insensitive to initial conditions–eventually it converges to the one and only attractor shown in Fig. 5(a) (at fixed ω).

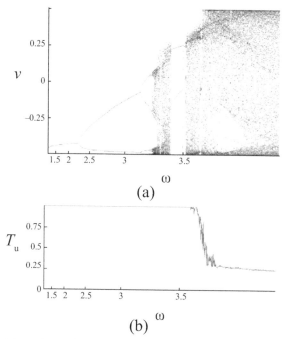

(a)

(b)

Figure 5 Transport of particles at increasing wave speeds (elastic collisions, viscous media over the surface). Reduced impact representation (a) shows the transition to chaos via a period doubling route. Note that impact heights are distributed in the interval [-0.5,0.5]. Nondimensional longitudinal particle's transport velocity T_u drops down at higherwave speeds due to the viscosity of themedia above the surface (b). System's parameters are $\alpha = 1$; $\beta = 0$; $\eta(x,t)= (2/3)\sin(\omega t - x)$; $\zeta(x,t)= (1/3)\cos(\omega t - x)$; $h= 0.1$; $m= 0.5$; $g=9.81$.

Figures 6(a)–6(d) show different dynamical behaviors of the transient processes for the elastic case ($\alpha = 1$) once we fixed the parameter values as follows: $\beta = 0$, $k=1$, $m=0.5$, $g=9.81$, b=0.5, and $a=b/1.5$. We obtain both, periodic and chaotic motions depending on the value of the parameter ω. Figures 6(a)–6(c) show for $\omega = 2$, $\omega = 3$, and $\omega = 3.45$ period 1, period 2, and period 3 processes, respectively. Chaotic bouncing for the value $\omega = 4$ is shown in Fig. 6(d).

The situation becomes different when collisions are inelastic, as shown in [22]. In the presence of inelastic collisons, a vanishing bouncing process takes place (complete chattering [23]) when the particle sets into the state of rest on a slope of the propagating wave. The term *complete chattering* is used in literature to describe the process when the time interval between inelastic bounces tends to zero and the ball finally "sticks" to the surface of the oscillating platform. As we mentioned previously, a complete description of this phenomenon is given in [22].

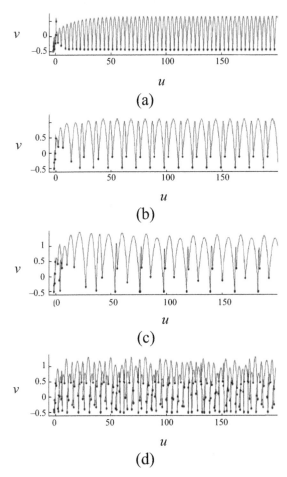

Figure 6 Transient processes for parameter values as follows: h=0.1, α =1, β =0, k=1, m=0.5, g= 9.81, b=0.5, and a=b/1.5. We observe the following behaviors: (a) period 1 process for ω = 2, (b) period 2 process for ω = 3, (c) period 3 process for ω = 3.45, and (d) chaotic bouncing for ω = 4, respectively.

The Sliding Particle Model

In this section we thoroughly analyze the case in which the particle is sliding on the surface instead of the case in which it is falling down on it, analyzed previously. Our motivation is the following. Conveyance of particles and bodies by propagating waves is an important scientific and engineering problem with numerous applications. Manipulation of bioparticles and gene expression profiling using traveling wave dielectrophoresis [14,25,26], segregation of particles in suspensions subject to ac electric fields [27], transport of sand particles and oil spills in coastal waters [28,29], powder transport by piezoelectrically excited ultrasonic waves [13,22], transportation of thin films in biomedical applications [21] are just a few examples of problems involving interaction between propagating waves and transported objects.

We now describe, as in the bouncer model, the equations of motions of our sliding particle model.

It is assumed that a mass particle is in contact with the deformed surface at a point (u,v) at a time moment t (Fig. 7). A point of the surface in the equilibrium state $(x,0)$ is translated to coordinates (u,v) at time moment t. This translation is sensitive to time t and coordinate x:

$$u = x + \eta\,(x,t),$$
$$v = \xi\,(x,t),$$
(4.1)

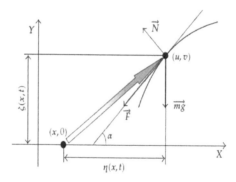

Figure 7 A geometric scheme of the dynamical system showing the particle sliding on the surface.

where $\eta(x,t)$ and $\xi(x,t)$ are predefined functions.

The condition that the particle is located on the surface leads to the following constraint:

$$v = \xi\,(x,t),$$
(4.2)

where x is to be found from the following algebraic equality (in which u is given and x is unknown):

$$x + \eta(x,t) = u.$$
(4.3)

Though the instantaneous shape of the oscillating surface cannot be described by an explicit function, the tangent to the surface at the point (u,v) can be expressed as

$$\tan\alpha = \frac{\xi'_x(x,t)}{1+\eta'_x(x,t)}. \tag{4.4}$$

Instantaneous velocities of the surface's point (u,v) in the direction of x- and y-axis can be expressed as follows:

$$\begin{aligned} u'\big|_{x=const} &= \eta'_t(x,t), \\ v'\big|_{x=const} &= \zeta'_t(x,t), \end{aligned} \tag{4.5}$$

where dots denote derivatives by t. When a mass particle slides on the surface, it does not necessarily move in contact with one point of the surface. Therefore x is no longer a constant. Thus,

$$\begin{aligned} u' &= x'(1+\eta'_x)+\eta'_t, \\ u'' &= x''(1+\eta'_x)+(x')^2\eta''_{xx}+2x'\eta''_{xt}+\eta''_{tt}. \end{aligned} \tag{4.6}$$

The condition that the mass particle continuously slides on the surface brings another constraint into force (the relative velocity in the normal direction to the surface at the contact point must be zero):

$$\tan\alpha = \frac{v'-\xi'_t(x,t)}{u'-\eta'_t(x,t)}. \tag{4.7}$$

Equation (4.4) with (4.7) in force yields

$$v' = \frac{u'-\eta'_t}{1+\eta'_x}\xi'_x+\xi'_t, \tag{4.8}$$

which together with (4.6) produces the following relationship:

$$v' = x'\xi'_x+\xi'_t. \tag{4.9}$$

Differentiation of (4.9) yields

$$v'' = x''\xi'_x+(x')^2\zeta''_{xx}+2x'\zeta''_{xt}+\zeta''_{tt}. \tag{4.10}$$

Then the relative sliding velocity of the particle on the surface v_{12} can be expressed as

$$\begin{aligned} v_{12} &= (u'-\eta'_t)\cos\alpha+(v'-\zeta'_t)\sin\alpha \\ &= \frac{x'}{\sqrt{1+\tan^2\alpha}}(1+\eta'_x+\zeta'_x\tan\alpha) \\ &= x'\sqrt{(1+\eta'_x)^2+(\zeta'_x)^2} \end{aligned} \tag{4.11}$$

The condition of dynamic equilibrium leads to the following system of equations:

$$mu'' + N \sin \alpha + F \cos \alpha = 0,$$
$$mv'' + mg + F \sin \alpha = N \cos \alpha,$$

(4.12)

where m is the mass of the particle; N is the reaction force at the contact point; g is the gravity acceleration; F is the friction force between the mass particle and the surface. The system of equations in (4.12) is in force when $N > 0$. Otherwise the particle jumps off the oscillating surface. It is assumed that the friction force is linear. Thus F can be expressed like

$$F = k v_{12},$$

(4.13)

where h is the coefficient of linear friction. Finally, the governing equation of motion can be derived from (4.12). Elementary transformations and substitutions lead to the following explicit differential equation:

$$B_1(x,t)x'' + B_2(x,t)x' + B_3(x,t) + B_4(x,t)(x')^2 = 0,$$

(4.14)

where

$$B_1(x,t). = m\left(1 + \eta'_x + \frac{(\zeta'_x)^2}{1 + \eta'_x}\right),$$

$$B_2(x,t). = 2m\left(\eta''_{xt} + \frac{\zeta'_x \zeta''_{xx}}{1 + \eta'_x}\right) + h\left(1 + \eta'_x + \frac{(\zeta'_x)^2}{1 + \eta'_x}\right),$$

$$B_3(x,t) = m\left(\eta''_{tt} + g\frac{\zeta'_x}{1 + \eta'_x} + \frac{\zeta'_x \zeta''_{tt}}{1 + \eta'_x}\right),$$

$$B_4(x,t). = m\left(\eta''_{xx} + \frac{\zeta'_x \zeta''_{xx}}{1 + \eta'_x}\right).$$

(4.15)

A major obstacle is eliminated, and direct numerical time marching techniques can be used for integration of (4.14)–computation of u and v is straightforward if the coordinate x is given at time t (4.1). Existence of a stability of the dynamic equilibrium can be analyzed explicitly.

But before proceeding with the analysis of dynamic equilibrium the following observation can be done. If kinematic relationships describing a traveling Rayleigh wave are in force, the change of variables

$$z = \omega t - kx,$$

(4.16)

transforms (4.14) to the following autonomous form:

$$C_1(z).z'' + C_2(z).z' + C_3(z) + C_4(z).(z')^2 = 0,$$

(4.17)

where

$$C_1(z). = -\frac{m}{k}\left(1 - ka\cos(z) + \frac{k^2 b^2 \sin^2(z)}{1 - ka\cos(z)}\right),$$

$$C_2(z). = \frac{h}{m}C_1(z),$$

$$C_3(z) = \frac{\omega h}{k}\left(1 - ka\cos(z) + \frac{k^2 b^2 \sin^2(z)}{1 - ka\cos(z)}\right) + mg\frac{kb\sin(z)}{1 - ka\cos(z)},$$

$$C_4(z). = -m\left(a\sin(z) + \frac{kb^2 \sin(z)\cos(z)}{1 - ka\cos(z)}\right).$$

(4.18)

An important conclusion can be done.

Dynamics of a particle sliding on the surface of a propagating Rayleigh wave cannot be chaotic. This is due to the fact that the governing equation of motion is a second-order autonomous ordinary differential equation with smooth parameter functions [30].

Equation (4.8) yields the dynamic equilibrium which represents a motion of the particle on a slope of the propagating wave with the velocity of its propagation:

$$u'' = 0,$$

$$u' = \frac{\omega}{k},$$

(4.19)

$$u = \frac{\omega}{k}.t - \varphi,$$

where ψ is a constant. Then, it follows from (4.4) that

$$x + a\sin(\omega t - kx) = \frac{\omega}{k}.t - \varphi.$$

(4.20)

The term $a\sin(\omega t - kx)$ is bounded, therefore (4.20) will be in force when

$$x = \frac{\omega}{k}.t - \theta,$$

(4.21)

where θ is a constant satisfying the equality $-\theta + a\sin(k\theta) = -\psi$. Moreover, conditions of existence of the dynamic equilibrium are similar in terms of x or u:

$$x' = \frac{\omega/k - \eta'_t}{1 + \eta'_x} = \frac{\omega}{k},$$

$$x'' = \frac{-x'^2\eta''_{xx} - 2x'\eta''_{xt} - \eta''_{tt}}{1 + \eta'_x}.$$

(4.22)

As mentioned earlier, the explicit governing equation is formulated in terms of x, not the coordinate of the contact point u. First, coordinates of the unstable saddle point are determined. Then coordinates of the same saddle point are calculated in the frame $(\omega t - ku)$; u' using the relationship in (4.3). Forward and reverse time marching techniques are used to construct basin boundaries of attractors when partial solutions of (4.14) are sought from the infinitesimal surrounding of the saddle point.

The described computational technique is used to construct basin boundaries of the system's attractors (Fig. 8). Solutions in terms of u (forward and reverse) are visualized only. It can be noted that two stable attractors can coexist–a stable equilibrium point and a stable limit cycle. Shaded regions in Fig. 8 correspond to a basin (attracting set of initial conditions) of stable equilibrium points, while white region corresponds to a basin of the limit cycle. The phase plane in Fig. 8 is periodic by 2π and can be visualized in cylindrical coordinates, but the plane representation is clearer.

A special attention should be paid to dashed line intervals on basin boundaries. Equation (4.14) describes a motion of a particle on the surface of a propagating wave. This governing equation of motion holds until the reaction force N in (4.12) is positive. Whenever N gets equal or lower than zero, the particle looses a contact with the surface and starts a free fly in a gravitational field until it bounces on the surface again. Therefore, the moment when the particle looses the contact with, the surface is detected, and the trajectory is marked by a dashed line. It can be noted that such motions occur only at relatively high particle velocities (Fig. 8).

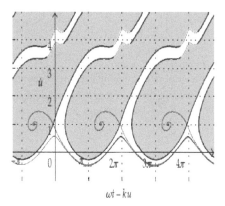

Figure 8 Basin boundaries at $b=0.5$; $a=b=1.5$; $m=0.4$; $h=0.1$; $\omega = k = 1$, shaded regions illustrate the basin of attraction of stable equilibrium points.

Conveyance of a particle by a propagating Rayleigh wave is a nonlinear problem, so such effects as the coexistence of stable attractors should not be astonishing. Stable equilibrium point type attractor in Fig. 8 corresponds to a surf-type motion on a slope of a propagating wave; stable limit cycle corresponds to a motion with an average velocity much lower than the velocity of the propagating wave. Coexistence of attractors (a stable equilibrium point and a stable limit cycle) enables development of motion control strategies based on a small external impulses which can bring the system from the regime of motion with small average velocity into motion with the propagating wave's velocity [21]. Such attractor control strategy is illustrated in Fig. 9 where the particle first oscillates in the limit cycle, and then a small external impulse kicks it to the basin of attraction of the stable focus point implemented only when the stable equilibrium point and the stable limit cycle coexist. Thus, it would be impossible to transport a sand particle with the velocity of the propagating wave by an acoustic surface Rayleigh wave. Nevertheless, such attractor control strategies could be implemented for transportation of biomedical objects on the surface of an undulation film [21]. The sliding particle model presented in this section also exhibits a very rich dynamics as in the case of the bouncer model. In particular, the sliding particle model should also have the sensitivity to the initial conditions for certain sets of parameter values as occuring in the bouncer model (see [1]). The sensitivity of transient processes to initial conditions takes place for both, the bouncer model and the sliding particle model.

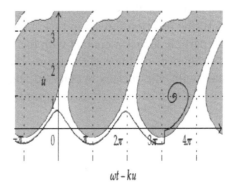

$\omega t - ku$

Figure 9 Illustration of the attractor control strategy: limit cycle is represented as a periodic trajectory in frame $\left((\omega t - ku), u'\right)$ small external impulse kicks the trajectory to basin boundary of stable equilibrium point where the particle eventually settles down.

Conclusions and Discussion

Transport of particles by surface waves is an important scientific and engineering problem, with numerous practical applications, including MEMS (micro-electro-mechanical systems) used to manipulate objects like particles or cells. We show that this problem is a modification of the classical bouncer model which is considered as a paradigm model in nonlinear physics. The formulations of our model are implicit, thus phase and velocity maps cannot be expressed in explicit form.

Chaotic dynamics of a conveyed particle is not an unexpected fact due to the complexity of the constitutive model. More surprising is the rich dynamical behavior in models comprising dissipative dynamics, elastic and inelastic collisions. It appears that the transition to chaos via a period doubling route is a universal property for bouncers and is observed in our model of particles transport in both, conservative and viscous media.

Moreover, the sensitivity to initial conditions can be useful for control techniques which can dramatically increase the effectiveness of particles transport by surface waves. These results are relevant in the sense that we have also found the sensitivity to the initial conditions for the sliding particle model, which may have important applications in practical implementations as powder transport by piezoelectrically excited ultrasonic waves, transport of sand particles, among others.

Though the numerical analysis was concentrated on the dimensionless system only, theoretical and experimental investigation of dry particle conveyance and its control is a definite object for future research.

Acknowledgments

This work was supported by the Spanish Ministry of Education and Science under project number FIS2006-08525 and by Universidad Rey Juan Carlos and Comunidad de Madrid under project number URJC-CM-2007-CET-1601.

References

1. G.M. Zaslavsky, "The simplest case of a strange attractor," *Physics Letters A*, vol. 69, no. 3, pp. 145–147, 1978.
2. L.D. Pustilnikov, "On Ulam's problem," *Theoretical and Mathematical Physics*, vol. 57, p. 1035, 1983.
3. M.A. Lieberman and A.J. Lichtenberg, "Stochastic and adiabatic behavior of particles accelerated by periodic forces," *Physical Review A*, vol. 5, no. 4, pp. 1852–1866, 1972.
4. E. Fermi, "On the origin of the cosmic radiation," *Physical Review*, vol. 75, no. 8, pp. 1169–1174, 1949.
5. B.V. Chirikov, "A universal instability of many-dimensional oscillator systems," *Physics Reports*, vol. 52, no. 5, pp. 264–379, 1979.
6. A.J. Lichtenberg, M.A. Lieberman, and R. H. Cohen, "Fermi acceleration revisited," *Physica D*, vol. 1, no. 3, pp. 291–305, 1980.
7. P. Pieranski and J. Malecki, "Noisy precursors and resonant properties of the period-doubling modes in a nonlinear dynamical system," *Physical Review A*, vol. 34, no. 1, pp. 582–590, 1986.
8. R.M. Everson, "Chaotic dynamics of a bouncing ball," *Physica D*, vol. 19, no. 3, pp. 355–383, 1986.
9. K. Wiesenfeld and N. B. Tufillaro, "Suppression of period doubling in the dynamics of a bouncing ball," *Physica D*, vol. 26, no. 1–3, pp. 321–335, 1987.
10. Z.J. Kowalik, M. Franaszek, and P. Pierański, "Self-reanimating chaos in the bouncing-ball system," *Physical Review A*, vol. 37, no. 10, pp. 4016–4022, 1988.
11. M.-O. Hongler, P. Cartier, and P. Flury, "Numerical study of a model of vibro-transporter," *Physics Letters A*, vol. 135, no. 2, pp. 106–112, 1989.
12. M.-O. Hongler and J. Figour, "Periodic versus chaotic dynamics in vibratory feeders," *Helvetica Physica Acta*, vol. 62, no. 1, pp. 68–81, 1989.
13. M. Mracek and J. Wallaschek, "A system for powder transport based on piezoelectrically excited ultrasonic progressive waves," *Materials Chemistry and Physics*, vol. 90, no. 2-3, pp. 378–380, 2005.
14. M.S. Talary, J.P. H. Burt, J.A. Tame, and R. Pethig, "Electromanipulation and separation of cells using travelling electric fields," *Journal of Physics D*, vol. 29, no. 8, pp. 2198–2203, 1996.
15. C.-F. Chou, J.O. Tegenfeldt, O. Bakajin, et al., "Electrodeless dielectrophoresis of single- and double-stranded DNA," *Biophysical Journal*, vol. 83, no. 4, pp. 2170–2179, 2002.
16. R.H. Plaut, A.L. Farmer, and M.M. Holland, "Bouncing-ball model of 'dry' motions of a tethered buoy," *Journal of Vibration and Acoustics*, vol. 123, no. 3, pp. 333–339, 2001.
17. J.D. Achenbach, *Wave Propagation in Elastic Solids*, Elsevier, New York, NY, USA, 1984.
18. L.D. Landau and E. M. Lifshitz, *Theory of Elasticity*, Pergamon Press, Oxford, UK, 1986.
19. K. Aki and P. G. Richards, *Quantitative Seismology*, Freeman, New York, NY, USA, 1980.

20. C.M. Flannery and H. Von Kiedrowski, "Dispersion of surface acoustic waves on rough anisotropic materials," in *Proceedings of the IEEE Ultrasonics Symposium*, vol. 1, pp. 583–586, 2001.

21. M. Ragulskis and K. Koizumi, "Applicability of attractor control techniques for a particle conveyed by a propagating wave," *Journal of Vibration and Control*, vol. 10, no. 7, pp. 1057–1070, 2004.

22. M. Ragulskis and M.A.F. Sanjuán, "Transport of particles by surface waves: a modification of the classical bouncer model," *New Journal of Physics*, vol. 10, Article ID 083017, 2008.

23. J.M. Luck and A. Mehta, "Bouncing ball with a finite restitution: chattering, locking, and chaos," *Physical Review E*, vol. 48, no. 5, pp. 3988–3997, 1993.

24. R. M. May, "Simple mathematical models with very complicated dynamics," *Nature*, vol. 261, no. 5560, pp. 459–467, 1976.

25. C.-F. Chou, J.O. Tegenfeldt, O. Bakajin, et al. "Electrodeless dielectrophoresis of single- and double-stranded DNA," *Biophysical Journal*, vol. 83, no. 4, pp. 2170–2179, 2002.

26. L. Cui and H. Morgan, "Design and fabrication of travelling wave dielectrophoresis structures," *Journal of Micromechanics and Microengineering*, vol. 10, no. 1, pp. 72–79, 2000.

27. A.D. Dussaud, B. Khusid, and A. Acrivos, "Particle segregation in suspensions subject to high-gradient ac electric fields," *Journal of Applied Physics*, vol. 88, no. 9, pp. 5463–5473, 2000.

28. W.N. Hassan and J.S. Ribberink, "Transport processes of uniform and mixed sands in oscillatory sheet flow," *Coastal Engineering*, vol. 52, no. 9, pp. 745–770, 2005.

29. S.D. Wang, Y.M. Shen, and Y.H. Zheng, "Two-dimensional numerical simulation for transport and fate of oil spills in seas," *Ocean Engineering*, vol. 32, no. 13, pp. 1556–1571, 2005.

30. R.C. Hilborn, *Chaos and Nonlinear Dynamics*, Oxford University Press, New York, NY, USA, 1994.

Chapter 6

Chaotic Behavior of the Biharmonic Dynamics System[†]

Vladimir S. Aslanov

Faculty of Aircraft Construction, Samara State Aerospace University. Theoretical mechanics chair, SSAU, 34, Moscovskoe shosse, Samara, 443086, Russia. Email: aslanov_vs@mail.ru.

ABSTRACT

Motion of a biharmonic system under action of small periodic force and small damped force is studied. The biharmonic oscillator is a physical system acting under a biharmonic force like: $a \sin \theta + b \sin 2\theta$. The article contains biharmonic oscillator analysis, phase space research, and analytic solutions for separatrixes. The biharmonic oscillator performs chaotic motion near separatrixes under small perturbations. Melnikov method gives analytical criterion for heteroclinic chaos in terms of system parameters. A transition from chaotic to regular motion of the biharmonic oscillator was found as the heteroclinic chaos can be removed by increasing the coefficient of a damping force. The analytical results obtained using Melnikov method has been confirmed by a good match with numeric research.

Introduction

Reduction of the Biharmonic Dynamics System to the Duffing Equation

As is well-Known, a nonlinear system can perform a chaotic motion under the action of periodic forces [1–4]. Frequently the Duffing equation is used to illustrate chaos [1–4], and the chaotic behavior of various forms of the Duffing equation [5], some of which exhibit two-frequency excitation [6] as well as the chaotic motion of Duffing system with bounded noise [7] have been investigated. However the Duffing equation in the expanded (generalized) form has many mechanical applications and it can be interesting to researchers.

[†]Reused with permission from: Vladimir S. Aslanov, "Chaotic Behavior of the Biharmonic Dynamics System," *International Journal of Mathematics and Mathematical Sciences*, vol. 2009, Article ID 319179, 18 pages, 2009. doi:10.1155/2009/319179.

In this paper, we will study periodically driven biharmonic dynamic system with a damping force:

$$\ddot{\theta} = a\sin\theta + b\sin 2\theta + \varepsilon\left(a_1\sin\theta + b_1\sin 2\theta + c\right)\cos\omega t - \delta\dot{\theta} \qquad (1.1)$$

where ε and δ are assumed to be small positive parameters; $\omega > 0$ is the frequency of the external force; a, a_1, b, b_1 and c are coefficients. The terms in ε and in δ equation (1.1) can be considered as small perturbations.

If $\varepsilon = 0$ and $\delta = 0$ then the periodic and the damping force are absent, and we have the conservative system describing the motion of the undisturbed biharmonic oscillator as

$$\ddot{\theta} = a\sin\theta + b\sin 2\theta \qquad (1.2)$$

We shall show, that the equation (1.1) is equivalent to the Duffing equation for small values of θ . It is known that

$$\sin x = x - \frac{x^2}{3!} + \dots \left(x = \theta, 2\theta\right) \qquad (1.3)$$

and the equation (1.1) can be written as

$$\ddot{\theta} - a\left(\theta - \frac{\theta^2}{3!}\right) - b\left(2\theta - \frac{(2\theta)^2}{3!}\right) = \varepsilon\left\{a\sin\left(\theta - \frac{\theta^2}{3!}\right) + b\sin\left(2\theta - \frac{(2\theta)^2}{3!}\right) + c\right\}\cos\omega t - \delta\dot{\theta} + \dots \qquad (1.4)$$

If the variable θ is a small of order ε , then we have the Duffing equation

$$\ddot{\theta} + \lambda\theta + \mu\theta^3 = \varepsilon\left(c - \lambda\theta\right)\cos\omega t - \delta\dot{\theta} + O\left(\varepsilon^4\right), \qquad (1.5)$$

where $\lambda = -\left(a + 2b\right)$, $\mu = \dfrac{a + 8b}{3!}$.

Mechanical Applications

The disturbed system (1.1) has applications in space flight mechanics when studying a problem of a spacecraft motion about its center of mass in the atmosphere. Atmospheric reentry is a critical phase for space vehicles. Dynamic stability issues play a crucial role for the success of their mission. For effective breaking while descending in the rarefied atmosphere of Mars or Titan we can compensate blunt-shaped spacecrafts by a small increase in length [8–11]. We shall consider blunt-shaped spacecraft. An aerodynamic restoring moment strongly influences the motion of the spacecraft relative to the center of mass. The aerodynamic restoring moment coefficient for an axisymmetric rigid body can be written as [12, 13]

$$m_\theta\left(\theta\right) = m_0\left(\theta\right) + c_n\left(\theta\right)\bar{x}_c, \qquad (1.6)$$

where θ is the spatial angle of attack (θ is defined as the angle between symmetry axis and the velocity of the spacecraft $\bar{V}$, Fig. 1), m_0 is the pitching moment coefficient concerning the leading edge body, c_n is the normal force coefficient, $\bar{x}_c = x_c / L$ is the nondimensional coordinate of the center of mass and L is aerodynamic reference length (shield diameter). Such spacecrafts can have three positions of equilibrium according to the angle of attack θ : stable position at the points $\theta_* = 0$ and $\theta_* = \pi$; and unstable in the third intermediate point $\theta_* \in (0, \pi)$ [12,13]. The restoring moment can be approximated as the biharmonic dependence (Fig.1)

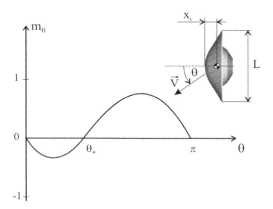

Figure 1 A blunt-shaped spacecraft.

$$m_\theta = a \sin \theta + b \sin 2\theta \tag{1.7}$$

The presence of the second harmonic in the expression (1.4) causes the possibility of appearance of an additional equilibrium position-saddle point on a phase portrait. For the considered spacecrafts position $\theta = 0$ is stable; therefore, a derivative of the function $m_\theta (\theta)$ with respect to the angle of attack θ at this point is negative

$$m'_\theta \big|_{\theta=0} = (a \cos \theta + b \cos 2\theta) \big|_{\theta=0} < 0 \tag{1.8}$$

or

$$2b < -a \tag{1.9}$$

and if there exists an intermediate position of equilibrium inside the interval of $(0, \pi)$, then

$$m_\theta (\theta) = \sin \theta (a + 2b \cos \theta) = 0, \tag{1.10}$$

which holds true, if

$$|2b| < |a| \tag{1.11}$$

It is obvious that expressions (1.5) and (1.6) are valid simultaneously when $b < 0$. Note that the dependence of $m_\theta (\theta)$ given in Fig. 1 satisfies conditions (1.5) and (1.6).

The stable position occurs not only in the point of $\theta = 0$, but also in the point of $\theta = \pi$ when condition (1.5) for the spacecraft is fulfilled. The motion of the spacecraft in a neighborhood of $\theta = \pi$ can not be allowed, because in this case the back part of the spacecraft will move towards to an approach flow.

The aerodynamic restoring moment can be written by

$$M_\theta = m_\theta (\theta) qSL, \tag{1.12}$$

where S is the reference surface and q is the dynamic pressure. Disturbance will simulate periodic change of position of the center of mass

$$\bar{x}_c = \bar{x}_{c0} + \varepsilon \cos \omega t, \tag{1.13}$$

where $\bar{x}_{c0}$ is an initial position of the center of mass. Using the expression (1.3) we represent the aerodynamic restoring moment as

$$M_\theta = A \sin\theta + B \sin 2\theta + \varepsilon (A \sin\theta + A \sin 2\theta) \cos \omega t \tag{1.14}$$

Thus, the planar motion of the spacecraft about the center of mass can be described by

$$I\ddot{\theta} = A \sin\theta + B \sin 2\theta + \varepsilon (A \sin\theta + A \sin 2\theta) \cos \omega t + M_d \tag{1.15}$$

where I is the transverse moment of inertia, $M_d = -\delta_1 \dot{\theta}$ is small damping moment. Obviously, the equation of spacecraft motion (1.8) corresponds to the biharmonic dynamics system (1.1) at $c = 0$ up to constants.

Let us observe, that the undisturbed equation (1.2) describes the motion of a known mechanical system—a heavy material point on a circle, rotating about a vertical axis [14]

$$\ddot{\theta} = a \sin\theta + b \sin 2\theta, \quad \left(a = \frac{-g}{l}, b = \Omega^2 > 0 \right), \tag{1.16}$$

where g is the gravitational acceleration, l is the radius of the circle and Ω is the angular velocity of the circle.

Aim and Structure of the Article

Let conditions (1.5) and (1.7) be satisfied, and let there be three positions of equilibrium. In this case, we find three regions in the phase portrait separated by separatrixes. Under the effect of disturbances the phase trajectory $\theta = \theta(\dot{\theta})$ can repeatedly intersect the separatrixes, thus moving from one an area to another, is accompanied by a jump change of the variable θ. We can observe chaos. It is an accepted fact in the theory of nonlinear dynamic systems that knowledge of the stable and unstable manifolds of hyperbolic equilibrium or hyperbolic periodic orbits may play a crucial role in understanding many issues of dynamics [1,2]. For many dynamic systems, the only general way of studying such stable and unstable manifolds is computing them numerically. However, in some cases we can obtain analytic solutions.

The aim of this paper is to analyze the motion of the disturbed system (1.1) near the undisturbed separatrixes and to define the boundaries of chaos. We will carry out the theoretical studies by means of the Melnikov method [15]. The Melnikov method is an analytical tool used to determine to first order, the existence of homo/heteroclinic intersections and so chaotic behavior. The Melnikov method allows us to obtain a necessary condition for the existence of chaos, therefore numerical simulation is needed to confirm the predicted behavior and to give a deeper understanding of the global dynamics of the system. We shall show that our theoretical results are in good agreement with the results of numerical calculations.

The present paper is structured in the following way. Section 2 gives the analysis of the unperturbed motion of the biharmonic dynamics system (1.1) and the phase portrait. On the phase portrait characteristic regions of the possible motions are found and for these areas the analytical solutions of the equation of the unperturbed motion are obtained. The main features of the phase space of the unperturbed system are defined. In section 3 the Melnikov criterion for the perturbed system is analytically calculated for various areas of the phase portrait with help of the theory of residues. In section 4 the Melnikov criterion is numerically calculated for various areas of a phase portrait for the case of disturbed motion of the spacecraft. By means of computer numerical simulations of the disturbed motion, we use several numerical techniques to check the validity of the analytical criterion for chaos obtained using Melnikov method.

The Unperturbed Solutions

The ε and δ terms in equation (1.1) are considered as small perturbations. If $\varepsilon = 0$ and $\delta = 0$ then periodic and dissipation forces are absent, and we have the conservative system (1.2). It is obvious, that if b=0 or a=0 (replacement of variables $\varphi = 2\theta$) we have the equation of a mathematical pendulum. However, if the conditions (1.5) and (1.7) are satisfied ($b < 0$), then biharmonic dynamic system has a more complicated phase portrait in comparison with the mathematical pendulum or with the heavy material point on a circle (1.9). The equilibrium positions of the system (1.2) are defined from the equation (1.6). If the conditions (1.5) and (1.7) are satisfied, then the undisturbed system (1.2) has four equilibrium positions at $\theta \in [-\pi, \pi]$: two stable—center type

$$\theta = 0, \pi \tag{2.1}$$

and two unstable-saddle type

$$\theta* = \pm \arccos\left(-\frac{a}{2b}\right), \tag{2.2}$$

where $b < 0$. The center $\theta_* = -\pi$ coincides with the center $\theta_* = -\pi$. At $\theta_* \to -\pi$ and at $\theta_* \to \pi$ the speeds $\dot{\theta}$ coincide, therefore we can say, that phase trajectories are closed on a cylindrical phase space. We shall consider the evolution of the cylindrical space in the range $\theta \in [-\pi, \pi]$. We will separate two regions A_0 and A_1, divided by the two saddles s_1 and s_{-1} (Fig. 2). It is necessary to note, that the region A_1 of the development of the cylinder undergoes a break at $\theta = \pi, -\pi$. From (2.2) it follows, that if the coefficient a is equal 0, the saddle s_1 is in the position: $\theta_* = \dfrac{\pi}{2}$. At positive values of the coefficient $a > 0$ the saddle s_1 belongs to the interval: $\theta_* \in (0, \pi / 2)$, and at negative values $a < 0$ the saddle s_1

belongs to the interval: $\theta_* \in \left(\pi / 2, \pi\right)$ (Fig.2). The following energy integral corresponds to the equation (1.2):

$$\frac{1}{2}\dot{\theta}^2 + a\sin\theta + b\cos^2\theta = E, \tag{2.3}$$

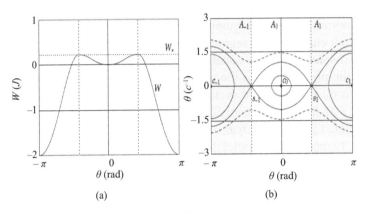

Figure 2 The potential energy $W(\theta) = a\sin\theta + b\cos^2\theta$ and the phase space for $a = 1, b = -1$.

where E is total energy. The biharmonic oscillator as well as the mathematical pendulum can perform oscillations and rotation. The shape of the phase portrait depends on the potential energy:

$$W(\theta) = a\sin\theta + b\cos^2\theta. \tag{2.4}$$

The centers (2.1) correspond to the minimum of the potential energy (2.4), and the saddles (2.2) - to the maximum of the potential energy (2.4). If $E > W_*$, where $W_* = W(\theta_*)$, then the motion is possible in the outer regions (Fig. 2). In the opposite case ($E < W_*$) the motion can occur in any of the inner regions, depending on initial conditions. The equality $E = W_*$ corresponds to the motion along separatrixes. In this case, the two saddles s_1 and s_{-1} are connected by four heteroclinic trajectories.

First of all, we will consider the separatrixes, limiting the region A_0. Separating the variables in the energy integral (2.3), and taking into account (2.2) and (2.4), the equation of the motion on the separatrixes can be written as in the integrated form

$$t - t_0 = \int_{\theta_0}^{\theta} \frac{d\theta}{\sqrt{2\left[W\left(\theta_*\right) - a\sin\theta + b\cos^2\theta\right]}}, \tag{2.5}$$

where

$$W\left(\theta_*\right) = a\sin\theta_* + b\cos^2\theta_* = -\frac{a^2}{4b}. \tag{2.6}$$

Substituting the variables:

$$x = \tan \frac{\theta}{2}, \tag{2.7}$$

we can rewrite (2.5) as:

$$t - t_0 = 2 \int_{x_0}^{x} \frac{dx}{\sqrt{\psi(x)}}, \psi(x) = Px^4 + Cx^2 + A, \tag{2.8}$$

where $P = 2 \left[W \left(\theta_* \right) + a - b \right]$, $C = 4 \left[W \left(\theta_* \right) + b \right]$, $A = 2 \left[W \left(\theta_* \right) - a - b \right]$..

The integral (2.8) can be simplified [16]

$$t - t_0 = \frac{2}{\sqrt{P}} \int_{x_0}^{x} \frac{dx}{x_*^2 - x^2} = \frac{2}{\sqrt{P}} \ln \left| \frac{x_* + x}{x_* - x} \right|_{x_0}^{x}, \tag{2.9}$$

where $x_* = \tan \frac{\theta_*}{2}$, $P = 2 \left[W \left(\theta_* \right) + a - b \right], = -\frac{(a - 2b)^2}{2b} > 0$ and conditions (1.9) and (1.11) are satisfied.

Finally, using the change of variables (2.7) and expression (2.9), the solution of equation (1.2) for the heteroclinic orbits, for the region A_0 (Fig. 2), can be written as

$$\theta_+(t) = 2 \arctan \left[\tan \frac{\theta_*}{2} \tanh \left(\frac{\lambda t}{2} \right) \right], \ \sigma_+(t) = \left(\right) = \frac{\lambda \sin \theta_*}{\cosh(\lambda t) + \cos \theta_*}, \tag{2.10}$$

$$\left[\theta_-(t), \sigma_-(t) \right] = \left[-\theta_+(t), -\sigma_+(t) \right]$$

where

$$\lambda = \tan \frac{\theta_*}{2} \sqrt{P} = \sqrt{\frac{a^2 - 4b^2}{2b}}. \tag{2.11}$$

Now we will consider the region A_1, including the center c_1. Let's make a substitution to the new variable

$$\beta = \pi - \theta \tag{2.12}$$

in the equation of undisturbed motion (1.2) and obtain the following equation

$$\ddot{\beta} = -a \sin \beta + b \sin 2\beta. \tag{2.13}$$

Computing as in equations (2.3)–(2.9), we obtain the solution of this equation in the form

$$\beta = 2 \arctan \left[\tan \frac{\beta_*}{2} \tanh \left(\frac{\lambda(t - t_0)}{2} \right) \right], \tag{2.14}$$

where $\beta_* = \pi - \theta_*$. Then, coming back to the variable θ with help of the substitution (2.11), we will receive the equation of the heteroclinic orbits, bounding the region A_1, including the center c_1 (Fig. 2):

$$\theta_+\left(t\right)=\pi-2\arctan\left[\cot an\frac{\theta_*}{2}\tanh\left(\frac{\lambda t}{2}\right)\right],\ \sigma_+\left(t\right)=\left(\dot\theta\right)=\frac{\lambda\sin\theta_*}{\cosh\left(\lambda t\right)-\cos\theta_*} \qquad (2.15)$$

$$\left[\theta_-\left(t\right),\sigma_-\left(t\right)\right]=\left[2\pi-\theta_+\left(t\right),-\sigma_+\left(t\right)\right].$$

Chaotic Motion The Melnikov Criterion

General Positions

Now we set the stage for our study of the disturbed system (1.1). The stable and unstable manifolds do not necessarily coincide and it is possible that they can cross transversally leading to an infinite number of new heteroclinic points. Then, a heteroclinic tangle is generated. In this case, because of the perturbation, the motion of the system (1.1), near the unperturbed separatrices, becomes chaotic. Inside this chaotic layer small isolated regions of regular motion with periodic orbits can also appear. The existence of heteroclinic intersections may be proved by means of the Melnikov method [15].

We present a more convenient form the for application of Melnikov method to the disturbed nonautonomous equation of the second order (1.1) as three differential autonomous equations of the first order [2]

$$\dot\theta=\sigma=f_1+g_1,$$
$$\dot\sigma=a\sin\theta+b\sin 2\theta+\varepsilon\left(a_1\sin\theta+b_1\sin 2\theta+c\right)\cos\omega t-\delta\sigma=f_2+g_2, \qquad (3.1)$$
$$\dot\phi=\omega,$$

where

$$f_1=\sigma,g_1=0,f_2=a\sin\theta+b\sin 2\theta,$$
$$g_2=\varepsilon\left(a_1\sin\theta+b_1\sin 2\theta+c\right)\cos\omega t-\delta\sigma. \qquad (3.2)$$

The Melnikov function [2] for system (3.1) is given by

$$M^\pm\left(t_0,\phi_0\right)=\int_{-\infty}^{\infty}\left\{f_1\left[q_\pm^0\left(t\right)\right]g_2\left[q_\pm^0\left(t\right),\omega t+\omega t_0+\phi_0\right]-f_2\left[q_\pm^0\left(t\right)\right]g_1\left[q_\pm^0\left(t\right),\omega t+\omega t_0+\phi_0\right]\right\}dt$$

$$=\int_{-\infty}^{\infty}\left\{f_1\left[q_\pm^0\left(t\right)\right]g_2\left[q_\pm^0\left(t\right),\omega t+\omega t_0+\phi_0\right]\right\}dt, \qquad (3.3)$$

where $q_\pm^0\left(t\right)=\left[\theta_\pm\left(t\right),\sigma_\pm\left(t\right)\right]$ are the solutions of the undisturbed heteroclinic orbits (2.10) or (2.13) for the areas A_0 or A_1.

3.2. Case 1 ($a_1=0,b_1=0,c=1$). The disturbed system (3.1) in this case takes the form

$$\dot\theta=\sigma=f_1+g_1,$$
$$\dot\sigma=a\sin\theta+b\sin 2\theta+\varepsilon\cos\phi-\delta\sigma=f_2+g_2, \qquad (3.4)$$
$$\dot\phi=\omega,$$

where

$$f_1 = \sigma, \ g_1 = 0, \ f_2 = a \sin\theta + b \sin 2\theta, \ g_2 = \varepsilon \cos\phi - \delta\sigma. \tag{3.5}$$

Substituting (3.5) into (3.2) gives

$$M^{\pm}(t_0, \phi_0) = \int_{-\infty}^{\infty} \sigma_{\pm} \{\varepsilon \cos(\omega t + \omega t_0 + \phi_0) - \delta\sigma_{\pm}\} dt$$

$$= \varepsilon \int_{-\infty}^{\infty} \sigma_{\pm} \cos(\omega t + \omega t_0 + \phi_0) dt - \delta \int_{-\infty}^{\infty} (\sigma_{\pm})^2 dt = M_{\tau} + M_{\delta}, \tag{3.6}$$

where M_{τ} and M_{δ} are the functions corresponding to both perturbations: the external periodic force $(\varepsilon \cos\omega t)$ and the damping force $(-\delta\dot{\theta})$, respectively. The Melnikov function describes the splitting of the stable and unstable manifolds of the disturbed hyperbolic fixed points defined on the cross-section. Thus, there are transverse intersections between the stable and unstable trajectories, if $M^{\pm}(t_0) = 0$.

   Firstly we consider the functions $M_{\delta}^{(0)}$ and $M_{\tau}^{(0)}$ for the area A_0, including the center c_0 (Figure 2). Substituting (2.10) into (3.6) gives

$$M_{\delta}^{(0)} = -\delta \int_{-\infty}^{\infty} (\sigma_{\pm})^2 dt = -\delta \sin^2\theta_S \int_{-\infty}^{\infty} \frac{dt}{\left[\cosh(\lambda t) - \cos\theta_*\right]^2}$$

$$M_{\delta}^{(0)}(t_0, \phi_0) = \varepsilon \int_{-\infty}^{\infty} \sigma_{\pm} \cos(\omega t + \omega t_0 + \phi_0) dt dt = \varepsilon\lambda \sin\theta_S \int_{-\infty}^{\infty} \frac{\cos(\omega t + \omega t_0 + \phi_0) dt}{\cosh(\lambda t) - \cos\theta_*}, \tag{3.7}$$

So using the tabulated integrals [16,17], for the integrals (3.7) and (3.8) we obtain the following expressions

$$M_{\delta}^{(0)} = -2\delta\lambda(1 - \theta_* \cos\theta_*),$$

$$M_{\delta}^{(0)}(t_0, \phi_0) = 2\varepsilon\pi \frac{\sinh\left(\theta_* \frac{\omega}{\lambda}\right)}{\lambda \sin(\theta_*) \sinh\left(\pi \frac{\omega}{\lambda}\right)} \cos(\omega t_0 + \phi_0) = M_{\varepsilon\max}^{(0)} \cos(\omega t_0 + \phi_0), \tag{3.8}$$

Similar expressions can be obtained for $M_{\delta}^{(1)}$ and $M_{\tau}^{(1)}$ for the region A_1, including the center c_1 (Fig. 2), using the solutions (2.15)

$$M_{\delta}^{(1)} = -\delta \int_{-\infty}^{\infty} (\sigma_{\pm})^2 dt = -\delta \sin^2\theta_* \int_{-\infty}^{\infty} \frac{dt}{\left[\cosh(\lambda t) - \cos\theta_*\right]^2},$$

$$M_{\delta}^{(1)}(t_0, \phi_0) = \varepsilon \int_{-\infty}^{\infty} \sigma_{\pm} \cos(\omega t + \omega t_0 + \phi_0) dt dt = \varepsilon\lambda \sin\theta_S \int_{-\infty}^{\infty} \frac{\cos(\omega t + \omega t_0 + \phi_0) dt}{\cosh(\lambda t) - \cos\theta_*}, \tag{3.9}$$

or

$$M_\delta^{(1)} = -2\delta\lambda\left[1 + (\pi - \theta_*)\cos\theta_*\right],$$

$$M_\delta^{(1)}(t_0,\phi_0) = 2\varepsilon\pi \frac{\sinh\left((\pi-\theta_*)\dfrac{\omega}{\lambda}\right)}{\lambda\sin(\theta_*)\sinh\left(\pi\dfrac{\omega}{\lambda}\right)}\cos(\omega t_0+\phi_0) = M_{\varepsilon\max}^{(1)}\cos(\omega t_0+\phi_0),$$

(3.10)

where $M_{\varepsilon\max}^{(1)}$ and $M_{\varepsilon\max}^{(2)}$ are measures of the maximum splitting of the stable and unstable manifolds, when the disturbed system (3.4) is only under the action of the one perturbation the external periodic force ($\varepsilon\cos\omega t$) for the regions A_0 and A_1 respectively.

Obviously, at $a = 0$ the undisturbed biharmonic oscillator (1.2) is transformed to the simpler system: $\ddot\theta = \sin 2\theta$. The regions A_0 and A_1 are equal. From (2.2), (2.10) and (2.13) we obtain

$$\theta_* = \frac{\pi}{2}, \lambda = \sqrt{-2b}.$$

(3.11)

Following the expressions (3.8)-(3.10), the Melnikov function becomes identical for the regions A_0 and A_1:

$$M^\pm(t_0,\phi_0) = M_\tau + M_\delta(t_0,\phi_0) = -2\delta\lambda + \varepsilon\frac{\pi}{\lambda}\operatorname{sec}h\left(\frac{\pi}{2}\frac{\omega}{\lambda}\right)\cos(\omega t_0+\phi_0)$$

(3.12)

From (3.8)-(3.10) it is easy to see that the conditions for the manifolds to intersect in terms of the parameters (δ,ε) is given by

$$\delta < \left[\frac{\pi\sinh\left(\theta_*\dfrac{\omega}{\lambda}\right)}{\lambda^2(1-\theta_S\sin(\theta_*))\sin(\theta_*)\sinh\left(\pi\dfrac{\omega}{\lambda}\right)}\right]\varepsilon \quad \text{(for the area } A_0\text{)},$$

$$\delta < \left[\frac{\pi\sinh\left((\pi-\theta_*)\dfrac{\omega}{\lambda}\right)}{\lambda^2(1+(\pi-\theta_*)\cos(\theta_*))\sin(\theta_*)\sinh\left(\pi\dfrac{\omega}{\lambda}\right)}\right]\varepsilon \quad \text{(for the area } A_1\text{)}.$$

(3.13)

Let us define a new parameter of the damping force, divided into amplitude of external force

$$\Delta = \frac{\delta}{\varepsilon}$$

(3.14)

then conditions (3.13) are given by

$$\Delta < \left| \left[\frac{\pi \sinh\left(\theta_* \frac{\omega}{\lambda}\right)}{\lambda^2 \left(1 - \theta_* \sin(\theta_*)\right)\sin(\theta_*)\sinh\left(\pi \frac{\omega}{\lambda}\right)} \right] \right| = \Delta_0 \text{ (for the area } A_0 \text{)},$$

$$\Delta < \left| \left[\frac{\pi \sinh\left((\pi - \theta_*) \frac{\omega}{\lambda}\right)}{\lambda^2 \left(1 + (\pi - \theta_*)\cot(\theta_*)\right)\sin(\theta_*)\sinh\left(\pi \frac{\omega}{\lambda}\right)} \right] \right| = \Delta_1 \text{ (for the area } A_1 \text{)},$$ (3.15)

Let's note, that θ_* and λ, according to (2.2) and (2.11), depend on coefficients a and b, therefore criteria (3.15) are functions of the parameters a, b and ω

$$\Delta_j = \Delta_j(a, b, \omega) \, j = 0, 1.$$ (3.16)

The criteria (3.16) define chaotic behaviour of the perturbed system (3.4) in the regions A_0 and A_1. In Fig. 3 we graph these criteria and the variable θ_* as functions of parameter a for the fixed parameter $b = -1$ and $\omega = 1$. Figure 4 shows the criteria (3.16) as functions of the frequency ω.

3.3. Case 2 ($a_1 \neq 0, b_1 \neq 0, c = 0$). The disturbed system (3.1) in this case takes the form of

$$\dot{\theta} = \sigma = f_1 + g_1,$$
$$\dot{\sigma} = a \sin \theta + b \sin 2\theta + \varepsilon (a_1 \sin \theta + b_1 \sin 2\theta) \cos \phi - \delta \sigma = f_2 + g_2,$$ (3.17)
$$\dot{\phi} = \omega,$$

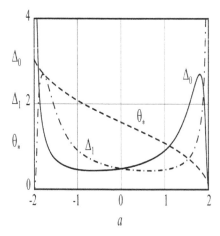

Figure 3 The criterions Δ_j and the variable θ_* of parameter a for $b = -1$ and $\omega = 1$.

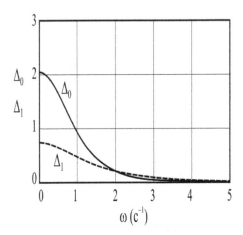

Figure 4 The criterions Δ_j as functions of the frequency ω.

where $f_1 = \sigma$, $g_1 = 0$, $f_2 = a\sin\theta + b\sin 2\theta$, and

$g_2 = \varepsilon(a_1\sin\theta + b_1\sin 2\theta)\cos\phi - \delta\sigma$.

$$M^{\pm}(t_0,\phi_0) = \int_{-\infty}^{\infty} \sigma_{\pm}\left[\varepsilon(a_1\sin\theta_{\pm} + b_1\sin 2\theta_{\pm})\cos(\omega t + \omega t_0 + \phi_0) - \delta\sigma_{\pm}\right]dt \qquad (3.18)$$

or

$$M^{\pm}(t_0,\phi_0) = M_\tau + M_\delta \qquad (3.19)$$

where

$$M_\tau = \varepsilon\int_{-\infty}^{\infty}\sigma_{\pm}\left[(a_1\sin\theta_{\pm} + b_1\sin 2\theta_{\pm})\cos(\omega t + \omega t_0 + \phi_0)\right]dt$$

$$(3.20)$$

$$M_\delta = -\delta\int_{-\infty}^{\infty}(\sigma_{\pm})^2 dt.$$

For the two regions A_0 and A_1 the functions (3.20) can be represented as

$$M_\tau^{(k)} = -\varepsilon I_{\pm}^{(k)}\sin(\omega t_0 + \phi_0),\ M_\delta^{(k)} = -\delta J_{\pm}^{(k)},\ k = 0,1, \qquad (3.21)$$

where

$$I_{\pm}^{(k)} = \int_{-\infty}^{\infty}\sigma_{\pm}^{(k)}\left[(a_1\sin\theta_{\pm}^{(k)} + b_1\sin 2\theta_{\pm}^{(k)})\cos(\omega t)\right]dt, \qquad (3.22)$$

$$J_{\pm}^{(k)} = \int_{-\infty}^{\infty} \left(\sigma_{\pm}^{(k)} \right)^2 dt. \tag{3.23}$$

It is obvious that using (3.8) integrals (3.20) can be rewritten as

$$J_{\pm}^{(0)} = 2\lambda \left(1 - \theta_* \cot\theta_* \right), \ J_{\pm}^{(1)} = 2\lambda \left(1 + \left(\pi - \theta_* \right)\cot\theta_* \right), \tag{3.24}$$

The improper integral (3.22) in view of solutions (2.10) and (2.15) is calculated numerically. For parameter (3.14) the conditions for the manifolds to intersect are given by

$$\Delta < \frac{I_{\pm}^{(0)}}{J_{\pm}^{(0)}} = \Delta_0, \ \text{(for the area } A_0 \text{)},$$

$$\Delta < \frac{I_{\pm}^{(1)}}{J_{\pm}^{(1)}} = \Delta_1, \ \text{(for the area } A_1 \text{)}. \tag{3.25}$$

Criteria (3.25) define behaviour of the perturbed system (3.17) in a vicinity of separatrixes.

Numerical Analysis

We have analyzed the evolution of the dynamical behavior of the disturbed system (1.1) as the parameters vary, studying the time histories of the variable θ and its derivative $\dot\theta$. Numerical techniques are based on the numerical integration of the equation of the disturbed motion (1.1) implementing a fixed step fourth order Runge–Kutta algorithm. For all numerical calculations the following biharmonic force parameters were used: $a=1$, $b=-1$ and the frequency of the perturbed force was $\omega = 1$.

For the numerical analysis of the disturbed system (3.1) we use the P Poincaré cross-section method, examining manifolds with plane sections, perpendicular to the phase axis ϕ in the two-dimensional space $\left(\theta, \dot\theta\right)$, divided with an interval of 2π. It allows us to study the disturbed system (3.1) using a discrete phase instead of examining the continuous dynamics of the system. At $\varepsilon = 0$, $\delta = 0$ the regular structure of phase space is observed, trajectories have no intersections, and Poincaré sections coincide with undisturbed phase portrait (Fig. 5).

Disturbances ($\varepsilon \neq 0$) result in the complication of phase space and the occurrence of a chaotic layer near the undisturbed separatrixes (Figs. 6–9). Figures 6,7 shows Poincaré sections for the case considered in section 3.2, and in Figs. 8, 9—in section 3.3. The growth of disturbances there leads to an increase in the width of the chaotic layer, and the new oscillatory modes determined by closed curves, uncharacteristic for the undisturbed case are observed in the presence of damping phase trajectories eventually tended to reach steady positions of equilibrium of the undisturbed system (Figs. 10, 11).

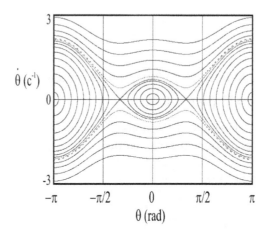

Figure 5 Poincaré sections for $\varepsilon = 0$, $\delta = 0$.

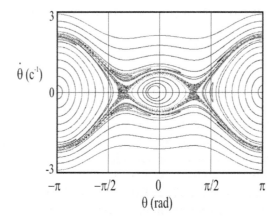

Figure 6 Poincaré sections in the case of $a_1 = 0$, $b_1 = 0$, $c = 1$ for $\varepsilon = 0$, $\delta = 0$.

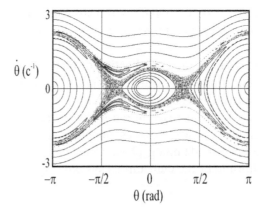

Figure 7 Poincaré sections in the case of $a_1 = 0$, $b_1 = 0$, $c = 1$ for $\varepsilon = 0.02$, $\delta = 0$.

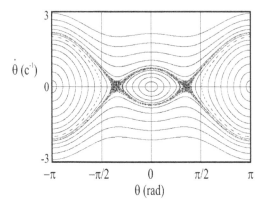

Figure 8 Poincaré sections in the case of $a_1 = 1, b_1 = -1, c = 0$ for $\varepsilon = 0.01, \delta = 0$.

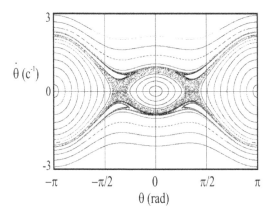

Figure 9 Poincaré sections in the case of $a_1 = 1, b_1 = -1, c = 0$ for $\varepsilon = 0.02, \delta = 0$.

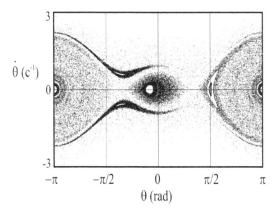

Figure 10 Poincaré sections in the case of $a_1 = 0, b_1 = 0, c = 1$ for $\varepsilon = 0.02, \delta = 0.0001$.

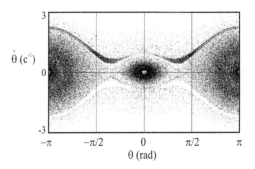

Figure 11 Poincaré sections in the case of $a_1 = 1, b_1 = -1, c = 0$ for $\varepsilon = 0.02, \delta = 0.0001$.

In order to check in a quantitative way the validity of the analytic criteria (3.15) we focus on the evolution of the stable and unstable manifolds associated to the saddle fixed points. Parameters $a = 1, b = -1$ and the frequency $\omega = 1$ in expressions (3.15) give a critical value for the regions A_0 and A_1:

$$\Delta_0 = 0.9115, \Delta_1 = 0.4530 \qquad (4.1)$$

or critical values of the coefficients of a damping force for $\varepsilon = 0.02$:

$$\delta_0 = \varepsilon\Delta_0 = 0.01823, \delta_1 = \varepsilon\Delta_1 = 0.00906. \qquad (4.2)$$

Figure 12 shows numerical simulations of the phase space with initial conditions close to the undisturbed separatrix ($\theta_0 = -1.0572, \dot{\theta}_0 = 0.01, \phi_0 = \pi/10$) for the region A_0. Now,

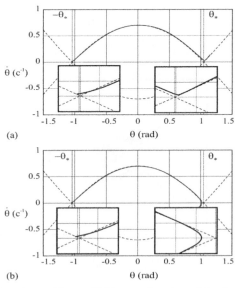

Figure 12 Evolution of stable and unstable manifolds as a function of the damping force coefficients δ for $a = 1, b = -1, \varepsilon = 0.02, \omega = 1$ and two different values of δ close to the critical value $\delta = 0.018$. (a) $\delta = 0.018$. (b) $\delta = 0.020$, with following initial conditions: $\theta_0 = -1.0572, \dot{\theta}_0 = 0.01, \phi_0 = \pi/10$ for the area A_0.

we reset the value of δ from $\delta_0 = 0.01823$ to greater ones (see Fig. 12). It can be observed clearly that, for $\delta < \delta_0$ ($\delta = 0.018$), the stable and unstable manifolds transversally intersect each other (Fig. 12(a)). However, when $\delta > \delta_0$ ($\delta = 0.020$), the invariant manifolds do not intersect (Fig. 12(b)). Similar results for the region A_1 are shown in Fig. 13 ($\delta_1 = 0.000906$).

For the following initialconditions: $\theta_0 = 0.9472$, $\dot{\theta}_0 = 0.2$, $\phi_0 = \pi$,. Figure 13(a), $\delta < \delta_1$ ($\delta = 0.009$). Figure 13(b), $\delta > \delta_1$ ($\delta = 0.0113$). Thus the description, based on numerical simulations for some certain parameter values, makes a good match with the analytic criteria (3.15) provided by Melnikov method.

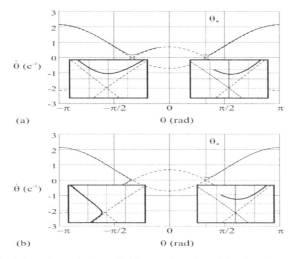

(a)

(b)

Figure 13 Evolution of stable and unstable manifolds as a function of the damping force coefficients δ for $a = 1, b = -1, \varepsilon = 0.02, \omega = 1$ and two different values of δ close to the critical value $\delta_1 = 0.000906$. (a) $\delta = 0.0090$ (b) $\delta = 0.0113$, with following initial conditions: $\theta_0 = 0.9472$, $\dot{\theta}_0 = 0.2$, $\phi_0 = \pi$ for the area A_1.

Conclusion

This work attempts to describe the transient cases occurring during a spacecraft descent in a planet atmosphere using methods of chaotic mechanics, in particular, the Melnikov method. We have suggested to introduce the concept of a biharmonic system (1.1) which reflects the behavior of the spacecrafts, and also, probably, more general mechanical systems. We have established the existence of transient heteroclinic chaos by means of the Melnikov method. Moreover, this method has provided an analytic criterion for the existence of chaotic behavior in terms of the system parameters. We have found a transition from chaotic to regular regime in the motion of the biharmonic oscillator, as the heteroclinic chaos can be removed by increasing the coefficient of a damping force. The analytic results given by the Melnikov method have been confirmed by a good match with the numeric research.

Acknowledgements

The author thanks Mr. Amany Qassem, Journal Publishing Editor for the invitation to the publication. The author would like to thank the Russian Foundation for Basic Research (Project no.06-01-00355a), which has supported this work.

References

1. J. Guckenheimer and P. Holmes, Nonlinear Oscillations, Dynamical and Bifurcations of Vector Fields, Springer, New York, NY, USA, 1986.
2. S. Wiggins, Introduction to Applied Nonlinear Dynamical Systems and Chaos, vol. 2 of Texts in Applied Mathematics, Springer, New York, NY, USA, 1990.
3. F.C. Moon, Chaotic Vibrations: An Introduction for Applied Scientists and Engineers, AWiley-Interscience Publication, JohnWiley & Sons, New York, NY, USA, 1987.
4. E. Simiu, Chaotic Transitions in Deterministic and Stochastic Dynamical Systems, Princeton Series in Applied Mathematics, Princeton University Press, Princeton, NJ, USA, 2002.
5. Z.-G. Li, W. Xu, and X.-Y. Zhang, "Analysis of chaotic behavior in the extended Duffing-Van der Pol system subject to additive non-symmetry biharmonical excitation," *Applied Mathematics and Computation*, vol. 183, no. 2, pp. 858–871, 2006.
6. I.I. Blekhman and P.S. Landa, "Conjugate resonances and bifurcations in nonlinear systems under biharmonical excitation," International Journal of Non-Linear Mechanics, vol. 39, no. 3, pp. 421–426, 2004.
7. W.Y. Liu, W.Q. Zhu, and Z.L. Huang, "Effect of bounded noise on chaotic motion of duffing oscillator under parametric excitation," Chaos, Solitons and Fractals, vol. 12, no. 3, pp. 527–537, 2001.
8. A. Wilson and A. Chicarro, Mars Express: The Scientific Payload. Vol. SP-1240, ESA, Noordwijk, The Netherlands, 2004.
9. R. Gershman, M. Adams, R. Mattingly, et al., "Planetary protection for Mars sample return," Advances in Space Research, vol. 34, no. 11, pp. 2328–2337, 2004.
10. J. Collinet, P. Brenner, and S. Palerm, "Dynamic stability of the HUYGENS probe at Mach 2.5," Aerospace Science and Technology, vol. 11, no. 2-3, pp. 202–210, 2007.
11. G. Colombatti, A. Aboudan, F. Ferri, and F. Angrilli, "Huygens probe entry dynamic model and accelerometer data analysis," Planetary and Space Science, vol. 56, no. 5, pp. 601–612, 2008.
12. V.S. Aslanov and A.S. Ledkov, "Features of rotational motion of a spacecraft descending in the Martian atmosphere," Cosmic Research, vol. 45, no. 4, pp. 331–338, 2007.
13. V. S. Aslanov, "Resonance at motion of a body in the Mars's atmosphere under biharmonical moment," WSEAS Transactions on Systems and Control, vol. 3, no. 1, pp. 33–39, 2008.
14. A.A. Andronov, A.A. Vitt, and S.E. Khaikin, Theory of Oscillators, Pergamon Press, Oxford, UK, 1966.
15. V.K. Mel'nikov, "On the stability of a center for time-periodic perturbations," Transactions of the Moscow Mathematical Society, vol. 12, pp. 3–52, 1963.
16. H. Bateman and A. Erdelyi, Table of Integrals Transforms. Vol. 1, McGraw-Hill, New York, NY, USA, 1954.
17. G.A. Korn and T. M. Korn, Mathematical Handbook for Scientists and Engineers, McGraw-Hill, New York, NY, USA, 1961.

Modeling, Chaotic Behavior and Control of Dissipation Properties of Hysteretic Systems[†]

J. Awrejcewicz[1] and L. Dzyubak[2]

[1]Department of Automatics and Biomechanics, Technical University of Łódź, 1/15 Stefanowskiego Street, 90-924 Łódź, Poland. Email: awrejcew@p.lodz.pl.
[2]Department of Applied Mathematics, National Technical University "Kharkov Polytechnic Institute", 21 Frunze Street, 61002 Kharkov, Ukraine. Email: ldzyubak@kpi.kharkov.ua.

ABSTRACT

In the present work, hysteresis is simulated by means of internal variables. Analytical models of different types of hysteresis loops enable reproduction of major and minor loops and provide good agreement with experimental data. Using an effective algorithm based on the analysis of wandering trajectories, an evolution of chaotic behavior regions of oscillators with hysteresis is presented in various parametric planes. A substantial influence of a hysteretic dissipation value on the form and location of these regions and also restraining and generating effects of the hysteretic dissipation on chaos occurrence are shown. It was demonstrated that for fixed parameters, which govern the shape of a hysteresis loop, the extent of pinch of this loop could be controlled by means of the amplitude and frequency of external periodic excitation.

Introduction

Hysteresis is caused by very different processes in nature and is characterized above all by the presence of an *output* delay with respect to *input* in an *input-output* correlation, energy dissipation, and memory in a system. The problem of hysteretic systems investigation occurs in many fields of science for mechanical, engineering, physical, and biological systems and even for sociological and economic systems.

The present paper is composed of five sections. Section 2 is devoted to the simulation of hysteretic systems. There is a lot of different phenomenological approaches to hysteresis modeling. General mathematical models of hysteresis are presented and discussed in

[†]Resued with permission from: J. Awrejcewicz and L. Dzyubak, "Modeling, chaotic behavior, and control of dissipation properties of hysteretic systems," *Mathematical Problems in Engineering*, vol. 2006, Article ID 94929, 21 pages, 2006. doi:10.1155/MPE/2006/94929.

the monographs [21,28]. Recent results of the analysis of hysteretic systems including modeling, experiments, dynamic response, and applications can be found in [27]. A large number of publications [7,9–15,17–19,22–26] are devoted to hysteresis simulation, since it is known that hysteretic systems are complicated to investigate and various difficulties occur when applying the existing models. The question of multipurpose and generally valid models describing the wide spectrum of hysteretic phenomena is still open. In the present work, hysteresis is simulated by means of additional state variables (internal variables).

In particular, the behavior of magnetorheological/electrorheological (MR/ER) fluids (which are known as smart materials and commercially available now) in a damper/absorber [24,25] is simulated. Another example is modeling of stress-strain hysteresis in a steel rope including minor-loop reproduction. The developed models are effective and contain principally less parameters than, for example, Bouc-Wen or Spencer models. Section 3 is dedicated to the analysis of chaotic behavior occurring in the hysteretic systems in various parametric spaces. The models describing systems with hysteresis are discontinuous and contain high nonlinearities with memory-dependent properties. The *output* is delayed with respect to the *input* and for every *input* there may be more than one equilibrium state. Investigation of these systems within the framework of an approximate analytical approaches as, for instance, slowly varying parameters or harmonic balance methods, results in a conclusion that irrespective of the values of control parameters occurring in the conditions of external periodic excitation, the hysteretic system has a stable symmetric asymptotic response. However, the recent publications and works [8,16,20], based on numerical and combined numerical-analytical techniques, present frequency response curves and bifurcation diagrams which indicate the presence of solutions and bifurcations mostly unexpected for hysteretic oscillators. At the same time, knowledge of the control parameter spaces is still insufficient. In this connection, the prediction of conditions for stable/unstable behavior of such systems is very topical. In the present work, a chaotic behavior occurring in the dynamic hysteretic system governed by a coupled differential set is investigated in various parametric planes using a methodology described in [1–6]. This methodology had been successfully applied already in particular to predict stick-slip chaos in a weakly forced oscillator with friction [3], in 2-DOF discontinuous systems with friction [2,4], and chaos in other smooth and non-smooth systems [4]. Section 4 is devoted to the control of dissipation properties of oscillators with hysteresis.

Hysteresis Simulation by Means of Additional State Variables (Internal Variables)

In modeling of systems showing hysteresis based on the parametric approaches, it is intended to use various linear and nonlinear elements simulating memory in the system as well as the energy dissipated in each cycle. In a physical sense, energy losses can be described by including into the model, for example, dashpots, friction elements, springs, and other mechanical elements. Thus, the classical Masing model of hysteresis merges the element of Hooke (elastic body model) and a number of St. Venant's elements (plastic body models) in parallel. The Biot model of hysteresis includes some number of Newton's elements (Newton flow models) instead of St. Venant's elements. Not infrequently successive joints of mechanical (physical) elements are also used as, for example, in the Spencer model which is an extension of the Bouc-Wen model. The parameters of each model can be defined during a parameter identification process with experimental data. The model validity is confirmed also by the experiment.

Let us consider the combination of N friction elements with the maximum friction forces $F_1, F_2, \ldots, F_N$ and springs $k_0, k_1, k_2, \ldots, k_N$ in parallel, as it is shown in Fig. 2.1. Here, x is the *input* (input signal) and z is the *output* (response) of the hysteretic system. The delay in the arrival of the output with respect to the input can be described with the aid of internal variables (forces) $y_1, y_2, \ldots, y_N$.

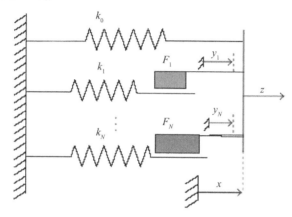

*Figure 1** Mechanical system with elastic-plastic properties for the hysteresis modeling (the Masing model).

If absolute value of the internal variable $|y_i|$ $(i=1,2,\ldots, N)$ is less than maximum friction force F_i, or if both this value is equal to F_i and the velocity of the input $\dot{x}$ and the force y_i have different signs (including the case when $\dot{x}=0$), then the evolution of force y_i in time is governed by the following equation:

$$\dot{y}_i = k_i \dot{x} \text{ for } |y_i| < F_i \vee \left(|y_i| = F_i \wedge \operatorname{sgn}(\dot{x}y_i) \le 0 \right), \; i=1,2,\ldots,N. \tag{2.1}$$

In all other cases this evolution is as follows:

$$\dot{y}_i = 0. \tag{2.2}$$

Equations (2.1) and (2.2) are generalized into the following equation:

$$\dot{y}_i = k_i \dot{x} \frac{1}{2} \left(1 - \operatorname{sgn}\left(y_i^2 - F_i^2 \right) - \operatorname{sgn}(\dot{x}y_i)\left[1 + \operatorname{sgn}\left(y_i^2 - F_i^2 \right) \right] \right). \tag{2.3}$$

Using the approximation

$$\operatorname{sgn}\left(y_i^2 - F_i^2 \right) \approx \left| \frac{y_i}{F_i} \right|^m - 1 \text{ for } y_i^2 \le F_i^2, \; m \in \Re^+ \wedge m \ge 1, \tag{2.4}$$

---

*Figure 2.1 in original source

equation (2.3) can be rewritten in the form

$$\dot{y}_i = k_i \dot{x} \left(1 - \frac{1}{2} \left(1 + \mathrm{sgn}\left(\dot{x} y_i\right)\right) \left|\frac{y_i}{F_i}\right|^m \right), \tag{2.5}$$

and then the response of the hysteretic system is

$$z(t) = k_0 x(t) + \sum_{i=1}^{N} y_i(t). \tag{2.6}$$

A generalization of (2.5) and (2.6) in the case of friction elements and springs with properties of a more complex character (in particular of nonlinear character) yields

$$z(t) = k_0(x) x(t) + \sum_{i=1}^{N} y_i(t),$$

$$\dot{y}_i = \left(A_i(x) - \left(\beta_i + \alpha_i \, \mathrm{sgn}\left(\dot{x}\right) \mathrm{sgn}\left(y_i\right)\right) \left|\frac{y_i}{F_i(x)}\right|^m \right) \dot{x}, \tag{2.7}$$

where $k_0(x), A_i(x) \geq 0, F_i(x) > 0$, the input signal belongs to admissible codomain $x \in [x_{min}, x_{max}]$, $\alpha_i, \beta_i \in \Re$; or it can be written in a more general form

$$z = p\left(x, y_1, y_2, ..., y_N\right),$$

$$\dot{y}_i = q\left(x, \dot{x}, y_i\right), \quad i = 1, 2, ..., N. \tag{2.8}$$

Here p and q are the nonlinear (in a general case) functions of their arguments. They are chosen depending on the properties of a hysteretic system and a loop form.

Note that, according to (2.7), the structure of differential equations describing the evolution in time of the internal variables is very similar to the structure of the differential equations of the Bouc-Wen hysteretic oscillator model.

The parameters of functions $p\left(x, y_1, y_2, ..., y_N\right)$ and $q\left(x, \dot{x}, y_i\right) (i=1, 2, ..., N)$ are determined via a procedure minimizing the criterion function

$$\Phi\left(c_1, ..., c_j, \alpha_1, ..., \alpha_k, ..., \beta_1, ..., \beta_l\right) = \sum_i \left(p\left(x\left(c_1, ..., c_j, t_i\right), y_1\left(\alpha_1, ..., \alpha_k, t_i\right), ..., y_N\left(\beta_1, ..., \beta_l, t_i\right)\right) - z_i\right)^2 \tag{2.9}$$

which characterizes an error between the experimental and calculated curves. Here z_i are the responses of a hysteretic system, which are known from the experiment and the values $p\left(x\left(c_1, ..., c_j, t_i\right), y_1\left(\alpha_1, ..., \alpha_k, t_i\right), ..., y_N\left(\beta_1, ..., \beta_l, t_i\right)\right)$ are obtained from the integration of the system which is described by model (2.8). To minimize the criterion function (2.9), the method of gradient descent is used. The step-by-step descent to the minimum of the criterion function is performed in the opposite direction to the criterion function gradient

$$\mathbf{grad}\Phi = \left\{ \frac{\partial \Phi}{\partial c_1}, ..., \frac{\partial \Phi}{\partial c_j}, \frac{\partial \Phi}{\partial \alpha_1}, ..., \frac{\partial \Phi}{\partial \alpha_k}, \frac{\partial \Phi}{\partial \beta_1}, ..., \frac{\partial \Phi}{\partial \beta_l} \right\}. \tag{2.10}$$

When solving the optimization problem

$$\Phi\left(c_1,...,c_j,\alpha_1,...,\alpha_k,...,\beta_1,...,\beta_l\right)\to\min \tag{2.11}$$

after a sufficiently good choice of the initial approximation $\left(c_1^0,...,c_j^0,\alpha_1^0,...,\alpha_k^0,...,\beta_1^0,...,\beta_l^0\right)$, the convergence of the approximation is reached. If in the step-by-step descent

$$\tilde{c}_i = c_i - h_{c_i}\left(\frac{\partial\Phi}{\partial c_i}\Big/\left|\mathbf{grad}\Phi\right|\right),\quad i=\overline{1,j},$$

$$\tilde{\alpha}_i = \alpha_i - h_{\alpha_i}\left(\frac{\partial\Phi}{\partial\alpha_i}\Big/\left|\mathbf{grad}\Phi\right|\right),\quad i=\overline{1,k}, \tag{2.12}$$

$$\dots\dots\dots\dots\dots\dots\dots\dots$$

$$\tilde{\beta}_i = \beta_i - h_{\beta_i}\left(\frac{\partial\Phi}{\partial\beta_i}\Big/\left|\mathbf{grad}\Phi\right|\right),\quad i=\overline{1,l},$$

the value of the criterion function Φ increases or remains unchanged, it is necessary to decrease the values of the steps $h_{c_1},...,h_{c_j},h_{\alpha_1},...,h_{\alpha_k},...,h_{\beta_1},...,h_{\beta_l}$.

The model presents the fast enough numerical convergence. Applications to different types of hysteresis loops confirmed that models with internal variables were appropriate to simulate hysteresis.

Consider two examples of such applications. As it was mentioned, hysteresis is widely found in nature. Among others, hysteretic behavior is peculiar to "smart" materials that more and more attract the attention of commercial structures. Electrorheological/magnetorheological (ER/MR) fluids belong to such materials. These fluids are used for construction of dampers, which act as interfaces between electronic control systems and mechanical systems. The dynamic characteristics of ER/MR dampers are strongly nonlinear that is reflected in numerous results of the experimental studies of MRF damper behavior [24]. Physically accurate and simple models of dampers are highly required by engineering community and they are necessary for producing effective samples. Due to the nonlinearities of hysteresis and jumps in damper behavior, difficulties arise while developing these models [25].

During analysis of seven known parametric models in [25], an attempt was made to find a compromise between simplicity, "economy" of the model, and its physical accuracy. However, it was concluded that the simplest involution model with two parameters could not be applied to simulate the damper behavior. An extension of the Bouc-Wen model proposed by Spencer enables the most accurate prediction of an actual MR damper behavior, but this model contains nine parameters.

We suggest here the developed model with a single internal variable

$$z(t)=y_1(t),$$
$$\dot{y}_1 = \left(c_1 - \left(c_2 + c_3\,\mathrm{sgn}(\dot{x})\,\mathrm{sgn}(y_1)\right)|y_1|\right)\dot{x}, \tag{2.13}$$

which contains only three parameters and simulates the actual hysteretic MR damper behavior in the damping force versus velocity plane. This analytical model is simple and provides a high degree of correspondence with experimental data. A comparison of the simulated loop with experimental data is presented in Fig. 2.2. Final values of the parameters used in model (2.13) for identification of the experimental data are given in Table 2.1.

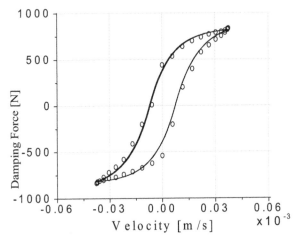

*Figure 2 Experimental (∘∘∘) and simulated solid line (as in ms) hysteresis loops for the magnetorheological damper filled with MRF-132LD (applied current 0.15 A, frequency 5 Hz).

**Table 1 Final values of the parameters used in model (2.13) for identification of the experimental data.

c_1	c_2	c_3
70000	80.7208	3.002

Further, we consider the stress-strain hysteresis in a steel rope. It happens that when a force acts upon a steel rope, some transient processes are reflected in minor loops reproduction. In this case, there are two (or more) ways to simulate these processes in hysteresis behavior of the steel rope. The first one is to increase the number of internal variables. But it leads to a complication of the model and increase in the number of parameters.

The model with five parameters, which contains a single internal variable and quite well describes the major stress-strain hysteresis loop, is

$$z(t) = c_4 x(t) + c_5 + y_1(t),$$
$$\dot{y}_1 = \left(c_1 - \left(c_2 + c_3 \operatorname{sgn}(\dot{x})\operatorname{sgn}(y_1)\right)|y_1|\right)\dot{x}. \tag{2.14}$$

It presents a fast numerical convergence. However, when the transient processes are taken into account, model (2.14) does not enable a proper reproduction of minor loops (Fig. 2.3(a)). They are not closed, though losses connected with the transient processes are presented. A quite accurate reproduction of minor loops is possible after introducing

---

*Figure 2.2 in original source
**Table 2.1 in original source

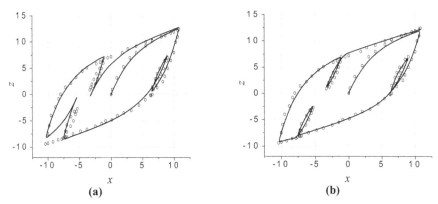

*Figure 3 Experimental (∘∘∘) and simulated solid line (as in ms) stress-strain hysteresis with transient processes for the steel rope (stress (N) versus strain (mm))—(a) simulation using model (2.14) with not properly reproducing minor loops; (b) simulation using model (2.15).

in model (2.14) the additional value z_c, which characterizes the geometrical centre of each minor loop

$$z(t) = c_4 x(t) + c_5 + y_1(t),$$
$$\dot{y}_1 = \left(c_1 - \left(c_2 + c_3 \operatorname{sgn}(\dot{x}) \operatorname{sgn}(y_1 - z_c)\right)\right) |y_1 - z_c| \dot{x}. \tag{2.15}$$

Model (2.15) enables a simulation of minor loops (Fig. 2.3(b)) and presents fast numerical convergence. The values of parameters in model (2.15), obtained during minimization of the criterion function (2.9), are given in Table 2.2.

**Table 2 Final values of the parameters used in model (2.15) for the experimental data identification.

c_1	c_2	c_3	c_4	c_5
2.22254	0.0010226	0.338787	0.387749	1.45286

Evolution of Chaotic Behavior Regions in Control Parameter Planes of the Masing and Bouc-Wen Hysteretic Oscillators

Consider classical hysteretic models such as Masing oscillator and Bouc-Wen oscillator. In both cases, an external periodic excitation with an amplitude F and frequency Ω acts on the mass m which oscillates along an inertial base. These oscillators possess hysteretic properties and it is supposed that there is a linear viscous damper with a coefficient 2μ. The following set of differential equations governs a motion of the Masing oscillator [16]:

$$\dot{x} = y,$$
$$\dot{y} = -2\mu y - (1 - v)g(x) - vz + F \cos \Omega t, \tag{3.1}$$
$$\dot{z} = g'\left(\frac{z - z_i}{2}\right) y.$$

---

*Figure 2.3 in original source
**Table 2.2 in original source

In the above $v \in [0,1]$; $g(x) = \dfrac{(1-\delta)x}{\left(1+|x|^n\right)^{\frac{1}{n}}} + \delta x$; $R = (1-v)g(x) + vz$ is the total restoring force

with nonlinear elastic part $(1-v)g(x)$ and with hysteretic part vz. The case $v=1$ corresponds to the maximum hysteretic dissipation and $v=0$ corresponds to the elastic behavior of the oscillator. The parameter δ characterizes a ratio between the post- and pre-yielding stiffness. The parameter n governs the smoothness of the transitions from the elastic to the plastic range. The couples $\pm(x_i, z_i)$ represent the velocity reversal points at $\dot{x} = 0$. According to Masing's rule which is extended to the case of steady-state motion of the hysteretic oscillator, the loading/unloading branches of a hysteresis loop are geometrically similar. So, if $f(x,z) = 0$ is the equation of a virgin loading curve, then the equations $f\left((x \pm x_i)/2, (z \pm z_i)/2\right) = 0$ describe loading/unloading branches of the hysteresis loop. In the case of non-steady state motion of the Masing's oscillator, it is supposed that the equation of any hysteretic response curve can be obtained by applying the original Masing rule to the virgin loading curve using the latest point of velocity reversal.

A motion of the Bouc-Wen oscillator [16] is governed by the following set of differential equations:

$$\dot{x} = y,$$

$$\dot{y} = -2\mu y - \delta x - (1-\delta)z + F \cos \Omega t, \tag{3.2}$$

$$\dot{z} = \left[k_z - (\gamma + \beta \operatorname{sgn}(y) \operatorname{sgn}(z))|z|^n \right] y,$$

where $R = \delta x + (1-\delta)z$ is the total restoring force; the parameters $(k_z, \beta, n) \in R^+$ and $\gamma \in R$ govern the shape of the hysteresis loop. The parameters δ and n have the same sense as in the case of the Masing model.

Describe briefly the approach based on the analysis of wandering trajectories in view of the state vector of systems (3.1) and (3.2) for $\mathbf{x} \in R^3$. In the chaotic behavior of nonlinear deterministic systems, wandering of the trajectories of motion around the various equilibrium states is assumed. They are characterized by unpredictability and sensitive dependence on the initial conditions. By analyzing the trajectories of motion of these systems, it is possible to find the chaotic vibration regions in the control parameter space.

The continuous dependence property on the initial conditions $\mathbf{x}^{(0)} = \mathbf{x}(t_0)$ of the solutions of set (3.1) or (3.2) will be used: For every initial condition $\mathbf{x}^{(0)}$, $\tilde{\mathbf{x}}^{(0)} \in R^3$, for every number $T > 0$, no matter how large, and for every preassigned arbitrary small $\varepsilon > 0$, it is possible to indicate a positive number $\delta > 0$ such that if the distance ρ between $\mathbf{x}^{(0)}$ and $\tilde{\mathbf{x}}^{(0)}$, $\rho\left(\mathbf{x}^{(0)}, \tilde{\mathbf{x}}^{(0)}\right)$, is less than δ, and $|t| \leq T$, the following inequality is satisfied:

$$\rho\left(\mathbf{x}(t), \tilde{\mathbf{x}}(t)\right) < \varepsilon. \tag{3.3}$$

In other words, if the initial points are chosen close enough, then during the preassigned arbitrary large time interval $-T \leq t \leq T$ the distance between simultaneous positions of moving points will be less than a given positive number ε.

For the sake of tracing chaotic and regular dynamics, it is assumed that, with an increase of time, all trajectories remain in the closed bounded domain of a phase space, that is,

$$\exists C_i \in R : \max_t |x_i(t)| \le C_i, \ i=1,2,3. \tag{3.4}$$

To analyze trajectories of sets (3.1) and (3.2), the characteristic vibration amplitudes A_i of components of the motion are introduced $A_i = \dfrac{1}{2}\left| \max_{t_1 \le t \le T} x_i(t) - \min_{t_1 \le t \le T} x_i(t) \right|$. Index number i runs over three values corresponding to three generalized coordinates x, y, z. $[t_1,T] \subset [t_0,T]$ and $[t_0,T]$ is the time interval, in which the trajectory is considered. The interval $[t_0,t_1]$ is the time interval, in which all transient processes are damped.

For the sake of our investigation, it seems the most convenient to use the embedding theorem and to consider a 3-dimensional parallelepiped instead of a hypersphere with the center in point $\mathbf{x}$. Two neighboring initial points $\mathbf{x}^{(0)} = \mathbf{x}(t_0)$ and $\tilde{\mathbf{x}}^{(0)} = \tilde{\mathbf{x}}(t_0)$ ($\mathbf{x} = (x,y,z)^T$ or $\mathbf{x} = (x_1,x_2,x_3)^T$) are chosen in the 3-dimensional parallelepiped $P_{\delta_x,\delta_y,\delta_z}(\mathbf{x}^{(0)})$ such that $\left| x_i^{(0)} - \tilde{x}_i^{(0)} \right| < \delta_i$, where $\delta_i > 0$ is small in comparison with A_i. In the case of a regular motion, it is expected that $\varepsilon_i > 0$ used in inequality $|x_i(t) - \tilde{x}_i(t)| < \varepsilon_i$ is also small in comparison with A_i. The wandering orbits attempt to fill up some bounded domain of the phase space. In instant t_0, the neighboring trajectories diverge exponentially afterwards. Hence, for some instant t_1, the absolute values of differences $|x_i(t) - \tilde{x}_i(t)|$ can take any value in closed interval $[0, 2A_i]$. An auxiliary parameter α is introduced, $0 < \alpha < 1$. αA_i is referred to as a divergence measure of observable trajectories in the directions of generalized coordinates and with the aid of parameter α one has been chosen, which is *inadmissible* for the case of the motion "regularity". The domains, where a chaotic behavior of considered systems is possible, can be found using the following condition:

$$\exists t^* \in [t_1,T] : |x(t^*) - \tilde{x}(t^*)| > \alpha A_x. \tag{3.5}$$

If this inequality is satisfied in some nodal point of the sampled control parameter space, then such motion is relative to chaotic one (including transient and alternating chaos). The manifold of all such nodal points of the investigated control parameter space sets up domains of chaotic behavior of the considered systems.

Motion stability depends on all parameters of the considered hysteretic models including initial conditions. We succeeded in tracing irregular responses of the Masing and Bouc-Wen hysteretic oscillators in the damping coefficient – amplitude (μ, F) and frequency—amplitude (Ω, F) of external periodic excitation planes after a coordinate sampling.

The Masing oscillator (3.1) is nonlinear both in the case of a pure elastic behavior without hysteretic dissipation ($v=0$) and in the case of motion with hysteretic dissipation ($v>0$). And at $v=0$ chaos has been found too. Figure 3.1 and 3.2 display the evolution of chaotic behavior domains with an increasing hysteretic dissipation value in the mentioned planes. The (Ω, F) and (μ, F) planes had been uniformly sampled by 100×100 nodal

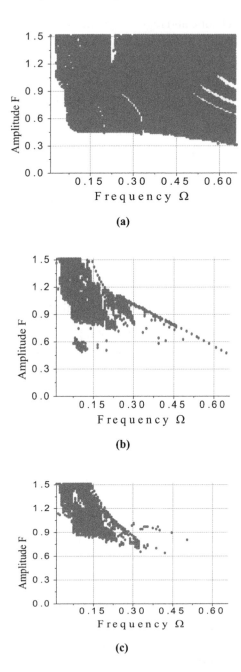

(a)

(b)

(c)

*Figure 4 Evolution of the chaotic regions for the Masing hysteresis model in the (Ω, F) plane with increasing hysteretic dissipation value (a) $v=0$; (b) $v=0.5$; (c) $v=0.8$. The parameters $\mu=0, \delta=0.05$, $n=10.0$, $x(0)=0.1$, $\dot{x}(0)$ $=0.1$, $z(0)=0$ are fixed for all cases.

---

*Figure 3.1 in original source

points in the rectangles ($0.01 < \Omega < 0.61$; $0.01 \leq F \leq 1.51$) and ($0.01 < \mu < 0.61$; $0.01 \leq F \leq 1.61$), respectively. The time period for the simulation T is of $\dfrac{300\pi}{\Omega}$ nondimensional time units. During computations, half of the time period T corresponds to the time interval $[t_0, t_1]$, where transient processes are damped. The integration step size is $\dfrac{\pi}{100\Omega}$. Initial conditions of the closed trajectories are distinguished by 0.5 percent with ratio to characteristic vibration amplitudes, for example, the starting points of these trajectories are in the 3-dimensional parallelepiped ($\left|x(t_0) - \tilde{x}(t_0)\right| < 0.005 A_x$, $\left|\dot{x}(t_0) - \dot{\tilde{x}}(t_0)\right| < 0.005 A_{\dot{x}}$, $\left|z(t_0) - \tilde{z}(t_0)\right| < 0.005 A_z$). The parameter α is chosen to be equal to 1/3. All domains are multiply connected. There are also a number of scattered points here. Such structure is characteristic for domains, where chaotic vibrations are possible.

One can observe the effect of *restraining* of the chaotic regions with the increasing of the hysteretic dissipation value in the (Ω, F) plane (Figures 3.1(a), 3.1(b), 3.1(c)). The "quickness" of the restraining decreases when v increases. So, in the case of maximum hysteretic dissipation value $v = 1$, the chaotic regions in the (Ω, F) plane are distinguished from the regions (c), Fig. 3.1, non-principally.

In the (μ, F) plane (Fig. 3.2(a), 3.2(b), 3.2(c)), the form and location of the chaotic domains are changed depending on the hysteretic dissipation of the nonlinear terms in set (3.1).

Figure 3.3 characterizes the obtained domains and demonstrates a different character of motion of the Masing oscillator as chaos and hysteresis loss (a), and periodic response (b).

Other situation occurs for the Bouc-Wen oscillator (3.2) which naturally is linear (when $\delta = 1$). So, addition of the hysteretic dissipation leads to chaotic responses that occur in this system. Figures 3.4 and 3.5 present the evolution of chaotic behavior regions with an increasing hysteretic dissipation value in the (Ω, F) and (μ, F) planes, respectively. One can observe a change in the form and location of the chaotic regions. Note that chaotic responses of the Bouc-Wen oscillator are not observed right up till $\delta = 0.2$ when the influence of the nonlinear terms becomes critical.

It demonstrates a *generating effect* of the hysteretic dissipation on chaos occurring in the hysteretic system, which appears after some critical value δ_{cr}. After δ_{cr}, both the form and location of the chaotic behavior regions are changed with the increasing hysteretic dissipation. In the case of a maximum hysteretic dissipation value (when $\delta = 0$), the chaotic behavior regions are practically the same as in case (c) in Fig. 3.4. A "friable" form of the chaotic regions (Fig. 3.5(a) and 3.5(c)) in the (μ, F) planes is conditioned by the fixation of the frequency $\Omega = 0.24$ and changing location of the corresponding domains in the (Ω, F) planes with an increase of the hysteretic dissipation. The straight line $\Omega = 0.25$ only slightly contacts to the chaotic behavior regions in cases (a) and (c), in Fig. 3.4.

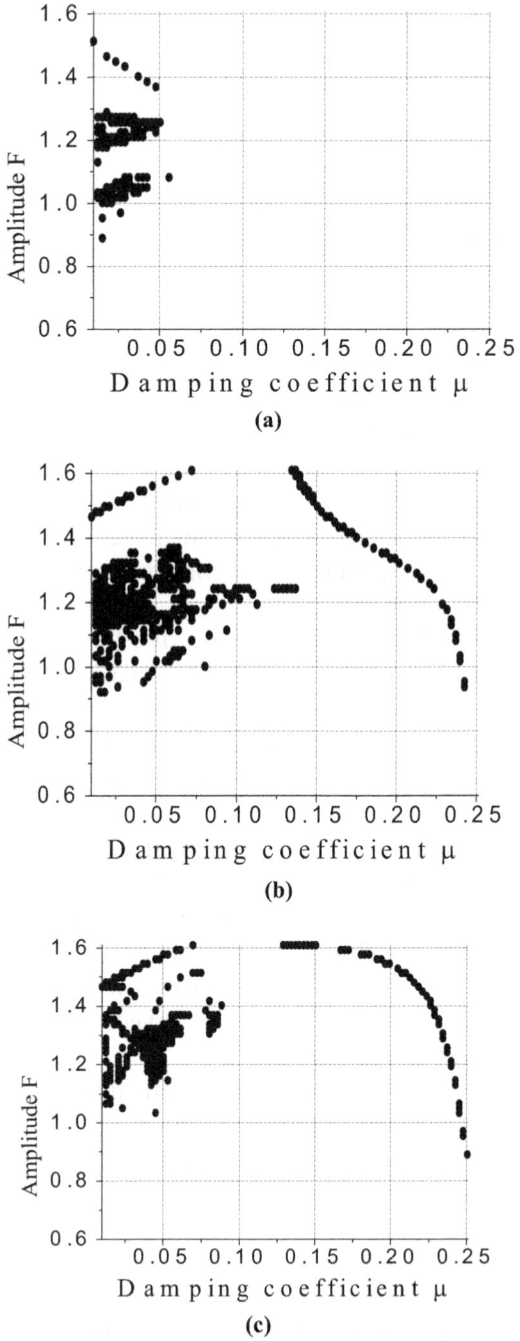

Figure 5 Evolution of the chaotic regions for the Masing hysteresis model in the (μ, F) plane with increasing hysteretic dissipation value (a) $v=0$; (b) $v=0.5$; (c) $v=0.8$. The parameters $\Omega=0.15$, $\delta=0.05$, $n=10.0$, $x(0)=0.1$, $\dot{x}(0)=0.1$, $z(0)=0$ are fixed for all cases.

*Figure 3.2 in original source

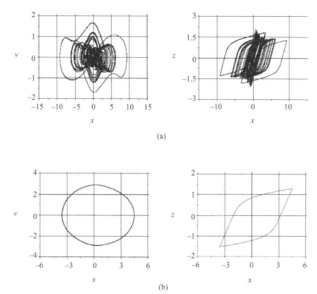

(a)

(b)

*Figure 6 Phase portraits and hysteresis loops of the Masing hysteretic oscillator in the cases of (a) chaotic (Ω=0.15, F=1.21, μ=0.026, v=0.5, δ=0.05, n=10.0, $x(0)$=0.1, $\dot{x}(0)$=0.1, $z(0)$=0) and (b) periodic (Ω=0.7,F=0.8, μ=0, v=0.5, δ=0.05, n=10.0, $x(0)$=0.1, $\dot{x}(0)$=0.1, $z(0)$=0) responses.

During simulation the (Ω, F) and (μ, F) planes had been uniformly sampled by 100×100 nodal points in the rectangles (0.01$\leq\Omega\leq$0.36; 0.01$\leq$F$\leq$2.05) and (0.001$\leq\mu\leq$0.04; 0.01$\leq$F$\leq$1.71), respectively. The time period for the simulation T is $\dfrac{300\pi}{\Omega}$. During computations, half of the time period corresponds to the time interval $[t_0, t_1]$, where transitional processes are damped. The integration step size equal to $\dfrac{\pi}{40\Omega}$ is chosen. As in the case of the Masing oscillator, the initial conditions of the closed trajectories are distinguished by 0.5 percent in comparison to characteristic vibration amplitudes. The parameter α is set equal to 1/3.

Figure 3.6 characterizes the obtained regions of regular/irregular motion and depicts various responses of the Bouc-Wen oscillator as chaos and hysteresis loss (a), and periodic response (b). Figures 3.7(a), 3.7(b) and 3.8(a), 3.8(b) with periodic pinched hysteresis also agree well with the obtained regions of regular/irregular behavior of hysteretic oscillators (see Figs. 3.1(b), 3.2(b), 3.4(a), and 3.5(a)).

Control of Pinched Hysteresis Phenomenon by Means of an Amplitude and Frequency of an External Periodic Excitation

Pinched hysteresis possesses reduce dissipation properties and are a consequence of residual phenomena diminution or "delay" reduction of the *output* relative to the *input* in a hysteretic system.

---

*Figure 3.3 in original source

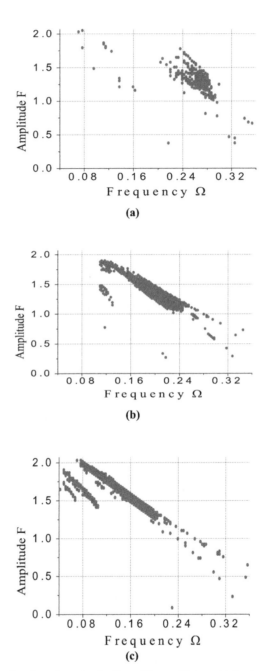

*Figure 7 Evolution of the chaotic regions for the Bouc-Wen oscillator with hysteresis in the (Ω, F) plane with increasing hysteretic dissipation value (a) $\delta=0.0476$; (b) $\delta=0.01$; (c) $\delta=0.001$ at $k_z=0.5$, $\gamma=0.3$, $\beta=0.005$, $n=1.0$, $x(0)=0.1$, $\dot{x}(0)=0.1$, $z(0)=0$ and $\mu=0$.

---

*Figure 3.4 in original source

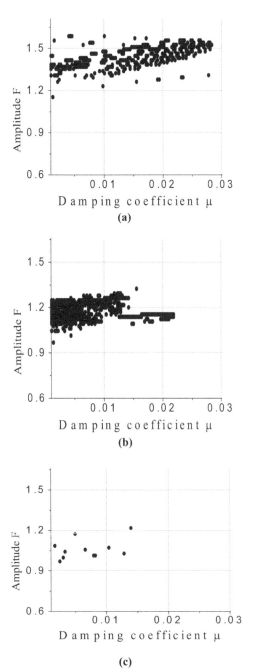

*Figure 8 Evolution of the chaotic regions for the Bouc-Wen oscillator with hysteresis in the (μ, F) plane with increasing hysteretic dissipation value (a) $\delta=0.0476$; (b) $\delta=0.01$; (c) $\delta=0.001$ at $k_z=0.5$, $\gamma=0.3$, $\beta=0.005$, $n=1.0$, $x(0)=0.1$, $\dot{x}(0)=0.1$, $z(0)=0$ and $\Omega=0.24$.

---

*Figure 3.5 in original source

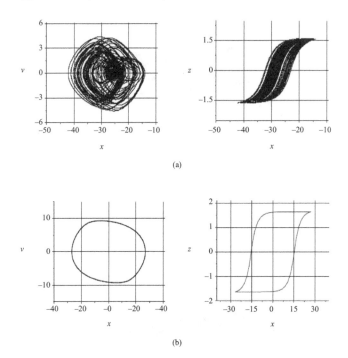

(a)

(b)

*Figure 9 Phase portraits and hysteresis loops of the Bouc-Wen hysteretic oscillator in the cases of (a) chaotic (Ω=0.24; F=1.1227, μ=0.00136, δ=0.01, k_z=0.5, γ=0.3, β=0.005, n=1.0, $x(0)$=0.1, $\dot{x}(0)$=0.1, $z(0)$=0) and (b) periodic (Ω=0.35; F=1.2, μ=0.0, δ=0.0476, k_z=0.5, γ=0.3, β=0.005, n=1.0, $x(0)$=0.1, $\dot{x}(0)$=0.1, $z(0)$=0) responses.

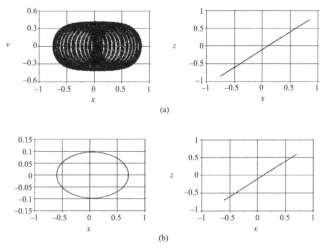

(a)

(b)

**Figure 10 Phase portraits and pinched hysteresis loops of the Masing hysteretic oscillator (a) (Ω=0.1114; F=0.43, μ=0, ν=0.5, δ=0.05, n=10.0, $x(0)$=0.1, $\dot{x}(0)$=0.1, $z(0)$=0) and (b) (Ω=0.15; F=0.634, μ=0.0856, ν=0.5, δ=0.05, n=10.0, $x(0)$=0.1, $\dot{x}(0)$=0.1, $z(0)$=0).

---

*Figure 3.6 in original source
**Figure 3.7 in original source

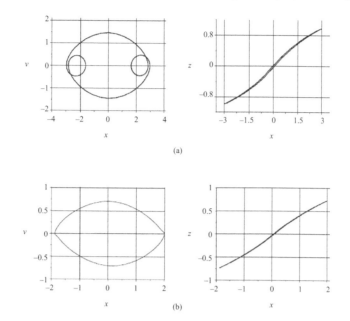

*Figure 11 Phase portraits and pinched hysteresis loops of the Bouc-Wen hysteretic oscillator (a) (Ω=0.2; F=1,μ=0, δ=0.01, k_z=0.5, γ=0.3, β=0.005, n=1.0, $x(0)$=0.1, $\dot{x}(0)$=0.1, $z(0)$=0) and (b) (Ω=0.24; F=0.724, μ=0.017257, δ=0.0476, k_z=0.5, γ=0.3, β=0.005, n=1.0, $x(0)$=0.1, $\dot{x}(0)$=0.1, $z(0)$=0).

In this connection, a control of pinching phenomena is of great importance for various practical applications. It was shown that for the nonzero fixed value of hysteretic dissipation (which, in changing from minimum to maximum, can produce behavior of the system with/without delay) and for the fixed parameters which govern the shape of a hysteresis loop, it is possible to choose an "appropriate" hysteretic process with desirable residual phenomena with the aid of an amplitude and frequency of an external periodic excitation:

i) Masing oscillator

$$\dot{x} = y,$$
$$\dot{y} = -2\mu y - (1-\nu)g(x) - \nu z + (F + i\Delta F)\cos(\Omega + j\Delta\Omega)t, \tag{4.1}$$
$$\dot{z} = g'\left(\frac{z-z_i}{2}\right)y, \ g(x) = \frac{(1-\delta)x}{\left(1+|x|^n\right)^{\frac{1}{i}}} + \delta x$$

ii) Bouc-Wen oscillator
$$\dot{x} = y,$$
$$\dot{y} = -2\mu y - \delta x - (1-\delta)z + (F + i\Delta F)\cos(\Omega + j\Delta\Omega)t, \tag{4.2}$$
$$\dot{z} = \left[k_z - (\gamma + \beta \operatorname{sgn}(y)\operatorname{sgn}(z))|z|^n\right]y,$$
$$i=1,2,\dots,n; \ j=1,2,\dots,m.$$

---

*Figure 3.8 in original source

The relation $\varepsilon_r/\varepsilon$ was chosen as a value that characterizes a pinch in the hysteresis loop. As shown in Fig. 4.1, ε is the distance between the projection of the velocity reversal points $\pm(x_p z_i)$ to the *input*-axis; ε_r is the distance characterizing the residual phenomena.

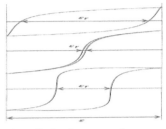

*Figure 12 Hysteresis loops with various dissipation properties.

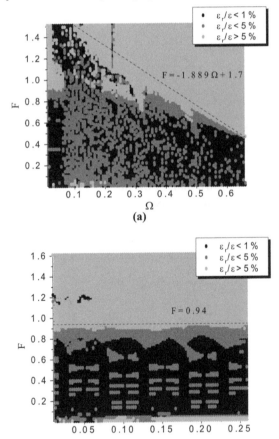

**Figure 13 Regions of hysteresis with various dissipation properties for the Masing model. The parameters $\delta=0.05$, $n=10.0$, $v=0.5$, $x(0)=0.1$, $\dot{x}(0)=0.1$, $z(0)=0$ and (a) $\mu=0$, (b), $\Omega=0.15$ are fixed.

---

*Figure 4.1 in original source
**Figure 4.2 in original source

It is shown that the thresholds of the regions with a "pinched hysteresis" in the planes frequency versus the amplitude of external periodic excitation in both cases of Masing and Bouc-Wen hysteretic oscillators (Figs. 4.2(a) and 4.3(a)) have an almost linear character. The behavior of Masing hysteretic oscillator with enhanced dissipation properties is possible above the straight line $F = -1.889\ \Omega + 1.7$. The analogous behavior of the Bouc-Wen oscillator is observed above the straight line $F = -4.8\ \Omega + 2.4$. Figures 4.2(b) and 4.3(b) indicate that a linear viscous damper μ almost does not influence the pinching and a hysteresis with the improved dissipation properties is observed above the lines $F = 0.94$ ($F = 1.4$) for the Masing (Bouc-Wen) hysteretic oscillator.

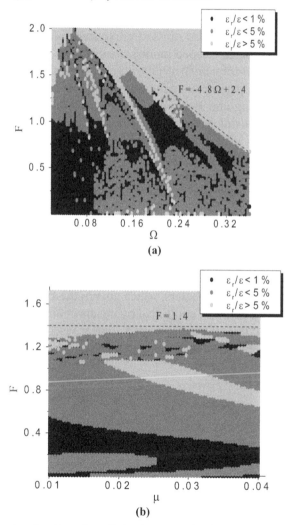

(a)

(b)

*Figure 14 Regions of hysteresis with various dissipation properties for the Bouc-Wen model. The parameters $\delta = 0.01$, $k_z = 0.5$, $\gamma = 0.3$, $\beta = 0.005$, $n = 1.0$, $x(0) = 0.1$, $\dot{x}(0) = 0.1$, $z(0) = 0$ and (a) $\mu = 0$, (b) $\Omega = 0.24$ are fixed.

---

*Figure 4.3 in original source

Figures 3.7 and 3.8 characterize the obtained regions and demonstrate pinching phenomena in the hysteresis loops. Figures 3.3(b) and 3.6(b) illustrating the enhanced dissipation properties of hysteresis loops are presented in the section devoted to the analysis of chaos occurring in the hysteretic systems. It is clear that there is no sense to discuss dissipation properties in the case of hysteresis loss or chaotic behavior of hysteretic oscillators (Figs. 3.3(a) and 3.6(a)).

Conclusions

In the present work, hysteresis is simulated by means of additional state variables (internal variables). In particular, the behavior of magnetorheological/electrorheological (MR/ER) fluids in a damper/absorber as well as stress-strain hysteresis with transient processes in a steel rope are simulated. The developed models are effective, enable production of minor loops, present fast numerical convergence, provide good agreement with experimental data, and contain principally less parameters than, for example, Bouc-Wen or Spencer models.

Highly nonlinear Masing and Bouc-Wen hysteretic models with discontinuous right hand sides are investigated using an effective approach based on the analysis of wandering trajectories. This algorithm of quantifying regular and chaotic dynamics is simpler and faster from the computational point of view comparing with standard procedures and allows us to trace accurately enough the regular/irregular responses of the hysteretic systems. The evolution of chaotic behavior regions of the oscillators with hysteresis is presented in various control parameter spaces: in the damping coefficient-amplitude and the frequency-amplitude of the external periodic excitation planes. A substantial influence of the hysteretic dissipation value on the possibility of chaotic behavior occurring in the systems with hysteresis is shown. The *restraining* and *generating* effects of the hysteretic dissipation on a chaotic behavior are demonstrated.

It was shown that for fixed parameters, which govern the shape of a hysteresis loop, the extent of pinch of this loop could be controlled by means of an amplitude and frequency of the external periodic excitation. It was found that a linear viscous damper did not influence the pinching. The regions of hysteresis with various dissipation properties for the Masing and Bouc-Wen models are presented.

Acknowledgment

This work has been supported by the J.Mianowski Foundation of Polish Science Support, and the Polish Ministry of Education and Science for years 2005–2008 (Grant no. 4T07A 031 28).

References

1. Awrejcewicz, J., Dzyubak, L., *Chaos exhibited by hysteretic systems*, Proc. 7th Conference on Dynamical Systems-Theory and Applications (Łódź, 2003), vol. 1 (2003) 123–130.
2. Awrejcewicz, J., Dzyubak, L., *Regular and chaotic behavior exhibited by coupled oscillators with friction*, University of Niš, Facta Universitatis. Series: Mechanics, Automatic Control and Robotics **3** (2003) 921–930.

3. Awrejcewicz, J., Dzyubak, L., *Stick-slip chaotic oscillations in a quasi-autonomous mechanical system*, International Journal of Nonlinear Sciences and Numerical Simulation **4** (2003) 155–160.

4. Awrejcewicz, J., Dzyubak, L., *Quantifying smooth and nonsmooth regular and chaotic dynamics*, International Journal of Bifurcation and Chaos in Applied Sciences and Engineering **15** (2005) 2041–2055.

5. Awrejcewicz, J., Dzyubak, L., Grebogi, C., *A direct numerical method for quantifying regular and chaotic orbits*, Chaos Solitons Fractals **19** (2004) 503–507.

6. Awrejcewicz, J., Mosdorf, R., *Numerical Analysis of Some Problems of Chaotic Dynamics*,WNT, Warsaw (2003).

7. Bernardini, D., *Models of hysteresis in the framework of thermomechanics with internal variables*, Physica B: Condensed Matter **306** (2001) 132–136.

8. Capecchi, D., Masiani, R., *Reduced phase space analysis for hysteretic oscillators of Masing type*, Chaos, Solitons and Fractals **7** (1996) 1583–1600.

9. Casoria, S., Sybille, G., Brunelle, P., *Hysteresis modeling in the MATLAB/power system blockset*, Mathematics and Computers in Simulation **63** (2003) 237–248.

10. Kádár, G., Szabó, G., *Hysteresis modeling*, Journal of Magnetism and Magnetic Materials **215-216** (2000) 592–596.

11. Ketema, Y., *Averaging and bifurcations in an oscillator with a history dependent restoring force*, Journal of Sound and Vibration **209** (1998) 187–198.

12. Kleineberg, T., *A generalized approach for modelling the nonlocal memory of hysteretic systems*, Journal of Magnetism and Magnetic Materials **166** (1997) 315–320.

13. Kolsch, H., Ottl, D., *Simulation des mechanischen Verhaltens von Bauteilen mit statischer Hysterese*, Forschung im Ingenieurwesen **4** (1993) 66–71 (German).

14. Koltermann, P.I., Righi, L.A., Bastos, J.P.A., Carlson, R., Sadowski, N., Batistela, N. J., *A modified Jiles method for hysteresis computation including minor loops*, Physica B: Condensed Matter **275** (2000) 233–237.

15. Ktena, A., Fotiadis, D.I., Spanos, P.D., Massalas, C.V., *A Preisach model identification procedure and simulation of hysteresis in ferromagnets and shape-memory alloys*, Physica B: Condensed Matter **306** (2001) 84–90.

16. Lacarbonara, W., Vestroni, F., *Nonclassical responses of oscillators with hysteresis*, Nonlinear Dynamics **32** (2003) 235–258.

17. Li, C., Yu, J., Liao, X., *Chaos in a three-neuron hysteresis hopfield-type neural network*, Physics Letters. A **285** (2001) 368–372.

18. Li, H.G., Zhang, J.W., Wen, B.C., *Chaotic behaviors of a bilinear hysteretic oscillator*, Mechanics Research Communications **29** (2002) 283–289.

19. Makaveev, D., Dupré, L., De Wulf, M., Melkebeek, J., *Dynamic hysteresis modelling using feedforward neural networks*, Journal of Magnetism and Magnetic Materials **254-255** (2003) 256–258.

20. Masiani, R., Capecchi, D., Vestroni, F., *Resonant and coupled response of hysteretic two-degree-of-freedom systems using harmonic balance method*, International Journal of Non-Linear Mechanics **37** (2002) 1421–1434.

21. Mayergoyz, I.D., *Mathematical Models of Hysteresis*, Springer, New York (1991).

22. Ortín, J., Delaey, L., *Hysteresis in shape-memory alloys*, International Journal of Non-Linear Mechanics **37** (2002) 1275–1281.

23. Ossart, F., Hubert, O., Billardon, R., *A new internal variables scalar model respecting the wipingout property*, Journal of Magnetism and Magnetic Materials **254-255** (2003) 170–172.

24. Sapiński, B., *Dynamic characteristics of an experimental MR fluid damper*, Engineering Transactions **51** (2003) 399–418.

25. Sapiński, B., Filus, J., *Analysis of parametric models of MR linear damper*, Journal of Theoretical and Applied Mechanics **41** (2003) 215–240.

26.   Smyth, A.W., Masri, S.F., Kosmatopoulos, E.B., Chassiakos, A.G., Caughey, T.K., *Development of adaptive modeling techniques for non-linear hysteretic systems*, International Journal of Non-Linear Mechanics **37** (2002) 1435–1451.

27.   Vestroni F., Noori, M., *Hysteresis in mechanical systems—modelling and dynamic response*, International Journal of Non-Linear Mechanics **37** (2002) 1261–1262.

28.   Visintin, A., *Differential Models of Hysteresis*, Applied Mathematical Sciences, vol. 111, Springer, Berlin (1994).

Nonlinear Dynamics and Chaos of Microcantilever-Based TM-AFMs with Squeeze Film Damping Effects[†]

Wen-Ming Zhang,[1,a] Guang Meng,[1,b] Jian-Bin Zhou[1,c] and Jie-Yu Chen[1,d]

[1]State Key Laboratory of Mechanical System and Vibration, School of Mechanical Engineering, Shanghai Jiao Tong University, 800 Dongchuan Road, Shanghai 200240, China;

[a]Email: wenmingz@sjtu.edu.cn.
[b]Email: gmeng@sjtu.edu.cn.
[c]Email: giantbean@sjtu.edu.cn.
[d]Email: jerrysmiling@hotmail.com.

ABSTRACT

In Atomic force microscope (AFM) examination of a vibrating microcantilever, the nonlinear tip-sample interaction would greatly influence the dynamics of the cantilever. In this paper, the nonlinear dynamics and chaos of a tip-sample dynamic system being run in the tapping mode (TM) were investigated by considering the effects of hydrodynamic loading and squeeze film damping. The microcantilever was modeled as a spring-massdamping system and the interaction between the tip and the sample was described by the Lennard-Jones (LJ) potential. The fundamental frequency and quality factor were calculated from the transient oscillations of the microcantilever vibrating in air. Numerical simulations were carried out to study the coupled nonlinear dynamic system using the bifurcation diagram, Poincaré maps, largest Lyapunov exponent, phase portraits and time histories. Results indicated the occurrence of periodic and chaotic motions and provided a comprehensive understanding of the hydrodynamic loading of microcantilevers. It was demonstrated that the coupled dynamic system will experience complex nonlinear oscillation as the system parameters change and the effect of squeeze film damping is not negligible on the micro-scale.

Keywords: TM-AFM; microcantilever; squeeze film damping; Lennard-Jones (LJ) potential.

[†]Reused with permission from: Wen-Ming Zhang, Guang Meng, Jian-Bin Zhou and Jie-Yu Chen, Nonlinear Dynamics and Chaos of Microcantilever-Based TM-AFMs with Squeeze Film Damping Effects, *Sensors* 2009, 9, 3854–3874; doi:10.3390/s90503854.

Introduction

Atomic force microscopy (AFM) has been developed to a nearly ubiquitous tool for studying physics, chemistry, biology, medicine and engineering at the nano-scale [1–5]. AFM could significantly impact many fabrication and manufacturing processes due to its advantages such as 3D topography of nano-fabrication and metrology for MEMS [2]. As a typical dynamic mode, the tapping mode (TM) is widely used in the operation of AFM where the cantilever is driven at a fixed frequency close or equal to the fundamental resonance frequency of vertical bending [4,6]; a schematic of the TM-AFM setup is shown in Fig. 1. Moreover, the vibration amplitude of the cantilever is much bigger than the equilibrium separation between the tip and the sample.

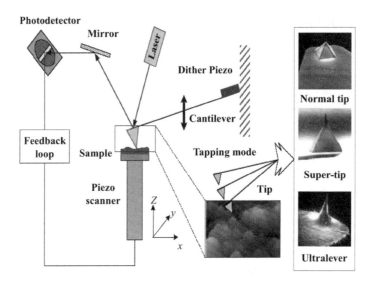

Figure 1 Schematic of the typical TM-AFM setup.

The TM-AFM has attracted extensive attention due to its ability to deal with compliant materials as well as to overcome adhesion forces.

The inherently and highly nonlinear tip-sample interaction will give rise to complex dynamics of the cantilever in TM-AFM [7]. Nonlinearity is essential in understanding the dynamics of cantilevers as there are many nonlinear forces in TM-AFM, such as the attractive van der Waals forces, the shortrange repulsive interactions, contact nonlinearities and capillary forces et al. [8–10]. Ashhab et al. [11] concluded that chaos in AFM depend on the damping, excitation and tip-sample distance, and suggested that a state feedback control can be used to eliminate the possibility of chaotic behavior. Using nonlinear analysis methods and numerical simulations, Basso et al. [12] found that the chaotic behavior may occur via a cascade of period doubling bifurcations. In their studies [11,12], Melnikov theory was used to predict the existence of chaos in AFM. The nonlinear dynamics to frequency sweeps in TM-AFM was simulated using the van der Waals forces and the Derjaguin-Muller-

Toporov (DMT) and Johnson-Kendall-Roberts (JKR) contact models [6,13]. Lee et al. [6] carried out numerical analysis using modern continuation tools for computational nonlinear dynamics and bifurcation problems where the tip-surface interaction was represented by the van der Waals and DMT contact forces. Nonlinear hysteresis and jumps in the dynamic response were examined as the tip approaches to or retracts from the sample at a fixed excitation frequency [14]. Zitzler et al. [15] considered the influence of hysteretic capillary forces in TM-AFM and studied the effect of the relative humidity on the amplitude and phase of the cantilever oscillation. By using the forward-time simulation and numerical continuation techniques [15], Hashemi et al. [10] investigated the nonlinear dynamics of a TM-AFM with tip-surface interactions, which include attractive, repulsive, and capillary forces. Hersam [3] conducted experiments to verify the importance of nonlinear dynamics in TM-AFM measurements. Rutzel et al. [9] used the Lennard-Jones (LJ) potential to model the tip-surface interactions and carried out a comprehensive investigation to the nonlinear dynamics and stability of the TM-AFM, and the results showed that considering the LJ interaction potential in modeling the dynamics of AFM could improve the qualitative prediction of the real system response.

Nonlinear dynamics and chaos of the cantilever for the TM-AFM remain a challenging and critical issue. After experimental observations, Couturier et al. [16] mentioned that the motion of the cantilever could become chaotic under instability conditions. When the cantilever is driven close to the surface of the sample, the squeeze film between the cantilever and the sample surface contributes significantly to the damping and gives rise to the complicated nonlinear behavior [17,18]. When studying the frequency response of AFM cantilevers in liquid media contained in a commercial fluid cell, Motamedi and Wood-Adams [19] found that such systems could exhibit complicated dynamics. However, a rational connection of the tip-sample-interaction and the nonlinear dynamics analysis of the cantilever where the coupled effects of squeeze film damping and hydrodynamic loading are considered has not been presented and addressed satisfactorily. To order to make the TM-AFM achieve good performance, it is necessary to identify and so as to eliminate the possible chaotic motion of the cantilever of the AFM. In this paper, the cantilever is modeled as a single spring-mass-damper system and a nonlinear dynamic model is developed to study the cantilever-sample interaction by using the LJ potential including the long-range attractive forces and short-range repulsive forces. A comprehensive investigation of the nonlinear dynamics and chaos of the TM-AFM is carried out.

The rest of the paper is organized as follows. Section 2 describes the mathematic model of the cantilever vibrating in air considering the effects of hydrodynamic loading and squeeze film damping and the physical model of the cantilever-sample interaction is established in Section 3 using the LJ potential. Numerical results and discussions of the quality factor and resonant frequency of the frequency response, and nonlinear chaos and bifurcation of the dynamic TM-AFM are presented in Section 4. Finally, we end the paper with our conclusions in Section 5.

Micro-Cantilever Vibrating in Air

Hydrodynamic Loading Effect

The gas flow around the cantilever is assumed to be incompressible and the Navier-Stokes equation is given by [20]:

$$\rho \frac{\partial v}{\partial t} + \rho (v.\nabla)v = -\nabla p + \mu \nabla^2 v \tag{1}$$

where ρ, v, p and μ are the density, velocity, pressure and viscosity of the gas, respectively.

It has been found that the wavelength of vibration greatly exceeds the dominant length scale in the flow [20,21]. The nominal width B is the dominant length scale in the gas flow and the appropriate Reynolds number Re can be expressed as [21]:

$$Re = \frac{\rho \omega B^2}{4\mu} \tag{2}$$

where ω is a characteristic radial vibration frequency and Re is a normalized Reynolds number which indicates the importance of viscous forces relative to inertial forces in the gas fluid. It can be found that the gas could be considered to be non-viscous in the limit as $Re \to \infty$, and the non-viscous gas model is applicable for practical cases where $Re \gg 1$ [21]. However, if ω is close to the resonant frequency of the cantilever in vacuum, a reduction in dimensions of the beam will result in a reduction in Re.

The surrounding viscous fluid medium plays an important role on the dynamics of the cantilever in AFM, Sader et al. [21,22] studied the effect of viscous fluid medium on the AFM cantilever and found that the shift in resonant frequency of the cantilever from vacuum to fluid (gas or liquid) was strongly dependent on both the density and viscosity of the fluid. For a rectangular cantilever beam [21], when the quality factor of the fundamental mode of the cantilever in gas exceeds 1, which is typically satisfied when the cantilever is placed in air, the relationship between the vacuum resonant frequency ωvac and the resonant frequency in gas ωgas can be written as [20]:

$$\frac{\omega_{gas}}{\omega_{vac}} = \left(1 + \frac{\pi \rho B}{4\rho_c H} \Gamma_r (\omega_{gas})\right)^{-1/2} = \left(1 + \frac{\pi}{4} \Pi \Gamma_r (\omega_{gas})\right)^{-1/2} \tag{3}$$

where ρ_c and H are the density and thickness of the cantilever, respectively, and the natural scaling parameter is $\Pi = \dfrac{\rho B}{\rho_c H}$, which is defined as the ratio of the added mass of the gas to the mass of the cantilever.

To study the effect of hydrodynamic loading on the dynamics of the cantilever, the hydrodynamic functions $\Gamma(\omega)$, which represents the real and imaginary pressure of the surrounding on the cantilever in two dimensions, is given by [20,21]:

$$\Gamma = \Gamma_r + j\Gamma_i \tag{4}$$

where $\Gamma_r = a_1 + \dfrac{a_2}{\sqrt{R_e}}$ and $\Gamma_i = \dfrac{b_1}{\sqrt{R_e}} + \dfrac{b_2}{R_e}$ and $a_1 = 1.0553$, $a_2 = 3.7997$, $b_1 = 3.8018$, and $b_2 = 2.7364$ are selected from [20].

The quality factor for the bending modes can be written as [20]:

$$Qgas = \frac{\dfrac{4\mu}{\pi\rho B^2} + \Gamma_r\left(\omega_{gas}\right)}{\Gamma_i\left(\omega_{gas}\right)} = \frac{\dfrac{4}{\pi}\cdot\dfrac{1}{\Pi} + \Gamma_r\left(\omega_{gas}\right)}{\Gamma_i\left(\omega_{gas}\right)} \tag{5}$$

For small amplitude of the normal vibration of the cantilever, the local interaction stiffness in the vertical direction k_n and damping coefficient C_n is given by [23]:

$$k_n = k_L\left(\frac{A_{n0}}{A_n}\cos\varphi - 1\right) \tag{6}$$

and

$$C_n = -k_L\frac{A_{n0}}{A_n\omega}\sin\varphi \tag{7}$$

where A_{n0} is the free amplitude in the normal direction, A_n is the measured amplitude, φ is the measured phase of the cantilever, ω is the drive frequency, and k_L is the normal bending stiffness and $kL = \dfrac{EBH^3}{4L^3}$, in which E, B, H and L are the elastic modulus, width, thickness and length of the cantilever, respectively. When the cantilever in AFM operates far below the resonance, the phase angle is close to zero, then k_n and C_n become:

$$k_n = k_L\left(K - 1\right), C_n = 0 \tag{8}$$

where $K = \dfrac{A_{n0}}{A_n}$.

Squeeze Film Damping Effect

When the cantilever is driven close to the sample, the squeeze film between the cantilever and the surface of the sample causes significant damping except for the fluid dissipation at the edges and above the cantilever. For the damping problems encountered at micro-scale [24], the squeeze film damping can be expressed by the nonlinear Reynolds equation:

$$\frac{\partial}{\partial\xi}\left(\rho\frac{h^3}{\mu}\frac{\partial p}{\partial\xi}\right) + \frac{\partial}{\partial\zeta}\left(\rho\frac{h^3}{\mu}\frac{\partial p}{\partial\zeta}\right) = 12\frac{\partial(hp)}{\partial t} \tag{9}$$

Under the basic assumptions including the gas in the gap has been regarded as a continuum and the gas undergoes an isothermal process, the squeeze film damping can be modeled as:

$$\frac{p_a h^3}{12\mu_{eff}}\left[\frac{\partial^2}{\partial\xi^2}\left(\frac{p}{p_a}\right)+\frac{\partial^2}{\partial\zeta^2}\left(\frac{p}{p_a}\right)\right]=\frac{\partial}{\partial t}\left(\frac{\xi}{h}\right)+\frac{\partial}{\partial t}\left(\frac{p}{p_a}\right) \tag{10}$$

where the pressure p is small compared to the ambient pressure p_a ($p_a = 1.013 \times 10^5 p$), μ_{eff} is the effective gas viscosity, and h is the gap between the cantilever and the surface of the sample. At microscale, when the ratio between the mean free path of the air λ and the film thickness h, i.e., Knudsen number $K_n = \frac{\lambda}{h}$, is not small, the no-slip boundary condition at the interface may be inadequate.

Considering the effect of slip flow, the effective viscosity μ_{eff} can be expressed as [25]:

$$\mu_{eff} = \frac{\mu}{1+9.638 K_n^{1.159}} \tag{11}$$

where μ is the viscosity coefficient at 1 atm.

For the normal motion of parallel plates, h and μ_{eff} are not functions of the position. Equation (10) can be simplified as:

$$\frac{h^3}{12\mu_{eff}}\left[\frac{\partial}{\partial\xi}\left(p\frac{\partial p}{\partial\xi}\right)+\frac{\partial}{\partial\zeta}\left(p\frac{\partial p}{\partial\zeta}\right)\right]=\frac{\partial(ph)}{\partial t} \tag{12}$$

Then, the damping pressure can be obtained by direct integration with the boundary conditions, i.e.:

$$p(\xi,t)=-\frac{6\mu_{eff}}{h^3}\left[\left(\frac{B}{2}\right)^2-\xi^2\right]\cdot\frac{dh}{dt} \tag{13}$$

The damping force on the cantilever is:

$$F_s = \int_{-B/2}^{B/2} p(\xi,t)Ld\xi = \frac{\mu_{eff}B^3 L}{h^3}\cdot\frac{dh}{dt} \tag{14}$$

and the coefficient of damping force is given by:

$$Cs = \frac{\mu_{eff}B^3 L}{h^3} \tag{15}$$

The Physical Model

The AFM is composed of an elastic cantilever and the achievable sensitivity and resolution of AFM depend largely on the geometry of the cantilever [11,26]. Considering only the first vibration mode, the cantilever can be modeled as a simplified spring-mass-damping system, as shown in Fig. 2. The tip is modeled as a sphere of radius R, and the cantilever-sample distance is characterized by z_0, which is the distance between the equilibrium position of the cantilever and the sample when only the gravity acts on it. The cantilever position is given by x measured from the equilibrium position.

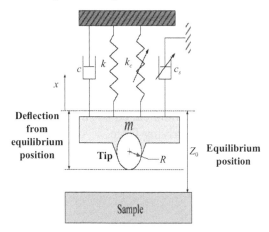

Figure 2 Schematic of the lumped spring-mass-damping model for the TM-AFM cantilever vibrating near a sample surface.

The LJ potential models the dispersive van der Waals forces as well as the short-range repulsive exchange interactions between two molecules [9]. The interaction between a cantilever tip and sample surface can be modeled as the interaction between a sphere and a flat surface. The tip-sample interaction is modeled by the LJ potential given by [9,11]:

$$U_{LJ}\left(x, z_0\right) = \frac{A_1 R}{1260\left(z_0 + x\right)^7} - \frac{A_2 R}{6\left(z_0 + x\right)} \tag{16}$$

where A_1 and A_2 are the Hamaker constants for the attractive and repulsive potentials, respectively. The Hamaker constants are defined as $A_1 = \pi^2 \rho_1 \rho_2 c_1$ and $A_2 = \pi^2 \rho_1 \rho_2 c_2$, in which ρ_1 and ρ_2 are the densities of the two interaction components, and c_1 and c_2 are the interaction constants, respectively.

When the cantilever is driven close to the sample, there are three different kinds of forces including the spring force, the van deer Waals attractive force which is proportional to the inverse square power of the distance between the cantilever tip and the sample, and the repulsive force which is proportional to the inverse eighth power of the distance between the tip and the sample. The LJ force can be defined as the sum of the attractive and repulsive forces and is expressed as [9]:

$$F_{LJ} = -\frac{\partial U_{LJ}\left(x, z_0\right)}{\partial\left(z_0 + x\right)} = \frac{A_1 R}{180\left(z_0 + x\right)^8} - \frac{A_2 R}{6\left(z_0 + x\right)^2} \tag{17}$$

During the AFM operates in the TM, a low-dimensional model reduction can provide an accurate description of the cantilever dynamics. As shown in Fig. 2, the cantilever is driven by the harmonic driving force, the tipample interaction force F_{LJ} (LJ force) and the force due to squeeze film damping F_s. The governing equation of the motion of the cantilever can be determined by:

$$mx'' + cx' + kx + k_c x^3 = F_{LJ}(x, z_0) + F_s(x, x', z_0) + f_0 \cos \omega t \tag{18}$$

where x is the instantaneous displacement of the cantilever tip measured from the equilibrium tip position in the absence of external forces with positive values toward the sample surface, and x' and x'' are the instantaneous velocity and acceleration of the cantilever tip,. m, k and c are the equivalent mass, spring stiffness and damping coefficients of the cantilever in air, respectively, and $k = k_L + k_n$ and $c = m\omega_1/Q + C_n$, in which $\omega_1 = \sqrt{k_L/m}$ is the first-order mode frequency, and k_c is the nonlinear cubic stiffness. f_0 and ω are the amplitude and angular frequency of the harmonic driving force.

The damping force $F_s(x, x', z_0)$ due to the squeeze film on the cantilever can be obtained from equation (14) and is given by:

$$F_s(x, x', z_0) = C_s x' = \frac{\mu_{eff} B^3 L}{(z_0 + x)^3} \tag{19}$$

In order to facilitate the investigation of the qualitative behaviors of the dynamic system, the equilibrium distance variable Z_s is defined and $Z_s = (3/2)(2D)^{1/3}$ [11], in which $D = A_2 R/(6k)$.

Introducing dimensionless variables:

$$\tau = \omega_1 t, y = \frac{x}{Z_s}, y' = \frac{x'}{\omega_1 Z_s}, d = \frac{4}{27}, \alpha = \frac{z_0}{Z_s}, \beta = \frac{k_c}{k} Z_s^2, \gamma = \frac{c_s}{m\omega_1},$$

$$\chi = \frac{\mu_{eff} B^3 L}{m\omega_1 Z_s^3}, F_0 = \frac{f_0}{kZ_s}, \Omega = \frac{\omega}{\omega_1}, \sigma = \left(\frac{A_1}{A_2}\right)^{1/6}, \Sigma = \frac{\sigma}{Z_s}, \xi = \frac{1}{Q} \tag{20}$$

Then, the LJ force F_{LJ} and the squeeze film damping force F_s can be rewritten as:

$$F_{LJ} = -\frac{d}{(\alpha + y)^2} + \frac{\Sigma^6 d}{30(\alpha + y)^8} \tag{21}$$

$$F_s = \chi \cdot \frac{1}{(\alpha + y)^3} y' \tag{22}$$

Therefore, the dynamic equation of the system can be given by:

$$y'' + \xi y' + y + \beta y^3 = -\frac{d}{(\alpha + y)^2} + \frac{\Sigma^6 d}{30(\alpha + y)^8} + \varepsilon \left(\Gamma \cos \Omega \tau - \eta \frac{1}{(\alpha + y)^3} y' \right) \tag{23}$$

where, $\Gamma = F_0/\varepsilon, \Delta = \gamma/\varepsilon, \eta = \chi/\varepsilon$, and ε is a small perturbation.

Results and Discussion

This section aims at numerically investigating the characteristics and nonlinear dynamics of a TMAFM cantilever-sample system driven by the harmonic excitation. The general properties of the cantilever and interaction properties with the respective sample are referred to [9], are listed in Table 1. The 4th order Runge-Kutta method is used to integrate the set of Equation (23). A small integration step ($2\pi / 200$) has to be chosen to ensure a stable solution and to avoid the numerical divergence at the points where derivatives of F_{LJ} and F_s are discontinuous. The effects of system parameters on the dynamic behavior of the cantilever vibrating system are investigated by using the bifurcation diagram, Poincaré maps, largest Lyapunov exponent, phase portraits, time histories and amplitude spectrum.

Table 1 Parameters of the silicon tip tapping and silicon sample used in the numerical simulations.

Description	Value
Length	449 μm
Width	46 μm
Thickness	1.7 μm
Tip radius	150 nm
Material density	2.330 kg/m^3
Young's modulus	176 GPa
Bending stiffness	0.11 N·m^{-1}
First-order resonant frequency	11.804 kHz
Quality factor	100
Hamaker constant (Repulsive)	1.3596×10^{-70} J·m^6
Hamaker constant (Attractive)	1.865×10^{-19} J

Effect of External Forcing Term Γ

The external forcing term is one of the most important parameters affecting the dynamic characteristics of the TM-AFM tip-sample system. Figure 3 and Fig. 4 show the bifurcation diagram and largest Lyapunov exponent map of the dynamic system where the amplitude of the external excitation is the control parameter and the small perturbation $\varepsilon = 0.1$ is added.

It can be seen from Fig. 3 that the system responses contain periodic and chaotic motions alternately at the interval of $0 < \Gamma < 80$. When $\Gamma = 6.8$, the vibration amplitude of the cantilever is small, the motion is synchronous with period-eight (P-8), and eight points are correspondingly displayed in the Poincaré map, as shown in Fig. 5(a). With the increase of the amplitude of the forcing term Γ, the motion becomes synchronous with period-two (P-2) at $\Gamma = 33.0$, as illustrated in Fig. 5(b). Moreover, the period-1 motion with only one isolated point in Poincaré map and one circle in phase portrait can be observed at $\Gamma = 65.0$ in Fig. 5(c), and the corresponding largest Lyapunov exponent becomes negative, according to Fig. 4. At $\Gamma = 75.2$, the chaos is shown in Fig. 5(d), the strange attractor has a fractal structure in Poincaré map, and it can be found in Fig. 4 that the corresponding largest Lyapunov exponent is positive. Therefore, as the amplitude of the forcing term increases, the changes of the system responses are very complex, with alternative periodic and chaotic motions.

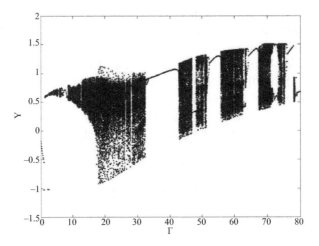

Figure 3 Bifurcation diagram of the amplitude of the external forcing term Γ.

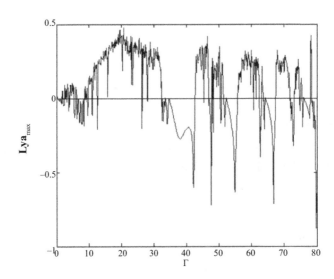

Figure 4 Largest Lyapunov exponent map of the amplitude of the external forcing term Γ.

Effect of Squeeze Film Damping η

At the micro-scale, the squeeze film damping coefficient and ratio are the key parameters for the dynamic responses of the micro-devices. The larger the squeeze film damping is, the higher the noise level results. The ratio of the fundamental resonant frequency in gas to

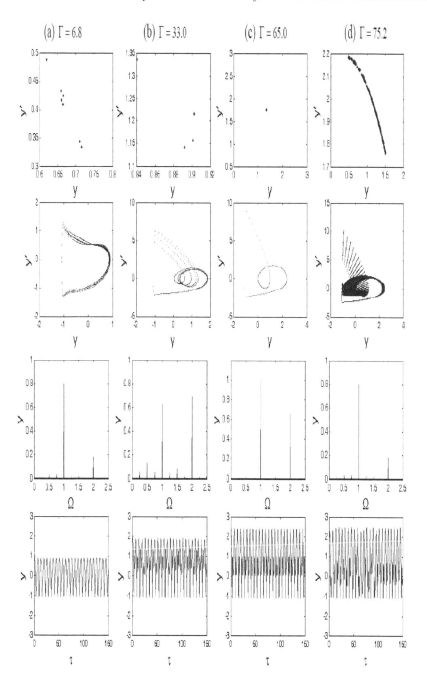

Figure 5 The Poincaré maps, phase portraits, amplitude spectrums and time histories of different Γ.

that in vacuum $\omega_{gas}/\omega_{vac}$ is numerically calculated from Equation (3). Figure 6 gives the ratio of resonant frequencies in vacuum and gas $\omega_{gas}/\omega_{vac}$ as a function of the Reynolds number R_e at different natural scaling parameter Π. It could be found that $\omega_{gas}/\omega_{vac}$ is increasing with the increase of Reynolds number Re and $\omega_{gas}/\omega_{vac}$ decreases with the increase of the scaling parameter Π. Figure 7 gives the quality factor Q_{gas} as a function of both natural scaling parameters, R_e and Π, which can be obtained directly from Equation (5). It is indicated that the quality factor Q_{gas} increases with the increase of the Reynolds number R_e and the decrease of the natural scaling parameter Π. However, when the natural scaling parameter Π tends to be a larger value, i.e., $\Pi = 10$, the quality factor Q_{gas} has small change. For example, when Π changes from 10 to 100, the quality factor Q_{gas} has very small change, as shown in Fig. 7.

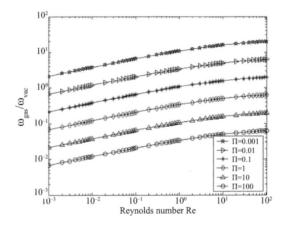

Figure 6 The relationship between the vacuum resonant frequency ωvac and the resonant frequency in gas ω_{gas} ($\omega_{gas}/\omega_{vac}$) as a function of the Reynolds number R_e at different natural scaling parameter Π.

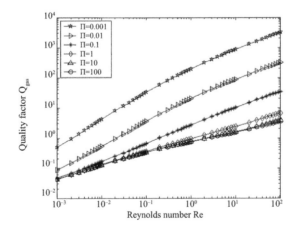

Figure 7 The quality factor Q_{gas} as a function of the Reynolds number R_e for the fundamental mode at different natural scaling parameter Π.

Microstructures undergoing motion transverse to a fixed plate exhibit damping effects that should be considered in the dynamics simulation. The heat transfer analogy is applied to obtain the damping and stiffness coefficients of the microstructures with PLANE 55 thermal elements in ANSYS 11.0 [28]. The effective damping and stiffness coefficients are determined by the finite element thermal analogy approach and theoretic analyses at different operation frequencies and the results are listed in Table 2. It can be seen that the damping coefficient decreases tardily with the increase of the operation frequency and the stiffness coefficient increases fleetly at the same time with slip and without slip. Meanwhile, the damping and stiffness coefficients with slip effect are smaller than those without slip. Figure 8 shows the pressure distribution of the damping component with slip at the resonant frequency. The pressure distribution is approximately parabola in the directions of length and width, and the peak appears at the center of the film.

Table 2 Effective damping and stiffness coefficients of the squeeze film at different frequencies.

Frequency f (Hz)	Damping coefficient			Stiffness coefficient		
	PLANE55		Analytic(slip) [27]	PLANE55		Analytic(slip) [27]
	No slip	Slip		No slip	Slip	
1	1.5199e − 4	1.3499e − 4	1.2529e − 4	9.2051e − 10	7.2610e − 10	6.2304e − 10
1 000	1.5199e − 4	1.3499e − 4	1.2529e − 4	9.2051e − 4	7.2610e − 4	6.2304e − 4
11 804	1.5197e − 4	1.3497e − 4	1.2528e − 4	0.1282	0.1012	0.0868
50 000	1.5163e − 4	1.3473e − 4	1.2509e − 4	2.296	1.8118	1.555
100 000	1.5056e − 4	1.3398e − 4	1.2449e − 4	9.117	7.206	6.190

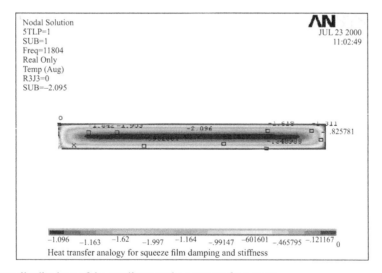

Figure 8 Pressure distributions of the cantilever at the resonant frequency.

Figure 9 displays the bifurcation diagrams of squeeze film damping ratio η in the range of $0.05 < \eta < 0.1$ for the coupling nonlinear dynamic system, from which it can be that the system response changes between periodic and chaotic motions alternately. Figure 10 shows the Poincaré maps and phase plane portraits of different squeeze film damping ratio η on the responses of the coupling system. The system response starts synchronous motion with period-4 at $\eta = 0.0525$ (shown in Fig. 10a), and then becomes synchronous motion with period-1 at $\eta = 0.0635$ (shown in Fig. 10b), and then leaves synchronous motion with period-1 and enters chaotic motion at $\eta = 0.07$, which can be seen in Fig. 10c. The strange attractor has a fractal structure and the corresponding largest Lyapunov exponent is positive. As indicated in Fig. 10d, with the increase of squeeze film damping ratio, the system response becomes synchronous motion with period-1 from chaotic motion. Therefore, the effect of squeeze film damping on the system response cannot be neglected for structures at the micro-scale.

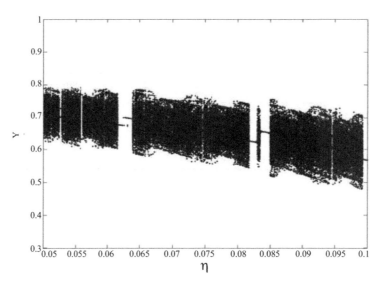

Figure 9 Bifurcation diagram of the squeeze film damping ratio η .

Effect of Material Property Parameter Σ

At micro-scale, the material properties of the AFM tip and sample play an important role on the surface force between them and, as a result, the dynamic response of the tip-sample system displays very rich nonlinear characteristics. Figure 11 is the bifurcation diagram of the material property parameter Σ for the coupling nonlinear dynamic system with different cubic stiffness ratios and squeeze film damping ratios. The system parameters are taken as follows: equilibrium parameter $\alpha = 1.2$, excitation frequency ratio $\Omega = 1$, integration step numbers $N = 200$ and the bifurcation step $\Delta\Sigma = 0.005$.

Figure 11a shows the bifurcation diagram of the material property parameter in the range of $0.3 < \Sigma < 0.42$ with the cubic stiffness ratio $\beta = 0.3$. The response of the coupled

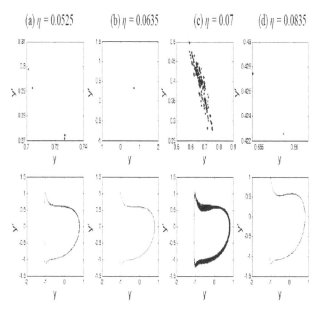

Figure 10 The Poincaré maps and phase portraits of different squeeze film damping ratios η .

nonlinear system undergoes a complete process from chaotic motion through period-2, period-1, period-4 and period-8 motions to steady-state motion with period-1 by the forms of period-doubling and anti period-doubling bifurcations. At the interval of $0.42 < \Sigma < 0.47$, the system response alters between chaotic motion with long time and periodic motion with short time. When Σ increases to $\Sigma > 0.47$, the system response becomes synchronous motion with period-1. With the change of the cubic stiffness ratio of the cantilever tip from $\beta = 0.3$ to $\beta = 0.5$, the response of the coupled system undergoes the process of chaotic and periodic motions alternatively. It comes into period-2 motion from chaotic motion with anti period-doubling bifurcation, and then enters period-1 motion in a large range of material property parameter ($\Sigma > 0.39$), as illustrated in Fig. 11b. In addition, it is found that the chaotic motion disappears at higher Σ with the increase of cubic stiffness ratio (from $\beta = 0.3$ to $\beta = 0.5$). Therefore, with the changes of the measured samples in experimental tests, the values of Σ vary accordingly, and as a result the response of the coupled system displays various nonlinear dynamic behaviors.

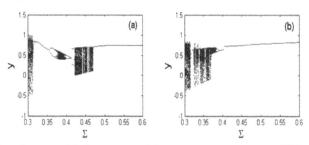

Figure 11 Bifurcation diagram of th 1 cm e material property parameter Σ at different cubic stiffness ratios: (a) $\beta = 0.3$; (b) $\beta = 0.5$.

Figure 12 is the bifurcation diagram of the cubic stiffness ratio β on the response of TM-AFM tipsample system at the interval of $0.3 < \beta < 0.6$ for various combinations of squeeze film damping ratios, material and equilibrium parameters, and the bifurcation step $\Delta\beta = 0.01$. It can be observed from Fig.12(a) that the response of the coupled system has a complete process from chaotic motion through periodic motion and chaotic motion to period-1 motion. At the interval of $0.3 < \beta < 0.43$, the system response enters periodic motion from chaotic motion, then it becomes chaotic motion again, and finally it comes into steady-state motion with period-1 in the range of $0.43 < \beta < 0.6$. With the increase of squeeze film damping η ($\eta = 0.14$), the system response changes noticeably and it mainly contains the periodic components, such as period-1, period-3 and period-6 motions, as illustrated in Fig. 12(c). As the equilibrium parameter α increases, the chaotic components of the system response decrease, while the periodic components increase and contain period-2, period-4 and period-8 motions with the case of $\alpha = 1.6$, as shown in Fig. 12(d).

The Poincaré maps and phase portraits for different cubic stiffness ratios are displayed in Fig. 13. When the cubic stiffness ratio is small, i.e., $\beta = 0.35$, the exhibited motion is period-6 motion, six points and circles can be seen in the Poincaré map and phase portrait

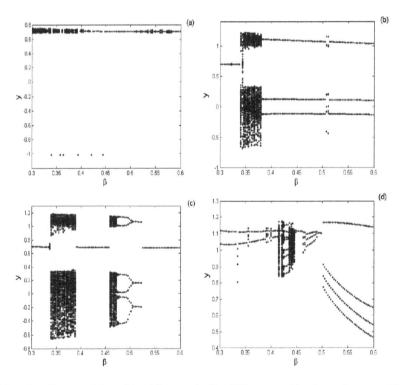

Figure 12 Bifurcation diagram of the cubic stiffness ratio β at different combinations of squeeze film damping ratios, material parameters and equilibrium parameters: (a) $\eta = 0.08$, $\Sigma = 0.3$, $\alpha = 1.2$; (b) $\eta = 0.08$, $\Sigma = 0.5$, $\alpha = 1.2$; (c) $\eta = 0.14$, $\Sigma = 0.5$, $\alpha = 1.2$; (d) $\eta = 0.08$, $\Sigma = 0.3$, $\alpha = 1.6$.

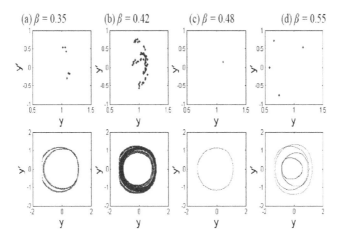

Figure 13 The Poincaré maps and phase portraits of different cubic stiffness ratios β.

respectively from Fig. 13(a). As the cubic stiffness ratio becomes larger, the closed circle is decomposed and the points in the Poincaré map gradually scatter. At $\beta = 0.42$, the system response comes into chaotic motion, and then the points of the attractor are decomposed again and finally converge to one point. When $\beta = 0.48$, a synchronous motion with period-1 can be observed. Then the system response comes into period-4 motion. It is indicated that the components of chaotic motions become wider with the increase of the squeeze film damping. Material properties of the tip and sample and equilibrium coefficient ratio play very important role in the nonlinear dynamics of the coupled system.

Effect of Equilibrium Parameter α

Equilibrium coefficient ratio α is one of the important parameters for determining the equilibrium position of the cantilever tip in AFM and it becomes the key factor to reflect the dynamic responses of the tip-sample model. The dynamic behavior of the coupled system depends on the value of α. Figures 14 and 15 give the bifurcation diagram and largest Lyapunov exponent map of the dynamic system with the control parameter of the equilibrium parameter α at $\Omega = 1$. It can be seen from Fig. 14 that the dynamic responses are very complicated, and the components contain periodic and chaotic motions at the interval of $1 < \alpha < 2$. The corresponding largest Lyapunov exponents are alternately positive and negative, as shown in Fig. 15.

To explain the dynamic responses of the system clearly, Fig. 16 shows the local bifurcation diagram and Poincaré maps of the dynamic system at the interval of $1.6 < \alpha < 2$. It can be found that the system responses exhibit the alternation of periodic and chaotic motions. The system response comes into steady-state synchronous motion with period-1 from chaotic motion, and enters period-2 motion from period-1 motion as the equilibrium parameter α increases, and then becomes chaotic motion with period-doubling bifurcation.

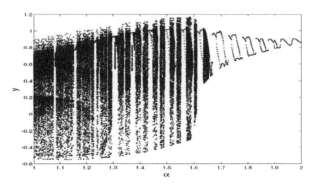

Figure 14 Bifurcation diagram of the equilibrium parameter α.

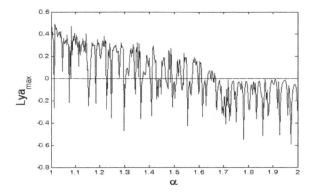

Figure 15 Largest Lyapunov exponent map of the equilibrium parameter α.

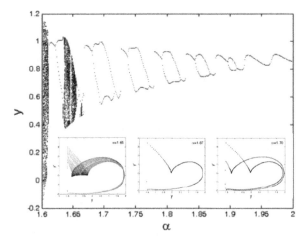

Figure 16 Local bifurcation and Poincaré maps of equilibrium parameter α.

Moreover, at $1.7 < \alpha < 1.9$, the system response changes between period-1 and period-2 motions alternately. When $\alpha > 1.9$, the system response comes into steady-state synchronous motion with period-1. These phenomena indicate that the dynamic responses of the coupled system are very complex. The cantilever tip can undergo a period-doubling cascade to possible chaos about the original equilibrium. It is demonstrated that, away from the surface, the net force on the tip is always in the downward direction and causes the tip to accelerate the sample until it passes the key point, where the repulsive force plus the spring force becomes larger than the van der Waals force, and then the tip is forced away from the sample.

Figure 17 shows the bifurcation diagram of the equilibrium parameter α on the response of TMAFM tip-sample system at the interval of $1.0 < \alpha < 1.8$, the squeeze film damping ratio is taken as $\eta = 0.008$ and the bifurcation step is $\Delta\alpha = 0.01$. It can be seen from Fig. 17(a) that the response of the coupled dynamic system comes into period-2 motion from period-1 motion and then enters chaotic motion with period-doubling bifurcation. With the increase of α, the system responses enters period-2 motion from chaotic motion, and it subsequently becomes period-1 motion again when $\Sigma = 0.3$ and $\Gamma = 2$. As illustrated in Fig. 17(b), the response of the coupled system has a complete process from chaotic motion through periodic motion to chaotic motion with the forms of period-doubling bifurcation and anti period-doubling bifurcations at the interval of $1.0 < \alpha < 1.8$ when $\Sigma = 0.3$ and $\Gamma = 4$.

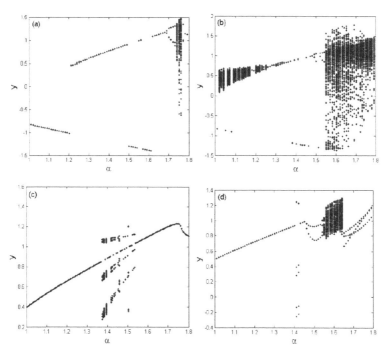

Figure 17 Bifurcation diagram of the equilibrium parameter α at different combined material parameters and external forcing terms: (a) $\Sigma = 0.3$, $\Gamma = 2$; (b) $\Sigma = 0.3$, $\Gamma = 4$; (c) $\Sigma = 0.5$, $\Gamma = 2$; (d) $\Sigma = 0.5$, $\Gamma = 4$.

With the increases of the material parameter coefficients, as shown in Fig. 17(c) and (d), the system response changes noticeably and it mainly contains the periodic components, such as period-1, period-2, period-3 and period-6 motions. Meanwhile, the chaotic components of the system response decrease and, on the contrary, the periodic components increase. In addition, the chaotic components of the system response shift to the smaller equilibrium parameter. It demonstrates that the components of chaotic motions become wider as the amplitude of external forcing term increases, and the material properties of the tip and sample and equilibrium coefficient ratio affect the nonlinear dynamics of the coupled system.

To illustrate the various motions, Fig. 18 shows the nonlinear characteristics of the coupled system with the plots of the Poincaré maps and phase portraits for different equilibrium parameters α at different conditions. The motion of the coupled system changes between periodic and chaotic motions alternately. At $\alpha = 1.2$, the motion with period-1 represented by a point in the Poincaré maps and characterized by a close curve in phase portraits is shown in Fig. 18(a). As illustrated in Fig. 18(b), the system response comes into period-6 motion at $\alpha = 1.42$ from synchronous motion with period-1 at $\alpha = 1.2$, as displayed in Fig. 18 (a), then leaves period-6 motion and enters chaotic motion at $\alpha = 1.6$, which can be seen from Fig. 18(c). The strange attractor has a fractal structure and the corresponding largest Lyapunov exponent is positive. With the increase of the equilibrium parameter coefficient, as shown in Fig. 18(d), the system response becomes periodic motion from chaotic motion again, and one can find the period-3 motion marked by three isolated points in Poincaré map and three circles in phase portrait at $\alpha = 1.47$. It is indicated that the components of chaotic motions of the coupled system increases obviously with the increase of the amplitude of the force term. In general, the effect of equilibrium parameter on the system response should be considered for the design of the TM-AFM.

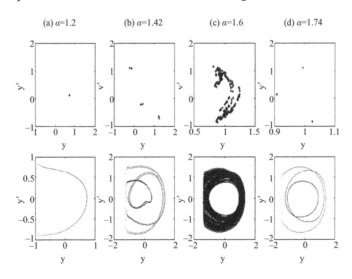

Figure 18 The Poincaré maps and phase portraits of different equilibrium parameters α.

Conclusions

The cantilever in tapping mode Atomic force microscope (TM-AFM) is one of crucial components and so it is very important to carry out a thorough dynamic analysis for the cantilever to further enhance the performance of the TM-AFM. In this paper, the cantilever-sample interaction and a harmonically forcing term in the TM-AFM have been considered. Numerical simulations have been used to investigate the nonlinear behaviors between the tip and the sample. The chaotic behavior appears to be generated via various system parameters including the amplitude of the external forcing term, equilibrium parameter, squeeze film damping and material property. The dynamic system responses display very rich nonlinear dynamic characteristics under the effects of these parameters and show an alternate changing process among periodic motion, quasi-periodic motion and chaotic motion.

The component of the chaotic motion in the system response increases with the increase of the amplitude of the external excitation, but will be weaken by the squeeze film damping. In addition, the increase of the equilibrium parameter coefficient would strengthen the component of the chaotic motion but weaken the component of the periodic motion increases in the system response. It is indicated that the external excitation and squeeze film damping are important for the design of dynamic TM-AFM. It is of significance to understand the cantilever dynamics in air/liquid and promote the development of the next generation of AFMs.

Acknowledgments

The authors would like to thank Prof. Z. K. Peng for his good suggestions. This work was supported by the National Natural Science Foundation of China under Grant No.10602033, the National Outstanding Youth Foundation of China under Grant No.10325209, and the Specialized Research Fund for State Key Laboratory of Mechanical System and Vibration.

References

1. Finot, E.; Passian, A.; Thundat, T. Measurement of mechanical properties of cantilever shaped materials. *Sensors* **2008**, *8*, 3497–3541.
2. Jalili, N.; Laxminarayana, K. A review of atomic force microscopy imaging systems: application to molecular metrology and biological sciences. *Mechatronics* **2004**, *14*, 907–945.
3. Hersam, M.C. Monitoring and analyzing nonlinear dynamics in atomic force microscopy. *Small* **2006**, *2*, 1122–1124.
4. Wei, Z; Zhao, Y.P. Growth of liquid bridge in AFM. *J. Phys. D: Appl. Phys.* **2008**, *40*, 4368–4375.
5. Yacoot, A.; Koenders, L. Aspects of scanning force microscope probes and their effects on dimensional measurement. *J. Phys. D: Appl. Phys.* **2008**, *41*, 103001.
6. Lee, S.I.; Howell, S.W.; Raman, A.; Reifenberger, R. Nonlinear dynamics of microcantilevers in tapping mode atomic force microscopy: a comparison between theory and experiment. *Phys. Rev.B* **2002**, *66*, 115409.
7. Zhao, X.; Dankowicz, H. Characterization of intermittent contact in tapping-mode atomic force microscopy. *ASME J. Comput. Nonlin. Dyn.* **2006**, 1, 109–115.

8. Yagasaki, K. Nonlinear dynamics of vibrating microcantilevers in tapping-mode atomic force microscopy. *Phys. Rev. B* **2004**, *70*, 245419.

9. Rutzel, S.; Lee, S.I.; Raman, A. Nonlinear dynamics of atomic-force-microscope probes driven in Lennard-Jones potentials. *Proc. R. Soc. Lond. A* **2003**, *459*, 1925–1948.

10. Hashemi, N.; Dankowicz, H.; Paul, M.R. The nonlinear dynamics of tapping mode atomic force microscopy with capillary force interactions. *J. Appl. Phys.* **2008**, *103*, 093512.

11. Ashhab, M.; Salapaka, M.V.; Dahleh, M.; Mezic, I. Melnikov-based dynamical analysis of microcantilevers in scanning probe microscopy. *Nonlin.Dyn.* **1999**, *20*, 197–220.

12. Basso, M.; Giarre, L.; Dahleh, M.; Mezic, I. Complex dynamics in a harmonically excited Lennard-Jones oscillator: microcantilever–sample interaction in scanning probe microscopes. *J.Dyn. Syst. Meas. Control* **2000**, *122*, 240–245.

13. Paulo, A.S.; Garcia, R. Tip-surface forces, amplitude, and energy dissipation in amplitudemodulation (tapping mode) force microscopy. *Phys. Rev. B* **2001**, 64, 193411.

14. Behrend, O.P.; Odoni, L.; Loubet, J.L.; Burnham, N.A. Phase imaging: Deep or superficial? *Appl. Phys. Lett.* **1999**, *75*, 2551.

15. Zitzler, L.; Herminghaus, S.; Mugele, F. Capillary forces in tapping mode atomic force microscopy. *Phys. Rev. B* **2002**, *66*, 155436.

16. Couturier, G.; Boisgard, R.; Nony, L.; Aime, J.P. Noncontact a tomic force microscopy: Stability criterion and dynamical responses of the shift of frequency and damping signal. *Rev. Sci. Instr.* **2003**, *74*, 2726–2734.

17. Zhang, W.M.; Meng, G. Nonlinear dynamical system of micro-cantilever under combined parametric and forcing excitations in MEMS. *Sens. Actuat. A: Phys.* **2005**, *119*, 291–299.

18. Zhang, W.M.; Meng, G. Nonlinear dynamic analysis of electrostatically actuated resonant MEMS sensors under parametric excitation. *IEEE Sens. J.* **2007**, *7*, 370–380.

19. Motamedi, R.; Wood-Adams, P.M. Influence of fluid cell design on the frequency response of AFM microcantilevers in liquid media. *Sensors* **2008**, 8, 5927–5941.

20. Basak, S.; Raman, A. Hydrodynamic loading of microcantilevers vibrating in viscous fluids. *J.Appl. Phys.* **2006**, *99*, 114906.

21. Sader, J.E. Frequency response of cantilever beams immersed in viscous fluids with applications to the atomic force microscope. *J. Appl. Phys.* **1998**, *84*, 64–76.

22. Sader, J.E.; Chon, J.W.M.; Mulvaney, P. Calibration of rectangular atomic force microscope cantilevers. *Rev. Sci. Instr.* **1999**, *70*, 3967–3969.

23. Matei, G.; Jeffery, S.; Patil, S.; Khan, S.H.; Pantea, M.; Pethica, J.B.; Hoffmann, P.M. Simultaneous normal and shear measurements of nanoconfined liquids in a fiber-based atomic force microscope. *Rev. Sci. Instr.* **2008**, *79*, 023706.

24. Bao, M.H.; Yang, H. Squeeze film damping in MEMS. *Sens. Actuat. A: Phys.* **2007**, *136*, 3–27.

25. Veijola, T.; Lahdenpera, J. The influence of gas-surface interaction on gas film damping in a silicon accelerometer. *Sens. Actuat. A: Phys.* **1998**, 66, 83–92.

26. Song, Y.; Bhushan, B. Atomic force microscopy dynamic modes: modeling and applications. *J. Phys.: Condens. Matter.* **2008**, *20*, 225012.

27. Andrews, M.; Harris, I.; Turner, G.A. comparison of squeeze-film theory with measurements on a microstructure *Sens. Actuat. A: Phys.* **1993**, 36, 79–87.

28. Ostergaard, D. *Using a heat transfer analogy to solve for squeeze film damping and stiffness coefficients in MEMS structures.* ANSYS Inc.: Canonsburg, PA, USA, January 23, 2003; Available online: http://www.ansys.com/assets/tech-papers/mems-thermal-analogy-fsidamping. pdf.

Mathematical Identification of Homogenisation Processes in Argon Stirred Ladle[†,*]

K. Michalek,[1] J. Morávka[2] and K. Gryc[1]

[1]VSB-TU Ostrava, FMME, Ostrava, Czech Republic.
[2]Material & Metallurgical Research Ltd., Ostrava, Czech Republic.

ABSTRACT

Mathematical models processed results of experimental investigation obtained during ladle gas argon bubbling realised by stir elements situated in the ladle bottom. Exact theoretical description of processes occurring at argon bubbling into steel would be very complex and it would lead to a system of non-linear partial differential equations describing transfer of momentum, heat, components, and with excitation function in the form of equation of so called deterministic chaos (argon bubbling). On the basis of pouring ladle model diagram and concentrations courses, the simplified linear physically adequate model was proposed, which described behaviour of steel concentration in pouring ladle during its bubbling. The analysed process was understood in the form of a cybernetic model.

Keywords: steel, Ladle, gas argon bubbling, stir elements, physical modelling, mathematical identification.

Introduction

Previously presented paper "Physical Modelling of Bath Homogenisation in Argon Stirred Ladle" was devoted to results of bath behaviour in the ladle model investigation during its gas argon bubbling realised by one or two stir elements situated in the ladle bottom. Study was performed with use of physical modelling method on a scale model 1 to 10. Development of homogenisation processes after start of bubbling was evaluated on the basis of electrical conductivity and temperature change, which were measured at three

[†]Reused with permission from: K. Michalek, J. Morávka, K. Gryc, Mathematical Identification of Homogenisation Processes in Argon Stirred Ladle, Metalurgija, 48(4), 219–222, (2009). Original Scientific Paper -Izvorni znanstveni rad.

The work was realised within the frame of solution of the grant project No.106/07/0407 under financial support of the Czech Science Foundation.

points of the ladle volume by conductivity and temperature sensors. The executed works were realised for conditions of 180 tons steel ladles.

It was found that the rate of homogenisation in the steel ladle is significantly dependent on the volume flow of argon. Mathematical method for the identification process was used for determining the optimum flow of argon.

Mathematical Description of Processes at Argon Bubbling

It was appropriate to create a physical-mathematical (physically adequate) model for the measured time courses of tracing substance concentrations in physical model of pouring ladle (PL) occurred during argon bubbling into the steel.

Description of Situation

Schematic representation of situation at argon bubbling into bath in the model of steel ladle (mLP) is shown in Fig. 1.

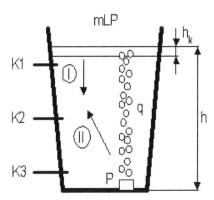

Figure 1 Argon bubbling into ladle model.

Argon bubbles flow from eccentrically situated stir element (P) in the bottom of mLP at constant volumetric flow (q). They "disrupt" the layer of concentration enriched and coloured water with thickness (h_k) and there occurs gradual mixing of enriched and clear liquid (steel, water). Two (pressure) forces II and I act basically against each other on molecules of water in proximity of the sensors.

Measured Data

Analysis and synthesis of mathematical models was realised with use of the data measured on physical model, where courses of the measured concentrations, (with period of sampling $\Delta\tau \approx 0.5$ s), had on the sensors K1, K2 and K3 are shown in Fig. 2.

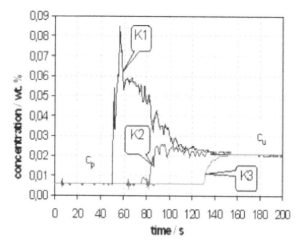

Figure 2 The time response of concentrations on the sensors K1, K2 and K3.

Several facts are obvious from the development of concentration in the sensors:

- start and progress of gas argon bubbling into steel can be approximately considered in the form of the Heaviside unit step function and it is therefore possible to consider the development of concentration as a unit step response,
- the sensors reacted only after elapsing of certain "dead time", which is proportional to the distance of individual sensors from the liquid level in the mLP,
- the overshoot of courses (apparently proportional to the magnitude of the force I or rather to the difference of the forces I and II) also descends with the distance from the liquid level in the mLP,
- steady-state (final) value of concentration is proportional to the proportion of volumes of pure water and water with enriched concentration.

Physical-mathematical Model

Exact theoretical description of processes occurring at argon bubbling into steel would be very complex and it would lead to a system of non-linear partial differential equations describing transfer of momentum, heat, components, and with excitation function in the form of equation of so called deterministic chaos (argon bubbling). On the basis of the mLP diagram and concentrations courses, the simplified linear physically adequate model was proposed, which described behaviour of steel elements concentration in PL during its bubbling. Analysed process was understood in the form of a cybernetic model, which can be transparently visualised by so called block diagram shown in Fig. 3.

It is series-parallel connection of three components, namely component of time delay and two parallel proportional (inertial) components acting against each other.

Two parallel and antagonistically connected simplest proportional systems with inertia of the 1st order (with transfers G_I a G_{II}) are assumed for the part of the model without time delay [1]:

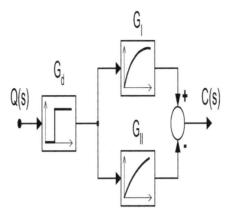

Figure 3 Block diagram of processes in ladle model (mLP).

$$G_I(s) = \frac{k_1}{\tau_1 s + 1}, \quad G_{II}(s) = \frac{k_2}{\tau_2 s + 1} \tag{1}$$

where:

k_1, k_2-coefficients of transfer (amplification) of systems / %·s/m³,

τ_1, τ_2-time constants of systems / s,

s-complex variable in the Laplace transform / s⁻¹.

For this part of the model it is then possible to compose on the algebra basis of transfer the following continuous L-transfers (for zero initial conditions):

$$G(s) = \frac{k_1}{\tau_1 s + 1} - \frac{k_2}{\tau_2 s + 1} = \frac{(k_1 - k_2) + (k_1 \tau_2 - k_2 \tau_1) s}{(\tau_1 s + 1)(\tau_2 s + 1)} \tag{2}$$

Wherefrom for the L-image and original of transfer function $H(s)$ and $h(\tau)$ we get the final expression for standard concentration transfer function, which can be used at the same time as non-linear regression model F1 with three parameters k_1, τ_1 and τ_2:

$$h(\tau) = c_n(\tau) = 1 - k_1 \cdot \exp(-\tau/\tau_1) + (k_1 - 1) \cdot \exp(-\tau/\tau_2) \tag{3}$$

Regression Model for all Concentrations

Figure 4 shows courses of the regression function based on the model F1, for standard concentration values in all three sensors K1, K2 and K3.

It is obvious from the results that the model F1 is suitable and usable for description of concentration course for all three sensors. It is also obvious that the transfer coefficient (gain coefficient) k_1 decreases with the distance of the sensor from the liquid level in the mLP, while the time constant τ_2 increases in this dependence.

Use of Model for Evaluation of Experiments

Proposed physically adequate model F1 was used for evaluation of the physical modelling results at argon bubbling in the ladle model. Its parameters were used for determination of other aggregated parameters, and all of them where used for definition of an optimum range of blown argon flow. The paper documents results of one stir element for position "A" in the ladle bottom.

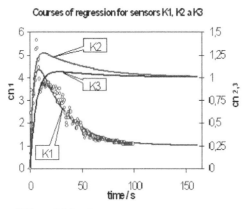

Figure 4 Courses of regression of F1 model for data.

Approaches to Solution

Three groups of parameters were chosen for characterisation of homogeneity "intensity" measure in respect to blown argon flow q: basic physical parameters of model (2 parameters -see the model in Fig. 3 and time delay, i.e., k_1, k_2, and τ_d), aggregated parameter defining derivative time constant (1 parameter, i.e., τ_D), and aggregated parameters defining maximum of model transfer function (2 parameters, i.e., τ_m, respectively τ_{max}, and h_m). Aggregated parameters were chosen in order to achieve simplicity, clarity and appropriate interpretability of homogenisation process.

Parameters of Physically Adequate Model

Graphical dependencies of foregoing mathematical model parameters on blown argon flow for the sensor K2 are shown in Fig. 5:

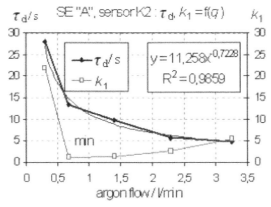

Figure 5 Dependencies of parameters τ_d, k_1 on argon flow for the sensor K2.

It is obvious from the Fig. 5 that physically adequate model, in conformity with the assumed behaviour of real system, has evaluated a trend of dependence of the transport delay τ_d on the argon flow rate. At higher flow rate homogenisation starts sooner and runs more rapidly.

Derivative Time Constant

Overall L-transfer of the defined model has character of so called real derivative member with a delay of the 2nd order [2], and derivative time constant τ_D is function of all three parameters. The constant τ_D can be expressed from the equation (2) with use of the relation:

$$\tau_D = k_1(\tau_2 - \tau_1) + \tau_1 \tag{4}$$

Lower values of this derivative time constant τ_D express faster reaction of the model on change of the input value, which is in this case represented by the volume flow of argon.

Graphical dependence of this time constant on the argon flow (for measurement at the sensor K2) is shown in Fig. 6.

It is obvious from the Fig. 6 that increase of the volume flow of argon to 1,4 l/min causes a significant step change of the value τ_D and therefore also more rapid changes in chemical composition of the model liquid (and development of homogenisation). The value of τ_D does not change significantly with further increase of the argon flow rate.

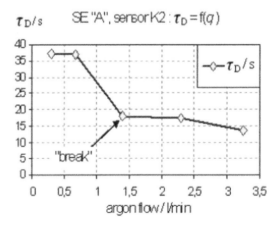

Figure 6 Dependence of derivative time constant τ_D on argon flow for the sensor K2.

Transfer Function Overshoot

This approach uses analytical relations for determination of transfer function position and transfer function maximum values. Graphical dependencies of the assumed aggregated maximum parameters of the blown argon flow for the measurement at sensor K2 are shown in Fig. 7 and Fig. 8.

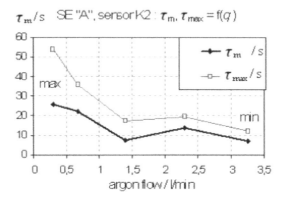

Figure 7 Dependencies of maximum times (τ_m, τ_{max}) of the flow of inert gas for the sensor K2.

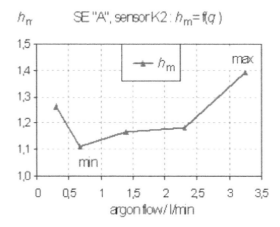

Figure 8 Dependencies of maximum values (h_m) on the flow of inert gas for the sensor K2.

Diagram in Fig. 7 shows development of times for achieving of maximum values in the impulse function in dependence on the argon flow rate (τ_{mx} on the axis y is a general expression of time parameters τ_m, which is a time of achieving the maximum regardless of transport delay τ_d, or τ_{max}). It is possible to obtain values of the parameter τ_{max} by adding the parameter τ_m to the transport delay τ_d. Gradual convergence of τ_m and τ_{max} is given by decrease of the transport delay τ_d with increasing volume flow of argon. Faster circulation of volume of liquid results in faster achieving the maximum values of the impulse function. This speed, however, steeply falls at argon flow rates below 1,4 l/min.

The Fig. 8 is devoted to a description of dependence of the maximum value of the impulse function (h_m–*maximum value of dimensionless concentration*) on the volume flow of argon. At its lowest flow rate (0,3 l/min) the injected substance gets to the sensor K2 in concentrated condition without being mixed. At increase of the argon flow rate to 0,68 l/min the faster dispersion of the marking substance at the expense of the speed of circulation of the volume of model liquid manifests itself the most from the viewpoint of this parameter

(h_m). Further increase in argon flow rate brings about again a gradual enforcing of speed of circulation. At the highest argon flow rate (3,25 l/min) the rate of circulation is too high and injected substance gets to the sensor K2 without being able to disperse in the volume of the liquid.

Conclusion

Physically adequate mathematical model was proposed and verified for the measured time response of substance concentrations in the pouring ladle model bath. This model was used for subsequent analysis of the input simulation parameters influence on its coefficients and aggregated parameters with consequences on appropriate (optimum) working mode setting for steel bubbling in the pouring ladle. Due to pre-set various flows of argon blown into bath three groups of (aggregated) parameters for homogenisation intensity characterisation were determined and verified. On the basis of the shape of their course, the argon flow optimum range was deduced—within limits from approx. 1,4–2,3 l/min (on the ladle model), which can be transformed with use of the appropriate volumetric flow scale factor to the realistic industrial equipment conditions, i.e., 243–400 l/min. It was established in the previous paper entitled "Physical Modelling of Bath Homogenisation in Argon Stirred Ladle" that after achievement of a certain "breakpoint" volume flow its further increase brings low efficiency from the viewpoint of homogenisation times. Parameters assumed in this work enable moreover a more detailed investigation of the character of homogenisation process, correct setting of which forms an integral part of the advanced secondary metallurgy [3,4]. For example magnitude of the maximum values of the impulse function (h_m) at individual argon flow rates gives evidence about behaviour, e.g., of alloying additions injected into the volume of real pouring ladle at the initial stage of homogenisation.

References

1. Víteček, L. Smutný, J. Kusyn, Teorie řízení I. Základní pojmy a řešené příklady z Laplaceovy transformace. III. [Management theory I. Basic notions and solved examples from Laplace's transformation. III.], VŠB-TU Ostrava, Ostrava, 1988.
2. S. Kubík, et al., Teorie regulace—I. Lineární regulace [Theory of regulation—I. Linear regulation]. SNTL, Praha, 1974.
3. J. Krajcar, Metalurgija 32 (1993) 3, 87–92.
4. V. Ferketić, J. Krajcar, A. Ivančan, Metalurgija 35 (1996) 1, 43–48.

Note: The responsible translator for English language is B. Škandera.

Analysis of a Nonlinear Aeroelastic System with Parametric Uncertainties Using Polynomial Chaos Expansion[†]

Ajit Desai[1] and *Sunetra Sarkar*[1,a]

[1]Department of Aerospace Engineering, IIT Madras, Chennai 600036, India.
[a] Email: sunetra.sarkar@gmail.com.

ABSTRACT

Aeroelastic stability remains an important concern for the design of modern structures such as wind turbine rotors, more so with the use of increasingly flexible blades. A nonlinear aeroelastic system has been considered in the present study with parametric uncertainties. Uncertainties can occur due to any inherent randomness in the system or modeling limitations, and so forth. Uncertainties can play a significant role in the aeroelastic stability predictions in a nonlinear system. The analysis has been put in a stochastic framework, and the propagation of system uncertainties has been quantified in the aeroelastic response. A spectral uncertainty quantification tool called Polynomial Chaos Expansion has been used. A projection-based nonintrusive Polynomial Chaos approach is shown to be much faster than its classical Galerkin method based counterpart. Traditional Monte Carlo Simulation is used as a reference solution. Effect of system randomness on the bifurcation behavior and the flutter boundary has been presented. Stochastic bifurcation results and bifurcation of probability density functions are also discussed.

Introduction

Fluid-structure interaction can result in dynamic instabilities like flutter. Nonlinear parameters present in the system can stabilize the diverging growth of flutter oscillations to a limit cycle oscillation (LCO). Sustained LCO can lead to fatigue failure of rotating structures such as wind turbine rotors. Hence, it is an important design concern in aeroelastic

[†]Reused with permission from: Ajit Desai and Sunetra Sarkar, "Analysis of a Nonlinear Aeroelastic System with Parametric Uncertainties Using Polynomial Chaos Expansion," *Mathematical Problems in Engineering*, vol. 2010, Article ID 379472, 21 pages, 2010. doi:10.1155/2010/379472.

analysis. Moreover, there is a growing interest in understanding how system uncertainties in structural and aerodynamic parameters and initial conditions affect the characteristics of such dynamical response. Uncertainty quantification in a stochastic framework with stochastic inputs has traditionally been analyzed with Monte Carlo simulations (MCSs). To apply this procedure one should use the distribution of the input parameters to generate a large number of realizations of the response. Probability density function _PDF_ and other required statistics are then approximated from these realizations; however, it is computationally expensive, especially for large complex problems. Hence, there is a need to develop alternate approaches which are computationally cheaper than direct MCS procedure. Perturbation method is a fast tool for obtaining the response statistics in terms of its first and second moments [1]. The statistical response is determined by expanding the stochastic parameters around their mean via a Taylor series [2]. The application of this method is, however, limited to small perturbations and does not readily provide information on high-order statistics [3,4].

The resulting system of equations becomes extremely complicated beyond second-order expansions as shown in the literature. Sensitivity method is a more economical approach, based on the moments of samples, but it is less robust and depends strongly on the modelling assumptions [5]. Another approach based on expanding the inverse of the stochastic operator in a Neumann series is also limited to small fluctuations only; even combining with the Monte Carlo method also seems to result in a computationally prohibitive algorithm for complex systems [4].

Polynomial chaos expansion (PCE) is a more effective approach, pioneered by Ghanem and Spanos [4], proposed first in the structural mechanics finite elements area. It is a spectral representation of the uncertainty in terms of orthogonal polynomials. The stochastic input is represented spectrally by employing orthogonal polynomial functionals from the Askey scheme as basis in the random space. The original homogeneous PCE was based on Hermite polynomials fromthe Askey family [6]. It can give optimal exponential convergence for Gaussian inputs [7]. A standard Galerkin projection is applied along the random dimensions to obtain the weak form of the equations. The resulting deterministic systems are solved using standard techniques to solve for each random mode [8]. Galerkin polynomial chaos expansion (Galerkin PCE) based approaches have been examined extensively with different basis functions to model several uncertain flow and flow-induced instability problems [9,10].

Galerkin PCE (also called intrusive approach) modifies the governing equations to a coupled form in terms of the chaos coefficients. These equations are usually more complex and arriving at them is quite often a tedious task for some choices of the uncertain parameters. In order to avoid these, several uncoupled alternatives have been proposed.

These are collectively called nonintrusive approaches. In a nonintrusive polynomial chaos method a deterministic solver is used repeatedly as in Monte Carlo simulation. The Probabilistic Collocation (PC) method is such a nonintrusive polynomial chaos method in which the problem is collocated at Gauss quadrature points in the probability space [11,12].

The deterministic solutions are performed at these collocation points. The nonintrusive polynomial chaos method proposed by Walters and coworkers [13,15] is based on approximating the polynomial chaos coefficients. A similar approach called nonintrusive Spectral Projection has been used by Reagan et al. [16]. Pettit and Beran [17,18] have also used a stochastic projection technique to compute the chaos expansion coefficients in an aeroelastic system. When multiple uncertain parameters are involved the collocation grids are constructed using tensor products of one-dimensional grids. Thus, the number of collocation points and therefore the number of required deterministic solutions increases rapidly. As an alternative, sparse grid collocation approaches can be implemented [19,21].

The intrusive and nonintrusive PCE approaches and their implementation to an aeroelastic model with structural nonlinearity are discussed in detail in the subsequent sections.

Nonlinear Aeroelastic Model

Figure 1 shows a schematic plot of the two degree-of-freedom pitch-plunge aeroelastic system and also the notations used in the analysis. The aeroelastic equations of motion for the linear system have been derived by Fung [22]. For nonlinear restoring forces such as with cubic springs in both pitch and plunge, the mathematical formulation is given by Lee et al. [23] as follows:

$$\varepsilon'' + x_\alpha \alpha'' + 2\zeta_\alpha \frac{\varpi}{U}\varepsilon' + \left(\frac{\varpi}{U}\right)^2 \left(\varepsilon + \beta_\varepsilon \alpha^3\right) = -\frac{1}{\pi\mu}C_L(\tau),$$

$$\frac{x_\alpha}{r_\alpha^2}\varepsilon'' + \alpha'' + 2\frac{\zeta_\varepsilon}{U}\alpha' + \frac{1}{U^2}\left(\alpha + \beta_\alpha \varepsilon^3\right) = \frac{2}{\pi\mu r_\alpha^2}C_M(\tau),$$

(2.1)

The above equations are shown in the nondimensional form. The nondimensional parameters are given below. The plunge deflection is considered positive in the downward direction and the pitch angle about the elastic axis is denoted positive nose up. Elastic axis is located at a distance $a_h b$ from the midchord where b is the half chord. Let us also use v as the wind velocity h as the plunge deflection. Among the nondimensional quantities, $\varepsilon = h/b$ = nondimensional displacement of the elastic axis point; $\tau = vt/b$ = nondimensional time;

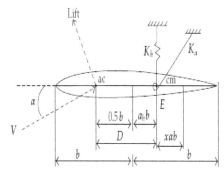

Figure 1 The schematic of a symmetric airfoil with pitch and plunge degrees-of-freedom.

$U = v/(b\omega_\alpha) =$ the nondimensional velocity (also called reduced velocity); $\varpi = \omega_\varepsilon/\omega_\alpha, \omega_\varepsilon$ and ω_α are the natural frequencies of the uncoupled plunging and pitching modes respectively. In the structural part, ζ_ε and ζ_α are the damping ratios in plunge and pitch respectively, r_α is the radius of gyration about the elastic axis, and μ is the airfoil mass ratio defined as $m/\pi\rho v$. β_α and β_ε denote coefficients of cubic spring in pitch and plunge respectively. For incompressible, inviscid flow, Fung [22] gives the expressions for unsteady lift and pitching moment coefficients, $C_L(\tau)$ and $C_M(\tau)$:

$$C_L(\tau) = \pi\left(\varepsilon'' - a_h\alpha'' + \alpha'\right) + 2\pi\left\{\alpha(0) + \varepsilon'(0) + \left[\frac{1}{2} - a_h\right]\alpha'(0)\right\}\phi(\tau)$$

$$+2\pi\int_0^\pi\phi(\tau - \sigma)\left[\alpha'(0)\varepsilon''(\sigma) + \left[\frac{1}{2} - a_h\right]\alpha''(\sigma)\right]d\sigma,$$

$$C_M(\tau) = \pi\left[\frac{1}{2} + a_h\right] \times \left\{\alpha(0) + \varepsilon'(0) + \left[\frac{1}{2} - a_h\right]\alpha'(0)\right\}\phi(\tau)$$

$$+\pi\left[\frac{1}{2} + a_h\right] \times \int_0^\pi\phi(\tau - \sigma)\left[\alpha'(0) + \varepsilon''(\sigma) + \left[\frac{1}{2} - a_h\right]\alpha''(\sigma)\right]d\sigma$$

$$+\frac{\pi}{2}a_h\left(\varepsilon'' - a_h\alpha''\right) - \left[\frac{1}{2} - a_h\right]\frac{\pi}{2}\alpha' - \frac{\pi}{16}\alpha''$$

(2.2)

The Wagner function $\varphi(\tau)$ is given by:

$$\varphi(\tau) = 1 - \psi_1 e^{-\varepsilon_1\tau} - \psi_2 e^{-\varepsilon_2\tau},$$

(2.3)

Values for the constants are, $\psi_1 = 0.165, \psi_2 = 0.335, \varepsilon_1 = 0.0455$ and $\varepsilon_2 = 0.3$ [24]. Introducing the following new variables $w_1, w_2, w_3,$ and w_4 [23], the original integrodifferential equations for aeroelastic system given by (2.1) are reformulated:

$$\omega_1 = \int_0^\tau e^{-\varepsilon_1(\tau-\sigma)}\alpha(\sigma)d\sigma, \omega_2 = \int_0^\tau e^{-\varepsilon_2(\tau-\sigma)}\alpha(\sigma)d\sigma, \omega_3 = \int_0^\tau e^{-\varepsilon_1(\tau-\sigma)}\varepsilon(\sigma)d\sigma,$$

$$\omega_4 = \int_0^\tau e^{-\varepsilon_2(\tau-\sigma)}\varepsilon(\sigma)d\sigma.$$

(2.4)

Now a set of autonomous differential equations of the form $X' = f(X)$ are obtained as, $X = \{x_1, x_2, x_3, x_4, x_5, x_6, x_7, x_8\} = \{\alpha, \alpha', \varepsilon, \varepsilon', w_1, w_2, w_3, w_4\}$.

Explicitly, the system looks like,

$$x_1' = x_2, x_2' = \frac{(c_0 N - d_0 M)}{(c_1 d_0 - c_0 d_1)}, x_3' = x_4,$$

$$x_4' = \frac{(-c_1 N + d_1 M)}{(c_1 d_0 - c_0 d_1)}, x_5' = x_1 - \varepsilon_1 x_5,$$

$$x_6' = x_1 - \varepsilon_1 x_6, x_7' = x_3 - \varepsilon_1 x_7, x_8' = x_3 - \varepsilon_2 x_8,$$

(2.5)

where

$$M = c_2 x_4 + c_3 x_2 + c_4 x_3 + c_5 x_3^3 + c_6 x_1 + c_7 x_5 + c_8 x_6 + c_9 x_7 + c_{10} x_8 - f(\tau),$$

$$N = d_2 x_2 + d_3 x_1 + d_4 x_1^3 + d_5 x_4 + d_6 x_3 + d_7 x_5 + d_8 x_6 + d_9 x_7 + d_{10} x_8 - g(\tau),$$

(2.6)

The coefficients $c_0 \cdots c_{10}$ and $d_0 \cdots d_{10}$ depend on the system parameters, and their expressions along with $f(\tau)$ and $g(\tau)$ are given in the appendix.

Uncertainty Quantification and Polynomial Chaos Expansion

It is increasingly being felt among the aeroelastic community that aeroelastic analysis should include the effect of parametric uncertainties. This can potentially revolutionize the present design concepts with higher rated performance and can also reshape the certification criteria. Nonlinear dynamical systems are known to be sensitive to physical uncertainties, since they often amplify the random variability with time. Hence, quantifying the effect of system uncertainties on the aeroelastic stability boundary is crucial. Flutter, a dynamic aeroelastic instability involves a Hopf bifurcation where a damped (stable response) oscillation changes to a periodic oscillatory response at a critical wind velocity. In a linear system the post flutter response can grow in an unbounded fashion [22]. System parametric uncertainties can significantly affect the onset and properties of bifurcation points. The importance of stochastic modeling of these uncertainties is that they quantify the effect of the uncertainties on flutter and bifurcation in a probabilistic sense and gives the response statistics in a systematic manner. The original homogeneous polynomial chaos expansion [4] is based on the homogeneous chaos theory of Wiener [6,25]. This is based on a spectral representation of the uncertainty in terms of orthogonal polynomials. In its original form, it employs Hermite polynomials as basis from the generalized Askey scheme and Gaussian random variables. Spectral polynomial chaos-based approaches with other basis functions have also been used in the recent past in various unsteady flow and flow-structure interaction problems of practical interest [8,26,27].

Classical Galerkin Polynomial Chaos Approach

In the classical Galerkin-PCE approach, the polynomial chaos expansion of the system response is substituted into the governing equation and a Galerkin error minimization in the probability space is followed. This results in a set of coupled equations in terms of the polynomial chaos coefficients. The resulting system is deterministic, but it is significantly modified to a higher order and complexity depending on the order of chaos expansion and system nonlinearity. After solving this set of coefficient equations, they are substituted back to get the system response.

As per the Cameron-Martin theorem [28], a random process $X(t,\theta)$ (as function of random event θ) which is second-order stationary can be written as

$$X(t,\theta) = \hat{a}_0\psi_0 + \sum_{i_1=1}^{\infty}\hat{a}_{i_1}\psi_1\left(\xi_{i_1}(\theta)\right) + \sum_{i_1=1}^{\infty}\sum_{i_2=1}^{i_1}\hat{a}_{i_1i_2}\psi_2\left(\xi_{i_1}(\theta),\xi_{i_2}(\theta)\right)$$

$$+\sum_{i_1=1}^{\infty}\sum_{i_2=1}^{i_1}\sum_{i_3=1}^{i_2}\hat{a}_{i_1i_2i_3}\psi_3\left(\xi_{i_1}(\theta),\xi_{i_2}(\theta),\xi_{i_3}(\theta)\right)+...,$$

(3.1)

where $\psi_n\left(\xi_{i_1},\xi_{i_2},...,\xi_{i_n}\right)$ denotes the Hermite polynomial of order n in terms of n-dimensional independent standard Gaussian random variables $\zeta = \left(\zeta_{i_1},\zeta_{i_2},...,\zeta_{i_n}\right)$ with zero mean and unit variance. The above equation is the discrete version of the original Wiener polynomial chaos expansion, and the continuous integrals are replaced by summations. For notational convenience equation (3.1) can be rewritten as

$$X(t,\theta) = \sum_{j=0}^{\infty}a_j(t)\psi_j\left(\hat{\mathbf{i}}(\theta)\right)$$

(3.2)

There is a one-to-one relationship between the ψ's and Φ's and also $\hat{a}_j$'s and a_j's in (3.1) and (3.2). In the original form, chaos expansion uses Hermite polynomials $(\psi_n s)$. The form of the one-dimensional Hermite polynomials is given as follows.

$$\psi_0 = 1,$$
$$\psi_1 = \xi,$$
$$\psi_2 = \xi^2 - 1,$$
$$\psi_3 = \xi^3 - 3\xi,$$
$$\psi_4 = \xi^4 - 6\xi^2 + 3,$$
$$\psi_n = \xi\psi_{n-1} - (n-1)\psi_{n-2}.$$

(3.3)

One can also use orthogonal polynomials from the generalized Askey scheme for some standard nonGaussian input uncertainty distributions such as gamma and beta [8]. For any arbitrary input distribution, a Gram-Schmidt orthogonalization can be employed to generate the orthogonal family of polynomials [29]. Any stochastic process $\alpha(t,\zeta(\theta))$, governed by Gaussian random variables ζ (ζ can always be normalized as a standard Gaussian one), can be approximated by the following truncated series:

$$\alpha(t,\xi(\theta)) = \sum_{j=0}^{\infty}\hat{a}_j(t)\Phi_j\left(\xi(\theta)\right)$$

(3.4)

Note that, here the infinite upper limit of (3.2) is replaced by p, called the order of the expansion. For n number of random variable and polynomial order n_p, p is given by the following [26]:

$$p = \frac{\left(n + n_p\right)!}{n! n_p!} - 1. \tag{3.5}$$

We demonstrate the Galerkin-PCE approach for a generalized dynamical system for a single random variable case, that is, with a random cubic stiffness. Let us write the governing equation with cubic nonlinearity in the following form [27]:

$$\pounds\left[\alpha\left(t, \theta\right)\right] + \beta_\alpha \left[\alpha\left(t, \theta\right)\right]^3 = 0, \tag{3.6}$$

here $\pounds$ is a linear differential operator.

Equation (3.4) is now rewritten for a single random variable ξ as

$$\alpha\left(t, \xi\left(\theta\right)\right) = \sum_{j=0}^{p} \hat{a}_j\left(t\right) \Phi_j\left(\xi\left(\theta\right)\right) \tag{3.7}$$

Here Φ_j's are now Hermite polynomials ψ_j as shown in (3.3). If the cubic spring constant β_α is assumed to be a Gaussian random variable with mean $\overline{\beta}_\alpha$ and standard deviation $\widetilde{\beta}_\alpha$, it can be characterized by

$$\beta_\alpha = \overline{\beta}_\alpha + \xi \widetilde{\beta}_\alpha = \sum_{l=0}^{1} \beta_{\alpha_l} \Phi_l, \tag{3.8}$$

with, $\beta_{\alpha_0} = \overline{\beta}_\alpha$ and $\beta_{\alpha_1} = \widetilde{\beta}_\alpha$.

Substituting the chaos expansion terms, (3.7) and (3.8) in (3.6),

$$\pounds\left[\sum_{j=0}^{p} \hat{a}_j\left(t\right) \Phi_j\right] + \sum_{l=0}^{1} \beta_{\alpha_l} \Phi_l \left[\sum_{j=0}^{p} \hat{a}_j\left(t\right) \Phi_j\right]^3 = 0. \tag{3.9}$$

The cubic nonlinear function can be expressed in the following form:

$$\left[\sum_{j=0}^{p} \hat{a}_j\left(t\right) \Phi_j\right]^3 = \sum_{i=0}^{p} \sum_{m=0}^{p} \sum_{n=0}^{p} \hat{a}_i \hat{a}_m \hat{a}_n \Phi_i \Phi_m \Phi_n. \tag{3.10}$$

Substituting (3.10) into (3.9) and simplifying, we get,

$$\pounds\left[\sum_{j=0}^{p} \hat{a}_j\left(t\right) \Phi_j\right] + \sum_{l=0}^{1} \beta_{\alpha_l} \Phi_l \left[\sum_{i=0}^{p} \sum_{m=0}^{p} \sum_{n=0}^{p} \hat{a}_i \hat{a}_m \hat{a}_n \Phi_i \Phi_m \Phi_n\right] = 0. \tag{3.11}$$

Using Galerkin projection on (3.11) by taking $\langle \, , \Phi_k \rangle$, for $k = 0, 1, \ldots, p$,

$$\pounds[\hat{a}_k\left(t\right)] + \frac{1}{\Phi_k^2} \left[\sum_{l=0}^{1} \sum_{i=0}^{p} \sum_{m=0}^{p} \sum_{n=0}^{p} \beta_{\alpha_l} \hat{a}_i \hat{a}_m \hat{a}_n \langle \Phi_l \Phi_i \Phi_m \Phi_n \Phi_k \rangle\right] = 0. \tag{3.12}$$

The expected value operator $\langle . \rangle$ called the inner product, is defined as,

$$\langle \Phi_1 ... \Phi_k \rangle = \int_{-\infty}^{\infty} \Phi_1 ... \Phi_k \omega (\xi) \, d\xi. \tag{3.13}$$

For Hermite polynomials the weighting function $\omega(\xi)$ is the Gaussian probability density function. For single random variable case it is given as

$$\omega(\xi) = \left(\frac{1}{\sqrt{2\pi}} \right) e^{-\xi^2/2} \tag{3.14}$$

The Hermite polynomials are orthogonal with respect to this weighting function in the Hilbert space. The polynomial chaos forms a complete orthogonal basis in the L_2 space of real-valued functions depending on the Gaussian random variables; hence the inner product of two orthogonal polynomial can be replaced by the identity

$$\langle \Phi_l \Phi_k \rangle = \langle \Phi_l^2 \rangle \delta_k \tag{3.15}$$

δ_k is the Kronecker delta function, given as:

$$\delta_{lk} = \begin{cases} 1 & \text{if } l = k \\ 0 & \text{otherwise} \end{cases} \tag{3.16}$$

The inner product terms in (3.12) $\langle \Phi_l \Phi_i \Phi_m \Phi_n \Phi_k \rangle$ and $\langle \Phi_k^2 \rangle$ can be evaluated analytically before-hand and substituted in the equation. The resulting system becomes a deterministic differential equation in terms of the chaos coefficients. Depending on the type of nonlinearity, the number of random variables and the number of expansion terms, evaluating the inner products could be tedious. In the present study, they are computed by numerical integration by using Gauss-Hermite quadrature rule and verified analytically by using symbolic mathematical solver *Mathematica*. Some typical nonzero inner-products are given in Table 1.

The Galerkin approach is also called the intrusive approach as it modifies the system governing equations in terms of the chaos coefficients. The modification results into a higher order and much more complex form. As a result, this approach may become computationally quite expensive.

Table 1

$\langle \Phi_1 \Phi_1 \rangle = 1$	$\langle \Phi_0 \Phi_0 \Phi_0 \Phi_0 \Phi_0 \rangle = 1$
$\langle \Phi_2 \Phi_2 \rangle = 2$	$\langle \Phi_0 \Phi_0 \Phi_1 \Phi_1 \Phi_2 \rangle = 2$
$\langle \Phi_3 \Phi_3 \rangle = 6$	$\langle \Phi_0 \Phi_1 \Phi_1 \Phi_1 \Phi_2 \rangle = 10$
$\langle \Phi_4 \Phi_4 \rangle = 24$	$\langle \Phi_0 \Phi_1 \Phi_2 \Phi_2 \Phi_3 \rangle = 48$

Nonintrusive Projection Method

A number of nonintrusive variants of PCE have been developed to counter the disadvantages of the classical Galerkin method. Stochastic projection is one of them [4,30]. In the present study, a stochastic projection-based approach is used to evaluate the chaos coefficients. Here, the chaos expansions are not substituted in the governing equations; instead samples of the solutions are used (using low-order deterministic simulations) to evaluate the coefficients directly using a projection formula. As a result, this approach can utilize the existing deterministic code and hence the name nonintrusive. The random process is approximated by a truncated series, as shown in (3.7).

The Hermite polynomials are statistically orthogonal, that is, they satisfy $\langle \Phi_i \Phi_j \rangle = 0$ for $i \neq j$, hence the expansion coefficients can be directly evaluated as

$$\hat{a}_j(t) = \frac{\langle \alpha(t, \xi(\theta)) \Phi_j \rangle}{\langle \Phi_j^2 \rangle}. \tag{3.17}$$

The denominator in (3.17) can be shown to satisfy $\langle \Phi_j^2 \rangle = j!$ for nonnormalized Hermite polynomials [31]. So the key step in projecting $\alpha(t, \xi(\theta))$ along the polynomial chaos basis is the evaluation of $\langle \alpha(t, \xi(\theta)) \Phi_j \rangle$. The inner product is given by the following integral:

$$\langle \alpha(t, \xi(\theta)) \Phi_j \rangle = \int_{-\infty}^{\infty} \alpha(t, \xi(\theta)) \Phi_j \omega(\xi) d\xi. \tag{3.18}$$

A Gauss-Hermite quadrature will be suitable for evaluating the above as the domain is $(-\infty, \infty)$ and the weight function is Gaussian PDF. The quadrature points are the zeros of the Hermite polynomials of chosen order. A number of deterministic runs are performed at the quadrature points which is much lower than the full Monte Carlo simulations. We refer to this step as a pseudo-Monte Carlo simulation approach. In the pseudo-MCS approach the samples of β_α are generated from the corresponding ξ values which are the Gauss-Hermite quadrature points. The realizations of the system response $\alpha(t, \theta)$ are then used to estimate the deterministic coefficients, $\hat{a}_j(t)$'s, in (3.17) using the Gauss-Hermite quadrature rule. It should also be noted that for each evaluation of the inner product integral a convergence study is done by gradually increasing the number of quadrature points.

Results and Discussions

The main focus of the present study is quantifying the effect of system uncertainties on the bifurcation behavior and the flutter boundary of the nonlinear aeroelastic system. A fourth order variable step Runge-Kutta method is employed for the time integration. The main bifurcation parameter in a flutter system is the nondimensional wind velocity, also called the reduced velocity. In a linear aeroelastic system, the response changes to an exponentially growing solution from a stable damped oscillation at some critical wind velocity, known as the linear flutter speed. Nonlinear aeroelastic system can stabilize the response at the post-flutter regime to limit cycle oscillations [23] and the critical point becomes a

Hopf bifurcation point. With a cubic nonlinearity, both supercritical and subcritical Hopf bifurcations are possible [32]. The latter case is observed for a softening cubic spring. Here in the stochastic analysis, we focus on the supercritical case. A deterministic bifurcation diagram with the following parameter values [23] is shown in Fig. 2: $\mu = 100$, $\varpi = 0.2$, $a_h = -0.5$, $x_\alpha = 0.25$, $\zeta_\alpha = 0$, $\zeta_\varepsilon = 0$, $r_\alpha = 0.5$, $\beta_\alpha = 3$, $\beta_\varepsilon = 0$. The variation of the limit cycle oscillation (LCO) amplitude is plotted with reduced velocities. Bifurcation occurs at the corresponding linear flutter speed of 6.285, and the observation match well with the earlier results [33]. At the post flutter velocities, limit cycle oscillations are observed and the amplitude of the LCOs increase as the reduced velocity increases.

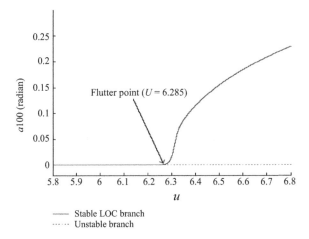

Figure 2 Deterministic flutter and bifurcation diagram with cubic nonlinearity (supercritical Hopf bifurcation).

We now consider random variations in the system parameters and investigate the influence on the overall dynamics. We consider only single uncertain parametric variation in this paper, that is, a single random variable model. First, the hardening cubic spring constant is considered to be a Gaussian random variable with mean $\overline{\beta}_\alpha = 3$ and standard deviation $\hat{\beta}_\alpha = 3$. All other parameters are assumed to be deterministic. Figure 3 shows bifurcation behavior with the cubic stiffness as random, it now has a range of possible LCO amplitudes for each reduced velocity and the onset of flutter is unaffected. The standard deviation, that is, the amplitude variation range increases as reduced speed increases.

A Galerkin PCE approach is used to quantify the propagation of this uncertainty on the response. The Galerkin approach modifies the 8th-order flutter system to an $8 \times (p+1)$ order system. It also involves calculating the complex fifth-order inner product terms as shown earlier.

As a result, the solution process is computationally intensive for the nonlinear system in question. After solving for the chaos coefficients, in the post processing stage, the coefficients are substituted back to the expansion form to get the stochastic response. Probability density functions (PDFs) and other required statistics can then be readily obtained. The time histories of the first few random modes in pitch are plotted in

Fig. 4. The zeroth-order mode is the mean; one can also see that the contribution of higher-order random modes is gradually diminishing. A representative PDF is shown in Fig. 5 for increasing order of chaos expansion terms. PDFs are calculated at time $t = 7000$ at which the solutions are well past their transients and stationary. The reduced speed considered here is $U = 6.42$, close to the deterministic bifurcation point. The figure also presents results from a standard MCS with 12000 samples as a comparison reference. One can see how increasing the order of expansion the CPU time for the solution is getting magnified. Results are presented up to the 12th order of expansion at which the PCE results match well with that of MCS. However, the simulation time also approaches to that of the reference MCS. While calculating the CPU time for the Galerkin-PCE approach the inner products computation and post processing of results are not taken into account.

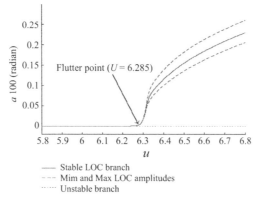

Figure 3 Uncertain nonlinear stiffness: stochastic bifurcation diagram.

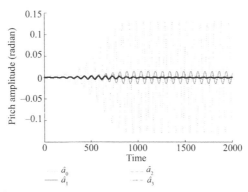

Figure 4 Galerkin-PCE: behavior of the first few random modes.

Now the nonintrusive projection approach is followed using a Gauss-Hermite quadrature. Galerkin-PCE and nonintrusive results are compared in terms of their accuracy and simulation time in Fig. 6. A good match with MCS is seen for the 12th order of expansion. Once again a standard MCS with 12000 samples is used as a reference solution (in other cases too we have used a standard 12000 samples MCS as reference).

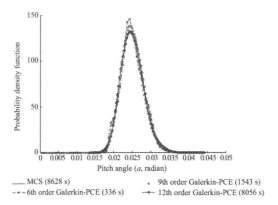

Figure 5 Galerkin-PCE: PDF comparisons for increasing order of chaos expansions, at t=7000 with U=6.42.

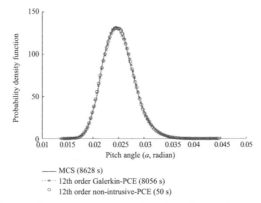

Figure 6 Galerkin-PCE: PDF comparisons for intrusive and nonintrusive PCE, at t= 7000 with U= 6.42.

However, nonintrusive approach is seen to be much faster than Galerkin-PCE for the same level of accuracy as indicated in Fig. 6. With this, the computational disadvantage of the conventional Galerkin based PCE for nonlinear systems is demonstrated. Henceforth, this approach will not be used for further simulations in this paper.

Now the nonintrusive projection approach is followed using a Gauss-Hermite quadrature. Galerkin-PCE and nonintrusive results are compared in terms of their accuracy and simulation time in Fig. 6. A good match with MCS is seen for the 12th order of expansion. Once again a standard MCS with 12000 samples is used as a reference solution (in other cases too we have used a standard 12000 samples MCS as reference). However, nonintrusive approach is seen to be much faster than Galerkin-PCE for the same level of accuracy as indicated in Fig. 6. With this, the computational disadvantage of the conventional Galerkin based PCE for nonlinear systems is demonstrated. Henceforth, this approach will not be used for further simulations in this paper.

The response realization time histories for a few samples of random variable ξ are plotted in Fig. 7. The response time histories show difference in amplitude but not in phase. A typical realization time history obtained with the 12th order PCE along with its deterministic counterpart is compared in Fig. 8. The match is perfect even at long time.

Amplitude response PDFs as a function of reduced velocities (bifurcation parameter) are shown in Fig. 9. They represent single peak monotonic behavior as all the realizations give finite amplitude LCOs. Effectively, the PDFs are not undergoing any qualitative change or bifurcations. Close to $U=6.4$ the PDF looks sharper and narrower as most realizations are going towards the same limit cycle amplitude. As the speed increases, the PDF is broader and less sharp, indicating that the realization amplitudes are spread over a wider band of amplitudes.

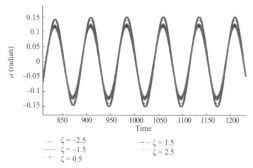

Figure 7 Uncertain nonlinear stiffness: five different realizations time histories at $U=6.42$.

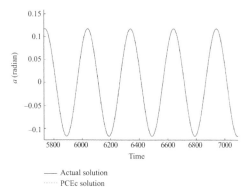

Figure 8 Uncertain nonlinear stiffness: comparison of a typical time history with 12th order PCE and MCS for $\xi = 1.5$ and $U= 6.42$.

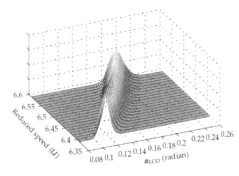

Figure 9 Uncertain nonlinear stiffness: amplitude response PDF as a function of reduced speed.

Next, we consider the viscous damping ratio in pitch (ξ_α) to be uncertain (in the earlier part damping was put to be zero) and all the other parameters deterministic. This case is potentially more interesting than the earlier one. The damping ratio is assumed to be a Gaussian random variable with mean $\xi_\alpha = 0.1$ and standard deviation $\xi_\alpha = 0.01$. Figure 10 shows the bifurcation behavior with random damping ratio. The firm line gives the deterministic bifurcation behavior. The other bifurcation branches are for the two different extreme realizations of the random damping. Thus they represent the boundaries of the possible random variations of the bifurcation behavior (stochastic). The major difference between the uncertain damping and the earlier considered uncertain stiffness is that, variation in damping can show phase shifting behavior in the response realizations. This is presented in Fig. 11 where five different realizations are shifted in phase from each other.

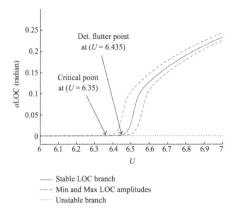

Figure 10 Uncertain viscous damping: stochastic bifurcation diagram.

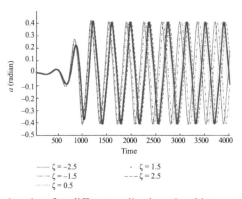

Figure 11 Uncertain viscous damping: five different realizations time history at $U= 6.52$.

This behavior becomes more pronounced as time increases. As a result, response PDFs can now show bimodal behavior especially at large times. A few representative PDFs are plotted now at different reduced velocities. Figure 12 shows the response PDF at $U=6.52$ and time=1400. Though the time level is past the transients it is not large and the phase shift is not yet very pronounced. The response PDF shows a single peak pattern. A reasonably good match with MCS is obtained within the 12th order of chaos expansion. However, at higher time levels as the phase shifting becomes stronger, the PDFs start to look distorted from their single peak behavior and goes towards a double peak pattern. Figure 13 shows the PDF at $U=6.52$ and time=5000 and 7800. In the first case, a double-peak bimodal PDF is just emerging as shown in Fig. 13(a). In this case a 12th order expansion is not sufficient to capture the response accurately; a 15th order expansion gives better accuracy. At higher time=7800, the response PDF is more towards a two-peak bimodal shape as is seen in Fig. 13(b). However, even a 15th order chaos expansion does not give the required accuracy. If one considers a different reduced velocity, the bimodal behavior can appear at some different time levels. The important observations from these figures are two-fold. First is the gradual double peak behavior with increasing time. The second one is the

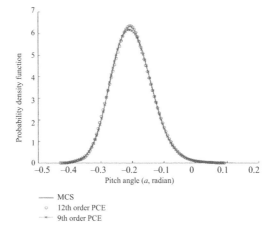

Figure 12 Uncertain viscous damping: comparison of the PDFs with increasing order of PCE at nondimensional time=1400 at $U=6.52$.

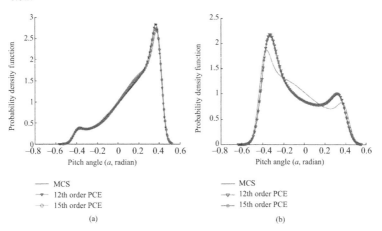

Figure 13 Uncertain viscous damping: comparison of the PDFs with increasing order of PCE at $U=6.52$ with nondimensional time (a) $t=5000$, (b) $t=7800$.

apparent mismatch between the MCS and PCE results which seems to be increasing again with time. The reason for the first is nothing but the increasing phase shifting between the realizations time histories. However, for the second, the reason for the mismatch is the long time degeneracy which is shown in Fig. 14. This mismatch can be improved by using higher order chaos expansions.

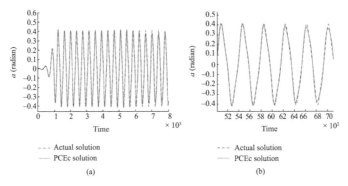

(a)   (b)

Figure 14 Uncertain viscous damping: a typical time history with (a) The 15th order PCE and MCS at $\xi = 2.3$ and reduced speed $U=6.52$. (b) A close-up of (a).

A typical realization time history with PCE along with its deterministic counterpart are presented in Fig. 14. One can clearly see a degeneracy in the time history which starts around time levels close to 6000. PCE can show such type of degenerate behavior in capturing LCO response [17], especially at large times. As a countermeasure, one can increase the order of the chaos expansion. However, this can only push the degeneracy to a later time but can not solve it entirely. Non polynomial based chaos approaches have been attempted in the recent past towards this end [17]. An unsteady adaptive stochastic finite elements method, developed by Witteveen and Bijl [34,36] has also been used successfully. This approach is based on time-independent parametrization. This achieves a constant accuracy in time with a constant number of samples. In this method interpolation of oscillatory samples is based on constant phase instead of a constant time.

The amplitude response PDFs for the uncertain damping case is shown in Fig. 15 for different reduced velocities. Here the LCO amplitudes are captured after the initial

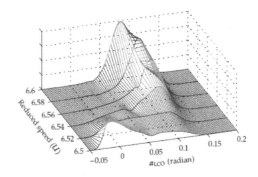

Figure 15 Uncertain viscous damping: amplitude response PDF as a function of reduced velocities.

transients have died down but before the time degeneracy has started. A nonmonotonic behavior is clearly indicated; some realizations are going to damped oscillation and others give LCO amplitudes scattered within the domain boundary. At $U=6.5$, the double-peak behavior of the PDF indicates the two different LCO amplitudes around which most of the realizations are concentrated. Towards $U=6.6$, all realizations give finite amplitude LCO, thus essentially they are of the same type. The PDF shows a single-peak monotonous behavior. Therefore, the PDFs of the response amplitude have clearly gone through a qualitative change here, in other words, a bifurcation.

For the uncertain damping case, we also see that the critical reduced velocity at which flutter can occur, has come down from its corresponding deterministic value. This value can be read off the bifurcation plot (Fig. 10) as $U=6.35$. This is the lowest extrema of the critical points. The cumulative distribution function (CDF) and the PDF of the critical points are shown in Fig. 16. The CDF can directly give the probability of flutter (in other words, probability of failure) at any given reduced velocity.

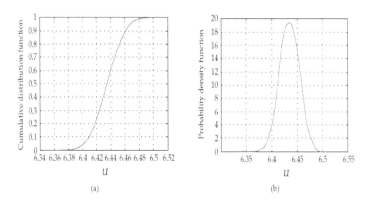

Figure 16 Uncertain viscous damping: (a) CDF, (b) PDF of the critical flutter point.

Conclusions

The bifurcation behavior of a nonlinear pitch-plunge flutter problem with uncertain system parameters has been studied. The problem is a simple model problem to understand the mechanism of nonlinear flutter in a stochastic framework. The parameters which have been assumed to be random could attribute their uncertainties to laboratory testing conditions.

Moreover, a cubic nonlinear stiffness is used for various sources of analytic nonlinearities; they often represent different control mechanisms and could face modeling uncertainties. The classical Galerkin Polynomial Chaos method and the nonintrusive Projection method are applied to capture the propagation of uncertainty through the nonlinear aeroelastic system. The focus of this work is to investigate the performance of these techniques and to see how the aeroelastic stability characteristics are altered due to the random effects. The Monte Carlo solution is used as reference solution. The computational cost of the Galerkin Polynomial Chaos method is seen to be very high and subsequently only the Projection method based on Gauss-Hermite quadrature is used for the analysis. The effect of uncertain cubic structural nonlinearity and viscous damping parameter are investigated.

Uncertainty in the cubic stiffness does not alter the bifurcation (flutter) point, it only affects the amplitudes of the periodic response in the post flutter stage. The PDF behavior also does not show any qualitative changes. On the other hand, uncertainty in damping affects the bifurcation point. It can lower the onset of flutter; the PDF of the response amplitude also undergoes a qualitative change. In other words, a bifurcation of the response PDF takes place. The results highlight the risk induced by parametric uncertainty and importance of uncertainty quantification in nonlinear aeroelastic systems.

The uncertain damping case by polynomial chaos suffer from long time degeneracy, as is also discussed in the literature. The degeneracy can be controlled by using higher order chaos expansions, though this cannot be a permanent solution. For the uncertain nonlinear stiffness, the problem of time degeneracy is not encountered.

Appendix

The coefficients introduced in Section 2 are used from [23] and are reproduced here for the sake of completeness:

$$c_0 = 1 + \frac{1}{\mu}, \quad c_1 = x_\alpha - \frac{a_h}{\mu}, \quad c_2 = 2\xi_\varepsilon \frac{\varpi}{U} + \frac{2}{\mu}(1 - \psi_1 - \psi_2),$$

$$c_3 = \frac{1 + 2(1/2 - a_h)(1 - \psi_1 - \psi_2)}{\mu}, \quad c_4 = \left(\frac{\varpi}{U}\right)^2 + \frac{2}{\mu}(\psi_1\varepsilon_1 + \psi_2\varepsilon_2),$$

$$c_5 = \beta_\varepsilon\left(\frac{\varpi}{U}\right)^2, \quad c_6 = \frac{2}{\mu}\left[(1 - \psi_1 - \psi_2) + (1/2 - a_h)(\psi_1\varepsilon_1 + \psi_2\varepsilon_2)\right],$$

$$c_7 = \frac{2}{\mu}\psi_1\varepsilon_1\left[1 - (1/2 - a_h)\varepsilon_1\right], \quad c_8 = \frac{2}{\mu}\psi_2\varepsilon_2\left[1 - (1/2 - a_h)\varepsilon_2\right],$$

$$c_9 = -\frac{2}{\mu}\psi_1\varepsilon_1^2, \quad c_{10} = -\frac{2}{\mu}\psi_2\varepsilon_2^2,$$

$$d_0 = \frac{x_\alpha}{r_\alpha^2} - \frac{a_h}{\mu r_\alpha^2}, \quad d_1 = 1 + \frac{1 + 8a_h^2}{8\mu r_\alpha^2},$$

$$d_2 = 2\frac{\xi_\alpha}{U} + \frac{(1 - 2a_h)}{2\mu r_\alpha^2} - \frac{(1 + 2a_h)(1 - 2a_h)(1 - \psi_1 - \psi_2)}{2\mu r_\alpha^2},$$

$$d_3 = \frac{1}{U^2} - \frac{(1 + 2a_h)(1 - \psi_1 - \psi_2)}{\mu r_\alpha^2} - \frac{(1 + 2a_h)(1 - 2a_h)(\psi_1\varepsilon_1 + \psi_2\varepsilon_2)}{2\mu r_\alpha^2},$$

$$d_4 = \frac{\beta_\alpha}{U^2}, d_5 = -\frac{(1 + 2a_h)(1 - \psi_1 - \psi_2)}{\mu r_\alpha^2}, d_6 = -\frac{(1 + 2a_h)(\psi_1\varepsilon_1 + \psi_2\varepsilon_2)}{\mu r_\alpha^2},$$

$$d_7 = -\frac{\left(1+2a_h\right)\left(\psi_1\varepsilon_1\left[1-\left(1-2a_h\right)\varepsilon_1\right]\right)}{\mu r_\alpha^2}, \quad d_8 = -\frac{\left(1+2a_h\right)\left(\psi_2\varepsilon_2\left[1-\left(1-2a_h\right)\varepsilon_2\right]\right)}{\mu r_\alpha^2}$$

$$d_9 = \frac{\left(1+2a_h\right)\psi_1\varepsilon_1^2}{\mu r_\alpha^2}, \quad d_{10} = \frac{\left(1+2a_h\right)\psi_2\varepsilon_2^2}{\mu r_\alpha^2}.$$

$$f\left(\tau\right) = \frac{2}{\mu}\left(\left(1/2-a_h\right)\alpha\left(0\right)+\varepsilon\left(0\right)\right)\left(\psi_1\varepsilon_1 e^{-\varepsilon_1\tau}+\psi_2\varepsilon_2 e^{-\varepsilon_2\tau}\right), \quad g\left(\tau\right) = -\frac{\left(1+2a_h\right)f\left(\tau\right)}{2r_\alpha^2}. \quad \text{(A.1)}$$

References

1. M. Kleiber and T.D. Hien, *The Stochastic Finite Element Method: Basic Perturbation Technique and Computer Implementation*, John Wiley & Sons, Chichester, UK, 1992.
2. T.D. Hien and M. Kleiber, "Computational aspects in structural design sensitivity analysis for statics and dynamics," *Computers and Structures*, vol. 33, no. 4, pp. 939–950, 1989.
3. R. Ghosh, S. Chakraborty, and B. Bhattacharyya, "Stochastic sensitivity analysis of structures using first-order perturbation," *Meccanica*, vol. 36, no. 3, pp. 291–296, 2001.
4. R.G. Ghanem and P.D. Spanos, *Stochastic Finite Elements: A Spectral Approach*, Springer, New York, NY, USA, 1991.
5. R.G. Hills and T. G. Trucano, "Statistical validation of engineering and scientific models: background," May 1999, http://citeseerx.ist.psu.edu/viewdoc/summary?doi=10.1.1.40.1901.
6. N. Wiener, "The homogeneous chaos," *American Journal of Mathematics*, vol. 60, no. 4, pp. 897–936, 1938.
7. R. Ghanem, "Stochastic finite elements with multiple random non-Gaussian properties," *Journal of Engineering Mechanics*, vol. 125, no. 1, pp. 26–40, 1999.
8. D. Xiu and G. E. Karniadakis, "Modeling uncertainty in flow simulations via generalized polynomial chaos," *Journal of Computational Physics*, vol. 187, no. 1, pp. 137–167, 2003.
9. D. Lucor and G. E. Karniadakis, "Predictability and uncertainty in flow-structure interactions," *European Journal of Mechanics B*, vol. 23, no. 1, pp. 41–49, 2004.
10. G. Lin, C.-H. Su, and G. E. Karniadakis, "Predicting shock dynamics in the presence of uncertainties," *Journal of Computational Physics*, vol. 217, no. 1, pp. 260–276, 2006.
11. I. Babuška, F. Nobile, and R. Tempone, "A stochastic collocation method for elliptic partial differential equations with random input data," *SIAM Journal on Numerical Analysis*, vol. 45, no. 3, pp. 1005–1034, 2007.
12. G.J.A. Loeven, J.A.S. Witteveen, and H. Bijl, "Probabilistic collocation: an efficient non-intrusive approach for arbitrarily distributed parametric uncertainties," in *Proceedings of the 45th AIAA Aerospace Sciences Meeting*, vol. 6, pp. 3845–3858, Reno, Nev, USA, January 2007.
13. S. Hosder, R. W. Walters, and R. Perez, "A non-intrusive polynomial chaos method for uncertainty propagation in CFD simulations," in *Proceedings of the 44th AIAA Aerospace Sciences Meeting*, vol. 14, pp. 10649–10667, Reno, Nev, USA, January 2006.
14. R.W. Walters, "Towards stochastic fluid mechanics via polynomial chaos," in *Proceedings of the 41st AIAA Aerospace Sciences Meeting and Exhibit*, Reno, Nev, USA, 2003.
15. S. Hosder, R.W. Walters, and M. Balch, "Efficient uncertainty quantification applied to the aeroelastic analysis of a transonic wing," in *Proceedings of the 46th AIAA Aerospace Sciences Meeting and Exhibit*, Reno, Nev, USA, 2008.
16. M.T. Reagan, H.N. Najm, R.G. Ghanem, and O.M. Knio, "Uncertainty quantification in reacting-flow simulations through non-intrusive spectral projection," *Combustion and Flame*, vol. 132, no. 3, pp. 545–555, 2003.

17. C.L. Pettit and P.S. Beran, "Polynomial chaos expansion applied to airfoil limit cycle oscillations," in *Proceedings of the 45th AIAA/ASME/ASCE/AHS/ASC Structures, Structural Dynamics and Materials Conference*, vol. 3, pp. 1975–1985, Palm Springs, Calif, USA, April 2004.

18. C.L. Pettit and P.S. Beran, "Spectral and multiresolution Wiener expansions of oscillatory stochastic processes," *Journal of Sound and Vibration*, vol. 294, no. 4, pp. 752–779, 2006.

19. B. Ganapathysubramanian and N. Zabaras, "Sparse grid collocation schemes for stochastic natural convection problems," *Journal of Computational Physics*, vol. 225, no. 1, pp. 652–685, 2007.

20. T. Gerstner and M. Griebel, "Numerical integration using sparse grids," *Numerical Algorithms*, vol. 18, no. 3-4, pp. 209–232, 1998.

21. D. Xiu and J. S. Hesthaven, "High-order collocation methods for differential equations with random inputs," *SIAM Journal on Scientific Computing*, vol. 27, no. 3, pp. 1118–1139, 2005.

22. Y.C. Fung, *An Introduction to the Theory of Aeroelasticity*, John Wiley & Sons, New York, NY, USA, 1955.

23. B.H. K. Lee, L. Jiang, and Y.S. Wong, "Flutter of an airfoil with a cubic nonlinear restoring force," *AIAA-98-1725*, pp. 237–257, 1998.

24. R. T. Jones, "The unsteady lift of a wing of finite aspect ratio," NACA Report 681, NACA, 1940.

25. N. Wiener, *Nonlinear Problems in Random Theory*, Technology Press Research Monographs, MIT Press and John Wiley & Sons, New York, NY, USA, 1958.

26. D. Xiu and G.E. Karniadakis, "The Wiener-Askey polynomial chaos for stochastic differential equations," *SIAM Journal on Scientific Computing*, vol. 24, no. 2, pp. 619–644, 2002.

27. O.P. Le Maître, O.M. Knio, H.N. Najm, and R. G. Ghanem, "A stochastic projection method for fluid flow. I. Basic formulation," *Journal of Computational Physics*, vol. 173, no. 2, pp. 481–511, 2001.

28. R.H. Cameron and W. T. Martin, "The orthogonal development of non-linear functionals in series of Fourier-Hermite functionals," *Annals of Mathematics*, vol. 48, pp. 385–392, 1947.

29. J.A.S. Witteveen, S. Sarkar, and H. Bijl, "Modeling physical uncertainties in dynamic stall induced fluid-structure interaction of turbine blades using arbitrary polynomial chaos," *Computers and Structures*, vol. 85, no. 11–14, pp. 866–878, 2007.

30. D.R. Millman, P. I. King, and P. S. Beran, "A stochastic approach for predicting bifurcation of a pitch and plunge airfoil," in *Proceedings of the 21st AIAA Applied Aerodynamics Conference*, Orlando, Fla, USA, 2003, Paper no. AIAA-2003-3515.

31. C.L. Pettit and P.S. Beran, "Spectral and multiresolution Wiener expansions of oscillatory stochastic processes," *Journal of Sound and Vibration*, vol. 294, no. 4, pp. 752–779, 2006.

32. D.S. Woolston, H.L. Runyan, and R. E. Andrews, "Some effects of system nonlinearities in the problem of aircraft flutter," Tech. Rep. TN 3539, NACA, 1955.

33. H. Alighanbari and S.J. Price, "The post-hopf-bifurcation response of an airfoil in incompressible two-dimensional flow," *Nonlinear Dynamics*, vol. 10, no. 4, pp. 381–400, 1996.

34. J.A.S. Witteveen and H. Bijl, "Higher period stochastic bifurcation of nonlinear airfoil fluid-structure interaction," *Mathematical Problems in Engineering*, vol. 2009, Article ID 394387, 26 pages, 2009.

35. J.A.S. Witteveen and H. Bijl, "An alternative unsteady adaptive stochastic finite elements formulation based on interpolation at constant phase," *Computer Methods in Applied Mechanics and Engineering*, vol. 198, no. 3-4, pp. 578–591, 2008.

36. J.A.S. Witteveen and H. Bijl, "Effect of randomness on multi-frequency aeroelastic responses resolved by unsteady adaptive stochastic finite elements," *Journal of Computational Physics*, vol. 228, no. 18, pp. 7025–7045, 2009.

Chaos Synchronization Criteria and Costs of Sinusoidally Coupled Horizontal Platform Systems[†]

Jianping Cai,[1,a] *Xiaofeng Wu*[2] and *Shuhui Chen*[1,b]

[1]Department of Applied Mechanics and Engineering, Zhongshan University, Guangzhou 510275, China.
[a]Email: mathcai@hotmail.com.
[b]Email: stscsh@mail.sysu.edu.cn.
[2]Center for Control and Optimization, South China University of Technology, Guangzhou 510640, China.
Email: wuxiaof@21cn.com.

ABSTRACT

Some algebraic sufficient criteria for synchronizing two horizontal platform systems coupled by sinusoidal state error feedback control are derived by the Lyapunov stability theorem for linear time-varying system and the Sylvester's criterion. The state variables are restricted in a sub-region in order to obtain easily verified criteria. The validity of these algebraic criteria is illustrated with some numerical examples. A new concept, synchronization cost, is introduced based on a measure of the magnitude of the feedback control. The minimal synchronization cost as well as optimal coupling strength is calculated numerically. The results are meaningful in engineering application.

Keywords: chaos synchronization, algebraic sufficient criterion, feedback control, synchronization cost.

Introduction

Horizontal platform devices are widely used in offshore engineering and earthquake engineering. Mechanical model for a horizontal platform system with an accelerometer is depicted in Fig. 1. The platform can freely rotate about the horizontal axis, which penetrates its mass center. When the platform deviates from horizon, the accelerometer will give an output signal to the torque generator, which generates a torque to inverse the rotation of the platform about rotational axis. The equation governing this system is

[†]Reused with permission from: Jianping Cai, Xiaofeng Wu, and Shuhui Chen, "Chaos Synchronization Criteria and Costs of Sinusoidally Coupled Horizontal Platform Systems," *Mathematical Problems in Engineering*, vol. 2007, Article ID 86852, 10 pages, 2007. doi:10.1155/2007/86852.

$$A\ddot{y} + D\dot{y} + rg \sin y - \frac{3g}{R}(B-C)\cos y \sin y = F \cos \omega t,$$

where y denotes the rotation of the platform relative to the earth, A, B and C are respectively the inertia moment of the platform for axis 1, 2, and 3, D is the damping coefficient, r the proportional constant of the accelerometer, g the acceleration constant of gravity, R the radius of the earth and $F \cos \omega t$ harmonic torque. More details about this model can be found in Refs.[1,2]. Such horizontal platform systems can reduce the swing of moving devices and keep the system close to horizontal position. They are used in modelling offshore platforms and earthquake-proof devices.

As shown in Fig. 2, the horizontal platform system has a double scroll attractor when its parameter values are $A = 0.3$, $B = 0.5$, $C = 0.2$, $D = 0.4$, $r = 0.1155963$, $R = 6378000$, $g = 9.8$, $F = 3.4$ and $\omega = 1.8$. It was numerically verified in Ref. [1] that two identical horizontal platform systems coupled by a linear, sinusoidal or exponential state error feedback control can achieve chaos synchronization. Analytic criteria for chaos synchronization have the advantage over numerical ones because they can reveal the relationship between the criteria and system parameters, and then they are convenient for design and analysis of the coupling controller [3–11]. Algebraic sufficient criteria for synchronizing the driving-response horizontal platform systems via linear state error feedback control were obtained in Ref. [12]. In this paper, some sufficient criteria for synchronizing the horizontal platform systems coupled

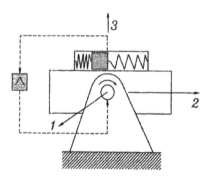

Figure 1 Mechanical model for a horizontal platform system with an accelerometer

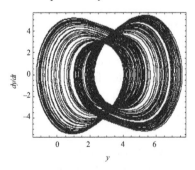

Figure 2 Double scroll attractor of the horizontal platform system.

by sinusoidal state error feedback control are further derived by the Lyapunov stability theory and the Sylvester's criterion. In order to obtain easily verified algebraic criteria, the state variables are restricted in a sub-region, which is different from Ref. [12]. Furthermore, a new concept of synchronization cost is introduced based on a measure of the magnitude of the feedback control. The minimal synchronization cost as well as optimal coupling strength is calculated numerically. Minimal cost means the lowest energy input, which is meaningful in engineering application.

Algebraic Sufficient Synchronization Criteria

Let $x_1 = y, x_2 = \dot{y}$ and $x = (x_1, x_2)^T$, and rewrite the governing equation in form of vector

$$\dot{x} = Mx + f(x) + m(t)$$

with

$$M = \begin{pmatrix} 0 & 1 \\ 0 & -a \end{pmatrix}, \quad f(x) = \begin{pmatrix} 0 \\ -b\sin x_1 + c\cos x_1 \sin x_1 \end{pmatrix}, \quad m(t) = \begin{pmatrix} 0 \\ h\cos\omega t \end{pmatrix},$$

$$a = \frac{D}{A} > 0, \quad b = \frac{rg}{A} > 0, \quad c = \frac{3g}{RA}(B - C), \quad h = \frac{F}{A} > 0.$$

A driving-response synchronization scheme for two identical platform systems coupled by a sinusoidal state error feedback controller is constructed as follows:

Driving system:

$$\dot{x} = Mx + f(x) + m(t) \tag{1}$$

Response system:

$$\dot{y} = My + f(y) + m(t) + u(t) \tag{2}$$

Controller:

$$u(t) = (k_1 \sin(x_1 - y_1), k_2 \sin(x_2 - y_2))^T \tag{3}$$

where $y = (y_1, y_2)^T$, T means transpose, k_1 and k_2 are constant coupling coefficients. Defining an error variable $e = x - y$, or $(e_1, e_2) = (x_1 - y_1, x_2 - y_2)$, we can obtain an error dynamical system

$$\dot{e} = M(x - y) - u(t) + f(x) - f(y) = (M - K(t) + N(t)\, e \tag{4}$$

with $K(t) = \begin{pmatrix} k_1 s_1(t) & 0 \\ 0 & k_2 s_2(t) \end{pmatrix}, s_1(t) = \frac{\sin(x_1 - y_1)}{x_1 - y_1}, s_2(t) = \frac{\sin(x_2 - y_2)}{x_2 - y_2}, N(t) = \begin{pmatrix} 0 & 0 \\ q(t) & 0 \end{pmatrix},$

and $q(t) = \dfrac{-b(\sin x_1 - \sin y_1) + c(\sin x_1 \cos x_1 - \sin y_1 \cos y_1)}{x_1 - y_1}$.

Our object is to select suitable coupling coefficients k_1 and k_2 such that $x(t)$ and $y(t)$ satisfy

$$\lim_{t \to +\infty} \|x(t) - y(t)\| = \lim_{t \to +\infty} \|e(t)\| = 0 \tag{5}$$

where $\|x(t) - y(t)\| = \sqrt{(x_1 - y_1)^2 + (x_2 - y_2)^2}$ denotes the Euclidean norm of vector. By the theory of stability, chaos synchronization of systems (1) and (2) in the sense of (5) is equivalent to asymptotic stability of the error system (4) at the origin $e = 0$. Taking a quadratic Lyapunov function $V(e) = e^T P e$ with P a symmetric positive definite constant matrix, then the derivative of $V(e)$ with respect to time along the trajectory of system (4) is

$$\dot{V}(e) = \dot{e}^T P e + e^T P \dot{e} = e^T [P(M - K(t) + N(t)) + (M - K(t) + N(t))^T P] e.$$

By the Lyapunov stability theorem for linear time-varying system (Theorem 4.1 in [13]), a sufficient condition that the error system (4) is asymptotically stable at the origin is that the following matrix

$$Q(t) = P(M - K(t) + N(t)) + (M - K(t) + N(t))^T P$$
is negative definite, denoting it by

$$Q(t) < 0 \tag{6}$$

For simplicity, we choose $P = diag\{p_1, p_2\}$ with $p_1 > 0$ and $p_2 > 0$, then

$$Q(t) = \begin{pmatrix} -2p_1 k_1 s_1(t) & p_1 + p_2 q(t) \\ p_1 + p_2 q(t) & -2p_2(k_2 s_2(t) + a) \end{pmatrix}.$$

By the Sylvester's criterion, $Q(t) < 0$ is equivalent to the following inequalities:

$$p_1 k_1 s_1(t) > 0, \quad 4 p_1 p_2 k_1 s_1(t)(k_2 s_2(t) + a) > (p_1 + p_2 q(t))^2 \tag{7}$$

Note that $s_1(t) > 0$ and $s_2(t) > 0$ if (x_1, x_2) and (y_1, y_2) are limited in the region $G = \{|x_1 - y_1| < \pi, |x_2 - y_2| < \pi\}$. So we conclude that under condition (7) the error system (4) is locally asymptotically stable at the origin in the region G. In order to get an easily verified algebraic condition, we further restrict the variables in the sub-region $G_0 = \{|x_1 - y_1| \le \frac{3\pi}{4}, |x_2 - y_2| \le \frac{3\pi}{4}\}$, then we have $\frac{2\sqrt{2}}{3\pi} \le s_1(t) \le 1$ and $\frac{2\sqrt{2}}{3\pi} \le s_2(t) \le 1$. Now a simple algebraic sufficient criterion for synchronizing the systems (1) and (2) can be obtained from (7) as

$$k_1 > 0, k_2 > \frac{9\pi^2 (p_1 + p_2(b + |c|))^2}{32 p_1 p_2 k_1} - a \tag{8}$$

in which the inequality $|q(t)| < b + |c|$ has been used as in Ref. [12].

The synchronization criterion obtained here only renders a sufficient but not necessary condition. It is natural to expect that a sharp criterion can provide more choices of the coupling coefficients.

To this end, we can minimize the lower bound of k_2 in inequality (8) by choosing $p = diag\{(b+|c|)p_2, p_2\}$ and obtain a sharper criterion

$$k_1 > 0, k_2 > \frac{9\pi^2(b+|c|)}{8k_1} - a \tag{9}$$

Similarly, if the controller is chosen as $u(t) = (k_1 \sin(x_1 - y_1), 0)^T$, the sufficient criteria associated with inequalities (8) and (9) become respectively

$$k_1 > \frac{3\pi(p_1 + p_2(b+|c|))^2}{8\sqrt{2}p_1 p_2 a} \tag{10}$$

$$k_1 > \frac{3\pi(b+|c|)}{2\sqrt{2}a} \tag{11}$$

The theoretical sufficient criteria are illustrated with the following examples. If we choose $p_2 = 1$ and $p_1 = (b+|c|)p_2 = 3.776615$, it is easy to verify that the coupling coefficients $k_1 = 5.6$ and $k_2 = 6.2$ satisfy inequalities (9). For this choice, the two coupled horizontal platform systems (1) and (2) can be asymptotically synchronized. The parameter values are chosen such that the system is in a state of chaos: $A = 0.3$, $B = 0.5$, $C = 0.2$, $D = 0.4$, $r = 0.1155963$, $R = 6378000$, $g = 9.8$, $F = 3.4$ and $\omega = 1.8$.

The result is shown in Fig. 3 with initial values $(x_1(0), x_2(0)) = (1,1)$ and $(y_1(0), y_2(0)) = (-1, -1)$, which are chosen arbitrarily in the region G_0.

In this paper, software *Mathematica* is applied to implement relative calculations and plots.

For the controller $u(t) = (k_1 \sin(x_1 - y_1), 0)^T$, inequality (11) should be $k_1 > 9.43706$. Chaos synchronization for $k_1 = 9.5$ is illustrated in Fig. 4, where p_1, p_2 and other parameter values are the same as above.

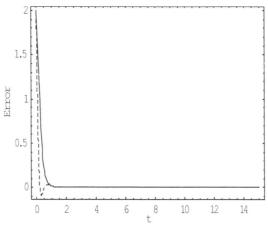

Figure 3 Error between the driving-response horizontal platform systems (1)–(3) with the coupling coefficients $k_1 = 5.6$ and $k_2 = 6.2$, solid curve for $x_1 - y_1$ and dashing curve for $x_2 - y_2$, initial conditions $(x_1(0), x_2(0)) = (1,1)$ and $(y_1(0), y_2(0)) = (-1, -1)$.

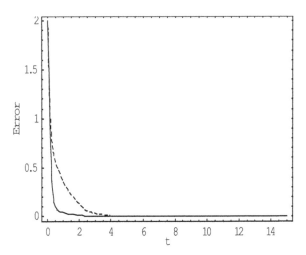

Figure 4 Error between the driving-response horizontal platform systems (1)–(3) with the coupling coefficients $k_1 = 9.5$ and $k_2 = 0$, solid curve for $x_1 - y_1$ and dashing curve for $x_2 - y_2$, initial conditions $(x_1(0), x_2(0)) = (1,1)$ and $(y_1(0), y_2(0)) = (-1,-1)$.

Synchronization Time and Cost

Firstly, we numerically investigate the behavior of synchronization time T_{syn} as a function of coupling strength k_1 and/or k_2. The synchronization time is defined as the initial time when the error measure $d = \sqrt{(x_1 - y_1)^2 + (x_2 - y_2)^2} < \varepsilon$ is satisfied and maintains in a long enough time interval $[T_{syn}, T_{syn} + L]$, where ε is the precision of the synchronization, and L is a sufficiently large positive constant. As shown in Figs. 5 and 6, the synchronization time T_{syn} gradually decreases with the increase of coupling strength, and approaches an asymptotic minimal value. This is a very interesting phenomenon, since one might think that the synchronization could be led as fast as desired if coupling strength is large enough. Figures. 5 and 6 confirm that very large values of coupling strength are not necessary to ensure the synchronization with approximately the minimum T_{syn}. Such phenomenon also occurred in synchronization scheme of single-well Duffing oscillators [14]. Generally, synchronizing two chaotic systems is not cost free. In order to evaluate what price must be paid to achieve synchronization, a new concept of synchronization cost for scheme (1)–(3) is introduced as follows,

$$\int_0^\infty k_1 \left|\sin(x_1 - y_1)\right| dt + \int_0^\infty k_2 \left|\sin(x_2 - y_2)\right| dt.$$

The meaning of this definition refers to the cost to achieve a certain degree of synchronization in the sense of (5). Note that the magnitude of $|x_i - y_i|$ is very small once synchronization is nearly achieved. So a good approximation of cost should be

$$\int_0^{T_{syn}} k_1 \left|\sin(x_1 - y_1)\right| dt + \int_0^{T_{syn}} k_2 \left|\sin(x_2 - y_2)\right| dt,$$

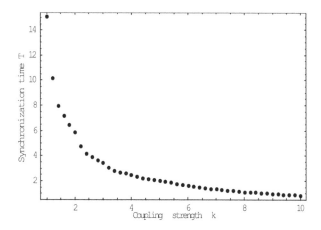

Figure 5 Synchronization time of systems (1) and (2) with sinusoidal controller $u(t) = (k\sin(x_1 - y_1),\ k\sin(x_2 - y_2))^T$, synchronization error measure $d < 0.001$, $L = 1000$, initial conditions $(x_1(0), x_2(0)) = (1,1)$ and $(y_1(0), y_2(0)) = (-1,-1)$.

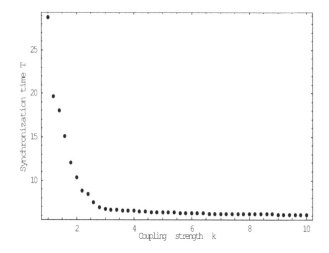

Figure 6 Synchronization time of systems (1) and (2) with sinusoidal controller $u(t) = (k\sin(x_1 - y_1),\ 0)^T$, synchronization error measure $d < 0.001$, $L = 1000$, initial conditions $(x_1(0), x_2(0)) = (1,1)$ and $(y_1(0), y_2(0)) = (-1,-1)$.

which will be adopted in the following simulations. Another definition of synchronization cost adopted in [15] for linear control is

$$\lim_{\tau \to \infty} \frac{1}{\tau} \int_0^\tau k_i |x_i - y_i| dt,\ i = 1,2,$$

which refers to the cost per unit time required to keep the synchronization going. The meaning is different from ours.

From the viewpoint of preventing from a useless increase of coupling strength, i.e., from an unavailing waste of input energy, the calculation of minimal synchronization cost as well as optimal coupling strength is of great practical interest. Synchronization cost versus coupling strength is simulated in Figs. 7 and 8 with different controllers. From these figures we can see that the synchronization cost decreases rapidly at first, then reach a minimal value and increases slowly with the increase of coupling strength at last. The explanation of this phenomenon is in agreement with the simulations of synchronization time shown in Figs. 5 and 6.

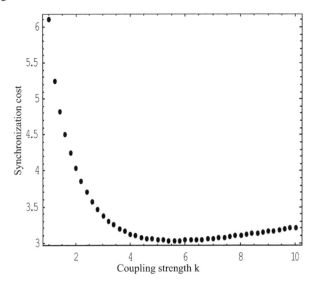

Figure 7 Synchronization cost of systems (1) and (2) with sinusoidal controller $u(t) = (k\sin(x_1 - y_1), k\sin(x_2 - y_2))^T$, with the initial conditions $(x_1(0), x_2(0)) = (1,1)$ and $(y_1(0), y_2(0)) = (-1,-1)$.

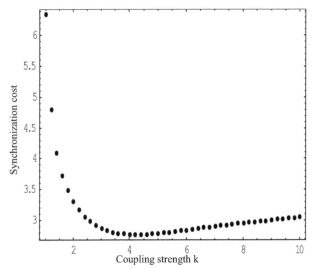

Figure 8 Synchronization cost of systems (1) and (2) with sinusoidal controller $u(t) = (k\sin(x_1 - y_1), 0)^T$, initial conditions $(x_1(0), x_2(0)) = (1,1)$ and $(y_1(0), y_2(0)) = (-1,-1)$.

The critical coupling strength with the minimal synchronization cost can be chosen as the optimal coupled strength in the sense of consumed energy. The optimal coupling strength and minimal synchronization cost are 5.6 and 3.03922 in Fig.7, 4.2 and 2.77078 in Fig.8 respectively. Although double variables coupled configuration (x and y-coupled) can lead to fast synchronization, its minimal synchronization cost is larger than that of single variable coupled configuration (x-coupled).

Conclusions

Some algebraic sufficient criteria for synchronizing driving-response horizontal platform systems coupled by sinusoidal state error feedback control are derived and their validity is illustrated with some numerical examples. Numerical simulations show that the synchronization time approaches an asymptotic minimal value with the increase of coupling strength. The concept of synchronization cost is introduced and the minimal synchronization cost as well as optimal coupling strength is calculated numerically. The minimal synchronization cost refers to the lowest energy input, which is of great practical interest.

Acknowledgements

The authors are grateful to the anonymous referees for their constructive comments. Research is supported by the National Natural Science Foundation of China under grant Nos. 10672193, 60674049 and 10571184, and the Foundation of Advanced Research Center of Zhongshan University under grant No. 06M13.

References

1. Z.M. Ge, T.C. Yu, Y.S. Chen, Chaos synchronization of a horizontal platform system, *Journal of Sound and Vibration*, 268 (2003) 731–749.
2. Chien-Lung Huang, Nonlinear dynamics of the horizontal platform, Master thesis, National Chiao Tung University, Taiwan, 1996.
3. J.A.K. Suykens, P.F. Curran, L.O. Chua, Master-slave synchronization using dynamic output feedback, *International Journal of Bifurcation and Chaos*, 7(1997) 671–679.
4. J.Lü, T.Zhou, S.Zhang, Chaos synchronization between linearly coupled chaotic systems, *Chaos, Solitons and Fractals*, 14(2002) 529–541.
5. G.P. Jiang, W.K.S. Tang, G.R. Chen, A simple global synchronization criterion for coupled chaotic systems, *Chaos, Solitons and Fractals*, 15(2003)925–935.
6. E.M. Elabbasy, H.N. Agiza, M.M.El-Dessoky, Global synchronization criterion and adaptive synchronization for new chaotic system. *Chaos, Solitons and Fractals*, 23(2005)1299–1309.
7. J.T. Sun, Y.P. Zhang, Some simple global synchronization criterions for coupled time-varied chaotic systems. *Chaos, Solitons and Fractals*, 19(2004)93–98.
8. Y.M. Lei, W.Xu, J.W. Shen, T.Fang, Global synchronization of two parametrically excited systems using active control, *Chaos, Solitons and Fractals*, 28(2006)428–436.
9. J.H. Park, Stability criterion for synchronization of linearly coupled unified chaotic systems. *Chaos, Solitons and Fractals*, 23(4)(2005)1319–1325.

10. Jiangen Wang, Yi Zhao, Chaotic synchronization of the master slave chaotic systems with different structures based on BANG-BANG control principle，*Chinese Physics Letters,* 22(10)(2005) 2508–2510.
11. Jianhe Shen, Shuhui Chen, Jianping Cai. Chaos synchronization criterion and its optimizations for a nonlinear transducer system via linear state error feedback control, *Chinese Physics Letters,* 23(6) (2006)1406–1409.
12. Xiaofeng Wu, Jianping Cai, Muhong Wang, Master-slave chaos synchronization criteria for the horizontal platform systems via linear state error feedback control, *Journal of Sound and Vibration，*295(2006) 378–387.
13. J. Slotine, W.P. Li, *Applied nonlinear control,* Prentice-Hall, New Jersey, 1991.
14. Y.C. Kouomou, P. Woafo, Stability and optimization of chaos synchronization through feedback coupling with delay, *Physics Letters A,* 298 (2002) 18–28.
15. C. Sarasola, F.J. Torrealdea, A. d'Anjou, M. Graña, Cost of synchronizing different chaotic systems, *Mathematics and Computers in Simulation,* 58 (2002) 309–327.

Index

1-D cellular automaton 24, 31
2-D cellular automaton 25, 26

A

Adams-type predictor-corrector 37, 44
Additional state variables 646, 664
Adnascent-type game model 215, 278, 279
Algebraic sufficient criterion 717, 720
Allometric growth 449–451, 453, 456, 459, 461, 466, 468, 469
Allometric scaling 355, 429, 443, 449–461, 463–470
Amplitudes 68, 232, 422, 423, 611, 612, 653, 655, 657, 706, 707, 709, 710, 712–714
Analog oscillator 345
Ancient philosophy 357, 372
Associative memory 29, 30
Astrobiological thought 93, 95
Astrocyte cellular automata 29
Asymmetry 339, 355, 487, 488–494, 497, 507, 508, 513–517, 523, 530, 533
Attractor 24, 25, 27, 45, 47, 50–53, 56–58, 61, 109, 153, 188, 189, 248, 249, 251, 262, 264, 265, 270, 275, 277, 288, 305, 308, 318–320, 322, 324, 345, 347, 353, 354, 369–372, 389, 393, 396–401, 403, 413, 416, 474, 475, 478–480, 491, 493, 521, 525, 528, 531, 617, 623, 624, 675, 680, 683, 686, 718
Attractor trajectory surrogates 474, 475
Attractor-based word modeling 397

B

Beauty 355, 365, 487, 488, 516, 517
Behavioral nutrition 1, 3, 14
Bifurcation 70, 124, 142, 169, 256, 257, 259, 262–264, 269, 275, 278, 279, 281–288, 317, 321, 347, 370, 405, 407–409, 413–417, 429, 430, 438, 440–444, 490, 515, 521, 523, 524, 528, 530, 533, 569, 570, 574–576, 616, 617, 646, 667–669, 675, 676, 680–685, 697, 701, 705–707, 709 710, 713, 714
Bifurcation theory 169, 279
Biharmonic dynamics system 545, 627, 630, 631
Bio-cybernetics 123
Biological evolution 145, 357, 359, 364, 369, 488, 516, 517
Bios 487, 489–493, 498, 501, 504, 506, 508, 509, 515, 517
Biosemiotics 123, 125
Bouc-Wen hysteretic oscillators 651, 653, 663
Boundary equilibrium 40
Bourbaki's mother 488, 515
Bush-Shaft system 545, 569

C

Cardiovascular applications 66
Catastrophic events 167, 169, 187
Central nervous system 22, 23, 31, 51, 135, 489, 513, 514, 516
Chaos anticontrol 278, 284
Chaos game representation 106, 108–110
Chaos synchronization 136, 545, 717, 718, 720, 721
Chaos theory 1, 3, 5, 6, 9, 11, 93, 147, 153, 167, 169, 215, 269, 272, 277, 315, 355, 376, 379, 384, 385, 387–389, 391, 397, 430, 438, 443, 521, 532, 533, 545
Chaotic model of a neuron 68
Chaotic view 1, 3
Chemical reactivity 545, 547, 548
Chromosome 106–108, 110–121, 145
Circuit realization 317, 348
Clinical psychology 520, 536, 537
Clock skew 405, 406, 408
Coarse-grained entropy 215, 304, 305, 309–312
Cobweb model 215, 256, 257, 262
Cognition 4, 34, 394, 395, 515, 520–522, 526, 529, 531, 532, 539

Cognitive science applications 521

Complex adaptive systems (CAS) 123, 134

Complex dynamics 210, 215, 250, 256, 262, 265, 278, 279, 286, 288, 345, 429, 430, 438–443, 614, 668

Complexity 7, 27, 52, 60, 93, 123, 125, 127–133, 137, 141, 143, 147–151, 153, 169, 171, 173, 175–177, 179, 181, 182, 184–188, 198–200, 203, 205, 237, 248, 251, 257, 288, 368, 370, 372, 373, 376, 384, 385, 387, 388, 390, 392, 394, 395, 401, 429, 439, 443, 444, 470, 473–475, 484, 485, 487, 490, 493, 507, 508, 516, 517, 538, 545, 584, 595–597, 599–605, 615, 624, 701

Complexity theory 123, 153, 373

Conservative dynamics 345, 346

Control 4, 10, 14, 77, 84, 87, 89, 123, 136 139–141, 148, 151, 152, 222, 244, 245, 256, 257, 265, 266, 271, 275–277, 284, 311, 334, 358, 378, 405, 406, 408, 409, 412, 414, 416, 426, 523, 530, 533, 535, 545, 570, 574–576, 589, 609, 614, 616, 623–625, 645, 646, 649, 651–653, 657, 661, 664, 668, 675, 683, 713, 717–719, 723, 725

Correlation dimension 50, 51, 53, 54, 60–63, 89, 169, 172, 175–177, 184–187, 189, 190, 192, 248, 250, 251, 305, 473, 474, 478, 481, 483–485

Correlation dimension algorithm 485

Correlation dimension estimation 169, 192

Creativity 149, 487–490, 507, 508, 514, 517, 529, 532, 537

Crisis 1, 14, 18, 240, 291, 312

Cryptography 328, 329, 341

CTM algorithm 51

D

Deflections 224, 394, 610

Delay feedback control 256, 265, 266

Delay times method 171, 172, 174, 185, 186, 188, 189, 192

Deteriorating items 215, 290–292

Determinism 51, 59, 138, 142, 153, 247, 306–308, 312, 372, 376–380, 382, 478, 481, 517, 538

Deterministic chaos 169, 196, 199, 210, 236, 238, 247, 249, 253, 270

Diabetes 85, 91

Difference equations 380, 405, 406, 426, 436

Discrete dynamics 217, 221, 256, 278, 290, 436, 437, 449, 520

Disruption management 290

Distribution of dynamic invariants 355, 473

Duffing equation 627, 628

Duration of memory 25, 27–29, 32

Dynamical systems 124, 140, 198, 201, 202, 265, 272, 318, 326, 345, 357, 371, 372, 379, 393, 394, 396, 397, 403, 425, 438, 443, 520, 595, 701

E

Economic science 215, 217, 223

Economic transition 215, 236–238, 241, 246, 247, 252

Economy model 236, 238, 252

EEG signal chaos characteristics 1, 50

Electronegativity 547–549, 552, 564

Electronics 315, 317, 328, 345, 346, 406

Emergence 24, 123–125, 134, 141, 145, 170, 174–178, 180, 186, 219, 226, 233, 270, 271, 414, 534, 591

Encroaching epidemic 14

Entropy 26, 28, 32, 50, 52, 53, 59, 60, 62, 63, 95–97, 99, 101, 148, 169, 173, 176, 177, 184–187, 190, 192, 215, 227, 304, 305, 309–312, 329, 361, 363–365, 372, 373, 393, 394, 473–475, 478, 480, 481, 483, 484, 487, 492, 493, 501–503, 508, 516, 536, 545, 547–550, 552, 554, 555, 562–564, 582, 585, 588, 590–592, 595–597, 599, 600, 601, 605, 606

Environment 6, 9, 10, 15, 16, 102, 125, 129, 133–136, 141, 144–147, 167, 196, 197, 199, 201–205, 208, 210, 224, 228, 230, 243, 244, 251, 312, 367, 370, 377, 384–389, 391, 405, 407, 409, 412, 414, 416, 417, 422, 423, 429, 449, 461, 526, 529, 532, 533, 591, 592

Environmental change 167, 196, 197, 201, 202, 208, 210

Equilibrium 37–44, 47, 182, 184, 185, 197, 198, 203, 220, 223, 224, 226, 228, 229, 232, 233, 256, 258, 260–263, 267, 278–286, 288, 304, 305, 320, 354, 362, 364, 379, 492, 508, 514, 516, 517, 584, 585, 588–592, 596, 610, 619, 621–624, 629–631, 639, 646, 652, 668, 673, 674, 680, 682–687

Equilibrium parameter 682–687

EU accession 236, 244, 252

Evolutionary geomorphology 167, 196–199, 206, 211

F

Feedback control 136, 256, 257, 265, 266, 284, 668, 717–719, 725

FIPS tests 328, 342

Flip bifurcations 278

Fractal dimension 84, 85, 87, 89, 183–186, 249, 251, 278, 279, 285–288, 368–371, 443, 449, 450, 452, 454, 456, 457, 459, 460–463, 467–470

Fractal patterns 6, 454

Fractal spectral analysis 170, 173, 181

Fractal structure 106, 109, 178, 179, 183, 262, 430, 449, 461, 469, 675, 680, 686

Fractional-order HIV model 1, 37

Free will 372, 376, 377, 380, 382, 538

Fusion 96, 582–584, 591, 592

G

Galerkin polynomial chaos approach 701

Gas argon bubbling 689, 691

Generalized entropy 595

Geomorphic systems 196–202, 204–207, 210

Geothermal 95, 99, 101–104

Global instability 169, 176, 177, 179, 180, 181, 183, 186

Growth 5, 99, 135, 200, 204, 215, 217, 219–222, 226, 233, 247, 284, 346, 355, 357, 368–371, 406, 429–431, 433–436, 439, 440, 442–444, 446, 449–456, 459–461, 463–466, 467–470, 528, 529, 639, 697

H

Hardness 547–549, 552, 553, 557, 558, 560, 564, 565

Health behavior change 3–6, 10

Healthy 1, 56, 84, 85, 89, 90, 134, 135, 184–187, 191, 312, 485, 539

Heartfailure 184

Hénon map 474, 477–480

High-level intelligence activity 1, 50, 51

Hindmarsh-Rose (HR) model 68

Human behavior 6, 9, 11, 203, 223

Human brain 1, 22, 27, 29, 50, 51, 59, 60, 61, 64

Human cellular automaton 27, 28

Human development 224, 528, 529

Human immune response against viral infections 37

Human photo-plethysmographic time series 355, 473

Hydrodynamic loading effect 670

Hydrogen atom 547, 548, 550, 554–564

hysteretic systems 545, 645, 646, 664

I

Information 7, 17, 22–25, 27–32, 37, 50, 51, 55, 56, 67, 95, 96, 99, 106, 107, 123–129, 135, 141, 143, 144, 147, 148, 151, 169, 171–174, 176, 187, 189, 209, 219, 228, 238, 240, 241, 247, 271, 275, 276, 307, 312, 328, 330, 338, 339, 346, 357, 361, 363–370, 372, 373, 384, 386, 389, 390, 391, 394, 395, 403, 450, 461, 462, 473, 492, 513–516, 522, 526, 533, 551, 590, 595, 606, 607, 698

Interior equilibrium 37, 38, 41–44, 47

International stock markets 215, 304

Inventory management 290

Irreversibility 180, 186, 218, 222, 224, 226, 227, 361, 364

K

Kalman filter 72

L

Ladle 545, 689, 690, 692, 696

Landform response 167, 196

Language contact 384, 385, 388–391

Language recognition 393, 401

Leadership emergence 534

Learning 29–31, 55, 257, 387, 390, 396, 520–522, 524, 526, 529, 530, 536

Lennard-Jones (LJ) potential 667, 669

Libertarianism 376, 377, 381

Linear events 4

Linear paradigm 4

Locally connected chaotic biological neurons 66

Logistic growth 429, 430, 433, 440, 443

Lyapunov exponent 50, 51, 54, 61–63, 89, 202, 248, 249, 251, 256, 262, 278, 279, 285, 286–288, 317, 320, 347, 370, 397, 398, 525, 533, 547, 548, 555, 569, 570, 573, 574, 576, 667, 675, 676, 680, 683, 684, 686

M

Market efficiency 304

Markov chains 109, 138, 153, 328

Mathematical identification 545, 689

Mathematical modeling 292, 406, 517., 570

Matrix frequency 93, 106–108, 110–112, 115–121

Measurement 1, 4, 6, 14, 17, 18, 66, 68–70, 73, 75–77, 79, 81, 87, 88, 127, 151, 174, 228, 304, 305, 309, 310, 312, 317, 327, 352, 379, 401, 449, 460, 475, 477, 484, 520, 526, 527, 529, 597, 669, 694

Melnikov criterion 631, 634

Mel'nikov's method 569, 570, 574

Memory in astrocytes 1, 22

Microcantilever 645, 667

Minimum entropy production principles 591

Modeling 145, 199, 204, 208, 256, 270, 292, 348, 355, 393, 395–397, 403, 405, 406, 408, 409, 429, 447, 468, 517, 524, 529, 539, 545, 569, 570, 591, 592, 645–647, 669, 697, 701, 713
Modified Bouncer model 610, 614
Multi-agent approach investigation 215, 226
Multi-fractality 184–186
Multi-lead correlation dimension 53
Music 223, 487, 490, 493, 497, 498, 501, 502, 504, 506, 511, 515, 517

N

Natural language processing 355, 393, 403
Natural phenomena 170, 357, 366, 397
Natural philosophy 357, 361, 363, 370
NDS in psychology 521
Neumannian economy 215, 226, 227
Neuro-capillary chaos 1
Neuropathy 84, 85, 88–91
Noise estimation 215, 304, 309, 310
Non linear dynamics 4, 9, 72, 123, 152, 153, 172
Nonintrusive projection method 705, 713
Nonlinear aeroelastic system 546, 697, 705, 713, 714
Nonlinear dynamics 1, 12, 37, 51, 53, 148, 189, 197–199, 201, 203, 205, 210, 217, 218, 248, 270, 278, 304, 305, 307, 310, 312, 355, 397, 520, 521, 532, 536, 539, 545, 583, 609, 610, 667–669, 675, 683, 686
Nonlinear filtering 1, 66
Nonlinear function synthesis 348
Nonlinear oscillator 70, 315, 317, 319, 324, 326, 345, 346
Nonlinear time series 51, 304, 309, 481
Nonlinearity 74, 167, 196, 198–201, 210, 217, 270, 272, 305, 306, 312, 318, 324, 346, 443, 552, 570, 598, 668, 699, 702–704, 706, 714
Normal forms 278, 283
Normalized Tsallis wavelet entropy 599, 605, 606
Nutrition behaviors 14, 18

O

One-virus model 37–40, 47
Orbit diagrams 405, 414
Oryza sativa (japonica cultivar-group) 93, 106, 107–111, 113, 115, 116, 121
Oscillatory measurements 1, 66

P

Particle transport 545, 609
Perception 395, 399, 403, 520–524, 528
Period-doubling bifurcations 405, 407, 408, 417, 681, 685
Persuasion slowly overcoming resistance 17
Phase graph analysis 56
Phase locked loops 405, 406, 408, 409, 423, 426
Phase space reconstruct technique 50, 51, 56
Phase space reconstruction 60, 174, 188–190, 304, 305, 307, 312
Photochemical 95, 101, 102, 104
Physical activity behaviors 3, 4
Physical modeling 689, 692, 696
Pidgin 384, 385, 390, 391
Pinched hysteresis phenomenon 657
Plantar pressure points 1, 84, 85, 91
Plasma 545, 582, 583–587, 589–592
Poetry 224, 487, 490, 493, 494, 497, 498, 501, 503, 504, 510, 511
Polynomial chaos expansion 546, 697, 698, 701, 702
Power spectra analysis 58
Prediction 5–7, 10, 18, 146, 149, 153, 170, 196, 199, 203, 204, 207, 209, 210, 269, 271, 277, 318, 363, 531, 577, 610, 646, 649, 669, 697
Predictive models 15, 16
Priority of the simple 515, 516
Psychology 355, 358, 514, 520, 521, 524, 526–529, 531–533, 536–539

Q

Quadratic iterated map 393, 397, 399, 400, 403
Quantum chaos 545, 547, 552, 595
Quantum event 3
Quantum fluid dynamics (QFD) 550
Quantum theory of motion (QTM) 550

R

Radiochemical 95
Randomness 7, 9, 127–129, 142, 147, 149, 151, 247, 249, 305, 310, 328, 329, 339, 341, 357, 360, 362–365, 371, 378, 380, 491, 600, 697
Rational-linear conceptualization 4
Real estate 256, 257, 267
Real estate market 256, 267
Reductionoxidation 95
Relaxation chaos 582

Resistance to chaos 9
Rising epidemic of obesity 14, 18
Rössler system 474, 478, 480–482, 485

S

Scaling exponent 178, 355, 443, 449–451, 453, 455–460, 462, 463, 465–467
Schwarzian derivative 278, 279, 282
Self-organization 67, 123–126, 134, 135, 141, 146–150, 153, 200, 201, 370, 388, 521, 525, 528, 529, 532–536, 545, 582, 583
Semiclassical theories 595
Sensitivity to initial conditions 5, 6, 14, 16, 18, 202, 248, 249, 251, 262, 266, 284, 305, 397, 539, 614, 624
Similarity of ion channels to cellular automata 23
Sliding particle model 618, 619, 624, 625
Small data sets method 54
SOM neural network 55
Split of water 95, 99–104
Spreadsheet modeling 405
Squeeze film damping 545, 667, 671, 676
Squeeze film damping effect 671
State space 78, 79, 198, 272, 318–320, 322, 345, 347, 348, 354, 396, 397
Statistical complexity 545, 596, 597, 600, 602–605
Steel 577, 646, 650, 651, 664, 689–691, 696
Stir elements 689
Supply chain management 290
Supremacy of the complex 515, 516
Sustainable life 93, 95
Symbolic dynamics analysis 141
Symmetry 161, 269, 271, 354, 355, 400, 414, 466, 487–493, 507, 514, 516, 517, 557, 585, 589, 628

Synchronization cost 717, 719, 722–725
Systems theory 123, 124, 135, 136, 149, 198, 520

T

Temporal redistribution 1, 84, 87
Tertiary education 405
Time series analysis 1, 64, 84, 189, 252, 253, 481
Transition path 236, 238, 242, 243, 247, 249, 253
Triangular map 278, 281
TRNG, PSoC mixed-signal hardware 328
Trophic ulcers 84, 85, 91
Turbulence 135, 208, 71, 582, 583, 585, 590
Two-population interaction 429, 431, 434–441, 443, 444
Two-virus model 43

U

UKF estimation 74
Unified approach 168, 169
Unified word feature 393, 401, 403
Uninfected equilibrium 40
Unpredictability 368, 371, 372, 391, 438, 652
Urban form 446, 450, 452, 459, 461–464, 467, 469, 470
Urban growth 449, 454, 463, 464, 465
Urban hierarchy 449, 453
Urban structure 449
Urbanization 355, 429–434, 436, 437–444, 446, 447

W

Wholeness 394, 395

Printed and bound by CPI Group (UK) Ltd, Croydon, CR0 4YY

18/10/2024

01776252-0011